PESTICIDES, POLLUTANTS, FERTILIZERS AND TREES:

their role in forests and amenity woodlands

FORESTRY SERIES

Series Editor: **Esmond H. M. Harris,** BSc., Dip.For., FICFor., CBiol., MIBiol.

1. Computers in Forestry: Use of Spreadsheets
   Roy Lorrain-Smith

2. Natural Management of Woods: Continuous Cover Forestry
   J. E. Garfitt

3. Pesticides, Pollutants, Fertilizers and Trees:
   their role in forests and amenity woodlands
   J. R. Aldhous

4. Wildlife Conservation in Managed Woodlands and Forests, SECOND EDITION
   Esmond Harris *and* **Jeanette Harris**

PESTICIDES, POLLUTANTS, FERTILIZERS AND TREES:

their role in forests and amenity woodlands

J. R. Aldhous, BA, FICFor

RESEARCH STUDIES PRESS LTD.
Baldock, Hertfordshire, England

RESEARCH STUDIES PRESS LTD.
15/16 Coach House Cloisters, 10 Hitchin Street, Baldock, Hertfordshire, England, SG7 6AE

and

325 Chestnut Street, Philadelphia, PA 19106

Copyright © 2000, by Research Studies Press Ltd.

Marketing:

Research Studies Press Ltd.
15/16 Coach House Cloisters, 10 Hitchin Street, Baldock, Hertfordshire, England, SG7 6AE

Distribution:

NORTH AMERICA
Taylor & Francis Inc.
47 Runway Road, Suite G, Levittown, PA 19057 - 4700, USA

ASIA PACIFIC
Hemisphere Publication Services
Golden Wheel Building, 41 Kallang Pudding Road #04-03, Singapore

EUROPE & REST OF THE WORLD
John Wiley & Sons Ltd.
Shripney Road, Bognor Regis, West Sussex, England, PO22 9SA

Library of Congress Cataloging-in-Publication Data
Available

British Library Cataloguing in Publication Data
A catalogue record for this book is available from the British Library.

ISBN 0 86380 199 4

Printed in Great Britain by SRP Ltd., Exeter

Editorial foreword

THE OBJECTIVES OF THE FORESTRY SERIES

Research Studies Press aims to make available specialist texts in developing areas of science and technology tailored to particular but limited markets. It is appropriate therefore that the Forestry Series should now include this comprehensive account of the present state of knowledge of the use of chemicals in forestry and the many issues associated with this.

Though primarily concerned with the situation in Britain, both the facts presented and the issues discussed are relevant throughout the world and will be of interest to foresters, land managers and tree growers everywhere. To meet wider needs, scientific facts and concepts are expressed as far as possible in terms understandable to the lay reader whenever obscure academic expression can be avoided without loss of accuracy.

THE OBJECTIVES OF THIS BOOK -

The use of chemicals in forests and woodlands has become both complex and controversial, presenting as it does many opportunities for efficiency but also threats. Who better to lead us through this complex field than John Aldhous, following his career in British forestry, both as a research scientist and as a senior manager with the British Forestry Commission.

In the 1950s and 1960s the author was involved in forest nursery work research, particularly the use of chemicals for weed control, nutrition and fertilizers, contributing to many Forestry Commission publications and at conferences. In particular, he was responsible for the first edition of Nursery Practice, the Forestry Commission Bulletin 43, which incorporated the wide experience that had been gained in Britain of growing trees from seed rather than naturally regenerating them in the forest.

In the 1970s John Aldhous moved into senior posts in forest management, being responsible for all aspects of wood production. With this background of lifelong experience of research and management, he has devoted three busy years of his retirement to putting together this comprehensive review of the chemicals available in forestry and the many opportunities for their use today. It is a factual and detailed account but running throughout is the theme of responsible use and an overall concern for the environment.

The first part of the book therefore appropriately starts with ethical issues before going on to laying out the chemical and legal frameworks and the safety system, summarising these in a chapter on integrated forest protection. In the core of the book, pests of all forms, nutrients and fertilizers are discussed in a way that leaves little of these complex subjects untouched, before going on to a comprehensive account of the present knowledge of pollutants and forest health.

In his final chapter the author returns to ethical issues and emphasises that if our woodlands and forests, as well as the ecosystems surrounding them, are to be sustainable the opportunities provided today by technological advances must be handled carefully, responsibly and with understanding, for which his text lays a comprehensive foundation.

This book therefore is particularly poignant at the present time when there is mounting concern about what is sometimes conceived as the irresponsible use of science. This sober, factual and masterful account of a complex subject, indeed a whole gamut of subjects, which offer both great opportunities and threats, is presented in a readable style easily accessible to forest managers and those with less scientific training but who are all concerned to grow trees well.

The text draws heavily upon and is well supported by over 2000 references, all meticulously quoted and thus the bibliography runs to 83 valuable pages and as such has never been produced as a whole before and is itself a significant contribution to present day forestry literature.

Esmond Harris
Editor, Forestry Series

Preface

This review describes the various ways in which man-influenced chemicals have impinged on woodland management in the United Kingdom over the last 50 years. 'Chemicals' in this context refers to pesticides and fertilizers selected and applied by woodland managers, and also the chemicals that affect trees and woodlands through atmospheric or soil pollution.

The ethos underlying the review is that with responsible management, use of pesticides and fertilizers is proper and necessary for commercial and amenity woodland and tree management. This ethos and its application is relevant not just in Britain, but can be extended to forests and amenity trees worldwide.

The following chapters provide a background to justify that stance. The benefits of well-managed forests and woodlands encompass production of timber for utilitarian purposes, wildlife habitat conservation, beauty both within woodlands and as part of a larger landscape, opportunities for recreation and opportunities for quiet and contemplation. Forests are not just timber factories.

The use of each of the various groups of chemicals has evolved during the past 50 years. For example, the need for and potential of fertilizers to improve tree growth on difficult sites commenced with empirical experiments on their use. Results from such work formed the basis for a deeper understanding of nutrient requirements and nutrient cycles, applicable to British conditions and more widely.

Studies of atmospheric pollution were initially aimed at finding the limits of tree growth in the vicinity of heavy industries discharging pollutants into the atmosphere. These became part of wider international studies on pollution from sulphur and nitrogen and have in the last ten years extended into global concerns about climatic change.

The role of pesticides has developed differently. Because of the small size of the forestry market for pesticides in Britain, materials available have been restricted almost entirely to those for which there are other markets, either in agriculture in the UK or internationally. Also, because of increasing criticism of what has been perceived as past cavalier promotion of dangerous chemicals for pest control, present-day pesticide practice is markedly more rigorous and circumscribed than at any time previously.

In the United Kingdom, establishment or refurbishment of forests, woodlands, parkland and street trees has very largely been by planting. Government policy over the last 70 years to re-establish woodland after earlier fellings have led to the UK holding a leading position in the silviculture of converting poor quality farm land back to woodland This programme has been the driving force behind many of the developments described below.

Acknowledgments

The author is deeply indebted to his many friends and former colleagues whose work is reported. Examination of the bibliography will show how numerous these are and how strongly represented are the staff of the Forestry Commission Research Division (now the Forest Research Agency) at Alice Holt, Farnham, Surrey, and the Northern Research Station, Roslin, Midlothian.

The work at many other centres of excellence has also been quoted extensively, in particular, the Macaulay Land Use Research Institute, Aberdeen, the Institute of Terrestrial Ecology, Penicuik, Midlothian, the Forestry Department, Aberdeen University, and Rothamsted Experimental Station, Harpenden, Hertfordshire.

Observations on individual chapters, much valued, were made by: Stuart Heritage, Lucy Sheppard, Jim Pratt and Malcolm Crosby. Throughout, the comments, encouragement and support from Esmond Harris have been greatly appreciated.

Thanks are also due to the publishers listed below, who have permitted the reproduction of figures or diagrams from their publications. In all cases, the author and year is shown under *Source* beneath each figure, full details of the source publication being given in the bibliography.

Academic Press, London: Figure 12.2;
Birkhauser Verlag, Basel, Switzerland: Figure 12.4;
Elsevier Science BV, Amsterdam: Figure 13.1;
Forestry Commission: Figures 8.1, 10.1, 10.4, 11.1, 13.2, 16.3;
Institute of Chartered Foresters, Edinburgh: Figures 10.2, 12.1, 12.3.
    14.2, 16.2;
Miller, Freeman UK Ltd. Tonbridge: Figure 10.3.
Stationery Office, Norwich: Figures 14.1, 16.1;

Contents

Part I - Contexts

Chapter 1   *Ethical issues: forests, food, man, land*

1.1   Introduction: ethical bases for management   3
1.2   Global trends: - the international debate   4
1.3   Population pressures   5
1.4   Global land use change   9
1.5   Global supply & demand for wood & wood products   10

Chapter 2   *The chemical framework*

2.1   Pesticides, fertilizers and pollutants   17
2.2   Pesticide availability and use   18
2.3   Pesticide nomenclature   21
2.4   Formulated products   26
2.5   Marketing new pesticides   27
2.6   Forestry's Green revolution   30

Chapter 3 .   *The legal framework*

3.1   Public safety   35
3.2   Regulatory schemes   35
3.3   Present provisions for safe use of pesticides   40
3.4   Consents   44
3.5   Protection of groundwater   47
3.6   Guidance on application of legislation   48

Chapter 4   *The safety system*

4.1   Safety - an international concern   51
4.2   Requirements for safe use of pesticides   58
4.3   Safety in use   65
4.4   Operator safety   69
4.5   Storage of pesticides   74
4.6   Safe disposal of pesticide wastes   75
4.7   Pesticide use and abuse   76

Chapter 5   *Integrated forest protection*

5.1	Minimising pesticide use	77
5.2	Integrating protection and forest management	78
5.3	Vigilance	81
5.4	Natural attractants	86
5.5	Biological control	89
5.6	Plants' defences and physiological state	92
5.7	Genetic engineering	94
5.8	Disintegrated pest control	94

Part II - Pesticides   95

Chapter 6   *Insect & other pests*

6.1	Introduction	97
6.2	Insect control & integrated forest protection	97
6.3	Chemical control	103
6.4	Integrated management of moth pests *Lepidoptera*	105
6.5	Integrated management of beetle pests *Coleoptera*	116
6.6	Integrated management of sawflies *Hymenoptera*	131
6.7	Integrated management of other pests	134
6.8	Vigilance against potential pests entering the UK	143

Chapter 7   **Fungal & other diseases**

7.1	Vital or virulent	145
7.2	Diseases acting through soil or roots	149
7.3	Vector-borne diseases	156
7.4	Diseases acting through stem or foliage	164
7.5	Cultural controls	167
7.6	Forest & hardy ornamental nursery stock	174
7.7	Virus diseases	178

Chapter 8   *Weeds*

8.1	Weeds not wanted	181
8.2	Weeding practice	192
8.3	Use of herbicides	197
8.4	Specific vegetation types	200
8.5	Forest nurseries	214
8.6	Other uses of herbicides	220

Chapter 9   *Other pests*

9.1	Introduction	223
9.2	Vertebrate control	223
9.3	Wood preservatives	238

Part III - Nutrients & Fertilizers

Chapter 10   Woodlands & fertilizers

10.1   Background to woodland cover   243
10.2   Fertilizers at and post-planting   249
10.3   Fertilizer need and stage of woodland development   253
10.4   Phosphorus (P)   255
10.5   Nitrogen (N)   260
10.6   Potassium (K)   267
10.7   Calcium (Ca), Magnesium (Mg), Copper (Cu) & Other Nutrients   269
10.8   Fertilizing pole-stage and older crops   273
10.9   Second rotations   278

Chapter 11   Fertilizers for tree nurseries, amenity planting & land restoration

11.1   Nutrition in forest nursery production   281
11.2   Organic sources of nutrients   297
11.3   Landscape & amenity plantings   304
11.4   Other techniques   315

Chapter 12   Nutrient cycles in woodland

12.1   Woodland cycles   315
12.2   Nutrients in tree and woodland growth   317
12.3   Mycorrhizas, heather-check and nitrogen   327
12.4   Effects of trees on soil profile development   332
12.5   Nutrient loss in drainage water   333

Part IV - Atmospheric & mineral pollutants

Chapter 13   Atmospheric pollutants

13.1   Atmospheric pollution   345
13.2   Air-borne ('fixed') nitrogen   349
13.3   Sulphur and acid rain   363
13 4   Ozone   373
13.5   Other air-borne pollutants   381

Chapter 14   Acid Rain, forest health & drainage water

14.1   Sources of concern within the UK   383
14.2   Water   386
14.3   Trees and atmospheric pollutant depositions   391
14.4   Acid rain and forest decline in Europe   397
14.5   Acidification and surface waters   405
14.6   Effects of forest practice   410

Chapter 15   Mineral and other pollutants

15.1   Mineral pollutants   413
15.2   Salinity   420
15.3   Other pollutant sources   425
15.4   Pollution and non-woodland trees   427

Chapter 16   Carbon dioxide and climatic change

16.1   Introduction   429
16.2   Carbon, carbon dioxide & climatic change   429
16.3   Other greenhouse gases   445
16.4   Renewable energy   446
16.5   Perspectives on climatic change and forests   455

Part V - Our future

Chapter 17   Woodland & environment for the 21st Century

17.1   Holistic living on a shrinking globe   459
17.2   Forestry in the UK in the 21st Century   462
17.3   Vigilance (again)   463
17.4   The challenge   466

Appendix I   *References to pesticides in literature*   469

Appendix II   *Pesticide tables - fields of use, toxicity etc.* 479

Appendix III   *Glossary*   487

Bibliography   491

Index   577

Part I

Contexts

CHAPTER 1

Ethical issues:
forests, food, man, land

'Only one world: our own to make and to keep' (UN 1992)

1.1   INTRODUCTION - ETHICAL BASES FOR MANAGEMENT

The choices made when managing woodland unavoidably reflect the ethical values underlying the objectives of management set by the woodland owners. At operational level, similarly, in discussing the options open to managers when considering the need for chemicals, the choices made may frequently be influenced by ethical values. This book is written from the standpoint that:

- mankind finds itself in a finite environment;
- mankind, no matter what colour, class or creed, has an obligation to steward global resources sustainably, deploying its talents and intelligence to the full to do so.

This ethic clashes with ethics based on unfettered exploitation of resources, wherever found. It also clashes with ethics based on national, cultural or religious supremacy.

Under the stewardship ethic, the aim is to provide man's needs for food and raw materials sustainably, and at the same time to maintain land productivity and protect the natural diversity of the living world.

These aims have to be achieved in a world that is not static in any meaningful way. Change is all around us:

- the world population has increased dramatically in the 20th century. The United Nations Population Reference Bureau estimated it to have reached six billion in October 1999. In the short term, the increase is likely to continue. The change is of fundamental significance because of the consequential need for food, and the call for energy and for raw material for housing and industry;
- world carbon dioxide levels have increased and will increase further if mankind's current activities continue as at present;
- understanding of the genetic codes embodied in DNA strands of many living organisms is making feasible biological changes previously believed to be impossible;
- micro-chip information technology is changing more rapidly and more conspicuously than any other current non-biological aspect of modern life. Its

influence is all-pervasive and is likely to alter life-styles and resource demands fundamentally.

Priority has to be given to addressing how to forecast the consequences of such changes and the actions that should be taken to minimise or exclude detrimental effects. Most of the developments in crop production and protection arise out of this priority.

Unforeseen and initially unquantifiable risks cannot be avoided in this or any other future scenario. It is not possible to apply any 'precautionary principle' seeking to know all the consequences of a novel action before taking it. That position is logically untenable, even in a static world.

At the other extreme, a completely *'laissez faire'* approach to risk is unacceptable to many on ethical grounds. Evidence since 1950 is overwhelming that unpredicted and unpredictable technical consequences of changes of practice can have devastating effects, *eg* accumulation of DDT in the food chain; side effects of thalilomide. Other evidence points to the damaging effects of over-complacent or inadequate management, *eg* the Chernobyl and Bhophal disasters.

A central thesis of this book is that continuing vigilance is essential, recognising that novel effects may occur and that it is prudent to be alert to such effects, noting them and taking appropriate action.

1.2   GLOBAL TRENDS - THE INTERNATIONAL DEBATE

Since 1990, a series of international conferences under the auspices of the United Nations, have caught up and focused concerns about global prospects for mankind.

The UN Conference on Environment and Development (the Earth Summit) in Rio de Janeiro in 1992 brought international agreement to proceed with:

- a 'Climatic Change Convention';
- a 'Biodiversity Convention';
- 'Agenda 21' - a worldwide programme seeking a more sustainable pattern of development for the next century;
- a statement of principles for the management, conservation and sustainable development of the world's forests.

The UK government's report on its strategy for sustainable development arising out of this conference quotes as a guiding definition :-

- *Sustainable development meets the needs of the present without compromising the ability of future generations to meet their own needs* (UK Govt 1994a).

The Rio Conference was followed in 1993 by a Ministerial Conference in Helsinki, on the 'Protection of European Forests'. Participating countries made commitments to implement guidelines for the sustainable management of European forests and for conservation of their bio-diversity (UK Govt 1994b).

At the UN International 'Population and Development' Conference in 1994 in Cairo, there was little dispute about past population increases and the fact that numbers are rising. However, there was less consensus as to future projections. See §1.3 below.

In 1996, a conference in Istanbul in June on human settlements, and the habitat for rural and urbanised man was followed later in the year by an International Conference in Rome on world food resources.

These conferences, and the programmes arising from them, have been received with varying degrees of approval in the developed world. There has, however, been dissent from under-developed countries. At the time of the Rio conference, developed countries were accused of wanting to evade the onus to rectify the consequences of their past lack of environmental concern and profligacy with fossil energy sources. Developing countries also accuse developed countries of attempting to impose environmental standards in order to maintain a competitive advantage. Under-developed countries' priorities are to alleviate widespread poverty of their peoples (Agarwhal 1992, Pearce 1992).

1.3   POPULATION PRESSURES

1.31   Population trends

The size of the world population, its location, health, wealth and prospects have been central to the issues of the 1990s international debates. The world

Table 1.1   *Estimates of population growth for developed and developing countries, and for the world: 1985, 2000, 2025, 2050*

Year	Developed countries Pop'n (Mill.)	Pop'n Change	Growth %*	Developing countries Pop'n (Mill.)	Pop'n Change	Growth %*	World Pop'n (Mill.)	Pop'n Change	Growth %*
1985	1179			3666			4844		
		+86	0.47		+1273	2.00		+1360	1.66
2000	1265			4939			6204		
		+71	0.22		+2139	1.45		+2211	1.23
2025	1336			7078			8415		
		-17	-0.05		+1638	0.84		+1620	0.83
2050	1319			8716			10035		

*% = compound rate of increase per annum over 25 year periods
Source   World Bank 1990

population in 1900 was about 2 billion, 3 billion in the 1950s and having doubled since then (Mitchell 1998, Pearce 1998b).

Forecasts in 1990 of population growth between 1985 and 2050 in *Table 1.1* show a further doubling of the world population over the period.

At the 1994 Cairo Conference, it was assumed that present trends in population increase would continue in the short term. The UN Population Fund report for 1994 pointed to a slowing of population increase through contraception; even so, their forecast was similar to earlier World Bank figures in *Table 1.1*, *ie* a world population increase to 8.5 billion (8.5×10^9) by 2025, and 10 billion by 2050. The Institute of Applied Systems Analysis (cited in Coghlan, 1996a) forecast approx 8.2 billion in 2025, and a peak of 10.6 billion in about 2080. A more recent estimate showed trends under 'high', 'medium' and 'low' predicted rates of population increase. For 2050, estimates based on these rates were about 12.5, 10 and 7.5 billion respectively (Pearce 1998a).

Population increase is predicted not to occur evenly across the globe. Populations in some African and Asian countries are predicted to more than double over the next 50 years, while populations of some European countries will remain static or will drop.

Such long-term forecasts are speculative in the extreme; they have to be monitored and continually updated for behavioural, health and demographic changes. Effectiveness and scale of use of contraception *etc.* may indeed alter trends in birth rate over a relatively short period (Webb 1996). Nevertheless, until trends have clearly changed, continuing population increase has to be a dominant factor in international strategic planning relating to land use.

Urbanisation of populations

In 1994, the UN Population Fund in its annual report focused on the distribution of world population between town and country, and drew attention to the urban growth taking place in developing counties (UNPF 1994). Another report, *An Urbanising World* from the UN Centre for Human Settlement (UN 1996), noted that the effects of urbanisation accentuated poverty, dispossession and limited access to food.

1.32   Food supply for an increasing population

Normal dietary intake for adults is of the order of 3000 kcalories per day (male) and 2200 kcalories per day (female). In a number of countries, mostly in Africa, the average daily intake is 2000 kcalories, the same as in pre-revolution France.

In 1990, globally, the area of land required to grow food crops averaged 2700 m² per person, ranging from 5600m² required per person in developed countries, to 2000m² in undeveloped countries. The difference is due to the higher meat content in the human diet in developed countries and the associated additional area needed for animal husbandry.

Malthus, the 18th century cleric whose name has become attached to his pessimistic views, noted that net family size had been increasing since about 1750. He predicted that population increase, if continued, would exceed 'the power in the earth to produce subsistence for man'. Malthus' predictions did not immediately come to pass. There were savage famines such as the 1840s' potato famine in Ireland but, overall, agricultural production increased and famine did not prevent population growth in the 19th century. Marx in 1840 observed that famine resulted more from unequal access to food than from failure to produce it, a situation that still applies.

Similar concerns have been expressed at intervals subsequently, for example, the Club of Rome in 'Limits to Growth' in the 1970s. Nevertheless, in the period between 1950 and 1984, while the world population increased at slightly less than 2%, global food production increased on average by 3% per annum as a result of the agricultural 'Green revolution'.

History is now repeating itself in relation to present day predictions of population and food production. The Worldwatch Institute, Washington, has predicted that the earth is close to its carrying capacity and has forecast that the rate of food production increase would flatten off (Brown & Kane 1994). Against this view, agricultural optimists point out how yields could be increased by a second 'green revolution', relying on further improvements from plant breeders and greater use of fertilizers, at the same time maintaining or improving present levels of control of weeds, pests and diseases.

There is no doubt that higher yielding varieties are 'in the pipeline'. Similarly, over much of the tropical and subtropical world, fertilizer use is now substantially less per hectare than in developed temperate countries. The science and practice of pest control is becoming continually more precise and selective. Yields could undoubtedly be increased, everything else being favourable. There are nevertheless, serious doubts as to whether these prospects can be achieved (MacKenzie 1994b). Some barriers are social and political; others, *eg* the probable development of resistant strains of insects, weeds and fungi, also cannot be ignored.

Members of the Consultative Group on International Agricultural Research have been responsible for introduction of many new high-yielding food crop varieties in tropical countries over the last 20 years. They expect comparable developments in breeding and biotechnology over the coming 20 years to be essential components of any continuance of the agricultural 'green revolution'.

In 1996, the Group's chairman expressed concern about biotechnology patents held by commercial companies in developed countries. He feared that genetically engineered seeds and plants could be too expensive for 80% of the peoples in developing countries, thereby preventing them from benefitting higher yielding varieties (Pearce 1996a). A analogous concern was expressed at a conference to develop the protocol for the Convention on Biological Diversity. Poor and underdeveloped countries fear that they will be used as test beds for genetic engineering; they also fear that multi-national companies will seek to evade liability for any ill-effects arising (Coghlan 1996d).

International support has been sought for endorsement of the principle of completely open access to the benefits of genetic engineering in food crops, in order to ensure maximum accessibility to all growers. Commercial biotechnology industries have resisted such moves.

1.33   Other factors influencing food supply - land

The importance of soil as a primary resource to sustain all mankind's activities is grossly under-rated. The lack of concern is of long standing; Satchel (1989) notes evidence for soil erosion in neolithic times in Cumbria. Perlin (1989) describes evidence for soil erosion following clearance of forest in Bronze Age times.

The cycle of land clearance, weakening of vegetation cover, soil impoverishment, physical loss, abandonment and slow revegetation, has been repeated with scarcely any interruption on a world-wide scale from those times to the present. Recent accounts of the process include a contemporary report of soil degradation by human activity in the drier lands of central and southern Europe (Thomas & Middleton 1994), while de Silva (1996) describes current concern in Australia. Fu (1989) noted continuing erosion in China.

In the UK, a Royal Commission on the Environmental Pollution Report (RCEP 1996) emphasised the importance in the UK of keeping good agricultural land in agriculture, observing that built-up areas and roads already cover one eighth of the surface area of England. The Commission urged that development should be directed to former industrial (brownfield) sites, tackling whatever residual pollution problems may be there. Allowing development on 'greenfield sites' on good land when other land is available, is viewed as an abuse of the land resource.

1.34   Other factors influencing food supply - water

Irrigation is estimated to have contributed as much to the higher yields of the 'green revolution' as increased fertilizer usage. It is claimed that most of the opportunities for increasing yields through additional irrigation have already been taken, and that 54% of presently accessible fresh water is currently being used by man. The practicable maximum is 70% and could be required by 2025. However, to harness such additional resources will involve either long distance transfer of water or construction of dams in difficult terrain. Such proposals are likely to controversial and may be opposed on environmental grounds (Mackenzie 1994b).

A study (MEDALUS) on behalf of the European Commission shows that in Spain, for example, farming currently consumes 60% of the water supply. In places near the coast, water has been drawn from underground aquifers more quickly than it has been replaced by fresh-water percolation from higher ground

(Spark 1996). At the coast, the direction of water flow has reversed; as a consequence, sea-water has entered wells previously supplying fresh-water for drinking, a situation repeated in other populated semi-arid coastal regions.

1.4   GLOBAL LAND USE CHANGE

As part of the storehouse of accumulated energy and fibre over large areas of the globe, trees in the natural forest have been the principal source of fuel and building material for human society for several millennia. Perlin (1989) outlines the often devastating exploitation of forests from Bronze Age times to the nineteenth century, emphasising the roles of wood both as a vital necessity for building and boat construction, and, as charcoal, the energy source for metal smelting. Perlin quotes widely from the earliest records from Mesopotamia and the eastern Mediterranean. He describes how the need to secure wood and charcoal as strategic and commercial resources significantly influenced the course of history. He also gives early examples of felling for timber being followed by heavy grazing so that the woodland did not regenerate.

The advent of iron and steel production based on coking coal, displaced wood as the key structural material for ship and other construction. Demand for charcoal also dropped. However, demand for land for food production and commodity crops has continued. Currently, 85% of land cleared of forest, is taken over for agricultural cropping.

Tree felling and forest clearance have frequently engendered social conflict, because few of the benefits of intervention have accrued to local forest peoples. Westoby (1987) emphasises the significance of wealth distribution between different levels of society, and the function of colonial societies in treating their colonies as sources of foodstuffs and timber to support the colonising power at the expense of the indigenous population. More locally, Hart (1995), in an account of the history of the Forest of Dean in south-west England, describes the power and extent of activity of 17th century iron-masters seeking fuel for their forges, and the three-way conflicts that arose from time to time between industrialists, local people and the Crown.

Loss of forest area continues. *Tables 1.2* and *1.3* show estimates of changes in world land use since 1980. More recent reports, *eg* Cohen (1996), Abramovitz (1998), indicate no reduction in the net global rate of deforestation, fewer fellings in temperate zones being offset by more in tropical regions.

If change of land use from forest to arable cropping continues at the rate shown in the tables, by 2025 the area of forest land will have fallen by 22 million hectares. This is approximately ten times the total productive woodland area of the UK.

Given that food production has to be increased, the choice will fall between continuing land clearance and more intensive crop production. If political,

Table 1.2   *World land use change*   *(Million ha)*

| | *Arable land & permanent crop land area* | | | *Forest area* |
	1975	1990	*Rate of annual change*	*Rate of annual change: 1975-90*
Developed countries	668	672	+0.03%	0.07%
Developing countries				
Africa	139	151	+0.56%	-0.37%
Latin America	126	151	+1.24%	-0.57%
Near East	84	85	+0.06%	-0.31%
Far East	375	383	+0.14%	-0.54%
Sub-total, devel. c'tries	726	772	+0.42%	-0.48%
World	1,394	1,440	+0.24%	-0.23%

Source   FAO 1992

Table 1.3   *Estimates of rate of global deforestation*   *(Million ha/annum)*

Year of deforestation	*1979*	*1976-80*	*1989*	*1981-90*	*late 1980s*
Closed canopy forest only	7.3	7.3	13.9	14.0	16.5
Closed & open canopy forest	-	11.3	-	17.0	20.5

Source   FAO 1988, 1989

commercial and technological conflicts can be resolved, increased crop yields could be achieved by:

- 30% improved varieties;   • 30% improved plant nutrition;
- the balance by improved cultural operations including pest control.

1.5   GLOBAL SUPPLY AND DEMAND FOR WOOD AND WOOD PRODUCTS

Global figures for wood supply and demand are summarised in *Table 1.4*. They illustrate the scale of wood production and the changes apparent in global annual production of wood raw materials and products over the last 30 years.

Between 1960/2 and 1985/7 wood raw material production rose by about 25%. About 70% of all wood for industry was harvested as sawlogs or veneer-logs. About 70% of logs harvested in 1985/7 were coniferous (Arnold 1991).

Table 1.4   *Global annual production of roundwood materials*   *(Million cu m)*

Wood raw material (roundwood)	1960-62 actual	1975 forecast	1985-87 actual
Sawlogs and veneer-logs	629	815	957
Pulping roundwood	226	493	395
Other industrial roundwood	188	185	(n/a)
Total	1043	1493	1452+other
Fuelwood	1088	1199	(n/a)
Source	- - - FAO 1967 - - -		FAO 1989

Table 1.5 summarises production of sawnwood products for the same period. There has been a significant change in product assortment over the 25 years from 1960/2 to 1985/7. While production of sawnwood has risen 40%, pulp products have doubled and panel products quadrupled.

Table 1.5   *Global annual production of wood products*   *(Million cu m)*

Wood raw material	1960-62 actual	1975 forecast	1985-87 actual	1986-2000 % increase
Sawnwood	346	427	480	+23%
Panel products	31	76	117	+70%
Pulp products	78*	162*	140**	+37%
Roundwood	188	185	(n/a)	(n/a)
Source of data	- - FAO 1967 - -		FAO 1989	FAO 1988
*million tons	**wood pulp only	(n/a)=no data given in quoted sources		

Forecasts of roundwood demands are precarious because of the cyclical nature of markets. In the short term, extrapolation of historic trends provides some guide. In the longer term, demand is dependent on economic activity, prosperity and technological changes in production.

Mould-shattering predictions are made from time to time, forecasting the obsolescence of wood as a result of substitution or synthesis. Nevertheless, because of its versatility and the relatively low amount of energy required to saw into planks, natural wood is likely to remain a dominant material in construction well into the 21st century. Panel products and wood fibre using wood wastes and smaller sizes of timber, have found increasing markets since they were first produced.

Importance of fuelwood

Fuelwood is an essential domestic requirement of many energy-poor countries; the quantity of fuelwood consumed approaches half of the global production of roundwood. See *Table 1.4*. While most economic forecasts have focused on sawlog and pulp production because of their significance for international commerce, fuelwood seldom enters international trade and is therefore under-recorded. Branchwood for animal fodder is even less noted. Nevertheless, both need to be recognised as important commodities which are essential to the well-being of local rural peoples and a significant element in global forest management.

1.51   Timber supply and demand in Britain

Great Britain, because of its high density of population and low woodland cover, has for centuries been dependent on imports of sawnwood and high quality logs. In the 20th century, this dependence has extended to wood-based pulp and paper products.

During the 1914-18 war, the flow of imports was disrupted. The over-riding need during those years was to draw on British timber reserves, regardless of longer-term plans. This vulnerability stimulated formation of the Forestry Commission in 1919 (Acland 1917) and its reinvigoration in 1943 with the objective of restocking wartime emergency fellings and increasing the area of productive woodland (FCms 1943). The Forestry Commission, since its creation, achieved a government-supported programme to restore in some measure, the area of forests cleared in preceding centuries (Aldhous 1997).

The following tables outline the main features of British forestry.

*   *Table 1.6* shows the productive area of forest as at 31.3.97 (FICGB 1998);
*   *Table 1.7* shows the volume of standing timber at the time of the 1980 woodland census (Locke 1987);
*   *Tables 1.8* and *1.9* give forecasts of UK annual roundwood and sawn wood production and demand for the period 1989 to 2050 (Whiteman 1991).

For Great Britain as a whole, conifers form about 70% of the productive woodland area (*Table 1.6*). This is not equally distributed between countries; conifers form over 90% of the area of Scottish productive woodland, about 70% in Wales but only 43% of the woodland area in England.

Because broadleaved woodlands in Great Britain have a higher than normal proportion of older trees, standing volumes of timber are not directly proportional to crop types. *Table 1.7* shows that the volumes for conifers and broadleaves forming the greatest part of the United Kingdom's timber reserves in 1980 were more evenly balanced than might be expected from the relative conifer and broadleaved woodland areas.

Table 1.6   *Woodland area   Great Britain as at 31.3.98*
              Total of Forestry Commission and private woodlands   *(000 ha)*

| Country | High forest | | Coppice | Total productive | Other |
	Conifers	Broadleaves		woodland	woodland
England	383	483	19	885	105
Wales	167	67	1	235	13
Scotland	989	120	0	1109	93
Great Britain	1539	670	20	2440	211

Source	Forest Industry Handbook (FICGB 1998).

The English landscape includes many older broadleaved woodlands originally planted in order to produce timber for markets now lost by substitution. The area of coppice now surviving is very small. It is a relic of former important underwood industries producing small roundwood for fuel, charcoal, tan-bark, turnery and other industrial, farming and domestic uses, markets that were recognised as obsolescent in the 1930s and earlier (Troup 1952). Oakwoods planted in the 19th century were in many areas originally intended to provide timber for ship construction. This reserve of standing timber is the predominant source of oak sawlogs in current production.

The figures in *Table 1.7* for timber volumes do not show volumes of elm lost as a result of Dutch elm disease which swept devastatingly through southern and central England in the early 1970s, killing over 20 million trees (Phillips & Burdekin 1982). See §7.31 and §6.54.

Table 1.7   *Standing volumes of selected species   Great Britain, 1980*
              (Million cubic metres over bark)

Species	Standing volumes	Species	Standing volumes
Scots pine	27.7	Oak	32.9
Corsican pine	6.0	Beech	15.2
Sitka spruce	28.2	Sycamore	8.0
Norway spruce	12.9	Ash	10.2
European larch	5.9	Birch	6.1
Jap/hybrid larch	11.6		
Douglas fir	6.1		
All conifers	106.3	All broadleaves	91.1

Source	Census of Woodlands (Locke 1987)

Timber production forecasts in the UK (*Table 1.8*) are dominated by conifers; because of their higher average yield and rapid early growth, rotation lengths maximising financial return seldom exceed 60 years.

For the period up to 2025, overall production of conifer sawlogs and small roundwood is forecast to increase. This forecast is reasonably reliable because the crops that will be harvested are already planted and growing. However, production of small roundwood and sawlogs will fall during the period 2025-2050 unless there is a surge of conifer planting in the next 10 years, or forests are managed so as to defer production from 2010-2025 to later decades.

Table 1.8   *UK forecast of annual production of roundwood*   *(Million cu m)*

Wood raw material	*1997-2001*	*2007-2012*	*2022-2025*	*2047-2050*
Conifer sawlogs	4.8	8.1	11.5	6.6
Conifer small roundwood	4.8	6.5	6.9	4.6
Broadleaf sawlogs	1.0	1.0	1.1	1.1
Broadleaf small roundwood	0.1	0.1	0.1	0.1

Source   Whiteman 1991

Table 1.9 shows that wood consumption forecasts predict increased consumption of most wood and wood products over the next 50 years.

Everyday experience is that demands are more susceptible to market changes, both in the long and short term. Some figures shown in *Table 1.9* are median values of alternative estimates, based on higher or lower consumption assumptions. Nevertheless, both high and low consumption forecasts showed a rise in consumption over the next 50 years, the differences being the size of that increase.

Table 1.9   *UK forecast of annual demand for wood and wood products*
                    (Million cu m, wood raw material equivalent)

Wood product	*1989* *actual*	*2000*	*2010*	*2025*	*2050*
				forecasts	
Sawnwood	18.8	16.8	17.0	17.2	17.7
Pulp products	20.8	28.4	32.9	40.5	49.2
Panel products	6.6	8.3	9.6	11.7	12.9
Other roundwood	0.4	0.5	0.4	0.4	0.3

Source   Whiteman 1991

Comparison of *Tables 1.8* and *1.9* show that wood production from UK woodlands is unlikely, during the next 50 years, to meet more than 20-25% of UK requirements. The UK is likely to remain dependent on imports of wood products of all kinds.

The Forestry Industry Handbook 1998 (FICGB 1998) gives fuller details of patterns of import and export for the period 1993-7. It also illustrates that exports from countries within the European Community provide over half the GB requirements for:

- coniferous sawnwood,
- fibreboard,
- particleboard,
- paper and paperboard.

Ethics of further increase of woodland cover in Great Britain

It has been argued that further planting in Britain is extravagant in resources. Nevertheless, from the global trends outlined above, there is a strong ethical case to ensure that wherever land is managed, it is managed to produce its best economically and ecologically sustainable yield. Heavily populated, developed countries such as the United Kingdom have no less a responsibility to contribute to the world supply of sustainable renewable resources than countries where forest is currently being removed on a large scale.

If there were not a prospect of great pressure on agricultural husbandry to produce increasing quantities of food and renewable raw materials, the pressure to clear forest for food and plant products would be less; similarly the pressure on foresters to maximise timber yield would be less. Reducing or even static demand for food and fibre is not a realistic prospect however.

Woodlands offer recreation, scenic beauty, wildlife conservation and a respite from the pressures of day to day living, uses which are widely appreciated but economically undervalued. These benefits, integrated around good quality wood raw material production, offer the best prospect of maintaining the whole portfolio of woodland, not just in the UK but worldwide.

In meeting such multiple objectives, the woodland industry has to bring together the most appropriate technological aids consistent with the objective of sustainable multi-purpose woodland - in contemporary terms, integrated and holistic management.

CHAPTER 2

The Chemical Framework

'One man's meat; another man's poison'

2.1   PESTICIDES, FERTILIZERS AND POLLUTANTS

'A pesticide is any substance ... or organism ... used to protect plants ... from harmful organisms' (MAFF Pst 1999).   This chapter outlines the broader contexts relevant to forestry arising from the use of pesticides - principally insecticides, herbicides and fungicides; these are discussed in greater detail in *Part II (Chapters 6-9)*.

The background to fertilizers as used in forests and in tree nurseries *etc.* are described in *Part III (Chapters 10 - 12)*, while pollutants and other chemicals to which trees and woodlands may be exposed during their life are covered in *Part IV (Chapters 13-16)*.

2.11   The agricultural 'Green revolution' and the role of pesticides

Agricultural production over the last 40 years has undergone a 'green revolution' with dramatic increases in yields of food crops world-wide.   The increases in yield have been achieved through the combined effects of:

- improved strains of crop species through plant breeding programmes;
- improved plant nutrition through added fertilizers;
- increased availability of water through irrigation and
- a reduced proportion of crop losses, through use of a wider range and larger amounts of pesticides.

The contribution to this revolution from pesticides has been due to rapid increases in the number, type and scale of their use.   Over the same period, the benefit due to fertilizers is through quantitative increases in use rather than change of material (see Chapters 10 and 11).

Table 2.1 shows potential and actual crop losses due to pathogens, animal pests and weeds, and the effectiveness of agricultural crop protection in different parts of the world.   Losses due to pests remain substantial.   While higher yields have been obtained in some regions because of fertilizer and improved crop varieties, without pest control, these would have been far less.   Potential losses everywhere represent over half of crops' potential yields. Actual losses are least where pesticide technology is most widely used.

The driving forces behind the continuing search for yield increases, comparable to those of the last 30 years, are the anticipated long-term demand for more food, and the commercial opportunity for crop yield improvement. Pesticides are seen as the means for protecting hard-won gains.

Table 2.1   *Crop losses due to pathogens, animal pests & weeds, & the effectiveness of crop protection in major farm crops in West Europe, North America + Oceania, and the rest of the world.*

Region	Loss scenario	Crop losses (%)				
		Pathogens	Animal pests	Weeds	Overall	Reduction in losses
Western Europe	*potential*	*18.7*	*17.3*	*21.4*	*57.4*	
	actual	7.3	8.9	6.4	22.6	*61%*
N. America & Oceania	*potential*	*13.0*	*15.5*	*27.8*	*56.2*	
	actual	9.9	10.2	11.5	31.6	*44%*
Other regions	*potential*	*18.2*	*24.0*	*30.4*	*72.6*	
	actual	14.3	17.1	14.0	44.4	*39%*

Source   Örke *et al.* 1994

2.2   PESTICIDE AVAILABILITY AND USE

2.21   Before pesticides

Before the advent of pesticides, effective control depended on the responsiveness of pests to man's direct intervention. Manual weed control was (and still is) laborious, but at least possible; larger animals, birds and mice could be deterred by barriers and scarers. Protection against a few insects could be achieved by trapping but there were no non-chemical cures for plagues of insects such as locusts or defoliating caterpillars. Similarly, there were very few means of reducing attack by fungi. Early uses of natural products as pesticides, including fumigation by sulphur, are recorded by classical writers. While small numbers of materials were used for control of insect and animal pests in the 19th century and the early part of the 20th century, only since the expansion of the bio-chemical industry from the late 1930s have pesticides played a major role in agricultural husbandry. Forestry was fifteen or so years behind agriculture in that respect.

2.22   Number of pesticides available

In the early 1940s, there were no more than about a score of agricultural pesticides on the market, mostly insecticides. However, several discoveries of biologically potent materials were being developed and began to become available from about 1943. Thereafter, there has been a continuing flow of new pesticides coming onto the market every year. This flow reached a peak in the 1980s and has subsequently diminished.

The most authoritative UK reference to world-wide availability of pesticides, *The Pesticide Manual*, lists 759 pesticides and biological agents believed to be in commercial production in 1996. A further 583 are listed as no longer manufactured or marketed for pest control (Tomlin 1997).

Table 2.2 shows, for the UK in 1978 and 1999, the numbers of active substances* available for use in land husbandry.

Table 2.2   *Pesticide active substances (a s's), single or in mixture, in products approved for use in the United Kingdom in 1978 & 1999.*

Year	Type of pesticide	Number of single approved a s's	Number of mixtures of approved a s's.	Total
1978	All	195 *(71%)*	78 *(29%)*	273
1999	All	353 *(52%)*	323 *(48%)*	676

Source	MAFF ACAS 1978, MAAF Pst 1999

All the pesticide active substances included in the table as single as's or in mixtures were at the time marketed in products conforming to the United Kingdom legal requirements. These are described in Chapter 3.

For any active substance, there may be more than one product available, depending partly on manufacturer's patent position, and partly on whether there are distinct markets for different formulations. In 1978, there were nearly 750 approved products based on the 273 active substances or mixtures. By 1999, there were over 3250 approved products for 676 active substances or mixtures, *ie* about 4½ times more products from 2½ times more as's or mixtures.

Single active substance products and mixtures
The relative availability of formulations containing a single or a mixture of active substances has changed between 1978 and 1999 The number of single active substance formulations has increased by about 50%, while the number of available mixtures has more than doubled.

Table 2.3 gives fuller details of availability of types of pesticide by single substance and mixture in 1999.

*   Until the introduction of EC Directive 91/414/EC, pesticide literature in the UK referred to 'active ingredients'. The EC Directive has substituted 'active substance'; that term being used in *Plant Protection Products Regulations 1995* (SI 1995/887) and some subsequent literature. The terms are, for practical purposes, synonymous.

Herbicides are the most numerous of the available pesticides.   Their total approaches the number of insecticide and fungicide products combined; this probably reflects the wide range of weed species and the breadth of the habitats where they occur.

There are appreciably more mixtures of herbicides and mixtures of fungicides than single active substance herbicides or fungicides.   For insecticides, the opposite applies.   For herbicides, mixtures are required where there is a wider range of weed species than can be controlled by one herbicide.   Insects tend to occur as single-species infestations where one insecticide gives adequate control.

For any one active substance or mixture, there may be any number of approved products.   See *Pesticides 1999* (MAFF Pst 1999).

Table 2.3   *Pesticides with single active substance (a s) or mixture of active substances in products approved for use in the UK in 1999.*

Type of pesticide	Number with single active substance	Number with mixture of active substances	Total	Mixtures as % of total by type
Fungicides	85	114	199	57
Herbicides	118	164	282	58
Insecticides	78	18	96	19
Vertebrate control	18	2	20	10
Other	17	5	22	22
Amateur*	68	64	133	45

*   *Many pesticides approved for amateur use contain materials already approved for professional use.   They are available in smaller quantities.   This category was not included in 1978 MAFF booklet.*

Source   MAFF Pst 1999

2.23   Government controls on pesticides

Governments in many parts of the world have introduced safety requirements in response to hazards perceived as arising from chemical manufacturers' development and market production programmes.

Initially, these were focused on protection of consumers of treated crops and the health of the people applying the pesticides.   Increasingly, however, additional requirements have been introduced covering pollinating insects, wildlife, water quality *etc.*, as evidence became available of the ramifications of the effects of pesticides throughout the environment.   Recent advice under the Government approved *Code of Practice for the Safe Use of Pesticides on Farms and Holdings* (MAFF/HSC 1998) requires users to consider, for the crop being grown:

- whether pest control operations can properly be integrated with good cultural practice to minimise pesticide use;
- what are the minimum quantities of pesticide that will achieve the control required; and,
- whether a less hazardous available product would have the same effect.

Legal requirements in the UK for safe use of pesticides and other hazardous substances are described in Chapter 3. Their safe implementation is reviewed in Chapter 4.

2.3 PESTICIDE NOMENCLATURE

There are four main 'types' of name by which a pesticide may be identified:

- full chemical name of the active substance;.
- common name of the active substance;
- development names or numbers;
- proprietary brand names of individual pesticide products available on the market.

Products may also be described in terms of their 'formulation' and may also be recommended for use in conjunction with 'adjuvants' (See §2.42).

2.31 Full chemical name of the active substance

As far as possible, names of established active substances conform to internationally agreed standards.

There are two standard systems in use, and a third which is applied to part of the first group of chemicals.

'IUPAC names' follow the rules agreed by the *International Union of Pure and Applied Chemistry* for systematic names for chemical compounds.

In addition, for pyrethroids, a system devised by chemists at Rothamsted Experimental Station (Hertfordshire, UK) is recognised as an alternative standard.

'Chemical Abstracts names' are the systematic names according to the rules of the 9th Collective Index period of the Chemical Abstracts service.

These names all describe the structure of the compound named. The differences between the names are mostly in the sequence of the component parts and the use of hyphens and brackets.

For the more complex molecules, names are lengthy. Where, in addition, there are stereo-isomers, names may occupy several lines of text. For example, flowers of *Pyrethrum cinerariaefolium* have been long known for their insecticidal properties. Extracts from the flowers contain six insecticidal constituents, the esters of natural stereo-isomers of chrysanthemic acid (pyrethrin I, cinerin I and jasmolin I) and the corresponding esters of pyrethric acid (pyrethrin II, cinerin II and jasmolin II).

The IUPAC-recognised Rothamsted name for pyrethrin I is :
   (Z)-(S)-2-methyl-4-oxo-3-(penta-2,4-dienyl)cyclopent-2-enyl(1R)-=*trans*-2,2
-di-methyl-3-(2-methylprop-1-enyl)cyclopropane=carboxylate.
   Names for the other pyrethrum compounds are of similar length. The full
chemical names of other current pesticides are often shorter, but seldom less
cumbersome or more easily memorised.
   Full chemical names of all currently used pesticides are given by Tomlin
(1997); they are not otherwise used in this text.

2.32   Common names of active substances

   As a preliminary to marketing a new active substance, manufacturers propose
a 'common name' for their compound. This often, but not always, has some
affinities with previously accepted common names of related compounds. The
aim is to get the name accepted by all the main standards organisations.
   The *British Standards Institute (BSI)* has played an important part in
standardising names but, more recently, has sought to bring its names into line
with the *International Standards Organisation (ISO)*. The latter organisation
runs two lists, one with English spellings, and one with French spellings and
accents. They are referred to as E-ISO and F-ISO lists of common names.
   Other regional standard lists of pesticide common names have been
developed by:

   • American National Standards Institute   (ANSI);
   • British Approved Name (by the British Pharmacopoeia Commission)
                                                    (BAN);
   • Entomological Society of America   (ESA);
   • Japanese Ministry of Agriculture and Forestry (JMAF);
   • Weed Science Society of America.   (WSSA).

France has its own list of common names.
   For the extract of pyrethrins from the flowers of *Pyrethrum* spp, the common
name 'pythrethrins' has been accepted for BSI, E-ISO, ESA and JMAF listings.
The common name 'pyrèthres' has been accepted for F-ISO listing.
   International common name usage is summarised in *The pesticide manual*
(Tomlin (1997) for each currently available active substance listed therein.

   Common names of active substances in use in the UK
   All pesticide products currently marketed in the United Kingdom are
required by law to include the common name of the active substance on the
product label.
   The MAFF/HSE Reference Book 500 *Pesticides, 19xx* is published annually
and contains the names of products approved at the beginning of that year.

   • Part A of the book lists products registered with the Pesticides Safety
   Directorate of the Ministry of Agriculture, Fisheries and Food (PSD) for use
   in agriculture, horticulture, forestry *etc*. Products are listed under

'Herbicides', *'Fungicides'*, *'Insecticides'*, *'Vertebrate control products'*, *'Biological pesticides'* and *'Miscellaneous'*.

• Part B lists products registered with the Health and Safety Executive as wood preservatives, surface biocides, insecticides for food storage practice and animal husbandry, and anti-fouling products for aquaculture, yacht and deep sea structures. Approved pesticide products are grouped under *'Wood preservatives'*, *'Surface biocides'*, *'Insecticides'* and *'Antifouling products'*.

• Products available for amateur use are so marked, or are listed separately.

For each of these main pesticide groupings, products are grouped together under the common name of the active substance(s) in the product. The names used are, wherever possible, taken from the relevant E-ISO or BSI list.

In this text, names of products and active substances are as given in the 1999 edition of the MAFF/HSE *Reference Book 500 'Pesticides'*, except where an older name is used in the title of a paper or reference book.

2.33 Development names or numbers

During the later stages of development and in particular for field trials, manufacturers usually allocate a number/letter code to the product for reference purposes. These may appear in publications of results of such trials; however, they are superseded by a product name as soon as this has been approved.

2.34 Proprietary brand names of individual pesticide products

Product names are assigned by manufacturers on purely marketing considerations, to assist sales. The name, as such, rarely indicates to the prospective purchaser which active substance the product contains. However, many manufacturers include their name as part of the product name.

While a pesticide active substance is under patent protection, there are usually only a few products on the market, all under the control of the patent holder. Once the patent has expired, or if there was no original patent, the number of products often increases substantially. As an example of the latter, in *Pesticides 1999*, there were 176 products containing 'pyrethrins' listed either under 'Insecticides' and 'Amateur products' in the MAFF/PSD section, or under 'Insecticides' in the HSE section.

The following is a selection of proprietary names for pyrethrin products from these lists:

Alfadex	Fortefog	PBI Anti-ant Duster
Bug Gun!	Keri Insect spray	Doff Greenfly killer
Aquapy	Dairy Fly Spray	Coopers Fly Spray N
Drione	Py Powder	Aquablast bug spray
Konk I	Pif Paf Fly Spray	Detia Pyrethrum Spray
Prevent	Home-base pest gun	Pyrematic Flying Insect Killer

Residex   B&Q Rose and Flower Insecticide Spray
Levington Natural Houseplant Insect Spray

In contrast, the active substance 'propaquizafop', first marketed in the late 1980s and within the life of initial patents, in 1999 had only six named products, three of which were approved for use in farm forestry (MAFF Pst 1999, Whitehead 1999).

Not only do the commercial names given by manufacturers to products offer little help to potential users; there are occasional examples of similar sounding names relating to markedly dissimilar products, *eg*:

Product name   Weedex   *Active substance*   simazine
                 Weedol   paraquat/diquat
                 Weedazol TL   amitrole

Potential users of pesticides must be well informed about the pest to be controlled and the alternative active substances that may be suitable. Only then should they start looking for products and should refer to the current MAFF/HSE *Pesticides* booklet or *The UK Pesticide Guide* for lists of approved products containing the active substance(s) they wish to use.

2.35   Groupings of pesticide active substances

From the outset, pesticide active substances have been grouped by chemical features in common, according to their known or assumed biochemical role. The choice of words for group or class names is somewhat arbitrary and the groupings are neither immutable nor infallible. With the increase in number of pesticides has come an increase in the number and size of classes into which the active substance may be grouped. The inclusion of biological agents has further expanded the classes. Altogether, 135 classes are listed in the 11th edition of the *Pesticide Manual* (Tomlin 1997).

Table 2.4 shows 32 of such classes consisting of 2 or more approved active substances. All the active substances shown in the table are constituents of approved products that may be used for one or other forestry requirement.

In addition, some pesticides used in forestry are in 'single compound' classes; *ie* they have no closely related compounds with which they can be grouped. These include asulam, chlorthiamid, dichlobenil and glyphosate.

Careful use of names

Pesticides are frequently referred to in the popular press and elsewhere by class groupings such as those in *Table 2.4*, and often in uncomplimentary terms.

Potential critics of pesticide use must recognise that the subject is complex and that the complexity is increasing rather than lessening. It is important that valid criticisms can be voiced and responded to. For criticisms to carry weight, however, they have to be related to materials and circumstances which can be objectively recognised and investigated. While classifications such as those in *Table 2.4* may provide a start in such situations, it is generally not sufficient to use a 'class' name as the basis for specific criticism. Site-, usage- and product-specific details are usually essential to be able to get to the core of a problem.

Table 2.4   *Examples of 'Classes of pesticides' and pesticides within them*

Pesticide class name	Approved* pesticides in class	Total in class	Type of pesticide
Alkylenebis(dithiocarbamate)			
	mancozeb, maneb, zineb	7	*fungicide*
Amide	isoxaben, propyzamide	3	*herbicide*
Aryloxyalkanamide	napropamide	2	*herbicide*
Aryloxyalkanoic acid	2,4-D, MCPA, mecoprop, triclopyr	15	*herbicide*
2-(4-aryloxyphenoxy)propionic acid			
	fluazipop-P-butyl, propaquizafop	10	*herbicide*
Azole	(imazalil, penconazole, prochloraz,		
	(propiconazole, triadimefon	30	*fungicide*
Benzimidazole	benomyl, carbendazim, thiabendazole	5	*fungicide*
Benzoic acid	chlorthal, dicamba	4	*herbicide*
Benzoylurea	diflubenzuron	10	*insecticide*
Bipyridylium	diquat, paraquat	2	*herbicide*
Carbamate	(chlorpropham	(	*herbicide*
	(carbosulfan, pirimicarb	(21	*insecticide*
Chloroacetanilide	metazachlor, propachlor	10	*herbicide*
Cyclohexanedione oxime	cycloxydim	5	*herbicide*
Dimethyldithiocarbamate	thiram, ziram	3	*fngcd/rplt*
2,6-dinitroanaline	pendimethalin	11	*herbicide*
Imidazolinone	imazapyr	6	*herbicide*
Methyl isocyanate precursor	dazomet, metam-sodium	2	*fumigant*
Organochlorine	dicofol, lindane	5	*insecticide*
Organophosphorus	(chlorpyrifos, diazinon, dimethoate,	)	
	(fenitrothion, malathion, trichlorfon)	71	*insecticide*
Pyrethroid	(cypermethin, fenpropathrin,		
	(deltamethrin, permethrin	37	*insecticide*
Pyridinecarboxylic acid	clopyralid, picloram	2	*herbicide*
1,3,5-triazines	(atrazine, cyanazine, simazine,		
	(terbuthylazine	14	*herbicide*
1,2,4-triazinone	metamitron	3	*herbicide*
Triazole	amitrole	1	*herbicide*
N-trihalomethylthio	captan	5	*fungicide*
Uracil	lenacil	3	*herbicide*
Urea	diuron	17	*herbicide*

*Only active substances commercially available in 1999 are included in this list.
Source   Tomlin, 1997 (Index 6)   MAFF Pst 1999

2.4   FORMULATED PRODUCTS

Under the various government schemes for approval of proprietary brands of agricultural chemicals, products have been listed under active substances. Nevertheless, at all times, it has been the product which has had approval rather than the active substance. This allows any effect of components other than the active substance which may influence the toxicity of the product to be taken into account.

From time to time when, for any reason, bans or restrictions of use are necessary, the mechanism for implementation has been amendment or withdrawal of approval of products containing the proscribed material. In practice, some flexibility has allowed specific minor uses of restricted active substances to continue for several years while alternative means of control of difficult pests are developed. Also, in most circumstances, existing stocks are allowed to be used up but not replaced.

2.41   Formulations

Almost all pesticide products sold currently include more than just the active substance which is toxic to the pest to be controlled.

Additional materials in formulations enable the active substance to be applied at the rate and in the appropriate physical condition to have maximum effect on the pest and least effect on the crop and the environment. Materials may include:

- *solvents* if the active substance is a solid at normal temperatures and is to be applied as liquid;
- *emulsifiers* if the active substance is not water-soluble and is to be applied diluted in water;
- *stabilisers* to ensure products remain in good condition and the components do not separate out while in store before use;
- *inert solid fillers/carriers* where the material is to be applied as a granule or wettable powder;
- *adhesives* to provide coherence for granules between the time of manufacture and application;
- *dispersing agent* to induce the active substance particles to spread in suspension;
- *anti-settling agents* to prevent fine particles of suspension concentrates settling;
- *encapsulating materials* where the active substance is intended for slow release after application;
- *adjuvants* to enhance the efficacy of the active substance.

The most widely used formulations are prepared for dilution with water, *ie*:

- soluble concentrates,
- emulsifiable concentrates,
- suspension concentrates,

- wettable powders,
- water-dispersible granules.

Pesticides may alternatively be applied as supplied, as dusts or granules.

2.42   Adjuvants

Adjuvants are materials included as part of a formulation or added to the spray tank and mixed with manufacturers' products at the time of application ('tank-mixes'). They are defined as 'substances other than water, without significant pesticidal properties, which enhance or are intended to enhance the effectiveness of a pesticide when added to that pesticide' (MAFF Pst 1996). They include:

- *wetting agents/surfactants* to reduce the surface tension of water diluent, improving spread of deposits of pesticide sprays, penetration of waxy barriers on foliage or skin, and lessening run-off from water-repellent surfaces;
- *adjuvant oils* to assist further the spread and penetration of pesticides;
- *sticking agents* to increase retention of pesticide on the treated surface. They are film-forming materials, however, and may reduce the activity of the pesticide by locking it into the film (Hance & Holly 1990, Makepeace 1996);
- *extenders/spreaders* which may also assist the spread and retention of pesticides.

Adjuvant products fall into two somewhat overlapping groups:

- *Inorganic ba*se

Non-ionic wetters)	reduce surface tension of droplets,
Tallow amine)	increase spread and give better coverage;
Glycol)	
Organo-silicones	super-wetters, rapid drying;
Emulsified mineral oils	mainly stickers.

- *Organic base*

Emulsified rape oil	extenders, reduce spray drift;
Acid lecithins	penetrants, used with dessicants;
Latexes	improve adhesion and retention;
Pinolene	filming agent, assists adhesion.

2.5   MARKETING NEW PESTICIDES

Between 1986 and 1995, no new product containing a pesticide could be marketed in the United Kingdom for use in agriculture, horticulture, forestry *etc.* without approval under the *Control of Pesticides Regulations, 1986* (SI 1986/1510). If a novel pesticide had not previously been marketed, full data had to be submitted to the competent authority (Pesticides Safety Directorate of MAFF, or the Health and Safety Executive).

In 1995, a more elaborate approval system was added under the *Plant Protection Products Regulations 1995* (SI 1995/887). This implemented EC Directive 91/414/EEC (as amended). Under these regulations, pesticide products can only be considered if the active substance is listed in Annex I of the Directive. Commercial companies may apply for a new active substance to be added to the list. Applications for approval of new products incorporating approved active substances have to meet criteria set out in Annexes II and III of the Directive, in respect of efficacy and safety. The evidence sought on safety is designed to expose the extent to which any new material could be hazardous in the context of the whole range of known ill-effects (§§ 4.14, 4.2).

If an extension of a previously permitted use is proposed, data relevant to the extension is required, together with any other new data on safety and side-effects that might affect its overall status.

For products introduced under the PPP Regulations, manufacturers are required to make data as listed in the Regulations readily available, including:

• physico-chemical data on the active substance and the product;
• summaries of results of tests to establish efficacy and harmlessness to humans, animals, plants and the environment.

Availability of evaluations

Any evaluations of approved products held by the *Pesticides Safety Directorate* (PSD) of MAFF are available on request (SI 1997/188, SI 1997/189). At the beginning of 1999, 183 evaluations were available from PSD for products approved under the 'old' system; none was yet available for products under the 'new' approvals completed under the PPP Regulations.

2.51   Costs of pesticide development

The cost and the time required to put a new pesticide on the market has increased substantially. Copping *et al.* (1990) describe in detail the process for a new herbicide extant in the late 1980s; they estimate the success rate at 1 product approved for sale and marketed out of 20 000 materials initially screened; they put the overall cost at £20 million and the time required at 10 years.

Some 20 years earlier, Galley (1968) estimated these costs to be £1 million. Other figures include:

• British Agrochemicals Association in the late 1980s estimated the costs to be £30-40 million;
• Örke *et al.*(1994) put the cost of each successfully introduced new pesticide at 200 million Dm, approximately equivalent to £80 million.

These costs include the costs of the approval process. As knowledge has increased about the indirect effects of pesticides, so additional tests have been added. Consequently, the costs of preparing evidence for approval have risen and the time required to assemble the data has lengthened. Until recently, the cost of securing approval for a new pesticide varied according to individual nations' requirements. Foulkes (1989) puts a 'core' cost at £2.35 million for EC countries, with additional per country costs of £10 000 - £130 000.

The first impact of the implementation of European Community Directive 91/414/EEC is mentioned above. The intention was to harmonise requirements where a pesticide is to be marketed in more than one EC member country and to save cost. First indications are, however, that the cost of the full package will be of the same magnitude as the figures quoted above (Anon 1996a).

Costs for registration or approval have to be put into the context of international expenditure on pesticides. Foulkes (1989) quotes figures for the value of agrochemical markets in the European Community in 1987 as £2665 million. By countries, the values (£million) were:

Belgium	£70	Denmark	£95	Eire	£20
France	£910	Germany	£370	Greece	£70
Holland	£120	Italy	£410	Portugal	£40
Spain	£220	UK	£340.		

These figures are not out of scale with estimates of expenditure world-wide; the British Agrochemicals Association (BAA 1987) estimated that $13 billion worth of pesticides were used annually in the mid 1980s. Dayton (1992) put expenditure by the Australian sheep industry on pesticides at A$300 million in 1991; UN/FAO estimate that $350 million is spent annually, world-wide, on locust control.

Pesticide use in the UK

The *British Agrochemicals Association* (BAA 1998) in their annual review give sales (UK use + export) at about 25 000 tonnes of active substance /yr over the period 1992-1997. The value of the sales rose from £1.16 billion in 1992 to £1.6 billion in 1997, two-thirds of which was accounted for by exports.

Sales in 1997 by pesticide type, by broad activity are given in *Table* 2.5.

Table 2.5　　*Value of pesticides sold in the UK in 1997*　　　　　*(£million)*

Use	Agriculture & horticulture	Industry & forestry	Garden & household	Other	Use totals	% of Gr total
Herbicides	£205.3	£12.1	£16.1	-	£233.5	*47%*
Insecticides	£35.5	£0.7	£5.7	-	£41.9	*8%*
Fungicides	£147.4	£4.3	£2.0	-	£153.7	*31%*
Other	£45.1	-	£3.7	£16.5	£64.9	*13%*
Totals	£433.3	£17.1	£27.5	£16.5	£494.0	

Source　　　BAA 1998a

'Other' in *Table 2.5* includes molluscicides, seed treatments and growth regulators. The table shows how agricultural and horticultural herbicides and fungicides dominate the market, constituting over 70% of the sales by value.

In these statistics, unfortunately, sales to forestry and industry are lumped together. Even so, taken together, they only account for just over 3% of the market.

2.6   FORESTRY'S GREEN REVOLUTION

Up to the 1940s, forestry techniques in the UK had much in common with land husbandry in less developed countries. Growth was dependent on the inherent fertility of the soil; fertilizers were rarely used. Pests and diseases in the forest were endured. Silvicultural techniques, especially timely thinning, were relied on to avoid situations where diseases might build up rapidly on weakened trees. Where newly planted stock could be expected to suffer attack from pine weevil, manual methods of trapping were the norm. Chrystal (1937) advocated natural control by birds, quoting successful moderation of a larch sawfly outbreak in northwest England by bird predation. He reported use overseas of aerial applications of arsenical dusts for control of defoliating forest insects, but not in the UK.

Extensive programmes of work were undertaken in the 1920s - 1940s to find means of establishing forest on upland heathland and moorland where trees had been lost to the combined effects, over centuries, of heavy grazing, repeated burning and encroaching peat. Zehetmayr (1954, 1960) describes how cultivation and, latterly, addition of phosphatic fertilizers, set the foundation for the post-1945 expansion of forestry. However, while heather *(Calluna vulgaris)* was recognised as a major impediment to crop growth, the possibility of control by herbicides was barely touched on.

The one area where UK foresters were at an advantage was in choice of species. Through the long-term interest of many foresters, both private and state, in new introductions discovered by explorers and plant collectors, there was a substantial corpus of observational data on rates of growth of introduced species for comparison with species native to the UK (Macdonald *et al.* 1957).

The national 'Post-war Forest Policy' set targets of replanting wartime fellings and increasing the area of woodland to 2 million hectares (FCms 1943). Under that stimulus, forestry in the UK from 1945 made many technical advances. The parallels with the international agricultural green revolution, while not exact, are similar in their effect of substantially increasing both the plantation area and the yield expectations from it.

Major factors have been:

• species choice, introduced conifers often yielding twice the volume of wood that would have been forthcoming from local native broadleaved and conifer species;

- phosphate applied at planting, enabling trees to be established and to grow where otherwise most would fail. Top-dressings of N and K have also been essential to sustain growth on the most infertile sites. Even so, total use of fertilizers in forests is, by agricultural standards, trifling. See also Chapters 10 & 11.

- the availability of heavy ploughing tractors, enabling cultivation and drainage of heathland, moorland and other sites where the soil had become compacted or was waterlogged. This provided a flow of nutrients to developing plants from decaying vegetation and organic matter, offsetting the need for fertilizers to supplement nutrients available on site;

On the more upland sites particularly, cultivation and drainage also kept the weeds down until plants were sufficiently established to grow away without need for further weed control. The benefit of absence of weed competition in the first three years following planting has become increasingly apparent (*eg* Davies 1987e);

- use of herbicides to supplement weed control achieved by site cultivation - see also Chapter 8;

- maintaining vigilance in respect of insect attack and occurrence of diseases. Routine use of insecticides and fungicides in the forest has been restricted to minimising the spread of conifer heart rot *(Heterobasidion annosum)* and protecting young planting stock against attack by the pine weevil and black pine beetles *(Hylobius abietis* and *Hylastes spp.)* - see §§7.21 and 6.51.

- Provision of healthy planting stock. In forest nurseries, under the stimulus to provide stock for a greatly increased planting programme, improved techniques during the same period halved the time required to produce good quality planting stock (Aldhous 1972a, Aldhous & Mason 1994). Nursery production systems, similarly, have been based on:

absence of weed competition ;   adequate nutrition;
appropriate soil texture and pH;   freedom from pests and diseases;
latterly, availability of irrigation to ensure an adequate water supply.

While these techniques can be itemised on a list, woodland managers and plant producers have long sought to integrate them cost-effectively. See §8.2.

2.61   Industry-specific pesticides for forestry in the UK

Independent development of pesticides specific to forestry has rarely been financially feasible anywhere in the world. This is partly because of the very high cost of testing products to meet requirements for approval, and partly because pesticides developed for agriculture have been able to be used effectively in the forest. Consequently, while there has been a time lag in development of pesticide use in forestry following initial introduction for use in agriculture, this has often been only a year or two. There have nevertheless been two developments pioneered in Britain, both relating to the control of colonisation of stumps by *Heterobasidion annosum*. Both involve treatment of the stump surface immediately after felling; one involved a chemical spray, the other an antagonistic fungus, *Phlebiopsis gigantea* (§7.21).

Table 2.6 sets out the number of herbicide active substances introduced into commercial use in the UK by decade for the period 1943-1989, the number relevant to forestry and the number of these still recommended for use in 1996.

The pesticides currently recommended for forestry include relatively fewer mixtures than are used in agriculture. The table also shows that almost half the herbicide products recommended for forestry during the period 1943-1969 have been superseded, whereas most of those subsequently marketed remain available for use. The table does not include herbicides specifically approved for farm forestry.

Table 2.6   *Numbers of herbicides in commercial use in the UK by decade since 1943, the number relevant to forestry & the number in use in 1996*

Number of new herbicides per decade over the period 1943 - 1989		*1940s*	*1950s*	*1960s*	*1970s*	*1980s*
Introduced into	Single product	6	12	41	31	14
commercial use	Mixtures	-	6	60	70	67
Relevant to	Single product	4	8	12	13	4
forestry	Mixtures	-	1	2	2	4
Relevant and	Single product	3	3	7	12	4
still available	Mixtures	-	-	2	2	4

Sources	Hance & Holly 1990	MAFF Pst 1996

2.62   Scale of international pesticide usage in forestry

Internationally, over the last 30 years, the control of spruce budworm in eastern Canada and adjoining areas in the north-eastern United States has been one of the largest and longest-continuing forest pest control operations. Moderate to severe annual defoliation of fir and spruce in eastern Canada alone has been in the range of 6 - 26 million hectares. While the peak of the outbreak appears to be past, trees died from repeated defoliation over millions of hectares during this period (Nigam, 1990).

Other major infestations overseas include the gypsy moth, introduced from Europe into North America, and Dutch elm disease, also an introduced pest (Carson 1963).

In the UK, the loss of elms from the lowland countryside is the most serious loss through tree pests on record. However, both the scale of the loss of elms and the scale of pesticide usage in forestry in the UK is extremely small in comparison with countries with more extensive forests and extensive pest problems. Williamson (1990a) estimated that 8.8 tonnes of herbicide active substance were used in 1988 by the Forestry Commission. This represented 0.1% of the amount of pesticides used by all industries in the UK.

Figures for usage in UK forestry for 1988 are compared with 1966 data in *Table 2.7.*

Table 2.7   *Pesticide usage by Forestry Commission in 1966 and 1988*

(000 kg active substance)

	1966*			1988		
	Herbi -cides	*Fungi -cides*	*Insecti -cides*	*Herbi -cides*	*Fungi -cides*	*Insecti -cides*
Nurseries	9.1	1.9	0.3	0.8	0.1	0.2
Forest	32.8	146.0	0.6	14.2	126.2	1.3

*Figures for 1966 include materials such as vaporising oil and creosote conventionally purchased by liquid measure rather than weight. Quantities have been converted on the basis of 1 gallon liquid measure = 4.5 kg.

Sources   Wood 1967   Forestry Commission report to MAFF (unpubl).

Figures for the two periods are not strictly comparable because of the changes in materials in use:

- forest nursery weed control in 1966 was dominated by mineral oils (vaporising oil and white spirit) whereas by 1986, pre-sowing treatment with diphenamid was the prevalent treatment;
- in the forest, 2,4,5-T was the dominant herbicide in 1966. It had fallen totally out of use before 1988, the most widely used herbicides then being glyphosate, atrazine and asulam;
- fungicide use has been dominated by the need to prevent colonisation of stumps by *Heterobasidion annosum.* In 1966, creosote and sodium nitrite were the two materials most widely in use. By 1988, both had been replaced by urea, with a small use of *Phlebiopsis (Peniophora) gigantea* (§7.21).
- Apart from *ad hoc* requirements to deal with outbreaks of defoliating caterpillars from the air, the continuing call on insecticides has been to minimise post-planting losses due to attack by pine weevil *(Hylobius abietis)* and black pine beetles *(Hylastes spp.).* In 1966, DDT was still in use; by 1988, it had been replaced by gamma-HCH (§6.51).

A more detailed comparison of changes in weeding practice in the 1980s is given by McCavish (1990); see also §8.11.

2.63   Application of chemicals from the air

Because of the small scale of most forest operations in the UK and the limited need to treat closed canopy woodlands, aircraft have had only a limited role in pest control.

The pattern of use described in 1969 is characteristic of UK circumstances (Aldhous 1969a):

- for insecticides, application only in relation to attacks threatening woodland survival;
- small scale aerial application of herbicides at time of planting to control bracken, heather and other woody weeds;
- nothing at all to control fungi;
- for fertilizers, substantial applications of N (urea) and lesser amounts of PK.

The total area of UK forest treated from the air with insecticides up to 1990 was 32 832 ha (Evans & Stoakley 1990); see also §6.33.

There are no comparable figures available for aerial application of herbicides to forest land in the UK. In the period 1960-68, in total approx. 3200 hectares of woody weeds in young woodland were sprayed (Aldhous, 1969a). Since that time, bracken has been the principle forest weed target for aerial spraying.

Since 1986, the use of aircraft has been stringently controlled under *Food and Environment Protection Act*, in particular, in respect of avoiding pollution of surface water supplies. Requirements for operators were originally set out in *Consent C(ii)* issued under part III of the Act. All consents have been amended and re-issued in closely comparable form (Tweedle-dum and Tweedle-dee!) as Schedule 4, both in the *Control of Pesticides (Amendment) Regulations* and the *Plant Protection (Basic Conditions) Regulations* (SI 1997/188, SI 1997/189).

The only products approved for aerial application to forests in 1999 contain as active substances, either asulam (for bracken control) or diflubenzuron (for insect control). Specific approval has to be sought in advance for any other aerial application to forests or woodlands (MAFF Pst 1999).

For bracken, the stringency of restrictions on use of aircraft has led to an increase in applications employing 'brush contact' methods, *eg* 'weed-wipers'.

CHAPTER 3

The Legal Framework

3.1 PUBLIC SAFETY

3.11 Legislation in the United Kingdom

The primary purpose of legislation regulating use of pesticides, fertilizers and disposal of wastes *etc.* in land husbandry in the UK has been to ensure:

- safe food supplies of high quality;
- public health in a sustainable natural environment.

The legislation in the UK exclusively applicable to the control of pests in forestry is restricted to very particular topics, *eg* defining the permitted use of warfarin for grey squirrel control (§9.24), requiring trees affected by watermark of willow to be destroyed (§7.52), restricting the movement of timber in order to minimise the risk of spreading insect pests (§7.31, §6.53). However, provisions made for safe use of pesticides apply wherever pests are encountered. They cover not only agriculture, horticulture, commercial and amenity forestry and arboriculture but also industrial sites, shipping (*eg* anti-fouling compounds) and food storage *etc.*

3.12 European Community Directives

Particularly during the last 10 years, the Commission of the European Community has issued Directives intended to establish Community-wide standards and to harmonise arrangements so as to facilitate trade. Recent directives relating to pesticides have necessitated amendments to UK regulations, several impinging on forests and woodlands. See §§3.31 & 4.17.

3.2 REGULATORY SCHEMES

3.21 Forerunners to current pesticide controls

Medicines, drugs and poisons

The control of medicines and drugs has a history going back to the Greeks and Egyptians of 1000 years BC. In the UK, the first controls were applied by the Guild of Pepperers early in the 14th century, seeking to avoid adulteration of imported spices and drugs and to maintain a standard of quality (Penn 1979).

During the reign of Henry VIII, the Royal College of Physicians was empowered to appoint inspectors of apothecaries' wares in the London area with the power to destroy defective stock. In Scotland, the Charter granted to the Faculty of Physicians and Surgeons in Glasgow the power to inspect and control drugs sold in the region.

In the 17th century, pharmacopoeias were issued first in London, later in Edinburgh, and in the 19th century, in Dublin. These contained approved lists of drugs and authoritative information as to how they should be prepared and used.

In 1864 the first edition of a British Pharmacopoeia was published, 'reducing to one standard, the processes and descriptions of three different pharmacopoeias and ... reconciling the various usages in pharmacy and prescriptions of ... three countries hitherto in these respects separate and independent'.

Legislation was introduced at about this time, to control the sale of poisons and the people selling them. This marked the beginning of the regulatory procedures affecting materials that could be used to control pests (MCA 1996).

The proliferation of pesticides synthesised by the chemical industry started in the late 1930s and early 1940s, following developments in medicine. In both pest control and medicine, synthesised products very largely supplanted naturally-occurring products. Nevertheless, the latter have acted as pointers to possible biological potency, a situation continuing to the present (Bell 1986).

A 20th Century landmark for control of poisons was the *1933 Pharmacy and Poisons Act* and *Poisons Rules* made under it. Nicotine and arsenical compounds were among the substances covered by these earliest rules, the list being added to as the chemical arsenal expanded. In 1975, the *Health & Safety (Agriculture) (Poisonous substances) Regulations* combined earlier provisions for poisons with the requirements of the 1974 *Health and Safety at Work Act*.

At this time, *Health and Safety (Agriculture) (Poisonous substances) Regulations 1975* (SI 1975/282) replaced *Agriculture (Poisonous substances) Regulations 1966* (SI 1966/1063) made under the 1952 Act.

Legislation current in 1999 includes the *Poisons Act, 1972*, and the *Poisons Rules, 1982* (SI 1982/218) and *Poisons List Order 1982* (SI 1982/217) (as amended) made under it.

A summary of the pesticide active ingredients subject to the poisons laws in 1999 is given in Annex F of *Pesticides 1999* (MAFF Pst 1999). Annex E in the same publication lists banned pesticides; it includes several materials on previous *Poisons Lists*.

The 1982 rules include general and specific provision for the storage, sale and supply of listed non-medical poisons. Poisons are segregated by 'list'.

The 'Part I list' includes poisons for moles and rabbits, and fumigants. Part I poisons are obtainable through registered pharmacists and registered non-pharmacy businesses. Poisons on the 'Part II list' include the more toxic insecticides, other poisons for vertebrates *eg* zinc phosphide, and also the fumigant formaldehyde and the herbicide paraquat; they are obtainable through pharmacists and listed sellers registered with local authorities (Whitehead 1999).

Veterinary products

Veterinary medicines during this century closely followed provisions for human medicine and drugs. Since 1968, under the 1968 and 1971 *Medicines Acts* and associated Veterinary Medicines regulations (SI 1993/2398; SI 1994/2157), products for professional use in animal and fish husbandry have required licensing. Proposals showing that a new product meets statutory criteria of safety, quality and efficacy for licensing are submitted to the *Veterinary Medicines Directorate* of the Ministry of Agriculture, Fisheries and Food.

The earlier Regulations were superseded by the *Marketing Authorisations for Veterinary Medicinal Products Regulations, 1994* (SI 1994/3142). These also implement European Community legislation and re-enacted most of the previously existing provisions. They provide for standardised safety, quality and efficacy testing within the European Community, analagous to arrangements for pesticides (VMD 1995a,b).

Sheep dip in this context counts as a veterinary medicine.

3.22 Pesticides

The first uses of chemical pesticides date from the end of the 19th century and the beginning of the 20th. The first herbicide was copper sulphate used for selective weed control in cereals. Prior to that, formulations of lead arsenate, lime sulphur and nicotine were in use as fungicides or insecticides. An industry-published specification for the latter group appeared in 1920 (Hance & Holly 1990).

In 1941-42, several herbicides became available for selective control of weeds in cereals. This led, in 1942, to the first involvement of government, other than through control of poisons legislation. A voluntary *Crop Protection Products Approval Scheme* for proprietary pesticides was agreed with manufacturers. Although by the end of 1943, DDT was being manufactured on a commercial scale in the UK, initially it was allocated for medical and public hygiene uses and was not available for agricultural use until 1945 (Mellanby 1992).

In the 1940s, incidents of operator poisoning when spraying arable weeds with DNOC and dinoseb, the introduction of highly toxic insecticides such as dieldrin, aldrin and endrin, and increasing concern about their side effects in the environment led in the UK to a series of reports to Government ministers:

- *Toxic chemicals in agriculture* (Zuckermann 1951),
- *Toxic chemicals in agriculture; residues in food* (Zuckermann 1953),
- *Toxic chemicals in agriculture; risks to wildlife* (Zuckermann 1957).
- *Toxic chemicals in agriculture and food storage* (Sanders 1961),

The first 'Zuckermann' report led to the passage of the *Agriculture (Poisonous Substances) Act, 1952* and regulations based on it. These laid down strict requirements for protective clothing and protocols for use when using the most toxic pesticides.

In 1957, a formal but voluntary *Notification of Pesticides Scheme* was set up to evaluate the toxicity hazards associated with the active ingredients of newly marketed products; this in 1964 was renamed the *Pesticides Safety Precautions Scheme* (PSPS). Under the scheme for each product, a sheet giving *Recommendations for safe use, agreed with Government Departments* was issued (MAFF 1966).

The *Crop Protection Products Approval Scheme* was also revised and re-launched in 1960 as the *Agricultural Chemicals Approval Scheme* (ACAS). A list of approved products was published annually (MAFF ACAS 19xx).

Concern about the persistence of DDT and the wide use of several persistent and highly toxic organochlorine pesticides led to a further series of reports:

- *Review of the persistent organochlorine pesticides* (Cook 1964a),
- *Review of the persistent organochlorine pesticides*; *supplementary report* (Cook 1964b),
- *Review of the present safety arrangements for the use of toxic chemicals in agriculture and food storage* (Cook 1967).

The PSPS and ACAS voluntary schemes ran smoothly; nevertheless, it was tacitly acknowledged within the industry that if manufacturers did not cooperate whole-heartedly, compulsion would follow quickly. There was also pressure on government from several quarters to introduce legislation regulating the safety and side effects of pesticides, similar to that being introduced for medicines (Watts 1968, Fryer & Evans 1968b).

Both schemes remained voluntary until the passage into law of the *Food and Environment Protection Act, 1985* (FEPA). However, the *Farm and Garden Chemicals Act, 1967* empowered ministers to make regulations about labelling and the need to carry indications of hazard and was the starting point for legislation on labelling now embodied in regulations under FEPA. See §3.32.

3.23   Health & safety in the workplace

Health and safety at work and accident prevention have been continuing themes in employment legislation with roots going back to the 19th century.

In 1946, safe working practices in many industries, including agriculture, were reviewed in *Health and welfare of employed persons not covered by Factories Act or Mines & Quarries Act* (Gowers 1949). While mainly concerned with other industries, this report included the statement:

> We think that the Minister of Agriculture and Secretary of State for Scotland should be given powers to prescribe by regulation the fertilizers, sprays and other chemicals whose use should make the supply and wearing of protective clothing compulsory.

The *Agriculture (Safety, Health and Welfare Provisions) Act, 1956* covered a number of issues raised in that report.

In 1974, the *Health and Safety at Work Act* (HSWA) was passed, consolidating and adding to earlier legislation, and covering all sectors of employment. Its major innovation was to extend the scope of health and safety legislation to all persons at work and also to any of the general public affected by work activities.

The 'Health and Safety Commission' (HSC) was formed under the Act, powers previously exercised by others passing to the HSC then or later; *eg* responsibility for farm safety was transferred from Agricultural Ministers to the Health and Safety Commission under the *Employment Protection Act, 1975*.

The Health and Safety Executive also regulates exposure to hazardous chemicals in uses unrelated to tree and woodland management practice, these forming the main focus of HSC's work.

Regulations and their revisions made under HSWA have increasingly influenced safe working practices, *eg*:

* *Control of Substances Hazardous to Health Regulations, 1988 (COSHH)* (SI 1988/1657), revised 1995, 1996, 1997 (SI 1994/3246, SI 1995/3138, SI 1997/11);
* *Management of Health and Safety At Work Regulations 1992* (SI 1992/2051); these implemented most provisions of the EC Workplace Directive 89/654/EC. They provide the framework for assessment of risk and hazard of pesticides under COSHH regulations; they were amended in 1997 to provide for young people and women of child-bearing age (HSE 1997b);
* *Reporting of Injuries, Diseases and Dangerous Occurrences Regulations, 1995 (RIDDOR)* (SI 1995/3163), (replacing SI 1985/2023);
* *Noise at Work Regulations* (SI 1989/1790); these apply to many sources of loud repeated noise including guns, chainsaws and harvesting machinery, bandsaws *etc.*;
* *Workplace (Health, Safety & Welfare) Regulations 1992* (SI 1992/3004);
* *Provision and Use of Work Equipment Regulations, 1992* (SI 1992/2932);
* *Personal Protective Equipment (EC Directive) Regulations as amended* (SI 1992/3139, SI 1993/3074, SI 1994/2326);
* *Health & Safety (First-aid) Regulations, 1981* (SI 1981/917);
* *Safety Representatives and Safety Committees Regulations, 1977* (SI 1977/500);
* *Agriculture (Safeguarding of Workplaces) Regulations, 1959* (SI 1959/428).

3.24 Protection of water supplies

For the larger part of this century, water authorities have been chiefly concerned with pollution of water from industrial and urban sources. Nevertheless, the potential hazard of pesticides had been recognised in the 1960s;

a *Code of Practice for the use of herbicides on weeds in water-courses and lakes* (MAFF 1967, revised 1985) has been available since that time. A early voluntary *Code of Conduct* (ABMAC 1968) included advice to:

- Avoid pesticides draining into water courses;
- Never wash or clean equipment in watercourses.

Since then, requirements have become progressively more stringent. Current legislation to meet drinking water standards is based on the *Water Resources Act, 1991*; disposal of surplus pesticides being covered by the *Groundwater Regulations, 1998*. Responsible practice is summarised in Appendix II of the Forestry Commission *Forests and water guidelines* (FC WG 1997), *The Water Code* (MAFF 1998b) and *The Groundwater Regulations SEPA Guidance Notes 1 and 3* (SEPA 1998, 1999). Practical implications are described in §4.29.

3.3   PRESENT PROVISIONS FOR SAFE USE OF PESTICIDES

3.31   European Community Directives

Current UK legislation has to be aligned with relevant European Community Directives. The most important is:

- 91/414/EEC Plant Protection Products Directive. Under this, plant protection products may not be placed on the market or used unless authorised by a member state under the Directive.

While harmonising arrangements for authorisation of plant protection products within the Community, product authorisation remains the responsibility of individual member states (MAFF Pst 1996).

Other directives affecting working practices include:

- 89/654/EEC Workplace Directive, also 92/85/EEC in respect of women of child-bearing age, and 94/33/EEC about young people;
- 89/686/EEC Personal Protective Equipment Product Directive, as amended by 93/68/EEC and 93/95/EEC,
- 90/679/EEC Directive on the Protection of Workers from the Risks related to the Exposure to Biological Agents at Work.
- 90/220/EEC Directive on the Deliberate release into the Environment of Genetically Modified Organisms;
- 90/219/EEC Directive on the Contained Use of Genetically Modified Organisms.

Biocidal Products Directive

A Directive (98/08/EC) intended to be implemented by May 2000 was published in 1998; it covers materials currently included by the Health & Safety Executive under the Control of Pesticides Regulations. Additionally, the Directive includes industrial preservatives, disinfectants, water biocides and pesticides other than for plant protection purposes. Technical Guidance drafts are expected to be circulated during 1999 (MAFF Pst 1999).

3.32 Current UK legislation - FEPA and HSWA

The *Food and Environment Protection Act, 1985* (FEPA) and the *Health and Safety At Work Act, 1974* (HSWA) remain the foundations of legislation on safe use of pesticides. Their most significant provisions (as at October 1999) are the:

• *Plant Protection Products Regulations 1995*, (SI 1995/887), supplemented by the *Plant Protection Products (Basic Conditions) Regulations 1997* (SI 1997/189);

• *Control of Pesticides (Amendment) Regulations 1997* (SI 1997/188) amending the *Control of Pesticides Regulations 1986* (SI 1986/1510);

• *Control of Substances Hazardous to Health Regulations 1994*, amended 1996 and 1997 (the COSHH Regs.) (SI 1994/3426, 1996/3138, 1997/11).

The first of these brings UK requirements into line with EC Directives. One consequence is that the definitive list of approved products when the Directive was introduced is listed in an Annex to the Directive. Similarly, conditions to be met when seeking approval for a new product are set out in other Directive Annexes. In the short term, there will be two types of approval, those under the 'old' system originally given by individual EC member states, and 'new' system approvals also given by member states but within the terms of the EC Directive.

The *Plant Protection Product Regulations* cover 'new' system approvals; the amended *Control of Pesticides Regulations* cover the 'old' system. Under both sets of regulations, evaluations leading up to approvals can be made available on request (for a fee representing the reasonable cost of the evaluation document), provided the recipient does not make commercial use of the information contained in them (SI 1997/188, SI 1997/189).

The 1994 COSHH Regulations replace the 1988 regulations of the same name (SI 1988/1657). The main effect of the 1994 version is to extend the range of 'hazardous substances' to include 'biological agents' (See *Glossary*).

Other relevant provisions related to safe use of pesticides include:

• *Chemicals (Hazard Information and Packaging) Regulations (CHIP), 1994* (SI 1994/3247, amended SI 1996/1092);

• *Personal Protective Equipment at Work Regulations 1992* (SI 1992/2966, HSE 1995);

• *Personal Protective Equipment (EC Directive) Regulations 1992* (SI 1992/3139), amended by SI 1993/3074 and SI 1994/2326;

• *Pesticides (Maximum Residue Levels in Crops, Food and Feeding Stuffs) Regulations 1994* (SI 1994/1985);

• *Reporting of Injuries, Diseases and Dangerous Occurrences Regulations, 1995 (RIDDOR)* (SI 1995/3163).

The *Pesticides Monitor* (previously *Register*) is issued monthly and gives details of recent pesticide approvals and of changes in regulations.

3.33   Disposal of pesticide wastes

Disposal of pesticide wastes fall under regulations mostly made under the

- *Control of Pollution Act, 1974,*
- *Water Act, 1989,*
- *Control of Pollution (Amendment) Act, 1989,*
- *Environment Protection Act, 1990* (EPA).
- *Groundwater Regulations* (SI 1998/2746) (See §3.51).

The *Environment Protection Act (Duty of Care) Regulations, 1991* (SI 1991/2839) relates to the disposal of controlled waste and replaces earlier regulations under the 1974 Act. A booklet *Waste Management, the Duty of Care, a Code of Practice* gives Government Departments' guidance to good practice under the 1990 EPA (D.Env 1991b, revised 1996).

Provisions for Scotland are included in various specifically designated Regulations. The underlying principles in both countries are the same.

Under the *Control of Pollution Act, 1974*, it is an offence to abandon waste, including waste from agricultural premises, on any land where it is likely to give rise to an environ-mental hazard (section 18), or to knowingly permit any poisonous, noxious or polluting matter to enter controlled waters (rivers, lakes, underground waters).

Guidance on safe disposal of pesticides used for non-agricultural purposes is given in the relevant HSE *Approved Code of Practice* (HSC 1995a). Where the ground where a pesticide is used is part of a farm or holding, guidance is given in the MAFF/HSC *Code of Practice for the Safe Use of Pesticides on Farms and Holdings* (MAFF/HSC 1998). See also §4.6.

Since April 1999, users have had to obtain formal authorisation from the Environment Agency or Scottish Environment Protection Agency to dispose of washings and unused dilute pesticide on non-crop land (Willoughby & Clay 1999).

3.34   Plant Health

Legislation has been used to :

- reduce the risk of introduction of pests - if pests are absent, the question of use of pesticides to control them does not arise!
- reduce the risk of pests occurring only in parts of Britain spreading over the whole of the country. Similarly, by minimising movement of infected plants, the potential need to use pesticides is reduced.

Minimising the risk of introduction of pests

The legislation in force in the 1950s, the *Destructive Insects and Pests, Great Britain Importation of Forest Trees (Prohibition) Order* (SI 1952/1929) constituted, for conifers, a broad embargo to protect against known and unknown pests. For broadleaves, prohibitions applied more specifically to oak and chestnut in respect of risks from *Endothia parasitica*, to elm on account of *phloem necrosis*, and poplar on account of potential canker-forming organisms.

Peace (1962) reviewed the dangers of introducing diseases and listed over 60 fungal species along with viruses, bacteria and parasitic plants (mistletoes) which are potential threats to forest and ornamental trees in the UK.

Earlier, when Dutch elm disease was known in Europe but not in Britain, the *Importation of Elm Trees (Prohibition) Order 1926* had been made in an attempt to prevent its entry. Nevertheless, the disease was discovered in England in 1927, and in 1928 was shown to be widespread in the south of England. These discoveries were not due to failure of legislation; the disease had probably been present for several years previously but had not been identified (Peace 1960). See also §5.3.

Phillips (1980b) in a review of international plant health controls pointed out the unreality of expecting to be able to keep all novel pests out in circumstances where there is massive import and export of plant and wood material.

On entry into the European Economic Community the United Kingdom came under EEC health controls and the philosophy that in order not to impede trade, provision should be made only against specific known pests (EC Directive 77/93/EEC and its many amendments) (Phillips 1978, 1980b). These changes and up-datings have been incorporated under the *Plant Health Act 1967* in the:

- *Plant Health (Great Britain) Order 1993* (SI 1993/1320) (revoking the earlier *Order,* SI 1987/1758 and its amendments),
- *Plant Health (Forestry) (Great Britain) Order 1989* (SI 1989/823) (revoking the *Tree Pests (Great Britain) Order 1980* (SI 1980/450) and the *Import and Export of Trees, Wood & Bark (Health) (Great Britain) Order 1980* (SI 1980/449) and their amendments).

The last-named order prohibits landing of trees, pests, wood bark, soil *etc.* where this is desirable as a protective measure against introduction of harmful organisms. It also provides powers to require phytosanitary certificates for incoming plants or wood, and powers to take appropriate action if requirements under the regulations are not followed.

3.35 Preventing spread of specific pests and diseases

Plant Passports

Under arrangements introduced in response to EC Directives, planting stock which is free from specified diseases can be issued with a plant 'passport' to allow such stock to be moved from country to country within the European Community. These arrangements are embodied in The *Plant Health (Great*

Britain) Order, 1993 (SI 1993/1320); Aldhous (1994) describes how these are applicable to forest nursery stock.

> *Legislation to minimise spread of specific pests*

See:

- §6.53 - *Great spruce bark beetle,*
- §7.31 - *Dutch elm disease,*
- §7.52 - *Sanitation felling for Watermark disease of cricket-bat willow,*
- §9.24 - *Grey squirrel.*

3.36   Containment of pesticide treatments

Section 4 in Schedule 3 in both the *Control of Pesticides Regulations (Amendments) Regulations 1997* (SI 1997/188), for the 'old' system approvals; and the *Plant Protection Product Regulations (Basic Conditions) Regulations 1997* (1997/189), states:

> 'Any person who uses a pesticide/pppp shall confine the application of that pesticide/pppp to the land, crop, structure, material or other area intended to be treated'.

3.4   CONSENTS

Under the *Food & Environment Protection Act* (FEPA), Ministers are required to give formal consent to the advertisement, sale, supply, storage and use of pesticides. Such activities are prohibited unless conditions in the Ministers' consents are met.

The original consents in the *Control of Pesticides Regulations 1986* have been replaced by twin sets of consents, one in the *Control of Pesticides Regulations (Amendments) Regulations 1997* (SI 1997/188) for the 'old' system approvals; the other in the *Plant Protection Product Regulations (Basic Conditions) Regulations 1997* (1997/189) for 'new' system approvals. The two versions differ very little except that where one refers to 'pesticides', the other substitutes 'prescribed plant protection products' ('pppp's).

In the both sets of regulations, there are four schedules giving:

> *Conditions relating to Consent to*:

- *Advertisement of pesticides* (Schedule 1),
- *Sale, supply and storage of pesticides* (Schedule 2),
- *Use of pesticides* (Schedule 3),
- *Use of pesticides by aerial application* (Schedule 4).

3.41　　　Fields of use

Agricultural use

The 'Consent' in both Schedules 2 includes the following :

"*a pesticide approved for agricultural use* means a pesticide/pppp ....
approved for use in one or more of:

- agriculture and horticulture (including amenity horticulture);
- forestry;
- in or near surface water, other than amateur, public hygiene or anti-fouling uses;
- industrial herbicides including weedkillers for use on land not intended for the production of any crop. "

Pesticides/ppppps approved for agricultural use may also be employed in other fields of use, in accordance with their approval, in:

- home garden (amateur gardening);　　• food storage practice;
- vertebrate control (*eg* rodenticides and repellents);
- domestic use;　　　　　　　　　• wood preservation;
- "other" as defined by Ministers.

Crop/situation within 'fields of use'

The 'crop' or 'situation' under which a pesticide may be used is defined for 'forestry' and associated fields of use. Willoughby & Clay (1999) tabulate 10 'crop/situation' terms. Most fall into the 'forestry', 'agricultural' or 'horticultural' fields of use. However, 'land not intended to bear vegetation' may come under the 'industrial' field of use; the 'use in or near water' situation comes within the 'in or near water' field of use. Godson (1996) describes how these terms were arrived at.

Requirements for training and competence

The 'Consent' in both Schedules 3, among other things, requires employers to ensure that any employee using pesticides has been given training to enable them to comply with the Regulations, and requires 'persons using pesticides in the course of a commercial service' to have obtained a certificate of competence unless otherwise exempted.

Certificates of competence

Users of 'pesticides approved for agricultural use' as defined in the Consent in Schedule 3 need a valid 'Certificate of Competence'; see §4.42.

Personnel responsible for storage of pesticides 'for sale or supply' also need to have obtained a certificate of competence; see §4.42.

3.42　　　'Label' and 'off-label' uses

Labels of approved pesticide products and ppppps are the most important single element in ensuring the safe use of pesticides in practice. They are required to be laid out and written so as to conform with conditions specified when the product was approved under FEPA. A 'model' label is shown and described in Annex E of the *Code of Practice for the Safe Use of Pesticides on Farms and Holdings* (MAFF/HSC 1998*).* Labels are usually set out with key information printed within a line frame or 'box'. They cover:

- field of use,
- crop or situation for which treatment is permitted,
- maximum individual dose,   • maximum number of treatments,
- maximum quantity or area that may be treated,
- latest time of application, or interval before harvesting,
- operator protection or training requirements,
- environmental protection requirements,
- any other specific restrictions relating to particular pesticides.

'Off-label' or 'extended use' approvals

The approval of specific pesticide products may be extended to cover uses additional to these approved and shown on the manufacturer's product label. Users either have to apply for extended use (off-label) approval or ensure that they possess a copy of the relevant existing 'extended use' approval.   Such specific off-label 'extensions of use' approvals may have additional conditions attached to them.   Use in these cases is at the user's choosing, and the commercial risk is entirely the users.

A copy of the revised long-term arrangements approved for 'extensions of use' is given in Annex C of MAFF Booklet 500 *Pesticides 1999*. No expiry date is given in the Annex.

Subject to specific restrictions set out in Annex C, pesticides approved for any growing crop may be used in forest nurseries on forest nursery crops prior to final planting out, and on ornamental crops where neither seed nor plant is to be consumed by animals or humans.

For 'farm forestry' schemes, herbicides approved for use on cereals may be used in the first five years of establishment in farm forestry on land previously under arable cultivation or improved grass.

3.43   Commodity substances

Certain chemicals are in widespread use commercially or occur naturally in large amounts in living organisms. At the same time, they may have useful functions in crop management or protection. These are termed 'Commodity Substances' (also, formerly, 'Commodity Chemicals') and special provision is made for them under *Control of Pesticide* regulations. Such substances may be *used* as pesticides under the terms of their designation as 'Commodity substances'. *Sale, supply, storage* and *advertisement* of a commodity substance *as a pesticide* is an offence unless specific approval has been granted for the substance to be marketed under an approved label.

Urea and formaldehyde are listed as commodity substances (MAFF Pst 1999). See also §7.21.

For some commodity substances, there is no 'date of expiry' of approval. For others, a date is set, sometimes coupled with a requirement for submission of data from interested parties by a set date if the approval is to be extended beyond the expiry date. Formaldehyde and urea are both in this category.

3.5   PROTECTION OF GROUNDWATER

3.51   Contamination by pesticide and fertilizer residues

Discovery in the 1970s of traces of pesticides and fertilizers in groundwater aquifers has led to an increasing concern for the quality of groundwaters which are potential sources of water for human consumption.

In 1980, the Commission of the European Communities issued a Directive on the permissible levels of residues in groundwater (CEC 1980). The actual levels chosen were influenced by political considerations as much as assessment of risk to consumers.

In the UK, *Groundwater Regulations* came into force in December 1998 (SI 1998/2746, SEPA 1998).

Under the regulations, authorisation is required for the disposal of certain waste agrochemicals; these are shown in two lists.

The substances in List 1 (the more stringent list) are:

- organohalogen compounds and substances which may form such compounds in the aquatic environment;
- organophosphorus compounds;   • organotin compounds;
- substances which possess carcinogenic, mutagenic or teratogenic properties in or via the aquatic environment;
- mercury and its compounds;   • cadmium and its compounds;
- mineral oils and hydrocarbons;   • cyanides.

Substances in List 2 include:

- the following elements and their compounds:

antimony	arsenic	barium	beryllium	boron
chromium	cobalt	copper	lead	molybdenum
nickel	selenium	silver	tellurium	thallium
tin	titanium	uranium	vanadium	zinc;

- biocides & their derivatives not in List 1;
- substances which have a deleterious effect on the taste or odour of groundwater, & compounds liable to cause the formation of such substances … and render it unfit for human consumption;
- toxic or persistent compounds of silicon …….. excluding those which are biologically harmless ………;
- inorganic compounds of phosphorus and elemental phosphorus;
- fluorides   • ammonia and nitrites.

3.52   Fertilizer application and water supplies

Wherever fertilizers are under consideration for use on any scale in an area where surface water is gathered for public consumption, the local water authority should always be informed at an early stage (FC WG 1993). See also §§3.33, 4.6.

3.6   GUIDANCE ON APPLICATION OF LEGISLATION

3.61   Statutory and non-statutory advice and guidance

Official guidance in the form of codes or notes on safe pesticide use have been prepared from time to time. MAFF has issued 'Codes of practice' and 'Guidelines' for example, on ground spraying, use of herbicides in water courses, safe use of poisonous chemicals on the farm, disposal of unwanted pesticides on farms *etc* (MAFF 1966, 1967, 1975a, b, c, 1980, 1983, 1989). Lists of products approved for farmers & growers have been issued annually since the 1960s; current product lists are contained in *Pesticides 1999* (MAFF Pst 1999).

The passing into law in 1974 of the *Health & Safety at Work Act*, and its wider provisions for safety, gave rise to a series of *Approved Codes of Practice (ACOPs)* with statutory status and to numerous notes, leaflets, guides *etc.* from the Health and Safety Executive. Important non-statutory codes and guidance notes have also been issued by government departments and others.

A desire in the 1960s to see a more broadly based consensus on pesticide usage led in 1968 to the preparation of *Pesticides, a Code of Conduct* drawn up and agreed jointly by industry, government departments, farm and trade representative bodies, conservation groups and agricultural unions (Moore & Evans 1968, ABMAC 1968). While it had no formal status, the code indicated both a need and a desire for guidance against yardsticks agreed by representatives of a wide range of interests.

The subsequent Government *Approved Codes of Practice* have been the subjects of extensive consultation prior to final publication and have lessened the need for such an informal 'Code of Conduct'.

3.62 ·   Statutory Approved Codes of Practice (ACOPs)

Statutory Approved Codes of Practice can only be issued if there is empowering legislation to do so. The Health and Safety Commission (HSC) under the 1974 Health & Safety at Work Act is so empowered in relation to health and safety, and has issued a substantial number of ACOPs.

These codes have special status. Each HSC code states 'If you are prosecuted for breach of health and safety law, and it is proved that you have not followed the relevant provision of the Code, a court will find you at fault, unless you can show that you have complied with the law in some other way' (HSC 1995a).

Under the Food and Environment Protection Act, the Ministry of Agriculture, Fisheries and Food was empowered likewise.

Codes for safe use of pesticides in forestry

There are no statutory approved codes of practice which apply exclusively to forestry. Forestry is included, however, in two codes which between them cover safe use of pesticides in amenity, farm and commercial woodlands, urban forest areas and forest nurseries. These are:

• *Pesticides; Code of Practice for the Safe Use of Pesticides on Farms and Holdings.* This was first published in 1990 (revised 1998), as a jointly approved code (MAFF/HSC 1990, 1998). It covers farm woodlands, defined as 'the use of land for woodlands where that use is ancillary to the farming of land for other agricultural purposes' (HSC 1995a).

• *The Safe Use of Pesticides for Non-agricultural Purposes,* approved by HSC, published 1990, revised 1995, covers forestry, amenity horticulture including forest nurseries, and vertebrate control (HSC 1991a, 1995a).

Other codes

Other more general statutory approved codes of practice include:

• *Management of health & safety at work* (HSC 1992a),
• *Workplace health, safety & welfare* (HSC 1996b),
• *Safety representatives and safety committees* (HSC 1988),
• *First aid at work* (HSC 1991b),
• *Code of practice for suppliers of pesticides to agriculture, horticulture and forestry,* published 1990, revised 1998 (MAFF 1990, 1998a),
• *Waste management; the duty of care* (D Env 1991b, revised 1996),
• *General COSHH ACOP, Carcinogens ACOP & Biological agents ACOP*, published 1989, revised 1995 and 1997 (HSC 1995b, revised 1997),
• *COSHH in fumigation operations* ACOP (HSC 1996d).

The first of these codes is targeted at all employers and workplaces; it aims to reduce the 30 million working days lost by industry and commerce due to accidents and ill-health.

While the regulations it amplifies were introduced several years after the COSHH regulations were first introduced, its wide scope requires that assessments under COSHH are seen as an element within of a firm's overall review of corporate health and safety risks, and not independent of it.

3.63   Other statutory lists *etc*

• *Occupational exposure limits* (HSE 1998a).

This publication lists *maximum exposure limits* (MELs) and *occupational exposure standards* (OESs). A number of pesticides have occupational exposure limits (see also § 4.24).

• *Information approved for the classification and labelling of substances and preparations dangerous for supply (4th edn)* (HSC 1998).

This list is the authoritative source document allocating toxicity ratings to pesticides (*ie* 'very toxic', 'toxic', 'harmful', 'irritant' etc.) (§ 4.23).

- *Pesticide (Maximum residue levels in crops, food or feeding stuffs) Regulations 1994* (see §4.28 - MRLs).

3.64 Non-statutory codes

- *Provisional Code of Practice for the Use of Pesticides in Forestry* (Anon 1989). This code contains eight guidance notes and two checklists all relating specifically to forestry operations.

3.65 Guidance on regulations

Recent HSC and HSE publications give guidance on regulations derived from HSWA 1974, stating: 'Following the guidance is not compulsory and you are free to take other action. But if you do follow the guidance you will normally be doing enough to comply with the law. H&S inspectors seek to secure compliance with the law and may refer to this guidance as illustrating good practice.'

Such guidance is contained in:

- *A guide to the Health & Safety at Work Act*, 1974 (HSC 1992b),
- *Personal protective equipment at work* (HSE 1995e),
- *Health risk management* (HSE 1995d),
- *Everyone's guide to RIDDOR 95* (HSE 1996a),
- *Staying healthy, a guide for workers in farming, forestry & horticulture* (HSE 1995a).

'Guidance' sections are also included in many 'Approved Codes of Practice'.

3.66 Other guidance notes

MAFF, HSE, the Forestry Commission and many other organisations have issued guidance booklets, leaflets and posters by the score, all aimed at informing current users, advisers *etc* as to good practice, new techniques, legislation, safety procedures *etc*. See 'Current publications lists' issued by HM Stationery Office, Ministry of Agriculture/Pesticides Safety Directorate, Health & Safety Executive, Forestry Commission and British Crop Protection Council.

3.67 Gaps and overlaps

There is not perfect coincidence of provision between various codes of practice, guidance notes *etc*. Farm forestry is covered by a joint code for farmers made under the Food and Environment Act (FEPA) and Health and Safety at Work Act (HSWA); non-farming forestry falls within the code made only under HSWA. The content of the HSWA element of the two codes is very similar. However, the joint FEPA/HSWA farming code carries additional information about environmental concerns and application procedures relevant to forestry operations and should be treated as an important aide-memoire to good practice.

CHAPTER 4

The Safety System

'One man's meat - another man's poison
One man's 'safety' - another man's 'risk'

4.1   SAFETY - AN INTERNATIONAL CONCERN

Pesticide safety is overtly an international issue. Safe use of pesticides in forestry in the UK has to be seen as part of an interrelated series of international working standards and practices alongside provisions for:

- *Human health*
    adequate amounts of nutritious and uncontaminated food;
    clean water;
    control of disease-carrying organisms, ectoparasites *etc* (*eg* mosquitoes, lice, ticks).
- *Animal health*
    nutritious and uncontaminated grazing and fodder;
    control of disease-carrying organisms and ectoparasites.
- *Food and fodder storage*; *environmental health*
    control of pests of food storage, in food preparation areas, in dwellings, *etc* (*eg* flies, weevils, cockroaches, rats, mice);
    control of pests on licensed waste tip sites, *etc*.
- *Industrial, commercial and urban site hygiene & safety*
    control of vegetation on roadsides *etc*, (*eg* to minimise fire risk).

The United Nations, in particular, FAO (Food and Agriculture Organisation) and WHO (World Health Organisation) have made important contributions to standards and practice on many of these issues (WHO 1970, 1991; Johnen 1990; FAO 1994).

The European Community has promulgated directives, seeking to harmonise standards Community-wide in respect of approvals for and uses of pesticides. Though in force, they are not to everyone's liking (Thomas 1990, Foulkes 1990).

4.11   Safety

While in everyday parlance, the words *safe, safety, risk, danger, hazard* are commonly used inexactly and ambiguously, provisions under the *Health & Safety at Work Act 1974*, and *Control of Substances Hazardous to Health Regulations*

(SI 1994/3246) have defined usages for certain terms.

Under the terms of the COSHH regulations, safety in any given circumstance is secured by identifying and evaluating separately:

- 'hazard' (*synonym* 'danger') and
- 'risk' (*synonyms* 'likelihood', 'probability') of exposure to a hazard and the extent of that exposure.

Where the safest of alternatives have to be chosen and circumstances allow, the *hazard* and *risk* for each alternative should be evaluated by the same criteria.

4.12   Hazards

A hazard is anything that has a potential to cause harm, *eg* overhead electricity lines, windblown trees, steep slopes, toxic chemicals.

Some hazards can be quantified, others cannot. In both cases, differences of judgement may lead to disagreement about the seriousness of the hazard, *eg* the relative toxicity of different pesticides.

For hazards that can be quantified, the extent of the hazard depends on the dose or exposure. Aspirin is a much quoted example of a hazardous toxic substance, widely used at low doses for headaches *etc*, but with the power to kill if taken at high doses. Electricity is another familiar hazard, in this instance, 'dose' being related to voltage.

4.13   Risk

The risk from a pesticide is the *likelihood* that it will do harm *in the actual circumstances of use* (HSE 1989b).

This clear statement particularly applies to assessment of hazard and risk for 'COSHH assessments'. However, 'risk' is also used more loosely in many situations, including other government publications. In this book, the HSE definition is followed.

Likelihood'

This concept, otherwise expressed as 'probability' or 'odds', is at the heart of divergences in judgement of risk based on quantitative values.

4.14   Pesticide hazards

Aspects of hazard considered when evaluating pesticides include:

- chemical properties of active ingredient,
- the formulation (§2.41),
- use and method of application,
- acute and chronic toxicity (§4.23),
- risk of damage to skin and eyes, risk of sensitisation (§4.22-4.26),
- carcinogenicity (risk of cancer) (§ 4.27),
- teratogenicity (risk of damage to the unborn child),

- mutagenicity (risk of damage to genes),
- operational exposure characteristics,
- behaviour and fate in plants, crops, soil and water,
- risk to non-target plants and animals, domestic and wild, (in particular, bees and other pollinating insects),
- efficacy and crop safety.

Numerical values can be assigned to expressions of severity of hazard, leading to quantitative scales for assessing hazard.

Manufacturers have to provide data covering such hazards and the associated risk of exposure when seeking approvals. Specific requirements for new active substances are given in Annex I of EC Directive 91/414/EEC as amended (SI 1997/189). Requirements for products containing approved active substances are given in Annexes II and III of the Directive.

4.15 Accessibility of data

Some data submitted for approval of new active substances and new products are readily available through published evaluations (see §4.17). This is a substantial improvement over the situation in the 1980s; however, much data is not readily available.

- Manufacturers' commercial interest is to ensure that the benefit of their investment in research and development is not lost to a competitor through premature release of commercially valuable data. Such information would be considered confidential to the company and the approving authority until published or covered by patent.
- At any point in the use of a pesticide, dispute may arise as to whether a use or action was harmful. If litigation is likely, directly related information not already made public may be considered confidential to the parties.
- An individual's medical records are almost always confidential between the individual and the medical advisor. Retrospective analysis of such records, *eg* in epidemiological studies of possible pesticide side-effects, may not be possible on this account.

4.16 Judgements on hazard and risk

Views on safety, whether of pesticide use or anything else, have to be seen as matters of *judgement*, based on *communicated knowledge* and the *ethical values* applied when considering hazard and risk. Sound judgements cannot be made if important blocks of information are not available to those who wish to form an opinion on an issue.

Interest groups

At least six major interest groups may express opinions (judgements) on current pesticide use (Summerscales 1990):

- pesticide producers, including their scientists, engineers, commercial & marketing staff, *etc*;
  - growers/pesticide users, and their suppliers, advisers and trainers;
- employees applying pesticides, a group often at greatest risk of contamination by pesticides;
  - consumers of treated foodstuffs, and users of other treated goods;
  - legislators/regulators with responsibilities for producers, growers, consumers and the environment;
  - environmentalists, neighbours, journalists and others interested in but having no direct responsibility for pesticide production or use.

Each interest group may form judgements about safe pesticide use based on the information that they have, their personal potential exposure to pesticides or their residues, the ethical values that influence their judgement, and their economic involvement.

In an ideal world, for any pesticide, there would be a comprehensive data base completely publicly accessible and widely known, combined with an agreed set of values and criteria by which to judge a new proposal. As outlined above, this is not the case in practice. It is therefore not surprising that differences in judgement of safety exist between the 'interest groups', arising out of different degrees of access to information and differing values.

In this situation, the authorities whose responsibility it is to make judgements on data not publicly available, must ensure that their impartiality and thoroughness are seen to be of the highest order.

Qualitative restriction of use

Qualitative judgements may be made either in the absence of a basis for a quantitative judgement, or may over-ride such judgements. Pesticides have been banned or restricted in several countries on qualitative, ethical grounds.

Recent qualitative judgements include:

- ethical judgements which dismiss probabilities as irrelevant because some practice is obnoxious, *eg* pharmaceutical tests using live animals;
- setting of arbitrarily low threshold values, *eg* of permissible pesticide residues. Current EC restrictions on use of triazine herbicides because of their presence in trace quantities in groundwater, are based on an ethical judgement, not on an analysis of the documented hazards (Henningsen 1990).
- The province of Quebec lost 30% of its forest growing stock between 1970 and 1985 through attack by spruce budworm. Under a strategy promulgated in 1994, a consultation with the general population showed a dislike of the use of chemicals; the province aims to have discontinued their use by 2001 (Spencer 1996).

Unquantified hazards

There remains the possibility of unforeseen hazards emerging, either previously unidentified, or viewed as low-risk and not to requiring attention.

While for any particular pesticide, the risk of unforeseen hazards diminishes with experience and time, the possibility recurs with every new active substance and new use. The need for vigilance is equally a recurring theme in safe use of pesticides.

For the UK, the identification, evaluation and early response to a new hazard can only come through reliable reporting to the Health and Safety Executive of all apparently inexplicable incidents, in addition to any action under RIDDOR. The HSE have a leaflet and report form available on request (HSE 1999).

4.17   UK practice

Most agricultural pesticides are used on food crops. Avoidance of inadvertent transfer and/or accumulation of toxins through food chains has therefore been a major element in framing safe application rates and working practices. In forestry operations, direct threats to the human food chain are trivial. Nevertheless, the framework of food safety, user safety and environmental checks developed for agriculture is directly applicable to forestry, with only minor modifications for the characteristics of trees and the land on which they grow.

In the UK, recommendations for safe use of pesticides are science-based and quantitative (Hollis 1990, Tooby & Marsden 1991). They are founded on assumptions that:

- a sufficiently broad and consistent scientific data-base exists and can be relied on;
- the subjects of such data-bases are adequately representative of all known classes of dangers and that results of tests on selected species and processes can be reasonably extrapolated to all other species or processes in the class;
- recommendations can be modified if evidence is forthcoming that either of the above assumptions is incorrect.

Approvals

The early development of the arrangements for approval of agricultural chemicals and the associated safety precautions schemes preceding the present arrangements are described in §3.2.

Requirements for new product approvals and reviews

Data and other information on the hazards listed in §4.14 are submitted by producers of new substances or products. Initially, they are treated as confidential, unless already published.

Since the late 1980s, summaries of the initial evaluations of new active ingredients and new products approved by the *Advisory Committee on Pesticides* (ACP) have been included in the *Pesticide Monitor* (formerly Pesticide *Register*). This is issued monthly and can be purchased, or inspected in reference libraries.

Reviews

It was recognised in the 1950s that regular reviews should be made of all pesticide products in use, because of the higher standards and greater knowledge accumulating over the years. However, during 1970s and early 1980s, the regulatory system carried a backlog of unreviewed older products, because of the continuing heavy flow of new pesticides and new products for initial evaluation.

With the introduction of the *Food and Environment Protection Act* (FEPA), the *Advisory Committee on Pesticides* started a new programme of evaluations of fully or provisionally approved products. Results have been published as completed, *eg* glufosinate-ammonium (MAFF Eval 1990), benomyl (MAFF Eval 1992a), simazine (MAFF Eval 1993b), *Phlebiopsis gigantea* (MAFF Eval 1998).

By the end of 1998, 183 such reviews had been completed; their titles are listed in Annex G of MAFF Booklet 500 (MAFF Pst 1999). The evaluations may be purchased from the MAFF *Pesticides Safety Directorate*, on condition that the recipient does not make commercial use of the information contained in them (SI 1997/188, SI 1997/189). See 'Access to Information of Pesticides' in introductory Chapter 2 of MAFF Booklet 500.

EC-coordinated evaluations

The EC Directive 91/414/EEC set out to harmonise the task of initial evaluation of active substances*, and to undertake a full review of all existing approved active substances.

To avoid duplication within the community, new evaluations and reviews were shared between its members, working to common standards. The reviews of materials on the market before July 1993 were to begin in 1995 and to be undertaken as part of a ten-year programme (MAFF Pst 1996).

Reviews already under way in member states would be completed. However, reviews of products on the EC list would take precedence unless an item on a national programme raised unresolved safety issues (ACP 1994).

For the first round of reviews, 90 active substances were selected and were listed in the *Pesticides Register No 8 1992*. Of these, the UK was allocated 12.

Because of this international programme, the UK programme was reviewed and work on active substances allocated to other EC countries was dropped.

At the beginning of 1999, while a small number of EC evaluations had been completed, none had been published.

Collaboration on analytical methods

The *Advisory Committee on Pesticides* drew attention to the role of its supporting 'Pesticide Analysis Advisory Committee' as a member of the UK branch of Collaborative International Pesticides Analytic Council (CIPAC). This council has world-wide membership and seeks to define and maintain

---

* '*active ingredient*' has been the term used for several decades in relation to pesticides in the UK; '*active substance*' is its equivalent in the text of the EC Directive. See §3.31. In some papers '*actives*' is used to include both phrases.

international standards in accepted methods of analysis appearing in CIPAC handbooks. The Council has links to the Food and Agriculture Organisation and to the World Health Organisation (ACP 1991). It seeks to ensure that analytical standards and procedures in the UK conform to best international practice.

Change of limits of use because of additional quantitative evidence

It is typical of new developments that the full limits to use are refined after a period of practical use. In some cases, uses are extended; in others, they are restricted or prohibited. The latter are normally based on additional quantitative data that became available only after an initial pesticide product approval had been issued.

The time taken to introduce restrictions has sometimes been criticised. Whatever the merits of such criticism, it is important that reviews leading to restrictions are thorough and objective.

Many more products have been withdrawn by manufacturers for commercial reasons than have been officially banned or restricted.

Reviews of safety of 2,4,5-T and 2,4-D

2,4-D and 2,4,5-T were in relatively widespread use in the 1960s (see *Table 8.8*). In 1975, the Forestry Commission commissioned an independent review of the safety of these materials following adverse publicity linked to its use between 1966 and 1970 in the war in Vietnam.

The report of the review gave full details of the problems arising when a pesticide is used in an unpopular military context and extrapolations are made from such usage to forestry and agricultural practices. The problems arising from dioxin impurities formed during manufacture were also described.

The conclusion from the survey was that 'well-made 2,4-D and 2,4,5-T are safe herbicides when used with care and forethought' (Turner 1977).

As part of its on-going monitoring, the Advisory Committee on Pesticides re-evaluated 2,4-D and its salts and esters. It concluded that all approvals should be allowed to continue, but that further studies should be undertaken. Among other requirements, its Investigation Service was asked to look for incidents of 2,4-D poisoning inconsistent with current evidence (ACP 1992, MAFF Eval 1993a).

Reviews of uses of lindane/gamma HCH

Four studies have reviewed aspects of the use of Lindane, predominantly in respect of its non-agricultural uses (MAFF Eval 1992a, b, 1996, 1997). These had no effect on forestry practice. In 1999, some restrictions were announced on the use of lindane as an agricultural seed dressing (ACP 1999b).

Review of off-label uses

The longer term requirement for off-label use was reviewed in 1994, 1995 and 1997. A current statement of the position in relation to 'off-label use' is given in Annex C of *Pesticides 1999* (MAFF Pst 1999). No restriction on future off-label use is stated. Land types on which off-label use is not allowed include amenity vegetation, parks, motorway verges and other non-cropped land.

4.2   REQUIREMENTS FOR SAFE USE OF PESTICIDES

4.21   Safety in use - the label on the container

'Read the label, especially the safety precautions' has for several decades been the first item of advice in official publications on safe use of agricultural chemicals.

With the advent of the *Control of Pesticides Regulations* 1986, the label on the pesticide container has become the key source of information about safe working practices.   Requirements on the label are intended to ensure that pesticides do no harm to:

*   the person applying them or to anyone handling the treated plant or adjacent soil afterwards;
*   the food chain if used on a food crop;
*   beneficial pollinating insects, especially bees;
*   wildlife, to water or to soil.

Other safety considerations that may be carried on the label are:

*   the *toxicity of the pesticide* product, shown on the product label using set phrases indicating the 'general nature of risk' (§§4.22, 4.23);
*   *safety of the user* requirements, giving instructions to the person applying the pesticide when to wear protective clothing (§4.43), and if necessary, to undergo periodic health monitoring for exposure to particular classes of pesticide (§4.45);
*   *safety of the ultimate consumer*, reflected in specification of 'maximum residue levels' (MRLs) in foodstuffs.   Achievement of such levels when treating food crops, is through observance of label requirements on 'maximum dose rates', 'maximum number of treatments' and specified 'latest timing' or 'harvest interval' between last treatment and first harvesting. Forest and amenity trees are not first choices as foodstuffs!   Even so, 'maximum residue levels' have been set for a few wild plants that may be found in and around woods and trees (§4.28).
*   specific guidance to ensure *safety of the environment* (§4.29).

4.22   'General nature of risk' classification

Under UK legislation, the toxicity classification for dangerous compounds in commerce is given in the HSC *Authorised and Approved List* (of) *Information approved for the classification, packaging and labelling of dangerous substances for supply and conveyance by road (4th edn)* - the CPLD list (HSC 1998).

The list covers industrial chemicals of all sorts; active ingredients of pesticides form only a small proportion of the substances listed.

Many pesticides are not listed because they are not sufficiently dangerous to warrant a hazard rating.

Symbols (black figures on an orange background in a square frame) to be used as visual indications of types of hazard are specified in *The Chemicals*

(Hazard Information and Packaging for Supply) Regulations 1994 (SI 1994/3247). Single letter abbreviations are also specified (see below).

In order to be considered 'dangerous' and included on the CPLD list, substances have to qualify for one or more of 48 'indictions of particular risk'. These are categorised according to the 'general nature of risk'.

The main categories in the 'Classification and indication of general nature of risks' are:

- Very toxic   T+
- Harmful   Xn
- Extremely flammable   F+
- Corrosive   C
- Oxidising   O

- Toxic   T
- Irritant   Xi
- Highly flammable   F
- Explosive   E
- Dangerous for the environment   N

The main 'indications of particular risk' relevant to forestry pesticides are:

- 'harmful...   )   (...in contact with skin',
- 'toxic...   )   (...if swallowed',
                    (...by inhalation',
- 'irritating...   ...to eyes',   ...to respiratory system' or ...to skin'
- 'may cause sensitisation...   ...by inhalation' or ...by skin contact'
- 'contact with water liberates toxic gas',
- 'flammable'.

There are 30 other 'indications' not relevant to forestry pesticides.

4.23   Toxicity ratings

When evaluating a pesticide, tests are carried out to determine its toxicity to a range of test organisms, both in the short term (acute poisoning) and in the long term (chronic poisoning). Tests normally cover intake by mouth and through the skin, and are carried out on small groups of test animals, insects, fish *etc.* as necessary.

Lethal dose - LD_{50}

The most widely used first expression of toxicity is 'LD_{50},' the amount eaten or entering through the skin required to kill half a population of test organisms. Rats or mice are frequently used, results being expressed in milligrams of active ingredient per kilogram bodyweight of the test organism. LD_{50} figures for pesticides classed as dangerous are given in Part VI of the CPLD list. It must be emphasised that tests on different organisms give differing results, both relatively and absolutely. LD_{50} (rats) figures are only an initial, crude guide to toxicity. Nevertheless, the World Health Organisation and the US Environmental Protection Agency both use LD_{50} values for rats in their pesticide toxicity classification, as shown in *Table 4.1*.

Table 4.1   *WHO and EPA toxicity classifications based on LD_{50} and LC_{50}*
             values for rats   *(mg/kg body weight)*

I   *World Health Organisation classification of pesticides by acute toxicity*
           (LD_{50}) values

		Oral		Dermal	
	Class	Solids	Liquids	Solids	Liquids
Ia	Extremely hazardous	≤ 5	≤ 20	≤ 10	≤ 40
Ib	Highly hazardous	5-50	20 - 200	10 - 100	40 - 400
II	Moderately hazardous	50 - 500	200 - 2000	100 - 1000	400 - 4000
III	Slightly hazardous	≥ 501	≥ 2001	≥ 1001	≥ 4001
Table 5 Products unlikely to present acute hazard in normal use		≥ 2000	≥ 3000	-	-

II   *Environmental Protection Agency (US) acute toxicity classification.*

Class	Oral LD_{50}	Dermal LD_{50}	Inhalation $LC_{50}*$	Eye effects	Skin effects
I	≤ 50	≤ 200	≤ 0.2	Corrosive; corneal opacity not reversible within 7days	Corrosive
II	50-500	200-2000	0.2-2.0	Corneal opacity reversible within 7 days; irritation persisting for 7 days	Severe irritation at 72 hrs
III	500-5000	2000-20000	2.0-20	No corneal opacity irritation reversible within 7 days	Moderate irritation at 72 hrs
IV	≥ 5000	≥ 20000	≥ 20	No irritation	Mild or slight irritation at 72 hrs

Source   Tomlin 1997   * LC_{50} = 'Lethal concentration' in air as mg/l
The oral LD_{50} (rats) for aspirin is 1200 mg/kg (Fryer & Evans 1968a).

4.24   Respiratory standards

'Inhalable' material is any airborne substance which enters the nose and
mouth during breathing; 'respirable' material penetrates into the region of the
lung where gas exchange occurs.   The distinction between 'inhalable' and
'respirable' applies to dusts and fibres, and to spray and other droplets. Inhalable

droplets are less than 15 μm (microns) diameter; respirable droplets are less than 5μm diameter.

Maximum exposure limits and *Occupational exposure standards* refer to the amounts of dusts, droplets and vapours hazardous to health that may be inhaled as a result of work activity. Particular limits are specified in the 1994 COSHH regulations (as amended); they impose requirements by reference to the current issue of HSE publication *Occupational exposure limits* EH40/98 (HSE 1998a), Tables 1 and 2 and Appendices 1 and 2. These therefore have legal status. Other text in that publication has 'guidance' status (§3.51). The publication also describes the background to setting and applying these limits and standards.

Exposure limits for substances used as active ingredients in pesticides refer to the specific active ingredient, not to the formulation as a whole.

Maximum exposure limits - MELs

There is a MEL for formaldehyde, a widely used industrial chemical. It is approved for use as a pesticide with 'Commodity Substance' status (§ 3.43), for treatment of soil and compost, and in greenhouse hygiene *etc.*

There are MELs also for inhalable hardwood dust and, since 1996, for softwood dust. These are relevant to any sawmill, especially if the production line also includes sanding. Otherwise, MELs appear not to be relevant to pesticide use in forestry. The case for extending the MEL to softwoods is given in a HSC Consultative Document (HSC 1996c).

Occupational exposure standards - OESs

OESs are less stringent than MELs and are assigned to substances where:

- there is evidence that high levels of exposure may be injurious and
- a lower level can be set at which there is no indication that the substance is likely to be injurious to health following exposure day after day at that concentration. This lower level may be based on a determination of a *'no observable adverse effect level'* (NOAEL).

There are approximately 450 entries in the 1996 OES list; entries range from 'cyanides' and 'warfarin' to 'marble' and 'gypsum' and include 'lindane', 'pyrethrins', 'o-acetylsalicilic acid' (asprin) and 'paracetamol'. 19 entries refer to substances listed as approved pesticides. Such entries are noted in pesticide tables in Appendix II. There is also one solvent, cyclohexanone, which is an essential component of the 'electrodyne' system of applying insecticide (§ 6.51).

Guidance as to appropriate measures to meet OES requirements in practice should be found on the approved pesticide product label or accompanying directions for use.

4.25   Exposure through the skin

For all pesticide users working out-of-doors, exposure to pesticides through skin contact is by far the most serious risk. This is especially so when applications have to be made by knapsack and other manual systems, rather than by tractor-mounted or aerial equipment.

Certain substances are able to penetrate intact skin and become absorbed in the body. The HSE *Occupational exposure limits* list annotates the substances for which this risk is recognised. These include chlorpyrifos, diazinon, lindane and nicotine. In Table 3 of the HSE booklet, a non-statutory 'guidance value' is given for lindane in blood samples, should biological monitoring be undertaken.

Pesticide absorption through the skin is minimised through good working practices, engineering controls and use of personal protective clothing.

4.26   Sensitisers

In addition to the possibility of suffering directly through the ill-effects of absorbing undue quantities of toxic substances, some substances are known to induce changes in the skin or respiratory system. Once these changes have taken place, further exposure to the substance, sometimes even in tiny quantities, can cause, in the case of respiratory sensitisation, symptoms ranging from a runny nose or nose-bleeding to asthma, or for skin sensitisation, reddening and swelling.

Not everyone who is exposed to sensitisers reacts; nor is it possible to predict which individuals are likely to become sensitised. Substances listed in *Occupational exposure limits* which may cause sensitisation are so marked. The only one relating to forestry is 'hardwood dust'.

More detailed information is given in *Preventing asthma at work: how to control respiratory sensitisers* (HSE 1994a).

4.27   Carcinogens, mutagens and teratogens

No pesticides thought to be carcinogenic, mutagenic or teratogenic are approved for use. The following wastes are included in the list of materials classed as potentially carcinogenic:

*   used engine oils,
*   hardwood dust (HSE 1998a).

4.28   Pesticide residues in food

Concern for quality in food is of long standing. In the UK, a report of 50 years of 'National Food Survey' was published (Slater 1990). In 1970, a report on a survey of legislation to control pesticides notes that a committee on pesticide residues first met in 1966, as part of the preparation of the joint FAO and WHO *Codex Alimentarius*. At that time, legislation in the UK had not proceeded beyond regulating lead and arsenic in food (WHO 1970).

Earlier legislation on food and food safety has to a considerable extent been subsumed into the *Food Safety Act, 1990*. Based on that Act, numerous Codes of Practice and Guides have subsequently been issued.

Since 1988, there has been a joint MAFF/HSE working party on pesticide residues in food, answering to the Steering Group on Chemical Aspects of Food Surveillance and the Advisory Committee on Pesticides. The working party's annual report is published (*eg* MAFF PRes 1995).

Three complementary concepts are used to ensure wholesome food supplies:

- No observable adverse effect level (NOAEL);
- Acceptable daily intake (ADI);
- Maximum residue limit (MRL).

The first two relate to the pesticide and the third to the foodstuff. Figures for NOAELs and ADIs are given by Tomlin (1997). Consumption of normal amounts of food containing no more than maximum allowable residues should result in a daily intake less than maximum ADI.

Current maximum residue levels for the UK are given in *Pesticide (Maximum residue levels in crops, food or feeding stuffs) Regulations 1994* (SI 1994/1985).

Maximum Residue Limits (MRLs)

There are 92 pesticides listed in two schedules *of the Pesticide (Maximum residue levels in crops, food or feeding stuffs) Regulations 1994*. These cover the full range of crops raised in the UK. Pesticides listed which might affect edible wild berries *etc*. occurring in forests are shown in *Table 4.2*.

Table 4.2 *Maximum residue levels for nuts, wild berries etc.*   *(mg/kg fresh wt)*

Pesticide	Nuts	Wild berries/fruits	Wild mushrooms
Aminotriazole	0.05*	0.05*	0.05*
Atrazine	0.1*	0.1*	0.1*
Cypermethrin	0.05*	0.05*	0.05*
Glyphosate	0.1*	0.1*	0.05 (?)
Paraquat	0.05*	0.05*	0.05*
Permethrin	0.05*	0.05*	0.05*

*Level at or about the limit of determination
'Wild berries' includes blackberries, raspberries, bilberries

4.29   Pesticide residues in water

Under the *Water Resources Act 1991* (or in Scotland, the *Control of Pollution Act 1974* as amended), it is an offence, without proper authority, to cause or knowingly permit any poisonous, noxious or polluting matter to enter controlled waters. These comprise all inland and coastal waters, including lakes, ponds, rivers, watercourses, groundwaters and underground and open drains (MAFF/HSC 1998, SOAFD 1997).

These provisions have been stimulated by the appearance in groundwaters of nitrate from farming operation and of traces of several pesticides. The latter have been used both in agriculture and for weed control in urban and industrial sites, railways *etc.* The EC Directive 80/778/EEC, in consequence, set limits at the level of 0.1 part per billion for any pesticide in drinking water, within an overall total of 0.5 ppb. Atrazine, in particular, occurs widely in groundwater, frequently at levels above those set. The UK Government White Paper *Our Common Inheritance* (UK Govt 1990) undertook to ensure that these standards would be met, approvals for non-crop uses of atrazine being subsequently withdrawn (ACP 1993).

Papers from a conference *Pesticides in soils & water: current perspectives* review the problems encountered when trying to balance pest control which relies on persistence of pesticide in the soil, with avoidance of pollution (Walker (Ed.) 1991, Fawell 1991). Methodological questions also arise, *eg* how to interpret a single figure standard against fluctuating levels which partly reflect seasonal weather and partly agricultural practice, when both operate over large tracts of land and periods of time. Tracing pesticides in water drawn from an aquifer back to point sources is not feasible (Hance 1989). Nevertheless, peak concentrations following heavy rain and surface run-off are reported, even though subsequently diluted or dissipated. Studies of surface- and groundwater flow during the late 1980s sought to establish base-line data and to model of water flow (Tooby & Marsden 1991, Lee-Harwood 1991, Hollis 1991, Foster & Chilton 1991).

The Advisory Committee on Pesticides reported in 1994 that out of over one million tests of water quality, 98.9% were within EC directive standards for pesticide load. This achievement resulted both from improved treatment and greater compliance (ACP 1994).

Surface-waters and forestry

Pest control in forests and woods in Britain is not on the scale to make any material contribution to groundwater pollution by any currently approved pesticide.

The situation is quite the opposite in relation to the potential for local contamination of surface-waters. In many parts of the uplands, forests not only have many surface streams but, not infrequently, water supplies for small communities and for individual farms come from take-off points within or close to woodland areas. Other stream systems may feed into water supply reservoirs (Hornung & Adamson 1991).

Prior to any pesticide operation in a new area, it is essential by local enquiry to find out whether a local surface-water supply is at risk and, if it is, to agree with the user and the local water authority, the appropriate course of action.

Buffer zones

The *Advisory Committee on Pesticides* (ACP 1991) considered that:

'as a general principal applying to all pesticides used, the protection of aquatic habitats by means of separation distances should be promoted ... as 'Good Agricultural Practice' (GAP) rather than a statutory condition of use.'

Where pesticides are to be used near water, Forestry Commission *Forest and Water Guidelines* recommend that spraying or granule application should not take place within 10 metres of a watercourse or 20 metres of a lake (FC WG 1997).

The *Code of Practice for use of pesticides on farms & holdings* advises 2-m buffers for hand-held applicators and 6-m buffers for tractor-mounted sprayers, unless a greater buffer width is specified in the product approval (MAFF/HSE 1998).

These requirements were modified in March 1999 by a need to undertake a *Local Environmental Risk Assessment for Pesticides* (LERAP) for some ground spraying operations. At the same time, buffer boundaries were redefined. For all spraying operations where previously a 6-metre buffer-zone adjoining streams and lakes had to be left unsprayed, the zone width is reduced by 1 metre, but on the riverside boundary, measurements have to be taken from the top of the bank of the watercourse, and not the water's edge. The same change is made for hand-operated equipment, where the unsprayed zone had previously been 2m. Also, the buffer zone for dry ditches for *all* sprayers is reduced to 1 m.

It is possible to reduce the buffer zone according to the width of the watercourse and the extent to which the spray equipment used is designed to produce minimum spray (MAFF 1999).

Weeds in or near water

A limited number of herbicides are authorised for use in or near water (MAFF Pst 1999, Whitehead 1999). Before use, the appropriate water regulatory body must be consulted. See also the MAFF *Code of Good Agricultural Practice for the Protection of Water* (MAFF 1998b), and SOAEFD Code on *Prevention of Environmental Pollution from Agricultural Activity* (SOAEFD 1997).

4.3   SAFETY IN USE

4.31   Safe working practices

For details of legislation covering safe working practices, see §3.3.

Safe working practices result from:

- correct identification of the pest;
- identification of the appropriate pesticide, and specification for use consistent with minimum essential pesticide use;
- identification of all foreseeable *hazards* (§4.12) associated with a task;
- recognition of the range of *risks* (§4.13) associated with the proposed work; selection of working practice such that the risk of any harmful consequence is acceptably low;
- use of operators who are competent and appropriately equipped.

COSHH assessment

Under the *Control of Substance Hazardous to Health Regulations* both in their original 1988 form and as re-enacted in 1994, 'an employer shall not carry on any work which is liable to expose any employees to any substance hazardous to health, unless he has made a suitable and sufficient assessment of the risks created by that work to the health of those employees and of the steps that need to be taken to meet the requirements of the Regulations' (SI 1994/3246).

Guidance Note 4 of the *Provisional Code of Practice for the use of Pesticides in Forestry* (Anon 1989) describes an approach to such assessments. While that *Guidance Note* was written in the context of the 1988 *Regulations*, it remains relevant to the 1994 *Regulations*. Other more general guides are *A Step-by-step Guide to COSHH Assessment* (HSE 1997b) and *Seven Steps to Successful Substitution of Hazardous Substances* (HSE 1994d).

All other things being equal, the less expensive options may be chosen. However, safety must never be sacrificed just to save money.

4.32   Droplets and drift

Nozzles and spinning discs used for applying pesticides produce droplets of a range of sizes similar to those in fine rain and mist.

Conventionally, the range of droplet sizes produced by a spray nozzle is characterised by the Volume Median Diameter (VMD) or the Number Median Diameter (NMD) of the droplets produced. For any VMD value, half the applied spray *volume* is in droplets equal or greater than the VMD figure, and half the volume is in smaller droplets. Clearly for any given VMD, the number of the smaller droplets must be proportionately larger to offset the volume of the larger droplets. For any NMD, half the *number* of droplets have a diameter equal or greater than the NMD figure, and half the number is in smaller droplets.

Spray droplets are formed:

• where pressure has forced a fluid column of pesticide + diluent in the spray line through a nozzle aperture so that it breaks up and spreads before striking its target. The range of droplet sizes so formed depends on the design and aperture size of the nozzle and operating pressure. The design of the nozzle will determine the spray pattern (BCPC 1994a,b);

• by spinning disc systems ('rotary atomisers' or 'controlled droplet applicators'). These fling droplets off the edge of a rapidly spinning disc, spray mixture (product+diluent) having slid in a thin film across the disc from its centre. Droplet size depends partly on the size and speed of rotation of the disc and partly on the physical properties of the spray liquid. The system generates a much narrower range of droplet sizes, and far fewer fine droplets liable to drift, especially if the disc is toothed. Droplet generation is disrupted if the flow rate onto the disc is excessive.

Spray application and droplet size

In practice, there is a conflict between ideal treatments for good pest control and the parallel requirements to operate safe systems, minimising drift *etc.*

For many weeds, fungi and insects, best control is obtained by cover from a large number of very small droplets rather than fewer larger drops. At the same time, droplets, must reach their target quickly and predictably, with minimum drift.

The British Crop Protection Council has published *Boom* and *Hand-operated sprayers handbooks* (BCPC 1994a, b). These include the BCPC Nozzle Code which characterises spray nozzles. The Handbooks:

- list five *Spray quality classes* - Very fine, Fine, Medium, Coarse, Very Coarse - and give typical uses for each class;
- describe fan, hollow cone and deflector designs and for fan nozzles the main variations in spray angle;
- provide manufacturers' data on nozzle output in terms of spray pattern and quality, and flow rate in litres per minute at specified pressure.

Under the BCPC system, a nozzle described as *F80/1.20/3* produces an 80° flat fan with an output of 1.20 litres per minute at 3 bar operating pressure.

Table 4.3 shows the relationship between spray quality class, spray volume rate and recommended uses for hydraulic nozzles and rotary atomisers.

Spray drift

Smaller droplets fall more slowly than larger ones; in the extreme, if the diameter of the droplet or particle is less than 15μm diameter, it may remain airborne, similarly to water droplets in clouds. In this state they may travel long distances and be deposited on any object lying in their path. Droplets may also be inhaled by operators not suitably protected. Droplets less than 5μm are respirable (§4.24).

The risk of damage by pesticide drift may be minimised by:

- using application systems and nozzles least likely to generate drift;
- using nozzles giving the coarsest spray pattern consistent with effective cover for the pest concerned;
- not spraying in high winds;
- in still weather, timing spraying to avoid warm sunny periods;
- not spraying if there are susceptible crops or other features that might be at risk in the vicinity and the wind is blowing towards them from the area to be treated.

Wind-assisted incremental spraying

This technique has been used for application of pesticides where there is both a dense cover of foliage and access is difficult. The pesticide is sprayed from a

Table 4.3   *Spray quality and associated attributes for hydraulic nozzles and rotary atomiser applicators*

Spray quality	Retention on difficult leaf surfaces	Uses	Potential drift risk	Rotary atomiser droplet range
Very fine	Good	Only exceptional circumstances	Highest	< 90 μm
Fine	Good	Good cover		91-200μm
Medium	Good	Most products and uses		201-300μm
Coarse	Moderate	Soil herbicides		301-440μm
Very coarse	Poor	Liquid fertilizers	Lowest	450μm +

Sources BCPC sprayers handbooks (BCPC 1994a, b)   Anon 1989

controlled droplet applicator (rotary atomiser) as droplets small enough to be wind-assisted to their target. Applications are in successive overlapping bands, so that a relatively even coverage of the whole area is obtained. Formulations are required which can be sprayed through the rotary atomisers at very low volumes to give droplets of the required size *eg* a VMD of 70 μm for control of bracken.

Applications have been either from aircraft for insect control, or from the ground for bracken, heather and woody weed control.

In 1999, active ingredients in products approved for use in the forest included only dimilin (for insects) and asulam (for bracken).

This practice is more dependent on favourable wind conditions for success than spraying using hydraulic nozzles.

Reduced volume spraying

For the less hazardous pesticides, the amount of diluent used per unit area may be reduced without reducing the amount of concentrate. However, explicit restrictions on reduced volume spraying may be stated on the product label.

Using less diluent results in fewer stops to refill and reduces the bulk of material that has to be brought to the site. At the same time, the smaller the volume applied, the finer the spray and the greater the risk of drift.

Table 3 in the *Code of Practice for the Safe Use of Pesticides on Farms and Holdings* (MAFF/HSC 1998) should be followed. See also Guidance Note 8 in *Provisional Code of Practice for the use of pesticides in forestry* (Anon 1989).

Low volume/low use techniques are seen as an important means of reducing pesticide usage in the medium term (Matthews 1999).

Adjuvants

Adjuvants are described in §2.42. They enhance or are intended to enhance the effectiveness of a pesticide when mixed with it (MAFF/HSC 1998). Over 200 adjuvants are given in the *List of authorised adjuvants* in Annex I in MAFF Booklet 500 (MAFF Pst 1999).

Use of the appropriate adjuvant can improve success of treatment and may enable smaller amounts of pesticide to be used. However, many products currently marketed already include adjuvants in their formulation. Expert advice should be sought to determine whether in particular circumstances, addition of an adjuvant is likely to be beneficial.

Any listed adjuvant not mentioned on the label of any particular product may be used with that product, but at the user's risk (Willoughby & Dewar 1995).

4.33   Bees and wildlife

Risks to honeybees have to be assessed if pesticides are being considered for application during plant flowering periods.

In forest areas, bees may be particularly at risk if areas of heather are to be sprayed during its flowering period, especially if hives have been brought to the area for that time. Local bee-keepers or their local spray liaison officer should be kept informed of plans and detailed timings (MAFF/HSC 1998).

Wildlife

Pests, and in particular, weeds may properly be part of the wildlife of a site. In assessing the effect of the use of a pesticide, the dynamics of pest populations have always to be evaluated.

When considering weed control operations, recognition has also to be given to the role of weeds as food-plants for many soil invertebrates which, in turn, may support larger creatures. Part of the lack of bird-life in extensive areas of cereal crops is due to the absence of herbaceous plants which supported insects on which birds rely.

Herbicides nevertheless have a place in the management of vegetation on sites where nature conservation is important, provided effects on all species present are evaluated against all the conservation objectives (Watt *et al.* 1988).

4.4   OPERATOR SAFETY

4.41   Occupational health

Concern for the health of users of pesticides has, from the outset, been a element of provisions for their use. The issue of *Recommendations Sheets* under the *Pesticides Safety Precautions Scheme*, briefly mentioned in §3.22, set out recommended precautions for approved uses, as agreed between manufacturers and Government departments. They remained in being until the introduction of the *Control of Pesticides Regulations 1986* and subsequent legislation.

The older recommendations were designed to ensure adequate protection for those applying pesticides. This was particularly important to forestry because of the relatively high proportion of manual operations compared with agriculture, and forest workers' closer contact with pesticides (Watts 1968).

The 1974 Health & Safety at Work Act marked the beginning of an increasing number of regulations seeking to improve safety in the work-place, both in industry and on the land. Concern that, in spite of these further provisions, there was more ill-health among farm workers than was being reported, led to a Parliamentary Select Committee considering the effects of pesticides on human health (Body 1987).

The outcome was an industry-wide requirement to consider the whole work-place and all its activities. This requirement was embodied in the *Management of Health & Safety at Work Regulations* 1992 (SI 1992/2051) and in Approved Codes of Practice on *Management of Health & Safety at Work* (HSC 1992a), and *Workplace Health, Safety & Welfare* (HSC 1996b). Among other aspects, greater consultation was required than formerly between employers and employees in respect of working practices and conditions.

Managing health and safety in forestry (HSE 1999b) provides overall guidance for forestry.

Health monitoring

While provision for health monitoring has been made in the legislation, the requirements have been specific to limited classes of products; there has been no requirement for routine monitoring of health of forestry workers.

Recent leaflets offering general guidance to good health include:
COSHH in Forestry (HSE 1999b),   *Farm Forestry Operations* (HSE 1995e),
Health & Safety guide for gamekeepers (HSE 1994c).

4.42   Operator competence

From the introduction of the *Control of Pesticides Regulations 1986*, operator training and competence have been stressed as essential components of safe use of pesticides in agriculture, horticulture and forestry. In this context, 'operator' includes employees and self-employed persons applying pesticides; it also includes people directly assisting such work *eg* loading a sprayer for another operator (MAFF/HSC 1990 & 1998).

Initially, operators required a certificate of competence unless they were born before 31st December 1964, or were working under close supervision from a certificate holder (MAFF/HSC 1998). Proposals were made in 1999 that all operators should have certificates of competence but no timetable was stated.

The subject matter of pesticide application has, for certification, been split into 12 modules. All operators have to pass the *Foundation Module* together with any other module directly related to their work.

Modules relevant to woodland and amenity tree pesticide treatment include:

* Tractor-mounted or trailed sprayers or granule applicators;

- Hand-held applicators; • Mixer/loaders; • Aerial application.

Schedules of the modules drawn up by the *National Proficiency Tests Council* (NPTC 1994) are available from:
- National Proficiency Tests Council, Avenue J, National Agricultural Centre, Kenilworth Warwickshire CV8 2LG
- Scottish Skills Testing Service, Ingliston, Edinburgh, EH28 8NE.

Training
Details of training for proficiency in safe use of pesticides and preparation for certification are available from local Agricultural Colleges, LANTRA (formerly ATB Landbase), Training Providers, Forestry and Arboriculture Safety and Training Council (FASTCo, 231, Corstorphine Road, Edinburgh EH12 7AT), or the Health & Safety Executive regional offices.

Other guides
The *Forestry & Arboriculture Safety & Training Council* (FASTCo) (formerly the *Forestry Safety Council*) maintains up-to-date safety guides.

Safety Guides 101, 102 and 202 cover insecticide dipping at planting, and application of pesticides in young crops (FASTCo 1990, 1994a, b). FASTCo also issues guides on other means of weed control (FASTCo 1994c, FSC 1981).

Marketing and storage competence
Under the *Control of Pesticides Regulations 1986* (amended 1997), personnel involved in 'sale and supply' of pesticides for agricultural use, or who are concerned with 'storage of pesticides for sale or supply', require certificates of competence. These are obtained by examination under the auspices of BASIS (Registration) Ltd., 34, St. John St., Ashbourne, Derbyshire DE6 1GH.

There is one certificate for storekeepers. However, separate certificates are issued for forestry, horticulture, agriculture and 5 other crop types for which pesticides may be supplied. A list is given in the *Code of Practice for Suppliers of Pesticides to Agriculture, Horticulture and Forestry* - 'The yellow code' (MAFF 1998).

The code also recommends that advisers not directly involved in sale or supply, also qualify for an appropriate BASIS certificate of competence.

4.43 Personal protective clothing & engineering solutions

The earliest recommendations for safe use of herbicides in Britain advised users to follow published *Recommendations for safe use* for each active ingredient (Aldhous 1965a); see also §3.22.

Engineering to avoid operator exposure
Reliance was initially placed on protective clothing to minimise contact between people applying pesticides and the chemical applied. From the 1970s, alternative approaches tried to minimise operator exposure by mechanising the heaviest work (Wittering 1974, Lane 1983, 1984, 1990a, b). Since 1990, more

than 20 reports and technical notes on aspects of pesticide application have been issued by the Forestry Commission Technical Development Branch (TDB 1997).

Over the period between the 1960s and the 1990s, other engineering and manufacturing developments also reduced risks to operators:

- the toxicity of newer pesticides has been markedly less than many of the materials first developed, so that the hazards to the operator are less;
- the design of equipment has improved, reducing risks of drift during application, and splash and drip during maintenance (*eg* Landers 1989, Matthews 1999);
- low-pressure non-drift systems *eg* the 'weed wipe' are now available;
- improved design of equipment for mixing concentrate and diluent and for rinsing containers has reduced the risk of splash during these operations.

From this and similar work, the principle has been established that wherever there is a risk to the operator, the first approach should always be to reduce exposure by 'engineering'. Protective clothing should be the back-up, not the first line of protection.

It is now a requirement under *COSHH Regulations* that wherever protective clothing is recommended on the product label, engineering control of operator exposure must in addition be used where reasonably practicable.

Protective clothing for use with forestry pesticides comes under the scope of the *1994 COSHH Regulations* (HSC 1995a), even though protective clothing for other aspects of forest work, for example, chainsaw felling is covered by *Personal Protective Equipment at Work Regulations* (SI 1992/2966, HSE 1996c),

Heat stress

As a side effect of manual application of herbicides, it was soon obvious that men working in warm conditions in difficult terrain were likely to become hotter and uncomfortably sweaty in the additional protective clothing (Aldhous 1969a).

Heat stress associated with protective clothing has been reviewed in the context of 'rest allowances' and standard time calculations for forest work. In practice, the 'forecast ambient air temperature' has been found to be a good guide to the risk of heat stress when waterproof or similar jacket, coat and hood, and face mask are required to be worn. If the forecast air temperature under still sunny humid conditions is 20°C or more, additional 'rest factors' are recommended to allow operators to cool down (Hodgkiss 1990, Chadwick 1992a, 1992b).

There should always be an effective system of maintaining personal protective clothing, to ensure it gives the protection for which it was designed.

Table 4.4 summarises requirements for protective clothing where engineering or other control measures do not provide adequate control of exposure; the requirements are supplementary to any stipulations on the product label.

Table 4.4   *Protective clothing for permitted uses of pesticides in forestry**

Hazard as shown on label	Operation				
	Preparing pesticide for any use	Standard application		Reduced volume Outdoors: Nozzle sprayers	Cleaning up or handling freshly treated material
Hazard letter		Liquids	Solids, baits, granules		
Toxic	gloves	gloves	gloves	USE	gloves
	apron	-	-		-
T	coverall	coverall	coverall	NOT	coverall
	boots	boots	boots		boots
	hood	hood	-	PERMITTED	-
	face-shield	face-shield	-		face-shield
	RPE**	-	-		-
Harmful	gloves	gloves	gloves	gloves	gloves
or	coverall	coverall	coverall	coverall	coverall
Irritant	boots	boots	boots	boots	boots
Xn or	hood	hood	-	hood	-
Xi	face-shield	face-shield	-	face-shield	face-shield
Not	gloves	gloves***	gloves	gloves	gloves
classified	coverall	coverall	coverall	coverall	coverall
on label	boots	boots	boots	boots	boots
No letter	face-shield	-	-	face-shield	-

Source   Table 1 *Approved Code of Practice* for the *Safe use of Pesticides for Non-agricultural Purposes* (HSC 1995a).

*   See Approved Code if corrosive substance fogs or mists are to be used.
** RPE = Respiratory protective equipment
*** Required only for hand-held applicators.

4.44   Dangerous and undesirable working practices

Apart from hazards arising directly from pesticide usage, a number of dangerous practices during application have been identified. They should be covered in operator training and certification and include guidance such as:

- Never blow through pipes or nozzles to clear them;
- Never continue work after being contaminated by a pesticide unless all affected skin has been washed and all contaminated items of clothing have been removed and replaced with clean ones;

- Never allow pesticide waste or surplus material to enter a water course, drain or sewer, or percolate into ground water, unless in accordance with the terms of a consent from the local Water Service Company, relevant water authority or environment protection agency;
- Never leave containers containing pesticides near areas used by children as playgrounds, *etc.*
- Do not attempt to burn any pesticide concentrate.

4.45   Medical checks on health

Some pesticide active substances affect body function to the extent that operators using them may require periodic medical checks.   These active substances fall mostly into two classes, organophosphorus compounds and carbamates which are classed as 'Toxic' or 'Very Toxic'. The product label will carry a warning phrase 'XXX is an organophosphorus (or anticholinesterase) compound.   DO NOT USE if under medical advice not to work with such compounds.' Records of the results of health surveillance checks have to kept for 40 years. See *Health surveillance under COSHH* (HSE 1995h, also HSE 1995g).

Active substances which can act as sensitisers may also need to be monitored.

Suspected pesticide poisoning

Whenever poisoning by pesticides is suspected, medical help should be sought immediately. If possible, the doctor or hospital should be given a copy of the product label or told the name of the product and the active ingredients in it.

The British Agrochemicals Association has distributed to emergency services a booklet, *Pesticide poisoning, notes for the guidance of medical practitioners*, giving specialist advice. The *National Poisons Units* and major manufacturing companies operate 24 hour emergency telephone services (BCPC 1996).

The names and addresses of the seven *National Poisons Information Service* hospitals are given in Appendix 2 of *The UK Pesticide Guide* (Whitehead 1999).

Reports of injuries, diseases and dangerous occurrences regulations

These regulations (RIDDOR) require employers to report any work-related injury or accident that for more than three days prevents an employee (incl. self-employed) from working or doing their full range of duties (HSE 1996a, 1999c).

4.5   STORAGE OF PESTICIDES

The amounts of pesticides kept in store should be kept to the absolute minimum, consistent with operational needs. This may include storage for short periods when application has been delayed by bad weather. For many smaller scale users, a purpose designed storage chest or bin should be adequate. Hazard warning signs need to be displayed.

In planning a store, provision has to be made, not only for day-to-day stock control, but also for emergencies, such as major spillage or leakage from a pesticide container. Impermeable floors should be installed and designed so that spillage drains to a sump or to where it can safely be retained by a low bund.

Guidance notes on storing pesticides for farmers and other professional users have been available from the Health and Safety Executive (HSE 1988, 1997a). See also:

- BS5502 'Buildings and structures for agriculture' part 81 (1989), *Code of practice for design and construction of chemical stores.*
- British Agrochemicals Association check-list: *Good Agrochemicals Storage* (BAA 1996).
- British Agrochemicals Association handbook *A Guide to the selection and use of amenity pesticides* (BAA 1998b) - covering the range of pesticide uses as well as storage and disposal of pesticides.

BASIS Certification is required for any storekeeper controlling pesticides 'for sale or supply'. Contractors expecting to use stored materials on behalf of a client do not require 'storekeeper' certification.

Moisture-activated gassing compounds

Materials used for rabbit control, which are normally activated by soil moisture, require a separate dry storage compartment made of metal or other fire-resistant materials. The store should be located away from direct sunlight and heat, and marked 'Gassing Compound - do not use water'.

4.6   SAFE DISPOSAL OF PESTICIDE WASTES

Wastes should be minimised by:
- using returnable containers or pesticides in water-soluble packaging;
- applying as much as possible of container rinsings and washings to the site as part of the intended application;
- purchasing no more concentrate than required in the short term.

The means of disposing of non-returnable cleaned containers should be agreed with the local waste regulation authority.

Residues from plant dipping operations should be disposed of through a registered waste carrier for disposal at an appropriately licensed facility.

Guidance on safe disposal of forest pesticides used for non-agricultural purposes is given in Appendix 4 of the *Approved Code of Practice* for the *Safe Use of Pesticides for Non-agricultural Purposes* (HSC 1995a).

Permission may be required under *The Groundwater Regulations 1998* before waste agrochemicals (including pesticides) may be disposed off to land *eg* by spraying over waste ground (SEPA 1998, 1999, MAFF 1998b). See the section on 'Disposal' in *Using pesticides* (BCPC 1999); see also §3.5 above.

4.7   PESTICIDE USE AND ABUSE

4.71   Surveys of use

The Advisory Committee on Pesticides reports annually on results of surveys of quantitative pesticide usage in sectors of land husbandry. Surveys initially covered England and Wales but have been extended to Scotland, and subsequently to Northern Ireland.

In most years, the surveys are not relevant to forestry or amenity tree management; a forestry survey was reported in 1991, figures being incorporated in tables elsewhere in this text.

4.72   Accidental and deliberate misuse

Accidental misuse

The Health and Safety Executive report annually the number of complaints received of accidental or deliberate misuse of pesticides in agriculture, horticulture and forestry. Over the period 1995-1998, complaints were of the order of 200 per year, about half of them alleging ill-health from by exposure to pesticides during a work activity. The HSE 'Pesticides Incidents Appraisal Panel' considered that overall, for about 10% of the complaints there was likely to be a link. Some successful prosecutions have arisen out of these complaints.

Reports on poisoning of wildlife

The Advisory Committee on Pesticides reports annually the scale of animal poisoning. In 1994, 763 incidents were noted. The cause was established in 212 cases, deliberate abuse occurring in 110 of them. Similar figures were reported in other years. The main targets have been foxes and crows.

In 1987, Agricultural Departments launched a 'Wildlife Incident Investigation Service' to investigate suspicious deaths and injuries to wild birds, livestock, pets and bees. A MAFF report on *Pesticide poisoning of animals 1988-1991*, showed links between approved uses of poisons and animal deaths:

Creature	Cause		Creature	Cause
• Foxes	rodenticides		• Hedgehogs	molluscicides
• Hares	paraquat		• Wildfowl	seed treatment
• Dogs	molluscicides		• Game birds	seed/eelworm t'ments

These investigations led to the *Campaign against illegal poisoning of animals* to publicise and try to reduce abuse (ACP 1991). It is supported by a range of organisations associated with animal welfare, nature preservation, field sports and gamekeeping. The campaign started in 1991 and continues.

Further information is available from *Pesticides Safety Directorate*, Room 317, Mallard House, Peasholme Green, York YO1 7PX (Whitehead 1999).

CHAPTER 5

Integrated forest protection

'Stop the world - I want to get off' is not an option

5.1   MINIMISING PESTICIDE USE

Historically, pests have accounted for substantial proportions of man's crops, food and other goods; if unchecked, they will continue to do so. Land husbandry in all its forms will not be able to meet world needs without continuing to control pests of all sorts. *Table 2.1* and §2.11 put the global position briefly.

The most recent UK Government approach to sustainable forestry, set out in *The UK Forestry Standard* (FC 1998), specifies that the use of pesticides should be no more than that necessary for the effective control of pests and compatible with protection of human health and the environment.   Integrated forest protection is an important means of implementing that policy (§5.2).

Fortunately, the coordinated integration of:

- changed techniques of husbandry,
- increasingly selective pesticides,
- greater knowledge of host and pest life cycles,
- greater knowledge of natural predators, parasites *etc*. on pests,

offers the potential in many specific circumstances both to minimise crop losses and to reduce pesticide use.

In recent years, such integration in agriculture and horticulture have come to be referred to as: 'Integrated Pest Management' (IPM) or more broadly, 'Integrated Crop Management' (ICM) according to circumstance (Wainhouse 1987, BAA 1994a, McKinlay & Atkinson 1995). For forestry, *'Integrated forest protection (IFP)'* seems an appropriate equivalent expression.

Minimising fertilizer use

The use of chemical fertilizers, the relationship between these and the nutrients needed to sustain forests growth, and the impact of pollutants, particularly compounds of nitrogen and sulphur, are reviewed in Chapters 10-14.

While the same principle applies, of using the least amount of fertilizer necessary for forest growth, the actual quantities used over the life of a woodland plantation are miniscule compared with agricultural or horticultural usage.

5.11   Forest protection strategy

Formulation of an IFP management programme, while important, is not, however, the first step.

Wherever trees are growing, they are likely to be the target of pests or diseases or be subject to competition from weeds *etc.* Some form of *Protection strategy* is desirable at both national and estate level, so as to ensure good growth of healthy trees, wherever trees and woodlands are under management. While individual circumstances will dictate the details, some principles should be common to all strategies.

Production of such a strategy should form part of preparations for either a business 'Health risk management assessment' (HSC 1992a, 1997b) or as the foundation for a 'COSHH assessment' in relation to pesticide use (HSC 1995a; MAFF/HSC 1990, 1998).

Three important components of an 'Integrated Forest Protection strategy' are:

• to recognise the need, and where appropriate to act, to prevent further addition to the list of introduced pests that might threaten trees or woodlands;
• to control movement of plants and timber so as to minimise the risk of spreading pests and diseases;
• to be aware of fluctuations in native and introduced pest populations and to know when and how to act to prevent widespread losses with minimal harm to the environment.

5.2   INTEGRATING PROTECTION AND FOREST MANAGEMENT

Integrating protection with management is not a novel concept to foresters. Wood (1968) pointed out:

'The forester is rarely able to provide blanket protection for his crops by chemical means, and the eradication of widely distributed insect or fungal pests from the forest environment is not regarded as biologically feasible. Forest protection is essentially applied ecology. Where the life habits of harmful organisms are sufficiently known, silvicultural and harvesting practices are adapted to avert damaging increases or concentrations of pest populations. Pesticides fulfil their most important roles when used at critical stages in the life of crops, or at specially vulnerable stages in the life history of important organisms. Such "stages" should be localized and accessible.'

5.21   Cultural   and   technical   means   of   minimising   pesticide dependence

Integrated forest protection (IFP) is *not* a 'single recipe cure-all'; it has to be tailored to pest species individually, according to their behaviour and potential effect on the local woodland. Forest and woodland management can often minimise dependence on pesticides if account is taken of the particularities of pest and crop; appropriate practices may be followed just as much in small-scale operations as in large, and are usually beneficial economically.

The various ways in which adaptations can be made are tabulated in *Table 5.1*. Fuller accounts are given in the subsequent chapters as indicated.

There will also be situations where IFP is not feasible.

Table 5.1   *Summary of opportunities for Integrated Forest Protection*

Management action	*Relevant pest &/or effect*	*Chapter/section*
No action		
None	Sooty bark disease *Cryptostroma corticale*	7.55
	Douglas fir wooly aphid *Adelges cooleyi*	6.74
Host plant management		
Select resistant variety or seed origin	Bacterial canker of poplar *Xanthomonas populi*	7.51
	Douglas fir needle cast *Rhabdocline pseudotsugae*	7.51
	Larch canker *Lachnellula willkommii*	7.51
Cut & destroy infected plants & propagating material	Watermark disease of Cricket bat willow *Erwinia salicis*	7.52
	Elm lightly infested with DED in isolatable areas	7.31
Dig out stumps and roots	Conifer heart rot *Heterobasidion annosum*	7.21
	Honey fungus *Armillaria* spp	7.22
Find disease-free isolated sites	Nursery production of Western red cedar free from *Didymascella (Keithia) thujina*	7.66
Avoid wide spacing	Achieve earlier crown closure and suppression of competing woody & herbaceous weeds	8.2
Site management		
Avoid burning lop & top	Group dying of conifers *Rhizina undulata*	7.56
Allow fallow period before planting	Reduction of attack by *Hylobius abietis*	6.51
Use fences, shelters, sleeves, collars	Exclude deer, rabbits, voles	9.22
	Hylobius abietis	6.51
Use mulches	Weed suppression	8.24
Plant on ploughed land, mounds *etc*	Quicker establishment at initial planting and at restocking; less need for weeding	8.23
		8.24

(Continued overleaf)

Table 5.1 contd   *Summary of opportunities for Integrated Forest Protection*

Management action	Relevant pest	Chapter/section
Regulate pH in forest nurseries	Maximise growth, reduce weeds	11.12
	Reduce risk of damping off	7.61
Remove hedges & belts of spp that may harbour pests	Avoid beech aphis, *Didymascella,* pine needle diseases, fire blight, plum pox	6.73

Operational management

Raise stock in containers	Avoid risk of *Phytophthora* on heavy texture soils	7.61
Keep 0.5m weed-free round young plants	Higher survival, quicker growth, less risk of attack by voles to newly planted trees	8.21 9.26
Consider trapping	Integrate to minimise use of warfarin in grey squirrel control programme	9.25
Observe controls on timber movement	Minimise spread of harmful insects	5.34, 6.2
'Hot log' pine	Minimise risk of blue stain	7.32
Consider wet storage	Prolong storage life of pine timber	7.32
Co-ordinate spraying with forecast weather	Avoid failure through wash off or excessive drift in windy conditions	4.32

Application management

Establish safety ethos among users	Create confidence that for *all* aspects of application, proposed safety procedures are appropriate	4.3, 4.4
Favour application techniques that minimise pesticide use	Placed sprays	8.2
	Electrodyne technique	6.51
	Rotary atomisers for aerial spraying	6.24
	Low drift nozzles	4.32
Consider slow release formulations	Extend persistence/availability	6.74
Favour rapidly bio-degradable materials	Pesticides; also, chainsaw oils; hydraulic fluids	15.33

5.3 VIGILANCE

I didn't know the gun was loaded ...

5.31 Introduction(s)

Pests and diseases seldom arrive to a flourish of trumpets; nor do weeds always announce their coming. Control of introduced pests whose arrival resulted from lack of vigilance accounts for a substantial proportion of global pesticide use. Internationally, Dutch elm disease, white pine blister rust, gypsy moth are a tiny selection of organisms which, introduced outside their native range, have caused havoc to trees and forests (Phillips & Bevan 1967).

National and international forestry plant health authorities have sought to establish a statutory framework to protect trees and forests against introductions of new pests (Phillips 1980a). Even so, commercial pressures to weaken or obtain derogation from regulations or to ignore them continue. Failure to resist them has been disastrous.

In the UK, past unwitting (if not witless) commercial activity resulted in the introduction of the Great spruce bark beetle *Dendroctonus micans* into the UK and accelerated the spread of Dutch elm disease (§6.53; §7.31). Fireblight (*Erwinia amylovora*, §7.52) and horse chestnut scale (*Pulvinaria regalis*) (§6.73) are recent examples of pests introduced by means unknown, first found here in 1957 and 1964 respectively (Phillips & Burdekin 1982, Wainhouse 1994).

Major threats remain, particularly *Ips typographus*, the eight toothed bark beetle (FC PH 1995, 1996, Evans 1996) (§6.55). This insect has been trapped in varying numbers at ports and, in 1997, at a pulpmill yard. Intensive searches have failed to find any source established in the UK; overall experiences indicate, fortunately, that *I. typographus* may not be able to become established in Great Britain as readily as had been feared (FC PH 1998a,b).

Asian longhorn beetle

At the end of 1998, the Forestry Commission invoked emergency powers against an Asian longhorn beetle, *Anoplophora glabripennis* (SI 1998/3109). Live adult beetles and larvae in packing material imported from China had been found in a number of places in England and Wales and a risk assessment showed that the insect could damage a wide range of forestry and amenity trees in UK.

Import controls have been imposed requiring all hardwood packing material originating in China to be bark and grub-hole free (Anon 1998, Perry 1999 (illustrated)). The insect has also become established in USA (FC NR 1998b).

5.32 'Keep 'em out' - Cross-frontier controls

It has been long recognised that legislation is required to control the import of plants and wood which are potential carriers of pests. Earlier, UK legislation (*eg* SI 1952/1929) looked to domestic needs on restriction of entry of pests. In the last two decades, European Community Directives have taken over. Current requirements applying within the European Community arise from Community

Directive 77/93/EEC and amendments (Burdekin & Phillips 1977, Phillips & Burdekin 1982, Gibbs 1984, Hansen 1985, Calderwood 1996).

The *Plant Health (Great Britain) Order, 1993* (SI 1993/1320) implemented the provisions of the EC directives. Forest tree plants intended for planting are regulated separately from 'forestry material' (definition on p 81) (FC PH 1994c).

The principle behind the Directives currently in force is that precautions against clearly identifiable threats must be allowed. Otherwise, restrictions on trade should be minimised lest they become a disguised form of trade protection.

Nevertheless, unintentional spread of pests through international trading has occurred repeatedly, usually at the expense of the grower rather than the trader.

While legislation can minimise risks, whole-hearted observance of requirements on international trade aimed at avoiding introductions of pests is an integral part of minimising use of pesticides. If the pest can be kept out, control measures do not have to be considered.

Imports from within the European Community

Plants: 'Plant passports'

Prior to the implementation of EC Directives, plant health inspections took place at customs posts at national borders or other agreed locations and often required the production of a 'phytosanitary certificate'. These arrangements have been replaced by preventative measures directed towards plant genera for which there are recognised risks to health if becoming infected. Listed genera are defined as 'controlled genera'.

For the production of planting stock of species or varieties of 'controlled genera' in the nursery, nurseries have to be registered and inspected. As long as no evidence of insect pest or disease *etc.* has been found, the nursery manager is allowed to trade in the species inspected (Aldhous 1994) but has to include the nursery's passport number on advice notes. The aim of the passport system is to be able to trace back to its source, any material which has been found inadvertently to be infested with an undesirable pest of disease. Genera for which passports are required when woodland trees and shrubs are moved include:

- *Crataegus, Malus, Prunus, Sorbus* (except *S. intermedia*), *Pyrus, Cotoneaster, Pyracantha, Chaenomeles, Cydonia etc.* (all potential host species for fireblight (*Erwinia amylovora*); also *Abies, Castanea, Larix, Picea, Pinus, Populus, Pseudotsuga, Quercus* and *Tsuga*.

Where a country or part of it is free from a particular pest that is present elsewhere in the community, the uninfested area may be given 'Protected zone' (PZ) status. Such areas are defined as being 'free from injurious pests of pathogens, even though conditions for their establishment are suitable'. All of Great Britain is a 'PZ' in respect of *Ips typographus*. Only parts of Britain are in a PZ in respect of *Dendroctonus micans* (FC PH 1993a, b).

In order to maintain PZ status, surveys have to be carried out to determine whether a zone remains free of the pest against which it is protected. The first survey of zones in Britain confirmed their initial status (Evans H.F. 1995).

Forestry material

Under the terms of the EC Directive, forestry material includes Christmas trees, wood, articles of wood, and isolated bark of one or more species or genera to which the plant health controls apply (FC PH 1992). All importers of wood are under an obligation to check all material received and to notify the Forestry Authority of any apparent pest infestation (FC PH 1993b, 1994c).

Passports have not been required for movement of wood and wood products within the domestic market in the United Kingdom (Calderwood 1996). At the same time, the possibility was considered for conifer wood moving into and out of the *'Dendroctonus micans* Control Area' (FC PH 1996).

Imports from areas outside the EC

Plants

Schedules 3 & 4 of the *Plant Health (GB) Order 1993*, (SI 1993/1320) detail restrictions and some complete bans on plant imports. These apply to the more important conifer genera and to some broadleaved genera including *Quercus, Ulmus,* and *Castanea*. Otherwise, imported stock has to meet requirements similar to those from EC countries, except that a phytosanitary certificate (*synonym,* Plant health certificate) is required rather than a plant passport.

Seeds

There are no formal requirements for tree and shrub seed imported into Britain to be covered by phytosanitary certificates at the time of import. There should, however, be valid seed germination test certificates.

Wood and wood products

Imports of all conifer wood and specified hardwoods require a phytosanitary certificate issued by the Plant Protection Organisation in the country of origin, except where alternative forms of certification, tailored to perceived threats, have been agreed.

'Oak wilt' is an example of a disease not yet known in Britain but which could cause damage similar to that wreaked by Dutch elm disease. Under the *Import and Export of Trees, Wood and Bark (Health) (Great Britain) Order, 1980*, oak wood from North America must carry a certificate that all bark has been removed and that either the moisture content of the wood does not exceed 20%, or the wood has been disinfected by hot air or hot water, or the wood has been squared so as to remove all of the natural surface (Gibbs 1984).

Another perceived threat to Europe as a whole is the Pinewood nematode (*Bursephelenchus xylophilus*) from North America. Timber imports are required to have been heated so that a core temperature of 59 - 60°C is reached for half an hour (TTJ 1992). Heat treatment certificates approved by plant protection authorities in importing counties are acceptable (FC PH 1993b). See also §6.81.

Export of plants and wood

Export of plants and wood within the EC has to follow the arrangements for issue and use of plant passports (FC PH 1994c - under revision).

For exports outside the EC, phytosanitary certificates are required. For plants, these have to be obtained through the MAFF 'Plant Health and Seed Inspectorate' or through the Scottish Office Agriculture and Fisheries Department (SOAFD).

5.33   'Fire brigade' action

Where a recent introduction of insect pests or diseases is still localised but is judged to be a major threat to woodlands if spread from the locality where it was discovered, the national protection strategy has been to contain the outbreak, on the analogy of a 'fire brigade' response. This has been done by:

 • reducing the scale of sources of infection by sanitation measures *ie* felling and burning, or spraying;
 • seeking to break the reproductive cycle of the pest;
 • containing the outbreak by preventing movement of potentially infective timber, bark or other plant material out of the infested area.

The success or otherwise of 'eradication' depends on timing and scale. The earlier an introduction can be tackled and the smaller the outbreak, the greater the prospect of success.

Eradication of the recent introduction of Gypsy moth (§ 6.44) seems to have a reasonable chance of success. *Dendroctonus micans* was too well established when first discovered in 1982 for eradication to have been a practicable proposition. Nor can one be confident that, had the *Dendroctonus* been discovered within a few years of its arrival, it would have been possible to eradicate it, because of the inaccessibility of some of the sites where it breeds.

A policy of eradicating affected host trees has been applied to the control of watermark disease in cricket bat willow. It has not eliminated the disease but has been effective in reducing its incidence to a low level (Preece 1977) (§7.52).

Contingency plans for control of rabies, should it be found in the wild in the UK, have included as an option the eradication of susceptible wildlife, especially foxes, that are at risk of infection. It is to be hoped that the need will not arise, but the prospect is daunting.

5.34   Don't spread them around

Where pests or diseases have become established in Britain but over a limited area, other provisions have been necessary to minimise spread within the UK.

Restricting the movement of timber

There have been two series of Orders restricting movement of potentially infested timber from areas known to have infestations into uninfected areas:

- *Dutch Elm Disease (Restriction on movement of wood) Order 1974* No 767 (SI 1974/767) and subsequent amendments, prevented movement of elm logs from within infested areas, unless debarked;
- the *Restriction on movement of spruce wood order 1982.* (SI 1982/1457) as amended, for areas affected by *Dendroctonus micans* (§6.53). This required more elaborate restrictions on movements.

In retrospect, the restriction order on elm had little visible effect on the ultimate scale of infestation of Dutch elm disease, though it may have slowed the rate of spread. Mature elms remain in the landscape only in those localities where movement restrictions, geographic isolation and a continuing 'fell and burn' sanitation programme have been possible, or where beetles are less active.

The spread of the area infested by *D. micans* increased only marginally after the initial natural spread of beetle populations. In this case, movement restrictions, eradication treatment to reduce insect numbers locally and the longer term control through the introduced predator, *Rhizophagus grandis*, appear to have been effective in avoiding further rapid spread.

Inadvertent carriage

The ubiquity of road transport and the widespread use of low quality wood as packing material in containers, dunnage *etc.* raises the probabilities both of spread of existing diseases and introduction of new ones, in spite of legislation. The nature of potential problems is illustrated by the egg masses of Gypsy moth (*Lymantria dispar*) shown in *Arboricultural Research & Information Note No. 124* (Winter & Evans 1994), and the recent Asian longhorn discovery (§5.31).

Requirements for dunnage are listed in *Plant Health Newsletter 6* (FC PH 1994a). While there are some exceptions, dunnage must meet the same specification in relation to avoidance of plant health risks, as for sawn wood of the same species from the same country of origin.

5.35 Monitoring pest populations

Integrated Forest Protection management programmes cannot be drawn up without knowledge of the local scale and more general relevance of threats of any particular pests.

In anticipation of this need, Integrated Forest Protection strategies have to ensure the availability of up-to-date information on significant changes in numbers of the more harmful pests, and occurrence of signs of damage to trees.

It is to be hoped that the Forestry Commission will continue to assess and publicise changes in major pest populations, and to monitor ports and sites of major timber importers for new introductions.

Nevertheless, forest managers and their staff fail in their duty if they also do not keep a continual watch for early indications of pest or disease attack and ensure that anything unfamiliar is properly investigated. Experienced forest staff are thin on the ground. If the responsible person does not check unexpected crop damage, it is unlikely that anyone else will be there as a back-up.

5.4   NATURAL ATTRACTANTS

5.41   Pheromones, kairomones & semiochemicals

The discovery of pheromones arose out of studies in the late 1950s of chemical communication between insects. The substances involved occur in extremely small quantities, similar to those occurring in scents. While a range of types of chemical compound were discovered, responses to them mostly related to reproduction. In a significant number of instances, release of a pheromone would bring large numbers of the opposite sex to aggregate round the pheromone source. In other instances, pheromones drew attention to a food source but with the same consequence of attracting other individuals of the species in the vicinity.

Other scents (kairomones) may attract a predator while 'semiochemicals' (signalling chemicals) may attract insects to a host food plant.

5.42   Monitoring major pests

Those insects which are thought to constitute threats to the survival of healthy woodland have been monitored regularly to detect build up of numbers. The changes in insect population numbers identified through such monitoring has provided the basis for decision-making in relation to timing and area of control operations. See §6.41 Pine looper, §6.42 Pine beauty and §6.53 Great spruce bark beetle for details of monitoring for these species.

Monitoring using biologically active compounds

It soon became apparent that one potential role for any pheromone compound which could be manufactured, was as a lure to attract insects to a trap, often a sticky substance. Pheromone traps are now widely used, the numbers caught under standard conditions indicating the level of population in the vicinity of each trap (Jones 1987, Ridgeway *et al.* 1992).

Traps used for monitoring are made up of two basic elements:

- a pheromone formulated in a device to control its release;
- an entrapment mechanism.

By placing traps at strategic locations, early warning can be secured of both casual introduction into the UK of new pests, and changes in populations of existing pests. However, there are two inherent problems:

- the concentration of attractant has to be set with some thought. If too high, traps may become quickly 'saturated' with insects when populations are high, and fail to give accurate estimates of population later in the monitoring period. Too low a concentration may not draw in enough insects. Some trials have included traps in pairs, one with high and the other with low initial pheromone concentrations.
- where the attractant is unprotected, dispersal is by natural volatilisation and removal by air currents. The rate of dispersal of attractant may initially

be high but may fall quickly, reducing the period during which the monitor is effective. Slow release systems are under trial.

Pheromone traps are complementary to other survey techniques such as sampling forest litter for overwintering pupae, egg counts on foliage.

5.43 Pheromone systems in use in forestry

Eight-toothed European spruce bark beetle *Ips typographus*

This species is considered the most dangerous of the European bark beetles. In central Europe, small populations developing on weakened or recently wind-blown timber, can increase sufficiently to attack extensive areas of live standing trees, mass aggregation pheromones acting as attractants to sites where beetles are already present and active.

Traps incorporating *I. typographus* aggregation pheromone have been deployed for some years at ports around the UK (Barbour 1987d). In 1993, 1994 and 1995 respectively, 3, 45 and 149 specimens of *I. typographus* were caught in pheromone traps at ports. Follow-up inspections in 1995 found larvae or adults on timber and dunnage, mainly on timber imported from Eastern Europe into ports on the east coast of England but also in one port in Wales (Evans 1996, Gibbs & Evans 1996).

Large larch bark beetle *Ips cembrae*

First found in the UK in 1955 in Morayshire, the beetle is thought to have been brought in on reparation timber in 1947. By 1975, it was reported at Glentress forest, Peeblesshire, and in 1988 it was recorded at Hamsterly forest in County Durham. Analyses and trials in 1976 identified the possibility of using pheromone-baited traps to reduce population numbers or to monitor populations (Stoakley 1977b).

Pheromone traps were used in succeeding years. Monitors established that *Ips cembrae* was present at 5 sites within 12 km of Hamsterly (Winter & Hendry 1993).

Gypsy moth *Lymantria dispar*

The role of pheromones in monitoring a small infestation of the gypsy moth in Essex is summarised in §6.44.

5.44 Pheromone systems under development

Pine beauty moth *Panolis flammea*

Sex-attractant pheromones of *Panolis flammea* were tested at Naver forest, Highland using fibres and microcapsules. Early successes were not sustained (Stoakley & Longhurst 1982, Stoakley *et al.* 1983a, Barbour 1987b).

Trials of a pheromone-based monitoring system based on a mixture of two *P. flammea* sex pheromones took place in 1981-84. Up to 2000 traps were deployed in pairs, carrying 5 or 25 μg of pheromone. Trap catches only partially reflected increases in population (Jones 1987).

Elm bark beetles *Scolytus scolytus & S. multistriatus*

Two active stereo-isomers of 4-methyl-heptanol are produced as sexual attractants by male *Scolytus scolytus* and female *Scolytus multistriatus*. Trials showed:

• these pheromones were effective attractants;
• they or synthetically produced mixtures of active and inactive isomers could be incorporated into traps, the latter working as effectively as naturally occurring pheromones (Blight *et al.* 1979, 1980).

Pine looper *Bupalus piniaria*

First trials of a sex pheromone of *B. piniaria* in the forest did not attract moths (Barbour 1987c).

Pine shoot moth *Rhyaconia buoliana*

Trials of mating disruption, comparing pheromone dispensed from fibres with ultra-low volume spraying with pheromone in micro-capsules showed promise. Both techniques reduced catches on traps, implying that the insects could not find their way to attractants in the presence of the disrupting materials (Longhurst & Billany 1982).

Aggregation pheromones for a conifer pinhole borer

Carter & Gibbs (1989) mention an aggregation pheromone for the conifer pinhole borer, *Xyloterus lineatus*, being developed for protection of logs stacked in timber yards.

5.45　　　　Kairomones for *Rhizophagus grandis*

Rhizophagus adults are believed to be attracted by specific attractants *(kairomones)* in the frass of *D. micans*. Wainhouse *et al.* (1993) report early screening work seeking a basis for a monitoring system on the success of *Rhizophagus* releases.

5.46　　　　Semiochemicals

Semiochemicals, compounds which are given off by certain plants, are possible inadvertent attractants for pests on outdoor ornamentals. For the vine weevil, they may work in combination with that insect's own aggregation pheromone in attracting other weevils to the site (Izat 1996).

5.5   BIOLOGICAL CONTROL

5.51   Introduction

Delfosse *et al.* (1996) recognize four types of biological control:

- 'Classical' or inoculative   (*eg* control of *Dendroctonus micans* by *Rhizophagus grandis* - §6.53)
- 'Augmentative' or inundative   (*eg* control of *Neodiprion sertifer* by spraying with nuclear polyhedral virus - §6.61)
- 'Conservation control' by enhancing natural indigenous enemies;
- 'Broad spectrum' - polyphagous plant eaters used specifically (*eg* controlled selective grazing).

'Classical' biological control in agriculture has been 'narrow spectrum', directed against specific pests, mostly introduced. Techniques were developed and practised in the first half of the 20th century in tropical and sub-tropical crops, where there were notable examples of successful application on introduced coconut and citrus pests. Burge *et al.* (1988) reported that, by 1980, there had been liberations, worldwide, of over 3000 potential biological agents At first, the overall success rate was 17%. However, this had risen to about 45% by the end of the period.

Waage (1996) reported 6000 programmes since 1888, only about 10% of which had been fully successful. The success of those that did work more than justified the effort lost in unsuccessful trials.

One of the leading organisations providing insect material for biological control worldwide in the 1930s was the laboratory of the Imperial Entomology Institute at Farnham Royal, Buckinghamshire (Thompson 1930). The Institute at that time investigated, among others, parasites for control of pine shoot moth (*Rhyaconia (Evetria) buoliana*) (Thorpe 1930). This moth had inadvertently been introduced into Canada and was causing widespread damage there. The Farnham Royal laboratory collected and exported to Canada three parasites of *Rhyaconia*, a braconid wasp and two ichneumonids, of which two are reported as having become established (Anon. 1936, Chrystal 1937).

Reviewing temperate forest pests of potential or actual danger to the UK in 1937, Chrystal mentioned *Rhizophagus* as a possible valuable predator, and noted that a naturally occurring virus caused rapid collapse of a pine sawfly infestation. He also stressed the length of time needed for studies of insect population fluctuations and the difficulties in working out specific roles for the various predators and parasites found in natural insect populations. Carter (1975) reiterated that point in the context of integrated control of forest aphids.

It is a reflection of such problems that in 1999, while the number of approved pesticide active substances has increased as shown in *Table 2.3*, there were only one bacterium and two fungus species approved for biological control of pests in agriculture, horticulture and forestry in the UK (MAFF Pst 1999).

The proceedings of a conference *Microbial insecticides: novelty or necessity* review the issues in producing and marketing microbial insecticides (Evans 1997). Many existing biological materials, while effective, are slower acting than corresponding insecticides. They are more expensive and also have a narrow spectrum of use. Such factors, coupled with limited availability of patent protection in many circumstances, has deterred commercial companies from pursing microbial insecticides energetically (Lisansky 1997, Harris 1997).

Predators & parasites

Living insects and allied organisms introduced as predators or parasites for biological control are additional to the tiny list of fungi and bacteria. Their release as biological control agents is regulated under legislation incorporated in the Wildlife and Countryside Act, 1981, Section 14. An *Advisory Committee on Releases to the Environment* gives opinions on such matters. The committee is also responsible for approval of genetically modified organisms that may be released into the environment.

There are some 15 approved predators and parasites currently commercially available in the UK for biological control. These have been of particular value for control of aphids, whitefly, red spider mite *etc.*, in glasshouse crops such as cucumber and chrysanthemum (Hussey & Scopes 1985, Scopes & Stables 1989).

It is claimed that pests are less likely to acquire tolerance or resistance to biological control agents than they are to pesticides.

In UK forestry, *Rhizophagus grandis* is now well established as a predator on *Dendroctonus micans* (§6.53). A naturally occurring parasitoid and commercially available entomopathogenic nematodes offer considerable potential for control of *Hylobius abietis* (Brixey 1997 - §6.51).

Vertebrates may be included in 'IFP' programmes, wherever silvicultural operations can be varied on account of populations of rabbits or voles *etc.* However, the past effects on rabbits of *Myxomatosis* and possible future effects of a new infection, a 'calcivirus' discovered in Australia, are not included below as elements of biological control for woodland management in the UK.

There is no self-evident potential biological control for the grey squirrel, however desirable that might be.

5.52   Biological control in British forestry

Several important forest pests in the UK have been controlled either by managed application of a naturally occurring predator, parasite or biological agent, or by spontaneous development of a biological agent. Others are under investigation.

Table 5.2 lists those biological control agents that need to be introduced and maintained by management action, and those where nature can be allowed to take its course, allowing that, occasionally, a natural system may need a 'kick-start'.

For pests not listed but requiring control, chemical pesticides are the only recourse. The method of control, however chosen, should nevertheless be part of an 'Integrated Forest Protection' management plan.

Table 5.2   *Forest pest species controlled by biological agents in Britain*

Pest species	Biological agent	Status	Section
Dendroctonus micans	*Rhizophagus grandis*	M***	6.53
Heterobasidion annosus	*Phlebiopsis gigantea*	M/S**	7.21
Neodiprion sertifer	Nuclear polyhedrosis virus	M**	6.61
Panolis flammea	Nuclear polyhedrosis virus	M**	6.42
Cephalcia lariciphila	*Olesicampe monticola*	S*	6.62
Gilpinia hercyniae	Nuclear polyhedrosis virus	S*	6.64
Hylobius abietis	Entomopathogenic nematodes	MD***	6.51
Hylobius abietis	*Bracon hylobii*	MD**	6.51
Hylobius abietis	*Beauvaria bassiana*	MD*	6.51
Ophiostoma nova-ulmi	Mycovirus	MD***	7.31
Tree wounds	*Trichoderma viride*	MD*	7.41
Willow rusts	Fungal hyperparasite	MD*	7.58

Status   M = Biological agent applied under management control
S = Biological agent occurs spontaneously to give control of pest
D = Under investigation; success not guaranteed.
*** ** *   relative (diminishing) woodland importance

5.53   Viruses for biological control

Viruses are most commonly thought of as disease-causing organisms capable of doing widespread harm in the animal and vegetable kingdom. Plant viruses are often only detectable by foliage discoloration, by altered plant morphology, by examination under a microscope or by biological test. Individual strains are reproducible but usually cause harm only to a narrow range of species.

The viruses involved in biological control affect insect pests, particularly, plant defoliators. It is their specificity that has enabled natural insect viruses to be used successfully and to be preferred over broader spectrum insecticides.

Their potential for use against forest pests arises partly out of their specificity and partly out of the speed with which killing infections of virus can spread. In Britain, for a period in the late 1980s, one virus product was approved for use. It was subsequently withdrawn because of lack of demand. Other trials exploring the use of baculoviruses for controlling leaf-feeding lepidopterous caterpillars have not yet progressed beyond feasibility trials, (Cory 1996); see also §§5.7 and 6.42.

5.6   PLANTS' DEFENCES AND PHYSIOLOGICAL STATE

Perennial plants rely on tolerance of pests as a first defence mechanism against attack. For example, caterpillars of over 100 species are recorded as feeding on oak leaves, conspicuous defoliation of oak usually being caused by high numbers of either the winter moth (*Operophthera brumata*) or the green tortrix moth (*Tortrix viridiana*). Oak is able to respond to early defoliation by immediate flushing of dormant buds that otherwise would be abscissed, and the ability to produce 'lammas shoots' (Gradwell 1974, Morris 1974.)

A second line of defence is by internal physiological change such as 'hypersensitive response' (Day 1993). By this response, cells around the point of attack die rapidly, depriving the invader of food; adjoining cells walls thicken and may become corky.

Bracken (*Pteridium aquilinum*), when threatened by imminent fungal invasion, produces a substantial physical barrier from lignified papillae (Burge *et al.* 1988).

Some internal resistance may be stimulated *eg* restriction of oviposition by *Panolis* on Lodgepole pine following defoliation (Leather 1987). As yet, however, such defence mechanisms appear to come into operation too slowly to be included in integrated forest protection management.

Defence responses at molecular level in response to fungal attack by *Heterobasidion annosum* are comprehensively discussed by Karjalainen *et al.* (1998) but indicate potential future lines of enquiry rather than short-term prospects for reduction of pathogen attack.

5.61   Nursery plant conditioning

Mason (1994) describes how plants may be manipulated by repeated undercutting (multicutting) to produce plants which have been subject to moisture stress in the season before lifting and as a result are better able to withstand the shock of subsequent transplanting into the forest. Better forest survival and early growth in plantation is likely to result in a shorter period when young trees are susceptible to weed competition and hence may need less use of herbicides.

5.62   Allelopathy

Allelopathy is the ability of plants to produce chemicals which prevent seed germination or seedling establishment in their immediate vicinity, thereby reducing potential competition for water and nutrients from the site. One of the best known allelopathic substances is juglone, produced by the black walnut *Juglans nigra*.

Rhododendron ponticum and *Calluna vulgaris* behave similarly (Rotherham & Read 1988).

Regrettably, many such organic compounds are general biocides and have not found a place currently as sources of pesticides (Copping *et al.* 1990).

5.63   Plant toxins

Many plants contain compounds in foliage fruit or bark which are toxic or repellent to potential predators.

Table 5.3 lists trees, shrubs and herbs which may be found in or on the edge of woodlands and waste ground, and which have been reported to cause poisoning symptoms to man or to livestock.

Table 5.3   *Woodland and woodland edge plants poisonous to man or livestock*

Alder buckthorn (*Frangula alnus*)	Beech (*Fagus sylvatica*)
Bluebell (*Hyacinthoides non-scripta*)	Bog asphodel (*Narthecium ossifragum*)
Box (*Buxus sempervirens*)	Bracken (*Pteridium aquilinum*)
Buckthorn (*Rhamnus cathartica*)	Cherry laurel (*Prunus laurocerasus*)
Cypress (*Cupressus* spp.)	Deadly nightshade (*Solanum nigrum*)
Dog's mercury (*Mercurialis perennis*)	Elder (*Sambucus nigra*)
False acacia (*Robinia pseudoacacia*)	Foxglove (*Digitalis purpurea*)
Giant hogweed (*Heracleum mantegazzianum*)	
Hellebores (*Helleborus foetidus, H. viridis*)	Herb paris (*Paris quadrifolia*)
Holly (*Ilex aquifolium*)	Horse chestnut (*Aesculus hippocastanum*)
Hemlock (*Conium maculatum*)	Laburnum (*Laburnum anagyroides*)
Lords & Ladies (*Arum maculatum*)	Mezereon (*D. mezereum*)
Mistletoe (*Viscum album*)	Oak (*Quercus spp.*)
Privet (*Ligustrum* spp)	Rhododendron (*Rhododendron ponticum*)
Snowberry (*Symphoricarpos albus*)	Spindle (*Euonymus europaeus*)
Spurge laurel (*Daphne laureola*)	Varnish tree (*Rhus verniciflua*)
Woody nightshade (*S. dulcamara*)	Yew (*Taxus baccata*)

Source: *Cooper & Johnson 1984*

A number of the plant toxins in these and other plants have been the base for medicinal products (*eg* digitalin from foxglove).   Many others have been screened for potential pesticide activity.

Dried *Pyrethrum* flowers had been first introduced into Europe from western Asia in the early 18th century to control household insects.   Extracts from flowers of *Pyrethrum cinerariaefolium* were found to contain a group of active compounds subsequently marketed as 'pyrethrins'.   Products containing pyrethrins are currently approved for sale as contact insecticides to control flies around the home and in dairy and other farm buildings;   at the same time, a whole family of pyrethroid-based insecticides has been developed by chemical modification of the pyrethrins.   See also §2.31 and Ch 6.

5.7 GENETIC ENGINEERING

There is scarcely any part of the realms of human health, human food supplies and raw material production which is not under enquiry in relation to potential 'improvements' from genetic engineering. Controversies abound and are likely at least to slow down the development of such products. Underlying them are fundamental disagreements about the ethics, hazards and risks involved (Coghlan 1994a, 1996c, Cory 1996. Tickell 1999).

Forestry in the UK is small in relation to the potential markets in international temperate agriculture for genetically engineered pesticides. Resistance to herbicides has been engineered into some arable crops but not yet to trees. In the short term, UK forestry is likely to remain outside the main stream of genetic engineering pressures, except where a forest pest is also either an important agricultural pest or is important in forestry internationally. See baculoviruses and control of Fox-coloured pine sawfly (*Neodiprion sertifer*) and Pine beauty moth (*Panolis flammea*) (§§6.61 and 6.42).

5.8 DISINTEGRATED PEST CONTROL

5.81 Resistance in pests

Integrated forest protection can be put into effect if it can be assumed that the pest remains susceptible to the pesticide. However, world-wide, a number of pest organisms of all sorts have developed resistances.

The society representing the interests of pesticide manufacturers:

- *Groupement International des Associations Nationales de Fabricants de Produits Agrochemiques (GIFAP)*

in the 1980s formed four 'Resistance Action Committees' to understand and manage resistance, so as to minimise impact on crop production (Tomlin 1997). Their concerns are most acute in relation to tropical agricultural crops mostly outside Europe.

The evidence of resistance developing in forest plants is so far limited to forest nurseries in the UK, where grounsel, willowherb and mayweed in some places have shown resistance to some herbicides (Williamson & Morgan 1994).

Dosage rate and resistance

To minimise development of resistance, if a crop on a site requires repeated treatment for a pest and there is more than one active substance which will give the necessary control, the active substances should be alternated.

As far as possible, the manufacturer's recommended dose should be adhered to. If there is insufficient material to treat the whole of an area selected, it is preferable to treat as much of the area/pest at full rate, leaving a proportion completely untreated. Treating the whole area at a reduced dose is likely to exert a stronger selective pressure favouring development of resistance than leaving part untreated.

Part II

Pesticides

CHAPTER 6

Insect and other pests

'For the locusts covered the earth and blotted out the sun;
they ate every plant, so that there remained not one living thing.'

6.1 INTRODUCTION

There are more than 20 000 species of insect listed for the British Isles
(Chinery 1973, Bevan 1987). Some 1500 are phytophagous insects feeding on
trees and are catalogued in FC Booklet 53 (Winter 1983). The large numbers of
small insects in even quite small areas of forests are illustrated, for example, by
775 species collected in various ages of native Scots pine woodland in Glen Tanar
and Abernethy (Young & Armstrong 1994).

Fortunately the number of species potentially harmful to woodlands and
ornamental trees is only a small fraction of the number of woodland insect
species. *Table 6.1* lists a selection of the more important pest insects. The status
of forest insects pests as in 1997 was reviewed in *Forestry* (Evans 1997c).

Table 6.1 includes:

- major insect pests subject to control operations using insecticides
applied from the air;
- major pests for which identifiable biological control has been effective;
- insects of local or specialist importance;
- others against which no action up to the present has been taken, either
because the threat has not developed into an infestation requiring action or
because no practical means of control has been found.

Almost all the pests described in this chapter are insects; however, sections
§6.71, §6.72 and §6.77 also mention the effects of specific mites and nematodes.

6.2 INSECT CONTROL & INTEGRATED FOREST PROTECTION

The current woodland area in Great Britain is 2.4 million hectares (*Table
1.6*, §1.51). 100 years ago, the corresponding woodland area was 1.1 million ha
(Aldhous 1997). Not only has there been a substantial increase in woodland area
but the area under introduced species has increased substantially, as may be
inferred from *Table 1.7*. Similarly, while there are only 35 trees native to the
UK, 500 introduced species can easily be encountered in parks and gardens, with
a further 1000+ in specialist collections and arboreta (Mitchell 1974).

Table 6.1   *Insects potentially damaging to woodlands, ornamental plants etc.*

English name	Latin name	Damage rating	Chapter reference
Aphids			
Beech woolly aphis	*Phyllaphis fagi*	x	§6.72
Felted beech coccus	*Cryptococcus fagisuga*	x	§6.73
Green spruce aphis	*Elatobium abietinum*	t	§6.74
Bark beetles on established trees			
Great spruce b.b.	*Dendroctonus micans*	xxx	§6.53
Large larch b.b.	*Ips cembrae*	tt	§6.55
8-toothed spruce b.b.	*Ips typographus*	ttt	§6.55
	Ips sexdentatus	t	§6.55
Large elm b.b	*Scolytus scolytus*)	vvv	§6.54
Small elm b.b	*Scolytus multistriatus*)		
Pine shoot beetle	*Tomicus piniperda*	ttt	§6.56
Bark feeders on newly planted stock			
Pine weevil	*Hylobius abietis*)	xxx	§6.51
Black pine beetles	*Hylastes* spp)		
Clay-coloured weevil	*Otiorhynchus singularis*	x	§6.57
Banded pine weevils	*Pissodes* spp	x	§6.57
Moth caterpillars			
Browntail moth	*Euproctis chrysorrhoea*	x	§6.45
Larch bud moth	*Zeiraphera diniana*	tt	§6.45
Gypsy moth	*Lymantria dispar*	tt	§6.44
Pine beauty moth	*Panolis flammea*	xxx	§6.42
Pine looper moth	*Bupalus piniaria*	xxx	§6.41
Pine shoot moth	*Rhyaconia buoliana*	tt	§6.43
Winter moth	*Operophthora brumata*	x	§6.45
Sawfly caterpillars			
Large larch sawfly	*Pristophora erichsonii*	x	§6.63
European spruce sawfly	*Gilpinia hercyniae*	xx	§6.64
Fox-coloured sawfly	*Neodiprion sertifer*	xx	§6.61
Large pine sawfly	*Diprion pini*	t	§6.61
Web-spinning larch s'fly	*Cephalcia lariciphila*	xx	§6.62
Pests of Christmas tree crops			
Pineapple gall woolly aphid	*Adelges abietis*	xx	§6.71
Brown spruce aphid	*Cinara pilicornis*	x	§6.71
Spruce bell moth	*Epinotia tedella*	x	§6.71
Gregarious spruce sawfly	*Pristophora abietina*	x	§6.71
Conifer spinning mite	*Oligonychus ununguis*	xx	§6.71

x minor/occasional pest	xx significant/frequent pest	xxx major/regular pest
v " vector	vv " vector	vvv " vector
t " threat	tt " threat	ttt " threat

Source Forest Insects. FC Handbook 1 (Bevan 1987).

Superimposed on this mixture of introduced and native tree and shrub species, there are both native and introduced insects, some of the latter being important pests.

In addition, there are further insects not yet established in Great Britain but which are major pests in Europe or North America and would threaten forests in Britain if introduced (Gibbs & Wainhouse 1986).

The vulnerability of British forests and ornamental trees in the 21st century is not entirely predictable because of the scale of use of introduced species on site types and in climatic conditions where they had not previously grown, and because of the continuing introduction and spread of non-native pests and diseases.

Most insect infestations requiring treatment have been either of native insect species attacking plant species growing outside their native range, or of introduced insect species on introduced tree species (Evans, H.F. 1996).

The most serious loss of countryside trees in Britain over the last 50 years has been due to Dutch elm disease, a fungal disease spread by an insect (See §§6.54 & 7.31). Some 22 million trees have been lost, virtually eliminating the many varieties of elm in hedgerows and woodland in parts of central and southern England, where they had dominated the landscape. There was no effective way of preventing the disease spreading, once established.

In view of the scale of introduction of so many species into Britain, it is perhaps surprising that losses have not been on a greater scale.

Infestations, however, do not always involve introduced species. For example, in north-eastern Canada and the United States, losses through a native pest, spruce budworm, have resulted in spraying programmes and tree mortality to native forest for over 20 years and involving spraying programmes ranging from 300 000 to over 9 million hectares annually (Chandra Nigram 1990).

Insecticides in the forest

Prior to 1945, only a very limited range of techniques to control insects were available to foresters in Britain; these could be summed up in the phrase 'forest hygiene' (Chrystal 1937).

From 1945, land husbandry has been transformed by the availability of increasingly sophisticated products of the pharmaceutical industry impinging on all elements of crop management and production. The scale and cost have been described in §2.5.

While 'biological control' has been sought for more than 100 years and has had some marked successes, it is only since the 1980s that the concept of 'Integrated pest management' has come to the fore, seeking to balance good crop protection with maintenance of a diverse natural environment and encouraging use of intensively raised natural predators for pest control. British forest entomologists were active in pursuing potential predatory or controlling agents in the 1930s and 40s and had several successes (Chrystal 1937, Hanson 1949,1950).

6.21   Toleration of damage

Whatever the circumstances, if woodlands or trees appear at serious risk of insect damage, judgements have to be made case by case, as to whether a loss from an insect pest should be borne and no action taken, or a control attempted.

Decisions depend on:

- whether there is a practical and economic method of control;
- whether there is an accepted 'threshold for action' (§6.22) in relation to predicted damage; if not, whether the wood or forest can be expected to survive with adequate stocking, even though some increment may be lost or some trees killed;
- the scale of any damage;
- for ornamental trees, whether an infestation or the damage caused is too unsightly to be acceptable, and whether felling is acceptable.

Examples of insect species which in woodlands are not normally treated, notwithstanding substantial defoliation or other damage, are given in *Table 6.2*.

For figures on possible loss of increment, see Gradwell (1974), Austarå (1987) and Carter (1977).

Table 6.2   *Examples of types of insect damage tolerated on woodland trees*

Type of damage	Pest	Host species
Defoliation by green spruce aphid; some depression in growth rate (Carter 1972, 1977).	*Elatobium abietinum*	Sitka spruce
Short-term reduction in leader growth	DF woolly aphid *Adelges cooleyi*	Douglas fir
Heavy white waxy covering on tree stems; some deaths	felted beech coccus *Cryptococcus fagisuga*	beech
Heavy defoliation; some loss of increment (Varley & Gradwell 1962)	oak leaf roller moth *Tortrix viridiana*	oak
Shoot (post-horn) distortion	*Rhyaconia buoliana*	Scots pine
Seed fly larvae feeding inside seed	*Megastigmus spermotrophus*	Douglas fir

6.22   Thresholds for action

Thresholds for action range between:

- a defined population level, *eg* of overwintering pupae determined through long-running survey programmes, which is exceeded,
- presence (as opposed to absence), as in the case of new introductions not yet established in the country.

Surveys and searches for threatening insects

Pupal surveys

Many forest insects overwinter in the surface layers of the soil as pupae. Surveys of threatening insects that pupate can be made while they are thus dormant. Using the number of pupae per sq. metre in susceptible crops as a yardstick, the potential emergence of adults in the spring following the survey can be predicted. If the pupal numbers appear threateningly high, an application of insecticide or biological agent may be required.

Before a final decision is made and to confirm that the population has not suddenly collapsed, a count of eggs is made after adults have emerged and mated. These counts are restricted to areas thought to be at great risk of heavy and widespread defoliation.

Pine looper (*Bupalus piniaria*) was the first species in Britain for which such counts were made. At that time, evidence from Germany was that a pupal density of 6 per sq. metre w as the critical figure (Bevan 1954). With experience, this was increased to about 15 pupae per m^2 for a forest as a whole or substantial sections of it. This was in turn replaced by 'Highest Compartment Mean', on the basis that individual compartment counts may be several times higher than the forest average, these occurring where possible control most needs to be focused.

The mean number of eggs per female (probable fecundity) was found to vary according to pupal weight, a regression equation of egg-laying fecundity in relation to pupal weight providing a basis for forecasting numbers of eggs (Bevan & Paramonov 1956). The confirmatory threshold for egg counts was taken as 2000 per tree (locally reduced to 1000 per tree) at Tentsmuir in 1957 (Crooke & Bevan 1958). In later surveys, the threshold was 3000 eggs per tree.

Cannock Chase was sprayed in 1963, when there was a mean of 19.5 pupae per m^2 (Brown 1973a) and a subsequent mean egg count of over 6000 eggs per tree (Bevan 1964).

Highest compartment means of pine looper pupae for 1991-1996 in 16 forests have been tabulated in the 1996 Forestry Commission Report on Forest Research (Gibbs & Evans 1996). The results indicated that larval numbers were threateningly high in several forests in north Scotland. Results of previous surveys were given in the appropriate year's Report on Forest Research.

For Pine beauty moth (*Panolis flammea*) on lodgepole pine on peat, sampling is more difficult because of the branchiness of crops of susceptible age and because the pupae penetrate more deeply into surface soil/peat than Pine looper larvae. At the outset, in the absence of previous experience, any block with a count averaging 15 pupae per m^2 or more was regarded as seriously at risk. Subsequent studies and experience confirmed the validity of that figure (Stoakley 1979a, b, 1985a, b).

While there is no regular survey for the web-spinning larch sawfly and no threshold figure, sampling gave peak numbers of overwintering larvae in excess

of 1000 per m² in an early infestation at Margam Forest, south Wales (Brown &
Billany 1973).

Search and control

As soon as it was realised that the Great spruce bark beetle (*Dendroctonus
micans*) was present in Great Britain and that sample surveying was not
appropriate to the situation, a widespread 'search and control' programme was set
up and a statutory *Restriction on Movement of Wood Order* imposed (FC AR
1983, SI 1982/1457). Plantation s of Norway and Sitka spruce within 20 miles
of known infestations were intensively surveyed, but with no better symptoms to
search for than characteristic resin bleeding from infested trees, and in more
extreme cases, dead tops to trees. It was great good fortune that there was in
Europe, a pre-existing programme of biological control involving *Rhizophagus
grandis* which could be brought into play, this predator being a far more effective
surveyor for the presence of *Dendroctonus* than any human being. Nevertheless,
vigilance has continually to be maintained during all inspections of potentially
susceptible woodlands, as well as by periodic surveys (§6.53).

The Asian strain of the gypsy moth *Lymantra dispar* has similarly been the
subject of 'search and control', the 'control' aim in this case being to eradicate
the insect before it has become widely established in the country.

In June, 1995, larvae were found on the eastern outskirts of London near
Epping Forest. An intensive survey and local spraying programme followed.
This was not completely successful, larvae being found in 1996 in a small
number of gardens where they were sprayed. Further surveys and spraying must
be expected until it is clear that the insect has gone (FA PH 1996, FC AR 1996,
Gibbs & Evans 1996). See also §6.44.

On August, 1997, twenty mature *Ips typographus* were discovered in a
Forestry Commission early warning trap set up close to incoming timber at the
Shotton paper mill. Immediate steps were taken to check all potential sources of
supply and to publicise the new risk through leaflets and journal articles. For
1998 the programme included:

- surveillance on 110 sites where timber was being harvesting;
- monitoring through pheromone traps, industrial premises where spruce
was being sawn or pulped, and ports through which spruce might be shipped;
- reappraisal of the relative susceptibility of Norway and Sitka spruce;
- continuing publicity about the need to report suspicious signs of beetle
attack (Burgess 1998). See also §6.55, *Ips typographus*.

Monitoring programmes for pests not in the country (§5.35) could at any
time give rise to similar operations to those for *Ips typographus* and gypsy moth.
It is only at the very earliest stages of colonisation by a new pest that eradication
is a feasible objective.

6.23 Minimising use of insecticides

Operational practices that minimise risk of insect attack include:

- maintaining a high level of vigilance towards perceived threats;
- managing each site to make it inhospitable to potential pests without spoiling it for beneficial organisms;

eg: minimise possible pest 'reservoirs';

mix or dilute plantings of potential host species;

favour habitats of known predators;

eliminate potential breeding material quickly, *eg* by removal and milling shortly after felling ('hot logging'), or burning infected material;

- use available biological controls;
- use advanced application techniques for best effect by insecticide.

6.24 Application techniques

In all insecticide spray applications, the aim is to ensure that the optimum quantity of insecticide is placed so as to come into contact with or be eaten by the target pest. That there is no single simple method is self-evident when the range of insect habitats is considered.

For spraying in the nursery and on plants in the first two or three years after planting in the forest, equipment similar to that used for horticultural and agricultural crops is suitable.

Specialist techniques which have been or are being developed for British forest crops include:

- developments of canopy treatment through ultra low volume (ULV) controlled droplet emission and turbulent airflow to achieve contact (§6.42);
- electrodyn electrostatic application of permethrin to nursery stock for forest areas that are being restocked (§6.51);
- slow-release carbosulfan granules mixed into soil at planting (§6.51).

6.3 CHEMICAL CONTROL

This book is **not** a manual of current recommendations.

A note on sources giving up-to-date prescriptions for pest control is given in *Appendix II Table 2*.

Individual insect species and insecticides for their control are discussed in §6.4 *et seq.* Literature references to insecticides are listed in *Appendix I Tables 1* and *2*. Background data for pesticides used or tested for forests and trees are listed in *Appendix II, Table 1*.

6.31 Insecticides used in British forestry practice

For the main blocks of forested land, insecticides are in use on a regular basis only to protect newly planted stock on replanting sites (§6. 51).

Otherwise, the use of insecticides in the forest has been restricted to:
- insect infestations where substantial blocks of forest appear at risk of being killed, necessitating treatment from the air (§6.32, §6.41 *et seq.*);
- local control of new introductions of insects (§6.22);
- protection of felled timber against beetle attack, where neither 'hot logging' nor wet storage are possible (§6.52).

Specialist Christmas tree crops may also be troubled by pests which affect foliage and shoots, and also require treatment with insecticide or acaricide (§6.71).

The production of planting stock in forest and hardy ornamental stock nurseries may require control of aphids, foliage-feeding caterpillars and soil-inhabiting organisms such as cutworm larvae, nematodes etc. Here, chemical control is often the only practical remedy (§6.72).

For ornamental trees in parks and gardens, the scope for insecticides is very limited because of difficulties of access and treatment and, in areas open to the public, requirements of public safety (§6.73).

Lists of insecticides used in woodlands and other land, 1950-1998

Appendix I lists keys to pesticide references in literature (mainly UK forestry sources). All the keys are in the form of 'one ' or 'two letters'. These can be linked to 'author(s) and year' through the lists in *Appendix I Table 2.* For each 'author(s) and year' entry, a full reference can be found in the main bibliography.

Appendix I Table 1a refers to insecticides used or tested experimentally in forests, farm-woodlands and amenity tree management or in forest nurseries since the 1950s. *Appendix I Tables 1b-d* cover insecticides, herbicides and other pesticides respectively.

Appendix II Table 1a lists most of the insecticide active substances in *Appendix I Table 1*, and shows:

- approximate year of first published description/patent/use;
- whether the active substance is listed in *Pesticides, 1999* (MAFF Pst 1999);
- the year of publication of any report by a *FAO Panel of experts on pesticide residues in Food and the Environment*, as described in *'The pesticide manual'* (Tomlin 1997);
- World Health Organisation and (United States) Environment Protection Agency toxicity classes for the active ingredient (see §4.23 and *Table 4.1*);
- The date of the most recent review or re-evaluation (if any) carried out by the Advisory Committee on Pesticides (§4.17).

The references listed in *Appendix I Table 1a* reflect the general level of interest in particular active substances. *DDT*, the subject of much attention in the 1950s, was not being reported by the end of the 1970s, following widespread restrictions on its use. *Lindane* is the only insecticide from the early period which is still in use. *Fenitrothion* and *diflubenzuron* came more into prominence in the 1980s while *permethrin* has been the most widely reported material in the 1990s. Current use notwithstanding, it is rumoured that permethrin will be withdrawn by 2002.

6.32   Quantities of insecticides used in British forestry

Details of quantities of insecticides used are difficult to obtain. *Table 6.3* gives figures for four periods in relation to control of the large pine weevil (*Hylobius abietis*). Estimates of possible private woodlands usage can only be obtained on the assumption that private woods and forest nurseries will use proportionate amounts. For areas of private planting, see *Tables 8.4 & 8.5.*

Table 6.3   *Quantities of insecticide active ingredients per annum used by the Forestry Commission in 1966/7, 1986/7, 1992/3 and 1996/8, in forest nurseries and for post-planting control of pine weevil and beetle*   *(kilograms)*

Active ingredient	1966/7	1986/7	1992/3	1996/8
DDT	590	nil	nil	nil
Lindane	70	1240	21	nil
Malathion	225	195	nil	nil
Permethrin	nil	9	845	599

Sources   Aldhous 1968a; unpublished data; Willoughby 1995.

6.33   Treatment from the air

A number of forests have suffered serious infestations by larvae, mainly of lepidopterous species. These have required treatment with insecticides from the air; areas treated are summarised in *Table 6.4.* Fuller details are given in sections §6.41 - 6.43 by insect.

6.4   INTEGRATED MANAGEMENT OF MOTH PESTS *Lepidoptera*

6.41   Pine looper *Bupalus piniaria**

The Pine looper moth overwinters as pupae, adults emerging between the end of May and the end of June. Larvae emerge in mid-late summer and commence feeding on older needles. Final instar caterpillars spin to the ground and pupate beneath the litter layer of the soil in the autumn, remaining there until the following spring. During the winter, populations in pine plantations thought to be at risk are assessed by counting the number of pupae on sample unit areas.

*The current scientific name for this insect is *Bupalus piniaria*. In some earlier publications, the specific name was given as *'piniarius'*.

Table 6.4   *Areas of forest treated by large-scale application of insecticide,*
              or biological control agent, by decade from 1950   (hectares)

Pest treated	1950-59	1960-69	1970-79	1980-89	1990-97	Total
Pine looper	2570	556	640	1100	-	4898
Bupalus piniarius	DDT	DDT	Tetvp	Diflub	-	
Pine beauty moth	-	-	8496	12079	5253	25828
Panolis flammea			Fentn	Fentn	Diflub	
Pine sawfly	-	-	-	7244	-	7244
Neodiprion sertifer				NPV		
Winter moth	-	-	-	340		340
Operophthora brumata				Fentn		
						38310

Principal active agent: Tetvp = tetrachlorvinphos   Diflub = diflubenzuron
                           Fentn = fenitrothion   NPV = nuclear polyhedrosis virus

The Pine looper moth first came into prominence in the British Isles in 1953 when more than 40 hectares of pinewoods, predominantly Scots pine, were defoliated by its larvae (Bevan 1954, Anon. 1954b).   Subsequent counts of overwintering pupae showed there to be extremely high populations in pine woods both in Cannock Chase, Staffordshire, and in Culbin forest, Morayshire.

Such surveys were the basis for deciding the first areas to be sprayed in 1954. Surveys have been repeated annually, results of being given in FC Reports on Forest Research.

Analysis of population fluctuations over 24 years indicated both density-dependent and delayed density-dependent relationships in pupal numbers (Barbour 1985); results enabled the scale and intensity of survey to be reduced.

In any year where high average pupal numbers per compartment have been revealed by winter survey, provisional plans to spray the areas affected have been made and confirmatory egg counts arranged for the following July, as described in §6.22.

Biological control

From the outset, attention was paid to the potential for natural biological control (Bevan 1954).   The 1953-4 pupal survey showed that at Cannock Chase, the proportion of larvae parasitised ranged between 11 and 33%.   The most numerous parasite, *Cratichneumon nigritarius*, was found to raise a succession of generations on one generation of host, more than had originally been appreciated. Nevertheless its effect as a natural control on *Bupalus* has not prevented heavy defoliation (Bevan 1966b). No other potentially effective biological measure to control pine looper has been found.

Insecticides

A programme of aerial spraying using DDT at 1.1 kg/ha was proposed for 1954, the first time insecticides had been applied from the air to forests in Britain. The areas designated for treatment amounted to a little over 1000 ha at Cannock and 1420 ha at Culbin (Crooke 1959).

Surveys in winter 1956/7 showed a small concentration of high numbers of pine looper pupae in Tentsmuir Forest, Fife. Subsequent surveys confirmed that the threat appeared substantial (Bevan *et al.* 1957). However, because only about 150 ha were considered to be at sufficient risk to warrant treatment, it was decided to apply insecticide by fogging (FC AR 1957).

Previous experience on the continent had shown the importance of still air and well marked temperature inversion just above the forest canopy for successful fogging. Experience at Tentsmuir confirmed these requirements and illustrated the practical difficulties. It was estimated that 27 hours fogging would be required, applying DDT at 1.1kg/ha. In the event, crews were standing by for 18 nights. Only on 5 nights were climatic conditions sufficiently calm for operations to proceed (Crooke and Bevan 1958).

In 1962/3, pupal counts for Cannock Forest reached an average of 19.5 over the whole forest. Subsequent egg counts (highest mean compartment count, 6000 per tree) showed that 556 ha of forest were immediately threatened; these were sprayed over 2 days with DDT at 1.1 kg/ha from fixed wing aircraft (Bevan 1964).

Surveys in winter 1970/71 showed high numbers of pupae in Wykeham Forest, Yorkshire. Egg counts indicated infestation level populations on 538 ha, the average number of eggs per tree exceeding 9500 (Bevan & Davies 1971). Tetrachlorvinphos was applied by helicopter at 0.6 kg per ha, timed in relation to the estimated egg hatch date.

In 1976/7, high winter pupal counts at Tentsmuir Forest led to egg counts being made. These, at 1926 eggs per tree for the 10 most heavily infested compartments, did not reach the 3000 eggs per tree level set as the indicator for spraying operations. However, because the trees were important to the amenity value of the area, 102 ha were sprayed by helicopter using tetrachlorvinphos at 0.56 kg/ha (Brown, & Barbour 1978, Bevan & Brown 1978). One compartment left unsprayed showed a population drop from 221 to 7 larvae per m², the parasite *Campoplex oxycanthae* accounting for approximately 70% of the *Bupalus* larval population (Barbour 1978). Radial increment of heavily defoliated trees remained depressed for 2 -3 years after the peak levels of infestation.

Counts of pupae in 1983/4 again revealed high numbers in Tentsmuir Forest and in Roseisle Forest (Moray). Egg counts were carried out in July, 1984, setting averages of 4000 eggs per tree as the minimum for insecticidal control.

These showed that at Tentsmuir, 1100 ha were at risk but that at Roseisle, spraying was not required, all the egg counts being less than the threshold figure. At Tentsmuir, diflubenzuron was applied, at 67.5 grams per ha, using a helicopter fitted with rotary atomisers and gave good control (Stoakley *et al.* 1985b). Diflubenzuron had first been tested on a small outbreak in 1979 on 72 ha and had been successful (Stoakley 1984a).

Between 1984 and 1995, pine looper populations remained low. However, in 1996, a localised increase in population in north east Scotland was reported as needing closer monitoring (Gibbs & Evans 1996). Since then numbers have fallen and in 1999 were very low.

6.42   Pine beauty   *Panolis flammea*

The pine beauty moth is native to most of Britain. Populations exist at low, stable levels in many Scots pine woodlands in northern England and southern Scotland (Barbour 1987a). Severe build up of numbers and subsequent serious defoliation, while reported intermittently in the 1800s in Bavaria, have not occurred on Scots pine in Great Britain.

Larvae immediately after emergence feed at the base of newly developing needle pairs, biting small holes and feeding within them. Fourth and later instars feed on older needles starting from the tip. In heavy attacks, both current shoots and older foliage are killed.

Pine beauty moth: its biology, monitoring and control (Heritage 1997b) gives details of the life-cycle and current recommendations for monitoring through pupal surveys, assessment of adults and control measures.

From the late 1950s, extensive areas of lodgepole pine (*Pinus contorta*) were planted on the poorest quality, unflushed peats of northern Scotland, outside what had been thought was the natural range of pine beauty. However, in 1973, *Panolis* larvae were found feeding heavily on 3 - 6m tall lodgepole pine on peat sites in Caithness and Sutherland. In 1976 and 1977, severe defoliation by *Panolis* larvae led to deaths of lodgepole pine on 240 ha of young pole-crop plantations in Naver forest, Sutherland, with significant defoliation in other parts of this and neighbouring forests (Stoakley 1977a,
b, 1979a, b).

Within its range as previously recognised, pupae of *Panolis* had been observed regularly in the course of surveys for pupae of the pine looper moth (*Bupalus piniaria*). Average pupal densities of 1 per 10m^2 were typical for more than 20 years; often 50% of pupae had been parasitised. These observations were remarkable if anything for the low numbers per m^2 and lack of year-to-year fluctuation (Barbour 1987b).

Nevertheless it was clear that the behaviour of the insect had changed and that large numbers were attacking lodgepole pine on deep unflushed peat, whether as hill peat or raised bog over Moine schists. On less infertile peat sites, attack was less (Barnett 1987).

Lodgepole pine in other parts of Scotland have also subsequently been attacked, but always on the poorest soils. It seems that the natural predators which otherwise would hold down the numbers of *Panolis* are not able to survive well on infertile deep peat and that lodgepole pine as a host tree is less hospitable to predators of *Panolis* than is Scots pine (Walsh 1991).

Seed origin of Lodgepole pine has also been found to be important; plants raised from seed of southern and central interior British Columbian origins were favoured for initial attack over plants of coastal Washington, Skeena river or Alaskan origin (Leather 1987, 1992, Leather *et al.* 1993).

Vigilance

Following the first reports of serious damage, surveys of northern areas of lodgepole pine were stepped up. As an initial criterion for the need for control, the threshold value of 15 pupae per m² was used. Within the range where infestations were occurring, densities of pupae up to a maximum of 1000 per m² were encountered. In areas potentially at risk, there was an egg-count to decide whether treatment would go ahead. By June 1977, it had become clear that a block in Naver forest was at high risk (Stoakley 1978).

Over the subsequent 20 years, following larval surveys and egg-counts, sprays have been found necessary on more than 25 000 hectares of forest, a larger scale of treatment than for any other forest insect in Great Britain (see *Table 6.4*).

Figure 6.1 shows the area sprayed annually over the period 1977-1997, and the main agents used. While *B. thuringiensis* and NPV were tried, fenitrothion was the mainstay of spraying between 1978 and 1988, and diflubenzuron, between 1991 and 1997.

Biological control

Trial of Bacillus thuringiensis

Bacillus thuringienis is effective against larvae of lepidoptera in vegetables, fruit *etc.* if ingested. Approved products have been available since the commencement of the present approval scheme.

In the first attempt to control *Panolis* in 1977, Dipel, an approved commercial formulation of *Bacillus thuringiensis*, was applied by helicopter to foliage during the early larval stage of *Panolis* but was ineffective. While an adequate dosage reached the foliage surface, to be effective, insects have to ingest the bacillus. It was subsequently concluded that as larvae in their early instars were feeding within the needle tissue, they were protected from the spray by the outer layers of the needles (Stoakley 1978, 1985b). A second trial carried out in 1985 showed that *B. thuringiensis* was less effective than insecticide or nuclear polyhedrosis (NPV) (Stoakley 1985a).

Nuclear polyhedrosis virus (NPV)

This virus is found in a range of lepidoptera; preparations may be identified by the host from which they were obtained. A preparation from *Panolis*

(PfNPV) was first tested against *Panolis* in 1981-3 (Entwistle & Evans 1987). In subsequent work, the virus was prepared from *Mamestra brassicae* (MbNPV). In forest trials NPV was applied from the air.   In 1985 and 1986, the combination of initial kill and secondary epizootics gave almost total mortality in each season (Stoakley 1987a, b).   In 1987 and 1988, further treatment was necessary, MbNPV being used successfully on 241 ha out of 2265 ha treated in 1987 and on 133 out of 774 ha in 1988.   The other areas were treated with fenitrothion or diflubenzuron.   Control using MbNPV took longer to take effect than a conventional insecticide (Leather 1986, Evans *et al.* 1991).

On the Isle of Lewis, just over 100 ha were sprayed with a virus formulation in 1992 (FBT 1992).

Heritage (1997b) concluded that the impact of both *Bacillus thuringiensis* and NPV is less than that of chemical insecticides and is insufficient to control a severe outbreak of *Panolis*.   He recommended that they should be used in the season before a full outbreak is expected.

Speeding up the effect of virus ingestion

The effect has been investigated of introducing into a baculovirus, toxin-producing genetic material from a North African scorpion. The aim was to speed up kill of the target larvae thereby substantially reducing feeding.   Licensed field trials were started in 1993, first to test toxicity, and subsequently, risk of affecting non-target species. These had been preceded in 1987-8 by small scale tests against *Panolis flammea* using the PfNPV virus into which an inert marker gene had been inserted. The addition of the marker gene did not inhibit the activity of the PfNPV (Coghlan 1994b, 1995).

Pheromone disruption

Pheromones emitted by one or other sex of an insect can attract others of the same species as potential mates or as colonists aggregating for a mass attack on a selected tree. Pheromones have been identified for pine beauty and in the period 1981-1984 were used by the Forestry Commission to monitor populations.   In 1984, pheromone traps failed to give advance warning of an outbreak.   The FC consequently discontinued their use in favour of pupal surveys.   Population monitoring using pheromones continued in the private sector and in 1991 the Forestry Commission renewed its interest in this monitoring system and the potential of a new trap design.   This is being used increasingly to replace pupal survey work.

In the mid-1960s, a small field trial examined the potential to disrupt mating by swamping natural pheromone plumes with high concentrations of synthetic pheromone.   In the following year, micro-encapsulated pheromone was sprayed from the air.   No reductions in egg numbers were observed subsequently; the studies were not continued.

Recent studies have modeled population changes using a spreadsheet. Population increases may be preventable if a sufficient proportion of the male population can be trapped prior to mating (Leather & Knight 1997).

Insecticides

With the failure in 1977 of *Bacillus thuringiensis* to control *Panolis*, and with evidence both of recent death of previously infested untreated blocks and the results of pupal surveys showing substantial additional areas at risk, spraying using an insecticide with good contact action was accepted as necessary. Fenitrothion was selected. Implementation, however, was not entirely straightforward compared with the first spraying against *Bupalus* in 1954 because:

- requirements under the *Pesticides Safety Precautions Scheme* had become much wider in their scope;
- the proposed insecticide, fenitrothion, had not previously been applied from the air to forests in Britain. It was not included in the 1978 'List of products approved for aerial application' in the *List of products approved for farmers and growers* (MAFF ACAS 1978);
- the insect was not a widely known pest species;
- the preferred application technique, Ultra-low Volume (ULV) spraying treatment was untested against forest pests in Britain.

Faced with the options of setting up a spray programme or facing the loss of several thousand hectares of young crops, it was decided to press ahead with a spraying programme. All aspects for which approvals were needed were identified and approvals sought and obtained. Programmes were then undertaken and were successful.

The 1978 and 1979 spraying operations were monitored and results reported. Aspects covered included effects on operators and equipment, forest birds, streams and aquatic wildlife. The short-term results indicated that ill-effects of spraying were small or not detectable. At the same time, it was accepted that longer-term studies were desirable (Holden & Bevan 1978, 1981).

One requirement of PSPS approval was that a comparison should be made between the results of ULV and Low Volume (LV) treatment. The ULV technique was shown to be superior and has formed the basis for application of insecticides to pests in forest canopies since. During the 1980s intensive work was undertaken to get best performance out of the spinning disc rotary atomisers (Stoakley 1987b, Evans *et al.* 1991). Heritage (1997b) describes the equipment used in 1997.

Figure 6.1 shows the areas sprayed to control pine beauty moth and shows clearly the cyclical nature of outbreaks.

In the period 1978 - 1988, fenitrothion was the predominant means of control (Stoakley 1978, 1985a, b, Holden & Bevan 1978). 142 hectares treated in 1984 were part of a study of the effects of fenitrothion on birds (Stoakley 1985a).

In the outbreak which started in 1985, in the first two years, the main treatment was fenitrothion, but small trial areas were treated with diflubenzuron (67.5 grams per hectare), nuclear polyhedrosis virus (MbNPV) and *Bacillus thuringiensis*. The results with diflubenzuron and MbNPV were sufficiently promising for them to be applied in 1987 on, respectively, 31% and 11% of the treated area; all gave 99 + % reduction in autumn pupal counts (Stoakley 1987a, 1988). In 1988, 784 ha were sprayed with fenitrothion or NPV.

An infestation in the Isle of Lewis developed, leading in 1992 to treatment again using MbNPV (Anon 1992d).

In May, 1993, areas requiring treatment were sprayed with diflubenzuron. The equipment used was the most advanced version of the helicopter-borne ULV equipment. This was an independent spray rig suspended below the helicopter; it delivered the product at the rate of 1 litre per ha in droplets of 70-80 μm volume mean diameter (vmd). With droplets of this size, a strong turbulent air flow is required to achieve effective impact of droplets. A wind speed greater than 2.5 m/sec. ensures adequate turbulence (Heritage *et al.* 1994a, b).

For the following 3 years no control was necessary; however, in 1996, a warning was issued of insect numbers increasing again in the north of Scotland (FC NR 1996a). Spraying with diflubenzuron took place in 1997.

Figure 6.1   *Area of lodgepole pine forest sprayed annually between 1977 and 1997 to control Pine beauty moth*   *(hectares)*

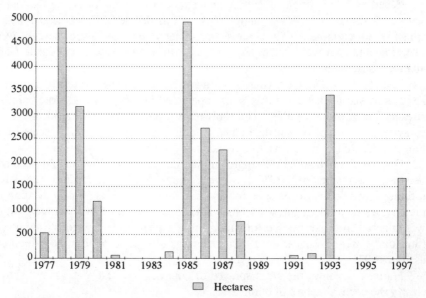

Source: Heritage 1999 *Pers Comm.*

6.43   Pine shoot moth   *Rhyaconia (Evetria) buoliana*

The pine shoot moth is native to a large area of temperate Europe and Asia. On the continent, it is recognised as a major pest of pine forests. Its presence is conspicuous in Scots pine because of the characteristic 'post-horn' response in branch-growth following larval damage to leading shoots in young plantations (Anon. 1936). In Britain, it is largely confined to southern and eastern England.

The pine shoot moth was inadvertently introduced into America at about the beginning of the 20th century. It quickly established itself as a serious pest and, as noted in §5.51, was the subject of an early application of biological control.

In the late 1930s, serious concern was expressed in the UK in respect of new plantations in eastern England. Chrystal (1937) noted that damage spread as the area of plantations increased. In 1936 and 1937, biological control operations were mounted, involving the release of 7400 egg parasites in pine forests at Wareham, Dorset, Rendlesham and Tunstall, Suffolk, and release of several thousand pupal parasites at Wareham (Hanson 1950). Subsequently, spontaneous biological control by naturally occurring predators and parasites seems to have kept the population in check.

High levels of pine shoot moth occurred in Wiltshire in a lodgepole pine seed orchard of Skeena River origin. Fenitrothion gave slightly better results than DDT (Winter & Scott 1977).

Scott (1972) states that general insecticidal control of *Rhyaconia* has never been practised in Great Britain and recommends insecticides only if localised control is required.

Bevan (1987) reports serious post-war problems in southern Ireland with *Rhyaconia* on lodgepole pine and on local plantings in the south of England of *Pinus muricata*. The latter was also studied by Straw *et al.* (1994).

6.44   Gypsy moth   *Lymantria dispar*

The gypsy moth, *Lymantria dispar* is a potentially serious forest pest capable of severe defoliation of a wide range of tree species. The insect, a native of central and eastern Europe, and across Asia to the Pacific coast, was inadvertently introduced into the United States in 1869. It initially spread at about 10 km/yr; causing heavy damage to oak woodland there; in 1971, 800 000 ha were defoliated. Attempts to prevent spread have been unsuccessful and the insect is ultimately expected to spread to all the oak regions east of the Great Plains. It is considered to be the most important defoliator of deciduous hardwoods in North America (Carson 1962, Gibbs & Wainhouse 1986, Booth 1997). During the 20th century, it has also spread within central Europe. Until recently, the last record of the insect in Britain was in 1907.

Booth (1997) concludes that broadleaves can tolerate one defoliation but may die back after two defoliations, conifers are killed by one severe defoliation, trees in the understorey or on poor sites are more quickly killed than vigorous trees.

There are two strains of the moth, European and Asian. In both strains, 1st instar larvae spread by 'ballooning' on silk threads. While this is the most widespread method of spread, most larvae travel no more that 200 metres in this way. Even so, about 0.1% of the total population may travel 5 km or more (Booth 1997).

In the European strain, the female is wingless and sedentary, whereas females of the Asian strain fly actively, thereby increasing the insect's potential for dispersal. Colour photographs of adults, larvae and egg masses are given in *Arboricultural practice, present & future* (Finch 1997).

It was discovered in 1993 that the Asian strain of the gypsy moth had become established in Germany and adjacent countries; egg masses were found on a ship returning military equipment to the USA from Germany. The Asian gypsy moth has also been intercepted around the Pacific Ocean (Winter & Evans 1994).

Traps baited with (+)-disparlure, a pheromone which strongly attracts male gypsy moths to the female, were set out in Britain in 1994 and 1995. In 1994, nine male moths were caught, thought to have been blown over from the continent.

In 1995, larvae were found in Essex in suburban gardens (Winter 1995) believed to be of the Asian strain. The incident was widely publicised (*eg* Drury 1995) because of the strain's greater mobility and ability to devastate, and absence of any 'community' tolerance that indigenous woodlands tend to acquire in relation to resident pests.

From the intensive follow up, it was concluded that the infestation was localised but was at least two years old. Larvae found were nearing pupation and were killed with insecticide. Pheromone traps subsequently placed in gardens locally and in nearby Epping Forest caught a further 36 moths, demonstrating that the infestation had not been entirely removed. Larvae were found in a small number of gardens in June 1996 and were sprayed (FC PH 1996).

Annual surveys included inspections for egg masses and larvae, as well maintaining pheromone traps. In 1998, no eggs or larvae were found. One adult was caught in a pheromone trap on the edge of Epping Forest, suggesting that the population is still present, but at a very low level (Heritage pers. comm.).

Biological control of Gypsy moth

There is a long history of attempts at biological control of gypsy moth, a strong stimulus coming from the United States seeking natural controls for a devastating European pest. Chrystal (1937) mentions the carabid beetle, *Calosoma sycophanta* L. as an important predator of larvae and pupae; he also records that a tachinid fly, *Compsilura concinnata* Meig. had been successfully introduced into the United States and had established itself as a successful parasite over a large area of the country. More recent studies have shown that, in sparse stable gypsy moth populations, about 80% of total mortality could be attributed to small predatory mammals. A fungus *Entomophaga maimaiga* has become an important biological control in North America.

Control options

Should infestations of gypsy moth require to be controlled by large-scale spraying, options include *Bacillus thuringiensis*, diflubenzuron, a material which interferes with chitin deposition in developing larvae, and a pyrethroid such as permethrin (Booth 1997). Earlier, a braconid parasitoid had also been thought to be a possible contender (Blumenthal *et al* 1979).

See also §6.22 Thresholds for action *Search & control.*

6.45   Other lepidoptera

Browntail moth   *Euproctis chrysorrhoea*

This moth has been known as an occasional pest since the 18th Century. It is widely distributed in central and southern Europe but in Britain is mainly confined to the southeast.

Numbers have fluctuated irregularly. Epidemics have occurred in the past, once or twice per century. In 1984, an outbreak was discovered in south and south east England (Sterling 1985, Greenwood & Halstead 1997).

When young, larvae feed as a clutch within a silken tent and prefer rosaceous trees and shrubs. They hibernate within the tent overwinter and recommence feeding in late spring, pupating in June. The insects feed voraciously in the spring, heavily infested trees often being completely defoliated. Also, the moths when touched, shed irritant hairs copiously, so that great care has to be exercised if attempting any control work.

Careful hand removal of tents with larvae during the winter is possible for small infestations (Carter & Gibbs 1989).

Recommendations for chemical treatment have depended on availability and product approvals. The biological agent *Bacillus thuringiensis* was available in early 1999, together with permethrin, diflubenzuron and pyrethroids (TAT 1999).

Larch bud moth   *Zeiraphera diniana*

This moth has occasionally caused serious defoliation to larch, Sitka spruce and lodgepole pine. No control has been necessary. Pheromones have been successfully tested as attractants in Britain (Bevan 1987, Evans 1997c).

Vapourer moth   *Orgyia antiqua*

The vapourer moth has a circumpolar distribution and is found throughout Great Britain. It is a very general feeder on trees and shrubs of all kinds, and is an occasional pest of forests and ornamentals. Adults emerge in mid- to late summer; the female is wingless and lays eggs on the silk cocoon from which she has emerged. Eggs are the overwintering stage. Larvae have conspicuous tufts of orange hairs on their backs; in their first instar stage, they are small and hairy and may be dispersed by wind from their hatching place (Chrystal 1937, Bevan 1987, Alford 1995, including illustrations).

A local defoliation of young Sitka spruce growing in *Calluna* in 1978 was reported from Carron Forest, near Cumbernauld, Central Scotland. The outbreak is thought to have been aggravated by previous drought and by particularly favorable weather during the period of egg-hatching. Spruce defoliation was more severe where *Calluna* was most prevalent; the *Calluna* was also heavily defoliated (Pinder & Hayes 1986).

Winter moth   *Operophthera brumata*

A general defoliator and a recognised pest of fruit trees and ornamentals, this moth's larvae normally feed on a wide range of broadleaved trees and shrubs; conifers may also be attacked. While the adult female cannot fly, young larvae spin silk threads and may be carried from tree to tree by the wind.

Damage was studied for 12 years in the west of Scotland and for a shorter period in southern Scotland.

In 1982, severe damage occurred on Sitka spruce on a private estate; 340 ha were sprayed in May 1983 using fenitrothion at 0.3 kg a.i. per ha from a helicopter fitted with rotary atomisers. The treatment was 80-85% successful (Stoakley 1984b).

6.5   INTEGRATED MANAGEMENT OF BEETLE PESTS *Coleoptera*

6.51   Large pine weevil   *Hylobius abietis*
Black pine beetles   *Hylastes* spp.

Hylobius abietis has been recognised for two centuries as the most important pest of European conifer regeneration. The Forestry Commission's first *Forest Leaflet* published in 1920 had the title *Pine weevils* (Anon. 1920).

The large pine weevil is a normal inhabitant of woodland, feeding in the forest canopy and breeding in conifer stumps, logs or branchwood left after thinning or final crop felling, natural mortality or storm damage. It is a relatively long-lived insect and is able to increase rapidly whenever material suitable for breeding occurs. In north east Scotland following a serious gales in 1954, areas of windblown pine were estimated to be carrying up to 180 000 adult pine weevils per ha within two years of the damage (Crooke & Bevan 1956).

The long-lived adults damage young plants by feeding on thin bark near ground level, often girdling them (Scott & King 1974, Bevan 1987). Average losses due to pine weevil attack on plants that have not had chemical or physical protection, is estimated at about 30% per year for the first two years *ie* 50% over the 2 year period. In extreme infestations, losses can exceed 90% (Heritage 1996).

The potential for future damage by pine weevil is likely to increase for the foreseeable future. While the annual rate of planting of unwooded land in Great Britain has fluctuated since 1970, the area being restocked following felling has slowly risen (Figure 1.1 in Aldhous 1994). In 1997, areas of new planting and restocking were, respectively, 16 900 and 14 300 ha (FC FF 1997). For the year 2000, the combined private and Forestry Commission restocking area has been forecast at 15 000 ha; for 2005, the Forestry Commission restocking programme was expected to rise by more than 60% to 12 500 ha (Heritage *et al.* 1989).

In the long term, as long as the net GB forest area under conifers increases due to new planting, the long term area of restocking will also continue to rise.

The total annual cost of controlling *Hylobius* on Forestry Commission land was estimated in 1997 to be £1-1.5 million (Moore 1997).

Black pine beetles Hylastes spp

While adult pine weevils feed above ground, *Hylastes* spp. may feed on roots of young plants in similar circumstances, also leading to losses in newly planted stock. Experiments showed that treatments against *Hylobius* could also be effective against *Hylastes* (Bevan & Davies 1970).

Vigilance

The pine weevil is so widespread that it should be assumed that protection measures are likely to be required. However, forest managers must be aware of the general level of background numbers, and whether in a given season, there are departures from the assumed level which should cause a review of pre-planting dipping, electrodyne treatment *etc* and provision for post-planting 'top-up' spraying.

Non-chemical means of control

Until the advent of insecticides, trapping by laying bark or billet traps was the only course of action. Traps on the soil surface were inspected daily and any weevils visible were destroyed by hand (Chrystal 1937). However, these were very laborious and not reliable. It has only been during the last fifteen years that alternative non-chemical means have been considered to merit trials.

Fitting a collar (Teno) to newly planted stock to act as a physical barrier to access by pine weevil was not effective (FC WS 1987). A cotton mesh stocking (Vinetta) over small container plants offered greater promise (Heritage & Stoakley 1988) but has not come in to general use. At the end of the 1990s, other physical barriers were, however, commercially available in Britain.

Delaying planting for several years has been suggested from time to time, but unless a replanted stand is completely isolated, weevils move in from adjoining crop areas and little reduction in damage is found (Booth 1990). Movement by adult weevils over distances of 0.5 km/day or more has been recorded.

Biological control

Investigations into natural parasitism of pine weevil have found five species of parasitic insect, two nematodes and various fungi, bacteria and yeasts. The possibility of maintaining sufficient populations of parasites and predators such that, in the long term, weevil populations may be significantly suppressed is being pursued. The two most promising lines, entomopathogenic nematodes (nematodes that feed on insects), and a braconid wasp, are complementary.

Neither appears to operate sufficiently quickly to be used as a substitute for an i3nsecticide where there has been no prior attempt at biological control. Nevertheless field trials have commenced testing whether in forests with a sustained felling programme, biological control agents can so reduce resident populations that intensity of attack by adults feeding on newly planted stock is reduced to a tolerable level (Brixey 1997).

Insect-feeding Nematodes

Indigenous entomopathogenic* nematodes, in the genera *Steinernema*, *Dirhabdilaimus* and *Heterorhabditis*, are effective biological agents. They can be used to control vine weevil and sciarid flies in protected and outdoor container-grown crops and are commercially available for that purpose; they also have considerable potential for a wider role.

In the first British field trials of such nematodes against adult and larval *Hylobius abietis* between 1988 and 1993, nematodes applied to buried conifer billets found and killed weevil larvae successfully. Subsequent trials also gave good results, in particular on lodgepole pine stumps. The infective juvenile nematodes act by searching for potential host larvae. Once found, the nematodes inject bacteria into larvae; the bacteria multiply, generating toxins which kill the larvae. The nematodes feed on bacteria and larval remains, mature and breed. When the parasitised larva is completely consumed, infective juveniles emerge to seek for other larvae. 100 000 juveniles can emerge from one larva (Heritage *et al.* 1990, 1993, 1995, Brixey 1997).

Good control of *Hylobius* has been more readily obtained when nematodes have been applied to two-year-old stumps, rather than freshly cut stumps. Results are better in high rainfall areas and wetter soils where there is less risk of nematodes drying out before encountering a target larva.

Parasitic wasp - Bracon hylobii

Parasitic wasps are one of the largest groups of insects, comprising over 5500 species. Physically, most are small, as are their hosts. Brixey (1997) lists four braconids two of which affect *Hylobius* larvae and two, the adult beetle.

The most promising braconid is *Bracon hylobii*. The limited data available suggest that at different times *Bracon hylobii* can destroy 30 - 90% of *Hylobius* larvae in pine and 50% in spruce; it is more host-specific than the entomopathogenic nematodes. However, populations studied show that peak numbers of the wasp lag behind the period of peak host availability. While mass rearing methods are known, it has yet to be found whether appropriately timed mass releases to augment wasp numbers would significantly reduce damage to young planting stock (Heritage *et al.* 1995; Heritage 1996, Brixey 1997).

Pathogenic fungi

Three strains of the pathogenic fungus *Beauvaria bassiana* have been isolated from *H. abietis* and are being cultured for initial laboratory trials (Heritage *et al.* 1995). *Sporotrichum globuliferum* has also been identified as a natural enemy of *H. abietis* (Brixey 1997).

Wood-rotting fungi may also compete with developing larvae. *Phlebia gigantea* has been noted as a possible candidate (Heritage et *al.* 1995).

---

*The terms *entomophagous*, *entomopathogenic* and *entomogenous* have been used by different authors to describe these nematodes.

Other agents

Other biological agents with potential to reduce numbers of *H. abietis* include an ichneumonid fly, bacteria and yeasts (Brixey 1997).

Insecticides

While biological methods are being developed, insecticides remain the mainstay of the protection of newly planted stock against attack by *Hylobius*. These have been under test or in use for over 50 years.

The first trials in the period 1948-50 were in the context of adding insecticides (DDT and benzene hexachloride) to bark and log traps. It was hoped that insects attracted to the traps would be killed by the insecticide, saving the work of having to inspect traps and manually remove and kill insects found there. Results, however, were not encouraging and led to consideration of application of insecticidal dusts to the plants themselves (Hanson 1951, Crooke 1953, 1954b, 1955b, Crooke & Bevan 1956).

Plant dips

Dipping shoots in 5% DDT emulsion shortly before planting was found to be successful as a protection for planting stock; it became the first recommended treatment (Bevan *et al.* 1957). Such dipping was also expected to control black pine beetles feeding in the same area. The recommendation remained in force until the worldwide recognition of the detrimental side-effects of DDT's persistence led to programmes to find alternatives.

Trials of dipping in a 1.5% dilution of lindane (Gammacol), either of shoots or complete plants showed promise compared with 5% DDT (Bevan 1966a), and led to adoption of lindane for this use. The main practical difficulty was that the insecticide deposit on plants had to have dried adequately after dipping before further handling.

In these trials, chlorpyrifos was also shown to be an effective alternative, given similar care in handling (King & Scott 1975b).

Results of trials with lindane in the forest repeatedly showed that pre-treatment by dipping prior to planting was usually better and very seldom worse than post-planting spot spraying of planted trees.

Further trials compared the effect of treatment before or after storage prior to planting. No ill-effect was found of treating before storage, provided insecticide deposits on plants had been allowed to dry properly (Heritage 1997a).

Lindane remained in use into the late 1980s (Scott & King 1974, Blatchford 1983, Hibberd 1986).

While not a regular occurrence, in ill-defined circumstances, lindane could be toxic to plants (Tabbush 1985a, Tabbush & Heritage 1987). In a comparison of spruce plants dipped in either permethrin or lindane and stored for 2 or 6 weeks, survival of lindane-dipped plants stored for 6 weeks was appreciably less than for other treatments (Tabbush & Heritage 1988). Corsican pine treated with permethrin suffered frost damage if lifted in November and stored and also if sprayed with permethrin before storage. Plants lifted in January and stored for three weeks once foliage had dried after treatment achieved 90% survival at the end of the first year after planting (McKay *et al.* 1998).

Pyrethroids

Programmes testing some of the newer and more persistent pyrethroids started in 1982; early results showed permethrin and cypermethrin were as persistent and less phytotoxic than lindane (Stoakley *et al.* 1984, 1985a, Stoakley & Heritage 1987). Early trials of permethrin were subject to close scrutiny for safety to the men treating and dipping the plants (Teasdale 1990b).

Dipping

Increasing concern about the safety of particular pesticides raised questions about lindane and other pesticides used in dipping programmes. For dips to be successful, the insecticide had to dry on the plant after application. It was of equal importance to ensure that all personnel involved in the dipping operation had minimal contact with insecticide. Achievement of these requirements, which apply no matter which insecticide is included in the dip, has called for considerable management skill, especially during periods of continuing wet weather when plants are due to be treated.

With increasing replanting programmes, the number of plants to be treated increased correspondingly; in one case, where dipping facilities served an increasing replanting area, the duration of exposure of workers treating plants prior to planting also increased, occasionally beyond acceptable limits (Teasdale 1990a). As a consequence, higher safety requirements have been introduced, along with a preference for 'engineering' systems. Such systems utilise equipment or mechanical devices to reduce substantially the scale on which people dipping or planting are exposed to pesticides (Heritage 1996, 1997a, b).

Electrodyn treatment

As an alternative to dipping, a system novel for forestry has been developed whereby electrically charged particles of insecticide are attracted to plants electrostatically, the 'Electrodyn' application system. Plants are fed onto a moving belt which carries them into a fully enclosed spray booth in which they are sprayed. Most of the insecticide is attracted to the plant foliage and stem, so that there is little wastage of insecticide. The carrier solvent evaporates during the time the plants are in the spray booth; special precautions have to be taken to ensure safe discharge of air from the spray booth. However, water is not involved so that, unlike dipping, there is no requirement to allow plants to dry after treatment. The target rate of treatment is 0.4 $\mu g/mm^2$. Use of the electrodyn method of application eliminates much handling of insecticide-charged dipped plants (Scott *et al.* 1990, English 1990, McKay *et al.* 1994).

By 1996, 4 million plants per annum were being treated in Wales under permitted field trials (Heritage *et al.* 1997b).

Carbosulfan

In the 1990s, trials were commenced using carbosulfan. This is a systemic insecticide, normally relatively non-persistent. However, for forestry purposes and for container-grown ornamentals, it is formulated as a polymer-based controlled-release granule to be mixed into the soil around the plant roots at the time of planting, persistence being achieved through slow release from the granule (Heritage *et al*.1997a).

Recent practice

Recommendations for practice in 1996 are summarised in Forestry Commission *Research Information Note* (RIN) 268 (Heritage 1996). This briefly describes *Hylobius'* life-cycle, the insecticides for which there is approval for use, the 'Electrodyn' equipment under trial and a note on the possible use of entomopathogenic nematodes.

Spraying after planting as an alternative or boost to dipping

Dipping is preferred to post-planting spraying because of the logistical difficulties of getting round all plants before weevils become active, and because of the difficulty on rough terrain of locating all the newly planted stock. Also, less insecticide is used in dipping compared with spraying.

If plants are to be treated only after planting, it is important to get treatments completed before the emergence of the first weevils (Heritage 1997a).

First season boost

Several studies showed that the concentration of insecticide on dipped trees diminishes after application, often before the plant has become robust enough to withstand weevil attack. Where there are exceptionally high infestations of weevil and continuing damage, it has often been found necessary to boost the effect of dips by spraying planted stock with lindane or permethrin.

Second season treatment

Dipping and 'Electrodyn' treatments only give protection in the first year after planting. A single post-planting spray gives 4 months protection against moderate attacks, provided that the insecticide has been able to dry on the plant and thereby become rainfast (Heritage & Johnson 1997).

Where weevil populations are high so that attack in the second growing season after planting has to be expected, a 'top-up' spray is recommended at the start of the second growing season (Heritage 1997a).

6.52 Bark beetles

Bark beetles are woodland insects, each restricted to a preferred range of host species. Many bark beetles cannot overcome the natural defences of thriving healthy trees but can invade and rapidly increase in numbers where stands have been weakened or damaged by drought, wind-blow, fire *etc*. Bevan (1987) groups *Dendroctonus micans, Tomicus piniperda, Ips cembrae* and *Ips typographus* as being able to attack green trees under stress through defoliation, drought or damage. He differentiates them from other bark beetles which, while mainly attacking dead, dying or felled trees, also have a period of maturation feeding on healthy trees. It is during this time that, for example, *Scolytus* spp and *Tomicus* spp may introduce a fungus which materially affects the health of the tree or the quality of its timber.

6.53   The great spruce bark beetle   *Dendroctonus micans*

Dendroctonus micans is the largest of the commonly occurring bark beetles (6-8 mm long). It was recognised in several central European countries in the mid- and late 19th century as a potentially destructive pest. Although not native to the United Kingdom, its ability to attack healthy Sitka and Norway spruce trees marked it as an important threat to be kept at bay by avoidance of imports of infested logs and timber (Brown, J.M.B. & Bevan 1966).

The beetle was inadvertently introduced into the United Kingdom in about 1972 near Ludlow, Shropshire, probably on timber imported from Germany following extensive windblow there. It became established and spread into adjoining forests along the England/Wales border. Its presence was not formally confirmed until 1982 (Bevan & King 1983, FC AR 1984).

Relative susceptibility to attack and subsequent mortality are shown below (FC PH 1998c). Ornamental spruces have also been attacked (Evans *et al*. 1984, Evans & King 1988).

	Susceptibility to initial attack	*Likelihood of mortality*
Higher	*P. abies, P. alba, P. omorika*	*P. pungens, P. omorika, P. orientalis*
Intermediate	*P. pungens, P. orientalis*	*P. sitchensis, P. alba*
Lower	*P. sitchensis*	*P. abies*

Dendroctonus is able to create brood galleries almost anywhere above ground on a tree stem and around the buttresses and bigger roots. The insect does not breed synchronously, so that on a heavily infested site, all stages of the life cycle may be present at one time. The duration of the life cycle is 10-18 months.

When the Ludlow outbreak was discovered, it was recognised that positive location of all infected trees was not feasible, total eradication was therefore not a reasonable option and that a 'containment strategy' was necessary. This was developed under the guidance of a *Dendroctonus micans working group*, a joint forestry Commission and private forestry sector advisory body which met regularly to review progress.

Containment strategy

The main features of the containment strategy have been:

* initial population reduction by 'sanitation' felling, debarking and spraying of all trees found to be infested with *Dendroctonus*;
* initial surveys to ascertain the extent of the outbreak, with follow-up surveys to determine the extent of reinfestation and spread;
* designation of a 'Scheduled Area' including all known infestations, plus a safety margin;

- legislation to control timber movement (SI 1982/1457):
  - felled timber to be processed at 'approved mills';
  - sawn timber only to leave the Scheduled Area if totally bark-free;
- an urgent programme to build up stocks of *Rhizophagus grandis* for mass release (King 1987, Fielding 1992). See p 124;
- cessation of the initial sanitation programme when it appeared that the introduction of *Rhizophagus* was succeeding.

Surveys

Surveys were immediately undertaken to discover the initial extent of infestation. Between 1982 and 1984, over 12 000 sites were examined, over 85 000 infested trees discovered and the area containing infested trees established.

Thereafter, annual surveys assessed changes in population size and range, seeking signs of the presence of *Dendroctonus* in plantations and other trees within a 10 km radius of the limits of the previously known infested area.

From assessment of sample trees, there was a rapid increase in numbers of *Dendroctonus* until 1986/7, after which time the rate of increase diminished. From 1991 the population decreased, reflecting the increasing effect of *Rhizophagus*.

Ten years after first discovery of the presence of *Dendroctonus,* and after the successful establishment of an effective biological control, its rate of spread was estimated to be 3 - 5 km per annum (FC AR 1988, 1992). It is expected that the *Dendroctonus* population will settle to a stable equilibrium level well below any threshold requiring more active control of its numbers. Fielding *et al.* (1991) give a full account of the early spread and results of surveys.

Restriction of spread through movement of timber

Once the area of the infestation had been defined by the initial survey, movement of spruce timber that could possibly carry the insect was controlled through the *Restriction on Movement of Spruce Wood Order 1982* (SI 1982/1457) and subsequent amendments.

In 1993, the control system was changed to bring it into line with the 'Single Market' passport mechanism and its associated provision of *Protected zones* (FC PH 1994c). New rules were set out in the *Plant Health (Forestry) (Great Britain) Order 1993,* (SI 1993/1283, as amended). Under this order, the area affected by *D. micans* was designated the *Dendroctonus micans Control Area* (DMCA). The rest of Great Britain was designated as a 'protected zone' (DMPZ).

Controls on the movement of spruce bark originating in the affected area were set out in the *Treatment of Spruce Bark Order 1993* (SI 1993/1282, as amended).

Arrangements under the 1993 orders were broadly similar to those previously existing. For full details see *Plant Health Leaflet No 8* (FC PH 1994c - under revision). Key points include:

- producers of 'forestry material' have to be registered unless exempted under 'local movement' provisions;
- movement from the DMCA to the DMPZ of spruce wood with bark is prohibited unless it has been kiln-dried;
- movement of non-spruce wood is controlled by means of passports and related health checks.

Other steps to minimise spread include:

- monitoring by surveys for up-to-date evidence of the occurrence of *Dendroctonus*;
- sanitation felling in peripheral zones and in localised outbreaks elsewhere;
- biological control by release of *Rhizophagus grandis* wherever called for (FC PH 1993b, 1996, 1998c).

The 'DMCA' was divided into:
- the main area of infestation, comprising most of the English/Welsh border forest and much of Wales (the 'Central Zone'). Once an area has been designated as infested, it retains that designation.
- an infested outlier at Bowland Forest, Lancashire;
- a 'Peripheral zone' approximately 10 km wide outside the known areas of infestation, and subject to regular survey to detect spread into it;
- an outer zone between the outer boundary of the peripheral zone and the boundary of the 'Scheduled area' (FC PH 1986, 1989).

In late summer 1996, a further outbreak was discovered in Kent, in southeast England (FC NR 1996b, FC PH 1998c). This outbreak is thought to be the result of an introduction on infested wood, separate from the Shropshire introduction. The same report mentioned another outbreak in Warwickshire, much closer to the first outbreak. Both were classed as requiring sanitation felling treatment.

Biological control and *Rhizophagus grandis*

For nearly 20 years before the bark beetle was discovered in England, biological control of *D. micans* had been practised in Russia and France. Adult predator beetle *Rhizophagus grandis* had been bred up in laboratories and successfully released. *R. grandis* adults located and laid eggs where *Dendroctonus* larvae were feeding, so that their larvae fed on and destroyed *Dendroctonus* larvae.

Populations of *Dendroctonus micans* build up relatively quickly in the absence of natural controls. When *Rhizophagus grandis* has been successfully introduced, increases become smaller and slower. Once both species have become established, *Dendroctonus* populations maintain themselves at a relatively low but stable level; population fluctuations are minor, a characteristic of 'immediate density-dependent' natural control of populations (Barbour 1985).

In Georgia, USSR, serious infestations of *D. micans* had appeared in extensive forests of Oriental spruce *(Picea orientalis)*. A programme of artificial release and breeding of *R. grandis* commenced in 1963 and continued into the 1990s. A second such programme began in 1983 in Belgium in collaboration with French forest authorities to control various outbreaks of *D. micans* in north east France and Belgium.

In Britain, a biological control programme was able to be put into effect quickly and has been effective. *R. grandis* adults imported from Belgium were the source of the 27 pairs released into infested spruce in July, 1983; in October, 1983, some larval progeny were recovered.

Annual releases in 1984-1988 were, respectively, 31168, 39392, 17604, 7100 and 7300, the last two years' releases being at new locations found in peripheral surveys (King *et al.* 1984, King & Fielding 1989, Evans & King 1989b, Evans & Fielding 1996). Between 1984 and 1991, over 131 000 *R. grandis* had been bred and released.

By 1992, up to 80% of all *Dendroctonus* broods studied were found to have been colonised by *Rhizophagus* (Fielding 1992). *Rhizophagus* was always to be found on sites containing over 100 infested trees (Fielding & Waters 1994).

In 1985, the entomopathogenic fungus, *Beauvaria grandis* was noted as a principal factor in 30% mortality in *R. grandis* between pre-pupal and adult stages (Evans *et al.* 1985a). Subsequent changes in breeding practice have minimised the threat from this fungus. See also §5.52.

The containment strategy for the Ludlow infestation has been successful. Overall, damage by *D. micans* is light and acceptable; where first found, it seems to have become a low level endemic pest (Evans & Fielding 1994).

Pesticide use was restricted to lindane sprays to 'clean up' infested trees in the first 'sanitation felling' stages. Spraying was discontinued in most areas in 1985; infested trees subsequently discovered were totally debarked, *Rhizophagus* being released to search out infestations in any nearby trees (FC PH 1986). Recommendations in 1998 were for felling and 'spot peeling to ensure removal of all beetle stages' (FC PH 1998c).

6.54   Elm bark beetles   *Scolytus* spp

Three species of *Scolytus* may act as vector for Dutch elm disease, *S. scolytus* (large elm bark beetle), *S. multistriatus* (small elm bark beetle) and *S. laevis*, the last only having been found recently in Great Britain.

These species breed under the bark of dead and dying elm. Shortly before leaving their breeding site, they may inadvertently pick up spores of the fungi which cause Dutch elm disease, *Ophiostoma* spp (§7.31). The newly emerged adult beetles then fly to the crowns of healthy elm and feed (maturation feeding), for a period on the bark of twigs. In so doing, they transfer *Ophiostoma* spores to those trees. The spores germinate, spread through the tree creating toxins which lead to the blocking of the water-conducting tissues of the sap-wood through formation of tyloses in vessels. Where all the vessels supplying whole limbs or branches are affected, they die through desiccation. Trees are thereby weakened, and become suitable targets for beetles seeking a breeding site.

The combination of beetle and fungus has resulted in the death of over 25 million of the 30 million elms in Great Britain (Brasier 1996). Losses have not, however, been equally spread within Britain.

- *S. scolytus* prefers *Ulmus procera* and *U. carpinifolia* (English and Wheatley elm) to *U. glabra* (wych elm); this preference probably accounts for the observed lower initial incidence of the disease in wych elm rather than any inherent greater resistance to the fungus.
- Wych elm is more prevalent in northern and western Britain;
- *S. scolytus* occurs throughout Great Britain up to the line of the highland boundary fault. The distributions of *S. multistriatus* and *S. laevis* overlap only to a small extent. The former occurs in Wales, central and southern England and the latter to the north. (See *figure 1* in Fairhurst & Aitkins 1987.) *S. laevis* while only recently discovered, is thought to have been present for a considerable time.
- A major factor restricting beetle breeding in wych elm is believed to be the fungus *Phomopsis oblonga* which is more common in the north and in wych elms (Webber 1981).
- Beetles disperse by flying; however, dispersal is temperature-dependent. In the forest, few beetles fly at temperatures below 21°C.

These factors account for observations that more elms, especially wych elms, have survived the onslaught of the 1970s epidemic in Scotland than further south.

Beetle control

Attention was given to the possibilities of controlling the vector beetle (Scott & Walker 1975). First attempts sought to control beetles during the period immediately after emergence, when they do their maturation feeding. Trials of DDT over three years in the early 1950s on trees 6 - 10 m tall, were at the time thought to be generally successful; however, treatment costs were too expensive for any but exceptionally highly valued trees (Peace 1954, 1960). Small-scale tests with methoxychlor, tetrachlorvinphos, chlorpyrifos and lindane were carried out in 1971 but were not pursued (Bevan & Davies 1972, King & Scott 1975a).

A trial of the effect of removing infested logs and spraying uninfected logs with *gamma*-HCH to prevent colonisation was considered to be effective but has not been widely applied (Gibbs *et al.* 1977).

Dutch elm disease was carried to the United States and Canada in 1930 and 1944, respectively, together with the beetle *Scolytus multistriatus* which has been an important vector there. By 1977, it was estimated that 60% of the 77 million urban American elms had been lost (Gibbs & Wainhouse 1986). Heavy handed attempts to control the disease by aerial spraying in the 1950s was one of the prime subjects of complaint in Rachel Carson's *Silent Spring* (Carson 1963).

6.55   *Ips* spp

The 8-toothed spruce bark beetle   *Ips typographus*

Forestry in the UK is vulnerable to introduced bark beetles. Several species which are not native to Britain exist on the continent and could inadvertently be introduced on imported roundwood or unbarked sawn timber. The most feared is *Ips typographus*. It characteristically infests damaged spruces, beetle populations quickly building up following wind-blow. Large populations can attack healthy trees. The beetle carries with it blue-stain fungi which help overcome trees' normally high resistance to beetle attack (Wainhouse *et al.* 1998).

Winter and Burdekin (1987) describe an outbreak in 1869-75 in Bohemia where two severely damaging storms in 1868 were followed by a similar storm in 1870. The consequent build up of beetles led to killing of trees over an area of 11000ha, the volume of timber involved amounting to several million (cubic) metres of timber. Chrystal (1937) also mentions catastrophic wind damage in Germany leading to bark beetle infestations. More recently in Europe where trees have been weakened by air pollution, *Ips* has moved in (see §14.43).

Occurrences in Britain

Considerable numbers of *I. typographus* were distributed around Britain on timber imported in the late 1940s (Laidlaw 1947, Hanson 1949). In this period, beetles were discovered in forests near Inverness, Gloucestershire and Devon and hibernating specimens in a Welsh timber yard (Winter & Burdekin 1987). The beetle is not known to have become established at that time.

For a number of years the Forest Authority monitored air in the vicinity of ports for the presence of flying beetles and other insects. In 1995, it was noted that the number of beetles trapped at major ports in the year reported rose to 45 from 3 in the preceding year (FC AR 1995). In the following year, 149 were caught (Evans 1996). In 1996, *Ips* infestations were reported to have been found on imported dunnage (Bradbury 1996). The Forestry Authority, through its *Plant Health Newsletter,* warned that imported consignments of infested timber would be refused entry to the UK (FC PH 1996).

In August 1997, pheromone traps monitoring air-borne beetles at the pulp mill at Shotton caught approximately 30 adult *I. typographus*. The number caught and the location were considered sufficiently important to make intensive efforts to check 110 potential sources of supply to Shotton by pheromone traps and to publicise to everyone involved in timber harvesting what to look for in the forest. The trapping programme was repeated in 1998 (FC PH 1998b).

In the event, no source was identified in the following 12 months (Burgess 1998, Wainhouse *et al.* 1998). Pheromone trapping was resumed in 1999 at major processing plants and ports of entry (FC PH 1999).

Biological control

Hanson (1949) also mentions the release of a hymenopterous parasite, *Ipocoelius seitneri,* against *Ips* in Rheola forest in south Wales in 1948.

Vigilance

The first legislative provision to reduce the risk of import of *Ips* spp was the 1961 *Landing of unbarked coniferous timber order* (SI 1961/656), prohibiting imports of unbarked spruce timber from central Europe. Subsequent legislation applied more generally to pests on imported timber, *Ips* spp always being included.

Ips sexdentatus

This non-native species is a pest of pines but is not considered as great a threat as the *Ips* spp that attack spruces (Bevan 1987).

In the 1940s and before, maritime pine pit props imported from France carried considerable numbers of *I. sexdentatus*. It became established as a breeding species in south Wales and parts of southern England, infestations occurring in several areas of south Wales in 1947. These died out naturally, no further infestations being reported in 1948.

Hypophloeus fraxini, a predatory species had been introduced with *Ips sexdentatus* and became numerous; it was considered to have played an important part in terminating the infestations in south Wales. The potential of other imported potential predators was also recognised (Hanson 1949).

The large larch bark beetle *Ips cembrae*

This beetle was first recorded in Britain in 1955 (Crooke & Bevan 1957). It is reputed on the continent to be a secondary species breeding in felled logs, windblown and dying trees, but to attack green trees on drought-susceptible sites. This has also largely been the case in Britain. The species has been found feeding on conifers other than larch, its galleries in Sitka spruce being similar to those of *I. typographus*. It has been found in its maturation feeding phase on Douglas fir shoots (Bevan 1987).

Trials of lindane and chlorpyrifos on log stacks for prevention and control of attack were successful, both materials giving protection at one or more of the rates applied (Stoakley 1976). Both were subsequently recommended for use in a paraffin or diesel oil diluent (Blatchford 1983). Carter & Gibbs (1989) recommended lindane in oil and chlorpyrifos in water.

As first preference, however, during the period April - September, potential breeding material should either be removed from the forest or should be debarked soon after felling.

Two natural pheromones generated by *Ips cembrae* have been identified and synthesised. In tests they showed promise, both as the basis for population monitoring and also for trapping (Stoakley 1977b).

6.56   Pine shoot beetle *Tomicus (Myelophilus) piniperda*

This bark beetle is endemic in pine areas. It feeds by boring into shoots, killing or weakening them and pruning the crowns of host trees; it may cause some loss of increment. For illustrations, see Strouts & Winter (1994), p201.

The beetle breeds under the thick-barked stem material. Healthy trees are rarely attacked except on marginal sites such as sand dunes. The insect can, however, act as an important secondary killer following initial weakening through defoliation, *eg* by *Bupalus* (Crooke 1954a, Bevan & Davies 1970, Bevan 1987).

Semi-mature ball-rooted trees in amenity plantings can be attacked and killed if subject to water stress and there is an endemic population of pine shoot beetle nearby (Winter 1991).

The pine shoot beetle is a carrier of fungal spores of the fungus causing 'blue stain' *Ceratocystis* spp (Savory *et al.* 1965, 1970, Dowding 1970). See §7.32.

Control of beetle and blue stain of timber

The first line of control is to ensure that, between April and August in any year, potential breeding material is removed within six weeks of felling, either by removing felled timber to sawmills within a few days of felling ('hot logging') or by removal of bark. This may not be possible, for example where there is extensive wind-blow. However, in the 1980s, a large volume of pine was successfully stored as 'wet logs' following severe windblow in Norfolk/Suffolk, and remained largely stain-free (Webber & Gibbs 1996); see also §§7.32 & 15.34.

Chemical spraying of logs to minimise blue stain is recommended only where it is economic, *ie* where the timber is in sizable stacks. It is preferable to apply treatments before or very shortly after beetle attack, before larvae can penetrate deeply into the bark.

Lindane was recommended as a spray in 1962 (Bevan 1962). For the 1970s and 1980s chlorpyrifos was added. Lindane was recommended to be applied in oil and chlorpyrifos in water (Davies & King 1973, Blatchford 1983, Carter & Gibbs 1989, Strouts & Winter 1994).

Hanson (1949) considered *Myelophilus minor* to be a more serious threat than *T. (Myelophilus) piniperda*. His concern has not been borne out by subsequent experience.

6.57   Other Coleoptera

Banded pine weevils *Pissodes* spp

The genus *Pissodes* contains several species of economic importance in Europe and America. Only *P. castaneus (=P. notatus)* and *P. pini* have been cause for concern in Britain, *P. castaneus* being considered more widespread and destructive than *P. pini*. In the 1940s, with *Hylobius abietis*, they were grouped among the most destructive forest pests. In appearance, they resemble smaller versions of *Hylobius*, and can attack young newly planted stock and pine plantations up to 5 m tall (Anon. 1952). They breed under the bark of dead and dying pine and felled material; they may also lay eggs in shoots, the larvae tunnelling in the pith.

Potential breeding material should be removed from the forest, following the same practice as for *Tomicus* (§6.56). For protection against attacks on newly planted stock, the same insecticide treatment as for *Hylobius* has been recommended (Blatchford 1983, Carter & Gibbs 1989)).

FC Leaflet 29 (Anon. 1952) lists six parasites of *Pissodes* weevils and comments that they are important factors in the control of *Pissodes*, especially where the weevils are breeding in leading shoots. Some parasites may be able to produce two generations in a season compared with the weevils' one.

The weevils' relatively low profile in recent years may indicate that natural controls have become increasingly effective as the older pine plantings have matured.

Clay-coloured weevil	*Otiorhynchus singularis*
Nut-leaf weevil	*Strophosomus melanogrammus*

Both these weevils are described as being general feeders on forest and ornamental trees and in the nursery, the adults being able to remove chunks from foliage and to ring-bark young plants (Bevan 1987, Alford 1995). Lindane has been recommended for control when damage is first evident (Blatchford 1983).

Striped ambrosia beetle   Xyloterus *(Trypodendron) lineatum*

While essentially an inhabitant of timber rather than the growing tree, populations are endemic wherever timber felling operations have left any substantial waste, or where there are wind-broken timber fragments. The species has an outstandingly powerful aggregation pheromone which draws adults into potential breeding material. Attack begins in April; timber felled between November and March is more susceptible to attack if not previously removed. The insects will find timber both in the forest and at the mill.

Stacked timber has been protected by lindane or chlorpyrifos in water sprayed before the start of the attack period (Bletchley & White 1962, Bevan 1987, Carter & Gibbs 1989).

Asian longhorn beetle   *Anoplophora glabripennis*

Towards the end of 1998, the Forestry Commission became aware that timber imported from China and used as packing material, could be infested with larvae and adults of this beetle, capable of attacking several important hardwood species, including commercial fruit trees. An amendment to the *Plant Health (Forestry) (Great Britain) Order 1993* added this insect to the list of prohibited pests, requiring that from 1st February 1999, all wooden packing material imported from China either shall be bark free and contain no grub-hole less than 3mm wide, or shall have been kiln dried (FC NR 1998b).

Illustrations of the adult beetle and feeding damage are given in *Horticulture Week* (Perry, 1999).

6.6   INTEGRATED MANAGEMENT OF SAWFLIES   *Hymenoptera*

Only a few of the four hundred or so species of sawfly found in Britain seriously affect plantation trees. Chrystal (1937) considered the pine sawflies and the large larch sawfly to be the greatest threats to British forests. By 1950, the large larch sawfly was no longer a threat; spruce sawflies were receiving greater attention (Hanson 1950). Subsequent experience showed that the species that caused serious local damage were the pine sawflies (on younger crops), European spruce sawfly and the web-spinning larch sawfly, a species first discovered in Britain in 1953 (Bevan 1987).

6.61   Fox coloured sawfly   *Neodiprion sertifer;*
Large pine sawfly   *Diprion pini*

Both these species of sawfly are pests of young pine forests. When present in large numbers, they are conspicuous through their aggregation into large feeding colonies. Larvae may be up to 25mm long; see illustrations and life cycles in *FC Field Book 17* (Winter & Carter 1998). *Neodiprion sertifer* has a palaearctic distribution while *Diprion pini* occurs in central and northern Europe.

In Great Britain, neither species has caused prolonged serious damage to its native host, Scots pine. However, severe outbreaks of short duration characterise the pest on the European mainland where losses are recognised as being of economic significance (Austarå 1987). Outbreaks are brought to a halt through collapse of the pest population. For *Diprion pini*, this is due to attack by parasitoids; for *Neodiprion sertifer*, population collapse follows widespread natural infection by nuclear polyhedrosis virus (NPV). *N. sertifer* populations seldom recover to their previous high level before host trees have grown beyond the susceptible stage (Anon. 1955, Bevan 1987, Entwistle *et al.* 1985). This pattern of pest population 'surge and collapse' illustrates a 'delayed density-dependent' response (Barbour 1985), in contrast to the low level endemic infection population pattern described above for *Dendroctonus* when controlled by *Rhizophagus*.

Before it was clear that populations would normally collapse, areas infested with pine sawfly were sprayed with nuclear polyhedrosis virus, the area being shown in *Table 6.4*.

Biological control of Neodiprion sertifer by Nuclear polyhedrosis virus

NPV belongs to the nuclear polyhedrosis group of the Baculoviridae. The major groups affected by baculoviruses are Lepidoptera and sawflies. They do not infect beneficial insects such as ladybirds, and have been recorded only on Arthropoda. In nuclei of infected cells, all NPVs produce large quantities of protein which is deposited around groups of the rod-shaped virus particles, to form 'polyhedral inclusion bodies'. These are thought to give some protection to the virus from adverse environmental factors and to adhere strongly to plant surfaces, including needles. The pine sawfly NPV replicates in the gut of larvae on a scale sufficient to kill the larva. The semiliquid contents of dead larvae are spread by natural means. Further infection is by ingestion of 'polyhedral inclusion bodies' while feeding on naturally infected pine foliage. The youngest instars are most susceptible to infection.

Pine sawfly was introduced accidently into North America and was first noted there in 1925. The pine sawfly NPV was deliberately introduced and first successfully tested there as a means of control. Subsequent control studies have been undertaken over a large part of the natural distribution of the sawfly. Up until the mid-1970s, all suspensions of 'polyhedral inclusion bodies' originated from infected larvae.

In 1977, it was clear that, while there could be a place for baculoviruses in pest control in the UK and elsewhere, the use of virus material obtained directly from insects not reared under sterile conditions, though effective, would not meet the increasingly stringent standards required by pesticide registration authorities. Purified virus material was therefore prepared and tested as an essential preliminary for any subsequent registration action.

Results showed that highly purified 'polyhedral inclusion bodies' of *Neodiprion. sertifer* NPV could give effective control if applied at the time of egg hatch. These trials were, in effect, a late stage initial screening of a potential commercial pesticide (Entwistle & Evans 1985).

In 1984, a stabilised formulation of NPV was marketed under the name 'Virox'. It was used successfully in the north of Scotland to control heavy infestations of *N. sertifer*. 3000 ha of young lodgepole pine plantations were treated that year at the rate of 1 litre product per ha., applied from the air at ULV. Similar applications were made to 2765 ha in 1985 and 310 ha in 1987 (Evans *et al*. 1985b, Stoakley & Heritage 1986, Stoakley & Patterson 1987).

Between 1984 and 1990, 'Virox' was an approved listed product containing pine sawfly NPV which could be recommended for general use in the UK (Carter & Gibbs 1989, MAFF Pst 1990). However, the NPV was not listed in 1991 (MAFF Pst 1991) and has not reappeared in a commercially available form, probably because of lack of demand. As long as there is no available approved product, control of *Neodiprion sertifer* by NPV in the UK has to occur spontaneously.

6.62　　Web-spinning larch sawfly - *Cephalcia lariciphila*

Spontaneous biological control

This sawfly was first encountered in the UK in 1953 but did not emerge as a threat to woodlands until 1972. The first outbreak was reported from Margam forest, south Wales. At that time, it was found to cause heavy defoliation on Japanese larch, *Larix kaempferi*. Over the following six years, sawfly attack spread to 21 woodland blocks at the pole stage or older, and spread through Wales and the west and central midlands of England (Billany & Brown 1980). European and hybrid larch were also attacked, no species preference having been observed. By 1978, the area subject to annual defoliation was 2243 ha, in many cases involving trees which had been defoliated each year for several years.

Repeated defoliation causes a reduction in height increment and needle size, and after 4 - 5 years' defoliation, trees may die.

The larval parasite *Olesicampe monticola* was found at Margam forest in 1975 (Billany 1977); it lays its eggs in young *Cephalcia* larvae in June and remains until its host dies at the prepupal stage in the following spring. The *Cephalcia* larvae then spin a flimsy cocoon from which adults emerge as adult wasps when first and early second instar larvae of *Cephalcia* are again present. Subsequent observations and trials showed that *Olesicampe* is an efficient parasite which, when established, maintains *Cephalcia* populations below the level causing defoliation (Bevan 1987). Trials have shown that the parasite can successfully be brought into sites newly infested with *Cephalcia* (*ie* 'augmentative release') and have an immediate effect on its host.

6.63   Large larch sawfly   *Pristophora erichsonii*

Spontaneous biological control

A native of Europe, this species had inadvertently been introduced into North America in the 19th century. Between 1880 and 1909, it killed 'between 50 and 100% of the mature larch over vast areas of eastern USA and southeastern Canada.'

In Britain, the insect attracted little attention until 1906 when a heavy infestation was reported in the Lake District in north west England; similar attacks were reported from north Wales and other parts of England. A subsequent survey showed that the most severe attacks were restricted to north Wales and the Lake District. While no clear figures are available, the number of trees killed or severely weakened was likely to have been of the order of several hundred thousands of trees; most were subsequently felled. In 1912, the Ministry of Agriculture registered the large larch sawfly as a 'scheduled pest'.

The infestations subsequently subsided, with reports from that time of a parasitic fungus and cocoons containing shrivelled remains following what was surmised as virus or bacterial disease. Many larvae were attacked by parasitic insects, the most frequent being an ichneumonid *Mesoleius tenthredinis* (Hanson 1950).

Against that background, the discovery in 1948 of a batch of larvae at Radnor forest (mid-Wales), led to a more widespread survey of the occurrence of larch sawflies. The insect was found to exist at low levels in all the areas where, 40 years previously, there had been high populations and tree losses. Elsewhere, populations were very small or non-existent.

The insect is scarcely mentioned in later publications.

6.64   European spruce sawfly   *Gilpinia hercyniae*

Spontaneous biological control

This sawfly is native to the mountainous areas of Europe and is also found in areas afforested during the 18th and 19th centuries. It was first recorded in

Hampshire in the south of England in 1906 and had spread into Somerset, Dorset, Buckinghamshire and Hertfordshire by the early 1930s. Twenty years later, it was found to be widely distributed.

Reports of persisting local outbreaks and defoliation were first reported from Hafren forest in mid-Wales in 1968; surveys between 1971 and 1976 showed the sawfly to be quite common in widely dispersed blocks of spruce throughout Wales, central and southern England. The insect was not found anywhere in Scotland. However, only in Wales did major outbreaks occur with resultant death of trees (Brown 1973, Billany 1978). Elsewhere, repeated defoliation brought about a reduction in height increment of the order of 25% but did not lead to tree mortality.

In 1974, many of the Welsh sawfly populations collapsed. While cold wet conditions in July that year were contributory factors, naturally occurring virus took its toll later in the year.

In mainland Europe and elsewhere, *G. hercyniae* larvae are very susceptible to a naturally occurring, host-specific NPV. This virus was found in the UK for the first time in 1971 at Hafren forest. Infected larvae are killed on the tree before pupation can take place. Several species of birds feed on *G. hercyniae* larvae, especially titmice. After such feeding, the birds pass the infectious virus in their faeces. Any falling into tree crowns may then be spread by rainwater drip or splash and may persist for several months. It appears that the virus spreads most effectively at high sawfly population levels. Once it appears, it spreads very rapidly; nearby less dense populations of larvae may also become infected.

Billany also reports the appearance of a parasitic ichneumonid wasp *Lamachus marginatus* in populations of *G. hercyniae* in East Anglia and in Wales. The parasite has been released in former infestation areas hoping that it will naturally control population levels, as in mainland Europe.

6.7   INTEGRATED MANAGEMENT OF OTHER PESTS

6.71   Christmas trees

The specialist commercial production of Christmas trees is more akin to forest nursery production than plantation silviculture. Features such as aphis galls which would be ignored in the forest can, if numerous, seriously impair the marketability of a Christmas tree.

Pests of Christmas tree crops include:
- Pineapple gall adelgids   *Adelges abietis,*
- •   *Adelges viridis,*
- Brown spruce aphid   *Cinara pilicornis,*
- Spruce bell moth   *Epinotia tedella,*
- Gregarious spruce sawfly   *Pristophora abietina,*
- Conifer spinning mite   *Oligonychus ununguis.*

Forestry Commission *Field Book 17, Christmas tree pests* (Carter & Winter 1998) details the commonly occurring pests on Norway spruce, pine, silver firs and other species prepared and sold as Christmas trees in Britain. More than 30 pests are recognised; 17 are aphids, 4 are moths, 3 each are beetles, mites or sawflies. These last are found on pines grown as Christmas trees. For many of these, *Field Book 17* contains:

- illustrations of larvae, adults or damage,
- diagrams of life-cycles, and
- indications of the most appropriate times to apply treatments.

In almost all cases, pests of Christmas trees are best (in some cases only able to be) treated when first symptoms of serious attack can be seen. A sharp eye, regular inspection and knowledge of vulnerable periods are essential if intervention is to be timely (Carter 1972).

Carter (1974) recommended that Christmas tree growing areas should be cleared periodically (*eg* every 7 years) to avoid build up of pests.

Carter and Winter (1998) note that attacks on spruce by aphids may be worse where there are older spruce stands in the vicinity; they also point out on the eastern side of Britain, Norway spruce is more likely to be subject to moisture stress and hence more susceptible to attack by mites and aphids.

Insecticides

Christmas trees may be treated with products with approval 'on-' or 'off-label' for use on ornamental trees and shrubs. Where treatment has to be repeated annually, it is recommended that products are alternated and should always follow evidence of infestation (Bevan 1987). Earlier recommendations include lindane, dicofol, malathion, pirimicarb and fenitrothion (Blatchford 1983, Carter & Gibbs 1989, Brooks *et al.* 1989). These are amplified or replaced by recommendations in *FC Field book 17* (Carter & Winter 1998).

6.72 Forest and ornamental nurseries

The aim in forest and hardy ornamental tree nursery production is to produce healthy plants fit for planting, in as short a time as economically practicable. There are no routine measures that need to be taken in outdoor nurseries other than soil treatment prior to sowing. For ornamental nurseries raising stock vegetatively under glass or polytunnel production of container stock, particular attention has, however, to be paid to potential pests such as red spider mite and vine weevil.

In forest nurseries in the 1950s, substantial changes in plant production occurred, associated with the development of heathland nurseries, improved plant nutrition and more rapid growth. The resulting shorter production cycles reduced the prevalence of damage to roots from soil dwelling insects.

The most commonly occurring insect and allied pests in forest nurseries are listed and briefly described in:

- *Forest nursery practice* (Strouts *et al.* 1994), and
- for ornamental trees and shrubs, *A colour atlas of pests of ornamental trees shrubs and flowers* (Alford 1995).

The latter is extremely well illustrated with colour photographs; it includes illustrations of almost all the species mentioned in *Forest nursery practice*. Neither publication specifies pesticides other than in general terms.

Table 6.5 lists operational practices to reduce risk of losses through pests.

Table 6.5   *Operational practices minimising risk of attack by insects and mites in forest and ornamental nurseries*

Pest	Action	Comment
Reduce numbers of potential host plants or habitat		
Black cherry aphid *Myzus cerasi*	Remove/do not plant cherry in hedges *etc* around nursery.	Avoid harbouring
Woolly beech aphid *Phyllaphis fagi*	Remove/do not plant beech hedges.	nursery stock pests
Cockchafers Ghost swift moth *Hepialus humuli*	Ensure ground regularly cultivated, minimise area of permanent grassy banks *etc* .	Larvae easily damaged when soil is cultivated mechanically.
Vigilance		
Aphids, mites and caterpillars	Examine both sides of foliage regularly from late spring for early indications of infestation.	Treat in good time to avoid build up.
Cutworms, esp. *Agrotis segetum*	Keep weeds down. Watch for seedling damage. Check if local pheromone monitoring system is available.	Treat while larvae are still small Consult about irrigation.

Soil-inhabiting nursery pests

Chafer beetles

These, especially the cockchafer (*Melalontha melalontha*), were in 1937 considered to be one of the most difficult problems faced by entomologists. Soil fumigation seemed the best option, carbon disulphide being one of the most commonly recommended substances (Chrystal 1937, Anon 1948a). Some of the earliest reported trials of *gamma*-HCH as an insecticide in forest nurseries were against cockchafer (Hanson 1949, Edwards & Holmes 1950).

Studies of biological control of cockchafers from two nurseries showed 35-48% infestation with a dipterous parasite. It was also noted that healthy larvae descended to the deeper layers of the soil for the winter earlier than parasitised larvae. The latter were more likely to be exposed by routine autumn digging. It was remarked on that the commonplace instruction to remove and kill all larvae found at that time would have reduced the predator population rather than the cockchafer population, the opposite of the effect intended.

As a by-product of studies of parasites, several thousand parasitised larvae were sent overseas to further biological control programmes (Hanson 1949).

In the 1950s, much forest nursery production was relocated onto to lighter texture, more acid heathland nursery soils. At the same time, annual cropping became commonplace; also, less ground was green-cropped and more left fallow and ploughed and rotovated. 2nd and 3rd year cockchafer larvae are easily damaged by soil disturbance and are likely to have been killed by the operations under the changed cultivation regime. They are now not frequently encountered as pests damaging plant roots. Nevertheless, undisturbed ground, whether as hedges or grass banks, may still harbour cockchafers.

Cutworms

Larvae of *Agrotis segetum* (turnip moth), *Agrotis exclamationis* (heart & dart moth) and *Noctua pronuba* (yellow underwing moth) are important horticultural pests. Their effect may be seen in mid-late summer in conifer seedbeds covered with sand or fine grit. Larvae make burrows about 4mm in diameter in the vicinity of which stumps of seedlings can be found; seedling tops may occasionally be found projecting from the burrows.

Alford (1995) comments that infestations are less likely on sites that are well-irrigated and weed-free.

Experiments on insecticides in the 1950s showed aldrin and dieldrin to be effective (Aldhous 1959). When these were withdrawn, further work showed lindane to offer control (Bevan 1968, Blatchford 1983). The recommendation was subsequently supplemented, carbaryl or cypermethrin being recommended as alternatives (Carter & Gibbs 1989).

In some parts of Britain, ADAS on a commercial service basis, provides advice to potato growers and others as to when risk of cutworm damage is high, and on cultural measures to minimise losses. Under suitable conditions, cutworm larvae may be controlled by irrigation.

Nematodes

Nematodes have seldom been a problem in forest tree and woody shrub nurseries. In contrast, nurseries raising herbaceous ornamentals and bulbs can suffer heavy losses through nematodes and good nursery hygiene is vital.

In the 1950s, one forest nursery in Dorset was found to be heavily infested with *Hoplolaimus (=Rotylenchus) uniformis*, the infestation being reduced by soil fumigation with formaldehyde solution (Goodey 1965). Similar infestations have not recurred. A local infestation of *Dolichorhyncus microphasmis*, a tylenchorhynch nematode, was discovered to be widespread a nursery in Sussex (Winter *et al.* 1984).

Weevils

Larvae of the strawberry root weevil *Otiorhynchus ovatus* feed on finer roots throughout summer months; they are not readily distinguishable from larvae of other weevils. Adult weevils also feed on shoots and needles. Against larvae, imidacloprid or chlorpyrifos are currently used, worked into infested soil before use. Entomopathogenic nematodes may also be used (Heritage *pers comm.*).

Vine weevils in container stock

The vine weevil *Otiorhynchus sulcatus* is a major pest of glasshouse and ornamental plants. While it goes through the normal stages of egg-laying, larval development, pupation and emergence, all adults are female (Greenwood & Halstead 1997).

The most important damage is caused by larval feeding on fine roots of container stock being raised in a peat compost medium. Severe infestations can develop in individual containers.

Biological control is possible using entomophagous nematodes; however, most strains operate best where the soil temperature is 10 - 15°C. Strains are being developed which are active at lower temperatures .

Alternatively slow-release insecticides can be incorporated in the rooting medium filling the container (Shall 1999).

Foliage and stem feeders

Aphids

Aphids may appear:

• on young expanding broadleaved foliage in spring, *eg: Myzus cerasi* and *Phyllaphis fagi* on cherry and beech respectively;

• on needles and shoot tips from August to October (*eg Aphis fabae* on various conifers);

• on pine roots on container and open-grown stock throughout spring and summer (*eg Prociphilus pini,* the pine root aphis);

• between July and May on spruce needles (*eg Elatobium abietum,* the green spruce aphis).

In practice, there is no respite from the threat of aphid infestation on one species or another, throughout the year. Foliage often becomes discoloured with the black sooty moulds that develop on sugary honeydew excreted by aphids.

While there are a number of natural predators on aphids, none is able to provide any significant preventative biological control. *Table 6.5* lists two species for which 'cultural control' can be exercised by minimising the opportunity for reservoirs of aphids to build up on hedges and older stock plants.

Recommendations to control aphids advise early treatment, before numbers have increased and before plants have sustained any damage or loss in growth. Insecticides that have been recommended include nicotine (Gray 1953), lindane or malathion (Aldhous 1972), malathion (Blatchford 1983), malathion, dimethoate, tar oil, lindane or pirimicarb (Carter & Gibbs 1989), aldicarb, cypermethrin, deltamethrin, dimethoate, lindane, malathion, nicotine or oxydemeton-methyl (Brooks *et al.* 1989).

Plants infested with root aphids may have to be burnt and the soil where they were growing treated with a soil insecticide such as lindane dust, or a soil drench with diazinon (Blatchford 1983, Carter & Gibbs 1989, Alford 1995).

Leaf beetles, sawflies, weevils

Poplar and willow stool beds and cutting lines have in particular been subject to attack by poplar leaf beetles, *Phyllodecta* spp. and *Chrysomela populi*.

Larvae of several species of sawfly also feed on foliage of poplar, willow and other broadleaves.

Adult weevils (*Otiorhynchus singularis, O. ovatus, Strophosomus melanogrammus, Baripeithes araneiformis* and *B. pellucidus*) may attack a great variety of trees, feeding on fine shoots, needles, leaves, hypocotyl of emerging seedlings or succulent shoots of recently inserted poplar cuttings.

Pine shoot beetles may also attack ball-rooted semi-mature pines (Winter 1991).

Insecticides that have been recommended include pyrethrum (Chrystal 1937), lindane powder (Gray 1953), DDT or lindane (Aldhous 1972), lindane (Blatchford 1983), lindane or carbaryl (Carter & Gibbs 1989).

Spider mites

Alford (1995) describes some 50 species of mite in 4 families. Many of these are responsible for galls familiar on various amenity trees, but merit no action. For mites of economic significance, either on fruit trees, in glass houses or Christmas trees, acaricides are recommended. Some resistance to acaricides has already been noted for the fruit tree red spider mite. For the prevalent glass-house red spider mite, *Tetranychus urticae,* biological control using a predatory mite, *Phytoseiulus persimilis,* is also possible.

6.73   Pests on specimen and ornamental trees and shrubs

By definition, ornamental trees and shrubs occur singly or in small groups and are often placed in locations freely accessible to the public. A substantial proportion of varieties are affected by the same pests that affect commercial timber and fruit trees; possible treatments can be based on recommendations for those varieties or species, making due allowance for greater public accessibility.

Guidance is also available in:

- *Pests & diseases* (Greenwood & Halstead 1997*)* and
- *A colour atlas of pests of ornamental trees, shrubs & flowers* (Alford 1995). The illustrations in this work are particularly comprehensive.

Aphids

These sap-feeding insects can be found on virtually all ornamental plants. Alford (1995) lists over 110 species as pests of ornamentals, not including adegids, scale insects and other hemiptera. With few exceptions, guidance for treatment has to be sought through extrapolations from the recommendations for forest nurseries and Christmas trees in §6.71 and §6.72 above.

Cypress, juniper, and American juniper aphids are widely distributed, affecting *Thuja*, Lawson and Leyland cypress and ornamental cultivars of these species.

They may damage foliage directly and also by releasing honeydew which is colonised by sooty moulds, causing blackening. Ants seeking honeydew may make an earth shelter over the lower stem down to the ground. Recommended control is by contact aphicide as soon as aphids are seen (Winter 1989, Alford 1995).

Scale insects

Scale insects are sucking insects where the female is protected by a covering of wax threads and cast-off nymphal skins; these may harden to a greater or lesser extent and give protection, or the waxy wool may remain unhardened. Alford lists 27 species.

Beech scale   Cryptococcus fagisuga

The felted beech coccus or beech scale is prevalent in many beech woods throughout the European range of beech.

C. fagisuga was introduced into North America in about 1890. It has spread at about 6-8 km/yr.

The adults have only vestigial legs and are immobile. Beech scale reproduces parthenogenetically (*ie* by reproduction from eggs without fertilisation). Eggs are laid in June-September; first instar larvae are able to crawl to other locations on a tree and may be carried short distances by wind.

Larvae and adults secrete considerable quantities of white waxen wool, deposits in extreme cases covering tree stems so that they appear as though whitewashed. The considerable tree-to-tree variation in level of infestation seen in many stands in Britain is thought to reflect genetically based inherent resistance within a woodland. Infestation is more uniform in North America (Gibbs & Wainhouse 1986).

Interaction with Nectria spp

Trees which have suffered heavy attack may subsequently suffer from beech bark disease in Britain caused by the fungus *Nectria coccinea* and in America, by a variety of the same fungus (Gibbs & Wainhouse 1986, Lonsdale & Wainhouse 1987). See §7.33.

Control of beech scale

There is no clearly operating natural biological control, although a number of predators are regularly identified when populations are studied (Parker 1974).

In ordinary woodland conditions, there is no practical justification for attempting to control the insect. If trees are in any way special and insects are unwanted, Lonsdale & Wainhouse (1987) recommended either scrubbing with mild detergent, or treating with tar oil winter wash at high volume while dormant. Earlier recommendations were for tar oils, DNC winter wash or lime sulphur (Anon. 1956a), tar oil winter wash, diazinon, malathion or dimethoate (Parker 1974), tar oil winter wash or diazinon (Blatchford 1983).

Horse chestnut scale   Pulvinia regalis

This scale, established in the UK since 1964, occurs widely on many common urban deciduous trees, including lime, elm, sycamore, maple, horse chestnut and magnolia. London plane is less susceptible.

Mature females are brown in colour. Before egg-laying, they migrate to main stems with their white egg masses and at that stage are conspicuous and are often considered unsightly. Eggs hatch in June-July and nymphs disperse to leaves and twigs. If control is sought, tar oil in the dormant period, or diazinon or deltamethrin to young insects, have been recommended (Brooks *et al.* 1989).

6.74   Other aphids

Green spruce aphis   *Elatobium abietinum*

This aphid is a major pest of ornamental and forest spruces. It frequently causes heavy defoliation, most apparent in early summer and affecting all but the current year's growth. Serious defoliations occur every 3-5 years. Such events, while they cause loss of increment, do not lead to tree death. Consequently, in the forest no attempt is normally made to control this aphis (Carter 1972, 1975, 1977, Bevan 1987).

A Danish study of the interaction between litterfall, spruce aphis infection and nutrient cycling is described in §12.26.

Green spruce aphis and spruce root aphids

In a study of reasons for decline in growth observed in Sitka spruce growing on sites on the South Wales coalfield, high levels of *Elatobium* infestation were associated with loss of increment. Root aphids were also consistently present and in combination were considered to be potentially significant contributors to the 'forest decline' syndrome (Carter 1995).

Five insecticides that had potential to control *Elatobium* were given a preliminary screening for phytotoxic effects on nursery transplants. At normal rates, no effects were observed from dimethoate, malathion and pirimicarb applied as foliage sprays or malathion and chlorpyrifos applied as soil drenches. Applications at twice the standard rate resulted in some damage (Straw & Fielding 1998).

Douglas fir adelges   *Adelges (Chermes) cooleyi*

Both host (Douglas fir) and pest, the adelges, were introduced from western North America in the 19th century. By the 1920s, this adelges was considered to be one of the more serious threats to plantation forestry and was the subject of one of the earliest advisory leaflets published by the newly formed Forestry Commission (Anon. 1921). It is particularly conspicuous on trees when between

3 and 7 m tall, when all foliage appears mottled and covered with aphids and length of the current leading shoot is less than in previous years. However, from the mid 1930s, it was apparent from the length of internodes between annual branch whorls in older Douglas fir plantations, that this is only a short-lived check to growth (Chrystal 1937) and that after 3-6 years, vigorous leader growth is resumed.

Paraffin emulsion or nicotine were recommended for control of outbreaks in the nursery (Anon. 1948b, Crooke 1960). Recommendations have not subsequently been thought necessary.

6.75   Pests of seed

Megastigmus spermatrophus

During the 1920s - 30s, because of the dependence of the expanding forestry programme on imported seed, attention was paid to the extent to which imported seed was infested with seed flies. In particular, Douglas fir, because of its rapid rate of growth and high quality timber was considered to be of great potential at that time.

Forestry Commission *Leaflet 8* (Anon 1961) reported seven species of seed fly in imported seed of nine exotic conifer species. The species most frequently found to be carrying seed flies were Japanese larch, Norway spruce and Douglas fir.

Megastigmus spermatrophus was found not only in imported seed, but also in the forest. It was first recorded in the UK in Scotland in 1906. Contemporary accounts were of heavy infestations in Douglas fir seed which previously had been pest-free. By 1950, out of 65 samples of seed from collections from all parts of Britain, 27 were infested, the average level of infestation being 36% (Hanson 1951). Chrystal suggested that infested seed should be fumigated for 48 hours using carbon disulphide. DDT and lindane were also proposed for the protection of Douglas fir seed orchards (Anon. 1961).

Two species of the parasitic chalcid fly, *Mesopolobus spermatrophus* and *M. pinus* were also identified in *Megastigmus spermatrophus* larvae.

In subsequent field and laboratory investigations into protection of seeds in the cone, malathion was more effective that DDT, lindane, diazinon, carbaryl, dinitrothion or azinphos (Stoakley 1965, 1967, Bevan 1966c, 1967, 1968). However, studies of oviposition and infestation levels in relation to size of seed crop, led to the conclusion that seed infestation was low in good crop years and only high in poor years. Provided full advantage could be taken of good seed years, there was no serious risk to Douglas fir seed sources except possibly in seed orchards.

Seed weevils

Curculio elephas, a weevil of acorns was found on imported acorns in 1955 and 1956. Acorns were sprayed with DDT in the seedbeds in an attempt to kill the weevil, but some survived and overwintered (Bevan *et al.* 1957).

6.76   Termites

Termites were discovered in 1998 to have become established and to be affecting ground and buildings in two adjacent properties in north Devon. The species, *Reticulitermes lucifugus*, is subterranean and is though to have been introduced up to 30 years before its discovery.

The colony is being treated, using a chitin inhibitor mixed in a bait. However, this is expected to have its effect only slowly (FC PH 1999).

6.77   Pinewood nematode (*Bursaphelenchus xylophilus*)

This nematode is a component of another 'vector + pest' association. The nematode is the causal organism of pine wilt disease, a major tree killer in Japan and China and is widespread in North America, where, however, it does not cause major tree mortality. It is feared that in the conditions of the warmer parts of Europe, infestations of this nematode would lead to tree losses similar to those of the Far East (Gibbs & Wainhouse 1986).

The nematode is transferred from tree to tree by ovipositing females of beetles in the genus *Monochamas*. Under warm conditions, nematodes can complete a life-cycle in four days so that numbers can build up rapidly. Phytotoxins are formed and xylem cells become blocked leading to wilt symptoms on affected trees. Trees usually die in the autumn and winter following infection.

In the 1980s, nematodes were found by several European countries in wood chips imported from North America (Anon. 1993b). Subsequent negotiations led to a European Union Directive dealing with coniferous wood originating outside the EU. Wood and wood chips must either have been heated to 56°C for 30 minutes, been fumigated or kiln dried and in addition must have been debarked and, for wood, be free from grub holes caused by the vector beetle (Fielding & Evans 1996).

6.8   VIGILANCE AGAINST POTENTIAL PESTS ENTERING THE UK

Additions to the list of insects and other pests found in the United Kingdom have been made regularly in the past and are likely to continue as long as international trade and travel continues. The UK continues to import 80% of its timber requirements.

Illustrations given above of the devastating effect of introduced pests on established natural and plantation woodlands give weight to the need for on-going vigilance.

Where a recognised new pest is found, there is seldom any option, if immediate eradication is the aim, other than to turn to rapidly acting pesticides or virus-based biological agents. Biological control, for example introducing *Rhizophagus grandis* to control of *Dendroctonus micans,* is too slow a process to be able to prevent the local establishment of an invading species; biological control is only relevant if it is accepted that the introduced pest has become established and can be contained.

Cultural measures such as debarking or kiln-drying that can take place in the country of origin to reduce the risk of transfer of wood-borne potential pests have to be encouraged.

To minimise the need to apply pesticides, the risk of introduction of new pests must be minimised. Recognition of potential risks and the appropriate form of vigilance to intercept are essential. International co-operation through the European Plant Protection Organisation (EPPO) and the operation of 'plant and timber passports' within the European Community are positive moves. At the same time, there is clear evidence that commercial pressure to try to get away with malpractice, *eg* placing sub-standard boards with bark on in the middle of packs of sawn boards, has occurred and is especially high at the peaks and troughs of the economic cycle *ie* both in times of glutted markets and timber shortage (*eg* Thompson 1996).

A shortcoming of the European Community Plant Health Directive approach is that it only lists known pests. There remains the risk of previously unknown combinations of native and introduced pests causing damage. For this, a more universal vigilance is required.

CHAPTER 7

Fungal & other diseases

7.1   VITAL OR VIRULENT

Fungi, while as ubiquitous as insects, differ in that their presence is often less obvious until they are well established. Like insects, fungi play essential roles in plant and soil ecology but, not able to photosynthesise, they must obtain their energy solely from living organisms (obligate parasitism), dead material (saprophytism), from both (facultative parasitism) or by exchange (symbiosis).

This chapter also includes diseases caused by other microorganisms, where these may be treated chemically or by appropriate silviculture or arboriculture.

That pests and diseases may develop resistant strains in response to exposure to pesticides is familiar. With diseases, equally important are:

- variations in resistance to disease displayed by the host species or genus, *eg* bacterial canker of poplar;
- differences in virulence of strains of the pathogen and potential rates of spread, *eg* Dutch elm disease fungi *Ceratocystis ulmi* & *novo-ulmi* (Peace 1960, Brasier & Webber 1987, Brasier 1996).

Future changes in relative susceptibility to diseases, independent of pesticide-induced resistance cannot be ruled out.

Over the last 15 years, the overall health of trees throughout north western Europe has been the subject of much concern in the context of damage to forests by air pollution and possible damage to the soil by plantation trees. As a result, the health of the more important forest trees has been surveyed in many countries as part of a co-ordinated international programme. This has not demonstrated any significant or novel worsening of tree health through fungal disease. The results of these surveys in Britain are described in §14.42.

7.11   Integrated crop protection

Against threat of fungal attack on trees and shrubs, the options are:

- avoidance or minimisation through modification of silvicultural or arboricultural practice;
- use of biological controls based on naturally occurring antagonists;
- application of fungicides;
- any combination of the above;
- acceptance that damage cannot be avoided.

There are two fungi which can be used as non-pathogenic antagonists to pathogenic fungi on woodland or ornamental trees. However, while worthwhile, the potential scale of their usage is small (§§7.21, 7.42).

Greatest emphasis has to be given to cultural techniques for minimising losses of trees and shrubs to fungal attack.

7.12 Important fungal and other diseases

Table 7.1 lists the more important tree diseases, the threat of which can be reduced by cultural action, or which can be controlled by use of a fungicide or biological agency in woodland or in the nursery.

Publications describing a fuller range of diseases of trees are listed in the introduction to the bibliography.

Table 7.1 *Fungi and other diseases potentially damaging to woodlands, ornamental trees or nursery stock*

English name	Scientific name	Damage rating	Section
DISEASES ACTING THROUGH SOIL OR ROOTS			§7.2
Conifer (*Fomes*) heart rot *Heterobasidion annosum* *		xxx	§7.21
Honey fungus	*Armillaria mellea* *	xx	§7.22
VECTOR-BORNE DISEASES			§7.3
Dutch elm disease	*Ophiostoma (Ceratostomella)* spp.	xxx	§7.31
Blue stain	*Ceratocystis* spp	xxx	§7.32
Beech bark disease	*Nectria coccinea*	x	§7.33
Oak wilt	*Ceratocystis fagacearum*	xxx	§7.34
DISEASES ACTING THROUGH STEM OR FOLIAGE			§7.4
Diseases entering through wounds		xxx	§7.41
Silver leaf	*Chondrostereum purpureum*	x	§7.42
Anthracnose & leaf spot of willow *Drepanopeziza sphaeroides*		x	§7.43
Bacterial canker of cherry *Pseudomonas mors-prunorum*		xx	§7.44
Scab & black canker of willow		xx	§7.45
CULTURAL CONTROLS			§7.5
Resistant varieties or origins			
Poplar bacterial canker	*Xanthomonas (Aplanobacter) populi*	xxx	§7.51
Douglas fir leaf-cast diseases *Rhabdocline pseudotsugae*		x	§7.51
	Phaeocryptopus gaumannii	x	§7.51
Larch canker	*Lachnellula (Trichoscyphella) willkommii*	xxx	§7.51
Sanitation			
Watermark disease of willow *Erwinia salicis*		xx	§7.52
Fireblight	*Erwinia amylovora*	xx	§7.52
Control alternate host			
Pine twisting rust	*Melampsora pinitorqua*	x	§7.53

(Table 7.1 contd)

English name	Scientific name	Damage rating	Section
DISEASES CONTROLLABLE BY CULTURAL MEASURES (Contd.)			
Needle rust of Scots pine	*Coleosporium* spp. (*C. senecionis*)	x	§7.53
Avoid planting			
White pine blister rust	*Cronartium ribicola*	xx	§7.54
Avoid disturbance			
Sooty bark disease of sycamore	*Cryptosoma corticale*	x	§7.55
Avoid fires			
Group dying of conifers	*Rhizina inflata*	x	§7.56
Seasonal restriction of operations			
Willow rust	*Melampsora amygdaline*	x	§7.57
Dilution of vulnerable varieties			
Willow and poplar rusts		xx	§7.58
Strict insistence on planting disease-free stock			
Phytophthora disease of alder	*Phytophthora* spp	xx	§7.59
FOREST NURSERY DISEASES			§7.6
Damping off	*Pythium* spp	xx	§7.61
	Phytophthora spp	xx	§7.61
Verticillium wilt	*Verticillium dahliae*	x	§7.61
Seed-borne disease	*Geniculodendron pyriforme*	x	§7.62
Grey mould	*Botrytis cinerea*	xx	§7.63
Leaf cast of larch	*Meria laricis*	xx	§7.64
Oak powdery mildew	*Microsphaera alphitoides*	x	§7.65
Keithia disease of western red cedar	*Didymascella thujina*	xx	§7.66
Needle cast of pine	*Lophodermium seditiosum*	x	§7.67
Birch rust	*Melampsora betulinum*	x	§7.68
Leaf spot on cherry	*Blumeriella jaapii*	x	§7.69
VIRUS DISEASES		x	§7.7

* Name formerly used for fungi now differentiated into two or more species
x, xx, xxx = slight, moderate, severe risk to trees, respectively

7.13 Fungicides used in woodlands, on ornamentals and in nurseries

For virtually all blocks of woodland, the only measures applied very widely and on a regular basis relate to *Heterobasidion annosum*, the fungus causing Fomes root and butt rot in growing conifers. Stump treatment, and occasionally more extreme measures to minimise spread of the fungus are in general use wherever there is considered to be a risk of a build-up in stumps of recently felled conifers (§7.21).

The vector-borne Dutch elm disease is the most important disease to have affected the British landscape this century, and also one of the most intractable to control. This and other vector-borne diseases are described in §7.3.

Diseases of stem and foliage that can to some degree be controlled by fungicides are described in §7.4.

Otherwise, for control of fungal diseases in the forest, variations in cultural techniques have to be relied on. These are summarised in §7.5.

Fungicides are used regularly:

• in nurseries producing forest and ornamental stock, where many plants are concentrated on small areas and the consequence of a heavy outbreak of a disease could lead to heavy losses or significant weakening of stock (§7.6);

• on ornamentals which may suffer disfiguring fungal attacks to foliage or which may have been physically damaged or which may have required tree surgery;

• on sawn pine timber which for any reason cannot be dried rapidly and while drying is vulnerable to blue stain fungi (§6.56 & §7.32).

A note on sources giving up-to-date prescriptions for pest control is given in *Appendix II Table 2*.

7.14   Scale of use of fungicides

Table 7.2 lists the fungicides and other materials used by the Forestry Commission to control fungi in three sample years since 1966. The first five materials listed have all been used to minimise the risk of spread of conifer heart

Table 7.2   *Quantities of fungicide active substances used in 1966/7, 1986/7 and 1992/3 in Forestry Commission woodlands and nurseries*

Active substance	Unit	1966/7	1986/7	1992/3
Sodium nitrite	kg	28 000	-	-
Creosote	litres	118 000	-	-
Polybor chlorate	kg	510	-	-
Urea	kg	300	126 250	149 100
Peniophora gigantea	sachets*	-	3500	1700
(Predominantly forest nursery use)				
Bordeaux mixture	kg	180	-	no data
Captan	kg	40	32	no data
Thiram	kg	45	-	no data
Zineb	kg	55	32	no data
Formaldehyde	kg	1600	80	no data
Dazomet	kg	nil	3800	no data

In addition in 1986/7, less that 25 kg ai each of: *benomyl* and *sulphur* were used.
* contents of sachets varied between 1 and 10 ml of spore suspension

Sources   Aldhous 1968a, unpublished data, Willoughby 1995.

rot (*Heterobasidion (Fomes) annosus*). They are the only fungicides or antagonists applied widely and on a regular basis to control any fungal pest.

Fungicides used in woodland or on amenity trees, 1950-1998

Appendix I Table 1b lists, by active substance, keys to references in literature to fungicides which have been used or tested experimentally in the UK in forests, farm woodlands, amenity tree management or in forest nurseries over the last 50 years. Full references are included in the main bibliography under author and year. Some 48 active substances are listed there. *Appendix II Table 1b* shows which of these were available in approved products in 1999.

7.2   DISEASES ACTING THROUGH SOIL OR ROOTS

7.21   Conifer heart rot   *Heterobasidion annosum*

Conifer heart rot is a disease of world-wide distribution with an extensive range of host species. It is considered to be the most important wood-rotting species of managed forests.

The fruit bodies of the fungus are typically bracket-shaped and occur on stumps, exposed dying roots *etc* at or just above ground level; they are often hidden by brash (Phillips & Burdekin 1982).

In the UK, it has until very recently been considered to be caused by a single species, *Heterobasidion annosum (Fomes annosus)*. However, accumulating evidence of differential susceptibility and 'intersterility' has led to proposals for three closely related species, *Heterobasidion annosum* 'sensu stricto', mainly in pine forests, *Heterobasidion parviporum* which mainly attacks spruce in Europe but is not present in Britain, and *H. abietinum* centred in Mediterranean countries (Niemelä & Korhonen 1998). Comprehensive reviews of the biology, ecology and control of *Heterobasidion annosum*, 'sensu lato' by Woodward *et al.* (1998) and Redfern & Ward (1998) places UK experience in international and historic contexts.

Losses thought decay, premature death and loss of increment

Decay caused by conifer heart rot reduces yields of timber on many sites, especially those in second or later rotations and on soils that are not acidic (Peace 1938, Risbeth 1951b). It is thought to account for about 90% of the decay in conifers in Britain (Anon. 1970). Spruces, larches and western hemlock are the most seriously affected by decay and staining. Pines are resistant to decay but prone to being killed on high pH soils; other species growing on very dry sites may also be killed. Noble fir, Douglas fir and grand fir are relatively resistant (Pratt 1974, Aldhous & Low 1974).

In a trial of underplanting under Scots pine on alkaline soils at Thetford Chase (Norfolk & Suffolk), losses attributed to *H. annosum* among the 18 species planted, were:

Leyland cypress   52%   Douglas fir   46%

| *Nothofagus obliqua* | 38% | Scots pine | 21% |
| red oak | 25% | hybrid larch | 24%. |

In an identical trial on acid soils at Thetford, mortality was low (Greig & Pratt 1973).

In Scots pine in East Anglia, in plots not initially given any protection against colonisation by *H. annosum* after thinning, volume loss was attributed partly to death of trees killed by *H. annosum*, and partly to loss of volume increment. Surviving trees did not compensate with increased increment following reduction in competition from the trees that died. While on an age/height basis, the Yield Class for the site was 12, growth of severely affected plantations over a 40-50 year rotation was expected to be reduced by 1-2 yield classes, *ie* 2 - 4 cu.m/ha/yr (Burdekin 1972).

In stands of Norway spruce, not only was growth slowed but the rate of spread justified premature clear-felling. The aggregate loss in revenue at a site in Devon was 53% of the maximum potential discounted revenue, while for a site in Aberdeenshire, the loss was estimated at 93% (Pratt & Greig 1988).

Infection

Infection in managed plantations can take place through colonisation by airborne spores of freshly exposed surfaces of stumps of newly felled or injured trees (Risbeth 1951a). While spores and mycelia are found in soil, invasion from such sources is very rare. The fungus first spreads through the stump; it may then invade adjacent growing trees, starting where their roots come into contact with those of the decaying stump. Once in the healthy tree, the fungus spreads up the affected root to the main stem and colonises the heartwood of the main trunk, often causing extensive rot to heights of 4 m or more.

This ability to attack healthy trees from adjacent infected stumps of thinnings or clear-felled trees is the attribute which makes the disease such a threat in developing plantations. It has the added potential to cause heavier loss in later conifer rotations than the first (Low & Gladman 1960, Pratt, 1979a, b, Redfern 1982, 1993, Greig 1985b, Redfern *et al.* 1997b, Redfern & Stenlid 1998, Stenlid & Redfern 1998).

Delatour *et al.* (1998) give a full review of factors affecting resistance of potential host trees to attack from *H. annosum*. Much data has been accumulated about tree responses under different ecological conditions; however, further work is required to clarify the extent to which natural host tree resistance can be managed or transferred to progeny.

Development of disease in forest stands has been modelled by several groups of workers world-wide (Pratt *et al.* 1998b). For Britain, studies include those by Burdekin (1972), Greig & Low (1975), Pratt (1979c), Pratt *et al.* (1989).

Control of *H. annosum*

H. annosum spores are usually plentiful in conifer woodland and can reach freshly exposed surfaces within hours of their being exposed. Subsequent colonisation can be prevented by applying chemicals. Treatments are aimed at making such surfaces inhospitable to germination of *H. annosum* spores and growth of hyphae for sufficient time to allow colonisation by other fungi.

Pine stumps can be sprayed with spores of the competing fungal species *Phlebiopsis gigantea*. This treatment enables the competitor to be quickly established, thereby inhibiting invasion by *Heterobasidion* (Risbeth 1952).

As a last resort, infected stumps of mature trees may be physically removed.

Chemical inhibitors of H. annosum

The chemical first recommended for stump treatment was creosote, applied by brush (Risbeth 1952, 1957, 1959a, Low & Gladman 1960). Treatments were applied in 1952 to 'high-risk' pine on highly calcareous sites in Thetford Forest in Norfolk & Suffolk. This was extended to all sites in that area in 1954 and subsequently came into general use throughout Britain.

Creosote continued in use until the late 1960s. Accumulated experience showed that complete control was not always achieved because of variations in quality of the creosote, and because creosote, acting as a wood preservative, hindered colonisation by non-damaging saprophytic fungi. Consequently, *H. annosum* could still invade if creosote-treated stump surfaces were subsequently broken by, for example, brash chopping machinery.

Alternatives had also been tested (Risbeth 1959b, Phillips & Greig 1970). Of these, sodium nitrite, urea and disodium octaborate showed most promise for widespread use. All were recommended as alternatives to creosote. A blue dye added to the solution to be applied was recommended, as a check that stumps had been fully covered (Anon. 1970).

Sodium nitrite was used as a stump treatment between 1960 and 1972 but was withdrawn because of concern about its toxicity to man and animals (Greig & Redfern 1974, Phillips & Burdekin 1982). Since that time, urea has been the only material in general use.

While from the point of view of risks to the operator and risks to the environment, urea is a desirable material, it is corrosive to steel in spraying equipment and is not always effective if applied on very wet stumps.

In order to have an alternative, information on the potential of disodium octaborate tetrahydrate has recently been reviewed and an application lodged with the *Pesticides Safety Directorate* for approval for its use as a stump treatment (Pratt 1996a, 1997, Pratt & Lloyd 1996).

Pratt *et al.* (1998a) have made a comprehensive summary of work in Europe testing fungicides to control *H. annosum*. An appendix in the same publication lists all the chemicals tried (Woodward *et al.* 1998).

Fungicides can be applied by mechanised harvesters using spray nozzles fitted below the chainsaw guide bar or in the harvesting head (Pratt *et al.* 1998a).

Urea as commodity chemical

Urea is a naturally occurring compound excreted by mammals universally; it is also an important component of many N-containing agricultural fertilizers. At the same time, by virtue of being used to control a harmful fungus, it comes within the scope of current regulations for the safe use of pesticides.

Currently, urea is included with other materials which have a wide spectrum of roles, of which use as a pesticide is a minor one. These are designated in the annual MAFF *Pesticides* listings as 'Commodity Chemicals'; most have clearance for use until 2001 or later. An application has been lodged for renewal (Pratt, pers comm.).

Treatment of stumps differs from most other applications of pesticides in forestry in that the person applying the material is either a chainsaw operator or a harvester driver, neither of whom would normally be expected to have undergone training and certification in the use of pesticides.

Urea is specifically classed as a 'Home Garden Fungicide' and is approved for amateur use. Materials with this approval may also be used by professional operators without certification (MAFF Pst 1999). In practice, this means that chainsaw operators do not require certification under the *Control of Pesticides Regulations* in order to apply urea to cut stumps.

Scale of use of chemicals against Heterobasidion

The little published data on the amount of material used on stump treatment is given in *Table 7.2*. Combining these with data for area treated (Aldhous 1968a), the average application of sodium nitrite, urea and borate was of the order of 1½-2 kg per ha for thinned areas and 5-6 kg per ha for clear fells.

Use of spore suspensions of *Phlebiopsis* was not expected to exceed 100 ml of spore concentrate per ha (MAFF Eval 1998).

Control using spores of *Phlebiopsis (Peniophora) gigantea*

Fungal inhibitors of H. annosum

Competing fungi may prevent colonisation of stumps by *H. annosum*. Risbeth (1951a) found that for pines in East Anglia, *Phlebiopsis gigantea* (previously known as *Peniophora gigantea*) was much the most important competitor. *Phlebiopsis* hyphae rapidly colonise and break down the woody tissue of stumps. Spores of *P. gigantea* are often present in sufficient quantity to colonise fresh stumps and prevent entry of *H. annosum* by making the surface unsuitable for spore germination and development (Greig 1984).

Some seasonal variation of infection of stumps was found, related to spore loads (Meredith 1959, 1960). *H. annosum* has the more consistent spore production throughout the year; *Phlebiopsis* spore production reaches an autumn peak and may fail if its sporophores become desiccated by drought or damaged by frost. Colonisation by *Phlebiopsis* is also restricted by any *H. annosum* infection already present (Risbeth 1963, Greig & Redfern 1974, Phillips & Burdekin 1982).

From Risbeth's pioneering work, one of the first practical biological control treatments for a plant pathogen was developed for pine in East Anglia (Greig 1976a).

Small sachets containing millions of asexual spores (oidia) of *Phlebiopsis gigantea* in a sucrose medium have been produced and can be stored. Immediately before use, the contents of each sachet are dispersed in water. Diluted spore suspensions are applied to stumps, either by hand using specially designed containers with a built-in brush (Webb 1973), or by devices attached to the cutting bar of chainsaws (Greig 1976b). These were later extended to applications through harvesting machines, increasing the size of sachets (to 10 ml) to provide for the larger volumes of treatment suspensions needed for the bigger machines (Pratt 1996b, 1997, MAFF Eval 1998).

Phlebiopsis spore suspensions have been used by Forestry Commission staff at Thetford forest and elsewhere for more than 25 years (Gibbs *et al.* 1996). In 1976, some 60 000 ha of pine crops were treated annually; in 1990 the area of thinnings and clear fellings treated had dropped to 30 000 ha (Booth 1990). By 1997, usage was confined to Thetford forest, area 18 000ha (Redfern & Ward 1998).

Phlebiopsis being a wood-rotting fungus, it is necessary to ensure that recently felled logs are also removed promptly and are not sprayed inadvertently if still at stump, lest they also be colonised.

Full approval for commercial use of a suspension of Phlebiopsis spores
A commercial product containing spores, 'PG suspension', was given full approval for use in the UK for controlling *H. annosum* in April 1998 and was included in *Pesticides 1999* in the section *Biological Pesticides* (MAFF Pst 1999). Pratt (1999) gives recommendations for its handling and application.

This 'approval' was accompanied by the publication of *Evaluation on PHLEBIOPSIS GIGANTEA* (MAFF Eval 1998), under the *Food & Environment Protection Act* 1986. The evaluation summarised the official view of the risks for each aspect of hazard considered before a pesticide product is approved. Because of the widespread natural occurrence of the fungus in woodlands, hazards and risks were viewed as low over the whole range of these attributes. Nevertheless, the applicant (in this case the Forestry Commission) when seeking approval, had either to produce relevant data or to make a case that the data was not necessary in the specific circumstances of the product, its use and the hazard in question.

A product of similar nature has approval for use in Finland, Sweden and Denmark, but no other EC country (See Korhonen *et al.* 1994)

A number of other species or genera have been studied as potential antagonists against *H. annosum*, including *Trichoderma* spp. and *Resinicium bicolor*. The cost of developing any of these to a stage where not only their practical utility is established but also they meet current regulatory requirements for pesticides has escalated to such an extent as to be a deterrent against such development, to the detriment of the forest environment (Holdenrieder & Greig 1998, Pratt *et al.* 1999).

Physical removal

Pine stump removal

Once infected, stumps may continue as infection sources for decades, depending on their rate of decay. In Thetford Forest, second rotation pine has suffered heavy losses when planted on sites where the previous crop had been seriously damaged. Experimental removal of stumps of infected trees reduced second rotation crop losses after 11 years from 54% to 20% (Greig & Burdekin 1970). Following extensive surveys, stump removal was adopted as a routine treatment prior to replanting the areas predicted to be most susceptible to serious losses (Greig 1971). The rate of destumping in 1966 was estimated at 100 ha/yr (Redfern & Ward 1998).

Economics of treatments

Twenty years after his pioneering work on the control of *H. annosum,* Risbeth (cited by Pratt 1998) identified the justification of disease control on economic grounds as one of the most intractable problems of intensive forest management, depending heavily on future projections of the scale and value of losses under alternative management options.

In Britain, as soon as it became apparent both that conifer heart rot was a potential cause of severe losses and that these losses might be reduced by stump treatment, the question was raised whether to introduce stump treatment as a general preventative measure. Low and Gladman (1960) recommended that all sites should be treated *except* those previously carrying a pure crop of hardwoods, and sites that were wet, acid and peaty. In the event, at least within the Forestry Commission, stump treatment has been applied universally to all felled conifers (Blatchford 1983).

With the benefit of 20 years of experience, a cost/benefit analysis taking account of different levels of stump infection and soil type showed that risks to Sitka spruce and benefits of stump treatment are greatest in regularly thinned crops on mineral soils. However, while stump treatment on peaty soils with low levels of infection may not pay in the short term, because low levels of stump infection cannot be avoided if treatment is discontinued, it was considered prudent to continue with treatments until there is stronger evidence about the build up of disease on the range of peaty soils (Redfern *et al.* 1994b).

Pratt (1998) reviewed attempts to estimate losses and to make appraisals in Britain and elsewhere. The results of formal appraisals in many cases imply that the costs of treatment are not always covered by future benefits. Plantations are treated in Britain, Ireland and Denmark, but only sporadically in areas of continental Europe and Scandinavia, possibly reflecting the difference between countries where recently established plantations dominate compared with countries with substantial areas of older forest.

7.22   Honey fungus   *Armillaria* spp

Armillaria species are very widespread in British woodlands and gardens and have been the subject of regularly updated Forestry Commission leaflets *etc.* (Peace 1962, Anon. 1958a, Greig 1967, Greig & Strouts 1983, Greig *et al.* 1991).

While earlier studies all referred to *Armillaria mellea,* already in the 1950s it was becoming clear that the fungus was highly variable in morphology and pathogenicity, implying the possibility of the existence of several closely related species. Risbeth (1987) mentions four species:

- *A. bulbosa, A. mellea, A. ostoyae* and *A. tabescens.*

Subsequent publications refer to five distinct species in Europe:

- *A. borealis, A. cestipes, A. mellea, A. ostoyae, A. gallica* (syns. *A. lutea & A. bulbosa*) (Gregory 1983, 1989, Greig *et al.*1991).

The species cannot be readily differentiated in the forest. However, they differ markedly in their pathogenicity (Redfern 1978). The most pathogenic species are recognisable from the fan-like sheets of white mycelium under the bark of dying trees. The 'boot-laces' or rhizomorphs by which honey fungus has been characterised in the past, are not diagnostic in relation to potential pathogenicity but reflect the saprophytic role of all the species in woodland conditions.

Armillaria species colonise plants principally by means of the rhizomorphs invading root systems. Colonisation of stumps by spores produced by its toadstools, or direct invasion across root contacts is possible but not common (Risbeth 1987).

Control of honey fungus

In hedges and gardens, the death of plants adjacent to others which have died previously is characteristic of how the pathogenic species of honey fungus spread.

There are no recommended chemical means of controlling the disease. Some derivatives of creosote have been used to kill rhizomorphs, but it has been predicted that the rhizomorphs will reinvade unless their source is eradicated by removing infected stump or root material (Pawsey & Rahman 1976, Risbeth 1976). Risbeth (1987) recommended killing of stumps of unwanted trees using ammonium sulphamate.

Removal of as much as possible of potential sources of infection is consistently recommended as the only practical means of keeping honey fungus in check (Phillips & Burdekin 1982, Greig *et al.*1991, Strouts & Winter 1994, Butin 1995).

Techniques for removal of trees, stumps and roots are described by Wilson (1981). Stump grinders or chippers are widely used for large stumps but usually cannot readily deal with lateral roots. Any fungus in the smaller wood chips mixed with soil from such operations quickly dies. It may, however, persist in substantial undisturbed root fragments outside the area of chipped roots.

Where digging out is not practicable, a vertical barrier has been suggested of PVC or similar material, between infected and uninfected or cleaned up ground, from the soil surface to a depth of not less that 45cm (Greig *et al.* 1991)

Rhizomorph development was observed in the laboratory to be inhibited by *Trichoderma viride* and *Pleurotus ostreatus* (Garrett 1956). These observations have not, however, been developed into any system of control.

Biological control

Phillips and Burdekin (1982) describes work in the tropics where trees to be felled were ring-barked before felling. The aim was to reduce the carbohydrate content of stump wood and make it less attractive as a substrate for honey fungus. Studies in Great Britain showed that ring barking hardwood trees before felling, or killing them with 2,4,5-T had no effect on colonisation of stumps by honey fungus but may have reduced the available food base in the longer term.

7.3   VECTOR-BORNE DISEASES

There are many diseases which are brought into contact with their host by a carrier, often an insect or other form of invertebrate. While at first sight, the involvement of a second organism offers the advantage of additional opportunities to control the disease, in many cases, this is not easily achieved.

7.31   Dutch elm disease *Ophiostoma (Ceratocystis) ulmi, O. novo-ulmi (Ceratostomella ulmi, Graphium ulmi)*

Dutch elm disease (DED) has been by far the biggest single pest-induced scourge to strike lowland woodlands and hedges in Great Britain. When account is taken of similar large-scale losses in Europe and in North America, the losses overall in the second half of this century constitute a disaster on an international scale.

Up to the present, Dutch elm disease has not been amenable to any form of large scale control.

Trees infected by DED die of desiccation following interruption of sap flow through the trees' vascular system. This is a result of fungal infection, the exact mechanism being still not fully explained. The essential features seem to be that the fungal spores are carried to the healthy twigs by newly emerged beetles which then feed (maturation feeding), usually at the junction of two twigs. Spores are transferred to the tree and germinate. The fungus transmutes to a small yeast-like cell form, in which condition it is able to be carried in the water-conduction vessels of the current year's sapwood. The form present in the 1930s and 40s seemed unable to cross from one year's rings to the next, so that repeated annual reinfection was necessary for trees to succumb. The more virulent forms are able to move from one year's wood to the next without reinfection.

Restriction of water movement is thought to be due partly to formation of tyloses (blockages caused by balloon-like cells) in conducting tissues, partly to damage to cell walls and partly through the formation of a toxin able to block pit membranes of vessels (Peace 1962, Butin 1995). The toxin is named as cerato-ulmin by Gibbs (1994a).

In the 20th century, the disease was first identified in France in 1918 and very soon after that, in Belgium, Holland and Germany. Subsequent studies and a realisation that earlier infections could be identified by characteristic markings in older growth rings, led to the conclusion that the disease appeared around 1910. Similarly, the first identification in England (Hertfordshire) was in 1927, but annual growth ring discoloration indicated infection at least 7 years earlier. The disease was first reported in the United States in 1930, Canada in 1944 and is believed now to be widespread throughout much of North America, most of Europe and into southwest Asia.

Rackham (1980, 1986) from 19th century and earlier documentary sources considers that the disease was present in Britain continuously from mediæval times. Evidence of a sudden decrease in elm pollen in prehistoric pollen profiles in early neolithic times has been taken as an indication of an earlier widespread attack by the disease.

Brief histories of the 20th century stages of the DED's spread and identification are given by Peace (1960, 1962), Gibbs, *et al.* (1977), Burdekin & Phillips (1982).

The discovery of two races of a more virulent strain of the fungus are described by Brasier (1979 and 1996) and Brasier & Webber (1987). Reports on the spread of the disease include Anon (1958c), Gibbs (1971, 1974, 1979), Gibbs & Howell (1972, 1974), Burdekin & Gibbs (1972, 1974), Gibbs & Wainhouse (1986). Gibbs *et al.* (1994) and Butin (1995) give recent summaries of the biology of the pathogen.

The first British outbreak was initially alarming, being characterised by rapid browning of affected foliage in mid-summer, and yellowing of foliage later. Often very substantial proportions of the crowns of many trees were affected in this way. More severely affected trees died. The peak of this first outbreak occurred in 1928-1930. From that time until about 1965 the disease was in decline. Fewer trees were affected and many of those that had suffered substantial loss of crowns recovered. Peace (1960), who had spent a considerable amount of his time surveying the progress of Dutch elm disease in elms in the 1930s and 40s concluded that ... 'The disease may long continue to be a minor nuisance, but unless its present trend of behaviour changes completely, it will never be the disaster once considered imminent.'

However, there was a complete change of behaviour.

Peace noted that from 1926, when Dutch elm disease was reported on the Continent until 1949, there was a complete prohibition on the entry of elms of any kind into the UK. From 1949, import was allowed under licence from

Europe, but not from countries outside Europe, principally to avoid importation of 'phloem necrosis', a serious disease widespread in the United States.

Ophiostoma novo-ulmi

In the late 1960s reports began to accumulate of a resurgence of Dutch elm disease in Britain with some outbreaks killing more trees than had been expected from previous experience. It became apparent that some infected trees were dying after infection whereas others similarly affected recovered.

By 1978, the disease had killed over 85% of the estimated 23 million elms over 6 m tall that had been growing in the southern half of Great Britain in 1970 (Gibbs 1979). Deaths elsewhere continued but over a longer period.

Mycological studies showed that a new 'virulent' strain of the fungus had appeared; its source was ultimately tracked down to importation of rock elm logs (*U. thomasii*) from Canada (Brasier & Gibbs 1973, Gibbs 1974). The new strain of the disease has been raised to species status and named *Ophiostoma novo-ulmi* (Brasier & Mehorota 1995).

This new lethal 'North American' strain spread rapidly and in regions where the landscape had previously been dominated by shapely hedgerow elms, effectively destroyed the feature which gave it depth and character.

A further complication was the discovery of a distinct 'Eurasian' form of the aggressive strain (Brasier 1979). The forms are able to hybridise (Brasier & Webber 1987).

Sanitation programme

Initially, a sanitation programme was set up, attempting to minimise spread out from areas where the virulent strain was established. Legislation was introduced through *The Dutch Elm Disease (Local Authorities) Orders* (SI 1971/1708 and later amendments). These empowered local authorities to take steps to organise control measures locally, including the power to destroy diseased trees. The order was revoked in 1973 (Gibbs *et al.* 1973, 1977).

The *Dutch Elm Disease (Restriction on Movement of Elms) Order 1974* (SI 1974/767 and later amendments) placed restrictions on the movements of elm in parts of the country where disease levels were slight, unless it had been debarked or treated with insecticide in an approved manner. Where attempts to avoid this order were discovered, legal proceedings followed. In 1981 for example, the Forestry Commission Annual Report mentions 8 convictions for contravention of elm movement orders and 13 cases pending (FC AR 1982).

Only in a limited number of towns has a sanitation programme persisted. The most well-known has been in Brighton and Hove, East Sussex. These adjoining coastal towns are isolated from most of the rest of the country by the chalky South Downs, providing an elm-free *cordon sanitaire*. As a consequence, there are elms still in that small part of southern Britain (Pinchin 1999a). Edinburgh, Scotland, has also pursued a rigorous sanitation policy in respect of trees in public parks and gardens so that there are also a substantial number of elms there also. However, losses in southern Scotland have occurred more slowly and have been spread over a longer period, compared with the south and central England, for reasons outlined in §6.53 (Greig 1992).

In many places, there has been a resurgence of the disease on 3-12m high root-suckers from trees the main stem of which had been killed in the 1970s (Brasier 1997, Pinchin 1999a).

Attempts to control DED

Attempts to avert the disease have investigated three strategies, targeting:
- the vector beetle while it is feeding in tree crowns;
- the development of the fungus within the tree;
- spread by root transmission

Control of insects

The most important vector for Dutch elm disease in the United Kingdom is *Scolytus scolytus*, the large elm bark beetle. While two other *Scolytus* species also act as vectors, they are much less important. See §6.54 for a fuller discussion of British beetle vectors of Dutch elm disease, including details of exploratory tests on insecticides undertaken in Britain.

In spite of much work, no environmentally acceptable means of large scale beetle control has been found (Burdekin & Gibbs 1974). Early concern about profligate use of insecticides in the United States (Carson 1963) arose out of attempts to avoid loss of American elms (*Ulmus americana*), planted as important and widespread town trees.

Phomopsis oblonga, a fungus which rapidly colonises bark of dying elms is thought to compete actively and to exert a strong natural control over beetle populations in the north and west (Brasier 1997).

Control of the fungus within the tree

Fungicides have been tested in attempts to control the fungus. The active substances have included benomyl, thiabendazole, imazalil, propiconazole and carbendozim (Gibbs *et al.* 1973, Meulemans & Parmentier 1983, Greig & Coxwell 1983, Greig 1985a, 1986, Brasier & Webber 1987). Of these, only thiabendazole injected into trees when only lightly and recently infested gives reasonable protection. Injections at low concentrations but high volume were tested experimentally between 1978 and 1982. 81% of trees injected survived compared with only 11% survival of untreated trees. However, the treatment is expensive and only viable for the most valued town and park trees treated shortly after infection. A design for a lance for injection is shown by Gibbs *et al.* (1973).

In 1995 there was a change in the product available for such injections, a new formulation being marketed as 'Storite' in place of 'Ceratotect' (Greig 1990, 1996).

Control of root transmission

Many instances are on record of DED attacking one tree from its infected neighbour, especially where a row or group of the trees has developed by growth of root suckers from one original planting, as is often seen in hedgerows. Mechanical severance of any roots linking infected and healthy trees has been recommended for locations where it is safe to dig. This can be achieved by cutting a trench 5-10cm wide to a minimum depth of 60 cm.

A barrier of dead roots created by pouring the soil sterilant metham-sodium into a line of augur holes 60 cm deep has also been recommended (Burdekin & Gibbs 1974).

Mycoviruses for attenuation of DED fungal infection

Studies attempting to reduce the pathogen's aggressiveness have arisen from the discovery in 1983 of a 'mycovirus' disease of the fungus. This is referred to as the 'd-factor' (Brasier & Webber 1987; Gibbs *et al.* 1994).

Preliminary trials showed that 50 times more spores were required to establish an infection in elm when using fungus affected by the d-factor mycovirus. Such a difference is significant in relation to the infectivity of the spore load carried by beetles when feeding in elm not affected by Dutch elm disease.

Brasier (1996) points out that this is a relatively unexplored area of research, it is long term and carries no guarantee of success. Nevertheless, the hope exists that through an agency such as the d-factor, an attenuation of the present outbreak may occur, similar in effect to the attenuation of the 1930s/40s outbreak. At the same time, the pathogen itself, because of its recent hybridisation between 'North American' and 'Eurasian' forms may give rise to novel forms that are resistant to the 'd-factor' and retain a high degree of virulence towards European elms (Tickell 1999).

Fungal antagonists

Trichoderma sp. and *Pseudomonas* sp. have been tested as antagonists that might inhibit the spread of Dutch elm disease within a tree. They only showed any promise with a very limited range of varieties (Brasier & Webber 1987).

Breeding resistant elms

It appeared in the 1930s that there might be strains of elm relatively resistant to *Ceratocystis*. Dutch workers were the leaders in this selection work (the 'Dutch' in the name of the disease arose in popular usage because of the Dutch interest in the disease and attempts to make selections of resistant varieties). The first Dutch release was in 1936 and the second in 1947 (Peace 1960). These, while showing some resistance to the less virulent stain of DED, had other shortcomings, such as susceptibility to *Nectria cinnebarina* or poor form. A third generation of Dutch selections in the 1970s coincided with the emergence of the more virulent strains of disease, and while further selections are in the pipeline, the prospects for the material then in hand is only moderate (Heybroek 1983).

In many areas, individual trees have been noted as having survived when all neighbouring trees have been killed.

Ulmus laevis, while a comparatively rare tree, has survived where others died (Harris 1996).

A complete break away from the approach of selection and breeding from promising resistant individuals is embodied in a programme investigating the possibilities of genetic manipulation set up in the early 1990s (Brasier 1996).

For elms, the best prospects of avoidance or resistance, albeit long-term and slender, are through techniques such as genetic manipulation of the host plant and exploitation of the virus-like 'factor-d' rather than use of insecticide or fungicide.

7.32   Blue stain   *Ceratocystis* spp and other species

Sapwood of pine may become discoloured by fungi in the genus *Ceratocystis* and related genera of *fungi imperfecti* (Dowding 1970, Phillips & Burdekin 1982). The discoloration ranges from pale grey or dull blue to almost black and is caused by partial reflection of light by fungal hyphae (Holtam 1966) rather than any pigmentation of the xylem. Hyphae live on the cell contents within sapwood, passing from cell to cell through the pits in cell walls. Cell walls are unaffected and retain their full strength but timber permeability is slightly increased. Blue-stains can also mask signs of brown stain from other wood-rotting species. The market value of stained timber is less than that for unstained wood, especially where the surface of affected timber may be seen in use.

Different fungi cause blue-stain in different circumstances. Butin (1995) considers that *Ophiostoma (Ceratocystis) piceae* is most frequent on coniferous round-wood, but that *Cladosporium* spp. are the main species involved in blue stain on sawn timber.

Following the extensive damage in southern England caused by the great storm in October 1987, studies were made of the spore load of adult *Tomicus piniperda,* the Pine shoot beetle. 17% of the beetles which had overwintered and were constructing breeding galleries in spring 1988 were carrying fast-developing *Leptographium* spp. More than half the adults emerging in June/July 1988 carried spores of these fungi. The proportion had fallen to 26% by the time of gallery construction in spring 1989.

Isolations from pine tissue around early brood galleries showed 25% were colonised also by *Leptographium* spp. Later-formed galleries showed over 50% colonisation. Four species of *Leptographium* were identified, principally *L. wingfieldii*. *Graphium* spp. were also quite common, as were two black yeasts (Gibbs & Inman 1991).

The blue stain in timber that becomes rewetted in use, *eg* garage doors, are caused by yet other species.

Spores of blue-stain fungi can only develop on exposed wood surfaces; stain spreads rapidly in warm conditions and is fastest when the temperature is between 22 and 30°C. A prerequisite for its development is a moisture content in the wood of 30 - 120% of dry weight (Butin 1995). Blue stain does not develop if the wood is soaking wet - see *Wet storage* below.

Two main routes for infection are by insects and by air.

In the forest, spores may be carried by the pine shoot beetle, *Tomicus piniperda* and other associated insect species (Dowding 1970), and are a principal cause of infection.

Spores may be carried by wind to infect ends of logs in the forest where intact bark protects their outer stem surfaces; airborne spores are the main source of infection of freshly sawn planks in sawmills.

Control measures

For control of blue-stain, proper timing of operations is crucial to the success of whatever techniques are used.

In the forest, measures to minimise risk of build-up of populations of potentially damaging beetles also reduce the risk of blue-satin:

- seasonal restriction of felling;
- 'hot logging';
- use of insecticides when beetles may breed.

See also §6.56.

At the sawmill, if uninfected timber can be sawn shortly after delivery and then kiln-dried so that its moisture content is maintained at less than 20% of oven-dry weight, it should remain free from infection. Where this cannot be done or there is a risk of timber being rewetted, timber after sawing may be dipped in fungicide, *eg* quaternary ammonium compounds, carbamates, octhilinone, borates, usually in mixtures (see *Surface biocides,* in *Pesticides 1999* (MAFF Pst 1999).

Wet storage

Blue-stain also does not develop in logs if a very high moisture content can be maintained (Findlay 1959). In the past, logs have commonly been stored in lakes and log ponds; more recently, experience has been gained of storing logs on dry land but keeping them wet with water from overhead sprinklers (Liese & Peek 1984, van dem Bussche 1993).

In 1987, 1.3 million m^3 of pine timber was blown down in October gales in the south and east of England (Grayson 1989b). This led to the first large scale trial in Britain of wet storage of timber. At its maximum, 70 000 m^3 of pine logs were stored in Thetford Forest, Suffolk. They were kept wet by water sprinklers, using recycled water (Webber & Gibbs 1996). Logs were stored for between 18 months and 3 years; samples being removed regularly to determine moisture content and to assess degrade (Thompson & Gibbs 1990).

Planks cut from logs which had been stored for six months were highly susceptible to blue-stain. With longer storage, susceptibility diminished; this was attributed to nutrient leaching under the water regime and presence of bacteria.

Overall, less than 2% of the wood was affected by blue-stain; any present in logs prior to storage was prevented from developing further. Porosity of stored logs increased sufficiently to affect the ability of sawn wood to accept surface treatments *eg* paints. Timber strength was slightly less, but not sufficiently to lower the stress grading category of stored timber (Bravery 1992).

About 1% of log ends showed signs of invasion by *Heterobasidion annosum*; there were also signs of colonisation by *Armillaria* spp but no decay (Webber & Gibbs 1996).

Influence of harvesting method

A comparison was made of development of blue-stain on Corsican pine where logs had been harvested by:

- mechanised harvester fitted with metal-spiked rollers;
- mechanised harvester fitted with rubber rollers;
- manually operated chainsaw felling and snedding.

The studies were made in Thetford forest (Norfolk/Suffolk borders) and near Inverness, Scotland on trees cut in September 1991. After felling, logs were stacked and sampled for spread of blue stain over the following 12 weeks.

Blue stain was overall more widespread at Thetford, considered to be due to the warmer temperatures during the storage period. At both sites, however, the amount of stain was least on logs felled using manually operated chainsaws and most on logs which had passed through spiked rollers. The incidence of stain closely reflected the amount of bark loosened during felling and conversion to log lengths (Lee & Gibbs 1996).

7.33   Beech bark disease and beech snap   *Nectria coccinea*

Trees may develop localised patches of dead bark infected by *Nectria coccinea* (Phillips & Burdekin 1982) following heavy infestations of beech scale *Cryptococcus fagisugi*, the disease colonising bark injured by the scale insect (Parker 1974; see plates 3 and 7 in that *Forest Record* for illustrations of *Cryptococcus* infestation and *Nectria* stem canker).

The disease and insect both occur over much of the European range of beech. Beech scale has also inadvertently been introduced into North America; attacks there have been followed by attack by *Nectria coccinea* var. *faginata,* the origin of which is not known (Lonsdale & Wainhouse 1987).

Trees may recover from light or moderate attacks of *Cryptococcus*. Nevertheless, substantial numbers of beech have died after exposure to the '*Nectria-Cryptococcus*' complex; several authorities consider it to be the most serious disease complex affecting beech.

Lesions caused by colonisation by *Nectria* can also follow where stems have been under stress through the effects of abiotic factors, in particular, drought, chlorosis, root disease (Lonsdale 1980, Lonsdale & Pratt 1981) or frost. None of these singly consistently induces *Nectria* infestation.

Once established, *Nectria* lesions may also be colonised by secondary wood-rotting fungi. These can invade quickly, weakening the stem of such trees so that they snap, commonly 3 - 6 metres above the ground, giving rise to the popular diagnosis of such damage as 'Beech snap'.

Treatments of *Cryptococcus* are discussed in §6.73.

In woodland conditions, removal of heavily infested trees as part of routine thinning is the only course of action recommended. There is no evidence that 'sanitation' thinning *ie* special intervention to remove heavily infested trees, has any significant effect on subsequent development of beech bark disease in the woodland. Trees showing signs of chlorosis should also be removed.

While natural enemies of *Cryptococcus* and *Nectria* have been observed, none seem to have the potential to be developed for biological control of either organism (Lonsdale & Wainhouse 1987).

7.34   Oak wilt   *Ceratocystis fagacearum*

This disease is not known in Britain but is native to North America. Like Dutch elm disease, its conidia are transported inside the tree as conidia by the transpiration stream in the outermost ring of sapwood. As a defence reaction, the tree forms tyloses; these disrupt the water conducting system and lead to leaf wilt over whole crowns or individual branches of oak (Butin 1995). Dispersal is by spores which may be carried by beetle vectors; spread may occur through root grafts also.

Because of the extent of oak forests in Europe, and presence of native beetles which could take over the role of vectors, oak wilt has been recognised as a disease against which steps should be taken to prevent introduction (Burdekin & Phillips 1977, Phillips 1980b). Legislation initially under the Plant Health Act, 1967 and periodically amended in the light of current information, requires that red and white oak timber from North America must be debarked and either dried to a moisture content of less than 20% oven-dry weight or fumigated with methyl bromide (Gibbs 1984, Gibbs *et al.* 1984).

7.4   DISEASES ACTING THROUGH STEM OR FOLIAGE

7.41   Entry through wounds

Treatment of tree wounds are of primary importance to arboriculturists, for many of whom tree pruning is a major, regular activity. Such wounds result from:

  • deliberate intervention to alter the shape of the tree crown by removing limbs;
  • removal of dead, diseased or decaying wood;

- damage limitation following breakage by wind *etc.*

Whether the reasons for such action are aesthetic, precautionary, or commercial is not relevant to the advisability to minimise the risks of introducing disease through the pruning wound and to take appropriate measures to hasten wound healing.

Peace (1962) gives a full review of arboricultural pruning practices in the preceding sixty years. More recent developments include a greater emphasis on 'Target' pruning *ie* cutting limbs just on the distal side of the 'branch bark ridge'. At this point, healing is likely to proceed more quickly than if cut 'flush' to the main stem (Lonsdale 1983a, 1984a, 1986c, 1987, 1992, 1993a, b, Mercer 1984a, Bradshaw *et al.* 1995).

Clifford and Gendle (1987) working principally on fruit trees, differentiate between 'fresh wound pathogens' able to colonise in the first 28 days following wounding, and 'mature wound pathogens' which are usually wood-rotting fungi. *Nectria galligena* and *Chondrostereum purpureum*, respectively canker forming and silver leaf disease fungi, are in the 'fresh wound' pathogen group.

Other approaches to wounds have come through observation of natural injury to trees, whether resulting from natural processes *eg* gale damage, or from human or animal activity. Some are beyond treatment because of size or inaccessibility; nevertheless, because of the risks arising if rot-damaged trees threaten persons or property, decay in trees has been taken as a factor affecting treatment of wounds in trees (Young 1977, Lonsdale 1984b).

Prevention of decay in pruning and other wounds

Treating new wounds with a fungicide and/or covering intended to resist entry of water has been recommended for many years. However, many such treatments have been unsuccessful (Mercer *et al.* 1983). Many older paint and tar preparations are no longer recommended.

To be successful, a wound treatment must:

- assist healing. Some earlier materials were too toxic, damaging the cambium and delaying occlusion of the wound (Young 1977, Lonsdale 1984b);
- reduce risks of entry of wood-rotting fungi;
- be part of a robust system which minimises the risk that the exposed tissues deteriorate and become an increasing hazard to the well-being of the plant;
- be applied to freshly pruned cut surfaces at the time of year most favorable to healing (Lonsdale 1993b).

Materials reported promising for short-term reduction of invasion by wood-rotting fungi include:

- thiophanate-methyl, beneficial in aiding callus development, (Mercer *et al.* 1983, Mercer 1984a);

- *Trichoderma viride* as a bio-control (Lonsdale 1984a, Mercer & Kirk 1984). For best effect, a non-fungicidal wound protectant should be applied after application of *Trichoderma* (Lonsdale 1983b);
  - ochthilinone (Strouts & Winter 1994);
  - a mixture of carbendazim and triadimefon. These were applied as a gel which allowed diffusion and retention of fungicides for up to 6 months and gave good control of the 'fresh wound ' pathogens (Clifford & Gendle 1987).

Use of any of these has to be linked to good pruning practice (Lonsdale 1987).

7.42   Silver leaf   *Chondrostereum (Stereum) sanguinolentum*

Chondrostereum sanguinolentum is an important 'fresh wound' pathogen affecting many ornamental and fruit trees, mostly rosaceous but including other families. It is the most serious disease of plums.

Pruning of infected trees is recommended to take place in June - August, treating fresh wounds with a paint containing *Trichoderma* or ochthilinone (Strouts & Winter 1994).

If confined to individual branches, these may be cut out; alternatively they may be left for a period as, on many species, affected branches recover spontaneously.

7.43   Anthracnose and leaf spot of willow   *Drepanopeziza sphaeroides (Marssonina salicicola)*

This fungal infection causes small brown spots on leaves and shoots. Affected leaves may fall prematurely. Infection on shoots develop into blackish cankers in which the fungus overwinters. It particularly affects weeping willow and is worse in wet seasons (Rose 1989a).

Fungicides recommended include Bordeaux mixture as leaves unfold and in the summer (Brooks *et al.* 1989), or repeated sprays from bud break to mid summer using benomyl, captafol, maneb, mancozeb or quinomethionate (Strouts & Winter 1994).

7.44   Bacterial canker of cherry   *Pseudomonas mors-prunorum*

This common and widespread disease of ornamental and commercial cherries enters twigs at the leaf scars at leaf fall. In the following spring, it may cause dwarf shoots or branches not to flush or to flush and then wilt. Affected trees produce copious flows of gum from affected bark. During the summer, bacteria may infect leaves causing dark brown spots which may then fall out (shot holes).

Different varieties show different degrees of resistance. Spread can be minimised in badly affected trees by pruning affected branches during the period June - August. It is recommended that any pruning outside that period should be treated with a protectant *eg* one containing ochthilinone. The bacterial load on foliage may be reduced by repeated spraying with Bordeaux mixture between August and November, a treatment more appropriate to orchard cherries than ornamental trees (Garrett 1982, Phillips & Burdekin 1982, Alford & Locke 1989, Strouts & Winter 1994).

Plums are extremely susceptible to infection of stems when trees have been staked and injuries have been caused by badly adjusted ties (Garrett 1982).

7.45   Scab and black canker of willow

Willow scab and black canker occur as shoot and leaf diseases in periods of persistent cool, wet weather during spring and early summer.   Infection can spread rapidly under favourable conditions.

Where willow is being grown commercially, scab may be controlled using penconazole (Rose 1989b, Strouts and Winter 1994).

7.5   CULTURAL CONTROLS

In the forest, there are a number of cultural measures that can be used to prevent or minimise the effects of disease when no fungicide or biological control is available.   These are summarised in *Table 7.1*.

7.51   Resistant varieties and seed origins

Poplar bacterial canker   *Xanthomonas (Aplanobacter) populi*

This is by far the most serious disease of poplars; it is widespread in Britain, Belgium, Holland and Northern France and extends into Poland and Russia.   It appears on stems or twigs of trees when about 6 metres or more tall.   The first signs of the disease are cracks in the bark of twigs, from which bacterial slime exudes.   Twigs if girdled die, but often, as the twig grows, adjacent issue may grow as a response to injury.   Further damage by bacteria and further injury response give rise to characteristic rough cankerous swelling on stems of affected trees.   The disease sometimes kills the tree but more often cripples its growth (Peace 1952a, 1962, Jobling 1990).

It was recognised in the 1930s that susceptibility varied according to species, or, where hybrids between species occurred, according to individual hybrids.

Because of the ease of propagating poplar from cuttings, substantial programmes have been undertaken in several European countries and in the United States, testing poplar clones for canker resistance.   This has been combined with selecting and breeding for vigour of growth and silvicultural characteristics enabling clones to be successfully grown on less favourable sites. This programme continues, with recent additions of fast-growing clones from Belgium (Potter & Tabbush 1991, Tabbush 1992b, Tabbush & Lonsdale 1999).

The importance of using clones of poplar tested for resistance to bacterial canker is reflected in the inclusion of *Populus* cultivars within the scope of the European Economic Community *Council Directive No 66/404 on the marketing of forest reproductive material* and its subsequent amendments.   This directive has been embodied into UK legislation as the *Forest Reproductive Material*

Regulations 1977 (SI 1977/891) and makes it a requirement that anyone selling poplar vegetative material for forestry purposes must use a poplar clone tested and approved under the regulations. While legislation imposes many restrictions on movements of forest plants and timber to protect plant health, this is the only example of a legal obligation to market only tested disease-resistant varieties when intended for forestry purposes (Aldhous 1994).

The *Directive on marketing of forest reproductive material* is currently under revision (CEC 1999). However, the provisions in the draft do not change the previous requirements for marketing tested disease-resistant poplar clones.

Where older poplar are affected by canker, current recommendations are to remove and burn badly cankered plants and to plant fully tested approved canker-resistant clones most suited to the site. See also §7.58 - vulnerability to rusts.

Douglas fir leaf-cast diseases   *Rhabdocline pseudotsugae* and *Phaeocryptopus gaumannii*

These needle cast diseases were introduced to Europe from North America in the early 1920s. *Rhabdocline* causes defoliation in pole crop trees within a year, while *Phaeocryptopus* induces needle shedding after 2-3 years. By the late 1930s, it had been noted that the effects of *Rhabdocline* can largely be avoided by growing 'green' Douglas fir originating from higher rainfall west slopes of the coastal and Cascade Mountains of USA and British Columbia rather than from lower rainfall interior regions (Anon 1956c, Peace 1962). *Phaeocryptopus* is found in Douglas fir stands particularly on the western side of Britain, but requires no control (Anon 1956c, Phillips & Burdekin 1982, Butin 1995).

Larch canker   *Lachnellula (Trichoscyphella) willkommii*

European larch has been planted in Great Britain for over 200 years; plantings during the nineteenth century being on a large scale. At that time, the significance of seed origin in relation to disease susceptibility was not understood and much seed was imported from high elevation alpine origins. These were subsequently shown to be particularly affected by die-back and canker, caused by a combination of susceptibility to late spring frosts and to colonisation by *Lachnellula (Trichoscyphella) willkommii* (Lines 1957).

Use of seed from carefully selected native seed sources from low elevation stands in the Alps or Tatra Mountains is likely to avoid such problems. At the same time, the development of a programme to produce *Larix* x *eurolepis,* the hybrid between European and Japanese larch, provides another alternative, the F1 generation of the hybrid not being susceptible to die-back.

7.52   Sanitation fellings

Watermark disease of cricket bat willow   *Erwinia salicis*

Tree willows are susceptible to 'watermark' disease, caused a bacterium *Erwinia salicis* (Day 1924). Infected trees die back severely; wood cut transversely appears wet and coloured by a reddish-brown stain which darkens on exposure to air. Of the tree willow species native to Britain *Salix alba* is by far the most susceptible, some clones being more prone to the disease than others.

Salix alba var. *caerulea* yields wood suitable for cricket bats and artificial limbs and on that account is highly valued. In 1992, it was estimated that there were 1.5 million bat trees, more than 50% of which were located in the counties of Essex and Suffolk (Turner *et al.* 1992). Unfortunately, it is highly susceptible to watermark disease, any tree with such wood being rejected by bat makers. The disease has been a major source of loss to the regional cricket bat industry which requires perfectly formed stem wood for the production of bat blades (Preece 1977).

The disease may remain latent in infected trees for several years. It spreads by natural means, such as birds or insects moving from infected to uninfected trees, often developing in sites of pruning wounds and other injuries. The disease may also therefore inadvertently be spread where sets are cut from symptomless but infected trees. Also, saws cutting diseased trees may carry disease unless sterilised every time the sawyer moves to a new tree.

Newly planted infected but symptomless trees may take several years to show symptoms of infection (Peace 1962, Phillips & Burdekin 1982).

The bacterium is not thought to be transmitted by root contact or through the soil so there is no objection to planting clean stock within a year or two of complete removal of infected trees. The currently recommended method of control is to fell and burn all above ground parts of infected plants or to grub out and burn.

From the 1930s onward, local authorities have been empowered to require felling or grubbing under a Watermark Disease (Local Authorities) Order (*eg* SI 1974/768). The first applied to Essex, but later amendments extended the area to most adjoining local authorities. A recent amendment added Wiltshire to the list of local authorities empowered under the 1974 order (SI 1992/44). This legal provision for control of a tree disease of cricket bat willow is the longest-standing of any such order.

Local authority inspectors are empowered to require infected trees:

- to be felled and burned, or,
- stumps to be grubbed up and burnt, or,
- in beds from which 'sets' may be supplied for planting, not to be so used for a specified period, which may be up t o three years.

A survey in 1972 showed that the eradication policy had reduced the disease to a low level (< 0.1% of trees affected) (Wong *et al.* 1974).

In Holland, willows have been planted and pollarded for many hundreds of years. While being cut on an annual or 2-3 year cycle, willow shoots show no signs of watermark disease. After 1945, the practice of pollarding became less common. At the same time, varieties of *Salix alba* were widely used in urban amenity and landscape plantings. By the 1960s when these trees were 15-20 years old, watermark disease started to appear and to cause losses on a massive scale. In the Netherlands, control of watermark disease appears to depend entirely on developing disease-resistant varieties (Turner *et al.* 1992).

Fireblight   *Erwinia amylovora*

Fireblight is a bacterial disease which, until about 1950, was only known in the United States and New Zealand. (It was first described in the US in 1794.) The first European record was in 1957 in Kent.

Fireblight attacks a wide range of roseaceous trees and shrubs, in particular, flowers as they are opening. It has caused significant damage to commercial apple and pear varieties and is also widespread on hawthorn and whitebeam (*Crataegus monogyna* and *Sorbus aria*).

The disease is carried in the spring by insects and by rain, and on susceptible varieties spreads throughout the plant, in extreme cases, leading to death within six months of attack (Phillips & Burdekin 1982).

Initially, fireblight was a notifiable disease; in 1993, the previous order was replaced by a requirement under the 'Plant Health Order 1993' (SI 1993/1320) for notification only when the disease occurs in a nursery registered under the order, or in an associated buffer zone (Strouts & Patch 1994).

Control is by sanitation felling or pruning. If caught early, infections can be arrested by pruning out affected branches. These must be cut well below where staining of the cambium indicates presence of the disease. Branch surfaces must be dry. It is essential between each cut to sterilize the cutting blade (Strouts & Winter 1994).

7.53   Control of alternate host

Pine twisting rust   *Melampsora pinitorqua*

This rust fungus alternates between Scots pine and *Populus* spp, aspen principally, also white or grey poplars but not black poplars. On pine, the disease can be quite serious; it attacks young shoots on plants up to about 10 years old. Infection can affect one side of a shoot and cause appreciable distortion. On the *Populus* host, it infects leaves. Peace (1962) stated that the disease has been recorded only in the southern half of England.

Recommendations for control are either to clear aspen or avoid planting within 200 (Murray) or 500 (Butin) metres of aspen or poplar. Alternatively, Corsican pine or lodgepole pine could be planted, both species being virtually immune (Murray 1955, Phillips & Burdekin 1982, Butin 1995).

Needle rusts of Scots pine   *Coleosporium* spp *(C senecionis)*

This group of rusts has as alternate hosts Scots pine and one of several *Compositae* and other herbaceous families. It can occasionally be troublesome on pine in forest nurseries where alternate hosts (groundsel and ragwort) are abundant. Removal of the alternate host as part of routine weed control operations prevents further infection (Murray 1955, Phillips & Burdekin 1982).

7.54 Avoid planting

White pine blister rust   *Cronartium ribicola*

This rust disease attacks five-needled pines. It is thought to have originated in Asia, alternating between *Pinus cembra* and *Ribes* spp. In the second half of the 19th century, white pines, especially *Pinus strobus* and *P. monticola* were being widely planted in Europe on account of their good form, vigorous growth and good quality timber. The rust was found in Europe on *Ribes* in 1854. It was found on *P. strobus* in Europe in 1887 and in England in 1892. Young plants of *P. strobus* were sent from Germany to the USA in 1900 and to eastern Canada from France in 1910. The American white pines are very susceptible to the blister rust. From these two introductions, the disease has spread widely; the cost of losses and of programmes slowing spread by eradication of the alternate host place it among the most serious American pests and diseases (Peace 1962, Phillips & Burdekin 1982).

White pine in Britain have also succumbed. While older plantings in Britain grew well, future planting has been dismissed as pointless until disease-resistant provenances or clones become available (Macdonald *et al.* 1957).

7.55 Avoid disturbance

Sooty bark disease of sycamore   *Cryptostroma corticale*

This disease was first observed in Britain in 1945. It causes the outer layer of sycamore bark to peel off revealing a dark brown powdery spore mass. The areas exposed are frequently extensive. It is assumed to have been introduced from North America where it has been reported on maples (Peace 1962).

Early outbreaks were restricted mainly to around the London area and occurred following the particularly hot dry summers of 1947 and 1949. Initially, it was feared that the disease could cause major losses throughout the country; however, after the initial surge, incidence of the disease declined.

By 1978, it appeared that the early pattern of infection was being maintained, the disease occurring only after particularly hot dry periods and mainly in the warmer parts of the south and east. Young (1978) forecast that the disease would only be prominent in years when the mean daily maximum temperature in two out of the three months June, July or August, exceeded 23°C. He also observed that as the 'sooty' spore-producing phase of the disease only occurs at a very late phase, any radical disturbance then or later, such as felling and burning would, if anything, assist dispersal of spores rather than minimise it.

Subsequent studies on the infection biology have shown that ... '*C. corticale* is an impressive example of a fungus which can live in an inactive state for many years in the secondary tissues of its host, only developing if the moisture regime of the tree is disrupted. This can occur under conditions of extreme drought stress, but it also happens if the tree is felled or blown down' (Gibbs 1997). Young's earlier observations on felling remain valid.

7.56   Avoid fires,   delay planting

Group dying of conifers   *Rhizina undulata* (syn. *R. inflata*)

This widespread fungus has long been recognised as being associated with the death of groups of trees. In the early 1950s, such deaths were identified as being linked to the site of small fires such as made when burning piles of brash or lop and top. Some fires were thought to have been lit by forest workers at meal breaks (Murray & Young 1961, Phillips & Burdekin 1982).

Avoidance of fires in plantations has reduced the incidence of losses. Because the fungus cannot decompose cellulose and lignin, it does not persist in the roots it has killed. Where burning has taken place, delay of planting for two years has also given local control of the disease (Gibbs *et al.* 1996).

7.57   Seasonal restriction

Basket willow rust   *Melampsora amygdaline*

The basket willow industry has contracted in scale for much of this century but nevertheless remains as a localised speciality industry in areas such as the Somerset levels. Baskets are made from shoots ('rods') cut annually in early spring from ground-level stools.

Regrowth from stools is liable to be seriously damaged by a monoecious rust *Melampsora amygdaline*. Phillips & Burdekin (1982) note a recommendation that cutting should be delayed as late as possible, or alternatively, that stoolbeds could be grazed or sprayed with a tar-oil winter wash to delay spring shooting and reduce risk of infection.

7.58   Dilution of vulnerable varieties

Willow rusts   *Melampsora* spp

Rust fungi are a serious threat to use of willows for high dry-matter biomass production. In 'single-variety' plots, infections of rusts can build up over a period of a few years and cause such defoliation that growth is substantially reduced, in some cases weakening plants to the extent that they die (DTI 1994a, Tabbush 1994, Tabbush & Parfitt 1996, Royle *et al.* 1996).

There is no suitable fungicidal treatment. To minimise build-up of rusts on biomass sites, 'polyclonal' mixtures *ie* several clones planted in alternating rows or similar mixtures, are recommended, omitting the known most susceptible varieties (Tabbush & Parfitt 1999).

Fungal hyperparasites

Included in the list of recommended clones is one known to suffer from a rust which is attacked by a fungal hyperparasite (Morris *et al.* 1994).

Poplar leaf rust   *Melampsora larici-populini*

Vigorous poplar hybrids used for timber production or for short-rotation biomass or energy crops have also been seriously affected by leaf rust. The rust has as alternate host *Larix* spp., where it reproduces sexually and in so doing gives rise to a range of new strains. Some of these have overcome the resistance to the rust previously observed in recently developed clones. For example, rusts develop on the clone *Trichobel* but are tolerated. In contrast, the clones *Beaupré* and *Boelare* are no longer recommended for short rotation coppice planting, because of risk of severe damage by rust when planted at close spacing.

To minimise risk, planting mixtures of poplar clones in small blocks or in intimate mixture now is recommended. Cutting material for new plantings should not be obtained from parent plants which were previously heavily infected with rusts. Revised lists of poplar varieties approved for commercial use, and of poplar and willow recommended for short rotation coppice were published in 1999 (Tabbush & Parfitt 1999, Tabbush & Lonsdale 1999).

Some new clones selected for rust resistance and under trial in 1999 have shown susceptibility to new rust strains (Lonsdale & Tabbush 1998, Tabbush 1999).

7.59   **Phytophthora diseases of alder and other broadleaves**

New disease causing alder die-back

This disease was not described until 1993; the fungal pathogen is considered to be a hybrid swarm between *Phytophthora cambivora* and a species closely allied to *Phytophthora fragariae*. Both fungi have been introduced to Europe (Gibbs 1994b, FC EA 1997, Gibbs & Lonsdale 1998, Pain 1999, Brasier 1999).

The disease has been found to be widespread and is thought to have been present both in the UK and in Europe for several decades. The disease is carried by water; even occasional inundation may allow the fungus to invade the bark at the base of the stem. Surveys carried out in 1994-6 in southern Britain showed 1.2% of trees in the survey area died between 1994 and 1995, and 2.2 between 1995 and 1996. If deaths continued at these rates, it would have a substantial effect on riverside ecology and landscape.

It is not possible to remove all infected stump and root material; felling is recommended only for safety or similar reasons. Planting of alder is not recommended on ground subject even to occasional flooding. Where alder is planted, the stock should be known to be disease free.

Poles cut from diseased trees or trees in their vicinity should not be used for river bank piling or other work near water.

Phytophthora pathogens and other tree species

Phytophthora spp, in particular *P. cinnamomi* and *P. cambivora,* have long been recognised as important causes of tree death and die-back of chestnut, beech, maple and some fruit trees. Their effect is spreading, worldwide, but as yet, there are no clear chemical or cultural preventatives or cures (Brasier 1999).

7.6   FOREST & HARDY ORNAMENTAL NURSERY STOCK

In the nursery, fungal damage can be locally severe at times. However, only a few diseases warrant routine preventative treatment. Fungicides should always be used with the awareness that repeated use may lead to a build-up of resistant strains of pathogens.

Pests and diseases of ornamental plants may occur equally on larger nursery stock as on smaller ornamental specimens. *A colour atlas of pests of ornamental trees, shrubs and flowers* (Alford 1995), *Pests & Diseases* (Greenwood & Halstead 1997) and *Collins guide to pests, diseases and disorders of garden plants* (Buczacki & Harris 1981) may provide relevant illustrations and guidance.

7.61   Soil-borne diseases

Damping off and root rot

Death of germinating seedlings before or during emergence occurs commonly in both broadleaves and conifers. Peace (1962) and Phillips & Burdekin (1982) give comprehensive summaries, discussing the range of soil fungi that may be responsible. Damping off problems are less on the more acidic soil types currently recommended for production of forestry planting stock, compared with nurseries on heavier agricultural soils.

In the 1950s, partial soil sterilisation with steam, formaldehyde solution or chloropicrin was widely tested in nurseries with poor yields and growth of seedlings. While these materials enhanced seedling growth, the need for them disappeared with the change to light acid soils for plant production (Benzian 1965, Edwards 1952, Faulkner 1957, Aldhous 1972c).

Where damping off is discovered, immediate treatment with captan has been recommended. If further attacks are expected, autumn treatment with dazomet as soil fumigant has been recommended (Phillips & Burdekin 1982, Brooks *et al.* 1989, Strouts *et al.* 1994).

Denne and Atkinson (1973) applied a 0.1% solution of captan to recently germinated seedlings of Sitka spruce, western hemlock and Scots pine. Whether grown in sand or compost, seedlings suffered a short period of check in growth, this being more severe where plants had been grown in sand.

Phytophthora cactorum

This fungus is associated with damping off or seedling blight in beech. Fungicides that have been recommended for application to check the disease include: Bordeaux mixture, zineb and maneb. Where beech has been affected, it is recommended not to raise it again on the same ground for at least 5 years (Phillips & Burdekin 1982).

Phytophthora spp can survive in poorly drained or wet land and attack trees suffering from waterlogging. They may also be inadvertently spread through nursery irrigation systems. Where *Phytophthora* spp are present on heavier or wetter ground and are regularly damaging stock, improved drainage or change of production technique to container production have been suggested as alternatives to soil fumigation (Strouts 1995a, Strouts *et al.* 1994).

Container-grown stock can be infected by *Phytophthora* spreading through capillary sand-beds.   Such spread has been reduced by using a proprietary copper-based chemical paint on the ground-cover matting.   In a trial where water draining from heavily infected plants was recycled through a slow sand filter, there was no re-infection (Labous & Willis 1997).

Verticillium wilt   *Verticillium dahliae*

This disease can cause substantial losses of *Acer* species and other trees, attacks on roots causing sudden wilting.   There is no cure for the affected plants; it is also thought that common weeds such as groundsel (*Senecio vulgaris*) may be symptomless carriers.

The risk of future attack can be reduced by not growing susceptible species on the affected ground for several years, keeping down weeds at the same time, or by treating previously affected areas with a soil fumigant.   Chloropicrin, dazomet, metham sodium or formaldehyde have been recommended but not all are available under current legislation (Pierce & Gibbs 1981, Phillips & Burdekin 1982, Brooks *et al.* 1989).

7.62   Seed-borne disease   *Geniculodendron pyriforme*

Commercial seed lots of Sitka spruce and other conifers may carry light infections with a seed-borne fungus *Geniculodendron pyriforme*.   This may substantially reduce seedling numbers, especially if seed is sown early or cold conditions follow sowing. While later sowing of seed treated before sowing by moist-prechilling improves seedling yield, moist pre-chilling combined with seed dressing with thiram or captan further reduced losses (Gordon *et al.* 1976, Phillips and Burdekin 1982).   The last two authors mention that the fungus was found in Canada to be the conidial stage of a discomycete *Caloscypha fulgens*.

Potentially damaging effects were minimised where seed had been treated with 50% thiram dust (Salt 1967).

7.63   Grey mould   *Botrytis cineria*

The grey mould fungus is a facultative parasite with a very wide host range. It occurs frequently in nurseries and can cause heavy losses of species which are slow to harden off in the autumn and are growing in cool moist conditions. Stock which is overcrowded is particularly susceptible.   Accounts of the fungus in relation to tree and ornamental species in Britain are given by Peace (1962), Pawsey (1964b), Buczacki & Harris (1981) and Butin (1995).

Losses can be minimised by avoiding over-dense sowing or use of fertilizer regimes likely to encourage late soft growth.   Pawsey recommends use of freshly prepared Bordeaux mixture applied at the first sign of fungal attack and repeated at intervals of three weeks until the end of September. Alternatively, for species known to be susceptible, routine spraying in late summer and autumn with thiram may be necessary (Phillips & Burdekin 1982).

7.64   Needle cast of larch   *Meria laricis*

Larch leaf cast is a fungal disease affecting only needles. It is widespread, occurring in many regions where larches are native or have been planted.

In Britain, European larch seedlings and transplants are markedly more susceptible than Japanese or hybrid larch.

Infection takes place through stomata and may appear within a few weeks of bud break. It can be distinguished from frost damage because symptoms are first seen in older needles of current shoots, whereas frost damage would be most conspicuous on the youngest, most tender part of the shoot tip (Peace 1936, 1962, Phillips 1963, Phillips & Burdekin 1982).

The only cultural aid is to minimise the risk of plants being infected by spores emanating from previous year's plants.

• Where possible, older European larch trees round the nursery site should be removed;
• within the nursery, plants should be raised so that they can be moved annually. One year's crops should be as far distant as practicable from the site of the previous year's larch beds.

Leaf cast can be controlled by spraying. Repeated applications of sulphur have been widely recommended (Peace 1962, Phillips 1963, Phillips & Burdekin 1982, Brooks *et al.* 1989). Zineb has been recommended as an alternative (Strouts *et al.* 1994).

7.65   Oak powdery mildew   *Microsphaera alphitoides*

Oak mildew is a member of a family of obligate parasites which, while forming a white felted mat of mycelium on the surface of their hosts plants, live on plant cell water and nutrients extracted by 'haustoria' which penetrate into plant epidermal cells.

Oak mildew lives mainly on foliage but may also develop on young shoots. Heavy mildew attack usually prevents foliage and shoots from reaching normal size and may cause some distortion. Height growth of heavily affected seedlings and transplants in nurseries may be less than expected at the end of the season but plants are not killed.

Oak mildew overwinters in Britain mainly as resting mycelium within bud scales and leaf primordia. First attacks on young leaves and shoots appear in May as cinnamon coloured spots. These spread and develop into white mycelium on which grow oidia and later, ascospores which are the source of secondary infections in late June. In the nursery, developing foliage, whether on one-year old seedlings or older plants, are at risk. Mildew can also particularly affect older trees recovering from insect defoliation or late spring unseasonal frost damage (Peace 1962, Phillips & Burdekin 1982).

Colloidal or wettable sulphur has been widely recommended as a control treatment, applied when symptoms first emerge and at 2-3 week intervals until the disease appears to be under control. Other materials recommended include benomyl and dinocap (Anon. 1956b, Peace 1962, Aldhous 1972, Phillips & Burdekin, 1982, Brooks *et al.* 1989).

Other mildew species affect many other plants, species which produce fresh young growth during the summer being most prone to attack. The same treatments as for oak mildew have been recommended where treatment can be justified for commercial or aesthetic reasons.

7.66   Keithia disease of western red cedar   *Didymascella thujina*

Western red cedar is a species that was widely used as a hedging plant in the 1940s and 50s but which has been supplanted in that role by Leyland cypress (*x Cupressocyparis leylandii*). It grows well in Britain as a forest tree but has not been grown on sufficient scale to have established a market as a timber producing species (Aldhous & Low 1974).

Didymascella thujina occurs naturally in North America on western red cedar. It was first noted in Ireland in 1918, in Britain in Sussex in 1919, and has since become widely distributed in most north European countries (Anon. 1967b).

The disease is most serious on nursery plants between three and five years old. On these, if allowed to build up, shoots can become so heavily infested that they become unlikely to survive planting into woodland.

Several cultural practices have been proposed for minimising the build up of infection in nurseries:

 • production should be planned so that western red cedar is produced on sites where there are no older trees or hedges of the species within a mile of the nursery (Peace 1958b);

 • plant production should be rotated between groups of three or four nurseries, or three or four widely separated nursery sections, so that in any one year, one nursery would have a complete break from raising the species (Pawsey, 1963);

 • plant production should be planned so that stocks are moved every year and ground cultivated as soon as possible after lifting, so as to minimise the risk of spore release from overwintering plant foliage (Aldhous 1972).

In trials, cycloheximide was effective against *Didymascella* (Pawsey 1964c). However, this fungicide is highly toxic; in the event, supplies have never been available commercially in Britain (Burdekin & Phillips 1970).

Prochloraz and benomyl have been recommended (Strouts *et al.* 1994) but with the *caveat* that the recommendation is based on work in France and is untested in Britain.

7.67   Needle cast of pine *Lophodermium seditiosum*

This widely occurring fungus was formerly misattributed to *L. pinastri* (Phillips & Burdekin, 1982) and is described under the latter name in earlier Forestry Commission and other publications. It is a potentially serious pathogen of Scots pine both in the forest and in the nursery.

Many serious outbreaks of needle cast have occurred where nearby Scots pine plantations, though only lightly infected, have acted as sources of infection for pines in adjoining nursery beds. Infection can also arise from diseased needles on lop and top as well as on growing trees (Pawsey 1964a).

Where it is not possible to raise plants where there are no nearby pines, several fungicides have been found to be effective. Bordeaux mixture was first to be recommended but has been supplanted by zineb, maneb and benomyl (Peace 1962, Pawsey 1964a, Aldhous 1972, Brooks *et al.* 1989, Strouts *et al.* 1994).

7.68   Birch rust   *Melampsora betulinum*

Brooks *et al.* (1989) state that young nursery stock are susceptible to attack on foliage by *Melampsora betulinum.* Where this is expected, repeated spraying from bud burst is recommended with materials including: zineb, mancozeb penconazole and Bordeaux mixture.

7.69   Leaf spot on cherry   *Blumeriella jaapii*

A previously insignificant leaf spot of *Prunus avium* became widespread in Britain in 1997. Foliage is infected in May and spreads during early summer. Affected leaves commonly turn red and fall prematurely. The disease could check growth of nursery stock and would require spraying; in France, mancozeb is used (Rose & Gregory 1997).

7.7   VIRUS DISEASES

7.71   Symptoms of virus infection

Viruses can cause blotching, patterning or yellow veining of leaves, distortions, cankers, pitting and stunting. However, while symptoms of viruses have been recorded on many genera, viruses are of no practical significance for the main groups of woodland trees raised from seed. The same applies to two classes of organisms which have in the past been confused with them, 'mycoplasma-like organisms' (causing some 'witches brooms') and 'rickettsia-like organisms'.

Virus particles are composed of two parts, a nucleic acid core and a protein-aceous coat. Although some plant viruses have complex shapes, most are either tubular or spherical. They are dispersed by:

- aphids, leaf hoppers, nematodes, mites, fungi *etc.*;
- parasitic plants;   • seed and pollen;

- contact, including natural root grafting;
- vegetative propagation including grafting and tissue culture;
- commercial trade in plants and seed.

Vector-borne viruses may be differentiated according to whether they persist in their host vector (persistent viruses) or disappear after an infective period (non-persistent viruses).

Means of control depend initially on securing virus-free material. Many virus-infected plants can be cleansed of virus if propagatable tips of plants are grown rapidly in conditions where the air temperature has been increased to 36-42°C.

Thereafter, the aim is to prevent reinfection by:

- strict plant health inspections beforehand and close inspection on arrival of any highly susceptible varieties, if purchased from another nursery;
- if soil-borne vectors are suspected, use of soil sterilised by methyl bromide or material with similar effect;
- if insects or mite vectors are suspected, control as appropriate, combined with removal of infection sources if known;
- if material is being propagated, meticulous attention to hygiene through rigorous exclusion of possible infected stock, use of sterile tools *eg* grafting knives, sterile media *etc*;
- in all circumstances, frequent inspection and rigorous 'roguing' *ie* removal and burning of any suspect plant, in otherwise virus-free stocks.

A full description of viruses is given in *Virus diseases of trees and shrubs. 2nd edn.* (Cooper 1993).

7.72   Viruses in trees

Forest trees are not considered to be prone to virus-caused diseases. However, many ornamental trees and shrubs may become infected with virus, and in a few instances may be propagated because of the virus patterning on their leaves.

Symptoms of virus infection have been noted in the following genera in Britain.

Acer	*Aesculus*	*Betula*	*Chamaecyparis*	
Cupressus	*Crataegus*	*Fraxinus*	*Ilex*	*Juglans*
Laburnum	*Malus*	*Picea*	*Populus*	*Pinus*
Prunus	*Sorbus*	*Ulmus.*		

For ornamental trees, especially species or varieties of *Prunus* and *Malus*, viruses can cause serious weakening and loss. Many genera of ornamental shrub are also affected (Cooper 1978, 1993).

Conifers seem largely to be unaffected by viruses.

In numerous instances, reported virus infection is associated with vegetatively propagated stock of broadleaved species, usually clonal ornamental varieties rather than populations of trees grown from seed. In a few cases, a virus appears to have been transferred from a rootstock of one genus to a scion of another genus.

Precautionary measures lie with nursery stock producers to ensure meticulous hygiene when propagating, and to take advantage of heat treatment to minimise virus infection where this is appropriate.

Cooper (1993) points out that with increasing use of vegetatively propagated stock through micropropagation, the opportunities for virus infection to become established increase. It is all the more important to be aware of the effects of viruses in debilitating plants and the means of avoiding such threats.

CHAPTER 8

Weeds

'And some seed fell on stony ground and the tares sprang up and choked them.'

8.1   WEEDS NOT WANTED

In this chapter, discussion is focused mainly on control of weeds to benefit crop trees. However, unless stated otherwise, comments apply equally to desired plants of all sorts, whether grown for consumption, physical utilisation or ornament.

The dictum, *a weed is a plant in the wrong place*, applies particularly to tree species that may be desired as woodland components in some situations and cut out as weeds in others.

8.11   Competition from weeds

Plants of all sorts benefit from being grown under weed-free conditions. From the beginnings of farming and cultivation of land, weeds have been recognised as undesirable competitors for light, soil moisture and nutrients. Control of weeds by cultivation or by hand cutting or pulling and occasionally by grazing, has been the norm. In woodland, however, *weeding*, *ie* cutting or treating competing ground cover following planting or natural regeneration, is normally differentiated from *cleaning*, *ie* later removal before canopy closure of unwanted species such as birch and willow that would compete for canopy space. Removal of trees from the established canopy is best considered as a *thinning* operation rather than weeding.

Early weed control

Freedom from competing weeds can be achieved for varying periods in most circumstances. In amenity and plantation forestry, nursery production and horticulture, freedom from weed competition in the early stages of growth is recognised as the best means of getting plants established quickly.

In many woodland planting schemes, treatment to control weeds is restricted to a circle 0.5m radius round each plant or 1 metre-wide strips centred on each row of plants. It may also be limited in time to the period starting at the time of preparation for planting or natural regeneration and ending when the desired young trees can overcome any competing weed growth without further intervention.

Weed 'control' may also include treatment of grasses and herbs so that their growth is checked, the plants remaining to provide low ground cover (§8.6, growth regulators). Repeated mowing or hand-cutting of weed stems and foliage may, however, place greater moisture stress on newly planted stock and is no substitute for local weed-free conditions.

More complete weed control, combined with use of fertilizers, may be justified for amenity plantings where quick, reliable plant establishment and growth are needed (Boylan 1988, Salter & Darke 1988, Putwain *et al.* 1988).

Figure 8.1 shows how control of herbaceous weeds boosted early dry matter production of oak and sycamore on roadside plots near Cambridge (Davies 1987e).

On another site, survival of cherry on a free-draining sandy soil was best on weed-free ground; where the vegetation had been selectively managed to create a predominately grass sward, survival was no better than on untreated ground.

Figure 8.1   *The effect of weeding on dry matter production three years after planting. Oak & sycamore on grassy roadside site in Cambridgeshire*

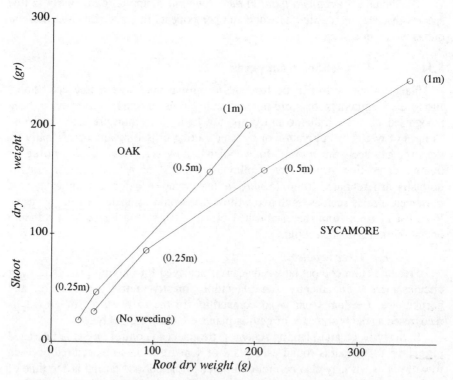

0.25m, 0.5m, 1m indicate the diameter of the patch kept fairly weed-free using paraquat and glyphosate between 1981 and 1984

Source   Davies 1987(e)

These growth and survival responses are primarily due to reduction in moisture stress and to a lesser extent, nutrient availability. They are obtained on lowland soils subject to a summer moisture deficit, wherever there is severe competition from herbaceous weeds and particularly with broadleaved species (Davies 1985, 1987b, 1987c, Davies & Colderick 1986).

On upland sites where moisture stress is not so acute, species do not always respond so strongly to freedom from weed competition. On some sites where plants responded to weeding, cutting overtopping weeds was as beneficial as killing them, suggesting that competition for light could be significant (Tabbush 1984b, Davies & Tabbush 1987, Nelson 1989). While need for light is implicit for all green plants, the less light-demanding species can survive and grow under partial shade without harm, albeit more slowly. At the extreme, however, plants can be killed by being smothered under fallen foliage of tall herbaceous weeds, especially bracken.

Once closed canopy woodland has formed, the competitive position of trees and ground flora is reversed, the canopy species dominate and it is the ground flora which is weakened by the severity of competition for moisture and nutrients. If the canopy is sufficiently dense, it may completely suppress the pre-planting ground vegetation, the ground remaining bare until breaks in the canopy are created by thinning or appear naturally as a result of within-canopy competition and death of individual trees.

Control of grass and herbaceous weeds at the beginning of a crop rotation is almost entirely restricted to the weeding period. Control of woody plants, however, may take place over a much longer period, including pre-planting site clearance, crop establishment, and cleaning and thinning. See also *Chemical thinning and respacing* (§8.46).

The one understorey woody species that is considered for treatment on any scale after canopy closure within a woodland is *Rhododendron ponticum* (§8.45).

At the end of a rotation if herbaceous weeds have invaded under the mature woodland canopy, they may also be treated shortly before clear felling if this seems likely to help establishment following replanting or natural regeneration.

Amenity plantings

In the more urban amenity planting sites, vandalism of young plants may be such that higher initial costs of weed-free sites and close spacing pay off in creating a dense thicket quickly. In these circumstances the aim has been to start, as far as possible, with a completely weed-free site and to maintain it in that state for the first five years (Salter & Darke 1988). If successful, this intensive treatment also quickly creates an attractive site.

8.12 Integrated crop establishment

Since 1945, in implementation of Government-promoted policy to increase the area of woodland in Great Britain alongside support for agriculture, much of the better quality agricultural land was reserved for food production. In the southern half of the UK, the land that became available for woodland planting was infertile acid heathland, sites with shallow soils over chalk, steeply sloping land, or felled broadleaved woodland, often on heavy clay soils. Former coppice areas, especially hazel, were also identified as 'derelict' and potentially well-suited to conversion to high forest or enrichment (Miller 1951).

In northern Britain, poor hill land, peats and heaths were the principal sites of forestry expansion.

Significant weed competition across such a range of sites ranged from heather and other moorland species on acid peats, to reservoirs of persistent dormant weed seeds on calcareous lowland sites and vigorous woody regrowth of coppice species.

In these circumstances, integrated crop establishment, while achievable, is complex because of the range of interactions of cultivation, drainage and weed growth with the local soils, topography and climate.

8.13   Methods of weed control

A review of weed control in 1967 noted that in the UK prior to 1955, weed control in the forest was a manual operation (Aldhous 1967a). Up to that time, it had been automatically assumed that any unwanted weed competition in the forest would be treated by hand. The tools commonly used were the sickle, short or long handled hook and axe (Wittering 1974). Hoeing was practised in small scale ornamental plantings and had been included in experimental treatments in the forest in the 1930s (Wood & Nimmo 1962, Evans 1984) but though successful, was too laborious on otherwise uncultivated soils.

With the advent of chemicals and sprayers able to kill weeds, machines able to carry flails, swipes or other cutting equipment into the wood, and other machines able to plough, cultivate and/or drain the soil, the integration of herbicides, cultivation and drainage became matters of everyday choice in creating or regenerating cost-effective and environmentally sustainable woodland.

8.14   Use of tractors and other equipment to clear vegetation

In the early 1950s, there was widespread interest in restoration of abandoned coppice, derelict woodland *etc.* to productive woodland. Attempts were made to harness heavy machinery developed for large scale civil engineering works to clear and remove unwanted woody growth. These were, however, not pursued beyond the trial stage (Miller 1955, Wood & Holmes 1958, Wood *et al.* 1960, Wood *et al.* 1967), partly on account of damage to soil and partly because of the inability at that time to deal with huge piles of stems, branchwood and roots.

The machines that subsequently came into use have been developed from smaller agricultural tractors strengthened and guarded to cope with woodland ground conditions and with a variety of attachments:

- front- or rear-mounted flails, swipes and rotary cutters,
- grass rollers.

Chipping machines were developed under the stimulus of whole-tree harvesting rather than weed control.

Small scale hand-operated powered weeding tools were also developed, such as:

- 'strimmers', hand-operated power-assisted nylon monofilament rotary flails, widely used for small to medium scale control of grasses and herbaceous weeds in amenity area;
- clearing saws, widely used on woody weeds, in cleaning operations and in tasks such as singling or respacing natural regeneration.

Spray and granule application equipment

Machine-mounted equipment

In parallel with mechanisation of cutting equipment, tractors and all-terrain vehicles have been developed to carry:

- spray booms, both conventional and controlled droplet;
- pressurised tanks containing diluted herbicide with hand-operated spray lances connected to the tank by a hose;
- contact applicators, or 'weed-wipes', where a deposit of herbicide is left where the applicator touches. Applicators may be in the form of rotating brush, a rope wick, sponge or similar porous material; they are fed with herbicide at low pressure so that spray drift is minimal. Weed wipes may be tractor mounted or fitted to hand-held lances.

Portable spraying equipment

Similarly, a wide range of portable equipment has been developed to apply pesticides in woodland conditions. These include:

- the pressurised knapsack sprayer; this has been and remains very widely used for herbicide application. Pressure in the tank is usually maintained by intermittent pumping during operation. Pressure at the nozzle is regulated by a valve usually located on the lance;
- knapsack-based small motors and fans driving air-assisted sprayers, mist-blowers and granule spreaders;
- hand-held rotary atomisers, or controlled droplet applicators driven by electric batteries located within the handle of the applicator (Rogers 1974, 1975, Brown & Thomson 1974). Models deliver droplets to 250 or 70 microns VMD specifications (Willoughby & Dewar 1995);
- the 'Forestry spot-gun', a hand-held short lance and manual trigger mechanism. A metering device attached to the lance enables a wide range of pre-determined quantities of herbicides to be applied to each treated patch;
- various designs of gravity-operated granule applicators. Some apply a predetermined volume of granules for each squeeze of the trigger. Others leave a continuous band of granules, the rate of flow being controlled by an interchangeable calibrated restriction tube;

- nozzles designed for tractor-mounted and knapsack systems ensuring desired spray patterns and droplet size spectrum for given operating pressures, and colour-coded to ISO 10625 (BCPC 1994a, b).

For comprehensive illustrations of equipment in use in the 1970s, see *Weeding in the forest* (Wittering 1974), *Chemical control of weeds in the forest* (Brown 1975), *Ultra-low volume spraying* (Rogers 1975) and *Control of heather by 2,4-D* (Mackenzie *et al.* 1976). Most of the equipment types shown are, with relatively minor modifications, still in use. Illustrations of spot gun, pepper pot and weed wiper are included in a journal paper *Weeding young trees - avoiding trouble* (Willoughby 1997a).

For a review of all-terrain vehicles (ATVs) as carriers for spraying equipment see *ATV sprayers on test* (Milward Forestry 1998).

8.15   Ground preparation and weeding

Ploughing bare land before planting
From the 1920s, the Forestry Commission, as part of its remit to extend the area of forest in Great Britain, set up research programmes to investigate how best to establish crops on the substantial areas of land considered marginal for planting at the time.

Techniques tested included ploughing using the best of the then available equipment, following local farming practices. However, comparisons were included to show the interaction of cultivation, added nutrients and species on sites which, by agricultural standards were at or beyond the limit for cultivation. Reports on tree planting on peat (Zehetmayr 1954), on upland heaths (Zehetmayr 1960) and chalk downland (Wood & Nimmo 1962) each include a section on early ploughing trials.

On peat, ploughing initially was seen as providing local drainage and a means of mechanising the production of turves onto which to plant. It followed the successful 'Belgian system' of turfing and draining on *Molinia* peat; additional weed control was not considered necessary. *Calluna* dominated peats were thought not to be capable of sustaining a tree crop.

On upland heaths, ploughing was seen as the essential means of cultivating soils that were indurated or had hard-pans or both; drainage remained a major pre-occupation. Weed control was viewed as a beneficial side-effect.

Cultivation to reduce weed competition was probably most important on the tight grassy swards of the chalk downland. Ploughing had to be sufficiently shallow not to bring up chalky subsoil; however, subsoiling was also desirable where there was any compacted surface layer of flint.

After 1945, at the same time that chemical and mechanical means of controlling weeds were developing, progressively more powerful tractors became available. These could draw ploughs through previously uncultivated land at increasing depths and with the option of deep tine attachments (Taylor 1970, Thompson 1978, 1984, Binns 1983).

The advantages of ploughing are:

- in turning over and burying surface vegetation, short-term weed competition is immediately removed;
- the plough furrow may act as a drain supplementing any more conventional system of drains that may have been installed. The drainage effect of spaced single-furrow ploughing underlaid its widespread adoption for forest planting on peaty upland soils;
- the plough ridge may offer immediate local shelter to trees planted off the ridge;
- some early release of plant nutrients can be expected from decay of buried vegetation and improved aeration of the plough ridge;
- on indurated soils, the addition of a tine to the plough may loosen the deeper soil sufficiently to assist drainage and improve soil aeration;
- the dark colour of peaty organic matter often exposed on the surface of plough ridges may absorb heat more readily in early spring weather and lead to earlier root growth in young plants.

Unfortunately, trees planted on wind-susceptible sites which were deep single furrow ploughed have blown down once they have reached a critical height. Sitka spruce on such sites can develop a root plate of adventitious roots spreading from the main stem shallowly through surface soil. These can, if not impeded, interlock with roots from adjacent trees to form a platform on which trees stand longer than where the development of such a platform is prevented by deep plough furrows (Coutts 1983).

Deep single furrow ploughing is not now widely practised. Shallow ploughing to obtain vegetation suppression combined with deep tining to break pans or induration is now preferred, any necessary drainage system being installed separately.

Screefing

The removal of surface vegetation at or just below ground level and loosening the surface soil is another long-standing method of site preparation to facilitate planting where weed removal is required, without the need to drain.

Planting into 'manually screefed patches' was very widespread. However, increasingly, as tractor-mounted equipment has been able to work over steeper and rougher land, machines have supplanted manual screefing. Tractors could pull ploughs turning over a thin turf ('continuous screefs') with or without tines (Thompson 1984, Tabbush 1988).

Site preparation on land previously carrying trees

On most older broadleaved woodland sites, the traditional practice was to heap and burn all the unsaleable lop and top from felling operations. This is not now so widespread due to:

- the availability of heavy tractors which are not unduly impeded by lop and top,

- a recognition of the desirability for deadwood to remain on site for the benefit of 'decomposer cycle' insects and fungi and to maintain site nutrient capital (§10.91).

On lowland private estates where shooting is important, felling debris may be cleared to a higher standard than strictly required for forestry operations, in order to allow access for dogs and beaters.

On many upland conifer sites, the volumes of branchwood and tops after clear-felling are substantial and seriously impede access to the soil for replanting. Effective burning of lop and top cannot be relied on in the wet climate of the uplands of Britain. Mechanical chopping using tractor-mounted flails mounted on a horizontal shaft was found to be effective (Neustein 1967). However, this work was overtaken by development of techniques using fully mechanised harvesters cutting shortwood for extraction by frame-steered forwarders.

On many of the wet peaty soils of upland Britain, forwarders run on 'mats' made using the brash from felled trees piled in rows, alternating with rows of pulpwood, sawlogs *etc.* from the site. Without such mats, they easily break through the layer of litter and surface soil layers and get bogged in the softer layers below. While the operational aim is to keep mat width to the essential minimum, after extraction, brash strips may cover up to 50% of the soil surface and obstruct replanting.

Site preparation equipment to facilitate replanting includes:

- disc-trenchers and mounders;
- boom-mounted rakes on tracked excavator bases or front-mounted raking blades;
- patch scarifiers.

Mounding

Mounds provide short-term local weed control, a drained planting site and warmer soil without adding to the windthrow risk (Tabbush & Ray 1988, Ray & Anderson 1990).

For new planting, mounds are made using back-acting diggers or specialised equipment such as the Maclarty mounder.

On areas being restocked, the difficulty of creating extensive new furrows and drains on stump and brash-covered land led to the development of:

- trailed or tractor-mounted heavy mounder-scarifiers capable of operating on brash-covered sites,
- tracked excavators fitted with buckets able to create a profiled drainage channel and at the same time distribute spoil from the drains over the site in mounds on which trees can be planted.

Illustrations of a disc trencher and several designs of patch scarifier/mounder are given in *Silvicultural principles for upland restocking* (Tabbush 1988).

8.16   Other non-chemical techniques for controlling weeds

Mulching is a long-standing technique for assisting early tree establishment by smothering weeds. Mulches were widely used in the 1950s and '60s for the establishment of wide-spaced poplars. Most sites where poplars have been planted have been moist and fertile, and weed growth has been luxuriant. Using cut weeds as a mulch around newly planted stock reduces local weed competition and supplies nutrients through decay of the mulch but is expensive (Peace 1952a, Jobling 1960, 1990). Dark coloured mulches increase the soil temperature around the planted tree.

Since then, trials have been undertaken of 'mulches' of plastic sheeting, straw, shredded bark and semi-rigid mats made of felt or other material. Sheet mulches have been effective in controlling weeds (Davies 1987c, d) but are not easy to maintain. Plastic sheets may be degraded by sunlight and are often torn. Substantial black polythene sheeting has given good weed control and has been more effective than thinner or lighter colour sheeting. Felt mats have found a place in small scale amenity plantings.

The area mulched per plant should not be less than 1 m^2 (Davies 1988a, b, Potter 1988).

Mulches are not recommended for sites where the soil is poorly draining as they increase the risk of anaerobic conditions developing around the surface roots and root collar (Davies 1987a).

In landscaping schemes in and around towns, bark mulches are widely used, being effective in controlling weeds, allowing rain to pass through into the soil, but not aggravating any pre-existing poor soil aeration. Finely shredded bark may be blown about in strong winds, However, mulches of coarser bark fragments are more stable, are seen as 'environmentally friendly' and a renewable resource, and are favoured in areas of high public access (Bradshaw *et al.* 1995).

Composted wood chips have also been recommended as mulches (Webber & Gee 1994, 1996), composting being recommended so as to reduce the risk of nitrogen deficiency following the decay of uncomposted wood chips.

See also §8.55 - mulches for container produced amenity stock.

Tree shelters, while in no sense a means of weed control, impinge on weed control practice:

- several important broadleaved species grow more rapidly in shelters and therefore require less weeding,
- shelters enable trees to be located quickly when weeding,
- shelters protect the lower part of the tree if a non-selective weed killer is being used.

Weeds should have been controlled before trees and their shelters are put in place, otherwise the weeds may benefit from the shelter as much as the trees (Tuley 1984, Potter 1991).

8.17   Conservation and weeds

Different forest management regimes over the last 40 years have had divergent views as to what are weeds. Species to be killed under one set of management objectives have been species to nurture under another.

Since 1970, increasing recognition has been given in the UK to the need to maintain and augment the range of wildlife habitats. At the extreme in some woodland conservation areas, it may be considered undesirable to interfere with existing natural undisturbed soil profiles so that cultivation is not an option for weed control. At the other extreme, herbicides are seen as having an important role in assisting the development of wildlife habitat (Marchington 1992).

Seedlings subsequently coming up in 'spot'-treated patches may differ markedly from the pre-existing vegetation, a diversification of species and a potential conservation bonus (Harris & Harris 1991).

The total suppression of former ground flora by developing even-age woodland is also sometimes criticised even though it is a widespread natural process. It has the advantage of removing the weed flora of the previous agricultural land use and may be necessary to allow recolonisation by a woodland flora able to survive under closed canopy conditions. In established deciduous woods, the ground flora is often 'pre-vernal', *ie* coming into leaf and flowering before the canopy trees break bud.

However, re-establishing the herb and shrub flora natural to woodland may be difficult where land has for a long time been used for arable or improved grass and there are no local seed sources of woodland plants. Often, very few herb species from earlier woodland cover remain. Also, not only may there be a substantial seed bank of aggressive agricultural weeds, the soil nutrient status may have been raised substantially and acidity neutralised by repeated applications of manures, fertilizers and lime ($CaCO_3$). Until their effect has been dissipated and seed sources of natural woodland species established, the signs of former arable husbandry will persist.

On unimproved upland grazings, similarly, centuries of grazing and burning has minimised the chances of survival of semi-natural woodlands and accelerated acidification, impoverishment of the soil nutrient reserve and formation of ericaceous or peatland plant communities. There is, however, no unanimity as to the appropriate wildlife conservation action when planting such areas.

Having taken into account all non-chemical options, if the least toxic and persistent herbicide is applied to the smallest area and at the lowest rate needed to achieve the required vegetation control, ill effects on wild life will be minimised.

8.18   Biological control of weeds

'Classical' biological control of weeds in agriculture is 'narrow spectrum' (see §8.43). Internationally, it has been mostly directed against specific introduced weeds, which, without the natural controls of their native habitat, spread rapidly. By 1982, 86 naturalised weed species had been the target of 200 organisms in worldwide control programmes. Control of *Opuntia* cactus (prickly pear) in Australia by the genus-specific moth *Cactoblastis cactorum* has been both successful and widely publicised (Wapshere *et al.* 1989).

In contrast, only 25 native weed species had been subject to biological control (Burge *et al.* 1988; Mortimer 1990).

To date in the UK, there have been no successful examples of the use of narrow spectrum biological agents to control weeds. However, bracken, *Rhododendron ponticum* and Japanese knotweed (*Polygonum cuspidatum*) have been identified as species for which effective biological control would be widely welcomed (Willis 1990). See also §8.43 (bracken).

The suppression of ground flora when a tree canopy closes can be considered a form of broad spectrum biological control. However, this is usually too late to be of value to forest management, except possibly through reduction of fire risk.

Weed control by fungus

In Holland, where the use of pesticides in forests is restricted, *Prunus serotina* is regarded as a serious weed problem on restocking sites. Spraying or brushing with 'silver leaf' fungus, *Chondrostereum purpureum*, has given good control (Bosman & Hoekstra 1990).

Animal grazing

Timely controlled grazing can occasionally be effective on herbaceous vegetation, if the aim is to reduce vegetation height and seed setting rather than outright kill. Any trees present must not be at serious risk from browsing or from bark stripping. Sheep grazing on land producing basket willows is an example where early grass competition is minimised and, by browsing willow shoots, risk of infection by rusts is reduced (§7.57). Controlled grazing of fire traces can reduce local fire risk if the crop is safe from browsing damage.

In areas where moderate scale deer populations have access to recently planted ground, it has repeatedly been observed that the deer will selectively graze naturally regenerating broadleaved species in preference to planted conifer. Fencing out deer has resulted in appreciable increases in the stocking of broadleaves within enclosures (See plate 12.4 in Pepper 1997).

Grazing by sheep has been identified as an option for the control of sea buckthorn on sand dunes (Baker 1996).

8.19   Striking a balance

Control of weeds around newly planted trees is only one component contributing to tree establishment. Weed control has to be integrated with the whole range of silvicultural and other requirements during this period. Woodland managers therefore have to decide, case by case, how best to achieve a cost-effective system of establishment that ensures a thriving developing woodland without losing desirable ecological and cultural features (Williamson & Ferris-Kaan 1990).

8.2   WEEDING PRACTICE

8.21   Scale of woodland weed control in Great Britain

The figures that follow come from Forestry Commission sources; the scale and type of use in privately owned woodlands has to be inferred by extrapolation from FC practice and the scale and location of planting.

Table 8.1 shows figures for areas weeded in Forestry Commission woodlands in 1966, 1982 and 1988, and the type of weed treated. The areas shown may have been weeded at any time from a few months to several years after planting.

The implication of the figures is that a substantial proportion of upland ground did not require weeding to establish trees and that weed control on such sites was achieved by cultivation treatments.

The Forestry Commission weeding programmes changed between 1966 and 1988. The increase in grass and herb spraying reflects the effectiveness first of paraquat and later of glyphosate. The decrease in treatment of woody weeds marks the change away from scrub conversion programmes of the 1960s (Stoakley 1962, Aldhous 1997).

Table 8.2 amplifies figures for 1988/9, showing methods of treatment. Herbicides were used on 67% of the area, weeds were cut by hand on 17% and a similar area was treated mechanically without herbicides. Tractor-mounted equipment, whether cutting or spraying, was used on 28% of the area.

Table 8.3 shows the areas of individual weed types treated with herbicides in 1988/9, both in new planting of bare land and restocking in recently felled areas.

Table 8.4 shows the scale of new planting and restocking for selected years in the period 1950-95 for Forestry Commission and private woodlands. The figures in *Tables 8.3 and 8.4* for areas planted and weeded show that much of the Forestry Commission area of new planting did not require weeding in 1988/9.

Table 8.1   *Areas treated for control of weeds; Forestry Commission wood-lands, 1966, 1982 & 1988 (Herbicide &hand/machine treatment)*

Year	Grass & b'lved herb	Bracken	Heather	Woody Foliar sprays	Stem & cut stump	Rhodo den- dron	weeds Gorse & broom	Total (000 ha)
1966/7 a	2.4	na	0.2	0.3	3.1	0.4	na	6.5
1982/3 a	11.9	4.8	1.2	6.1	0.5	0.4	0.5	25.4
1988/9 f	*9.2*	*3.4*	*2.0*	*4.3*	*0.5*	*0.1*	*0.1*	*19.9*
a	11.4	1.5	0.3	1.1	0.4	0.3	0.1	15.2

Sources   Aldhous 1968a   McCavish 1990
          a = actual: *f* = forecast   na = data not available

Table 8.2   *Forestry Commission woodland area weeded - 1988-9*

Method of treatment	Area (000 hectares)	% of total
Herbicide (hand spray/spread)	7.7	51
Herbicide (tractor mounted)	2.4	16
Hand cutting	2.6	17
Hand-held clearing saw	0.6	4
Tractor-mounted mechanical cutting	1.8	12
Total area treated	15.1	

Source   Booth 1990

Table 8.3   *Areas treated with herbicide - Forestry Commission 1988-9*   *(ha.)*

Weed type	New planting	Restocking	Total	% of total
Grass & herbs	1 214	6 574	7 788	76
Bracken	188	701	889	9
Woody weeds	14	486	500	5
Bramble	-	504	504	5
Heather	294	27	321	3
Rhododendron	124	122	246	2
Gorse and broom	12	25	37	<1
Total herbicide area	1 846	8 439	10 285	
% of area weeded (ex Table 8.2)				67

Source   Williamson 1990

Table 8.4   *Scale of FC & private planting - 1950-1995*   *(000 ha)*

| Year | Forestry Commission | | | Private | | | Combined |
	New plant	Restock	Total	New plant	Restock	Total	Area
1950	15.1	6.6	*21.7*			*6.1*	27.8
1960	16.6	8.4	*25.0*			*14.9*	39.9
1970	18.5	4.0	*22.5*			*19.3*	41.8
1980	15.8	5.7	*21.5*	8.3	2.9	*11.2*	32.7
1985	5.1	5.9	*11.0*	16.0	3.1	*19.1*	30.1
1989	4.1	8.5	*12.6*	25.4	4.9	*30.1*	42.7
1990	4.1	7.9	*12.0*	15.6	6.3	*21.9*	33.9
1995	0.9	7.9	*8.8*	18.5	6.2	*24.7*	33.5

Source   Forestry Commission Annual Reports

Table 8.5 shows planting and restocking by FC Conservancies for the 12 months to 31.3.89 and is typical for that period. It shows that FC new planting was virtually all in Scotland, half being in northern Scotland, where peat soils predominate.   On such sites, initial cultivation would be expected to give sufficient suppression of ground vegetation so that weeding would not be required.

For the areas restocked in contrast, the area to be weeded in one year is similar in order of magnitude to the area restocked. *Table 8.5* shows that these sites are distributed throughout Britain.   Nevertheless, the figures include 2nd and 3rd year weeding.   The implication is that, because of lack of weeds under the previous crop, site preparation post-clear felling has on many sites been good enough to avoid the need for weeding in the first year after planting.

Table 8.5   *Areas planted and restocked in 1988-9 by Forestry Commission Conservancies and private woodland owners*   *(000 ha)*

Country Conservancy	New planting		Restocking	
	For Comm.	Private	For Comm.	Private
Scotland				
North	1.96	na	1.28	na
Mid	1.03	na	1.26	na
South	0.92	na	1.31	na
Total	*3.91*	*22.43*	*3.85*	*2.24*
England				
North	0.11	na	1.78	na
East	0.01	na	0.64	na
West	0.01	na	0.48	na
Total	*0.13*	*1.75*	*2.90*	*2.33*
Wales *Total*	*0.06*	*0.92*	*1.74*	*0.26*
GB *Total*	*4.10*	*25.10*	*8.49*	*4.83*

Source   Forestry Commission Annual Report 1989.   na = data not available

Conflict of interest - nesting birds and early summer weeding
Both on private estates where game bird production is part of the estate's activities, and in areas of special conservation interest, agreement has to be reached over timing of weed control operations so as to minimise disturbance to nesting birds.

8.22   Costs of weed control in forests

Table 8.6 shows figures from the Forestry Commision's Annual Reports between 1959 and 1972, during which costs for weeding were stated for FC operations. It illustrates how significant weed control has been as a silvicultural cost. Figures for later years have not been published.

Table 8.6   *Establishment costs (prep ground, planting, beating up & weeding)*
             & weeding costs   *(All figures excluding overheads)*

Year	Area planted (000ha)	Total estab cost up to & incl. weeding (£000)	Weeding cost (£000)	Weeding as % total
Forestry Commission, 1959-1972				
1959	22.3	£3188	£1053	33%
1963	22.5	£3373	£1264	37%
1967	21.8	£3818	£1035	27%
1972	26.2	£3294	£794	24%
Private Woodlands, mean for 1989-1992				
Conifers (12.2)		£1680	£115	7%
Broadleaves		£2753/3764	£154	4-6%

Sources   FC Annual reports   Mitchell *et al.* 1994

The Forestry Commission figures for 1959 and 1963 very largely represent costs of hand or machine weeding. The reduction in weeding costs as a percentage of overall establishment costs for 1967 and 1972 is largely due to the impact of herbicides. The Private Woodland figures reflected partly well-established herbicide practice with cost-effective materials and partly the disappearance of the heavy scrub-covered woodland sites that the Forestry Commission was tackling in the late 1950s.

In 1964, Forestry Commission weeding cost per ha averaged £15 in Scotland, £65 in Wales and £100 in England; this reflected the disposition of weedy sites in the planting programmes of the time *ie* peats predominating in Scotland, while in England, programmes included sites where woody weeds were prevalent (Binns 1964).

Table 8.7 shows comparative costs in 1988 and 1997 for alternative weeding regimes in lowland Britain.

In a survey in 1993/4 of first year costs of establishment in farm woodland, the cost of herbicides ranged from £9-41 per ha., averaging about 2% of the total (Britt *et al.* 1996).

Costs and output guides

Because of the high cost of plantation formation, the Forestry Commission Work Study unit extended its range of investigations to include silvicultural operations. At that time, weeding costs largely arose from hand work.

Table 8.7   *Comparative weeding cost in lowland Britain*   *(£ per ha)*

Costs in 1988	Manual cutting	Manual chemical (atrazine)	Tractor cutting	Tractor chemical (atrazine)
Labour (incl oncosts)	325	35	14	6
Machinery	-	-	12	4
Materials	-	15	-	23
Total	*325*	*50*	*26*	*33*
No of treatments for establishment	5	3	5	3
Total weeding cost	*1625*	*50*	*130*	*99*

Costs in 1997	Mechanical cutting/ploughing	Plastic mulches	Alternative ground cover	Herbicides
	750 [1]	950 [2]	150-1000+ [3]	370-700 [4]

[1] difficult among planted trees; cutting by itself often detrimental.
[2] 1m diameter spot; effective, not always durable; organic mulches more costly.
[3] cost for between-rows nurse crop; in addition to, not instead of, other techniques.
[4] effective in most situations; figures for intensive regime on weedy site.

Sources   Williamson 1991a,   Willoughby 1997a

From the mid-1960s as 2,4,5-T for control of woody weeds came more widely into use, work-studies influenced, in particular, organisation of diluent supplies to reduce the time and cost of carrying knapsacks of diluted pesticide. Cost-savings were thereby achieved, in addition to those directly resulting from use of the herbicide (Dannatt & Wittering 1967).

By 1970, sets of standard times or output guides were available for the more important weeding operations and included comparison of costs of alternative materials (Wittering 1974, Brown 1975, Crowther 1976, Edwards *et al.* 1994). Appendix 1 of *Weeding in the Forest* (Wittering 1974) gives a good summary of the output guides available at that time, and their use. For current figures in more abbreviated form see Section 11 of *FC Field Book 8: The use of herbicides in the forest* (Willoughby & Dewar 1995).

Costs of chemicals and treatments as at November, 1995 are given in *Herbicide update* (Willoughby 1996b,c) and are updated in *Herbicide update* (Willoughby & Clay 1999).

8.3   USE OF HERBICIDES

The background to the development and use of herbicides is described in Chapter 2, particularly in §§ 2.2 and 2.6.

Recent development of use of herbicides for the forest has had to take place observing a wide range of constraints. Robust techniques have had to be found, enabling herbicides to be used safely and effectively under the range of conditions found in woodlands,

- minimising the cost of weeding operations in establishing young trees on newly planted land and land being restocked after clear felling,
- minimising the risk to operators applying herbicides and anyone visiting treated land,
- minimising risks to the environment - plant, animal, water, soil and air.

A review of weed control in forestry in 1967 commented that, in 1955, virtually all forest weeding was still by hand (Aldhous 1967a).

Trials of herbicides had begun in 1949 but recommendations for practical scale forest trials of 2,4-D, 2,4,5-T and ammonium sulphamate to control unwanted trees and woody growth were not made until 1957 (Holmes 1952, 1957a, b). However, from that time to the present, the Forestry Commission has maintained a programme of research and development into the application of herbicides, at the same time ensuring that up-to-date recommendations have been readily available.

Summaries of recommendations for forestry were included in editions of the *Weed control handbook* from the 1st edition in 1958 to the 8th edition (Anon 1958b, Hance & Holly 1990). The first Forestry Commission leaflet was *Chemical control of weeds in the forest*. This ran to three editions (Aldhous 1965a, 1969b, Brown 1975). *The use of herbicides in the forest* replaced Brown's 1975 edition, first as Booklet 51 (2 editions) running to 110 pages (Sale *et al.* 1983a, 1986). The third edition appeared as *Field book 8*, a 150 page booklet (Williamson & Lane 1989). This in turn has been replaced by a fourth edition comprising over 300 pages (Willoughby & Dewar 1995), the greater length resulting from the inclusion of copies of 'off-label approvals' for many of the products listed. The scope of recommendations was extended by the issue of *Herbicides for farm woodlands and short rotation coppice* (Willoughby & Clay 1996).

Progress reports on the supporting research are given in the Forestry Commission's *Annual Reports on Forest Research*, *Research Information Notes* and in numerous journal papers.

Herbicides used or tested in woodland or on amenity trees, 1950-1998
Appendix I Table 1c lists by active substance, references in literature (mainly UK forestry sources) to herbicides which have been used or tested experimentally in the UK in forest, farm woodland and amenity tree management or in forest

nurseries over the last 50 years.   All the references are included in the main bibliography under author and year.   Some 60 active substances are listed there. *Appendix II Table 1c* shows which of these were available in approved products in 1999.

8.31   Scale of use of individual herbicides

Figures giving quantities of individual herbicides used in forestry are not readily available.   *Table 8.8* brings together a few figures for Forestry Commission use; quantities used in private woodlands are likely to be similar but with a higher proportion of herbicides controlling grasses and herbs. These are modest amounts when compared to the estimate of herbicide expenditure in agriculture annually during the 1980s of more than £100 million (Mortimer 1990).

Comparison of quantities in *Table 8.8* shows that usage in the 1960s was dominated by ammonium sulphamate and 2,4,5-T.   These were used for killing woody weeds at a time when treatment of former coppice and scrub was at its peak.   By the 1980s, there was little call for coppice conversion; if anything, coppicing was subject of a small resurgence on wildlife conservation grounds.

Some 2,4,5-T and ammonium sulphamate was used for control of *Rhododendron* for which there are now several alternatives (§8.45).

Herbicide usage in the 1980s and 1990s is dominated by materials giving broad spectrum control of grasses and herbaceous weeds.

As an indication of the place of forestry in the scale of use in Britain, *Table 8.9* gives figures from the MAFF series, *Evaluation of fully approved or provisionally approved products* for the few products related to forestry for which such evaluations are available. In these examples, usage in forestry is small compared with other uses.

Market availability

Herbicides remain on the market as long as there is an approval in force and the manufacturer has taken a commercial judgement that continuing production is likely to be profitable.

None of the herbicides used in forestry has been withdrawn through loss of approval for use specifically in forestry.   Several products or active substances have, however, been withdrawn by manufacturers.   These include 2,4,5-T, atrazine/cyanazine mixture, atrazine/terbuthylazine mixture, atrazine/dalapon mixture, hexazinone, oryzalin and diphenamid (Palmer 1993).

8.32   Damage

Herbicides misapplied, whether through carelessness or accident, can damage the trees or shrubs which were meant to benefit from the control operation.

Peace (1962) described herbicide damage to trees adjoining areas treated with sodium chlorate, by root contact with poisoned trees, and to foliage and shoots by hormone weedkillers.   Phillips and Burdekin (1982) describe symptoms of

Table 8.8   *Quantities of herbicide active substances used per annum, 1967 1998 in Forestry Commission woodlands & nurseries*

Active substance	Unit	1964/6	1966/7	1986/7	1992/3	1996/8
Ammonium sulphamate	kg	25 000	17 000	1200	15	-
2,4-D	kg	*2000	*980	1440	750	779
Dalapon	kg	2000	590	*465	5	-
Paraquat	kg	21 000	1270	168	-	-
2,4,5-T	kg	*20 000	*12 930	-	-	-
Atrazine	kg	-	-	*2765	*1630	1832
Asulam	kg	-	-	2050	1745	2324
Fosamine ammonium	kg	-	-	66	19	-
Glyphosate	kg	-	-	4400	4475	831
Hexazinone	kg	-	-	17	130	-
Imazapyr	kg	-	-	-	56	-
Napromide	kg	-	-	25	-	-
Propyzamide	kg	-	-	630	510	468
Terbuthylazine	kg	-	-	-	*196	-
Triclopyr	kg	-	-	89	195	322

(Predominantly forest nursery uses)

Mineral oils						
Vapourising oil	litres	-	69 100	-	-	-
White spirit	litres	-	16 800	230	-	-
Simazine	kg	-	540	162	no data	no data
Diphenamid	kg	-	-	145	no data	no data

*   Of these quantities, some or all was in mixture with other active substances.

Sources   Aldhous 1968a   Brown 1969   FC unpublished data   Willoughby 1995

Table 8.9   *Quantities of pesticide active substances used per annum in the 1980s in agriculture and on non-agricultural land   (kg)*

Active substance	2,4-D	Atrazine	Simazine
England & Wales			
Agricultural land	47 742	47 000	31 510
Non-agricultural uses	-	* 135 100	78 500
Scotland			
Grassland (none on cereals)	4 000	no data	no data

* Prior to revocation of approvals for non-crop uses

Source   (MAFF Eval. 1993a,b,c)

damage from 10 herbicides, including simazine damage in the nursery, and chlorthiamid, glyphosate and 2,4,5-T in the forest. Strouts & Winter (1994) describe six types of damage caused by chemicals and name 9 potential causes; they also include road salt as a cause of serious harm to plants (§15.23).

Damage may also occur to plants outside the area treated through herbicide drift, mis-direction or volatilisation. See also §4.32.

Manufacturers, in the early years of marketing growth regulating herbicides such as 2,4-D, found that short-chain ester formulations could volatilise and be carried as vapour onto susceptible agricultural and horticultural crops. Longer chain non-volatile esters were consequently substituted.

Suspicion that volatilisation of 2,4,5-T had occurred in warm weather in 1967 led to experiments testing an iso-octyl ester in comparison with the previous standard N-iso-butyl ester. Volatilisation damage recorded was considerably less on the low-volatile iso-octyl ester plots (Brown & Mackenzie 1969).

Mechanical damage

While misapplied herbicides can cause serious damage inside and outside woodlands, far fewer amenity trees and shrubs suffer from herbicide damage than are damaged by mechanical injury. 'Sheffield blight' was the colloquial name for trees accidentally damaged when weeding by hook; trees usually recovered by growth from a bud below the injury. However, hand-cutting has mostly been replaced by machinery, either hand-operated or tractor-mounted. Strimmers can also damage plants by removing or severely injuring the bark if the flailing nylon monofilament comes into contract with the stem. Damage is usually at just above ground level and may not be obvious until several months later when the foliage begins to look unhealthy. If the area of bark is too wide to heal over naturally, the plants die. Damage following impact by mowing or other ground maintenance machinery is also commonly observed (Patch & Denyer 1992).

8.4   SPECIFIC VEGETATION TYPES

Weeds may form a ground cover dominated by one species *eg,* bracken or *Calluna,* or grow as intimate mixtures of species. The section following discusses them under:

mixtures of species:

- pre-and post-planting control of grasses and herbaceous weeds (§8.41),
- woody plants (§8.44),
- farm forestry (§8.47),

and species-specific treatments for:

- *Calluna* and heathers (§8.42),   • bracken (§8.43),
- *Rhododendron* (§8.45).

Chemical thinning of crop trees is discussed in §8.46.

8.41   Pre- and post-planting control of grasses and herbaceous weeds

Herbicide trials and development of recommendations

Trials of herbicides to control grasses in woodlands started in the late 1950s and included dalapon, amitrole and simazine for pre- and post-planting control of grasses. These were shortly followed by trials including paraquat and diquat (Wood & Holmes 1958, 1959, Wood *et al.* 1961, Aldhous 1965c). Trial work in Scotland also included grass control prior to clear felling, to facilitate establishment of the following crop (Neustein 1966).

The first edition of *Chemical control of weeds in the forest* included recommendations for the use of herbicides for grass control in young plantations. These were based on dalapon and paraquat but had to impose seasonal limitations, emphasising that for dalapon, grasses should be growing fairly vigorously while paraquat sprays had to be applied to green tissues, at the same time avoiding crop plants. Figures for usage in 1966 in *Table 8.8* show that recommendations for paraquat were taken up more than for dalapon. Provisional recommendations for chlorthiamid were added for the 2nd edition, based on previous trials (Aldhous 1964c, 1967b, 1969b).

Other broader spectrum materials subsequently under trial, *eg* atrazine (Aldhous *et al.* 1968a, Brown 1970, Brown & Mackenzie 1972), led to much fuller recommendations for grass control *etc.* in the 3rd edition of *Chemical control of weeds in the forest* (Brown 1975). Table 2 and Figure 1 of that booklet gave a good overview of the range of important grass weeds and the different degrees of susceptibility of individual species in relation to atrazine, chlorthiamid, dalapon and paraquat. The booklet also gave notes *'for information'* on glyphosate and propyzamide.

Tests on new materials continued *eg* dichlobenil/dalapon and atrazine/cyanazine mixtures, dichlobenil, (McCavish & Smith 1976), hexazinone and triclopyr (McCavish 1978), along with more extensive trials of glyphosate and propyzamide (Biggin & McCavish 1980). Recommendations based on results to date were summarised by Tabbush (1982) and included in the first edition of *The use of herbicides in the forest* (Sale *et al.* 1983a).

In practice, between 1966 and 1986, usage of dalapon and paraquat dwindled (*Table 8.8*); by the end of the period, glyphosate, atrazine and propyzamide by weight collectively accounted for about 60% of total forest herbicide usage. The same relationship continued to 1992.

In the next period, new materials reported on included terbuthylazine /atrazine mixtures and granules (Williamson & Tabbush 1988, Nelson & Williamson 1989), and imazapyr (Winfield & Bannister 1988).

The 1989 edition of *The use of herbicides in the forest* (Williamson & Lane 1989) listed for control of grasses and herbaceous weeds:

- for spring application, atrazine by itself or in mixture with dalapon, cyanazine or terbuthylazine,
- for summer and autumn treatment, glyphosate,

- for winter treatment, propyzamide.

Further testing continued, both of newer materials, *eg* terbuthylazine (Tracy & Nelson 1991) and methods, *eg* over-all spraying glyphosate onto dormant trees (Garnett & Williamson 1992).

At the same time, political concern was being expressed about non-crop uses of several triazine based herbicides. This led to withdrawals in 1992 of approvals for *supply* of simazine and atrazine for use on non-cropped land and in 1993 of *use* on non-cropped land (Carter & Heather 1996). Consequently, manufacturers withdrew from the market several products containing atrazine by itself or in mixture, their minor role in forestry not being sufficient to justify marketing specifically for this purpose.

The 1995 edition of *The use of herbicides in the forest* (Willoughby & Dewar 1995) added recommendations for:

- clopyralid for spring and summer control of broadleaved herbaceous weeds,
- glufosinate-ammonium for spring and summer control of grasses and broadleaved herbaceous weed mixtures,
- triclopyr or a mixture of 2,4-D, dicamba and triclopyr for summer control of broadleaved weeds,
- isoxaben for winter control of broadleaved weeds.

Tables 4 and 5 of that edition give the susceptibility of 18 grass and 90 broad-leaved herbaceous weed species to approved herbicides available at the time of publication.

Testing of new products or formulation continues to be reported *eg* comparisons of two adjuvants and three formulations of glyphosate (Willoughby 1997b), the effects of dormant season application of broad spectrum herbicides (Willoughby 1996e). These have enabled periodic 'updates' also to be published (Willoughby 1996c, Willoughby & Clay 1999). The latter includes extended recommendations for the control of thistles in new plantings of farm woodlands and comments on rainfastness of glyphosate products.

Recommendations for amenity trees are available in *eg Herbicides for sward control among amenity broadleaved trees* (Willoughby 1998).

Post-felling weed control

Actual and potential weed control programmes required when establishing a successor crop should be an integral part of the planning *preceding* the decision to fell, taking into account:

- the immediately preceding felling programme in respect of timing of the operation and any subsequent treatment of the lop and top,
- site preparation techniques prior to replanting,
- potential weed competition, including any effect of delay in replanting (Lund-Hoie 1988),

- for conifer crops, the scale of risk of damage from pine weevil (*Hylobius abietis*). See also §6.51.

Where woodland can be managed to full rotation age, there is a case for maintaining as heavy a canopy as possible until the time of felling, in order to maintain a relatively bare forest floor.

Natural regeneration of woodland is most likely to be successful, if felling follows shortly after a heavy seed fall and the ground is relatively weed-free.

8.42　　Heather

Heather (*Calluna vulgaris*) has been the most widespread single weed species encountered in upland afforestation and afforestation of lowland heaths; it often grows in association with *Erica cinerea* and *E. tetralix*. *Calluna* is exceptional in that it is able to cause certain species growing in its vicinity not to grow with the vigour possible in the absence of *Calluna*. Both Norway and Sitka spruce are subject to such heather 'check' or allelopathy (Weatherell 1953, Read 1984). Handley (1963) attributed the effects to the ability of heather to inhibit mycorrhizal formation in tree roots and thereby restrict nutrient uptake and in particular nitrogen, the 'check' being an 'induced' deficiency.

The need to control heather can often be avoided by:
- planting non-susceptible species,
- burning heather before ploughing,
- restocking felled areas immediately after felling, before heather has time to invade,
- planting spruces in mixture with pine or larch (Sale *et al.* 1983a, b).

If initial establishment fails to ensure satisfactory spruce growth and an area suffers heather check, this can often be rectified either by killing the heather or by adding fertilizer to correct nutrient deficiency or both. Everard (1974) in *Fertilizers in the establishment of conifers in Wales and southern England*, gives recommendations for application of phosphorus fertilizers to enable plants to increase their take up of nitrogen as an alternative to spraying. In 1974, the area of *Calluna* needing spray was estimated at 8000ha/yr (Mackenzie 1974b).

Taylor and Tabbush (1990) in *Nitrogen deficiency in Sitka spruce plantations* categorise sites where Sitka spruce is growing, according to whether the spruce is likely to respond to *Calluna* control, added N or both, recognising that on some of the coldest upland sites, slow rate of mineralisation may be as much a cause of nitrogen deficiency as heather check. See also §12.32 *Heather check*.

Recommendations for treatment

Where heather is considered to be the cause of actual or potential check, overall control of heather has to be the aim. Patch treatment suitable for grass control is not appropriate.

Heather was found to be susceptible to ester formulations of 2,4-D from the earliest trials in 1954 (Holmes 1956b, Everard 1974, Mackenzie *et al.* 1976). From that time, 2,4-D has been recommended for heather control in all editions of *Chemical control of weeds in the forest* and *The use of herbicides in the forest* up to and including the current (1995) edition.

Lack of suitable spraying weather may limit the opportunities for safe application. One study showed that weather was suitable for spraying operations only for 25% of the nominal season of susceptibility (Malcolm 1975).

Glyphosate came on the market in the 1970s and after tests has been recommended as an alternative since 1983 (McCavish 1979, Biggin & McIntosh 1981, Sale *et al.* 1983a). Cyprazine showed some initial promise (McCavish 1980) but only imazapyr has been added as a further recommendation (Willoughby & Dewar 1995).

Heather has been sprayed from the air, principally with 2,4-D (Aldhous 1969a). It is second to bracken as the forest weed species most widely sprayed from the air. Glyphosate has also been applied from the air on a trial basis (McIntosh 1980). However, there are no products currently approved for control of heather by aerial application.

Bees

Because heather dominates large tracts of moorland and is an attractive nectar source for bees, bee keepers put out hives during the flowering period. The recommended spray period for control includes when the heather is in flower. It has been long established that local bee keepers must be informed in good time when spraying is intended, to enable them to move hives and minimise the risk of bees visiting sprayed flowers.

Water

2,4-D can be detected by taste in water at low concentrations. In the 1970s, water authorities would not permit concentrations more than 0.01 parts per million. Immediately after application in the drains leading from a treated site, higher concentrations have been identified (Aldhous 1967c).

In response to concern expressed by certain water authorities in Scotland about the risk of taint through spraying where surface water is collected for public water supplies, trials were undertaken. These showed that the actual run-off was slight and that spraying should not give rise to any problem of taint (McIntosh 1980).

Nevertheless, restrictions were recommended to limit the proportion of a catchment sprayed at one time, so as to allow adequate dilution of any run-off should heavy rain fall shortly after treatment. Initially, as a rule of thumb, a limit was proposed of 1/1000 of the catchment sprayed at any one time (Brown 1975). More recent practice is to discuss and agree the acceptable scale of treatment with the local water authority before any large scale treatment in a surface-water catchment is undertaken.

See also Brown and Mackenzie (1969) for a report of a study of water quality following use of 2,4,5-T in diesel oil for scrub clearance.

8.43 Bracken

Bracken (*Pteridium aquilinum*) grows on every continent except Antarctica and is considered to be one the five most common plants on earth. It is widespread on drier soils in the UK. During the 20th century, it is estimated to have extended its area annually at an average rate of 1%, encroachments on remote, ungrazed commons and rough grazings rising to 3%. The area under bracken in Britain in the late 1970s was estimated as approaching 400 000 ha (Cooper & Johnson 1984). Taylor (1986) estimated bracken cover in England and Wales in the mid-1980s at 672 000 ha.

The whole plant contains toxic constituents, some of which remain if live fronds are cut and dried. In Britain, cattle, horses, sheep and pigs have suffered acute and chronic poisoning. However, dry brown bracken cut in winter is safe for use as animal bedding.

In parts of the world where bracken is part of human diet, it is associated with increased incidence of tumours. The part of the plant usually eaten is the uncurled frond which contains high concentrations of carcinogenic agents; indirect exposure through cows milk is also possible (Cooper & Johnson 1984).

Pattern of growth

The growth pattern of bracken is distinctive. It forms a massive repeatedly branching rhizome capable of sending up fronds from multiple growing points. Fronds commonly reach 1-1.5 m in height and may reach over 2 m. In densely stocked areas there may be 60 - 80 fronds per m^2 of ground surface. From mid-summer to the first frost of autumn, they cast heavy shade. In winter, the mass of accumulated dead fronds may smother smaller plants, especially if combined with thick snow. For these reasons, the '1 m^2 treated patch' found effective for control of grass weeds is inappropriate for bracken. Complete kill or very substantial weakening are the only reasonable targets for chemical control; manual or mechanical methods have to seek the severance or suppression of the main annual frond mass no later than the time it is first fully developed.

Biological control

While superficially an obvious target for biological control programmes, there is a dearth of candidate pests. Two foliage-feeding moths have been identified in South Africa and first steps taken to evaluate them as biological agents. Two fungi have been found attacking pinnae or causing leaf curl. In field trials even in combination, these have not been potent enough to overcome bracken's ability to produce a physical barrier to fungal attack, though they might have a place 'mopping up' weakened fronds emerging after treatment with herbicide *eg* asulam (Burge *et al.* 1988).

Cultural control

Traditionally, bracken has been killed by exhaustion. The massive underground rhizome system of any patch of bracken can replace individual fronds either within the season if damaged early in the year, for example by frost, or in the following year. It was only by single-mindedly organising mechanical damage to fronds two or three times a year for several years, that farmers in the past could reduce the scale of bracken infestation on their land. Mechanical damage could be by cutting, whipping, crushing/trampling by cattle, rolling or any other means having the effect of prevention of expansion of fronds. Such treatments had to be sustained so as to exhaust the rhizome system.

Herbicide treatment

First trials of dicamba for control of bracken (Aldhous 1964b, 1965b) showed it to be a potent translocated herbicide; it drastically reduced bracken growth in the season following application but also was toxic to trees rooting in the area. It is applied to the soil and is washed down to the rhizomes by rainfall. Provisional recommendations for its use before planting appeared in the 2nd edition of *Chemical control of weeds in the forest.* Chlorthiamid and picloram had been less promising (Aldhous 1966b, Aldhous & Atterson 1967).

Trials of asulam started in 1971 (Brown & Mackenzie 1972). It is applied to fully expanded foliage and is translocated to the rhizomes where it prevents development of the following year's fronds. It soon became apparent that this material was preferable to dicamba, being cheaper and usable before and after planting. Recommendations for its use appeared in 1975 (Brown 1975).

Subsequent tests included fosamine-ammonium (McCavish 1978) and glyphosate (McCavish 1980). The latter is also taken up by foliage and translocated to the rhizome; it causes eventual death of fronds and prevents regrowth from the rhizome (McCavish 1979, 1981). Recommendations in 1983, 1986 and 1989 for control of bracken included only asulam and glyphosate (Sale *et al.* 1983a, 1986c, Williamson & Lane 1989).

Further trials in 1985 tested dicamba, hexazinone and dichlobenil/dalapon applied shortly before frond emergence as a narrow band between planted rows (Sale *et al.* 1986a, Palmer 1988). Imazapyr is a more recent additional candidate for use to control bracken. It is applied when fronds are fully developed (Williamson *et al.* 1987, Winfield 1988). Of other recent chemicals tested, sulfonylurea compounds caused less damage to broadleaved trees than asulam (Lawrie & Clay 1994a).

The most recent recommendations for the control of bracken include asulam, dicamba, glyphosate and imazapyr.

'On sites with a mixture of bracken and other weeds, glyphosate or imazapyr may be the best choice due to their broad weed control spectrum. Dicamba will give good pre-plant control where bracken is the only weed present. Asulam may be the most appropriate where bracken is the primary weed problem.' (Willoughby & Dewar 1995).

Dicamba treatments can be applied in early spring when the previous year's bracken fronds are largely fallen and flattened, whereas asulam, imazapyr and glyphosate have all to be applied to foliage so that they can be taken up and translocated to rhizomes. Foliage treatment from the ground may require special access racks to be cut, especially if hand-held rotary atomisers are to be used.

The one commercial product containing asulam available in 1999 is approved for aerial application where this is appropriate (Whitehead 1999).

8.44 Woody plants

The circumstances requiring removal of competition from woody plants can be split into:

- conversion from coppice to high forest of non-coppice species;
- scrub clearance (usually on land recently neglected);
- control of gorse and broom and other woody shrubs and climbers;
- cleaning *ie* removal of species shortly before canopy closure;
- chemical thinning;
- respacing;
- post-planting competition.

In most of these situations, the threat to be averted is of the unwanted plants becoming established in the canopy of the developing wood to the detriment of the favoured plants.

The use of the words 'unwanted' and 'favoured' highlights the situation that species like birch and oak can be weeds or desirable woodland trees according to management intentions. The concept of chemical thinning using herbicides instead of cutting tools puts further emphasis on the role of management in deciding the target for use of herbicides rather than the plant species as such.

Control of gorse may be required partly to enable plants to get established but also, because of its characteristic of forming dense patches of impenetrable vegetation which can burn fiercely, partly to remove a fire hazard.

Rhododendron ponticum is in many places an unwanted woody shrub. However, its characteristics are sufficiently distinctive for it to be considered separately in §8.45.

Physical removal of trees and/or their stumps may be required rather than reduction of competition. While it is easier to remove a stump if it is dead, there are several other options including winching the tree using the trunk for leverage. Nevertheless, if it necessary to kill the stump before removal, current recommendations for herbicide treatment of stumps should be followed.

Control methods for woody plants fall into three categories:
- treatments of the stump;
- stem treatments, including basal bark sprays, stem injection and frill girdling;
- foliage treatment.

These alternatives were recognised from the outset in trials to control unwanted trees and woody growth. While often the same herbicides could be used, each control method requires its distinctive method of application and/or formulation and/or dilution.

Stump treatment is most effective if applied to freshly cut wood and adjoining bark surfaces. The technique requires least amounts of herbicide because the treated surface areas are relatively small compared to stem or foliage treatment, even though concentrations may be higher. Imazypyr and glyphosate have taken over the role previously dominated by 2,4,5-T.

Use of *stem treatments* may depend on the conservational, environmental and aesthetic acceptability of standing dead trees. If these are accepted, then basal bark, stem injection and frill girdling can be considered.

Foliage treatment currently has to be considered in the context of birch, willow *etc.* invading stands of other species within a few years of planting, after any post-planting weeding has been done. Germinating woody plants appearing in the first two or three years of growth after planting should be considered as part of the grass and broadleaved weed competition and treated with these weeds (§8.41).

Aerial application of herbicides to foliage of pole crop trees was practised in the period 1960 - 1969, the area sprayed over the 9 year period totalling about 3200 ha (Aldhous 1969a). However, from the early 1970s, there was no longer any call for this sort of vegetation control; there is no current approval for aerial application of any herbicide suitable for woody weed control.

Woody weed control in coppice conversion and scrub clearance

In the late forties and early 1950s, there were extensive areas of abandoned coppice and felled woodland requiring clearance and replanting. Early attention was therefore given to new materials that offered the possibility of assisting the costly task of initial clearance and subsequent weeding (Holmes 1952).

In the first trials in the 1950s, ammonium sulphamate and 2,4,5-T were immediately found to have considerable potential to kill a wide range of deciduous woody species. 2,4-D was consistently less effective except on heather. While species susceptibility varied according to the method of treatment, out of 27 woody species tested, 24 were susceptible or moderately susceptible to 2,4,5-T in one form or another and only 3 were resistant or moderately resistant (Holmes 1957a).

2,4,5-T and ammonium sulphamate, and to a lesser extent 2,4-D, formed the basis for recommendations for woody weed control for the following 20 years (Aldhous 1965a, 1969b, Brown 1975). Much detailed work was done on the safety and efficiency in use of these materials. *Weeding in the Forest* (Wittering 1974) brought together not only a comprehensive summary of techniques and materials, but also a pervasive awareness of working conditions and concern for the safety of the operator.

Other materials were tested when available *eg* :

• picloram and cacodylic acid - tree injection (Brown & Mackenzie 1971);

- glyphosate as a herbicide for grasses and herbaceous weeds; its potential to control woody weed compared with fosamine-ammonium, hexazinone, triclopyr and 2,4,5-T (McCavish 1978, 1979);
- triclopyr tested for phytotoxicity; compared to glyphosate, fosamine-ammonium and 2,4,5-T on cut stumps (McCavish 1981, Sale *et al*. 1982).

More recent materials to come into use following trials have been imazapyr and, as a replacement for 2,4,5-T, a mixture of triclorpyr, dicamba and 2,4-D (Valkova 1988, De'Ath 1988, Darrall 1988, Palmer *et al*. 1988).

Recommendations for control of woody weeds have been regularly updated, according to most recent trial results and the products available on the market in successive editions of *The use of herbicides in the forest* (Sale *et al*. 1983a, 1986c, Williamson & Lane, 1989, Willoughby & Dewar 1995). Between major revision, there have been periodic 'up-dates', the most recent appearing as Forestry Commission *Technical Paper 28* (Willoughby & Clay 1999).

Gorse and broom
A number of shrubby species were not as susceptible to ammonium sulphamate or 2,4,5-T as the various tree species. For a time, separate recommendations were made for gorse and broom. However, in trials, triclopyr was found to be effective against both species (McCavish 1980) and recommendations for its use to control both species are included in the main table for woody weeds in the 1995 edition of *The use of herbicides in the forest*.

Clematis vitalba
Traveller's joy or Old man's beard is found as a rampant climber over plants and walls on chalk and limestone soils; it can spread over thicket and young pole crops and can cause physical malformation through the weight of foliage and the pervasive entanglement of its stems. For many years no recommendation could be made for its control. Trials starting in 1992 showed that imazapyr gave best control when applied to a hedgerow overgrown with *Clematis* and to cut stumps. Triclopyr also had some effect. However, there are as yet no recommendations for control of *Clematis* (Clay & Dixon 1996a).

Bramble
Bramble (*Rubus fruticosus* agg.) is widespread as a perennial shrub in lowland broadleaved woodland. In established woodland it can form a light but extensive ground cover, with characteristically arching stems which, where they touch ground, root at the tips. Where woodland is clear-felled and replanted, bramble present can respond by vigorous growth, potentially smothering newly planted trees within 3 - 4 years.

Early trials showed brambles to be susceptible to 2,4,5-T as a foliage spray, but moderately resistant to 2,4-D (Aldhous 1965a).

A trial of the effect of spraying foliage of fruiting bramble with 2,4,5-T showed that ripening fruit discoloured and withered. Sprayed ripe fruits became mouldy after 3-4 days, but if collected before moulds were visible, contained about 100 ppm (freshweight) of 2,4,5-T (Brown & Mackenzie 1971). 2,4,5-T is no longer approved for use in Britain.

Most recent recommendations give a choice of several herbicides, including triclopyr, glyphosate and fosamine-ammonium (Willoughby & Dewar 1995). As bramble is frequently one of a mixture of weeds present on a site, choice should be influenced by the susceptibility of other weeds present.

If ripening fruits are present, warning signs are recommended advising the public that brambles have recently been treated and should not be collected, the signs remaining as long as any fruits appear wholesome.

Bramble is also a favoured feed for deer. Killing bramble too soon after planting has led to severe deer browsing on planted trees (Harris & Harris 1991, Harris personal communication).

8.45   Rhododendron

Rhododendron ponticum is native to southern Europe and Asia minor. It was introduced to Britain first in 1763 (Salisbury 1964) and at various times subsequently. It is thought that of the sources of introduction, 'Black Sea' origins may predominate over 'Atlantic' sources, but that there may be hybrids both between the main races and also with other *Rhododendron* species (Stace 1997). There is no evidence of any need, as far as its status in British woodland is concerned, to differentiate between sources of introduction.

The species grows under full light or under partial shade but in either situation, *Rhododendron* thickets cast heavy shade; it produces an acid litter and totally suppresses other ground flora through the exudation of toxic organic acids into its rooting zone (Rotherham & Read 1988).

Its growth habit is sprawling, the evergreen foliage and dense network of branches forming a barrier that is difficult to penetrate. The thick evergreen leaves are toxic and unpalatable to grazing animals and their semi-glossy waxy surface makes them difficult to wet.

Rhododendron was initially planted as decorative shrubbery. However, it grows vigorously on acid soils. Its branches layer and it sets viable seed; consequently, it has spread and become widely naturalised in many parts of the British Isles. It does not sucker (produce adventitious shoots from its roots). While its masses of spring flowers continue to be admired by some, others see it as a serious threat to native flora (Gritten 1994), or as a costly obstruction to the establishment of productive woodland (Tabbush & Williamson 1987).

Rhododendron clearance has always been an expensive operations. Manual or mechanised cutting and clearance remain an essential first step in *Rhododendron* control. Initially, hand tools and chainsaws were used. In the 1990s, advances in tractor capability on rough terrain enabled tractor-mounted flails to be used to break up sizeable *Rhododendron* bushes, markedly reducing the initial cost of clearance (Murgatroyd 1993, 1996, Edwards *et al.* 1993, Gritten 1994, Edwards & Morgan 1996, 1997).

The first herbicide trials in 1949 tested a range of compounds on cut stumps and coppice regrowth. In these tests, ammonium sulphamate was the only fully successful material against *Rhododendron*, the next best being 2,4,5-T (Holmes 1956a, 1957b). While further studies were carried out on both materials, the first recommendations for control only referred to ammonium sulphamate (Aldhous 1965a). However, 2,4,5T was included in 1969 and 1975 (Aldhous 1969b, Brown 1975).

As they became available, new herbicides, *eg.* picloram, chlorthiamid, glyphosate, triclopyr, hexazinone, were included in further rounds of trials (Aldhous & Hendrie 1966, Sale *et al.* 1983b , 1986b, Tabbush & Sale 1984, Sale 1985). These led to recommendations for glyphosate and triclopyr being added in 1983 and 1986, respectively.

At this time, the addition of a surfactant was also recommended both to enhance the effectiveness of glyphosate applied as a foliage spray and to enable a lower rate of herbicide to be used (Willoughby 1997b). A full description of work on *Rhododendron* at that time is given in *Rhododendron ponticum as a forest weed* (Tabbush & Williamson 1987) and was updated by Stables & Nelson (1990).

In the early 1990s, imazapyr was found to be more effective than glyphosate, especially as a foliage treatment of larger bushes (Clay *et al.* 1992a, Edwards *et al.* 1993, Edwards & Morgan 1996, Willoughby 1997b). Imazapyr was also successful against *Rhododendron* that had been cut down using a tractor-mounted flail (Edwards & Morgan 1997).

Herbicides recommended for control of *Rhododendron* in *The use of herbicides in the forest* (Willoughby & Dewar 1995) are:

- ammonium sulphamate,
- glyphosate,
- triclopyr.
- 2,4-D/dicamba/triclorpyr,
- imazapyr,

8.46   Chemical thinning and respacing

While conventional weeding and cleaning remove competing species, it is considered good silvicultural practice to remove crop trees by thinning, once they have reached a height of about 10m and the thinnings are marketable. The primary aim of thinning is to stimulate increment on the biggest and highest quality trees in the crop by providing space in the canopy for their crown to develop. Contemporary line-thinning and harvesting methods requiring tractor or forwarder access racks have undoubtedly increased wind turbulence over the canopy at tree-top height and have increased the risk of premature windblow.

Where there is no market for thinnings or it may be undesirable to break the canopy, chemical thinning has been attempted. Trials in Scotland in crops of Norway and Sitka spruce included injections with 2,4-D, picloram, picloram+2,4-D and cacodylic acid. The most effective material, however, was also translocated to untreated trees (Brown & Mackenzie 1970, 1971). Trials of glyphosate in the 1980s both by a forestry management company and by the Forestry Commission led to a working recommendation for treatment with glyphosate (Ogilvie 1984, Tabbush 1987a, Woolfenden 1988).

Respacing natural regeneration

In many upland areas where Sitka spruce plantations have been felled, natural regeneration has colonised sites. It has not been found possible to regulate stocking density during regeneration and means have been sought to thin out the denser patches to a spacing nearer what would result from conventional planting.

The clearing saw is used where stems are less than 6cm in diameter. For bigger stems, the chainsaw becomes the more suitable tool. Tractor-mounted vertical or horizontal shaft flails are under trial (Harding & Adam 1994). Earlier trials in 1984 and 1985 investigated possibilities of applying foliage sprays or treatments to tree stumps through an attachment to clearing saws (Tabbush 1984a, 1985b). Further trials considered herbicides, herbicide mixtures and adjuvants (Nelson 1990, Lawrie *et al.* 1992). These led to recommendations for the management of Sitka spruce natural regeneration (Nelson, 1991).

8.47   Farm forestry

Since 1988, a Farm Woodland Scheme and later a Farm Woodland Premium Scheme have been promoted jointly between the Forestry Commission and Agricultural Departments. A feature of these schemes was that higher rate grants could be paid for planting on former arable or improved grassland.

In parallel with that development has been an increased interest in short-rotation coppice on former farm land as a possible source of biofuels, pulpwood or chipwood.

Such farm sites are likely to be both more fertile and weedier. Trials covering cultivation, fertilization and intensity of weed control were set up (Davies *et al.* 1988). At the same time, herbicides widely used on arable farm crops were screened for their effects on young conifers and broadleaves. Specific off-label approvals have been obtained for the most promising herbicides (Williamson 1991, Tabbush 1992a, Willoughby 1996c, d, Willoughby & Clay 1999).

While providing trees with a weed-free square metre patch or 1 metre-wide strip has been the norm, on the most fertile farm sites, the growth of the vegetation in the unweeded strip has been found to be extremely vigorous, reducing growth of young planted trees. This reduction may be less if the unweeded strip is mown or put down to kale. The latter as a side effect could possibly improve the farm shooting (Williamson 1992).

Herbicides for farm woodlands and short rotation coppice

Willoughby & Clay (1996c, d) list all the herbicides for which specific off-label approval is available for use in farm woodland. Copies of the relevant approvals are included as part of their publication.

The *Herbicide update* Willoughby & Clay (1999) includes new recommend-ations for control of germinating perennial weeds and in particular, thistles.

Some farmers have attempted direct seeding as a method of establishing new woodland. While it can be expensive in seed, Forestry Commission Research Information Notes 285 (Willoughby *et al.* 1996) and 286 (Willoughby 1996d) describe early trials of techniques and a list of available herbicides. *Herbicide update* (Willoughby & Clay 1999) includes recommendations for post-sowing, pre-emergence herbicides where seed has been direct sown and the site is still weed-free.

A separate leaflet, *Noxious weeds* (Willoughby 1996a) describes and gives recommendations for the control of undesirable weeds that may be encountered in woodland or farm woodlands. Noxious weeds under the *Weeds Act 1959* are:

- spear thistle *(Cirsium vulgare)*,
- creeping thistle *(Cirsium arvense)*,
- curled dock *(Rumex arvense)*,
- broadleaved dock *(Rumex crispus)*,
- common ragwort *(Senecio jacobea)*.

Under the *Wildlife and Countryside Act 1981*, two further species may not be planted or otherwise caused to grow:

- giant hogweed *(Heracleum mantegazzianum)*,
- Japanese knotweed *(Polygonum cuspidatum)*.

8.48   Short Rotation Coppice

Dry matter production from short-rotation coppice for biomass can build up to 12+ tonnes /ha/yr. However, for the greatest success, crops need to be grown in a regime where weed competition is minimal from the outset.

Many short rotation sites have been located on arable soils where there is a substantial reservoir of arable weed seeds in the soil, and where climatic conditions are more favourable to early weed germination and vigorous growth. Willows and poplars are sensitive to some herbicides; nevertheless, the most successful short rotation plantations have been established following:

- pre-planting herbicide treatment to control perennial weeds;
- use of post-planting residual herbicides for medium-term control of annual weeds;
- subsequent treatment with foliar-acting herbicides to kill weeds emerging later.

Early stage trials with clorpyralid, glufosinate, glyphosate and simazine *etc.* offer effective regimes to maintain the weed control sought in short-rotation coppice (Clay & Dixon 1996b). Other trials have tested non-triazine herbicides for weed control in newly planted farm woodlands (Britt & Smith 1996).

In the first year, three applications must be anticipated, with a fourth 'foliar + residual' after first year shoots have been cut back. Weed control costs in early 1994 averaged £280/ha for chemicals and application (ETSU 1994b).

Short rotation coppice is discussed at greater length in §16.4, in the context of its role in 'renewable energy' programmes.

8.49   Motorway and railway vegetation

Establishment techniques for landscape plantings along motorways are similar to those for amenity and urban landscape schemes but with a greater risk of damage from road salt (see §15.23).

On railway lines in the era before electrification, there was always the risk of railway verge vegetation catching fire from embers shed from passing trains, and spreading beyond the railway fence onto vulnerable heath of forest land.   Fire breaks and fire patrols were the norm in periods of high fire risk.

Today the question is more of maintaining an aesthetically acceptable setting for the safe passage of high speed trains. This requires total weed control on the track and diminishingly rigorous control of vegetation from the track edge to the boundary fence where trees and shrubs often occur naturally (Fisher *et al.* 1990). Sargent (1984) listed selective herbicides in use in 1983 as part of a much wider discussion of management of verge vegetation.

8.5   FOREST NURSERIES

Forest nurseries in Britain have had the benefit of the sustained interest and activity of the Forestry Commission in producing high quality plants at low cost to sustain planting programmes of the last 50 years both for itself and for private planting. There have been two editions of *Nursery Practice*, a technical bulletin covering the main aspects of forest nursery work (Aldhous 1972a, Aldhous & Mason 1994) and several shorter papers, each with specific recommendations for control of weeds in forest nurseries (*eg* Aldhous 1962a, Biggin 1979, Williamson & Mason 1989, 1990a, b, Williamson *et al.* 1993).

The size and production of forest nurseries has been entirely dependent on the scale of state and private woodland planting, which is shown in *Table 8.4*.

Since 1980, there has been a very substantial reduction in the demand for plants as the Forestry Commission new planting programme has diminished. Plant demand for private planting has fluctuated, reflecting responses to the grant and taxation provisions at the time; in most recent years, the proportion of broadleaved species required has increased substantially compared with conifers.

The layout of seed-beds and transplant lines is a historical consequence of the use of small farm tractors to straddle them.

Weed control in nursery transplant lines is similar to weed control in many agricultural crops in that plants are close-planted in defined rows, almost everywhere being in beds of 5 or 6 continuous long lines.

Inter-row tractor-based cultivation or spraying is feasible, given appropriate tools and materials. Successful overall spraying depends on presence of susceptible weeds and resistant crop plants.

Seedbeds for forestry stock have for many years been 0.9-1.3m wide, with alleys 0.2-0.4m between. Originally, bed width was determined by the need to reach across to the middle of the bed to remove weeds by hand. For many years, however, bed and alley widths have been defined by the ability of tractors to work over them.

Seed may be sown broadcast on beds or sown in long drills. Broadcast sowing dominated during the 1960s to '80s but precision sowing associated with a carefully controlled regime of undercutting and side-cutting has increased in scale in recent years (Mason 1994).

The earliest record of weed control other than by hand or mechanical tool is the use of a blowlamp to scorch newly germinating weeds before the sown seedling crop emerged. The technique was introduced in the 1920s (Alexander 1954) and is illustrated by Wood (1974) plate 3. While it has fallen out of use in forestry, the technique has been promoted in the 1990s as an alternative to herbicides on hard surfaces in public places (Fryer & Stevens 1996).

Technical changes since 1950

The period in the late 1940s was one where nursery plant production was transformed by the discovery that acid heathland soils with adequate added nutrients were more suitable for production of young conifers than the heavier texture, less acid soils previously in use. The sowing and ground preparation techniques were scaled up to create what became known as *heathland nurseries* (Wood 1965). The nutritional background to these nurseries is described in §11.1.

Possibly the most important consequence for conifer nursery stock husbandry was the discovery that plantable stock of many tree species could be produced in two years rather than the three or four years formerly the norm. A less protracted period over which weed control was necessary was a by-product of that programme.

A second by-product of that programme was the recognition that soil sterilisation techniques, included as part of investigations into the potential role of pathogenic fungi and nematodes in preventing plant development, also controlled weeds. Subsequent development of partial soil sterilants has continued to maintain both improved plant growth and effective weed control (§8.51).

Keeping on top of weeds

Many heathland sites, because they had not been cultivated for many decades, if at all, had no load of dormant weed seeds from previous agricultural cropping. This was immediately recognised as a feature to protect by sowing seed in heathland nurseries and not importing seedlings.

'Experience has proved that many weed seeds and even growing weeds are introduced with seedlings from outside; thereby, one of the tremendous initial advantages of the heathland nursery is lost (by such transfers). With strict nursery hygiene and a relentless search for weeds, the annual cost of weeding can be kept down to £60 per hectare. In less fortunate nurseries, weeding regularly costs £250 to £500 per hectare' (1970 values).

Nursery practice (Aldhous 1972a):
Comparison in the 1950s of the time taken to hand weed conifer seedbeds showed that, where annual weeds were dominant, it took less time in aggregate to weed at monthly intervals than if weeding frequency was left longer (Aldhous 1961a). The increasing time was due to a few weed species, *Poa annua* in particular, producing viable, quick-germinating seed in little more than 4 weeks from germination. Consequently, weeds removed when weeding at 6 week intervals or more included weeds germinated from seeds from parent plants also present as weeds.

While herbicides and mechanical weeding have almost entirely superseded hand-weeding, the principle that weeds should as far as possible never be allowed to set viable seed remains valid. In practical terms, some form of control on a monthly basis, whether manual, mechanical or chemical, minimises the risk of build up of weed seed. See §8.52 last two paras.

Scale of use and availability of herbicides
Table 8.8 includes figures for the quantities of herbicides used in Forestry Commission nurseries in 1966 and 1986.

The forest nursery market for herbicides is small, even more than the forestry market. If a product is withdrawn because of reducing agricultural or horticultural demand, growers have little choice other than to seek alternative products. Diphenamid and oxyfluorofen are in this category. Products may also be withdrawn for other reasons. The withdrawal of approvals for use of atrazine and simazine on uncropped ground in the context of the European Community Drinking-water Directive (ACP 1993) has previously been mentioned in §4.16. Simazine remained approved 'on-label' for use in forest nurseries in 1999.

8.51   Pre-sowing soil partial sterilisation

As part of the post-war investigations into growth in forest nurseries, partial soil sterilisation was tested to determine the extent, if any, that soil-inhabiting pathogens might be limiting growth. Techniques included steaming, injection of chloropicrin or drenching with formaldehyde. In many of the nurseries where the soil pH was 6.5-7.0, substantial growth and yield responses were found (Benzian 1965, Aldhous 1972c). In addition, weed growth was very substantially reduced (Edwards 1952, Faulkner & Holmes 1953).

Out of these early methods, formalin was the recommended material; however, it was costly and only justifiable if there were benefits through better growth of seedlings as well as weed control.

Trials with dazomet began in 1967. It was soon shown to be more effective in reducing weeds than formalin (Low 1974). It was also easier to apply, not requiring the volumes of water needed for drench treatments. Recommendations for the use of dazomet were included in *Nursery Practice* (Aldhous 1972a), and have remained in force (Williamson & Mason 1989, Williamson *et al.* 1993, Williamson & Morgan 1994).

For production of *Alnus glutinosa* seedlings, investigations showed that if dazomet was needed, the rate of application should be reduced by 25% in order not to inhibit development of *Frankia* root nodules when inoculated from crushed nodules (Moffat 1994b).

Methyl bromide has been included in the list of partial soil sterilants based on its role in horticulture. However, it has to be applied by specially trained and qualified staff (Williamson & Mason 1990b, Whitehead 1999).

8.52   Seedbed herbicides

The first investigations into herbicides in the 1940s tested materials then in use in agriculture including sulphuric acid, ferrous sulphate, MCPA and pentachlorphenol (PCP). While these damaged crop trees too extensively to be usable as contact herbicides, mineral oils in the form of two readily available commercial products showed promise.

Vaporising oil (a tractor fuel) was found to be effective as pre-emergence spray on seedbeds, taking advantage of the longer germination time for dry-sown conifer seed compared with resident weed seeds. Vaporising oil was too phytotoxic to be used post-emergence. White spirit (a hydrocarbon solvent with many uses, *eg* as a paint diluent) was less phytotoxic and was used as a post-emergence spray on seedbeds and as an inter-row spray on transplants. However, it was not effective against anything other than the most recently germinated weeds (Edwards & Holmes 1949, 1950, Holmes & Ivens 1952, Faulkner & Holmes 1953). Vaporising oil and white spirit were the only two materials recommended for seedbed sprays in *Weed control in forest nurseries* (Aldhous 1962a). *Table 8.8* shows the scale of use of oils in the 1960s.

Trials of alternative materials showed paraquat to be a possible alternative to vaporising oil as a pre-emergent spray (Aldhous 1964a). While paraquat is an effective weedkiller offering few environmental risks, it is highly toxic to mammals. Consequently, alternatives were sought. Simazine had been shown to be effective as a pre-emergence spray for drill-sown large seeded hardwoods and was included in *Nursery practice* (Aldhous 1972a); however, the scope for its use was limited.

Screening trials of new materials started in 1971 (Brown 1972b, Low & Brown 1974). These and trials of other new materials led to recommendations for diphenamid, propyzamide and chlorthal-dimethyl in *Forest nursery herbicides*

(McCavish 1981, Sale & Mason 1985, Clay *et al.* 1988, Mason & Williamson 1987, 1988, Williamson & Mason 1988, 1989).

Further searches for more effective and selective post-emergence and inter-row transplant sprays led to screening of 13 contact herbicides (Clay 1992, Clay *et al.* 1992b). Some were tested further to determine the effect of post-spraying rainfall (Clay *et al.* 1996).

The concept of repeated doses of post-emergence herbicides at 6 week intervals at half the dose rate that would otherwise be appropriate was tested and found to be less damaging than full rate treatment (Williamson *et al.* 1990). This could be thought of as the contemporary means of achieving a frequency of weeding comparable to that described above in §8.5 under *Keeping on top of weeds*.

Seedbed herbicides recommended in *Forest nursery herbicides* (Williamson *et al.* 1993) and *Forest nursery practice* (Williamson & Morgan 1994) are:

• *Pre-sowing*	dazomet, methyl bromide,
• *Stale (unsown) seed-beds*	glufosinate-ammonium, glyphosate,
• *Seedbed pre-emergence*	chlorthal-dimethyl, diphenamid, glufosinate-ammonium, napromide, simazine
• *Seedbeds post-emergence*	diphenamid, propyzamide, simazine,
• *Seedbed repeated low-dose regime*	metamitron, metazachlor, napromide, propyzamide.

In the mid 1990s, it appeared that diphenamid would be withdrawn by the manufacturers; in a screening trial of some alternatives in the Republic of Ireland, oxyfluorfen was the most promising (O'Carroll & O'Reilly 1997).

Heather in seedbeds

In some heathland nurseries, late germinating heather has become a significant weed in seedbeds. Dixon & Clay (1996) identified a number of materials for further testing in 'repeat low-dose post-emergence' treatments.

8.53   Transplant line herbicides

The 1950s programme that led to the use of mineral oils for weed control in seedbeds also showed that white spirit could be applied between rows of transplants. However, only the youngest of weeds were controlled in this way.

Trials in the late 1950s showed simazine to be very well suited as a weedkiller for application to transplant lines shortly after lining out, and led to recommendations for use (Aldhous 1961b). *Table 8.8* shows that there had been an appreciable take up by 1966.

Simazine residues

Being a residual weedkiller, there was concern that simazine might accumulate in the soil on site. Trials of repeated cropping on the same land for eight years showed no evidence in terms of plant crop growth or health from residues. A bio-assay and chemical analysis of soil cores from two sites showed, after applications in three consecutive years of approx. 18 kg active substance per hectare per annum, the maximum detectable residue was equivalent to no more than approx. 1.1kg per hectare (Aldhous 1966a).

In the early 1990s, soil samples were taken from two forest nursery soils in England and tested for diphenamid and simazine. No detectable residues were found at any depth. It was concluded that there had been no build up and that it was unlikely that there had been any significant downward movement because the materials are relatively strongly adsorbed.

Simazine resistance; alternative herbicides

Over time, certain weeds, especially grounsel (*Senecio jacobea*) and American willowherb (*Epilobium ciliatum*), were found to be showing resistance to simazine (Ross 1988, Syme 1988). Alternative residual herbicides were investigated and several found to give useful control of a spectrum of weeds (Sale & Mason 1986, Mason 1986, Williamson & Mason 1990a, b, Williamson *et al.* 1990, Mason & Williamson 1992). Sulphonylurea herbicides also show promise (Lawrie & Clay 1994b).

Transplant herbicides recommended in *Forest nursery herbicides* (Williamson *et al.* 1993) and *Forest nursery practice* (Williamson & Morgan 1994) are:

- atrazine
- chlorthal-dimethyl
- clopyralid
- cyanazine
- diphenamid
- fluazifop-P-butyl
- glufosinate-ammonium
- glyphosate
- isoxaben
- lenacil
- metamitron
- metazachlor
- napromide
- oryzalin
- oxadiazon
- paraquat
- pendimethalin
- propyzamide
- simazine.

8.54 Fallow

In any year, a proportion of nursery land should be fallow or under a green crop. Green cropping is now only occasionally practised. A fallow period provides the opportunity to control any of the more difficult perennial weeds if these have become established, and by regular light cultivation to encourage germination of weed seeds in the soil. Herbicides currently recommended for use on fallow are glufosinate-ammonium, glyphosate and sodium chlorate.

8.55 Vegetatively propagated stock and plants grown in containers

Containers used in forest and amenity plant production range over the whole scale of available sizes. Container growing-media usually start with the advantage of being free from weed seed. Nevertheless, liverworts occur commonly and air-borne seeds of bittercress and willow herbs can germinate and threaten plants (Clay *et al.* 1990). Annual meadow grass and chickweed may also be prevalent.

Growers should reject supplies of peat contaminated with weed seed and should keep and stock piles free from weeds.

Willoughby & Clay (1999) report trials to control weeds in propagation beds in polytunnels but could not make any firm recommendations.

Checking root growth in containers

In container plant production, if containers rest on the ground, roots commonly grow out of the pot into the ground and may become difficult to lift. If growth is vigorous, roots quickly reach the sides of containers and follow them in a spiral. Rooting from the bottom of the container may be checked by suspending pots off the ground (air-pruning); copper net or paint containing hydroxides of copper have been used to inhibit growth of roots when they reach the sides of containers thereby preventing spiralling of roots and growth out of the container (Morgan 1997). Copper-painted matting also reduced infection by *Phytophthora cryptogea* (Labous & Willis 1997).

Control of liverworts and moss on containers

Withdrawal of products from the market has made it more difficult to find appropriate herbicides to control casual weeds and liverworts in container plants. Mulches have been proposed as 'container toppers' shading out any potential weed growth. Hewson (1996) found the herbicides oxadiazon and isoxaben were cheaper than any mulch; the least expensive treatment of those included in trials was bark.

A physical layer of a sprayable starch-based product has also been tested as a barrier to moss and weeds on containers (Anon 1996b).

8.6   OTHER USES OF HERBICIDES

8.61   Chemical debarking

In the 1960s, small hardwood roundwood could be used for pulping if the bark was removed before pulping. Straight material could be processed by debarking machinery, but not crooked lengths. Holmes (1961) described tests of diquat, 2,4,5-T and sodium arsenite for their effect on ease of bark removal of crooked small hardwood roundwood.

Results showed that of the materials under trial, diquat and sodium arsenite were the only materials that were consistently effective in loosening bark; however, sodium arsenite was considered too toxic for such use. 2,4,5-T showed promise but was not consistent. Treatments were effective in loosening bark on all straight and crooked material. Crooked material could be pre-treated to facilitate debarking at roughly twice the cost of machine debarking of straight material; however, this was commercially unacceptable (Semple 1964).

8.62   Epicormic shoots

The market value of the bole of large oak trees is commonly devalued, though otherwise sound, by the presence of the knot clusters left by epicormic branches. Varieties of lime, willow poplar and elm may also develop profuse epicormic shoots and may for that reason be considered unsightly in amenity areas.

2,4,5-T was found to kill oak, elm and poplar shoots but gave no lasting suppression. Also, when applied to shoots more than one year old these remained as dead twigs on the tree. There was little evidence of abcission on the affected branches (Holmes 1962a).

In trials 20 years later, glyphosate, maleic hydrazide and naphthylacetic acid were compared with physical removal and bark paints and wraps. Maleic hydrazide gave control lasting up to 2 years but it was concluded that the best approach is to use species and varieties not prone to develop epicormics (Evans 1986c, 1987, Patch *et al.* 1984, 1989).

8.63   Fire traces and fire breaks

Fires remain a threat to forests world-wide. In Britain, they have been a subject for regular study and report (Charters 1961, Connell 1967, Aldhous & Scott 1993). Analysis of the scale and nature of fires in British forests since 1950 shows that fires have predominantly been on recently planted land covered with herbaceous vegetation. The fires may have arisen within the forest boundary or may have spread from adjoining land.

Fire traces play a vital role in facilitating fire control. Between 6 and 9 metres wide, they may be bare land or carry non-inflammable low vegetation. They may

- provide access,
- when created on forest boundaries, act as 'cut-offs' for fires on adjoining land, especially where there is 'controlled burning' on adjacent land,
- within the forest, provide lines from which fire can be controlled.

Fire breaks have been maintained by mowing, swiping, regular ploughing or, in a few instances, grazing.

Fire breaks have also widely been created using close-spaced larch; the tree does not burn easily, is fast-growing and makes an effective barrier within 10 years after planting (Parsons & Evans 1977).

Pruning of lower branches and removal of branchwood on woodland adjoining roads can also create a fire break, provided that the canopy remains sufficiently dense to prevent invasion by a ground flora of woodland grasses and bracken which would carry a ground fire through the belt.

The earliest trials of herbicides as alternatives to mowing or cultivation to control fire break vegetation tested monuron, diuron, sodium chlorate, sodium tetraborate, 2,4-D, MCPA and TCA, either alone or in mixture; one of the more promising treatments, monuron + sodium tetraborate, left a vegetation cover almost entirely composed of non-inflammable broadleaved herbaceous species (Wood & Holmes 1959, Holmes & Fourt 1960).

Later trials included the use of paraquat for fire traces (Holmes 1962b, Connell & Cousins 1969). This was particularly effective against purple moor grass, *Molinia caerulea*. It has been used to control regrowth or reinvasion following fire trace clearance by bulldozer or may be used to desiccate vegetation so that it can be burned safely in the summer and leave a non-inflammable trace for the following spring (Fryer & Evans 1968a).

Mono- and di-ammonium phosphate, borates and bentonite clays were tested as fire retardants (Wood *et al.* 1961). However, though promising, they were overtaken by chemical supplements to improve the efficiency of water used for fire fighting. Sodium alginate has proved effective in thickening water so that six times more water is retained on sprayed foliage than if plain water is used (Connell & Cousins 1969); it is most effective in post-fire 'damping down' operations.

Low and medium expansion protein foams have been in regular use for holding and stopping fires; best effects are obtained if applied at least 5 minutes and not more than 1 hour before the arrival of a fire front (Ingoldby & Smith 1982).

8.64   Power lines

Under power lines, tall-growing woody species need to be eliminated and a low-growing, mixed herbaceous ground cover sustained. Such cover protects the soil, minimising the risk of colonisation by woody species; also, it is desirable environmentally, both for wild life conservation and for visual amenity. In the 1960s, 2,4,5-T was widely used.

Trials in the 1980s of herbicide alternatives to 2,4,5-T included dicamba, triclopyr, fosamine-ammonium, glyphosate and triclopyr/dicamba/2,4-D mixture. Applications were ruled out if treated plants were unsightly. Acceptable treatments, properly timed included cut-stump and foliage treatments with fosamine-ammonium, triclopyr and glyphosate (Darrall 1984, 1988).

8.65   Growth regulators to check growth of roadside vegetation

Maleic hydrazide alone or in mixture with 2,4-D retards growth of roadside grasses for 12-14 weeks (Fryer & Evans 1968), thereby reducing the need for mowing. Applications repeated over many years led to a short turf dominated by fine grasses (Willis 1988, 1990).

More recent trials indicate that other growth retardants, melfluidide and paclobutrazol, may be even more effective (Parr 1988).

CHAPTER 9

Other pests

If you go down in the woods today, you .... might get eaten alive

9.1   INTRODUCTION

This chapter describes a number of pests of woodlands and trees for which chemical control might be relevant when considering their overall integrated management (§9.2).   There is also a short section on those wood preservatives commonly used in the forest and in outdoor estate work (§9.3).

9.2   VERTEBRATE CONTROL

9.21   Damaging species

Woodlands and woodland edges are the natural habitat of a very high proportion of the terrestrial wild animals and birds of Britain.   In the 1950-60s, a series of leaflets described several woodland birds and animals (Campbell 1958, 1964, 1967, Palmar 1958a, b, 1962, Rogers Brambell 1958, Chard 1962). Subsequently, considerable efforts have been made to maintain and improve woodland habitats and encourage their wild-life (Steele 1975, Smart & Andrews 1985, Carter & Anderson 1987, Harris & Harris 1991).

Only a limited number of vertebrate species, mostly mammals, do enough material damage to trees to necessitate some form of control:

- deer (the native red, roe, and introduced sika, fallow and muntjac),
- rabbit (an introduced species),
- grey squirrel (also introduced).
- In the nursery, seedbeds need protection from mice and birds.

Illustrations of damage to trees by deer, rabbits, squirrels, dormice and voles are given in the *Wildlife Rangers Handbook* (Springthorpe & Myhill 1985, 1993).

While outside the scope of this book, requirements of estate shooting programmes are an important consideration when planning any integration of woodland protection against damage by squirrels, deer and rabbits.

Foxes are no direct threat to trees; where they are not controlled, rabbit numbers may be reduced.   Nevertheless, whether to meet estate sporting requirements or to maintain good relations with stock-rearing farming neighbours, foxes have to be included in the range of vertebrates to be monitored and, where necessary, controlled.

Control of red deer and other herbivorous vertebrates

As with all other pests, control measures have always to be integrated into the overall plan for the whole land management unit. Where, as in the case of red deer and grey squirrels, animals can cross property borders at any time, co-ordination of plans between groups of estates may be required to ensure best returns for the effort put into control work.

With some species, there is often a strong seasonal pattern of behaviour with a period of spread and seeking 'territories' and a period after the breeding season, when animals aggregate into flocks or herds and may also migrate between summer and winter feeding grounds.

Two widely recommended rules are:

• be vigilant about the size, location and movements of the population of any animal or bird that could damage trees or require control;

• focus control measures initially on the trees most at risk, and the season when most damage occurs. If possible, while leaving enough time for control to be achieved, start controlling as near as practicable to the time when damage is anticipated. If damage occurs normally in early summer, reductions in pest numbers in winter months may be nullified by invasions of territory-seekers in early spring.

Control using poisons is legal only for rabbits and mice. For deer, only repellents may be considered. For squirrels, the use of warfarin is restricted to defined regions where red squirrels are not at risk (§9.24). It is illegal to lay poison baits for foxes or other predators in Great Britain (Chadwick *et al.* 1997).

Import regulations backed by a competent port inspection service minimise the risk of rabies being imported into the UK. Should this occur and rabies reach any local fox populations in UK woodlands, a poison/eradication programme is one of the options that would have to be considered.

9.22   Deer

There are no chemicals that can legally be used to kill deer. If numbers permanently resident within any woodland unit have to be kept down, then shooting is the only option. Otherwise, plantations may have to be protected from incoming animals by a full deer-fence. Tree shelters protect lower stems but cannot prevent browsing of foliage above the level of the shelter (Hodge & Pepper, 1998, Pepper & Tee 1972, 1987, Pepper *et al.* 1984, Potter 1991, Pepper 1992a, 1999).

Electric fences have been tried; they are not as effective as a line wire and netting or mesh fence and, in the long run, not appreciably cheaper (Roe & Tee 1980, Pepper *et al.* 1991. Roe, fallow and muntjac are undeterred by them (Mayle,1999).

Various materials have been tried, either as deterrents or repellents, predominantly against roe deer.

- branchwood over trees, brash barriers, old netting (Stewart & Neustein 1961), naturally occurring bramble thicket;
- for individual tree protection, wire netting well staked, 4 feet high (Chard 1964);
- for plantations up to approx 3 hectares, 1.2 or 1.8 m tree shelters (Potter 1991);
- for small areas, chemical repellents: thiram (Wood *et al.* 1961), bone oil, 'Fowikal', (Neustein 1964), ziram (Rowe 1967a) and more recently, denatonium benzoate, a bitter-tasting compound (Wright & Milne 1996).

Physical protection by branches, brash or netting was not reliable in keeping roe deer off. Of the repellents, ziram (product name 'Aaprotect') has given protection over-winter for those parts of the plant that have been treated, and has been the yardstick against which other products were compared (Rowe 1977). It was on the list of 1999 Approved Products (MAFF Pst 1999).

Because repellents are costly, protect only the parts treated and need to be reapplied each year, they are recommended only for small scale use (Pepper 1978, Pepper *et al.* 1996).

Pepper (1999) describes new specifications of temporary and reusable fencing for use against the smaller deer species (roe, fallow, muntjac).

9.23 Rabbits

Rabbits were introduced into Britain by the Normans in the 12th century as a species to be farmed for their meat and fur. From that beginning, they have become a most serious and damaging vertebrate pest. At the beginning of the 1950s, the national rabbit population was estimated at about 60 million animals. The viral disease *Myxomatosis* was introduced in 1953; in the two years following its introduction, numbers in England were down to 1% of their mid-century peak (Tittensor & Lloyd 1983). Over the years, surviving rabbits became progressively more resistant, mortality in 1996 being estimated at about 20% of rabbits infected. Similarly, numbers have increased to between a third and a half of the 1950 level. Crop losses in Scotland through rabbit damage in 1990 were estimated at £12 million (Pepper 1976, 1998, Barclay & Foster 1996).

A second viral disease affecting rabbits, a calcivirus causing rabbit haemorrhagic disease, was discovered in central Europe in 1984, and was first reported in Britain in 1992. On the continent and in Australia, (where annual losses due to rabbits are put at over £280 million), the virus appears to be even more lethal and quickly spreading than myxomatosis; however, in Britain, 75% of rabbits appear to be immune to it (Barclay & Foster 1996, Anderson & Nowak 1997, Pepper 1998).

Legal obligations

Under the Pests Act, 1954, Agriculture Act 1947 and Agriculture (Scotland) Act 1948, the occupier is under a legal obligation to control rabbits on his land to whatever extent necessary to prevent them doing damage of economic importance to neighbouring farm crops or woodlands.

Silvicultural need

Rabbits, if present, constitute an immediate threat to any new woodland planting; there is no escape from the need to exclude rabbits from new plantings by fencing and to clear any present at the time of enclosure or subsequently found within enclosed areas, by shooting, gassing, or netting.

In older woodland in severe winter weather, rabbits can inflict substantial damage on pole-stage and larger trees, especially broadleaves. Bark-stripping of butts by rabbits is illustrated by Springthorpe & Myhill (1985), Pepper (1998) and Hodge & Pepper (1998). In practice, such damage cannot be totally avoided but may be minimised by an on-going overall programme to keep down the number of rabbits in woodlands.

Integrated rabbit control

Any plan has to take into account whether changes can be made to any features which might favour rabbits, *eg* patches of dense shrubby cover, broken down boundary banks, piles of stumps from a destumping operation, excessive fox control *etc*. It has also to take note of wider circumstances, *eg* rabbits from neighbouring land are damaging trees, or if rabbits in woodlands are damaging farm crops.

Fencing and, latterly, tree shelters have in almost every situation, been the first line of defence (Pepper & Tee 1972, 1987, Pepper *et al.* 1984, Potter 1991, Pepper 1976, 1992a, 1995b, 1998).

Repellents have been tried; ziram is effective whilst in place, but it gives no protection for young spring growth and has to be re-applied annually (Pepper 1976, 1978, 1998, Hodge & Pepper 1998).

Of the older methods of control, shooting and trapping remain feasible. They are labour-intensive and have been recommended for clearing up smaller populations as a 'mopping up' operation. Traps have to conform to the *Spring Traps Approval Order 1975*. Snaring is not recommended, being considered unselective and inhumane. The *Wildlife and Countryside Act 1981* requires snares to be visited daily and prohibits the use of self-locking snares.

Gassing

Pesticides available for killing rabbits by gassing are highly toxic and can only be used by fully trained and qualified operators (HSE 1994c, 1997c).

The pesticides in use are in solid form; they are placed in burrows which are immediately sealed. In contact with damp earth, the solids decompose to produce either hydrogen cyanide gas or phosphine gas.

Sodium cyanide powder was the first to come into use for control of rabbits on agricultural land. It may be applied by spooning into burrows or may be blown as a dust through a burrow system by mechanical or hand pump (Pepper 1976).

Cartridges giving off phosphine gas when ignited were introduced in the mid-1970s and recommended as an occasional alternative to spoon gassing for small burrow systems.

Reformulation as tablets in the mid-1980s allowed aluminium phosphide and magnesium phosphide to be delivered down a tube to within the mouth of rabbit burrows (Mayle *et al.* 1984, Rowe *et al.* 1985, Pepper 1982, 1987). Magnesium phosphide was subsequently withdrawn from the market.

The availability of tablets and applicator for aluminium phosphide has led to increasing use of this method of treatment (Pepper 1995b, 1998).

9.24 Squirrels

The grey squirrel has been introduced into Britain several times, first in the 1860s and again later. From an early date, its unwelcome behaviour of stripping bark from stems of pole stage and older trees was recognised. By the 1930s, occurrences of bark-stripping and squirrel numbers were increasing to such an extent that the continuation of broadleaved woodlands as a timber resource in many areas was considered to be threatened (Middleton 1931, Davidson & Adams 1973, Gill 1992b, c).

Not all trees are equally vulnerable. A list of the relative susceptibility to grey squirrel damage of sixteen of the most commonly planted broadleaves and conifers is given by Rowe and Gill (1985); it is based on several surveys carried out between 1954 and 1983. Sycamore and beech consistently were shown to be the most vulnerable to grey squirrel. Of other broadleaves, oaks, ash, birches and sweet chestnut were also commonly attacked. Attack on conifers as a whole has been less than on broadleaves, the most susceptible species being western hemlock, lodgepole and Scots pine, and Lawson cypress.

Mercer (1984b) studied squirrels in beechwoods in the Chiltern Hills, Buckinghamshire, which they colonised in about 1930. Trees most frequently attacked were 30-40 years old, damage being worst near the bases of trees. Trees in the 30-60 year age groups sustained most damage. On 70% of damaged trees, staining and decay was confined to the outermost two rings of stemwood.

Mountford (1997) reported significant loss of canopy potential in beech in Lady Park Wood, on the England-Wales borders in the Wye Valley and considered the grey squirrel to be a considerable 'disturbance agent' in the development of natural beechwood. Others would consider that to be a major understatement of the risk to woodland associated with grey squirrels.

While squirrel damage can occur on any age of tree, serious economic damage is mainly to pole-crops 10-40 years old. Any rapidly growing woody stem between 10 and 30 cm diameter is vulnerable, attack taking place in any year, mainly between April and August. Attack is sporadic; damage does not occur every year in a vulnerable stand, nor are all trees attacked (Rowe 1984). Stands previously unaffected may be attacked following thinning.

A tree that is severely bark-stripped is likely be substantially weakened, leading to breakage or rot or both and is unlikely to grow into a well-formed and sound mature woodland tree. Damage to beech and sycamore tends to be to the bark on the main bole from the butt upwards (See cover illustration *FC Report on Forest Research 1973*). On oak, because the bark on its lower stem is thick, attack occurs in the crown where bark is thinner and commonly results in ring barking and subsequent breakage at the point of damage.

Causes of debarking

Bark damage usually occurs in May-July when bark is easily peeled off. After removing and dropping the hard bark, squirrels scrape off and eat the sappy phloem tissues which lie beneath. They also suck the sap.

Many reasons have been proposed as to why squirrels remove bark from trees. Studies starting in 1978 at the Institute of Terrestrial Ecology (ITE) and sponsored by the Royal Forestry Society (RFS), sought initially to identify causes of damage and how to ensure that control operations were focused on the most susceptible areas and species (Kenward 1982).

Several hypotheses were dismissed from the outset:

* preventing incisor overgrowth,
* use for lining dreys,
* uncontrolled gnawing reflexes.

* water shortage,
* genetic mutation.

Hypotheses consistent with data available at the start of the study were:

* mid-summer food shortage,
* agonistic (territorial/display) behaviour,

* trace nutrient deficiency.
* liking for sweet sap.

Over a three year period, radio-tagged animals were monitored by repeated 'catch-and-release' trapping in mature oak/ash woodlands, and in adjacent young beech/oak/spruce plantation mixtures.

Observations showed that between August and February, radio-tagged squirrels spent on average 80% of the time foraging on the ground. During April to June, ground foraging time fell to between 30 and 50% of the total forage time, much of the difference being taken up on this site by foraging in the crowns of mature broadleaved trees. In April-May, the potentially vulnerable young plantations were not visited as much as tops of mature oak and ash woodland trees; in June-July, numbers visiting the young woodland rose while the numbers in older trees fell. This period coincided with the period of range expansion of individual adult males, a peak in gonad development and mating activity.

Squirrels were shown to lose body-weight in the summer if food supplies were inadequate. Weights of residents in woodlands from which 66% of the population had been removed increased while animals lost weight where there had been no removal. In neither wood was there any damage in the year of observation; however, other data showed a trend for damage to be worse on sites where squirrels also lost body weight.

In studies in other damage-prone woods, trees most highly damaged were those yielding greatest amounts of sap. Glucose and sucrose were the main nutrients in sap; however, there was no correlation between damage severity and sugar concentration.

To satisfy a squirrel's energy requirements (130 kcal/day) with a sap supply at 0.5 litres/m², a squirrel would have to remove 0.5 m² of bark a day. This could take 2-3 hours. In practice, it is rare for damage to be more than would sustain two squirrels for a couple of days (Kenward 1982). Phloem thickness was found to be closely correlated with sap supply and, in subsequent years, was the feature assessed, being easier to measure.

Studies between 1981 and 1983 showed a close relationship between phloem thickness and damage. Economically important bark stripping was restricted to trees with 0.3 mm or more thickness of phloem.

Other evidence showed that for the areas in the study:

- damage was greater, the greater the density of juveniles,
- if there had been appreciable damage on a site in one year, squirrels tended to take bark from more trees the following year. Only very 'sappy' trees were repeatedly attacked,
- damage was frequently unexpectedly severe around locations where baited traps had been set and in pheasant feed areas. In a study area where additional food had been supplied, juveniles spent less time foraging but gained more weight than in comparable woodland without an extra food supply. Damage to young beech was worse in the wood with the extra food.
- where wheat fields adjoined woodlands, immature heads of wheat in the 'milky' stage also formed a substantial food source (Kenward *et al.* 1988a).

Overall, 53 stands in Dorset and in the Midlands were studied, some for five years, some for up to nine years. A survey in 1985 assessed the cumulative squirrel damage in these woods; *Table 9.1* gives some results of this survey.

The table shows that the presence of mature trees increased damage and was the most significant of all the variables analysed. This effect was attributed to their role as a shelter and as an abundant source of food in good mast years to stimulate squirrel breeding. Damage was proportional to the number of mature trees within 200 m of the susceptible woodland edge.

While self-seeded trees in the study were less damaged than planted ones, this resulted from the fact that virtually all the self-seeded trees in the study had grown at close spacing. When thinned, damage commenced (Harris 1997, Pepper 1997b).

The lack of effect of pheasant feeding areas, notwithstanding previous comments, arises from the observation that almost all the pheasant feeding areas in the study were in mature woodland and that feeding ceased at the end of the shooting season, contributing little to squirrel breeding. Breeding was enhanced where pheasant feeding continued into the spring (Kenward *et al.* 1988b).

Table 9.1   *Site factors associated with accumulated squirrel damage in
53 beech and sycamore woods, 1985*

Variable Comment	Regression coefficient	Statistical Significance
Presence of mature seed-bearing trees	+4.7	v. highly significant
Presence of mature trees increased damage		(P <0.001)
Warfarin used / not used	-3.1	highly significant
Where warfarin was used, damage was less		(P< 0.01)
Sycamore / beech	+2.9	highly significant
Sycamore was worse affected than beech		(P< 0.01)
Average phloem width	+2.9	highly significant
The thicker the phloem the worse the damage		(P< 0.01)
Self-seeded / planted	-2.5	significant
Damage was less in stands that had regenerated naturally		(P< 0.02)
Pheasant feeding / none	+0.0	not significant
Presence of pheasant feeding areas did not affect incidence of damage		

Source   Kenward *et al.* 1988b

Salt licks/trace elements

Salt licks including trace elements were placed where they would be accessible to caged or penned squirrels. The availability of salt licks did not prevent bark stripping of beech by these animals which were otherwise adequately fed and watered (Tee 1984).

Red squirrel conservation

The native red squirrel has declined in numbers and has disappeared from large parts of its natural range in Britain over the past 50 years. This is partly due to habitat fragmentation and disease, but mainly to displacement by the grey squirrel, as the grey's numbers have increased.

The red squirrel also can cause damage to trees, especially Norway spruce (Sinclair 1832), but rarely on the scale done by the grey squirrel.

Tittensor (1975) describes the red squirrel life cycle. A Forestry Commission *Practice Note* describes in more detail, habitat preferences and management recommendations; it also gives a current distribution map for red squirrels (Pepper & Patterson 1998).

The areas occupied by grey and red squirrels in 1967 are illustrated in *Grey Squirrel and its Control in Great Britain* (Rowe 1967b). Maps of the distribution of red and grey squirrels at 1974 and 1985 in Forestry Commission forests are given by Gurnell & Pepper (1988). Use of warfarin against grey squirrels is illegal in red squirrel areas. See also *Differential trapping where red and grey squirrels are both present* below.

Under part of the UK Biodiversity Action Plan, a *Species Action Plan* has been published by the British Government (Anon 1995a); a supporting strategy for red squirrels has been developed by the Joint Nature Conservation Committee (JNCC 1996).

9.25   Grey squirrel control

As a mammal, the squirrel is exceptional in its ability to exploit habitat from the highest tree tops to the surface layers of the soil. This coupled with its adaptability makes grey squirrel control exceptionally difficult.

A campaign to control the grey squirrel was mounted in the 1930s and was reactivated with considerable publicity in 1952. A 'shilling a tail' incentive was offered (two shillings in 1958), expecting that the main means of achieving control would be both by shooting and trapping. Leaflets describing suitable traps were distributed. At that time, the declared intent was to 'encourage the grey squirrel's destruction and eventual extermination' (FC AR 1953). The bonus scheme was discontinued in 1958.

Over the period 1952-1956, the Forestry Commission Annual Reports record 1¼ million squirrels killed. Nevertheless, in spite of the scale of kill, at the end of 1957, squirrel numbers were similar to those in 1953 (Visoso 1967). It was clear that the adaptability and fecundity of the squirrel were such that extermination in the circumstances in Britain was not achievable. Quite apart from woodland populations, such a substantial proportion of the squirrel population was living in urban peripheral and amenity woodland habitats that these would always remain a reservoir from which animals could spread.

Studies in the 1950s conducted jointly between staff of the Infestation Control Laboratory of the Ministry of Agriculture and the Forestry Commission initially focused on establishing the distribution and rate of spread of grey squirrels, their basic life cycle and techniques for control (Nimmo 1955, Visoso 1957, Lloyd & Taylor 1960, Shorten 1962). Details of breeding success *etc.* are given in *Squirrel Populations and their Control* (Visoso 1967). Subsequently, squirrel damage and estimated changes in numbers were assessed annually by the Forestry Commission (*eg* Pepper 1992b).

From the 1960s, work by the Forestry Commission concentrated on control techniques.

Cage and spring trapping

Techniques initially available for control were winter shooting and trapping. Winter shooting was often in co-ordination with drey poking to rouse and expose the animals. All spring and cage traps recommended or used had to be of designs approved under the *Spring Traps Approval Order, 1975* and its predecessors.

Under the *Pests Act, 1954* and the *Protection of Birds Act 1954*, spring traps may only be set in some form of tunnel so that no birds or other animals are at risk (Anon. 1962a, Rowe 1980). The Forestry Commission's policy is not to recommend the use of spring traps (Pepper & Currie 1998).

Warfarin for grey squirrel control

The first preliminary cage tests of warfarin as an anti-coagulant poison to reduce squirrel population took place in 1960; they were followed up by the first field trials in the same year. Because of differences in the law, these and subsequent six years' trials were restricted to Scotland. Studies sought to find out the best:

- timing and duration of baiting in relation to times of damage and site recolonisation (Pepper *et al.* 1994),
- design and siting of hoppers to ensure access only by squirrels,
- rate of application and carrier bait (wheat, maize, *etc.*) for warfarin.

Studies of recolonisation after trapping and killing revealed that areas cleared of grey squirrels could be recolonised within two or three months. In consequence, poison-baiting had to be timed to have taken effect just at the beginning of the period when trees were most susceptible.

The conclusion in 1967 was:

'Squirrels can be killed at any time of the year by one method or another. However, since at most times of year no significant damage is done and sporadic control has little influence on population fluctuations compared with natural factors, the basis for economic management must be intensive control of numbers in and immediately around vulnerable crops just before and during the damage period. Killing squirrels at other times is wasting effort since reproduction and immigration result in filling any vacuum so produced' (Rowe 1967b).

Tests also included comparisons of alternative poisons, and analyses of small animals and birds for traces of squirrel bait (Lloyd & Taylor 1960, Lloyd *et al.* 1961, 1962, Rowe 1966).

The Scottish results for warfarin were the basis for further trials between 1968 and 1971 under licence from the Ministry of Agriculture (Rowe 1968-1971). They resulted in the making of the *Grey squirrels (Warfarin) Order 1973* (SI 1973/744) under the *Agriculture (Miscellaneous provisions) Act 1972*. The order specifies the poison, bait, design and dimensions of hopper, and the parts of England and Wales where warfarin may be used. Areas were excluded where the red squirrel was believed to continue to survive; these included substantial parts of the north and east of England and parts of Wales. The legislation did not cover Scotland at all. The period during which warfarin could be used was limited to between mid-March and mid-August (Pepper 1990b). Forestry Commission monitoring data showed that warfarin poisoning appeared to be effective (see *Figure 7* in Gurnell & Pepper 1988).

Subsequent studies

The hopper design was modified in the late 1980s by adding a door flap hinged to the roof inside a square-section entrance tunnel. This has made it more difficult for mice, voles *etc.* to gain access to bait (Pepper 1988, 1989). The amount of bait taken per modified hopper has fallen markedly. However, studies showed that the flap doors had not deterred grey squirrels; they did eliminate almost entirely the amount of warfarin-treated bait taken by small rodents and small birds (Pepper & Stocker 1993). Hoppers with circular-section entrance tunnels could not be adapted for use with any flap door.

Warfarin bait-poisoning remains the main means of control (Pepper 1990a, b, Pepper & Currie 1998); it may be used where red squirrels (and also pine martens) are absent. Outside these areas, cage trapping remains the recommended means of control.

An extension to the area where warfarin can be used was announced in 1996 but this extension was less than had been hoped for. The extension applied mainly to England and Wales, with only small scattered areas in Scotland. None of the areas had any recent history of red squirrel.

Premixed bait has been available commercially; more recently it has had to be mixed by hand by appropriately trained personnel (Pepper 1995a).

Effectiveness of controls

In 1983, grey squirrel bark-stripping damage was surveyed in a sample of stands greater than 0.3 ha in area and between 10 and 40 years old. Sites were chosen so as to cover the six most susceptible tree species and to get a good spread across England and Wales.

Table 9.2 shows the methods of use found by the survey. The difference in scale of use of warfarin is very clear. Equally, although considered to be ineffective because of the ease with which squirrels recolonise land where winter control operations have been carried out, it was clear that shooting and drey poking continued in widespread use in private and Forestry Commission woodlands (Rowe 1984).

Table 9.2   *Methods of control used against squirrels - survey in England and Wales, 1983*   *(Number of survey sites)*

	Poison	*Cage traps*	*Spring traps*	*Drey poking*	*Shooting*
Private woodlands	27	16	3	28	74
Forestry Commission	140	40	18	26	62

Source   Rowe 1984

A study on radio-tagged squirrels was undertaken in Dorset under the auspices of the Institute of Terrestrial Ecology. It was found that where warfarin had been placed at one hopper per 2 hectares throughout a wood which, in the previous year had suffered bark-stripping, all the tagged squirrels died within 25 days (Kenward *et al.* 1996).

Damage prediction

In the ITE studies described above, two main factors accounted for at least two thirds of the variation in long-term damage:

- damage increased with the number of mature seed-bearing trees within 200 m of a susceptible woodland;
- the prevalence of ground cover up to 30 cm tall was inversely related to severity of damage. Only one site out of 12 had significant damage where there was more than 50% cover, compared with 18 out of 41 with less than 50% cover (Kenward *et al.*1996).

From these and observations previously noted, Kenward & Dutton (1996) proposed a 'minimum intervention' three-phase damage risk prediction system summarised as:

- *Summer*   Survey in July all areas at risk for signs of bark stripping, both 'trial' stripping and larger areas of bark removed.
- *Autumn/winter*   Identify where there is abundant winter food supply for squirrels, *ie* heavy nut, mast or other seed crops, pheasant feeding areas *etc*. Divide areas at risk into two groups, higher and lower priority.
- *Spring*   Place hoppers in early March on higher priority sites, pre-bait and subsequently fill with warfarin-coated wheat; leave hoppers out for 4 weeks and move to lower priority sites.

Following the above scheme, hoppers are only put out to protected stands where risk appears high.

An analagous approach proposed by Gurnell and Pepper (1988) is to assess woodlands in terms of '*damage vulnerability*' and their potential for holding high densities of squirrels ('*potential high density areas*'). Assessment of winter numbers is not used as it takes no account of subsequent population movements before the damage period. Instead, the breeding potential is assessed. If winter food supply is good, breeding will start in mid-winter and there will be high recruitment of young in the following spring and summer.

Poisoning should be centred on areas with a potentially high density of squirrels (*ie* high breeding status and plentiful food supplies), within 1 km of damage-vulnerable areas. As part of this approach, winter trappability is being tested as a means of assessing food availability. This is on the premise that well-fed squirrels do not readily enter traps, whereas hungry ones do. Trials of predictions based on damage vulnerability, potential high density areas and winter trappability have been under way since 1990 (Pepper *et al.* 1994) and are continuing (Pepper & Hodge 1996).

Representations have been made to allow use of warfarin from mid-January in high density areas when winter food is in short supply.

Costs

Cost of control using warfarin was found to be scale-dependent. For larger woodland areas, it averaged £2 per hectare for materials and labour but ranged from 0.21p to over £5 (Kenward *et al.* 1988a, Pepper 1992b). In the mid 1980s, costs in the Chilterns were of the order of £5/ha/yr (Tilney-Bassett 1988); however, ten years later, for a 14 ha small beechwood in the same area, costs in a year of 'heavy' squirrel pressure were recorded as £32 per ha. Over a period of 7 years, uptake of bait ranged from 1.1 to 4.1 kg/ha. In 1997, uptake was 3.6 kg/ha (Morris & Whipp 1998).

By careful analysis of risk and applying a 'minimum intervention' prescription, costs using warfarin on 13 woods in Dorset were kept to about 40% of what annual control would otherwise have cost (Dutton 1993a, b, Kenward & Dutton 1996).

Losses

A stand of beech, yield class 8, aged 85 was estimated to have lost £1700/ha or 15% of its value as stocked woodland, the effect of squirrel damage increasing the proportion of timber sold for firewood by 20% at the expense of more valuable logwood (Kenward & Dutton 1996).

In a study of the effect on revenue of squirrel damage to Corsican and Scots pine thinnings in a Cheshire woodland, revenue reduction was estimated at only 2%. In spite of between 30 and 50% of trees being affected, much of the damage was in the crown and had less effect on the marketable section of the trees than anticipated (Tee & Rowe 1985).

While the scale may be relatively small, uncontrolled squirrels can completely destroy young crops, giving no return to the owner at all. Such damage is evident in group plantings in the Chilterns, 10-15 m tall in 1997.

Alternatives to warfarin

Alternative anti-coagulants and mixtures of warfarin with toxophene were tested during initial trials of warfarin but were not pursued.

The potential for reproductive inhibitors was reviewed in 1982 (Rowe 1982). Investigations into possible immunosterilants commenced in 1995. While first indications are promising, at least 5 years development work was anticipated (Pepper & Hodge 1996).

Differential trapping where red and grey squirrels are both present

Notwithstanding efforts to control it, the grey squirrel has continued to spread and the red squirrel to diminish in range and abundance (Gurnell & Pepper 1988). This diminution and restriction of range led to recommendations on how to manage conifer forests to maximise the competitive advantage of red squirrels (Gurnell & Pepper 1991).

Cage-trap designs in use up to the end of the 1990s operated unselectively, capturing both red and grey squirrels. While red squirrels can be released and the greys killed, a proportion of red squirrels died from the stress of being trapped. Alternative designs based on *hopper+flap-door*, which admitted grey but excluded red squirrels were tested in 1992. Results led to applications for licences for further trials.

In the 1990s, studies commenced on the use of warfarin to control grey squirrels so as to reduce pressure on red squirrels, rather than to protect trees against debarking. (Pepper 1993, 1995c, Pepper & Hodge 1996).

Red squirrels are also the subject of a conservation study jointly with English Nature, testing the effects of supplementary feeding, grey squirrel removal, red squirrel release and habitat improvement. Food hopper designs were sought which could be used for supplementary feeding of red squirrels but not greys in the same area. A hopper with a section of the entrance tunnel floor able to swing like a seesaw has been tested. Under the weight of a fully grown grey squirrel, it swings down ejecting the animal. It remains in place under the lighter red squirrel allowing it access to food in the hopper (Pepper 1993, Pepper *et al.* 1994, Gurnell & Pepper 1998).

A full summary of the Forestry Commission research programme into squirrels is given in *Squirrel population and habitat management* (Pepper & Hodge 1996)

Grey squirrels in Europe
Grey squirrels were introduced in 1948 to northern Italy. They have spread slowly but are reported to be within a short distance of an important hazel nut production area. In Britain, the grey squirrel's liking for immature hazel nuts is such as to have effectively eliminated the hazel nut crop in many coppice areas in the south of England (Bonner 1996).

9.26   Voles and mice and other vertebrates (not in nurseries)

Field vole (Microtus agrestis)
The field vole is the commonest member of the British vertebrate fauna. Essentially a grassland animal, its populations exhibit cycles of abundance to the point of plague followed by rapid decline. They are most noticed by foresters when in abundance on grassy planted areas before canopy closure. Young trees may sustain severe damage to bark at and just above ground level, often being girdled (Rogers Brambell 1958, Gill 1992b, c). A small mammal population indexing technique was developed in 1982 for upland predator studies and for predicting cyclic patterns in relation to damage (Rowe 1982).

On grassy sites, maintaining weed control around individual trees by spot weeding reduced the incidence of vole damage, through reduced grass cover (Rowe 1974, Davies & Pepper 1990).

While many raptors, particularly tawny owls, regularly take field voles, no raptor appears able to increase in numbers sufficiently quickly when vole populations surge, to act as stabilisers for vole populations. The converse effect is more apparent, tawny owl breeding success being poor when numbers of voles are low (Petty 1983, 1987). Artificial perches erected where vole damage is severe can reduce vole populations significantly (Harris *pers. comm.*).

In the 1960s when the opportunity presented itself, warfarin was successfully tested as a bait for vole control using alternative cereal carriers and rates of application (Rowe 1964, 1965) and was subsequently recommended for vole control (Hibberd 1984).

Plastic guards were tested in the mid-1980s on motorway sites where it is not practical to use warfarin. Over a two year period, spiral guards and split tube guards reduced damage by 90% (Pepper & Tee 1984, Pepper *et al.* 1986). 200 - 300 mm high spiral guards have been recommended for 'urban forestry' situations (Hibberd 1989, Hodge 1995). Short, point-ended tree shelters ('quills') pushed into the soil to a minimum depth of 50 mm have also been used.

Bank vole (Cleithrionomys glareolus)
While this vole from time to time strips bark of young trees up to a height of 2 to 3 metres, damage is sufficiently infrequent not to require chemical control.

Other vertebrates
Other denizens of the woods may need management or control if they become numerous or persist in unwelcome activity, *eg*:

- foxes,
- badgers, to ensure use of badger gates in fences,
- starlings, for avoidance or dispersal of excessively large roosts.

None of these requires the use of chemicals, however. Illegal use of poisons against foxes can only be deplored (Lloyd & Hewson 1986, Chadwick *et al.* 1997).

Moles do not damage forest trees but very occasionally may disturb nursery seedbed areas or grassland peripheral to the forest. If control is required, conventional techniques have been recommended (Hibberd 1986). While strychnine is approved for control of moles underground, it may only be supplied to holders of authority to purchase issued by government agricultural departments and used where there is restricted public access (Whitehead 1999).

The edible dormouse (*Glis glis*) was released in 1902 in Hertfordshire on the edge of the Chiltern hills; it has become established in that area, locally causing severe damage (Jackson 1994). In one study of more than 14000 trees, nearly 15% had been gnawed; damage was uneven but could reach 70% of trees locally. This dormouse also explores and damages human habitations and stores. It is a protected species under the Berne Convention and for the moment all that can be done is to study population trends and behaviour. A study of the suitability of nesting boxes for such work was started in 1996 (Morris *et al.*1997). Subsequent work showed that inexpensive nest boxes could be made using tubular tree guards (Morris & Temple 1998).

9.27　　　Forest nursery vertebrate pests

The principle attraction in the nursery is seed. Whether awaiting sowing or after sowing, seeds are at risk.

Formerly, conifer seed was dressed with red lead, believing it to act as a deterrent against birds and mice after sowing (Ackers 1947). This view became discredited, partly because lead compounds became recognised as environmentally undesirable and partly because, by observation, red lead was not very effective. In the search for deterrent seed dressings, anthroquinone was tested but found to be ineffective. Thiram was tested (Rowe 1967a) and later recommended as a bird deterrent if netting was not used. If a colorant was required purely to ensure an even distribution of seed when sowing, lithofar red was recommended (Aldhous 1972).

Losses of conifer seed through bird damage have been substantial even though the exact quantities have been difficult to identify (Rowe 1971, Tee & Petty 1973, Rowe 1974). Nevertheless, from the early 1970s, physical barriers have been adopted as the means of deterring birds. Initially plastic mesh netting was spread over beds. In many nurseries, cloches and floating mulches have come into use and, as a beneficial side-effect, provide protection against birds for the species under them.

With advances in nursery technology, and in particular, moist pre-chilling of conifer seed and mechanised sowing, seed is sown without any dressing (Mason 1994).

Seed of broadleaved species, in particular oak and beech, has been difficult to handle because of the necessity to maintain an adequate moisture content in the individual large seeds (Aldhous 1972). Whether stored in an aerated shed, in stratification or sown in the autumn, steps have had to be taken to protect against mice and birds. As for conifer seed, however, these steps have been physical rather than chemical.

Mice, where they occur, may be excluded by fine mesh netting, or trapped with conventional break-back traps or, within buildings, may be poisoned using warfarin (Gray 1953, Aldhous 1972, Mason 1994, Jinks 1994, Strouts *et al.* 1994).

9.3　　WOOD PRESERVATIVES

This section considers only those uses of wood preservatives within the forest for fencing and outdoor estate work and treatment of logs before reaching the sawmill. The availability and type of materials in use have largely been determined by agricultural and urban needs.

It is indicative of their national status that wood preservatives in *Pesticides 1997* under *Control of Pesticides Regulations* are listed under 'Health & Safety Executive registered products', rather under 'Pesticides Safety Directorate registered products' with insecticides, fungicides *etc.* for use in agriculture, horticulture, forestry *etc.*

British Standards have been issued and periodically updated for:

- creosotes (BS 144: 1990, BS 913: 1973, BS 3051: 1972)
- copper-chrome and copper-chrome-arsenic mixtures (BS 3452:1962, BS 4072:1974, BS 5589:1989).

Creosote has for many decades been the standard dressing for outdoor woodwork (Small 1960, Spencer 1999). A by-product of coke and coal-tar, it has been a variable commodity, hence the various BS specifications. However, with the demise of domestic coal mining and replacing of town gas with natural gas, previous sources of creosote are much diminished. At the same time, greater concern for environmental and operator safety and comfort have raised questions about the desirability of some former constituents of creosote. Even so, the UK was estimated to have produced 62 000 m³ of creosote-treated wood in 1993 (Palfreyman *et al.* 1995).

In 1957 fencing stake trials were set up in Scotland, Wales and England comparing the relative performance of a range of species treated with creosote and a fluor-chrome-arsenate (FCA) water-borne wood preservative (Richards 1957, 1961, Aaron 1962).

After 15 years,

- all the creosoted posts remained intact.
- results for FCA-treated posts were dependent on species:
for conifers, best was Scots pine (10% failure) >European larch >Japanese larch >Sitka spruce (35% failure);
for broadleaves, best was Sweet chestnut (38% failure) >birch and sycamore >oak >ash >elm (95% failure).
- for untreated posts, more than 50% of all conifer species failed except larch; for broadleaves, all failed completely by 15 years except for sweet chestnut.
- Peaty sites had less failures than sites with loamy soils - clays were intermediate. Failures were worse on drier (=warmer) sites than wetter ones.
- Charring ends of posts before driving them in hastened deterioration

Fluor-chrome-arsenate was subsequently replaced by pressure-treated copper-chrome-arsenate (CCA) as the standard alternative to creosote (Clarke & Boswell 1976).

Recent changes in legislation arising out of European directives are likely to restrict the use of creosote mainly to transmission poles, railway sleepers and some marine uses. At the same time boron-based compounds may play an increasing role. European (EN) standards will replace British Standards (Eason 1996).

Use of Sitka spruce for power transmission poles

While Norway spruce is used on the continent for power transmission poles, in Britain, neither Norway nor Sitka spruce has been accepted because insufficient preservative could be incorporated into the wood after felling to provide the required service life for the poles.

Investigations started in 1978 into whether water storage or spraying would stimulate bacterial activity throughout the sapwood and make it permeable to preservatives. By 1983, ponding had been found to be the most successful treatment, increasing permeability three-fold (Aaron 1978, 1983).

At about this time, sap displacement was tested. Freshly felled logs were placed in a pressure cylinder and a cap under suction fixed on the butt of each log. The cylinder was then filled and pressurised, maintaining suction through the cap on the butt. It was found that sap could be removed from the log in this way. When copper-chrome-arsenate preservative was applied, it was rapidly fixed in the log. However, there was a significant difference between species; the sapwood of Norway spruce was usually wide enough to take adequate quantities of preservative salts. With Sitka spruce, logs had to be examined after felling, only those with a wide sapwood being sent for treatment (Aaron 1982).

Sap-displacement for Norway and Sitka spruce poles was included in permitted techniques under BS 1990. Consequently, work on improving permeability under water was concluded (Hands 1985, Aaron & Oakley 1985).

Disposal of CCA-treated timber

With increasing attention to the problems of past accumulations of industrial wastes, the question has been raised how best to dispose of CCA-treated timber, because of its chromium and arsenic content. In Japan, current production is of the order of 300 000 cu.m/annum; figures for the UK lump together all water-borne wood preservatives at 919 000 cu.m in 1993. Leaching from the transmission poles *in situ* is insignificant. Alternatives include dump to land fill, incineration and metal recovery, and leaching with the aid or micro-organisms (Palfreyman *et al.* 1995).

See also §7.32 under *Blue stain - Wet storage.*

Part III

Nutrients & fertilizers

CHAPTER 10

Woodlands and fertilizers

10.1   BACKGROUND TO WOODLAND COVER

10.11   Introduction

From the beginning of the post-glacial period until the late stone age, the forest cover of the British Isles reflected the nutrient status and history of the soils and their underlying parent geology.

Mankind, once beyond the hunter-gatherer stage, began to exploit the land and the conserved energy and nutrients in the natural vegetation and soil.

In the British Isles, over the last 5-6000 years, settlers spread and multiplied, managing land primarily for arable crops and the maintenance of introduced domestic livestock species.

From Neolithic times until the late 19th century, farming practices relied for growth of crops on the nutrient capital of sites, together with any release of nutrients due to weathering, woodland being removed in the process.

Often in addition, both heathland and any surviving adjoining woodland remained a continuing source of litter for bedding, turf, wood and peat for fuel, and branchwood for browse. While dung and crop residues may have been returned to farm fields, amounts returned to the woodland were small. The widespread practice of coppicing, whether for wood or for charcoal, also reduced the nutrient status of many woodland sites.

On uncultivated heathlands, moorlands and peats, the periodic burning of heather and grass to provide succulent shoots for sheep and cattle accentuated the long-term nutrient drain from such sites. The leaching of soluble salts following fire supplemented the mineral nutrient removal in body tissues of the grazing animals (Fraser, 1933). On dry southern heathland, losses due to burning were replenished by nutrients brought in by rainfall, in respect of sodium, potassium, calcium and magnesium. However, there were net losses of phosphorus and nitrogen (Chapman 1967). Up to 80% of nitrogen in burned *Calluna* may be lost in the smoke, and if the temperature of the smoke exceeds 600°C, appreciable amounts of phosphorus, potassium and iron may also be lost (Allen 1964, Evans & Allen 1971).

Infertile ground which has been cleared and has been subject to such nutrient depletion has often required more nutrient supplement from fertilizers than land remaining under woodland.

Latterly, by industrial processes drawing on the geological reserves of minerals and hydrocarbons, mankind has mobilised nutrient sources far beyond the natural reserves in subsoils accessible to natural weathering processes.

One consequence of industrial activities and population increase has been the generation of toxic or noxious polluting wastes which by their quantity, concentration or composition cannot be assimilated by the natural recycling system without undesirable side-effects.

Up to about 1970, the dominant effect as far as trees were concerned was the harmful effect of sulphur dioxide and its derivatives (§13.3 & §14.4). Since 1950, however, nitrogen-containing gases have had an increasing effect on trees and on the rural environment, not only as primary pollutants, but also as inadvertent but highly significant nutrient sources (§13.2).

10.12　　　Stimuli to reafforestation

Woodland cover in Great Britain, initially of the order of 70-80%, had decreased to approximately 5% of the land area by the beginning of the 20th century. Lowland riparian high forest had largely disappeared and many other types of woodland were reduced to genetically depleted remnants. At the beginning of the 20th century, the woodland area in Britain was one of the smallest for the size of the country anywhere in Europe. Fortunately, it held enough reserves to sustain the economy during the 1914-18 war, but these were only mobilised through widespread untimely fellings.

In 1919, the Forestry Commission was formed with the brief from the government of the day to restore the wartime fellings of conifer and broadleaved woodlands and increase the overall woodland area to 2 million hectares (Aldhous 1997). The impetus to increase the area of commercial woodland in Britain was renewed from the end of the 1939-1945 war.

During the period 1945-1975, it was also Government policy to raise agricultural production towards national self-sufficiency in basic foodstuffs. This placed a limitation on the type of ground available for forestry; proposals for the conversion of farmland to woodland had to be approved in principle by Government agricultural departments before an owner could himself proceed with planting, or sell it for planting. Consequently, whether for private landowner, potential investor or state agency, most land available for planting during this period was marginal to farming *ie* poorer land at higher elevations, infertile lowland heath, or land too steep for agricultural machinery.

Introduced conifers were the species mainly favoured because of their fast rates of growth. However, while there had been many pioneering species trials (Macdonald *et al.* 1957), silvicultural knowledge about the nutritional requirements of many introduced species was limited or non-existent, especially for the more difficult sites.

Investigations over the last 70 years into how new forests could be created and managed for vigorous growth can be grouped under three main headings:

- *empirical experimentation* into the role of fertilizers in ensuring that the species introduced under the afforestation programme would be quickly established and would grow well,
- *within-forest nutrient cycling* aiming to broaden and generalise the results of empirical experiments,
- *interaction of forest practice with environmental and atmospheric processes*. This has become increasingly important since the 1960s, partly because of the links between forests and the quantity and quality of water draining from the forest and partly because of fears originating principally in central Europe, that forests were declining in health.

The earlier approaches to use of fertilizers through experimentation in the forest and the forest nursery *etc.* are reviewed in this chapter and in *Chapter 11* respectively.

Chapter 12 reviews nutrient cycles in woodlands, and includes a short section on the interaction of mycorrhizas and conifer growth. (See also §11.13.)

Chapters 13 & 14 outline links between forest plantations, atmospheric pollutants and stream-water quality.

Chapter 15 describes effects of other pollutants (salt, heavy metals *etc*) affecting trees and woodland, while *Chapter 16* discusses CO_2 and global climatic change.

10.13 Integrated silviculture using fertilizers

From the appearance in 1933 of the first edition of *Forestry Practice*, (Anon. 1933) contemporary guidance has been available to the best silvicultural practice of the time for growing vigorous, healthy trees. In respect of use of fertilizers, this has been supplemented periodically by journal articles and bulletins giving more depth and background to such recommendations (*eg* Binns 1966, Atterson & Davies 1967, Thompson 1972, Everard 1974, Mackenzie 1974, Mayhead 1976, Blatchford 1978, McIntosh 1981, 1982, 1983a, b, 1984a, b, Taylor 1986, 1987b, 1990a, b, 1991, Miller 1966, 1981, Miller *et al.* 1986, Taylor & Tabbush 1990, Taylor & Worrell 1991).

All the results of the studies of N, P and K and their interaction with *Calluna* have been embodied in recommendations for the integrated use of fertilizers, herbicides and cultivation.

In *Fertilizers in the establishment of conifers in Wales and southern England,* Everard (1974) created a new level of integrated silviculture in linking together:

- crop species,
- vegetation and its control,
- soil type,
- region,
- nutrient need through foliage analysis,
- geology,
- depth of peat,

as the basis for prescribing fertilizers supplying N, P or K at planting and where necessary up to canopy closure. This has been replaced by *Forest fertilization in Britain* (Taylor 1991), which includes more detailed lists of lithologies and extends recommendations to Scotland. Otherwise it follows the principles established in the earlier work.

General appreciations of the background to best practice for use of fertilizers in British forests are given in *Forest fertilizing: some guiding concepts* (Miller 1981), *Forest fertilization in Great Britain* (Taylor 1986), *Dynamics of nutrient cycling in plantation ecosystems* (Miller 1984c) and others (Hagner 1966, Miller *et al.* 1986, Miller & Miller 1980, Malcolm 1997). Bonneau (1995) gives a 'state of the art' review of mineral fertilization in temperate forests.

10.14   Financial justification for fertilizers at planting and on checked crops

The prospective profitability of use of fertilizers &/or herbicides to bring trees out of severe heather check appears self-evident. However, throughout experimental programmes on the poorer sites over most of Britain, crops on some sites have responded more than others. Everard (1974) propounded the view that because fertilizers often advanced the height of crops, a conservative approach was to assume no long-term change of yield class but to use the 'years saved' or number of years of advancement to calculate the discounted benefit of using fertilizers. He developed this approach into a series of tabular decision trees for Wales and southern England, taking into account vegetation and geology, and depth of peat, if present.

Miller & Cooper (1973), reviewing growth following application of nitrogen, also concluded that 'there is little evidence to contradict the idea that fertilizer response can adequately be described using the simple analogy of an accelerated time scale'.

In the context of £2 million annual expenditure by the Forestry Commission on fertilizers in 1988, Whiteman (1988) reviewed growth responses from 30 experiments on the Moine schists of northern Scotland. Using standard NPV discounting procedures and a discount rate of 5%, he found that only on the least fertile soils was growth sufficiently boosted to cover the cost of the fertilizer.

Taylor (1991) pointed out that all appraisals of the benefit of fertilizers depend on assumptions about future growth patterns. His preference was to use final crop yield class but recognised that this will not be feasible until long-term predictive data are available from fertilizer trials. In the shorter term, he opted for conventional NPV assessments in preference to the 'years saved' approach.

10.15   Scale of use of fertilizers in British forestry

Figures for the use of fertilizer in British forestry have never been collated on an industry- or country-wide basis. However, *Table 10.1* summarises figures from various sources. They are incomplete in that they refer almost exclusively to Forestry Commission activities. Also, no attempt has been made to estimate the scale of application of phosphate by hand in the period 1930-1970.

In 1985, the cost of the year's programme approached £2 million; it was predicted that in later years, the Forestry Commission area treated would fall but

Table 10.1   *Scale of use of fertilizers in forestry in Britain*

	Year	N	P	PK	Total	Method of application
a	1957		3 200		3 200	mainly by hand
b	Pre 1960		+		327	from air
b	1961		c 8 800	c 175	c 8 975 + b 50	"
b	1962				475	"
b	1964				325	"
b	1965				125	"
b	1966				982	"
b	1967				2 675	"
b	1968				6500 est.	"
d	1960/1	0	6 900	300	7 200	n
d	1969/70	450	23 000	2 700	26 200	n
d	1971/2	200	13 600	9 200	23 050	n
d	1972/3	+	25 650	14 650	40 400	n
d	1973/4	650	32 050	18 300	51 000	n
e	1974/5				45 100	mainly from air
e	1975/6				27 100	"
e	1976/7	1000	17 000	7050	25 000	"
e	1977/8				26 100	"
e	1978/9				32 352	"
	1980/1	1650	16 450	23 050	41 150	"
f	1981-6	2000	25 000	29 000	56 000	"
	1982/3	1350	10 350	24 150	35 850	"
g	1983/4				27 304	"
h	1982/3*	4+0	84+83		171	"
h	1983/4*	95+11	109+0		215	"
h	1984/5*	5+2	800+27		834	"
h	1985/6*	14+0	177+122		313	"
h	1986/7*	235+0	422+68		726	"
i	1989/90				16 000	By air
i	1995/6	2000	1100	2200	5300	"
i	1998/9	900	340	2000	3340	"

Sources: a   Edwards & Stewart 1958: incomplete figures for FC in Scotland
b   Aldhous 1969a   c   Binns 1966   d   Binns 1975b   e   McIntosh 1981
f   Taylor 1991 (Av. annual for period)   g   Farmer *et al.* 1985 (+pers.com.)
h   Moffat & Bird 1989   i   Pers. com.   n = no data available
*   where two figure are separated by + , they refer to England and Wales respectively.

that rates of application would increase. The area of privately owned forest treated would become of similar magnitude (Farmer *et al.* 1985).

Fertilizer usage in N. Ireland was reported to include K 'refertilized' on 13000 ha of poorly growing Sitka spruce in 1991-96 (Wright 1996). The Northern Ireland Forest Service Annual Report for 1997/8 tabulated usage of fertilizers for forest improvement, showing approximately 1000 tonnes used annually in 1994 and 1995 and 2000 tonnes in 1996 and 1997; K was applied to about 62% of the area treated (DANI 1998).

10.16   Method of application of fertilizers to plantations

Prior to 1959, most fertilizer applied at time of planting was spread by hand. In 1959, fertilizer was first applied from the air to checked crops. This was at Wilsey Down, Cornwall, and Halwill forest, Devon, as part of investigations into the effect of N, P and K to relieve check (Holmes & Cousins 1960). Ground methods using tractor-mounted spinners previously tested could not easily tackle either the roughness of the terrain or the obstruction of the checked trees. The first use of helicopter to spread fertilizer was in 1967 in Kilmory forest, Ayrshire (Davies 1967). Where hand application of fertilizer, placed around newly planted trees was feasible, trees grew more quickly in the first few years than where similar land had been treated broadcast by helicopter.

Nevertheless, from these early trials, aerial application was quickly taken up (Atterson & Davies 1967). Any response by weeds rarely added to local weeding costs. *Table 10.2* shows figures for the scale of different methods of application between 1960 and 1973. A very high proportion of the fertilizer use shown in *Table 10.1* also was applied from the air.

Table 10.2   *Means of applying fertilizers to plantations & area treated (ha)*

	Year	Manual	Helicopter	Fixed wing aircraft	Total
a	1960/1	7 100	0	100	7 200
b	1962-1968	n	. 11 100*	.	11 100
a	1971/2	3 500	18 500	1000	23 050
a	1972/3	4 500	33 050	2 850	40 400

Sources	*a*	Binns 1975b	b	Aldhous 1969a	n = no data available
	*	*aircraft type unspecified*			

See also §12.54 *Nutrient losses in first-rotation crops* and §14.62 *Forest practice and water quality.*

10.2   FERTILIZERS AT AND POST PLANTING

10.21   Early trials

At the beginning of the 20th century, many landowners were well aware of the difficulties in establishing trees on their poorer land. Zehetmayr (1954) reviewed attempts in the 19th century to establish woodland on peat; these included comparison of species and use of planting on turf. However, these failed and it was not until 1907 that Sir John Stirling Maxwell, when planting on peat moorland on his estate at Courrour, Invernessshire, first introduced the use of fertilizers for forest planting in combination with planting on peat turfs. He used 'basic slag' at planting and also applied some as a top-dressing to checked plantations (Stirling Maxwell 1925).

From the outset after its formation in 1919, the Forestry Commission investigated the potential of new species and employed the most advanced establishment techniques of the time.

Full accounts of the work over the thirty years up to the 1950s on afforestation of upland heaths and peatland soils were given by Zehetmayr (1954, 1960). Wood and Nimmo (1962) described work over the same period on chalk downland afforestation and (*ibid* 1952) gave a shorter account of experiments into establishment of trees on lowland heathland at Wareham, Dorset.

In all these studies, the emphasis was on identifying the best 'integrated silvicultural package' of species, cultivation, drainage, weed control and nutrient supply that could be incorporated into current forest practice.

While in §§10.4 to 10.7 following, individual nutrients are discussed, from the earliest reports and studies, mineral nutrition has usually been seen holistically. All potential nutrients, their interactions, their relative proportions (balance), and their relation to site variables such as peat quality or mineral soil characteristics, were to be taken together (Atterson 1966b, Wright & Will 1958).

At the same time, aids to effective use have been sought through studies of soils (§10.22), through identification of deficiency symptoms (§10.23) and through foliage analysis (§10.24).

10.22   Soil type and lithology

Soil type and underlying lithology have increasingly been important characteristics in refining nutrient requirements of forest trees on fully developed mineral soils.

Description and classification of soils of sites for afforestation developed in parallel with investigations of nutrition. The *Guide to site types in forests of North and Mid-Wales* (Pyatt *et al.* 1969) linked soil characteristics and cultivation with use of fertilizers at planting and as top dressings. *Soil groups of upland forests* and the associated soil classification (Pyatt 1970, 1982, Pyatt *et al.* 1979) provided a framework for *Forest fertilization in Britain* (Taylor 1991).

At a more general level, Pyatt's *Ecological site classification for forestry for*

Great Britain (Pyatt 1995, Pyatt & Suarez 1997, Wilson *et al.* 1998) has at its core, soil nutrient status, its links with lithology and its ability to support woodland.

For peaty soils, peat types, their diagnostic vegetation and their relationship to tree nutrition and growth, have been recognised and described with progressively greater refinement by Fraser (1933), Zehetmayr (1954), Pyatt *et al.* (1979), Pyatt & Suarez (1997).

For planting on reclaimed sites, chemical analysis of the nutrient content of soil-forming materials is recommended. See also §11.3.

10.23   Deficiency symptoms

Diagnosis of symptoms of mineral deficiency in agricultural and horticultural crops developed during the 1930s. *The diagnosis of mineral deficiencies in plants by visual symptoms* (Wallace 1943) has gone through several editions, *(eg* 5th edn., Bould *et al.* 1983) and remains the classical reference for growers, farmers and market gardeners.

The first coherent account of deficiency symptoms in forest plants in Britain arose out of investigations between 1945 and 1965 into the nutrition of nursery plants (Benzian 1965) and related to 1-year-old Sitka spruce seedlings. In the historical introduction to Benzian's report, Wood mentions that, in 1923, Laing had described the symptoms of magnesium deficiency in spruce seedlings grown in nutrient culture. This early observation was not, however, followed up.

Distinctive colours or death of tissues of seedling foliage were shown to be due to deficiency of nitrogen, potassium, magnesium or copper as appropriate. Symptoms on Sitka spruce transplants were similar to those of seedlings. While only a limited range of species were included in the nursery nutrition experiments, Sitka spruce showed symptoms far more clearly than Scots pine, lodgepole pine, western hemlock or western red cedar (Benzian 1965).

In the forest, the foliage of many poorly growing plants can be seen not to have the colour normally associated with healthy plants. Causes of discolorations have been identified through experimental programmes applying fertilizers, through foliage analysis, and through experiments designed to induce deficiencies (Wood *et al.* 1961, Binns & Aldhous 1965, Binns & Mackenzie 1969).

Colour illustrations of deficiency symptoms of N, P, K, Mg and Cu on established Scots and lodgepole pine, Sitka and Norway spruce are given in *Nutrient deficiencies of conifers in British forests* (Binns *et al.* 1980), *Forest fertilization in Britain* (Taylor 1991) and for N and Sitka spruce in *Nitrogen deficiency in Sitka spruce plantations* (Taylor & Tabbush 1990).

Circumstances leading to deficiencies and options for treating them are also reviewed by McIntosh (1983b).

Reproducibility of colours indicating deficiencies

Phillips & Burdekin (1982) show foliage affected by nutrient deficiencies in lodgepole pine from the same originals that had been used by Binns *et al.* (1980). Differences in the two publications between the colour values for the same symptoms are a reminder of the problems of documenting and communicating colour symptoms which themselves vary both within the growing season and from year to year.

Symptoms of nutrient deficiency on seedlings in the nursery were matched to standardised 'Munzell' colours (Benzian 1965).

10.24   Foliage analysis

In the 1950s, analytical techniques to determine concentrations of major nutrients in foliage became standardised in respect of position on tree and season of sampling (Leyton & Armson 1955, Tamm 1955, Ovington 1956, Leyton 1957, 1958, Wright & Will 1958, Everard 1973).

In 1957, joint investigations into relationships between foliage nutrient content and conifer crops were started by the Forestry Commission and the Macaulay Institute for Soil Research (Wright 1958, Edwards & Stewart 1958, Binns 1960).

From this work, foliage analysis became the basis for determining need for fertilizers on crops in check or at risk of going into check.   Standardised procedures, selecting a shoot from the top whorl of possibly nutrient-deficient trees, were part of annual programmes of several thousand foliage analyses to determine nutrient status (Binns *et al.* 1970, Everard 1973).

For most nutrients, there was a direct relationship between foliage concentration in the plant and availability of nutrients. Only with nitrogen and phosphorus were there indications of interaction between nutrients, implying a need to consider not only levels but the balance between them.

Table 10.3 gives values for nutrients in foliage from the forest and the forest nursery. Values are similar but not identical. Figures in columns headed *Def* (deficient) indicate that the plants would benefit from added fertilizer. Figures in *Sat* (satisfactory) columns indicate a satisfactory level, crops with such levels being unlikely to respond to fertilizer. Values intermediate between the two are 'uncertain', necessitating further on-site consideration of the trees in relation to others in the vicinity and their history.

Nutrient concentrations of seedlings and transplants

Published nutrient content of foliage of trees in the forest are normally based on analyses of foliage. Figures for nursery plants are based on whole shoot or whole plant samples.

In addition to the figures in *Table 10.3*, data are available for nutrient concentrations in 'ordinary' plants not exhibiting deficiency symptoms for:

• whole plants (N, P & K) for Sitka spruce seedlings and transplants from main treatments of long-term fertility experiments (Benzian *et al.* 1972);

Table 10.3 *Nutrient concentrations in foliage of forest trees & nursery stock*

		N Def %	N Sat %	P Def %	P Sat %	K Def %	K Sat %	Mg Def %	Mg Sat %	Ca Def %	Ca Sat %	Cu Def %	Cu Sat %
Spruces	a	1.2	1.5	0.14	0.18	0.5	0.7	0.03	0.07	n	n	(1.5	2.4)*
	b	1.2	1.6	0.16	0.18	0.5	0.7	0.06	0.08	0.10	0.15	3	5
LP &	a	1.1	1.4	0.12	0.14	0.3	0.5	0.03	0.05	n	n	n	n
SP	b	1.5	1.8	0.16	0.18	0.6	0.7	0.07	0.10	0.06-0.10		3...5	
Corsican	a	1.2	1.5	0.12	0.16	0.3	0.5	0.03	0.05*	n	n	n	n
pine	b	1.5	1.8	0.16	0.18	0.6	0.7	0.07	0.10	0.06	0.10	3...5	
Western	a	1.2	1.5	0.25	0.30	0.6	0.8	n	n	n	n	n	n
hemlock	b	1.6	1.8	0.18	0.20	0.7	0.8	0.10	0.12	0.15	0.20	3...5	
Douglas	a	1.2	1.5	0.18	0.22	0.6	0.8	0.04	0.06*	n	n	(1.1	1.5)*
fir	b	1.6	1.8	0.18	0.20	0.7	0.8	0.10	0.12	0.15	0.20	3...5	
Larches	a	1.8	2.5	0.18	0.25	0.5	0.8	n	n	n	n	n	n
	b	2.0	2.5	0.20	0.25	1.0	1.2	0.10	0.12	0.20	0.25	3...5	
Other conifers	c	1.6	1.8	0.18	0.20	0.7	0.8	0.10	0.12	0.15	0.20	3...5	
Alder	c	2.5	2.8	0.16	0.18	0.7	0.9	0.08	0.13	n	n	n	n
	b	2.3	2.8	0.18	0.25	1.0	1.2	0.10	0.15	0.15	0.20	3	5
Birches	c	2.5	2.8	0.19	0.22	0.7	0.9	0.08	0.13	n	n	n	n
	b	2.3	2.8	0.18	0.25	1.0	1.2	0.10	0.15	0.15	0.20	3	5
Oak	c	2.0	2.3	0.14	0.16	0.7	0.9	0.08	0.13	n	n	n	n
	b	1.7	2.3	0.14	0.20	0.7	1.0	0.15	0.20	0.20	0.30	3	5
Ash &	c	2.0	2.3	0.19	0.22	0.7	0.9	0.08	0.13	n	n	n	n
N. maple	b	2.3	2.8	0.18	0.25	1.0	1.2	0.10	0.15	0.15	0.20	3	5
Beech &	c	2.0	2.3	0.14	0.16	0.7	0.9	0.08	0.13	n	n	n	n
Sw ch'st't	b	1.7	2.3	0.14	0.20	0.7	1.0	0.15	0.20	0.20	0.30	3	5
Cherry lime & willow	b	2.3	2.8	0.18	0.25	1.0	1.2	0.10	0.15	0.15	0.20	3	5
Other broad-leaves	c	1.7	2.3	0.14	0.20	0.7	1.0	0.15	0.20	0.20	0.30	3	5

Sources   a Binns *et al.* 1980   b Proe 1994   c Taylor 1991
* = *limited data.*   *a* and *c* = nutrient content in needles (forest trees);
b = nutrient content - shoots for conifers; - foliage for broadleaves (nursery stock).

- needles (N, P, K, Mg) of 1-year seedlings of 7 common conifers grown in the south of England (Aldhous 1972a);
- tops and roots of seedlings of five conifers grown in research experiments in the south of England, together with figures from Norway, the United States, Canada and Czechoslovakia (Benzian & Smith 1973);
- whole plants (transplants) sampled after 15 years of cropping in a *forms of N x forms of P* factorial experiment (Benzian *et al.* 1974);
- whole plants (N, P, K, Mg, Ca, Mn) from naturally occurring wild populations of Sitka spruce in south east Alaska (Farr *et al.* 1977).

Table 10.4 gives values for trace-elements in young Sitka spruce. The figures show the range encountered in healthy trees. There is no basis for knowing how far the smaller value is from 'deficiency' level; the smaller values themselves do not indicate deficiency.

Table 10.4   *Trace elements in foliage of healthy Sitka spruce*

Observed range of nutrients in foliage of 7-12 year-old plants from 10 soil types

Al ppm	B ppm	Ba ppm	Co ppm	Cr ppm	Fe ppm	Mn ppm
100-830	13-29	8-62	0.11-0.47	1.6	20-294	205-1180

Mo ppm	Ni ppm	Pb ppm	Sr ppm	Ti ppm	V ppm	Zn ppm
0.07-0.33	1.9-15.2	0.9-7.6	5-125	1.9-21.4	<0.1-2.3	28-83

Source   Everard & Mackenzie 1974

Figure 10.1 shows the relationship from an experiment in mid-Wales showing how increment over a 15 year period reflected the initial differences in the P content of Sitka spruce foliage. (See also *Figure 10.2* in §10.54, *Pretreatment N levels in foliage in relation to subsequent response to N.*)

10.3   FERTILIZER   NEED   AND   STAGE   OF   WOODLAND DEVELOPMENT

10.31   Woodlands

In Britain, response to fertilizers has been investigated principally for conifers on sites where lack of nutrients appeared to be limiting,

- on afforestation sites between planting and closure of canopy, §10.4 - §10.7.
- in pole-stage and older crops, §10.82 - 10.84,

Figure 10.1   *Foliar phosphorus and subsequent height increment in Sitka spruce*
Tarenig, Ystwyth Forest, Montgomery

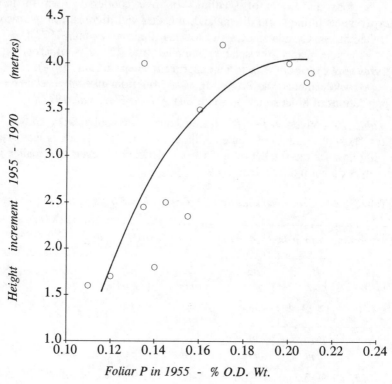

Source   Binns *et al.* 1972

- for successor crop establishment, §10.85.

The nutrients most widely tested have been phosphorus, nitrogen and potassium. Magnesium, copper and sulphur have also been investigated but are rarely lacking.

Calcium has been applied from time to time as limestone and is included as a component of several phosphatic fertilizers. It has also been used as a means of reducing the effects of acidity in streams (§14.61) but not as a direct nutrient for tree growth.

Broadleaved woodland

Most commercially managed broadleaved woodland in Britain is on relatively fertile sites and has not been thought to require fertilizers to supplement the nutrients available from the site (Miller 1984d). However, on some sites nutrient deficiencies may occur:

- lowland heaths - phosphorus;

- restored man-made sites - nitrogen;
- chalk downland - nitrogen and potassium;
- long-worked coppice woodland - phosphorus (Evans 1984, 1986b).

Notes on individual species are sparse, reflecting the overall low expectation of fertilizer responses by broadleaves. Most relate to use of fertilizers on established woodland or coppice, rather than during establishment, and are summarised in §10.85.

Uneven-aged woodland

While there is contemporary interest in uneven-age woodland systems, they are dominantly closed canopy woodland systems. They are closest to the Miller's 'Stage 2' of plantation woodland, *ie* with sufficient nutrient capital on the site to maintain the woodland, accommodating the more localised and smaller scale of felling associated with uneven-age woodland management (Miller 1981).

10.32   Ornamental trees

Trials of the effect of fertilizers on broadleaved species of a range of ages in landscape schemes showed that poor soil structure and weed competition were more detrimental to good establishment than lack of nutrients.

There were no detectable effects of added fertilizer on avenues of lime, horse chestnut or London plane. For younger trees, there were limited increases in height growth when fertilizers were applied in conjunction with herbicides. However, herbicides by themselves brought about the biggest improvement in growth, implying that competition for water was more critical on these sites (Davies 1987g).

10.4   PHOSPHORUS (P)

10.41   Forms of phosphatic fertilizer

Five main sources of phosphate have been applied to forest plantations during the present century,

- basic slag   • bone meal   • ground mineral phosphate
- superphosphate   • triple superphosphate.

Basic slag was a by-product of steel-making. For forest use, it was ground to a fine black powder and applied in this form. The amount of available phosphate was determined by its solubility in citric acid. The most suitable basic slag had the highest proportion (c. 80%) of its P content as 'citric soluble phosphate'. As the scale of the steel-making industry decreased and the technology involved changed, less basic slag was produced and at the same time it became less suitable for use in the forest. While formulation as a small granule overcame difficulties in handling when in dust form, basic slag fell out of use, being replaced by rock phosphate. Basic slag also had neutralising effect on acid soils, its lime content being equivalent to approximately 36% CaO.

Bone meal is a product of animal carcass processing. Bones after being degreased are ground to a fine meal texture for use in horticulture. Bone meal contains 3-5% nitrogen. It was used successfully in many early trials, particularly those on the Dorset heathland. It gave results similar to those obtained from basic slag (Wood and Nimmo 1952) but was discarded on grounds of its much greater cost.

Rock phosphate or *mineral phosphate* occurs as a mineral deposit in the form of apatite in several parts of the world, notably the Pacific island of Nauru, Florida (USA) and in Tunisia (Gafsa). It is the starting point for the manufacture of commercial water-soluble phosphatic fertilizers. It is imported as 'unground' material, similar in consistency to a dusty coarse sand. Initially it was applied as a finely ground powder (ground mineral phosphate or GMP) so specified that >80% passes a '100-mesh' sieve. GMP has also been available in granular form. Trials in the 1950s and 60s, however, showed that unground rock phosphate (URP) gave results similar to those using GMP; URP is now the most commonly used material.

The quality of rock phosphate may also be assessed by its 'citric soluble' phosphate content. For most sources, less that 40% is citric soluble, a lower level of solubility than for basic slag. The material from Gafsa has had the highest citric-soluble P content and has been the source of rock phosphate most widely used in British forestry.

Superphosphate and *triple superphosphate* provide the quickly available water soluble phosphate required in agriculture and horticulture. Through industrial treatment with sulphuric acid, rock phosphate is converted into water-soluble phosphates. Superphosphate contains about 45% $CaSO_4$. Triple superphosphate contains very little calcium sulphate and provides the same amount of available P from about a third of the weight of material, reducing handling and distribution costs accordingly. Both are commercially available as granular fertilizers, either singly or more often in combination with nitrogen and potassium (Binns 1975b, McIntosh 1984b).

Potassium metaphosphate has been available as a fertilizer; it was a 'low salt' alternative to conventional agricultural PK fertilizers and is not water soluble (Binns 1965).

10.42   P on peats

Through the Belgian turf system introduced by Sir John Stirling Maxwell, phosphorus, as 'basic slag', was the first mineral nutrient successfully applied to forest plantings in Britain. It was used for planting on *Molinia* peat moorland; the system combined drainage, spreading of peat turfs and application of basic slag, and gave rise to good growth of young plants. In contrast, planting into notches in undrained moorland generally failed. Anderson in 1925 laid down Forestry Commission phosphatic fertilizer trials; Fraser (1933) reported small scale experiments on peat from the same period, using N, P K and Ca fertilizers at agricultural rates. These were the beginnings of the substantial programme of experimentation on peatlands in Britain; they led to the extensive successful planting of poorer peats previously considered unplantable (Anderson 1997).

Neither basic slag nor mineral phosphate were manufactured to be marketed as a fertilizer; their effectiveness for newly planted young trees depended on P content and solubility. In an experiment in western Invernessshire set up in 1937, Japanese larch was treated with equal gross weights of various grades of basic slag and a standard form of mineral phosphate. After 13 years, best results were obtained from highest grade materials, because, on a weight for weight basis, they contained more phosphate. For equal amounts of P, there was little to choose in the responses obtained. The most cost-effective application amount was about 5-6 grams P per tree (Edwards 1959, figure 5).

By the 1990s, while regimes could be defined to ensure the satisfactory growth on Sitka spruce and lodgepole pine on any peat that could be drained, concern for the peatland habitat and in particular the 'Flow Country' of northern Scotland had led to virtual cessation of planting on extensive tracts of deep peat (Anderson *et al.* 1995, Anderson 1997). Active raised bogs and active blanket bogs have come under the EU Habitats and Species Directive (FC AR 1998).

10.43   Heaths

After peatlands, the second major target for conifer planting in the first years of the Forestry Commission was upland and lowland heathland. A full account of the distribution of heathlands in northwest Europe and their relation to British heathland is given in *Ecology of heathlands* (Gimmingham 1972). *Heathlands* (Webb 1986) contains a more detailed general account of British lowland heaths.

As with peat, first forestry plantings of trees on untreated ground did not do well. Subsequent research programmes soon showed that for upland heaths, critical factors were soil compaction and impeded drainage caused by the presence of peaty pans and iron pans. Cultivation improved survival and growth; in many sites, the larger the volume of soil disturbed, the larger was the initial growth response.

At the same time, phosphorus, as basic slag, ground mineral phosphate, superphosphate or a mixture of the latter were all tried in combination with cultivation treatments. Superphosphate by itself or in mixture placed under the turf before planting reduced plant survival in drought conditions, both on heathland and peatlands. However, subsequent work showed that if spread over the ground surface, losses were much less severe (Edwards 1958).

The application of 27-54 grams (1-2 ounces) of high grade high solubility basic slag or ground mineral phosphate per tree consistently improved early growth on many poorer heathlands. (The rate is equivalent to the 6 g P per tree quoted above for Japanese larch on peat.)

Table 10.5 gives results after 25 - 27 years, from two experiments laid down in 1929 and 1933 on upland heathland. The beneficial effect of P has persisted. Differences in stem number reflect thinning differences. The overall effect on growth is shown by height and total basal area figures.

These and other experiments showed that on the poorest upland heaths characterised by *Calluna-Trichophorum* vegetation, phosphate is essential for the establishment of species other than pines. For less demanding sites, while it is not essential, it is often useful where rapid early growth is required, the effects persisting. On the most fertile heaths, there is no response to added phosphate if there has been adequate ground preparation.

In the southern half of Britain, heathland is less extensive. Nevertheless, attempts were made in the 1920s to establish maritime pine (*Pinus pinaster*) by direct sowings on the infertile sands and gravels on the Surrey-Hampshire border, and in Dorset on the heathland extending between Dorchester and the New Forest. The failure of these led to a long series of experiments into alternative establishment techniques to assist direct sowings.

Table 10.5   *Long-term effect of phosphate applied at planting on upland heathland*

Experiment species and treatment	Age	Top height m*	Stems per hectare	Stem diameter cm	Total crop basal area sq m / ha
Teindland 41 P 29	27				
Lodgepole pine		7.8	5500	8.1	29
" + 57 gr (2 oz) basic slag/tree		9.0	5100	9.7	42
Wykeham 26 P.33	25				
Lodgepole pine		9.4	3100	10.5	37
" + 57 gr (2 oz) basic slag/tree		9.6	2800	10.9	39
Scots pine		8.5	3200	10.5	30
" + 57 gr (2 oz) basic slag/tree		9.0	2700	11.3	34
Corsican pine		8.2	3200	11.3	37
" + 57 gr (2 oz) basic slag/tree		9.1	2600	21.8	47
Japanese larch		9.1	1750	11.3	23
" + 57 gr (2 oz) basic slag/tree		11.6	1430	21.8	35

Source   Based on Table 50, Zehetmayr 1960

*Top height = mean height (metres) of 250 largest diameter trees per hectare

At Wareham, as with upland heaths, it soon became apparent that the combined provision of a source of phosphate and cultivation were necessary for good germination and growth of direct sown seedlings (Wood & Nimmo 1952).

10.44   1950s to the present

Since 1950, a wide range of experiments have tested refinements, alternatives or extensions to current practice. These included:

• differential rates of phosphate in species mixtures to even up growth rates (Zehetmayr 1951);

• tests of triple superphosphate (Wood & Zehetmayr 1955, Edwards & Stewart 1958);

• trials of heathland nursery seedlings in place of transplants (Green & Wood 1956). This trial showed that one-year old seedlings could not be relied on to survive and grow as well as transplants but also indicated that phosphate responses were only obtained on mineral soils where the amount of P in the subsoil (upper C horizon) was less than 1500 parts per million;

• relief of checked spruce on Cornish heathland,
      by triple super-phosphate (Wood & Zehetmayr 1955);
      by aerial application of P (Wood & Holmes 1958, 1959);

• relief of poor quality spruce in mid-Wales (Wood *et al.* 1960);

• relief of poor growth of DF on acid and impoverished Wealden soils in Kent (Wood *et al.* 1960);

• additional P and/or K on deep peat sites (Edwards *et al.* 1960); forms of P (Binns *et al.* 1970);

• relief of checked spruce using triple superphosphate in Devon (Soussons, Dartmoor). Applications from the air brought part of the Sitka spruce crop out of check. Over the remainder, N top-dressing improved growth for three years (Binns & Aldhous 1966);

• P compounded with K as potassium metaphosphate compared to mixtures as 'potassic superphosphate' on deep peats (Atterson 1966b). Results after three years showed potassium metaphosphate to be less effective than potassic superphosphate (Binns & MacKenzie 1969);

• sources of rock phosphate, ground and unground on peat soil. Ground rock phosphate applied immediately around each tree slightly reduced survival of lodgepole pine in proportion to the rate used, but not if broadcast (Binns & Atterson 1967);

• forms and rates of P on Sitka spruce on deep peat in Northern Ireland (Dickson 1971). Broadcast application initially gave better results than application as a ribbon along the plough ridge. Both were superior to application under the ridge;

• production of generalised response curves between concentration of P in foliage and growth for a range of species and sites (Binns *et al.* 1972);

• interaction of P fertilization and climate, indicating reduced sensitivity to temperature effects and greater negative response to wind where P supply was adequate (Malcolm & Freezaillah 1973);

• review of nutrient requirements for trees on peat (Atterson & Binns 1975);

- interaction of clonal selections to rates of phosphate, and indications of clonal differences (Taylor 1988);
- PK top-dressing to 18-year-old crops on ground given different intensities of cultivation. There was no interaction, the PK giving a boost to trees on all treatments (Wilson & Pyatt 1984);
- phosphorus uptake by 30-year-old birch came largely from the top 10 cm of the soil; less than 15% came from 50 cm depth (Harrison *et al.* 1988).

A review of 35 experiments since 1960 showed that response to P at planting reflected the lithology and soil type; it was not related to rainfall or elevation (Taylor & Worrell 1991).

10.5   NITROGEN (N)

Nitrogen differs from phosphorus or potassium in that nitrogenous compounds can be brought to a site in any of three ways:
- in solid form as a conventional fertilizer (§10.51);
- via transfer from nitrogen-fixing plants, *eg* leguminous plants, alders and other plants growing on site; these, however, in turn may also require adequate supplies of essential nutrients to be able to become established and grow (§10.52);
- by deposition from the atmosphere of nitrogen oxides and ammonia from the atmospheric pollutant load (§13.2).

Nitrogen can also be lost from a site in soluble nitrogen compounds in drainage water (§12.55). Also, fixed nitrogen can be lost from soil by denitrification.

10.51   Forms of nitrogen fertilizer

Urea has been the N fertilizer most widely applied to forests. It has a higher content of nitrogen (46%) than ammonium nitrate (34.5%) or ammonium sulphate (20.6%). Urea is stable when dry, but may break down and lose ammonia to the air in warm, moist conditions in the absence of rain. Once dissolved, urea is converted by hydrolysis into ammonium compounds. When washed into the soil, these are held firmly in the soil by organic matter (Binns 1975b). However, if applied to saturated land following heavy rain, some of the ammonium from hydrolysis may not pass through the soil but be carried by surface water drainage into water courses. Areas treated have to be set so as to minimise the risk of causing the levels of ammonium in the water to exceed the maximum concentration set by the *European Commission Water Directive (78/659/EEC)* (Nisbet & Stonnard 1995).

Ammonium nitrate is a strong oxidising agent; it is also available mixed with calcium carbonate as an agricultural fertilizer but in neither form is it used routinely in the forest. Nitrate applied to the soil in fertilizers is readily leached by heavy rain.

Ammonium sulphate has probably the longest history of use in forestry of any inorganic nitrogen fertilizer. However, while the ammonium component is commonly taken up relatively quickly, the sulphate remains in the soil and acidifies it. Ammonium sulphate used in forest experiments in the 1930s caused plant losses.

Hoof and horn and *dried blood*, both organic forms of nitrogen, have in the past been used in horticulture and in forest nurseries. They have largely been superseded by slow-acting fertilizers. None of the organic forms of N is used in the forest.

During the first half of the present century, the potential for additions of nitrogen in developing forests on difficult peat and heathland sites was not well understood. Applications of soluble nitrogen fertilizers were included in small scale trials in the 1920s and 30s but these were not followed up. Fraser (1933) reported that trees planted on peat to which nitrogenous manures had been applied were a better colour; however, growth was not appreciably increased and the colour advantage was generally lost in the second year. Zehetmayr (1954) gives figures for higher losses as a consequence of application of ammonium sulphate to plants on peat. Use of ammonium sulphate applied to new plantings on upland heath at Wykeham forest, Yorkshire, also caused heavy losses (Zehetmayr 1960)

This experience of damage or losses following use of soluble fertilizers damped enthusiasm for further enquiries until the 1950s.

10.52   Nitrogen-fixing plants

The value of nitrogen-fixing plants is widely recognised as a mainstay of many agricultural cropping systems, and also when reclaiming disturbed land. The two main groups are leguminous species and alders.

Trials of leguminous species

In legumes, root nodules develop in symbiosis with bacteria in the family Rhizobiaceae. Species of *Rhizobium* ('rhizobia') colonise many temperate legumes, some being exclusive to their host. Cultures of selected strains are commercially available for agricultural crop species.

In the early 1930s, broom was tested as a nurse for direct-sown seedlings and newly planted stock on soils of low nutrient status. Trials were sited on heathlands where difficulties were being encountered in establishing vigorous young plantations.

More than 70 experiments were established, continuing over a period of 30 years; they examined the potential benefit mainly of common broom (*Cytisus scoparius*) but also of 10 other leguminous species (Nimmo 1952, Nimmo & Weatherell 1961).

Experiments with common broom showed that crop trees grew most strongly when vigorous broom was grown about 0.9 m from the crop trees either in parallel rows or in mixture. The broom had to be kept back by frequent cutting so as not to damage the nursed species. Trees failed to respond where the number of planted broom was brought down to 1000 per ha. Broom patches were fertilized with phosphate at planting at the same rate per patch as the patches where trees were planted; the site (and the crop overall) thus received twice as much P as a crop planted without a broom nurse. However, it was subsequently shown that there was a continuing benefit where broom had been sown, over and above any improved growth due to extra phosphate.

The increase in growth due to broom has been attributed to the continuing release of nitrogen from root nodules. Nevertheless, the treatment was found to be too expensive and was superseded by application of N or NP fertilizers to checked crops where required.

Alders

Alnus species form a symbiotic relationship with actinomycete fungi in the genus *Frankia*. On alder roots, nodules form; within them *Frankia* spp. fix nitrogen.

In the nursery, alder beds can be inoculated with *Frankia* by watering with a coarsely filtered extract of crushed nodules (Wheeler *et al.* 1991, Walker & Wheeler 1994). Cultures of selected *Frankia* are also available. Where nursery seedbed soil has been sterilised with dazomet, response to inoculation depends on the source of inoculum, best results having been obtained with cultures (Moffat 1994b).

Nitrogen top-dressings markedly decreased nodulation; when planted on open-cast reclamation sites, the first year increment of top-dressed plants was markedly less than plants which had higher proportions of root nodules when planted out (McNeill *et al.* 1989).

Alder spp. have been widely and consistently recommended for mineral soil reclamation. Growth has generally been good (Wilson 1985a, Jobling 1987, Jobling & Carnell 1985, Malcolm *et al.* 1985). Vann *et al.*(1988) describe early growth of four alder species, *Alnus glutinosa, A. rubra, A. viridis* and *A. sinuata* on former ironstone workings, commenting that the shrubby alders, in particular *A. viridis*, being bushy and giving good ground cover, offer an additional element when designing landscapes involving land reclamation. Binns & Fourt (1981) recommend that alders or *Robinia* should form 50% of planting on restored sites which have no organic matter or top soil. *Alnus incana* has been disappointing on thin chalk soils (Wood & Nimmo 1962).

Comparison of strains of *Frankia* inoculants on six alder species showed that one strain usefully enhanced nodulation on five species but had no significant effect on common alder (Moffat & Roberts 1987).

See also §11.31 *Landscape and reclamation*.

Alder in mixtures on immature soils

Young, naturally regenerated Sitka spruce colonising glacial outwash grew more strongly where associated with *Alnus sinuata*. The N, P, K and Mg content of the young spruce growing with the alder was close to that of nursery transplants and was appreciably higher than in those growing without alder (Farr *et al.* 1977).

Other leguminous and non-leguminous species for urban planting

For use in urban areas, Bradshaw *et al.* (1995) recommend buckthorn, alders and *Robinia*; they also mention clover and lupins.

10.53 Conifer nursing mixtures and Calluna

Towards the end of the 19th century and following years, on the better forest soils, Norway spruce was commonly planted in mixture with European larch and Scots pine. The larch was removed in early thinnings for fencing, the spruce and pine remaining in intimate mixture for many years. On drier sites, the final crop was predominantly pine; on damper sites, spruce predominated. However, by the 1950s, planting in mixtures was no longer widely practised. (Macdonald *et al.* 1957).

Where Sitka spruce had been planted pure on *Calluna*-dominated heathlands, it was observed that it commonly went into 'check' after a promising start. Spruce growing in close proximity to Scots pine or Japanese larch, however, was less affected. If planted in intimate mixture, Sitka spruce or other 'heather-sensitive species' could be stimulated by the growth of these nurse species (Macdonald 1936, Macdonald & Macdonald 1952, Weatherell 1953, 1957, O'Carroll 1978, Taylor 1985).

Additions of litter of nurse species to a checked stand of Sitka spruce led to increases in annual shoots and increase in foliage content of nitrogen, indicating a possible mechanism for the nursing effect (Leyton & Weatherell 1959).

At Inchnacardoch Forest, Invernessshire, in a mixture of Sitka spruce with lodgepole pine, spruce growth was correlated to the amount of nitrogen in spruce foliage and the degree of suppression of heather by the pine (Binns 1960). At the same forest, spruce benefitted from being grown in mixture with larch, in spite of *Calluna* being a minor component of the ground vegetation (dominated by *Molinia caerulea*) and the larch being so stunted by deer browsing and frost that its beneficial effect on spruce could not be ascribed to vegetation suppression (Miller *et al.* 1986).

In the Irish Republic, studies showed that:

- nitrogen concentrations in spruce on mixed plots were considerably higher than in spruce in pure plots;
- the extra nitrogen came from the soil; the soil in the mixture plots showed a corresponding decrease in total nitrogen;
- there was a more rapid turnover of N in the mixtures compared with the pure spruce plots; nitrogen in throughfall and in litterfall was higher in

mixture plots. The quantities were not sufficient to account for the improved growth;

 • there were substantial differences in the fungal flora associated with spruce mycorrhizas in mixed and pure stands. More work was required to determine whether this difference accounted for the improved growth. Nevertheless observational evidence of the spatial arrangement of mixtures (better growth from mixtures on the same plough ridge compared with mixtures separated by plough furrows) strengthened the view that root activity was a key issue (Carey *et al.* 1988).

Relative rates of growth in some mixtures has been unbalanced; in particular, fast-growing lodgepole pine provenances have so outgrown inadequately fertilized spruce that the latter has been suppressed. Because of the poor stem form of the pine, the immediate reaction was to plant Sitka spruce pure and to rely on additions of fertilizer to sustain its growth. However, in the Irish Republic, it was recognised that the slower growing origins of lodgepole pine could play a valuable role on peat soils, where neither Scots pine nor larch thrived (Carey *et al.* 1988).

Reviews of experiments in Great Britain confirmed the value of mixtures on the growth and nutrient status of Sitka spruce on poor sites. In mixture with lodgepole pine, Scots pine or Japanese larch, spruce was markedly less affected by the effects of nitrogen deficiency than when grown pure (McIntosh 1983b, Taylor 1985). *Figure 10.2* shows how, at Inchnacardoch, after 18 years with either a nurse or nitrogen fertilizer application, Sitka spruce was nearly twice as tall as the single species unfertilized control (Taylor 1985).

The ratio of nurse species to spruce ranged from 33% to 75%, the latter predominating. Spruce grown as 2 rows of spruce to 1 Scots pine grew almost as well as where a higher proportion of nurse species had been used (McIntosh 1984c).

To avoid the suppression previously experienced when using the more vigorous lodgepole pine of coastal Washington origins, Alaskan or Skeena River origins are recommended, using equal numbers of nurse and nursed species.

The benefit of nursing species could be a consequence of pine or other nurse species having mycorrhiza-forming fungi in common with the spruce so that both species benefit from the additional nutrient supply, the nurse being the main source of carbohydrate for the fungi (Read & Finley 1985, Walker 1987).

In 1984, a joint project between the Macaulay Institute for Soil Research, Aberdeen and the Forest & Wildlife Service of the Republic of Ireland was mounted to study the nutrient cycles in mixture crops (Miller 1984a, Miller *et al.* 1986, Carey *et al.* 1988).

See also §12.3 *Mycorrhizas, Heather check and Nitrogen.*

Investigations directed towards maximising yield rather than analysing nutrition set down mixtures of Sitka spruce/western hemlock, Douglas fir/western hemlock, and Scots pine/red oak (Wood *et al.* 1960).

Figure 10.2 *Height growth of Sitka spruce grown pure, with and without nitrogen, or in mixture with lodgepole pine or Japanese larch*

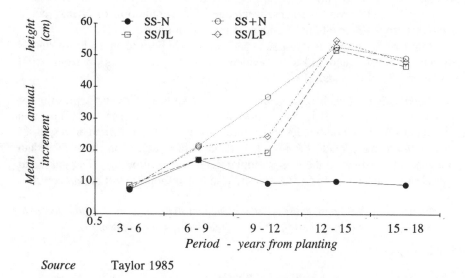

Source Taylor 1985

10.54 Nitrogen fertilizer investigations since 1950

Research programmes in the 1950s and 60s followed the complementary paths of investigating chemical means of killing heather in young plantations (see §8.42) and the effects of nitrogen top-dressing in checked or potentially checked crops. Studies included:-

- use of N to relieve checked SS (Zehetmayr, 1952);
- forms of fertilizer, comparing salts of weak acids with those of strong acids (Wood & Holmes 1958, Hinson and Reynolds 1958);
- slow release N (also K) through large size granulation (Wood & Holmes 1959);
- N top-dressing on checked crops on Dartmoor, Devon (Binns & Aldhous 1966);
- effects of rates of N fertilizer in upland Britain on spruce (Malcolm 1972, Taylor 1987b) and lodgepole pine (McIntosh 1982). Responses were short-lived.
- studies of the interaction of NPK fertilizing and heather control with 2,4-D, seeking to identify sites where N may not be necessary (Mackenzie 1974a);
- nitrogen requirements of Sitka spruce on low nutrient status peat in Northern Ireland (Dickson & Savill 1974, Dickson 1977);
- comparison of root responses of young spruce and sycamore to added N (Mackie-Dawson *et al.* 1995).

From these investigations, it became clear that on the less fertile colder soils, in spite of adequate P, growth of Sitka spruce becomes limited by shortage of available N. Use of herbicides (2,4-D) to kill *Calluna* increases its availability temporarily. However, the subsequent rate of mineralisation of N may be so slow that crops become N-deficient and growth is limited.

Many of the results from these experiments have been integrated into studies of nutrient cycles and linked to evidence of nitrogen deposition from aerial pollution. See §12.3 and §13.2.

Figure 10.3 shows height increment of Sitka spruce following application of N fertilizer on five sites in Scotland of differing responsiveness. *Calluna* was also controlled using herbicides; the figure also shows height increment responses to this treatment. While the response to herbicides is less on these sites than obtained by fertilizers, there is a similar inter-site relationship. The sites that were less responsive to N fertilizers were also less responsive to *Calluna* control.

Figure 10.3   *Responsiveness of Sitka spruce on five sites to added nitrogen & to heather control, shown by relative height increment over 6 years*

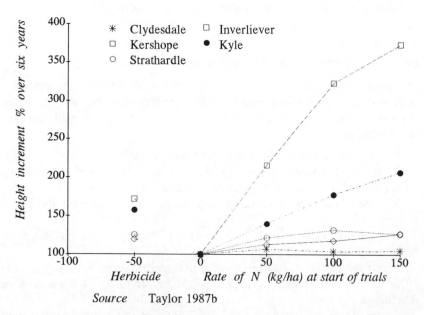

Source   Taylor 1987b

Nitrogen and peat

Peat contains substantial reserves of nitrogen in organic residues of peatland plants *etc.* Mineralisation proceeds in proportion to availability of oxygen, nitrogen accumulating in undrained peat.

When cultivated and drained, mineralisation increases, the rates depending on peat type, local climate and the extent of disturbance. *In situ* study of mineralisation is not possible because of uptake by plants and micro-organisms.

In a laboratory study of peats in northern Scotland, mineralisation rates were found to increase in the order, raised bog < unflushed bog < flushed blanket bog, reflecting the characteristics used to evaluate peatland for afforestation purposes. The top 300mm of peat on three flushed blanket bog sites averaged 7100 kg N/ha, whereas a tree crop, age 47, Yield Class 10 was estimated to contain 280 kg N/ha, *ie* less than 4% of the nitrogen in the upper part of the profile (Williams *et al.* 1979). See also §16.26.

Nitrogen deposition in lowland Britain

While atmospheric 'fixed' nitrogen is discussed more fully as a pollutant and nutrient in § 13.2, the scale of nitrogen deposition as reported for the site of the long-term continuous wheat experiment at Rothamsted, Hertfordshire is worth noting. Here deposition of atmospheric nitrogen has increased from about 10 kg/ha/yr in 1843 to 45 kg/ha/yr in 1997 (Goulding *et al.* 1997). Woodland canopies may have intercepted more.

Effect of complete weeding on foliage nitrogen

After two years of relative freedom from weeds, foliage of 7 species of conifers showed small increases in foliage N, compared with trees on unweeded ground (Brown & Mackenzie 1969). No growth effect was subsequently reported.

10.6   POTASSIUM (K)

Evidence of any benefit through use of potassium fertilizers on peat or on upland heathlands was almost completely absent in any of the work on these sites up to the 1960s. Fraser (1933) concluded from trials on peat that potash manures (potassium chloride, wood ashes, potassium sulphate) may produce a small temporary improvement in the second year after planting, or they may have no effect. In his summaries of the work on peats and upland heaths, Zehetmayr (1954, 1960) could produce nothing to show a need for K.

Planting on chalk downland showed very little response to any fertilizer at the time of establishment. Wood & Nimmo (1962) noted a benefit from potash to young Scots pine transplants on sands overlying chalky boulder till but considered the response of no practical significance as untreated plants grew adequately.

Stimulated by the clear benefits of inorganic fertilizers in nursery production programmes, potassium fertilizers were included in experiments on sites thought to be limiting such as slowly growing crops in dense grass, dwarf gorse, bilberry and heather on steep slopes in mid-Wales (Wood & Zehetmayr 1955, Edwards & Holmes 1956). One of the first favourable reports of the use of combined PK fertilizer was from experiments on soils overlying serpentine, local to the Lizard in west Cornwall. However, further experiments on the Cornish sites testing the two nutrients separately and in combination showed no growth effects from K but benefits following P application (Wood & Holmes 1958).

The role of foliage analysis and the identification of deficiency symptoms are described in §10.23 and §10.24 above. Recognition of such symptoms and the availability of foliage analysis led to the realisation that it was on peat soils that K was frequently deficient and that regular supplements might be necessary (Binns 1960, 1964, Atterson 1964).

Work and reports developing this hypothesis included:

- slow release N and K through use of large size granules (Wood & Holmes 1959);
- large scale trials of lodgepole pine on deep peat to follow up need for K applications after planting (Edwards *et al.* 1960);
- factorial PxK experiments and foliage analysis on suspected K deficiency on peats in Wales (Binns 1965);
- selection of Towy, mid-Wales as a site for a long-term experiment, being more K deficient and less N deficient than expected (Binns 1966);
- potassium applied on deep peat in northern Scotland eight years after planting. It stimulated growth for four years but was insufficient to sustain good growth for a longer period (Atterson 1966b);
- P & K at Mabie forest, Dumfriesshire on raised lowland bog enhanced growth when applied at planting, though the highest rates of K caused scorch (Binns & Atterson 1967);
- potassium uptake was best following fertilizer application in May-July rather than at other times of year (Binns *et al.* 1973);
- re-emphasis of the importance of potassium as well as phosphorus on sites where peat is over 25 cm deep (McIntosh 1979);
- trials of types and rates of fertilizer, potassium chloride being effective while a potash-felspar was not (Dutch *et al.* 1990);
- review of 23 experiments since 1960 showing K response mainly occurred on deep peats and was not related to rainfall or elevation (Taylor & Worrell 1991);
- In Northern Ireland, in the 1990s, a marked decline in the health of 20-30 year-old Sitka spruce was observed across many plantations on oligotrophic peats. Growth of drain-side trees was poorer than that of trees between drains. 4 years after experimental treatments of N and K, basal increment had increased in response to K but not N. There was no interaction between fertilizer treatment and tree position in relation to drains (Schaible 1998).

Table 10.6 shows the effects of K when applied in addition to P on a range of peats (McIntosh 1981). It illustrates the variability in response and shows broad differences according to lithology.

Table 10.6   *Sitka spruce on peats; % increase in mean height due to K,
6 years after application of P or PK*

Forest	Lithology	Peat bog type	% increase in height
Clydesdale	Carboniferous	*Molinia*	84
Kershope	Carboniferous	Unflushed blanket	79
Minard	Dalradian	*Molinia*	28
Clydesdale	Carboniferous	Unflushed *Sphagnum*	21
Minard	Dalradian	*Molinia*	21
Arecleoch	Ordovician	Unflushed *Sphagnum*	19
Castle O'er	Ordovician	*Molinia*	19
Cymer	Carboniferous	*Molinia*	18
Carron Valley	Basalt	*Molinia*	18
Glentress	Ordovician	Unflushed blanket	17
Tywi	Ordovician	*Molinia*	14
Shin	Moine	*Molinia*	11
Shin	Moine	Unflushed blanket	8
Corrour	Moine	*Molinia*	8
Shin	Moine	*Molinia*	2

Source: McIntosh 1981

10.7   CALCIUM (Ca), MAGNESIUM (Mg), COPPER (Cu) & OTHER NUTRIENTS

10.71   Calcium

Calcium has a multiple role in woodland nutrient cycling. It is an essential nutrient and also, where present in soil as calcium carbonate, has a dominant effect on soil acidity both in respect of soil surface organic matter and as a potential buffer against acidifying materials in the woodland environment.

Calcium as a nutrient

As limestone, gypsum or slaked lime, calcium was applied in several early trials. There was no significant lasting effect on peat (Fraser 1933, Zehetmayr 1954). Zehetmayr (1960) reported a small response by Japanese larch and certain hardwoods, notably grey alder on two heathland sites.

In studies of nutrient reserves in poor heathland soils in north-east Yorkshire, and their interaction with cropping, Rennie (1954) expressed concern that rapidly growing conifer crops could so deplete the calcium reserves of the site as to lead to reduced production.

As a constituent of rock phosphate and superphosphate, calcium is applied wherever these materials are used. It is absent from triple superphosphate; however, lime was included in a series of trials comparing triple superphosphate with rock phosphate but produced only negligible effects (Edwards & Holmes 1956).

In trials of up to 10 tons per hectare of lime, four sites were each planted with Scots pine, Sitka spruce and Japanese larch. Only larch at one site, Allerston, Yorkshire, increased in mean height in response to the amount of lime applied (Atterson 1964).

On deep peat, application of calcium-containing fertilizer was associated with increased levels of N, a short-lived effect previously observed in several localities. An experiment was set up at Borgie, Sutherland to differentiate Ca as nutrient from its possible effect of ameliorating the acidity of the peat and increasing microbiological activity (Atterson & Binns 1968).

In Northern Ireland, ground limestone was included in a fertilizer trial on peaty gley and surface water gley soils (Adams *et al.* 1970). After 9 months, the effect of lime had been to reduce N and P content of foliage (Adams & Dickson 1973). Assessments after six years showed this effect still to persist. Growth of the spruce was not affected but there were marked effects on the litter (Adams *et al.* 1978).

At Culbin, where needle foliage content and tree growth responded quickly to added nitrogen (as ammonium sulphate), foliage calcium at first was little affected but over the seven year period for which figures are given, calcium as a % of needle weight in the plots given highest N fell to less than 50% of the calcium level in the untreated tree foliage. (The treated needles were also bigger.) Miller & Cooper (1973) considered this to be a 'dilution' effect. The implication is that calcium, as reflected in pine foliage nutrient content, is not a critical nutrient for Corsican pine.

Calcium and humus

There is a substantial quantity of literature going back to the 1880s on the significance of the type of humus formed under woodland. In *Mull and mor formation in relation to forest soils*, Handley (1954) described a series of studies on material from heathland soils in Yorkshire and elsewhere. He concluded that mor formation results from the formation of nitrogenous compounds resistant to natural breakdown processes but that the constituents from the litter of some species associated with the development of mull humus can cause mor humus to revert to mull.

Dimbleby (1952) found that the presence of birch on moorland soils led to mull humus-forming processes. Increasing birch cover on former *Calluna* heathland led to several important changes in soil surface characteristics, including a rise in exchangeable calcium (Miles 1981, Miles &Young 1980).

Adams *et al.* (1978) in the trial described above, found that additions of 5 and 10 tons of ground limestone per hectare altered the pH of the litter from 4 to 6-6.5. After six growing seasons, the number of actinomycetes, bacteria, fungal fruiting bodies and earthworms on the limed plots had increased significantly and the mycorrhizal associations had changed. The lime in these trials had been applied to the soil surface.

Studies of the effects of liming on forest soils have been undertaken in several central and northern European countries. These have considered the presence of mull humus as more desirable than mor and have stressed the importance of incorporating lime into the soil to have the best prospect of achieving this aim.

Contemporary ploughs and tines are able to cultivate or rip indurated sub-soil material and iron or humus pans, physically disrupting, aerating and often mixing the natural humus layers at the same time. Natural processes of humus formation can only recommence when the soil has settled and a vegetation cover re-established.

The severity of such changes only reinforce the need for a sound understanding of the role of the humus complex in the nutrient cycle, its response to disturbance and its relevance to contemporary silvicultural practice.

See also *Lime-induced chlorosis* under *Other deficiencies* below and §14.61 *Liming in acid catchments*.

10.72   Magnesium (Mg)

In the forest, magnesium deficiency is rare; it has only been identified on soils in southern Britain well supplied with N, P and K but with poor physical structure, often associated with waterlogged soils. Magnesium deficiency shows up late in the season as bright golden yellowing of the needles at the base of the current year's shoot. Foliage analysis will distinguish between N, K and Mg deficiency if there is any doubt (Binns *et al.* 1980).

Magnesium deficiencies in Germany and the USA have been noted (Binns & Grayson 1967). The former have subsequently been strongly linked to 'Type 1 forest decline'. Characteristic discoloration is found on older foliage, as Mg is depleted to support growth of new foliage (Roberts *et al.* 1989). See §14.43.

Magnesium deficiency has been found in young plants in forest nurseries on acid ground; a characteristic 'hard' yellow develops in late summer on spruce seedlings (Benzian 1965 - with illustrations). Treatment with magnesium sulphate has normally cured the deficiency; if the soil has become too acid, dolomitic limestone may eliminate the magnesium deficiency and reduce soil acidity at the same time (Pyatt 1994).

10.73   Trace elements

Hewitt (1966) in a review of nutrients in forest trees, mentions copper, boron, zinc, manganese and iron as causing deficiencies in various parts of the world. Proe (1994) also gives a summary of potential deficiencies associated with trace element deficiency, but is more oriented to production in forest nurseries.

Stockfors *et al.* (1997) describe the effects of applications of nutrients to a Norway spruce stand sustained over a period of 33 years. Stems needles and branches were analysed for 5 micronutrients (B, Fe, Mn, Cu and Zn). No deficiency symptoms had been observed. However, needle concentrations of boron were considered to be approaching deficiency level in plots given N and P but no additional trace elements.

Figures for trace element nutrient concentrations in foliage of thicket-stage Sitka spruce are given in *Table 10.2.*

Copper

In the forest, occurrence of copper deficiency is uncommon. It has occurred when tree shoots are elongating at their maximum rate and manifests itself as a failure of the leading shoot and upper side branches to straighten (Wood & Holmes 1958, Binns *et al.* 1980 with illustrations, Taylor 1991).

It has also been seen in a number of species on second rotation sites where cultivation + P has resulted in rapid mineralisation of organic nitrogen and very fast tree growth. In an early trial, reduction of available N was more effective in reducing symptoms than applying copper sulphate (Atterson & Binns 1968).

Further examples of copper deficiency were found on several sites throughout the uplands (McIntosh & Tabbush 1981, Taylor 1991). Trees subsequently recovered, whether or not they had received some form of copper (McIntosh 1984c).

Copper deficiency in the nursery

Copper deficiency symptoms on conifers in Britain were first identified as a browning of tips of needles on Sitka spruce seedlings in the forest research nursery at Wareham, Dorset (Benzian & Warren 1956a, b). In the nursery, this deficiency has been restricted to the area of the Dorset heaths. Only western hemlock seedlings have exhibited similar symptoms. Young foliage of freshly rooted poplar cuttings also suffered from copper deficiency at Wareham; the symptoms were of marginal leaf scorch and intervenal yellowing; treatments with copper sulphate or Bordeaux mixture prevented further symptoms.

Copper deficiency symptoms on Sitka seedling are shown by Benzian (1965).

In the nursery, tip-burn on needles of Sitka spruce seedlings caused by copper deficiency increased with added P, an observation consistent with appearance of symptoms on rapidly growing fertilized crops. Tip-burn was reduced with added potassium and had at one time been thought to be a symptom of K deficiency.

Use of organic matter including hopwaste usually eliminated tip-burn; subsequent analyses showed that the copper content of spent hopwaste from the brewery was ten times that in fresh hops, implying that the hops had picked up copper during the brewing process (Tables 74-76 in Benzian 1965). Benzian later reported that an instance had been found of tip-burn having developed where hopwaste used had been supplied from a brewery fitted with stainless steel vats (Benzian *et al.* 1972b).

Suspected deficiencies
Lime-induced chlorosis

Peace (1962) reviewed chlorosis encountered on limestone and chalk soils and pointed out that this may be due to induced deficiency of iron or of manganese, linking the former with beech on chalk soils and the latter with pine on oolitic limestone.

Lonsdale & Pratt (1981) noted that chlorotic beech occurs on chalky soils where the soil contained microscopically divided chalk and concluded that lime-induced chlorosis and chlorosis induced by Beech bark disease operated independently.

Needle fusion

Peace described and illustrated 'needle fusion' observed on lodgepole pine at Wareham forest, Dorset. He suspected unknown deficiencies and probable pathogenic influences.

Shoot dieback

Buds and young shoots of Corsican pine at Culbin Forest, Morayshire died following repeated high rates of application of N fertilizers. The effects looked unlike those normally associated with fertilizer scorch; trace element deficiency (other than copper) was suspected (Craig & Miller 1966).

10.8   FERTILIZING POLE-STAGE AND OLDER CROPS

10.81   Introduction

At the same time as the needs of newly planted and checked trees were being investigated, observations on nutrient uptake and soil reserves of major plant nutrients raised the possibility that inadequate nutrition would limit growth of closed canopy woodland on the poorer heathland soils (Rennie 1955). Scandinavian work also indicated benefits from late rotation applications of nitrogen to dry boreal forest (Binns & Grayson 1967, Hagner 1966).

In Britain this led to:
- short-term trials of the effects of fertilizer to established crops after canopy closure but before first thinning (*ie* pole crops) §10.82;
- short-term trials of N applied later in the rotation §10.83/84.
- investigation of nutrient cycling in pine and spruce §12.21.

10.82   Pole-stage crops

Experience at the start of the research programme into the effects of pole-stage fertilizing indicated that responses were likely to be short-lived.

First trials started in 1958 and tested N, P, K Ca, Mg and trace elements on poor quality pine on southern heathlands. In Dorset, Scots pine failed to respond to any treatment. The same species in Hampshire gave a 12% increase in basal area over seven years in response to added N. In Devon a similar response was obtained to added P (Wood & Holmes 1958, Binns & Grayson 1967).

In 1959 and 1960, trials of fertilizers were extended to pole crops of Sitka spruce in Wales and south west England, Douglas fir and Corsican pine in Kent, and to Norway spruce and Scots pine in Scotland (Wood & Holmes 1959, Edwards *et al* 1959, 1960). Results after 5 years showed no consistent substantial increases. There were no responses to K, Mg or Ca on any site.

N applied to Corsican pine on sand dunes in south Wales resulted in both an initial increase in foliage N content and substantial increases in basal area increment. The increase in foliar N was short-lived but the basal area increment response was observable for seven years.

While the response at first was barely statistically significant, Norway spruce in Devon continued over a seven year period to maintain a 16% increase in basal area increment following application of N. The foliage of the spruce showed no detectable effect, an unexpected observation in the context of the persistence of enhanced growth (Binns & Atterson 1967).

On other crops and sites, N either had no effect or tended to depress basal area increment. P did not depress increment anywhere; there were significant increases in increment of Douglas fir; an increase in increment of Sitka spruce was obtained only on one site out of five reviewed (Binns & Coates 1965).

In the 1960s, on poor gravelly sand, Scots pine basal area responded to N, even though foliage analysis suggested P was more limiting than N (Binns *et al.* 1970).

In three experiments in Scotland applying NPK to Scots pine, all increased in basal area, these increases being paralleled by increases in needle weight and needle N concentration. At Speymouth, measurements summarised in *Table 10.7* below, showed denser crowns, less light and more cones with increasing fertilizer (Atterson & Binns 1968).

Results after 3-6 years from other trials started in 1964-69 on northern sites showed N responses on some sites but not on others (Binns *et al.* 1972).

Table 10.7   *Crown and cone response to fertilizer, Speymouth Forest*

Treatment	Visual score of crown density	Light intensity (lumens /sq ft)	No of cones per tree
No fertilizer	2.2	3.76	68
NPK (low rate)	4.4	3.02	127
NPK (high rate)	4.8	2.84	143

Source: Atterson & Binns 1968

The results from 55 fertilizer experiments in established Scots pine, Sitka spruce and Norway spruce stands in upland Britain have been summarised in *Fertilizer experiments in established conifer stands* (McIntosh 1984a).

The *Scots pine* crops ranged in age from 23 to 48 (and one of 85 discussed under §10.84). Results showed:

- application of nitrogen consistently increased basal area increment in Scots pine for up to 5 growing seasons, the maximum response occurring at about 200-250 kg N per hectare;
- response to nitrogen decreased as stocking density increased;
- foliar nitrogen concentrations gave a reliable indication of the degree of response to be expected (see *Figure 10.4*).

Figure 10.4 *Scots pine basal area increment in relation to pre-treatment foliar nitrogen concentration*

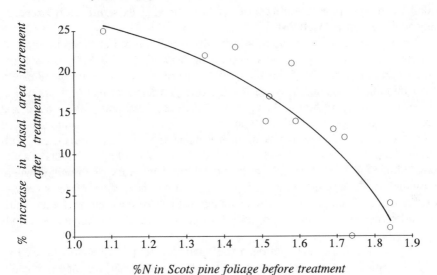

%N in Scots pine foliage before treatment

Source   McIntosh 1984a

*Spruce*s covered a similar age range to the pine, but included a higher proportion of plots 20 to 30 years old. Responses from the spruce experiments were variable:

- although responses to N, P and K were recorded, there was no consistent pattern in relation to site factors or foliar nutrient concentrations.
- Where there was a response, for N (5 out of 16 trials), the response persisted for not less than 5 years;

for P (6 out of 31 trials), responses persisted for 8-10 years;

for K (3 out of 10 trials), responses were associated with soils with over 30cm depth of peat.

Profitability of pole-stage fertilizing

Calculation of the economics of pole-stage fertilizing showed it to be less profitable than stimulation of growth shortly before felling. Also, because of the period of time between treatment and felling, the uncertainties of timber values are greater (Binns & Grayson 1967). For Scots pine, foliage nutrient levels before treatment appear to be a reliable guide to expected response; for spruces, there is no alternative but to seek expert local knowledge.

10.83   Interaction between fertilizer and thinning

Reports from the United States and New Zealand of benefits of fertilizing combined with thinning led to four thinning/fertilizer interaction experiments, three in Sitka spruce and one in lodgepole pine. Fertilizer significantly increased basal area increment in the one Lodgepole pine experiment and in two out of three of the spruce. With spruce, the effects of thinning and fertilizer were additive; for the pine, the effects of combined treatments significantly exceeded any additive effects (McIntosh 1984a).

In Sweden, crops have been both fertilized and thinned at the same time. A survey in 1991 of company woods occurring between 63°and 64°N showed the species present to be 70% Scots pine, 20% Norway spruce and 10% broadleaves. First thinning is at age 40. Since 1980, a nitrogen fertilizer had been applied at time of thinning, 33 500 ha being treated between 1980 and 1990. Wind and snowbreak damage occurred in about 1150 ha in 1989-91. It is quite accepted both that thinning increases risk of such damage and that damage increases with stand elevation above sea level. This survey showed additionally that, while post-thinning damage normally occurred within four years of thinning, where thinning was combined with fertilizer application the period of damage was extended by 2-3 years. This was attributed to the additional crown foliage built up following fertilization and its ability to retain more snow (Valinger & Lundquist 1992).

At Culbin, Morayshire, nitrogen was applied to four sample plots thinned so that the ratio of stems per hectare at the start of the trial was approximately (most lightly thinned) 4:3:2:1 (most heavily thinned). After two years the volume increment on all plots had at least doubled (Atterson 1965).

10.84   Stimulating increment in older crops

Highly profitable increases in increment following fertilization of stands more than 70 years old have been reported from central Europe and Scandinavia (Binns 1966).

The two oldest tests of the response of older conifers to applications of fertilizer in Britain were started in 1960 and 1963. An 80-year-old Scots pine crop at Alltcaileach Forest, Aberdeenshire showed 50% more girth growth both at 1.3m and 7.5 m height in response to added N, compared with untreated trees.

The response persisted for six years and was larger when N and P were applied (Binns *et al.* 1970).

Valuing the added increment at the then ruling price for timber and using a 3½% discount rate, the return was equivalent to 1½ x the cost of the fertilizer used; an alternative calculation put the discounted cost of the extra wood at 67% of its market value (Binns & Grayson 1967).

N and P were applied to 47-year-old Sitka spruce in Devon on a podsolised brown earth derived from granite.   After two years, there was a response to P but not to N (Binns *et al.* 1970).   However after 7 years, the response was no longer statistically significant (McIntosh 1984a).

Cores were taken by increment borer from 46-year old Corsican pine at Culbin Forest on the site of experiments into the effects of ammonium sulphate added 7-9 years previously. Fertilizer treatments could be seen to have increased ring width and decreased summer-wood percentage, effects lasting for over five years.   There were related changes in anatomical characteristics of tracheids and resin canals.   There was a suggestion that weather influenced the effect of fertilization on wood properties (Smith *et al.* 1977).

10.85   Fertilizers applied to broadleaved woodland

The following summarises the very limited amount of work on the potential of broadleaves to respond to fertilizers:

• Oak did not respond to added fertilizer whether as a pole crop or as coppice;   it was 'site demanding' rather than 'nutrient demanding' (Evans 1986a, b).

• In a review of silviculture of beech and oak in France, there was no mention of nutrient additions at all except for a reference to trials of fertilizers to induce seed production of beech (Oswald 1982, Le Tacon & Oswald 1977).

• A review of the silviculture of cherry (Pryor 1985, 1988) makes no mention of nutrients or fertilizers but comments that trees do best on permeable silty clay or clay loam soils.

• Ash responded to nitrogen and potassium fertilizer applications by increased diameter increment three seasons after treatment. Ash is said to be 'site demanding'; Kerr (1995) quotes figures for quantities of nutrients removed in nursery plants and considers its nutrient demands may be heavy.

• No response by young ash was observed over a six year period to additions on N, P, K or Ca in the first four years following planting on a fertile clay loam soil (Culleton *et al.* 1996).

• Sweet chestnut stem diameter increment responded positively to added phosphorus but was slowed by added lime (Evans 1986b).

• Birch fails on low-nutrient heathland soils where Scots pine survives (though growing slowly). See also §12.24.

10.9   SECOND ROTATION

The long-standing assumption has been that, if a woodland has been established and has grown satisfactorily up to the time of felling, then any successor crop, whether planted or regenerated naturally, should not require any additional nutrient supply.

Nevertheless, not all crops are healthy when clear-felled.   Site nutrient capital could also be reduced by fire *etc.*

In the early 1970s, questions were being raised as to whether any aspect of clear-felling might have adverse effects on the environment, and in particular on stream-water drainage from felled areas. For example,

- is the nutrient loss through removal of timber or other causes, ever sufficient to jeopardise the early growth of the successor crop?
- is there any loss of material from the site that could adversely affect the environment, in particular, in water draining from the site?
- to what extent is nutrient loss either by physical removal in plant material at the time of felling, or loss via drainage water affected by more intensive biomass removal such as in whole-tree harvesting?

Also, at this time, wood panel products were developing rapidly. Advances in the technology of chipping whole trees opened the possibility of accepting a lower quality of chip which included twig and foliage material. The consequent need for guidance on the silvicultural and environmental viability of supplementing wood raw material supply with crown and branch wood and foliage added a degree of urgency to these questions.

10.91   Fertilizers, brash removal and replanting

At Kielder Forest, in an area where nutrient release following conventional felling was being studied (§12.55), part of the site was planted as it was, *ie* with alternating brash zones and clear zones and without fertilizer; a similar part was given added NPK every three years and part had all brash removed while needles were still green, to simulate the effect of whole-tree harvesting.

Assessments after ten years showed that height growth of second rotation trees on the whole-tree harvesting plot was 17% less than trees on the unfertilized conventionally felled site. A fuller stem analysis after 14 years confirmed the height reduction and showed a similar reduction of basal diameter increment. On the conventionally clearfelled plot with added fertilizer, trees were similar in height to unfertilized trees. For a period 6-9 years after planting, basal diameter of fertilized trees was greater than that of the unfertilized trees; subsequently the effect disappeared. The case for adding fertilizer on conventionally cleared plots on this site is weak. However, the use of fertilizers on the whole-tree harvesting plot seemed more easily justified (Proe & Dutch 1994, Proe *et al.* 1996).

In USA and Scandinavia, as well as being a source of chips for particle board, crown material was also being used as a fuel for community heating plants.   Anderson (1985) concluded that if whole-tree harvesting became commonplace, upwards of £2 per tonne of brash would have to be held back from the sale revenue to pay for replacing the nutrients removed.

10.92   Models for nutrient requirements in areas for restocking

An alternative approach to direct experimentation is to use methods developed from Miller's work on nutrient cycling and modelling in Corsican pine (Miller *et al.*, 1980, Table 8). See also §12.21. From figures for biomass and nutrient accumulation, weights per annum have been derived for major nutrients by yield class for the main structural elements of the tree, *ie*,

- foliage
- stem wood
- stump
- live branches
- stem bark
- roots.

These, if multiplied by the age of the crop at felling according to its Yield class (from *Forestry Commission Yield Tables,* Hamilton & Christie 1971), give the predicted potential losses from a site due to conventional harvesting of tree stem wood.   Biomass values vary over a relatively small range, showing an increase if stumps are removed and a smaller increase if, at first thinning, whole trees are chipped.

For the same age of felling, for individual major nutrients, losses where whole trees are harvested are between 145 and 250% greater than where only the stem wood is removed, the differences depending on the nutrient.

The effect on biomass loss of conventional felling at rotation age compared with whole-tree chipping on 'no thin + premature clearfell' sites is less than might be expected.   The increase due to whole-tree removal is more than off-set by the overall reduction in biomass because of the shortened rotation.

In *Dynamics of nutrient cycling in plantation ecosystems,* Miller (1984c) summarises the mean concentrations of N, P, K and Ca for 14 species grown as plantation crops in cool and warm temperate regions. They show great variation and no consistent basis on which to group species. Nevertheless for each species, they provide the starting point for calculating changes in nutrient balance for alternative harvesting options.

Overall, the conclusion was that only very intensive harvesting is likely to remove more than the expected atmospheric inputs of most nutrients, with the exception of P, for which atmospheric inputs never match harvesting removals.

10.93   Interactions with other operations

Cultivation and nutrition

In parallel with the work on nutrition and vegetation control, great efforts were made in the 1950s-1970s to develop means of improving soil aeration and drainage by cultivation (Thompson & Neustein 1973).   The aim of cultivation was to improve the rootability of the soil both to benefit growth and crop stability. On Speyside, Scotland, one major experiment continued for 30 years on a podzolic ironpan soil with heavily indurated material below the iron pan.

Comparisons included spaced furrow and complete ploughing at different depths, with and without tining. For the first 8 years, the greater the volume of soil disturbed, the greater the initial advantage in crop growth. However, for the second decade of the experiment, the stimulus to tree growth from the more complete cultivations was reversed. During the third decade, there was no clear pattern of height growth in relation to original treatment, and only an indication that basal area was higher on the more completely cultivated plots. Cultivation clearly improved soil aeration. However rooting depth was limited by the undisturbed indurated subsoil. Nutrients were applied overall to this experiment according to the best practices of the time. Data for foliage nutrients for lodgepole pine and Sitka spruce show no evidence of interaction of nutrient status with cultivation treatment (Wilson & Pyatt 1984).

Peat drainage and nutrition

In a trial of the effect of controlled water tables in deep peat, tree height growth 10 years after planting was most clearly related to drainage depth and depth of well aerated peat. While nutrient levels for N, P and K in all treatment were on the low side, foliage nutrient levels for N and P were highest on the two best drained plots (Boggie & Miller 1976).

Brown *et al.* (1966) concluded for a raised bog at Kirkconnell Flow, Kircud-brightshire, that plants failed to grow because of nitrogen deficiency. The key feature of the site was that lack of aeration in the undrained peat reduced the ability of roots to grow and to take up nitrogen. The amount of 'available' ammonium was unaffected by lack of aeration.

High input establishment

Trials testing the effects of heavy complete fertilizing with complete weed control showed that, while early growth was good, for routine forest practice, the additional early expenditure could not be justified (Binns & Aldhous 1965). However, the technique has found a place in establishment of crops in amenity plantings in towns (Hodge 1990, 1991a, b), many ornamental broadleaves responding to nutrients and weed-free conditions on lowland sites. See §11.3.

10.94   Shortcomings of treatments

Not all attempts to apply fertilizer have been successful (Taylor 1990a). Control of machines, whether wheeled or airborne is difficult on much of the terrain to be treated. Uneven application can in particular arise through ineffective monitoring of the operation of spreading equipment (Farmer *et al.* 1985). Various criteria for adequacy of distribution, including a 'distribution quotient' have been considered (Binns *et al.* 1971).

A second class of failures arise through incorrectly identifying deficiency symptoms so that deficient crops are not treated and *vice versa*.

CHAPTER 11

Fertilizers for tree nurseries amenity planting & land restoration

'And from the seeds, tall plants grew ...'

11.1   NUTRITION AND FOREST NURSERY PRODUCTION

11.11   Introduction

Trees have been grown from seed following horticultural practices from time immemorial. It was formerly a common tradition for the larger private estates to raise forest trees for their own woodland planting, along with other plants required for vegetables, fruit and flowers. Landowners not wishing to grow their own young trees could purchase from established nurserymen in various parts of the country.

In the 1920s, the Forestry Commission set up nurseries to produce stock on sites located near areas where planting was expected, using the cultural practices of the time. Broadleaved species had been widely grown. However, the Forestry Commission in promoting the use of conifers on more difficult sites, required these on a very large scale.

Experimentation in the 1920s and 30s focused on cultural techniques for improving seedling yield, evaluating grades of plants, and examining alternative means of transplanting. Yields from conifer seed sowings, it was found, were higher when seedbeds were covered with silt-free coarse sand. While the importance of maintenance of nursery fertility was recognised and the use of manured green crops favoured (Guillebaud 1937), experiences with inorganic fertilizers had discouraged any strong belief that the latter had any important role when applied directly to seedbed or transplant ground (Wood 1965, 1974).

The production time for most stock was three or four years - two years in seedbeds and one or two years as transplants. At the same time, the quality of plants produced in some of these nurseries was considered to be deteriorating.

Rayner's work in the late 1930s and 40s, using composts to assist growth of seedlings on direct sowings on the very acid soils in Wareham Forest is described in §12.3. In the mid-1940s, the first heathland nurseries were opened at Wareham, Dorset, and at Harewood Dale, Yorkshire; outstanding seedlings and transplants were raised on composts in two years, rather than the 3-4 years necessary in the traditional nurseries of the day.

In 1944, in expectation of an expanded post-war conifer planting programme, a sub-committee of the Forestry Commission's *Advisory Committee on Forest Research* was formed 'to assess the problems underlying the nutrition of forest trees in nurseries and to suggest any desirable extensions of investigations already in progress'. It was arranged that Crowther of Rothamsted Experimental Station* should play a leading part, along with Rayner. The sub-committee was widely referred to as the 'NNC' or Nursery Nutrition Committee.

Under Crowther's leadership, a very substantial body of research was planned and organised under three main headings:

• to analyse problems of soil fertility and plant nutrition, and develop manuring and other methods for maintaining output of stock that perform well in the forest;

• to investigate how far the proven value of certain composts on acid soils depends on the available nutrients in the composts and whether there are benefits to be ascribed to soil physical and microbiological changes;

• to discover the cause of failure of Sitka spruce and other species in production nurseries.

The results of the programme between 1945 and 1956 are fully described in *Experiments on nutritional problems in forest nurseries Volumes I and II* (Benzian 1965). Work in progress is described both in the Forestry Commission's annual *Report on Forest Research* under the authorship of Crowther or Benzian and in the annual reports of Rothamsted Experimental Station.

All aspects of nutrition were tackled in the early programmes, some in single nutrient tests of rates and forms, others in factorial combination, designs being developed which allowed flexibility to add treatments without loss of experimental precision.

Few nurseries had a piped water supply. While the detrimental effects of drought on growth were noted as they occurred, irrigation was not a serious option for forest nurseries at that time.

In Scotland, since the 1930s, a modest programme on aspects of nursery production had been undertaken in collaboration with staff at the Macaulay Institute for Soil Research in Aberdeen. This included studies of the long-term effects of greencropping (Faulkner & MacDonald 1954) and the effects of sterilisation of soil using steam or latterly, formalin (Edwards 1952), the studies of sterilisation being coordinated with the work in England.

---

*Rothamsted at this time was the leading institution in agriculture experimental design and statistics. Forest nursery seedbeds were very well suited to compact replicated experimentation. Fertilizers themselves in their mode of action were also ideal subjects for 'factorial' design experiments. This whole programme became an exemplar of good field research practice.

11.12   pH

One of the earliest areas of investigation was soil pH and its relation to conifer seedling growth. From the first surveys in 1946, there appeared to be a case that the tallest and heaviest Scots pine and Sitka spruce seedlings were to be found on the more acidic soils. Work over the following decade reinforced and refined the general proposition that all species were sensitive to soil pH and that growth in the nursery of the large majority of conifer species was best under acid soil conditions. *Figure 11.1* shows the relative height growth of seedlings of four conifer species in response to change in soil pH (as measured in water). For Sitka spruce, a variation of one pH unit either way from the optimum reduced height by 25%. Seedlings on near-neutral soils (pH 6.5-7) were only half as tall.

Norway spruce, western hemlock and lodgepole pine all responded to soil pH in the same way as Sitka spruce, *ie* a preference for acid conditions and a marked fall-off in growth on more neutral soils. For Japanese larch and for western red cedar, the pH optima were 0.5 and 1 pH unit less acid respectively. Scots pine, Corsican pine and Douglas fir were more tolerant of a range of soil acidity; their optimum was similar to that for Japanese larch, but unlike other species, growth was only a little less within a range of +/- one pH unit.

Figure 11.1   *Relative height of conifer seedlings in response to soil pH at Wareham (W) and Kennington (K) nurseries*

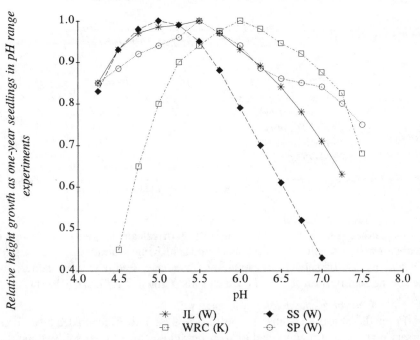

Source   Redrawn from Benzian 1965 (diagrams 10-12)

Other work showed that most of the common broadleaved seedlings grew well under the same conditions as conifers, but ash seedlings and poplar cuttings needed slightly acid soil (pH 5.5-6.4) and did not tolerate very acidic conditions.

The role of seedbed grits and liming

The most important cause of the observed decline in conifer seedling performance in the late 1930s and 1940s was inadvertent reduction of soil acidity. This came about either by use of seedbed grits including calcium carbonate, or by following green-cropping regimes which required application of lime to ensure good growth of the greencrop, or by using farmyard manure in which lime had been incorporated. Plates 1 and 2 in *Experiments on nutrition problems in forest nurseries* illustrate one of the pH range experiments and the effects of calcareous seedbed covers in reducing seedling growth.

With the success of the first heathland nurseries, almost all nurseries formed since 1944 were on acid soils. However, former cultural practices continued, including the use as seedbed cover of whatever local sand or grit was readily available. At Harwood Dale nursery, the second heathland nursery ever to be opened, a calcareous grit had been used when required from the outset. After four years, it was noticed that there was some depression in conifer seedling growth apparently related to recent past treatment. Reductions in growth were found to be related to applications of calcareous grit, as shown in *Table 11.1*. The depression in growth following use of calcareous grit was considered to typify the depression of seedling growth in older 'Sitka sick' nurseries.

Table 11.1   *Soil pH values and height of Sitka spruce*
             seedlings, Harwood Dale, 1950

Calcareous cover applied	pH*	Height (cm)
None since 1944	3.7	4.8
1944-1945 only	5.4	3.6
1948-1950 only	6.1	3.8
1944-45 and 1948-50	6.6	2.5

Source   Benzian 1965   * pH measured in $CaCl_2$

Soil acidification

In parallel with comparison of the effects of calcareous and non-calcareous seedbed grits on seedling growth, means of acidifying neutral soils were tested. Whether by use of sulphur or ammonium sulphate, seedlings grew better, the greater the acidification achieved on such soils (Crowther & Benzian 1951).

Control of the acidity of nursery soils

From the above and other supporting results, it was concluded that forest nursery management regimes had to keep very close control over soil pH and that use of limestone, or materials containing limestone, should be exceptional. Recommendations to this effect were included in *Forestry Practice* starting with the 3rd Edn (Anon 1951) and subsequently in more detail in *Nursery Practice* (Aldhous 1972a) and its successor *Forest Nursery Practice* (Aldhous & Mason 1994).

Response of weeds to pH

The ground providing the data for *Figure 11.1* carried several common nursery weeds. These also responded to soil pH, *Poa annua* showing a sharp maximum at pH 5.5, with indications that other common weeds preferred less acid conditions (Aldhous 1972d).

11.13   Mycorrhizas in forest nurseries

When the first heathland nurseries opened, seedbeds were prepared following cultivation of the cleared soil.   Composts were added following Rayner's recommendations.   No fertilizers were added. From her work in Dorset, Rayner believed that the composts did not act as nutrient sources or direct stimulants to growth. She considered that the prevailing poverty in nutrients in heathland soils was not an effective limiting factor to growth of young trees, but that qualitative changes in the organic constituents following directly upon modification of the biological activities of humus were responsible for the observed inhibition of growth (Rayner & Neilson-Jones 1944).

While seedlings grown on cultivated heathland ground were, in time, well provided with mycorrhizal rootlets, at first, it was not clear whether such rootlets were a prerequisite for good growth or whether both good growth and development of mycorrhizal rootlets were the result of adequate supplies of nutrients. (See §12.3 for description of mycorrhizas.)

As part of the programme undertaken under the auspices of the 'Nursery Nutrition Committee', meticulous examination of plants from a range of nursery treatments established that the primary cause of good seedling growth was a plentiful supply of major nutrients in acidic conditions and that while mycorrhizal rootlet formation usually followed, these frequently were not apparent until late in the year. There was no consistency between development of mycorrhizal rootlets and nutrient supply in the nursery; neither organic nor inorganic regimes showed any advantage (Levisohn 1965).

After comparison of seedling growth in long-term comparisons of production under compost or inorganic fertilizers, Benzian *et al.* (1972a) concluded 'we have no evidence that any factor other than shortage of nutrients interfered with growth at Wareham'.

11.14   NPK

The work in England up to 1962 investigating the use of fertilizers supplying major nutrients to conifer seedlings is comprehensively described in *Experiments on nutrition problems in forest nurseries* (Benzian 1965). Full descriptions of most materials and associated experimental treatments mentioned below are given there.

On soils where the soil acidity (pH) was well suited to the crop, forest nursery seedlings and transplants responded well to supplies of the three major nutrients, nitrogen, phosphorus and potassium. On a limited number of soils, magnesium was also required.

Potassium chloride, superphosphate, and a mixture of ammonium nitrate and calcium carbonate marketed initially under the brand name 'Nitro-chalk', were found to be perfectly adequate to supply plants' requirements. (Nitrochalk was designed to supply N without altering the soil pH.) These compounds prepared as granules, were first recommended for general forestry nursery use in 1951 and have continued to be the basis for maintenance of production of outdoor seedbeds and transplant lines (Anon 1951, Aldhous 1968b, 1972a, Proe 1994).

Slow-acting fertilizers

A wide range of more slowly acting forms of fertilizer have also been also tested, the majority as part of the 1950s research programme. They include:

Nitrogen compounds
- crushed hoof,
- dried blood,
- formalised casein
- formalised urea,
- isobutylidene urea (Benzian 1966, Benzian & Freeman 1967, Aldhous *et al.* 1967)

Benzian also mentions that shoddy, a by-product of the textile industry, being largely small fragments of wool fibres, had been reported in use as a nitrogenous fertilizer in the 1940s.

Phosphorus compounds
- rock phosphate
- basic slag

Potassium compounds
- glauconite

Multiple nutrient compounds
- potassium metaphosphate (Benzian *et al.* 1965, 1969, Benzian 1966, 1972, Aldhous *et al.* 1967)
- magnesium ammonium phosphate (Benzian *et al.* 1965, Atterson 1966a, Benzian 1966, Aldhous *et al.* 1967, 1968b, Aldhous 1972a, Low 1975).

While several of these gave results similar to those obtained with soluble fertilizers, most were more expensive and could not be justified in preference to the widely available granular compounds based on potassium chloride, super phosphate and 'Nitro-chalk'.

Nutrient residues in nursery soils

It is normally assumed that while K and N applied in fertilizers are soluble and readily leached from the soil, P persists in the ploughed layer in the soil.

In an experiment at Wareham nursery, Dorset, on a soil derived from Eocene sands, superphosphate, Gafsa rock phosphate and basic slag were applied annually for 14 years.

Sampling after 7, 11 and 15 years showed that in the 8 years between the first and last sampling, 93, 33 and 59% of the P in superphosphate, rock phosphate and basic slag respectively had been lost from the top 150mm of soil. Rock phosphate and basic slag residues were recovered from sand and silt fractions of the soil (Mattingly 1965).

On the same site, examination of soil horizons down to 440mm showed that the P losses in the A_1 and A_2 horizons accumulated in the B_1 and B_2 horizons; after removal through plant uptake is allowed for, there was very little loss of P from the site.

About 70% of the applied K had leached from the profile (Bolton & Coulter 1965).

Nutrient losses to surface- and ground-water

Forest nurseries have to observe the same limitations on use of soluble fertilizers as other nurserymen and growers in the same area. In the UK, the scale of forest nurseries is small so that the overall effect of the industry is insignificant. Nevertheless, nursery managers must know the requirements of the local water undertaking about use of fertilizers and surface run-off or flow through the soil into ground-water. Where irrigation is practised, the conditions under which nutrients can be added need also to have been agreed.

11.15   Green cropping

At the beginning of the 20th century, estate nurseries practised a three-year rotation, with green crops in the first year and seedbeds or transplants in the ground for the second and third years. Fertilizers or manures were spread and incorporated in the ground before sowing the green crop, which also might be top-dressed during the year.

No fertilizers or manures were applied directly to seedbeds or lines in the second and third years; seedlings or transplants were expected to benefit from residues from the green crop together with whatever manures had been applied to them in the first year.

11.16   Long-term maintenance of fertility

Since the 1930s, concern had been expressed about how to maintain the fertility of forest nurseries. The first two editions of *Forestry Practice* (Anon 1933, 1937) recommended that one year in three in a forest nursery should be under potatoes, a green-crop or summer fallow, partly to maintain growth, but also to reduce the potential for weed growth.

In the late 1930s and 1940s, interest in green-crops was particularly strong in Scotland.

Inchnacardoch trial

Faulkner & MacDonald (1954) describe a long-term trial of soil fertility maintenance, set up in 1938 at Inchnacardoch nursery, Invernessshire, jointly between the Forestry Commission and the Macaulay Institute for Soil Research, Aberdeen.

The trial area was divided in the first year into areas receiving 'no manures', 'artificial manures', or 'farmyard manure + artificials' plots. Each of these plots was divided into sub-plots on which were grown:

- oats and tares*,
- rye grass and clover,
- potatoes followed by rye,
- oats,
- tares sown in Feb, crushed and treated with calcium cyanamide in late June, followed by rye grass sown in late July, ploughed under in early winter.

For comparison, two other treatments were included :
- fallow
- continuous tree crops.

The 'artificial manures' consisted of applications of superphosphate (and in the first year, steamed bone flour), sulphate of potash and hoof meal. Green-crops received top-dressings of sulphate of ammonia or 'Nitrochalk' or calcium cyanamide. Farmyard manure was applied at 20 tonnes per ha.

Ground limestone was applied at the beginning of the trial to equalise the soil pH across the site.

After 12 years, 2 year-old seedlings in 1950 ranged from 7 to 14 cm in height, those on plots given the most heavily manured green-crops being on average about 70% taller than those on ground which had not received fertilizer or farmyard manure.

* The original publication does not specify what comprised 'tares'. Bentham & Hooker (1924), Salisbury (1964) and Stace (1997) all state that races of *Vicia sativa* had been widely cultivated for green fodder and forage. *V. tetrasperma* and *V. hirsuta* are also called 'tares' and are associated with past cropping (Godwin 1975).

NNC Long-term maintenance of fertility experiments

By 1950, the early results of the 'Nursery Nutrition Committee' programme, had clearly demonstrated the importance of soil pH and the potential role for inorganic fertilizer. At the same time, Rayner had demonstrated the benefits of composts, maintaining that their beneficial action was through other mechanisms than supply of nutrients (§11.13), while in Scotland, the practice of green cropping was well established.

These continuing and potentially conflicting interests led to the establishment in 1951 of two long-term experiments under the auspices of the NNC. These used Sitka spruce seedlings and transplants to test:
- whether any untoward long-term effects could be attributed to the continued use of the newly developed regimes using inorganic fertilizers,
- growth with inorganics compared with the use of the best of the compost regimes identified from Rayner's work,

- whether there was any identifiable advantage using any particular regime of green-cropping compared with fallowing or continuous cropping. The green-crops tested were rye grass, lupins and rye, all able to grow on soils with a pH suited to Sitka spruce seedlings.

The two experiments were located at Wareham nursery, Dorset and at Kennington nursery, Oxford and ran for 15 years. They were, with little doubt, the most complex and, in the event, the most meticulously conducted experiments in British forestry.

Progress was noted annually in the Forestry Commission *Reports on Forest Research*, and in the reports of Rothamsted Experimental Station. A full account is given in *Comparison of crop rotations and of fertilizer with compost, in long-term experiments with Sitka spruce in two English nurseries* (Benzian *et al.* 1972a, b).

Conclusions may be summarised:

- On the slightly acid sandy loam soil at Kennington, inorganic fertilizers sustained as good or better seedling growth than was obtained with composts, the differences, however, being small.

At Wareham, on the acid sandy heathland soil, plants raised on inorganics in the first years grew slightly taller than those on organics but did not maintain this position. Subsequently, neither fertilizer nor composts fully supplied plants' needs, growth being best where both were supplied.

Under no treatment at either site was there anything arising from the use of fertilizers which would indicate any serious long-term problem in their use.

- No benefit emerged from the use of composts which could not be attributed to their nutrient content. There was a consistent reduction of the order of 8% in the number of seedlings on compost plots.

- At neither nursery was there any evidence over the period of the experiment, of any benefit to yields or growth of Sitka spruce seedlings or transplants from including a fallow year or a green crop in the rotation.

These intensive experiments were to be backed up in production nurseries by five large scale comparisons of fertilizer, bracken/hopwaste compost, both or neither. In the event only two of these were successfully established, one on a heathland nursery soil at Bramshill Forest, Hampshire, (Appendix IV in Aldhous 1972a) and the other in Scotland at Teindland nursery, Moray in a site cleared in a pole-stage pine plantation (Low & Sharpe 1973). The former ran for 13 years and the latter for 20 years.

Both supported the conclusion that there were no obvious long-term harmful effects from using inorganic fertilizers. On the infertile soils of both sites, best growth was obtained where both inorganic fertilizers and bulky organics such as hopwaste had been applied.

Low and Sharpe point out the inherent dilemma in managing such trials, where nutrient availability and practice changes over time. Should the original prescriptions be adhered to rigorously, or should regimes be modified in the light of later experience, but at the risk of invalidating the long-term value of the comparisons being made? Smaller more frequent top-dressings would undoubtedly have resulted in larger seedlings on inorganic plots. Use of irrigation now available in many nurseries would have removed a good deal of the year to year variation caused by differences in summer rainfall reported in the more intensive trials (Benzian *et al.* 1972b, *Figures 4 and 5*).

Benzian, in discussing the results of the two major long-term experiments, quotes Guillebaud, the Forestry Commission's Research Officer in 1943 at the time nursery research strategy was being developed. He wrote:

'It has been a common, if not universal, experience that our forest nurseries deteriorate after a certain number of years working ... often ... after about 6 to 8 years ...'

From the experimental results and the subsequent experience in practice, Guillebaud's concern can be allayed in respect of properly applied contemporary techniques using inorganic fertilizers.

Physical effects of partially decomposed organic matter on the soil

Organic matter in the partially decomposed form applied as composts, raw hopwaste or ploughed in green crops, also has a physical effect. Benzian commented that soils on compost plots at Wareham were 'fluffier' and more difficult to consolidate because of the organic content. This is probably at least partly the cause of the small reduction in yield of seedlings in seedbeds receiving heavy applications of organic matter.

Binns & Keay (1962), Lowe & Sharpe (1973) and Proe (1994) quote figures for the Teindland long-term fertility experiment, where the soil organic matter content immediately after site clearance was 17%. After 20 years cropping, the organic matter on untreated and hopwaste-treated soils was 7% and 20% respectively. The micro-aggregate stability of the hopwaste-treated soil was 30% higher than for the untreated soil. See also §16.28.

On heavier textured soils, a late-sown crop of rye grass has been favoured; in the autumn when lining out under wet conditions, it provides a working surface which is less smeared or compacted. Alternatively, if ploughed in shortly before lining out, the fibrous roots keep the soil more open.

Soil organic carbon and nutrient residues

Table 11.2 shows changes in soil nutrients and organic carbon following the long-term comparisons of fertilizer and compost at Kennington and Wareham.

At both sites, continued use of compost increased the soil organic carbon content. Even so, the lowest organic carbon content at Wareham was higher than any at Kennington.

Benzian (1965) in Table 6 of Bulletin 37 gives soil organic carbon % for 13 nurseries, almost all in England; these ranged from 0.7% to 8.9%.

In the heathland nursery at Bramshill Forest, Hampshire, after 13 years continuous cropping, the organic matter content on organic-treated plots was 4.2%; on plots given inorganics, it was 3.1% (Aldhous 1972a Appendix 4).

At Teindland (figures quoted in the previous section), the use of organics slightly boosted the original soil organic matter level.

Table 11.2   *Soil nutrient levels before and after 15 years cropping*

Nursery and treatment	% organic C	Total % N	Total ppm P	Exchangeable ppm K	Exchangeable ppm Mg	Exchangeable ppm Ca	CEC me/ 100g*
Kennington							
Before cropping	0.7	0.07	200	42	29	798	
After cropping							
Fertilizer	0.6	0.06	586	92	59	629	4.8
Compost	1.0	0.09	515	171	68	474	5.8
Wareham							
Before cropping	1.4	0.05	33	11	19	80	
After cropping							
Fertilizer	1.6	0.05	66	9	11	334	3.3
Compost	2.6	0.16	92	26	44	334	5.2

* Cation exchange capacity in milli-equivalents per 100 grams

Source   Benzian *et al.* 1972

Use of bulky organic manures on nursery ground is consistently associated with higher content of soil organic matter on any one site. The fall on the plots not given organics represents the consequence of regular cultivations and the associated soil mixing, aeration and mineralisation.

The level of soil organic carbon appears not to have been directly related to the growth of seedlings or transplants.

Interaction of Cation exchange capacity (CEC) and nutrient regime

In *Table 11.2*, at the two sites, the difference in cation exchange capacity in response to treatment probably reflects the associated differences in organic carbon content. However, when comparing sites, the higher CECs on the inorganic fertilizer plots at Kennington reflect the higher clay content in the sandy loam there, compared with the corresponding plots on the Wareham sand.

11.17   Fertilizer damage to crops

Fertilizers may cause direct and indirect damage.

Direct damage

In the late 1940s, conifers had acquired the reputation of being 'salt-sensitive'. This seems more to have come from experience in the forest than documented evidence in the nursery. Nevertheless, Benzian (1966) commented that 'when E.M. Crowther was first asked (in 1945) to study the nutrition of young conifers, it was impressed on him that these crops were especially susceptible to damage by fertilizer salts'.

In the subsequent NNC research programme, Sitka spruce seedlings and transplants were subjected to a thorough-going range of treatments using soluble agricultural fertilizers. No damage to seedlings was encountered. When introduced into general nursery production practice, fertilizers were found to be safe for seedbeds if properly used.

Damage has been observed on seedbeds which can be attributed to misuse of fertilizers. The more visible forms of damage have usually been associated with overdosing with top-dressings of nitrogenous or potassic fertilizers. Over-doses have led to very obvious scorch, *ie* browning of the tips of needles; this is not uncommonly visible at the ends of beds where dosage has increased as tractors have slowed. More widely distributed damage has come about when individual fertilizer granules have lodged on foliage, causing localised browning or 'burn'. This is normally avoided by brushing plants to dislodge granules and not spreading fertilizer when foliage is damp.

Spreading inorganic fertilizer on top of seedbeds and rolling the surface immediately before sowing may lead to local concentration of salts that could damage emerging radicles, especially where pre-chilled seed is sown and the weather is dry (Mason 1994).

Scorch may also occur when top-dressings are applied in drought conditions or if very dry or warm weather follows within a short time of a top-dressing and it has not been possible to use irrigation to wash the fertilizer in.

Transplants have not been so free from damage by fertilizers as seedlings. Benzian (1966) described 'rust-red scorch' on Norway spruce transplants which developed in July after a cold spring and dry weather in June. All the plants affected had been lined out into plots which had received potassium chloride in late winter, several weeks before planting. Subsequent work showed that in the foliage of Norway spruce transplants treated with potassium chloride, the concentration of chloride ions was about twice as much as that found in Norway and Sitka spruce seedlings or Sitka spruce transplants. There was also an indication that height of transplants was 10-20% reduced where materials supplying chloride had been used.

Abies procera transplants were also subject to browning of foliage, possibly to a greater extent than Norway spruce, on plots given double the recommended rate of 'potassic superphosphate' - a granular mixture of potassium chloride and superphosphate (Aldhous 1964d).

Subsequent recommendations for the use of fertilizers for plant production included separate provisions to minimise the risk of scorch on transplants for Norway spruce and *Abies* spp (Aldhous 1972a).

In the course of raising a yellow lupin greencrop in the long-term fertility experiment at Wareham, lupins on the fertilizer plots were often discoloured and had distorted leaves. The injuries were associated with high concentrations (up to 2%) of P in the foliage; these did not occur when the amount of potassic superphosphate applied was reduced (Warren & Benzian 1959).

Induced susceptibility to frost

There are three components of plants' resistance to cold:

- resistance to winter cold by fully hardened-off plants;
- late-summer or early-autumn 'early' frost causing damage before plants have ceased growth or have acquired full winter hardiness;
- late-spring or early-summer 'late' frost, damaging plants which have broken winter dormancy and have commenced new season's shoot growth.

Winter hardiness

Lack of resistance to the extremes of winter cold limits the choice both of species and origins of plants for planting in Britain (Wood 1957; Cannell *et al.* 1985, Murray *et al.* 1986, Blackburn & Brown 1988). There is no evidence of added nutrients affecting winter hardiness. However, both onset and loss of full winter hardiness may be affected both by weather and by plant nutrient status.

Cannell *et al.* (1990) reported on analyses of Sitka spruce and Douglas fir transplants for: root growth potential (RGP), non-structural carbohydrate, and bud dormancy measured by the amount of cell division in buds. These were related to storage and lifting treatments and resultant frost hardiness. Hardiness was assessed by placing plants in a controllable freezing chamber. The study related nursery storage and lifting practice to other treatments that could influence frost hardiness and methods of its assessment.

Autumn frosts

Damage to plants by frost occurring while plants are actively growing is an ongoing hazard, both in the nursery and in the forest.

There has been long-standing concern that nitrogenous fertilizers applied late in the summer, may prolong plants' growth period and increase the risk of damage to autumn frost.

Redfern & Low (1972) reported widespread damage to nursery and plantation trees following a series of severe frosts in October 1971. Circumstantial evidence suggested that plants of low nutrient status were particularly susceptible.

Summer and early autumn applications of nitrogen before growth has ended for the season can prolong the growing season or induce a second flush of growth ('lammas' growth) and increase the risk of frost damage in such plants (Malcolm & Freezaillah 1975).

Atterson (1966a) considered that the death in the autumn of tips of plants suffering from acute magnesium deficiency was due to frost rather than tissue necrosis.

Redfern and Cannell (1982) describe and provide illustrations of natural and induced autumn frost damage on Sitka spruce. They also describe occasions where damage has been observed in the nursery and in the forest. They quote an instance of damage in an experiment where NxPxK was applied factorially, six years prior to the frost. While damage was less on plants given P, the possibility that this was due to trees being taller than untreated plants is not discussed.

Assessment 'scores' for autumn frost damage to four origins of Sitka spruce in a provenance experiment are given in *Table 11.3*; they show a strong relationship between latitude of seed origin and extent of frost damage. The greater frost susceptibility is attributed to the longer growing season of plants from the more southerly origins.

Red alder also shows a similar interaction between frost-hardiness and latitude of origin (Cannell *et al.* 1987).

Table 11.3   *Autumn frost damage to Sitka spruce of different origins*

Origin	Mean score for frost damage*
Oregon, USA	2.12
Washington, USA	1.96
Queen Charlotte Isles, Canada	1.21
Alaska, USA	0.50

* *Scores 0-5   0 = no damage;   5 = all current year shoots brown*

Source   Redfern & Cannell 1982

Spring frosts

Unseasonal severe late spring frosts have from time to time caused heavy losses especially in nurseries on sites susceptible to pooling of cold air under radiation frost conditions. Any practice which brings forward date of bud break exposes the affected plants to a greater risk of damage, should a frost occur. However, all new tissue is equally frost sensitive; differential damage to plants may more reflect differences in date of flushing rather than tissue sensitivity.

Late-season top-dressings

There has been extremely little critical work on the interaction of nutrient levels, time of fertilizer application and frost susceptibility.

In one year, severe ground frosts occurred in November on seedbeds given late-season top-dressings and intended to be lined out in the spring following. Of four species treated in September with N, Sitka spruce and western hemlock seedlings were less frost-damaged on plots given N than on untreated plots. Norway spruce and Corsican pine were not damaged. On Sitka spruce seedlings which without top-dressings were K-deficient, late-season K reduced frost damage on Sitka spruce more than did N (Benzian & Freeman 1967).

In trials of late-season fertilizing on transplants to be planted out on several forest sites, N and K applications were timed to be too late to influence current season's growth of Sitka spruce, Norway spruce, lodgepole pine, grand fir and western hemlock. The nitrogen levels resulting from treatment were within the range for healthy plants and were mostly 10-15% higher in the plants given the late-season dressings.

Assessments after planting in the forest showed that nitrogen treatments brought forward bud-break by up to a little less than a week; the more nitrogen that had been applied, the earlier the bud break. Trees given late-season nitrogen were up to 5% taller at the end of the first season on most sites.

Frost damage occurred on several sites; at two, while spring frosts damaged plants, previous nitrogen treatments did not affect the degree of damage. The same applied to autumn frost damage at a third site. At a fourth site, between 30 and 60% of plants were damaged by a frost in June of the second growing season. Unexpectedly, plants which received N some 21 months previously were more heavily damaged that those which did not.

Attempts to obtain similar contrasting nutrient levels for potassium failed (Benzian *et al.*1974a).

In a trial of late-season N and K top-dressings on one-year seedbeds of nine conifer species, foliage analysis showed that N was taken up by all except lodgepole pine. Uptake of K was more variable. No frost damage was observed but a few spruce and hemlock seedlings were killed following nitrogen top-dressing. This was attributed to fertilizer lodging in damp foliage just above soil level in fully stocked seedbeds (Aldhous *et al.* 1967).

See also §13.36 Sulphur & nitrogen interactions.

11.18   Broadleaves

Seedlings and transplants of broadleaved species have been raised on a large scale following the principles outlined above for conifers. The most important aspect of production is to apply the appropriate seed treatment to ensure adequate germination, and to ensure that the soil pH is suited to the species.

For birch and alder, smaller quantities of N top-dressings are recommended (Proe 1994).

Brookes *et al.* (1980) showed that seedlings from germinating acorns of *Quercus robur* and *Q. petraea* could draw stored nutrients from cotyledons independent of nutrients available from the soil, up to the production of the second flush of leaves. If nutrients were not used immediately, cotyledons could act as a reservoir for two years (Jones 1959, Ovington & MacRae 1960).

11.19   Other nutrients, alternative systems of plant production etc

Sulphur

For the period between 1950 and 1985, sulphur was a obligate additional nutrient, being included as part of the composition of commonly used materials, in particular, ammonium sulphate, sulphate of potash and superphosphate. In addition in the period 1950-1970, with every rainstorm came involuntary additions of sulphate from air pollution. Proe (1994) notes that sulphur deficiencies are not known in the range of soils normally encountered. However, with the diminution of sulphur deposition from air pollution (§13.33), and the tendency for fertilizers to have higher contents of N, P and K by using compounds with less S, the need for S applications may become as necessary for trees as it is already for some arable crops (MacKenzie 1995).

Container production

The possibility of reducing early death of newly planted trees by using container-raised stock came under consideration in the mid 1960s, following developments in Ontario, Canada where small seedlings were being raised in eight weeks in plastic tubes. Investigations tested the system's suitability for planting on peat (Aldhous *et al.* 1968a, Low 1971, 1975). Tests of other container-based systems considered:

- Japanese paper pots,
- 'Nisula' rolls,
- Moulded plastic or polystyrene pots and cell-strips of various sizes and designs (Brown & Low 1972).

Japanese paper pots and various designs of injection-moulded pots continue in use. 'Nisula' rolls and the Canadian plastic tube system were not taken up as production systems in Britain. 'Rootrainers' - strips of cells formed by bring together two preformed plastic sides are also in established use in Britain.

By 1990, it was estimated that 8-10 million seedlings were being produced in containers of one form or another, this constituting about 7% of the total forest planting stock production (Mason & Jinks 1990).

Such systems brought with them use of liquid feeds and slow release fertilizers, (Low & Brown 1972, Aldhous 1972a, Biggin 1981, Hollingsworth & Mason 1989).

Most liquid feeds contain N, P and K; many also include Mg, Ca, S and micronutrients. It is important to balance ammonium and nitrate N sources and to take account of any nutrients in the water supply (Jinks 1994).

Slow release fertilizers depend on the granule size and durability of coating to sustain nutrient release for periods specified.

Morgan (1994) showed that acceptable size planting stock can be raised using lower rates of fertilizer than recommended for conifers by Hollingsworth & Mason (1989).

Vegetative propagation

From the early 1980s, large-scale vegetative propagation of genetically improved Sitka spruce was increasingly undertaken (Gill 1983, Mason & Gill 1986). The associated production systems also utilised liquid or slow release fertilizers (Biggin 1981, Mason & Jinks 1994).

Carbon dioxide enrichment

A greenhouse experiment tested the effects of increases in CO_2 concentration, nutrient supply and day length on Sitka spruce, Corsican pine and lodgepole pine seedlings. Under an increased nutrient supply and a fourfold increase in CO_2 the seedling height achieved without treatments in sixteen weeks was achieved in 14-15 weeks. The effects were additive (Canham & McCavish 1981).

Electrolyte leakage

The ability of cell membranes to control the rate of ion movement in and out of cells has been used as a test of the health of tissues including seeds, stems, roots bulbs, needles *etc.*

The technique has been developed for forest nursery stock; it requires samples of roots to be taken at whatever time an assessment is required. It has come into widespread use as a means of assessing the quality of forest nursery plants at the time of despatch from nurseries (McKay *et al.* 1994, McKay 1998).

11.2   ORGANIC SOURCES OF NUTRIENTS

'One man's meat is another man's poison'

11.21   Sources

Accumulations of bulky organic matter from a wide range of sources have from time to time been considered as nutrient sources. These include:

Woodland	• forest litter,	• leaf mould,	
	• tree bark,	• sawdust;	
Farming	• broiler house litter,	• cattle slurry/farmyard manure,	
	• pig slurry,	• sugarbeet washings,	
	• straw;		
Horticulture	• mushroom compost;		
Arboriculture	• chipped wood & foliage from pruning, thinning etc.;		
Industry	• hopwaste,	• shoddy, textile wastes etc.;	
Urban	• sewage sludge,	• pulverised refuse fines,	
	• land fill leachate,	• paper mill effluent;	
Open land	• bracken,	• peat.	

Apart from bracken, peat, forest litter and leaf mould, these may all be process by-products or wastes*. Supply may therefore fluctuate, fall away or change in composition. For example, in the 1940s-1960s, farmyard manure, a mixture of droppings of yarded cattle with straw and, usually, lime, has largely been replaced by cattle slurry from larger animal units.

The more nutrient-rich wastes, including farmyard manure, have in the past been considered for use in plant nurseries and in amenity planting. The less easily broken down wastes have been found uses as mulching for weed control. For the more liquid materials, poplar and willow woodland has been used, overseas, as a biological filter to remove heavy organic matter load, reducing

*Waste Management Licensing Regulations 1994 under the *Environment Protection Act 1990* include detailed schedules of materials for which licensing is required. Waste soil, compost, wood and other plant materials are exempt (Moffat 1994a).

potential biological oxygen demand in drainage water and also absorbing available N and P (Hansen *et al.* 1980).

All bulky organic wastes are of relatively low unit value; costs of haulage, handling and distribution mostly limit use to sites near the point of production.

11.22   Fermentation or 'composting' of organic materials

Most bulky organic materials are capable of being broken down by aerobic or anaerobic bacteria. For most, if left piled up, bacterial activity will commence spontaneously, the inside of the heap warming in the process.

The traditional process of 'composting' creates conditions favourable to a rapid two-stage mesophilic and thermophilic aerobic decomposition. Considerable heat is generated in the process, temperatures reaching 60 - 70°C. If the compost is properly managed so that all the material being processed is held at this temperature for a period of 2-3 days or more, other organisms are killed. For bulky organic matter containing weed seeds, these also are most likely to be killed if high temperatures are sustained during compost making.

For fermentation to succeed, the unfermented material has to be of fairly high moisture content and may need to be wetted; there also has to be an adequate supply of nitrogen from the material itself or from external sources, to allow rapid build up of bacterial numbers. The 'turning' process in composting improves local aeration and ensures that material initially on the outside of the heap is subject to the fermentation process.

During composting, various toxic constituents of a raw material may disperse or be broken down. Composting bark eliminates volatile monoterpenes (Aaron 1976). Composting may also speed disintegration of the woody tissues, making them more suitable for subsequent use (Webber & Gee 1996).

If material is too compacted or otherwise not well aerated, anaerobic decomposition may take place, giving rise to different break-down products including foul odours, and less predictable properties of the compost.

Bark and sawdust

Fresh tree bark and sawdust from the sawmilling industry contain a high ratio of carbon to nitrogen (C/N ratio 150:1). Sawdust can be composted but requires substantial amounts of added nitrogen for the process to go well (Webber & Gee 1994). Sawdust from larger sawmills is most commonly utilised by wood fibre-processing industries.

Shredded bark has proved to be a valued mulch for weed control in urban areas and gardens (§8.24). Bark has also found a place in horticulture as a rooting medium, but must first be composted (Aaron 1976).

Chipped arboricultural waste

Chipped foliage, branch and stemwood waste from arboricultural operations has also been recommended as a mulch or, if nitrogen supplements are added, chipped waste may be composted for horticultural or amenity use (Webber & Gee 1996).

Conifer litter for nursery production

Freshly gathered spruce litter was used briefly as a growing medium for nursery seedlings under the *Dunemann* system (Aldhous 1962b). Subsequent studies on nutrient cycling show that the forest litter has too important a role in the nutrient cycle of the forest to justify practices necessitating large-scale removal of forest litter (§12.24).

Bulky organics as a nutrient source

There has been little advantage from using bulky organic manures in forest nurseries (§§11.15, 11.16). Benzian (1965, table 82) lists the N, P and K content of seven bulky organic materials.

Materials have to be viewed circumspectly and their nutrient content determined before use. For tree nurseries, farmyard manure and mushroom compost have both been considered potentially unsuitable sources wherever there is any likelihood of lime having been added; the former also had the disadvantage of containing many weed seeds. Fresh broiler house litter may cause damage to adjacent growing plants through emissions of ammonia (Aldhous 1972a).

Sewage sludge is discussed below (§11.23).

Bracken

Bracken is one of the most widespread plants in the world (Taylor 1986). Its foliage is unpalatable and toxic to grazing animals (Cooper & Johnson 1984).

While it has been used in the past for animal bedding, surface mulching and as ashes as a fertilizer, in Britain, these practices are now not widely followed. It is controlled by farmers and foresters mechanically or by herbicides (§8.43).

Bracken has been used by gardeners and nurserymen in composts and potting mixtures; it was one of the materials used by Rayner in her trials of composts on direct sowings (§11.1), and was widely used in the late 1940s as one of the two ingredients of compost specified for heathland nursery nutrition. In the early 1950s, it was replaced in general practice by the cheaper uncomposted hopwaste.

Bracken may spread rapidly in upland grazings. In the early 1990s, its continuing spread was considered a threat to the upland environment; recommendations were made to restrict it by repeated cutting (Lowday & Marrs 1992). The availability of a waste material with a known horticultural value has re-opened the possibility that collecting and composting cut bracken may create a viable alternative to peat for container-raised ornamental plants (Pitman & Webber 1998).

The most effective physical regimes for reducing vigour of bracken are those where stems are cut or broken before fronds are fully expanded. Nutrient levels are highest in July; bracken cut and fermented at that time will give the most nutrient-rich compost. Composting dying or dead bracken is no deterrent to the further spread of bracken; more nitrogen will be required than when younger less woody fronds are cut and the resulting compost will be less nutrient-rich.

11.23   Sewage and sewage sludge

The increase in human populations and simultaneous adoption of higher environmental standards have together increased the quantities of sewage to be disposed of, and created problems in ensuring the safety of the disposal process.

Sewage sludge is defined as 'residual sludge from sewage plants treating urban and domestic waste waters' (DEnv 1992). It has been disposed of on land but because of its potential content of micro-organisms and heavy metals, disposal of raw sewage in agriculture is subject to legislation, *The Sludge (Use in Agriculture) Regulations 1989* (SI 1989/1263) being current in 1999.

Raw sewage sludge consists of a carrier of nutrient-rich water in which is suspended fine organic material equivalent as dry solids to between 2 and 8% of the gross volume. On arrival at the processing plant, it is given a preliminary coarse screening and may be put through a fine screen to remove smaller fragments of plastic *etc.* that are not removed in the preliminary screening.

It may be used in its raw screened state; alternatively it may be processed by:

* 'digestion', *ie* subject to microbiological breakdown;
* 'pasteurisation' *ie* having been heated for a prescribed period to between 55 and 70°C, followed by mesophilic anaerobic digestion at 25 - 35°C for a further prescribed period;
* 'dewatering' into a water phase and a solid 'cake' phase containing between 20 and 50% dry solids. Sludge may be dewatered before or after digestion;
* drying. Screened, digested, dewatered sewage sludge may be further heated and dried into granules. This granular form of sludge is likely to meet present and future public health requirements for the safe use of sludge on land, as far as these can be foreseen (Wessex Water Authority, personal communication).

Disposal of sewage

In coastal towns and areas with access to the sea, untreated sewage was commonly discharged into the sea by pipe, or in the larger conurbations by boat. The *Oslo Convention on dumping waste at sea* gave rise to the *Control of Pollution Act 1974.* The latter set limits to the use of sludge on farm land and required licences for its disposal through long outfalls to the sea, a requirement that was renewed by the *Food & Environment Act 1985. Table 11.4* shows the sludge disposal routes in the UK in 1989. Disposal to sea at that time constituted nearly 30% of the UK total sludge disposal.

European Community Directives

The European Community Council of Ministers, concerned with matters of water quality and the environment, in 1986 issued Directive 86/278/EEC. This required that the practice of discharging sewage wastes to sea be discontinued. A timetable was set by Directive 91/271/EEC providing that disposal of sewage wastes to sea by any means must cease after 31 December 1998.

Table 11.4   *Sewage sludge disposal in the UK in 1989*   *(ktonnes)*

Country	To:	Farm	Landfill	Sea	Incineration	Total
England & Wales		507	151	234	66	958
Scotland		16	12	67	1 (+10 'other')	107
Northern Ireland		12	4	11	0	27
Total UK		535	167	312	67	1092

Source   Anderson 1992

As a consequence, over the whole of Britain, local authorities responsible for sewage disposal undertook reviews of the alternatives for sludge disposal:

- disposal to farmland,
- use in land reclamation,
- incineration,
- disposal to forest land,
- development as a growing medium,
- dumping as landfill.

Comparative costs to the disposal points in Holland in 1981 were (van Voorenburg & van Veen 1992):

Undigested sludge for injection in farmland	1
Composting	3
Landfill dumping after conditioning and dewatering	2.7
Incineration	3

The cost of drying and granulation is higher than any of these but, as with composting, some expenditure can be recovered by sales of the product.

Mineral and organic matter content of sewage sludge

Sewage sludge contains desirable plant nutrients and undesirable heavy metals, the latter being from domestic and industrial sources. The balance between these govern the use of sludge. *The Sludge (Use in Agriculture) Regulations 1989* (SI 1989/1263) allow farm land to receive the benefit of the nutrient content of sludge, monitoring closely the heavy metal content so that land is not rendered unfit for food production.

Table 11.5 gives the nutrient content of three main forms of sewage sludge. Differences are partly due to removal of water and its soluble nutrients, and partly reflect losses of nitrogen during the digestion and drying processes.

Table 11.6 summarises the limits set on the amount of zinc, copper, nickel, cadmium, lead and mercury that can be applied annually to farm land. While no rate limits are required by statute to be applied in forestry, the agricultural limits are recommended (Wolstenholme *et al.* 1992). See also §15.12 *Heavy metals*.

Table 11.5 *Nutrient content of sewage sludge products*

Product type	% dry substance	% organic matter	%N	%P	%K
Digested liquid	2 - 8				
% in dry substance		50 - 60	0.9 - 6.8	0.5 - 3.0	0.1 - 0.5
% per cu. m of liquid		1 - 5	0.018 - 0.54	0.01 - 0.24	0.002 - 0.04
Digested cake	20 - 50				
% dry substance		50 - 70	1.5-2.5	0.5-1.8	0.1 - 0.3
% per fresh tonne of cake		10 - 35	0.3 - 1.2	0.1 - 0.9	0.02 - 0.15
Dried granular sludge	90 - 99				
% dry substance		50 - 60	3.1 - 4.6	1.4 - 2.9	0.09 - 0.32
% per tonne of granule		45 - 59	2.8 - 4.6	1.3 - 2.9	0.08 - 0.32

Source Bradshaw *et al.* 1995 Wessex Water - personal communication.

Table 11.6 *Maximum permissible rates & concentrations of potentially toxic elements (PTEs) in soil after application of sewage sludge*

Potentially toxic element (PTE)	Average annual rate of PTE addition over preceding 10 years kg/ha	Maximum permissible concentration of PTEs mg/kg soil*
Arsenic	0.7	50
Cadmium	0.15	3
Chromium	15 (provisional)	400 (provisional)
Copper	7.5	80
Fluorine (as fluoride)	20	500
Lead	15	300
Mercury	0.1	1
Molybdenum	0.2	4
Nickel	3	50
Selenium	0.15	3
Zinc	15	200

Source Department of the Environment (DEnv 1992)
* Soil samples taken to depth of 15 cm including litter layer.

Sludge in forestry

Safe disposal of sewage is sought internationally, many systems not involving forestry at all. However, the effectiveness of using the established root systems of vigorous young trees to filter out organic material and pathogenic micro-organisms and to absorb nutrients has been tried using willow, poplar,

Eucalyptus and other species in America, Australia and Europe (*eg* Stewart *et al.* 1986, Bramryd 1980, Bayes *et al.* 1987, McCary 1989).

In Britain, the first test of sewage sludge in forestry occurred in 1945 with the inclusion of sewage sludge in forest nursery experiments Benzian (1965). She referred to material used as 'well tried in agricultural experiments' testing alternatives to farmyard manure during the 1939-45 war.

The composition of sewage sludge varied according to its regional source. In the more industrialised areas, it could contain excessive levels of heavy metals. Sewage sludge was not recommended for use in forest nurseries unless the supplier could guarantee absence of heavy metals (Aldhous 1972a, Proe 1994).

While the *Control of Pollution Act 1974* laid down procedures for use of sewage on agricultural land, use of sewage sludge in forestry was permitted for 'the purpose of fertilizing or otherwise beneficial conditioning of the land' without other restriction.

Trials during the 1970s sought to determine the effect of digested liquid sewage on nutrient-poor sandy soils. In reality, they had first to find solutions to the practicalities of spreading liquid sewage sludge among trees. High pressure rotary hoses operating from the forest road and rack systems were effective (Taylor 1991, Wolstenholme *et al.* 1992). Most experiments using sewage sludge showed it had a beneficial effect on trees either in improved growth rates or higher foliar nutrient levels (Bayes *et al.* 1987, 1991).

Moffat & Bird (1989) assessed the potential for using sewage sludge in forestry in England and Wales. They concluded that over 33 000 hectares of Forestry Commission land were suited to application of cake or liquid sludge and that, if available at little cost to the recipient, sludge had the potential to replace inorganic fertilizers. Taylor (1987a) considered that in Scotland, the greatest potential was in the east and northeast.

Sewage, forestry and the EC Directives

The speed and magnitude of the change required to meet the EC requirement of no further discharges to sea by the end of 1998 has resulted in most major sewage disposal undertakings opting for incineration rather than engage in the uncertainties of securing access to land. For Scotland, the additional area required would have been five times what was already in use.

The change of sewage disposal methods has not led to greater use in forestry in practice. While disposal to forestry remains possible, the pressure for increased public access to woodlands operates against uses of non-sterile sludges that necessitate consequential exclusion of the public. Nevertheless, there appear to be no grounds for restrictions on access following use of granular heat-dried sludge, and the material is becoming increasingly available.

Land reclamation

Because of the restrictions to public access while land restoration is in progress and heavy machinery is working, there is greater choice in the forms of processed sludges to provide starter nutrients and organic matter when seeking to establish woodland on newly restored land. Several successful operations have been reported *eg* Page (1993), Wolstenholme & Dutch (1995), Salt *et al.* (1995).

An initial application of 1000 kg/ha of organic N per ha is recommended for spoils lacking any organic matter (Bradshaw *et al.* 1995); to achieve this would require approximately 5-600 m³/ha of liquid sludge, or 150 tonnes of cake or 30 tonnes of granules. *Table 11.5* illustrates the variability of the nutrient content of liquid and cake sludge; rates in any particular instance should be adjusted according to the nutrient content of the material available.

Peat

Granular dried sewage sludge is under trial in the Republic of Ireland for the reclamation of cut-away bogs (White 1998).

11.3   LANDSCAPE & AMENITY PLANTINGS

The practice of landscape and amenity planting is rooted in antiquity, applying to ornamental and amenity plants, techniques of plant production and planting developed for horticulture. Major movement of soil for amenity rather than military, agricultural or industrial purposes was restricted to landscaping for wealthy land-owners because of the labour involved and lack of mechanical aids. Movement of large ball-rooted trees is nevertheless long-established practice. On farmland from the time of enclosures, maintenance of hedgerows and provision of hedgerow trees was thought of as part of good estate management.

In town parks and gardens in many situations, ornamental plantings of trees and shrubs have grown well, notwithstanding that they were often natives of distant lands. As a general rule, the fertility of such sites has been at least adequate and standard horticultural practices have sufficed.

At the same time, over the last 100 years, standards of amenity set for parks and gardens and the best 'townscapes' have been sought for land disturbed by industry and transport where the soil may have disappeared totally or may have been seriously damaged.

11.31   Landscaping and reclamation

From the late 19th century, attempts have been made locally to cover the unattractive raw surfaces created by engineering works, especially spoil heaps and roadsides. In a survey of colliery spoil in 1955, a few plantations dating from the end of the 19th century were noted, and many more that were between 40 and 50 years old (Wood & Thirgood 1955). During the 1930s the 'Roads Beautifying Association' successfully campaigned for amenity planting beside new or improved roads.

At the end of the 1939-45 war, many industrial waste heaps were recognised as eyesores, especially where the associated mines were exhausted and buildings abandoned. Through the *Mineral Workings Act (1951)*, the *Opencast Coal Act (1951)*, the *Industrial Development Act (1966)*, the *Derelict Land Act (1982)* and much other legislation, local authorities and others have been given powers to require or to assist land restoration, past, present and future.

Table 11.7 summarises results from a survey of mineral works in 1974. The areas covered by spoil heaps is one third of the area covered by excavations, sand and gravel workings constituting the largest single type of mineral working.

Table 11.7   *Areas of mineral workings in England in 1974*   *(000 ha)*

Mineral	spoil	*Area currently affected* excavations	total	*Potential future workings*	*Total current + future*
Sand & gravel	1.2	17.0	18.2	15.2	33.4
Clay & shale	0.8	4.6	5.4	6.5	11.9
Limestone	0.5	5.0	5.5	5.2	10.7
Ironstone	-	1.2	1.2	4.9	6.1
China clay	0.8	1.0	1.8	3.2	5.0
Silica molding sand	0.1	1.9	2.0	1.9	3.9
Chalk	0.1	2.1	2.1	1.6	3.7
Igneous rock	0.4	0.9	1.3	1.4	2.7
Sandstone	0.3	0.8	1.1	1.1	2.2
Gypsum, slate, etc	0.9	0.6	1.5	2.2	3.7
Total - England	12.3	36.8	49.1	45.3	94.4

Source   Binns & Fourt 1981, citing Department of the Environment

Derelict former industrial land is additional to the area of mineral workings. Derelict land surveys in 1988 found 40 500 ha in England, and 7400 ha in Scotland (DEnv 1991c, Moffat & McNeill 1994, citing the Scottish Office).

Nearly ten years later, derelict and disturbed land in England was estimated at 175 000 ha for England and 20 000 for Wales (Handley & Perry 1998). In Scotland, there were 12 000 ha of land 'derelict, vacant or contaminated' in the Central Belt (the region between Edinburgh and Glasgow) and lesser areas elsewhere (MacGillivray 1998).

Together, unproductive waste and polluted sites include:

Coal, petrochemical and power generation wastes (§11.33)
- coal deep mine spoil,
- oil-shale 'bings',
- opencast workings,
- pulverised fuel ash.

Mineral extraction (§11.33)
- ironstone workings,
- gravel workings,
- metalliferous mine workings
- china clay waste,
- limestone waste,
- other mineral extraction.

Waste disposal/polluted sites
- landfill (municipal domestic waste),
- capped noxious wastes.

Other sites
- civil engineering sites,
- derelict industrial sites,
- urban and rural roadsides.

11.32   Integrated approach to sites in restoration schemes

Disturbed land is commonly in the form of *heaps, holes, banks* or *rubble. Heaps, holes* and *rubble* are dominated by the nature of the industrial works which created them.   Notes of the characteristics and experience on a range of former industrial sites are given in §§11.33-11.34.

Banks and *cuttings* have been widely created as part of the 'cut and fill' process when constructing roads, railways and canals with appropriate gradients. The hospitability of the sloping surfaces so created has depended partly on the geology of the underlying material, and partly on the scale of provision of topsoil to cover them.

Integration of operations

Because of the great range in physical and chemical properties of material available for planting in land restoration schemes, cultural operations have to be integrated and tailored to sites individually.   Nevertheless, all restoration schemes should integrate, as relevant, provisions to avoid or alleviate:

- *over-compaction*, causing loss of aeration and poor drainage.   It can be minimised by avoiding all unnecessary transits of heavy machinery across land and may be alleviated by ripping.   Compaction of top-soil can be kept to an acceptable level by 'loose tipping' of the uppermost layers of spoil or soil, and keeping subsequent traffic to an absolute minimum (Fourt 1984, 1985, Ramsey 1986, Moffat & Roberts 1989b, Dobson & Moffat 1993);
- *soil texture deterioration*. Fresh shales and mud-stones weather into poorly draining fine-textured silts and clays, potentially leading to long-term drainage difficulties.   Alleviation may be possible during land restoration by mixing in a substantial proportion of coarser, more slowly weathering material to assist long-term soil aeration and site drainage (Wilson 1985a);
- *nutrient deficiency*.   Nitrogen deficiency is common on sites where there is little organic matter.   A nutrient capital of 1000 kg N per ha has to be secured for any site, by one or a combination of:

replacing the original top-soil,

adding sewage sludge or other bulky organics,

using nitrogen-fixing plants as 50% of the site planting mixture,

adding NPK fertilizers based on foliage analysis if available, or based on soil analysis (Bradshaw 1981, Binns & Fourt 1981, Moffat *et al.* 1989, Page 1993, Bradshaw *et al.* 1995);

- *toxic or excessively acidic materials* by ameliorative treatment, *eg* liming, or protective treatment by capping;

- *accumulation of gases eg* from landfill, by safe venting;
- *pollution from site drainage waters*, by filtration or recycling;
- *waterlogging* through inadequate natural drainage.   With the availability of heavy machinery, a large scale 'ridge and furrows' (or 'riggs and furrs') surface pattern can be imposed on flat land forms to create slopes between 10% and 20% (Wilson 1985a, Moffat & Roberts 1989b).

Recommendations for planting on disturbed and reclaimed land are given in *Reclaiming disturbed land for forestry* (Moffat & McNeill 1994), *A guide to the reclamation of mineral workings for forestry* (Wilson 1985a) and *Trees in the urban landscape* (Bradshaw *et al.* 1995).

Comments in the following sections are supplementary to such an integrated approach

11.33   Colliery spoil, coal open-cast workings and oil shale heaps

Mining for coal and minerals probably reached its peak in the late 19th and early 20th century.   In 1954, coal was being deep-mined at more than 1000 collieries; by 1980 there were about 250 collieries; in 1998 there were scarcely any.   In 1980, 12000 ha of land was covered with colliery spoil, the area at that time increasing by approx. 250 ha/annum (Jobling 1981).

Colliery spoil consists predominantly of shales and mudstones which weather into clay loams; however, materials derived from sandstones may also be present depending on the local geology.   Burned tips break down to coarser soil.   All such soils are capable of supporting growth of trees that can tolerate the initial low nitrogen status of the sites.

There may also be small localised intractable residues of furnace ashes, and washery waste.   The latter is fine-grained, high in carbon content, and exceptionally acidic.   Other spoils contain iron pyrites and may also be very acidic.   Both may better be buried under other spoil (NCB 1967).

Fresh spoil is mostly neutral or slightly alkaline (pH 7.0 - 8.0) but usually acidifies on weathering; acidification is more severe if pyrites is present. Although many spoils weather relatively quickly, soils remain stony and difficult to work on that account.

Deep mining spoil heaps

The accepted mining norm was to bring out spoil, heap it and abandon it. Trees have been planted on colliery spoil tips since the 19th century and there were a number of successful 'improvement' schemes at the beginning of the 20th century.   In Cumbria, for example, 82 year-old ash and beech on colliery waste reached heights of 19 and 17 m respectively (Richardson 1993).   However, reshaping was rare until the middle of the century (NCB 1967).

A survey in 1954 of 'unimproved' colliery spoil heaps, reviewed their potential for tree growth in the context of the obligations for restoration imposed by the *Opencast Coal Act, 1951*, and subsequently, the *Opencast Coal Act, 1958*.

The two main conclusions were:

- localised atmospheric pollution was considered one of the most serious restrictions on choice of species, the coalfields of the central belt of Scotland, Lancashire, the Black Country and North Stafford being the worst affected;
- natural colonisation of spoil was commonplace, with birch an important pioneer. Vegetation slowly developed on favourable sites in a succession towards oak woodland. Alder was noted as being specially suited to these sites. There was no evidence of failure because of lack of nutrients. Lack of available water in the first year was often the main cause of death (Wood & Thirgood 1955).

Opencast workings and land restoration

Opencast working of coal commenced in Britain in 1942 to supplement war-time production from deep mines. Initially, spoil heaps were left *in situ*. At the same time, improved and more powerful earth-moving equipment that made opencasting possible was also available to restore land, either to something close to its original form, or, on flat sites with poor natural drainage, a more undulating and better draining form.

From 1951, the initial requirement where opencast working had been completed was to restore the land to its previous use, usually agriculture. As long as it remained national policy to encourage British farming, restoration to grass was the accepted norm. However, increasingly, restoration of opencast land to woodland has been preferred.

The policy to restore opencast spoil and reshape colliery waste initially resulted in sowing or planting into unweathered or partially weathered material, the original top-soil having been mixed randomly with other spoil. In the more recent workings, under requirements imposed by planning authorities, topsoil has had to be removed and re-used, because of its nutrient and organic matter content (Moffat & McNeill 1994). Careful storage of topsoil is also necessary to ensure that the soil does not lose its physical structure or aeration through being heaped too high or being too heavily compacted.

Similar techniques were applied increasingly to heaps of deep-mined spoil.

In reviews of planting practice after nearly 30 years of reshaping and replanting colliery spoils (Jobling & Stevens 1980, Jobling 1981, Binns & Fourt 1981), the most important factors identified were:

- *excessive spoil compaction.* In the early years, restoration and reshaping were seen purely as engineering operations where long-term land stability and easy machine operation were the foremost criteria. The weight of machines and the frequency of their movement over any piece of ground led to very heavily compacted soils, impeding planting, and inhibiting root development and access to soil water reserves (Jobling & Carnell 1985, Fourt 1984, 1985). Fourt considered a 0.5 -0.75 m depth of rootable soil was necessary to provide the 150 - 200 mm of reserve moisture needed for successful growth of plants on disturbed land.

Where ground has been restored after opencast mining, the bulk-density of surfaces layers can increase by slumping and consolidation if there are no stable channels from deep-burrowing worms and roots. Installation of drains with permeable backfill has been suggested as a means of minimising waterlogging through impeded drainage through such denser surface soils.

• *excessive acidity.* Iron pyrites (FeS_2), wherever present and exposed to moist air, oxidises through a chain of chemical reactions. The final products are sulphuric acid which acidifies drainage water from the site, and ferric hydroxide, the flocculent orange deposit characteristic of many streams draining colliery spoil. The extent and speed of acidification depends on the particular structure of the pyrites, but the surface of severely acidifying deposits can be pH < 3. Limestone and limestone-containing wastes can be used to neutralise acidity. However, because spoils vary in their potential to acidify, this needs to be characterised in order to ensure adequate initial liming. Highly pyritic spoils may require liming at 100-400 t/ha to control pH down to a depth of 45 cm (Costigan *et al.* 1981, Gemmel 1981).

• *salinity.* This is a problem particularly in north east England, being worse in areas of low rainfall. Time should be allowed for rainwater to leach out salts.

• *need for adequate organic matter.* This, on the sites surveyed, varied according to the extent that top-soil or organic supplements had been used and their quality, and reflected the N available on the site.

Shaping of reclaimed spoils heaps into large-scale ridge and furrow landforms combined with subsoil cultivation and loose-tipping has been carried out successfully in areas such as the Forest of Dean. It has provided controlled drainage, easy access and good plant growth (Moffat & Roberts 1989b).

Interaction with surface vegetation

Grass seed and clover mixtures were widely used up to the mid-1970s to stabilise soil surfaces as part of restoration to agriculture. Trees were planted two years after sowing. In spite of shallow ploughing in preparation for sowing, by the time trees came to be planted, soil had often become recompacted.

In trials of ground cover and cultivation to relieve compaction, tree survival was worse on sites with complete vegetation cover, whether grass or legume. Of five legume covers chosen to test their ability to fix N, tree survival was worst on lucerne plots, which were also the most vigorous.

In the very early post-planting period, competition for water by any form of ground vegetation appeared to be more relevant to plant survival than any potential benefit from N (Jobling 1987).

Mine drainage water

Colliery spoils containing iron pyrites left a legacy of iron-charged acid drainage water, even though the spoil heap surface supported plant life.

Oil-shale

The 19th/early 20th century oil-shale industry in Britain was confined to the eastern end of the 'Central Belt' of Scotland *ie* the land lying between Edinburgh and Glasgow. The local shaley carbonaceous deposits of Carboniferous age differed from most coals in that they could be distilled to yield liquid hydrocarbons of a range of melting and boiling points. Orange-red tips ('bings') of burnt oil-shale characterised the landscape. The angle of repose of the material is steep and natural colonisation is slow.

The freshly burned shale resembled unburnt colliery spoil in its texture but weathered more slowly. Its pH initially is slightly above 7.0 but falls below 7 within a few years. Reshaped spoil heaps have been successfully planted with a range of species, with application of NPK fertilizer at time of planting. At that time, the material had little vegetation or soil organic matter. Alder species have grown well; so also have two pioneering conifer species, lodgepole pine and Japanese larch (Neustein & Jobling 1969, Aldhous 1995).

11.34   Tree establishment on other mineral workings and wastes

Sand and gravel workings

Sand and gravel quarries are most frequently associated with recent glacial drift and alluvial deposits or the younger geological formations.

Table 11.7 gives area of actual and approved potential workings for 1974 in England; by 1988, permissions for extraction had been granted over 29000 ha. From these in 1987, 82 million tonnes of sand and gravel were extracted, compared to >3 and <3 million tonnes for Scotland and Wales respectively (Moffat & McNeill 1994).

Workings of flinty gravel are widespread, particularly in the parts of the Thames basin near chalk. The sites are usually acidic and P deficient (Fourt & Best 1983). Many workings when exhausted are used for land fill. However, where gravel waste is being restored, nutrient and pH treatment have to be integrated with the provision of adequate drainage through ridge and furrow surface landform, and the avoidance of compaction by tining and loose tipping (Fourt 1979, 1980).

See also §11.35 *Landfill and other capped material.*

Ironstone

Ironstone occurs in the Midland counties of England in the Northamptonshire sand, a stratum of the Jurassic period occurring below overburdens of Oolitic limestone, sands, clays and silts. About 8000 hectares have been worked since Roman times. Of these, the 6500 hectares worked before 1940 have mostly been restored to agriculture. Deposits were more heavily and more deeply quarried in the 1939-45 war, leaving nearly 1500 hectares of long ridges and furrows ('hill and dale'). The *Mineral Workings Act 1951* obliged mineral operators to restore such ironstone workings.

In the early 1950s, growth of trees on older ironstone workings was surveyed; larches and pines were found to be growing adequately on the better drained soils. Of the broadleaves, sycamore, poplar and alders grew well, other species thriving only on sites where substantial amounts of top soils had been included in the restoration. Most species failed when planted into clay-dominated spoils. No species had received any external nutrient supplement (Pinchin 1951).

Ironstone has been mined in the central valley of Scotland from deposits in Carboniferous Coal Measures. The spoil is extremely acidic because of the iron pyrites present. Individual reclaimed heaps have either been capped or treated with lime (Aldhous 1995).

Limestone waste
Richardson (1993) reported that for three sites in Durham on magnesian limestone waste, nothing grew well on unimproved waste. However, the addition of topsoil or topsoil + fertilizer improved survival and growth of most species tried. *Alnus glutinosa* and *A. incana* were among the more vigorous species.

China clay waste
China clay is widely used in paper finishing, potteries, cosmetics and as a carrier for pharmaceutical products. 7 tonnes of coarse micaceous quartz sand is the by-product of every tonne of china clay. Heaps have a steep angle of repose and in west Devon and around St Austell, Cornwall, tower above the surrounding countryside (Fourt 1985). The wastes naturally are acidic (pH 3.9 - 4.8). Repeated liming has been applied to promote grass growth to control surface erosion. Climatically, the upper slopes are severely exposed.

The sand is very low both in nutrients and in available water capacity. Recolonisation by natural vegetation has been slow. Experimental tree planting using *Pinus radiata* and *P. muricata* showed only small responses to adjacent legumes (*Lupinus arboreus* and *Lathyrus sylvestris*); air-borne additions to the site of 6 - 8 kg N per ha in 1986 barely maintained plant growth (Moffat & Roberts 1989a, Moffat & McNeill 1994).

Power station wastes
Pulverised fuel ash (PFA) from combustion of coal, and gypsum from 'flue gas desulphurisation' (FGD) are materials continuing in production from contemporary power stations and may be dumped on land.

While some PFA is used for building blocks for the construction industry, currently approximately 5 millions tonnes are produced annually additional to that use. Fresh PFA has a pH of 11 - 12 and a high boron and soluble salt content. A period of leaching by weathering or holding in a lagoon is necessary before tree planting can be contemplated. Thereafter, the same integrated approach as for other materials is required, dealing with compaction/ consolidation and drainage, ensuring an adequate nutrient supply by additions of top soil or organic matter ameliorants or fertilizers, and establishment of nitrogen-fixing and other species capable of growing on neutral or slightly alkaline conditions. Moffat and McNeill (1994) list seven species tolerant of PFA. These include poplar varieties, common alder and (surprisingly!) Sitka spruce. Beech, ash and sycamore are classed as 'sensitive'.

FGD gypsum is probably the most recent industrial waste currently being produced in quantity. Gypsum is used in the manufacture of plasterboard. Some FGD gypsum may be a substitute for mined anhydrite. However, production of FGD gypsum is likely to be substantially more than can be taken up in that way so that disposal to land will be required. FGD gypsum comes from the same power stations that produce PFA and disposal of a mixture is possible. FGD gypsum is neither alkaline nor acid; it contains no plant nutrients and initially has a high soluble salt content. In short-term trials, poplar, alder and *Robinia* grew well if adequately supplied with added nutrients (Moffat & McNeill 1994), but there are as yet no long-term trials of trees.

11.35   Landfill and other capped sites

Dumping into excavations made during mineral workings has been and in the short term is likely to continue to be the most widespread means of disposing of domestic and commercial wastes. Incineration and recycling may make some inroads into the scale of the operation but will not replace it.

National and European standards for waste control have risen markedly since the beginning of the 1980s; practices previously accepted have been challenged. Control of waste disposal in Britain is governed by various *Waste Management Licensing Regulations* (SI 1994/1056, SI 1995/288, SI 1996/972, SI 1996/2019).

The objectives in landfill are to restore land to a beneficial use, preventing pollution and achieving a landscape that is aesthetically acceptable. The extent to which special provisions have to be made depend on the hazardous nature of the deposited material and the risk that it may cause damage if not contained.

Five categories of waste recognised are:

- inert waste - unaltered material from land excavation, brick and concrete waste and similar material,
- household waste - domestic refuse and civic amenity waste,
- commercial waste from shops and offices,
- industrial wastes,
- special wastes - toxic materials, acids and alkalis, cyanides, phenols and other organic and inorganic compounds.

All except inert waste are defined by legislation as 'controlled waste'.

The two primary requirements in waste deposition are that

- any water leaving the site does not harm surface or ground water supplies,

- any gases generated through decomposition or chemical reaction within the fill are discharged in a controlled manner.

In 1993, over 90% of the 136 million tonnes of 'controlled waste' was used for landfill on over 4000 sites. For these, present standards require that the waste is fully encapsulated by impervious materials.

The essential feature of a cap is that it is heavily compacted so that it presents a barrier to root penetration and has a very low oxygen permeability. The question remains as to whether large scale clay caps could actually be constructed to maintain a consistent bulk density without less dense parts, and the extent to which the integrity of a cap is lost if the waste subsides through decomposition of its organic constituents.

The selection of species that can be established on capped sites has been controversial. This is related to the question, if trees are planted on caps, whether their roots can weaken the integrity of the cap and allow increased water penetration.

In 1986, the *Department of the Environment* recommended that trees should not be planted on containment landfill sites (DEnv 1986). Detailed subsequent investigation into tree roots and landfill cap structure showed that trees are typically relatively shallow-rooted, 90% of all roots and virtually all larger roots are found in the top metre of soil. Compacted cap material presents a real barrier to roots. Trees growing on landfill sites with 1 metre depth or more of drained rootable soil over a cap are no more at risk of windblow than trees on similar soils on undisturbed sites, nor are they a threat to the integrity of properly constructed landfill caps (Dobson & Moffat 1993).

Most recent work has examined the use of multilayer caps of high density polyethylene or 'geotextile' membranes made of non-woven felted polyethylene and polypropylene fibres. A polyethylene membrane and a geotextile into which the herbicide trifluralin had been injected each successfully prevented root penetration (Bending & Moffat 1997).

As a result of this work and following a consultation paper circulated by the Environment Agency in 1996, the 1986 restriction on trees was amended to allow trees in belts and small blocks. Tree establishment practices now follow the same principles as for other wastes, but with the additional restraint that species choice should be limited to those species not likely to threaten the integrity of the cap.

Tree establishment on landfill sites (Bending & Moffat 1997) includes lists of trees and shrubs considered suitable for planting on capped sites.

Landfill gas

Domestic waste contains much decaying material ('putrescible waste'). When compacted and sealed in a capped landfill, any oxygen present is rapidly used up; anaerobic decomposition follows and may last for several decades, producing 'landfill gas'. Its composition depends on the waste present but it is typically roughly 2:1::methane:carbon dioxide. Hydrogen sulphide is normally only present as a trace but exceptionally has reached 35% by volume of the landfill gas.

Landfill gas seeping through cap material harms plants primarily through its ability to displace oxygen from the soil in the plants' rooting zone. Bacteria able to oxidise methane in the soil may also be present, making a further demand on soil oxygen. Trees affected by landfill gas typically show yellowing of leaves, wilting and stunted growth. Severely affected trees may die (Dobson & Moffat 1993).

Toxic mining and industrial wastes

While such wastes are less likely to generate landfill gas, water seeping through the material may become charged with soluble toxic compounds. Such leachate may require special treatment. Toxic wastes may be encapsulated as for domestic landfill.

Hard rock spoil from metalliferous ore workings have been used as forest road-building material. Surface wash-off from such roads have been toxic to plants in the immediate vicinity. Such materials should be avoided in road construction.

Rubble

Rubble from demolition of buildings is common in urban sites. On many sites it is innocuous, but where former industrial premises have been knocked down, there may be toxic manufacturing residues requiring specific treatment

11.4   OTHER TECHNIQUES

11.41   Injection of nutrients into trees

Trials of injection of nutrients into trees were not successful in delivering nutrients. The wood was heavily discolored and the stem injured around injection points (Hodge 1993).

CHAPTER 12

Nutrient cycles in woodland

Dust to dust and ashes to ashes?

12.1   WOODLAND CYCLES

12.11   Cyclic processes

The forest over the British Isles shortly before the arrival of the first land-clearing farmers some 5500 years ago was characterised by:

- woodlands of all ages from recent regeneration to senescence;
- a climatically defined tree line as part of a primary zonation of plant associations directly related to differences in elevation;
- within any elevation zone, different associations of vascular and non-vascular plants, reflecting the local combinations of climate, land form and the mineral content of subsoil geological formations.

At this point, the main cyclic processes affecting woodland were:

- *energy* capture through photosynthesis:
    partitioning within the plant;
    removal or loss in litter-fall, radiation, evaporation of water,
        consumption of energy-rich plant material by predators;
    deprivation through reduction in efficiency of leaf photosynthesis
        caused by toxins or lack of essential nutrients or drought;
- *nutrient* release by soil weathering, fixation by micro-organisms and
        deposition from the air and by rain:
    uptake to create and maintain photosynthetic tissues and their
        structural supports, and the organism's reproductive system;
    internal storage and translocation to renew energy capture tissues;
    removal from the living organism in litter-fall, tissue leaching,
        consumption by predators *etc.*;
- *water* supply through rainfall, redistribution through ground water:
    uptake to maintain tissue turgor and photosynthesis;
    loss by evaporation of intercepted and transpired water;
    drainage to surface streams and underground aquifers;
- *carbon* cycling through capture by photosynthesis:
    translocation and storage within the plant and as soil organic matter;
    losses by litter fall, decay, and by respiration in plant and in soil.

Cycles and models

During the last 50 years, cyclic processes have increasingly been subject to detailed study. Research programmes investigated the structure and flow of cyclical processes, attempting to identify the salient features of cycles and their chemical and physical components. Increasingly, quantitative information about scale of components, speed of reactions and rates of flow of reactants enabled figures to be assigned to parts of cycles previously only identified qualitatively.

With the availability of computers, 'mathematical models' have been constructed, attempting to describe cyclic and other processes in numerical form. The yardstick for their success is the extent to which 'process-model'-based predictions reflect field observations. Models developed by Miller (Miller *et al.* 1980) for within-forest nutrient cycles were early examples of modelling that was strongly 'forest-oriented'.

Many older studies of nutrient flow and of response to fertilizers were designed when neither the present knowledge of depositions nor the modelling technology was available. In many cases, the inputs due to dry deposition and cloud-water deposition are likely to have been underestimated.

Sections in this chapter cover:
nutrient cycles in tree and woodland growth (§12.2);
interaction of *Calluna vulgaris*, mycorrhizas, nitrogen nutrition and tree growth (§12.3);
the effects of trees on soil (§12.4);
nutrient loss in drainage water (§12.5).

The wider issues of atmospheric pollution are described in the following chapter.

12.12   Nutrient availability and need

At the beginning of this century, because of the past widespread distribution of natural woodland, forests were thought not to require nutrients additional to those already present in the soil as a result of natural weathering processes. However, a number of early attempts to re-establish woodland on agriculturally unproductive land failed, and it became apparent that on sites where nutrient capital had been depleted or soil lost, addition of fertilizers might be essential to establish a tree canopy.

Two broad lines of enquiry followed. One, described in the two preceding chapters, investigated empirically the extent to which poorly growing crops would respond to added nutrients supplied through fertilizers. The other sought to establish the role of nutrients in woodland growth, and is considered in this chapter. While distinctive, the two lines were often run with close collaboration between staff of the institutions involved. Together, they created the foundation of knowledge on which current silvicultural practice is based.

The principles underlying nutrient cycling in woodland apply equally to ornamental and amenity trees.

12.2   NUTRIENTS IN TREE AND WOODLAND GROWTH

12.21   Nutrient cycles

Driving the woodland ecosystem is a series of interlocking nutrient cycles which maintain the intact woodland canopy as the mechanism for capture of solar energy. Where trees have been removed or destroyed and that canopy has to be recreated, establishing whether there are adequate quantities of available nutrients to do so has been a major component of woodland research programmes over the last 50 years.

While quantitative studies of the nutrient content of plants began on the continent in the 19th century, nutrient studies in British woodland date from the 1950s (Rennie 1955, Ovington 1956, 1962, Ovington & Madgewick 1959).

From the 1950s, work on nutrient cycles has been based, partly on sites where empirical fertilizer trials had given striking results and partly on independent work.

Results of earlier nutrient cycle studies are described below.   Interaction between the nitrogen nutrient cycle and deposition of 'fixed' nitrogen (NO_x and NH_y) from the atmosphere are discussed in §13.2. Effects of sulphur compounds are considered in §13.3.

12.22   Culbin studies - nitrogen, other nutrients and pine

In northeast Scotland, crops of Scots pine and Corsican pine have been planted on sand dunes at the Culbin sands (Morayshire). The nutrient status of the soil is low (Ovington 1950). The area has cool summers and limited summer heat and would be considered marginal for Corsican pine for that reason (Macdonald *et al.* 1957).

Over a 20 year period commencing in 1956, studies based on Culbin and supplemented by other experiments, provided the basic data leading to the formulation of models of the role of nitrogen and other nutrients, first in Corsican and Scots pine and then in other species. The first of these studies assessed the N, P, K, Ca and Mg content in needles, branches, bark and wood in crops of three ages, and compared samples taken in June and December (Wright & Will 1958, Binns 1960).

In 1964, also at Culbin, a long-term experiment was set up in 36-year-old Corsican pine stands. The crop selected had grown adequately for the first ten years, *ie* up to canopy closure. For the second decade, growth was slow, reflecting the very low nitrogen content of the soil. In the following period, growth deteriorated as N was increasingly immobilised either in the tree or in developing 'mor' humus which was not releasing sufficient N through decay to keep up with plant N requirement (Miller 1966, Miller & Cooper 1973).

At the time the experiment was started, the crop appeared acutely nitrogen deficient; needles were small and chlorotic. Plots were treated with N as ammonium sulphate, at x½, x1, x2 or x3 of the 'standard' rate of 168 kg N per ha annually, these treatments being repeated for three years. Sample trees were taken for analysis before treatments were applied, after three years (*ie* after the completion of the fertilizer applications) and after seven years. Litter and humus were sampled at intervals and incoming precipitation and its chemical content measured (Miller & Mackenzie 1964).

Cycle of nutrients and growth

There followed a series of analyses each contributing to a facet of knowledge of the nutrient cycle or of growth response.

• After 2 years, foliage N had increased almost linearly in relation to the rate of application. Litter fall was initially reduced by up to 22% on the plots given highest rates of N because of improved needle retention, but this effect was subsequently reversed with the fall of larger needles resulting from the higher N levels (Craig & Miller 1966).

• The combined nitrogen on the site amounted to 1950 kg per ha, 13% of which was in the trees, 16% on the forest floor (Litter+Fermentation+Humus layers) and 70% in the sand below the humus layer (Miller & Williams 1968, Williams 1972).

• At the lowest rate of N application, the additional amount of N found in the nutrient system equalled the amount supplied *ie* all the additional application had been retained on site; for the highest application rate, the recovery fell to 45%. The greater part of this extra nitrogen was located in the tree crop rather than in the litter (Miller & Williams 1969).

Table 12.1   *Nutrient flux sinks & sources for N, P & K in a Corsican pine stand at Culbin, Morayshire*   (kg/ha/year)

Flux	N	P	K
Sinks			
new needles	92	9.4	45
structural tissue	46	4.7	15
replace leaching losses from crown	0	0.0	6
Total requirement	*138*	*14.1*	*66*
Sources			
uptake from soil	69	6.0	28
recovery from needles before death	61	7.2	32
recovery from other tissues	8	0.9	6
Total of all sources	*138*	*14.1*	*66*

Sources   Miller & Proe 1986, Miller 1984b   Av. needle retention 2.4 years, Stand vol. 230 cu.m per ha.   Current vol. increment 20 cu.m per ha.

• Potassium, calcium, magnesium and sodium from dust and dissolved salts in rainwater constituted an important supplement to potassium and magnesium obtained by recycling from material either leached or reabsorbed from the tree crowns (Miller & Cooper 1973, Miller & Miller 1976a, Miller & Williams 1974, 1976).

• Basal area increased proportionately to the amount of added nitrogen, this stimulus persisting for seven years but tailing off at the end of the period. Response was unaffected by tree size. The continuing response was attributed to the storage in plant tissues of nitrogen from the treatments (Miller *et al.* 1976b).

• Height growth was increased by all treatments and continued for the period of measurement. The biggest response was for 168 kg N per year and was less for the x2 and x3 rates. Trees in the middle of the height range responded more than the biggest trees.

Figure 12.1 *Pattern of cycling of potassium in Corsican pine*

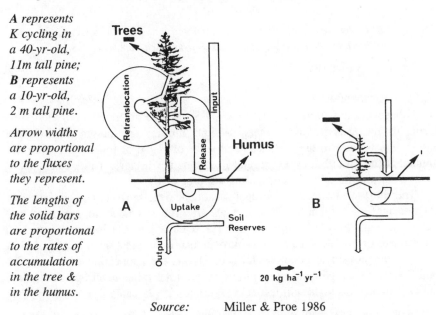

A represents K cycling in a 40-yr-old, 11m tall pine; B represents a 10-yr-old, 2 m tall pine.

Arrow widths are proportional to the fluxes they represent.

The lengths of the solid bars are proportional to the rates of accumulation in the tree & in the humus.

Source: Miller & Proe 1986

Table 12.1 above gives 'net' values for nitrogen, phosphorus and potassium circulating in the pine on the research plots at Culbin.

Figure 12.1 above illustrates the quantities of potassium circulating within the pine. Short horizontal bars in the figure illustrate that the K stored in the tree is markedly greater than the K stored in soil humus layers.

Figure 12.2 below shows the quantities of biomass and N in Corsican pine accumulating over the period between canopy closure and economic rotation age.

Figure 12.2   *Patterns of accumulation of biomass & nitrogen in managed plantation of Corsican pine*

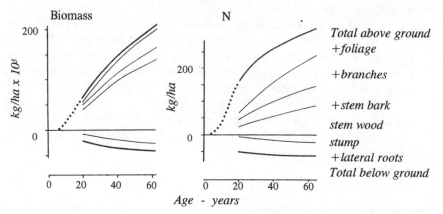

The top and bottom lines indicate total accumulated biomass; the spaces between the intermediate lines indicate the contribution from each component listed.

Source   Miller 1984c

Scots pine

Other work at Culbin included tests on Scots pine of the effect of alternative forms of nitrogen fertilizer (urea, calcium nitrate, ammonium sulphate and ammonium nitrate) in the presence or absence of P and of lime. The latter was applied at the rate of 2500 kg ground limestone per hectare, a rate calculated to change the soil pH as well as providing a source of Ca.

The short-term effects of nitrogen treatment were fairly large; subsequently, foliage nutrient content levels and rates of growth tended to return to pre-treatment conditions. N supplied as urea remained in the humus longer than other forms of N used. The form of nitrogen had little effect on N mineralisation.

The biggest single effect on form of N was the application of lime. This both increased the percentage of nitrate in the total mineral nitrogen present in humus and also made the soil less acid (Williams 1972, Table IV).

Scots pine was also studied at Devilla, Fife. Here, a study of nutrient circulation and translocation was carried out over two years in a 46-year-old Scots pine plantation. Sample trees were dismantled separately into 1st, 2nd and older branches and needles, bole, bark cones and wood. Good regressions of tree components on girth at 1.3 m were obtained. All plant tissues were analysed for major nutrients, as were rainfall and litter fall (Lim & Cousens 1986a, b). *Table 12.2* gives figures for circulation of major nutrients in the stand.

In the same stand, a study was made of litter fall over a 12 year period. The crop was thinned early in the period. After an initial fall, total litter production and needle biomass reached pre-thinning levels in 3-4 years (Cousens 1988).

Table 12.2 *Cycling of N, P, K, Ca and Mg in the above ground parts of a Scots pine stand at Devilla, Fife* *(kg/ha)*

	N	P	K	Ca	Mg
Nutrient capital in trees	340	35.5	217	142	47
Annual requirement	109	13.2	74.9	20.2	7.9
translocation	60	8.5	42.9	- 11.2	1.2
uptake	49	4.7	32.0	31.4	6.7
Annual litter fall	37	2.2	7.6	12.1	2.0
leaching loss from trees	0.7	0.2	11.3	8.1	3.1
Annual input from rain	2.6	0.4	4.3	7.6	1.4
net aerosol input	0	0	13.5	0	0.3

Source Lim & Cousens 1986b

Sitka spruce

Studies on nutrient cycling were extended in 1972 to Sitka spruce, with a study of the relationship between tree growth and nutrient movement in pole-stage plantations. The studies were was based on application of fertilizer on six sites covering a range of soil types from brown forest soils to peats. The work was linked to glass-house studies of responses by seedlings to nutrient regimes.

The main comparison was of 0 v NPK treatments (*ie* with and without NPK). It was supported by smaller experiments to ascertain the role of individual nutrients. Techniques developed for Corsican pine were applied also to Sitka spruce with slight modifications *eg* to capture the greater volumes of rainfall flowing down stems.

A comprehensive analysis after 11 years growth showed there to be no significant response in terms of height growth or basal area increment to added fertilizer. Taken together with results of other forest fertilizer trials (see §10.82), they support the assertion that nutrient cycling in Sitka spruce stands 25 to 35 years old is sufficient to maintain tree health and growth and that responses to fertilizers are unlikely at this phase of plantation growth (Miller *et al.* 1992).

Another study on pine, in this case *P. radiata*, emphasised that within-tree recirculation of nutrients between needles, growing points, roots and woody tissues during the growing season is an important process; translocation is not just a means of recovery of nutrients from older needles (Namibar & Fife 1991).

Comparison of pine and spruce

Studies of nutrient cycles showed that the weight of foliage in closed canopy of even-aged pine and spruce of comparable rates of growth was 7-14 tonnes/ha for pine and 15-20 t/ha for spruce. The differences were due mainly to differences in the duration of needle retention. In Corsican pine at Culbin, needles more than 2 years old constituted 20% or less of the foliage of 30-year-old trees, whereas for Sitka spruce, older needles could amount to 33-50% of the foliage weight. From models, the weight of first year foliage predicted for Sitka spruce and Corsican pine of similar rates of basal area increment was the same. The nutrient requirement necessary to maintain annual crown foliage replacement reflects primarily the rate of growth of the crop, differences in nutrient content of foliage being relatively minor (Miller 1984a,c, Miller & Miller 1987).

12.23   Nutrient flux models for plantation crops

The outcome of the studies on pine and spruce was a comprehensive understanding of distribution, balance and change of nutrients with time, and the partitioning of dry matter and nutrients within the plantation ecosystems (Miller 1984c). Three stages could be identified:

-    *Crown development* During this phase, any shortage of nutrients through infertility, immature soil *etc.* is rectifiable by additions of appropriate fertilizer.

-   *Full-crown closed canopy* The natural nutrient cycle is normally able to sustain itself; drastic loss of foliage through thinning, insect defoliation *etc*, may require a short-term boost to assist recovery.

-   *Ageing crops* associated with immobilisation of nutrients, especially N in humus. Such crops may respond to N.

N immobilisation has been most frequently encountered on cold dry sites in 'stagnating' boreal forest types but was also suspected to be operating in pole-crop pine at Culbin and in 80-year-old Scots pine in Thetford.

Nutrient content of forest crops

From this work, it became possible to integrate crop nutrient data with timber volumes and so, through volume yield tables, to estimate the nutrient content of forest crops according to age and yield class. Miller *et al.* (1980) give a worked example for the detailed nutrient content at 10 year intervals for Corsican pine yield class 12; data for other yield classes are available.

Table 12.3 shows weights per hectare of stem, branch and foliage components for five conifer species and birch; it highlights the relatively greater proportion of foliage in spruces compared with pines, larch and birch.

These data were further linked with data for nutrient inputs to a site, both natural and through fertilizers, to predict the consequences of alternative felling regimes on the nutrient capital of the site. It was concluded that:

-   phosphorus removal was not likely to be replaced from natural sources;
-   for nitrogen, input from airborne natural sources was likely to exceed removal;
-   for most other nutrients, natural deposition and weathering input were considered to be adequate (Miller *et al.* 1980, Miller & Miller 1991).

The method has been used to model other species and areas (Miller 1979a, Miller 1983, Miller *et al.* 1986), including crops on the water catchment study at Balquhidder (Miller *et al.* 1993).

Table 12.3   *Stem, branch & foliage weight of trees at clear felling   (tonnes/ha)*

Species	Age	Stem	Branches	Foliage	Total
Birch	40	84	15	2	101
Eur. larch	63	91	13	2	112
Scots pine	48	118	8	16	146
Cors. pine	50	136	9	22	175
Norway spr.	50	202	22	27	254
Sitka spruce	50	237	22	34	300

Source   Miller & Miller 1991

Other approaches

In predicting productivity of Sitka spruce from site factors, soil categories contributed significantly to the precision of predictions (Worrell 1987, Worrell & Malcolm 1990). Predictive computer models have also been created showing growth response to application of fertilizer, in relation to soil type and lithology (Taylor 1988).

12.24   Nutrient cycle in broadleaves and on lowland sites

There are only a limited number of studies of broadleaf nutrition and fertilizer response in Britain.

Birch

Following Dimbleby's work in North Yorkshire, Miles showed that birch invading *Calluna*-dominated heathland led to changes in soil pH from 3.8 to 4.9, a rise in exchangeable calcium levels and a substantial increase in the populations of earthworms (Miles & Young 1980, Miles 1981).

Harrison *et al.* (1988) using radio-isotope phosphorus, found that birch took up P predominantly from the surface layers of soil (0-100mm) and that the proportion of P from 0.5 m depth was less than 1%.

Nutrient cycle data for birch in Britain have been constructed using Forestry Commission yield table growth data, and biomass data for *Betula papyrifera* from North America (Miller 1984e). Values for birch were compared with what data was available from other broadleaved species. Nutrient cycling in birchwoods appeared similar to that of other species with similar rates and patterns of growth. While birch has a reputation as a 'soil improver' (Gardiner 1968), there were no clear indications why birch should improve soil. Miller concluded that his evidence does not support the hypothesis that birch promotes a more rapid circulation of calcium through the nutrient cycle.

Lowland sites

Anderson (1986a, b, 1987) reports results of soil nutrient measurements made in 1951 and repeated after 23 years. Samples were taken from 10 conifer and 9 broadleaved species growing in species trials, mostly at one of three sites of contrasting soil type. Data for total soil N, P, K and Ca are given. Only Corsican pine had been planted on all three sites.

On all sites, most nutrient capital changes after 23 years were of a similar order of magnitude. On the two non-calcareous sites, amounts of available N and Ca fell slightly and P and K increased. On the calcareous site, all levels increased. For most nutrients and sites, soil nutrient reserves were positively correlated with growth rate as indicated by Forestry Commission yield classes.

Coppice

Short-rotation coppice for the production of biomass for fibre or energy is an intensively managed plantation-type crop. Good growth requires fertile farm land and freedom from weed competition; even so, nutrients removed in stem wood alone by the act of coppicing are likely to exceed the inputs from weathering and atmospheric deposition (Cannell 1982).

In France, nutrient cycles were studied in three types of coppice stand:- sweet chestnut, mixed broadleaves and short rotation poplar. The first two crop types were on rotations up to 41 years.

Reliable relationships were found between biomass and nutrient content of the main stem and branches. While the proportion of leaves to stemwood was measured, leaves were not included in analyses of nutrient content. No data for foliage nutrient content is given. However, equations were derived for the relationships between soil properties and both dry matter and stem nutrient content (Ranger & Nys 1996).

12.25   Crop growth rates and timber quality

Corsican pine in three of the experimental treatments at Culbin were sampled to study the effects of varying nutrient levels on wood quality. While the proportions of early-wood increased and late-wood diminished with corresponding changes in ring width and wood bulk density, effects were considered to be no greater than could arise through natural variation in climate on the site, in particular in relation to early summer moisture stress (Smith *et al.* 1977).

Brazier (1977) reviewed the effects of forest practices on timber quality. He dismissed any effects of fertilizers at planting as the volume of wood produced in the first few years is small. During the late establishment phase, fertilizer could affect both the amount and character of juvenile wood. Applications to older crops consistently increased early wood at the expense of late wood, with a consequent reduction in wood density.

12.26   Litter

Litter and soil surface organic matter, being easily accessible components of the woodland, have a long history of study and comment. Many workers in the 1950s sought to understand the role of litter, its breakdown and the consequences for the trees (Ovington 1950, 1956, Dimbleby 1952, Handley 1954, 1963, Owen 1954, Ovington & Madgewick 1957, 1959).

The '*International Biological Programme*' in the 1970s renewed interest in litter of natural and plantation woodland as an element in overall site productivity and nutrient cycling.

Litter under pine

In investigations of the nutrient cycle in Corsican pine at Culbin (§12.22), while there was little correlation between nitrogen content of foliage with the litter-fall at the end of the year, there was a good relationship between foliage nitrogen and N in litter-fall in the following year (*Figure 12.3*). Where pine crops 20 m or more tall appeared to be at risk of nitrogen deficiency, systematic monitoring of nitrogen in freshly fallen litter was proposed as easier than sampling foliage from the top whorl (Miller & Miller 1976b).

Figure 12.3   *% N in top whorl foliage of CP in one year in relation to % N in needle fall collected in the subsequent year*

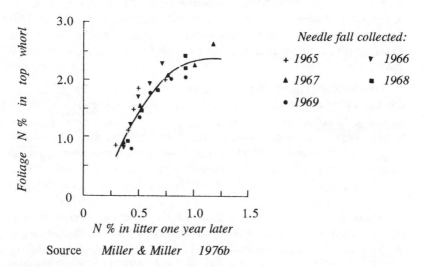

Source   *Miller & Miller   1976b*

Johansson (1995) describes a Scandinavian study on Scots pine, Norway spruce and *Betula pubescens* (white birch) on 17 sites over a latitudinal range of 10°. Nutrient content of litter quantitatively was pine < spruce < birch. No significant relationships were found between litter nutrient content and site index. The report contains a substantial bibliography of continental and American studies on nutrients, litter, tree growth, and nutrient cycles.

In a 4-year study at Bramshill Forest, Hampshire, litter under Scots pine was treated with different N fertilizers and analysed. The organic matter was described in terms of pH, ash, total N, polysaccharides, amino acids *etc* (Tinsley & Hutcheon 1965).

Litter under spruce

In Northern Ireland, annual litter-fall under Sitka spruce on surface water and peaty gleys differed little between soil type. Litter-fall quantitatively reflected vigour of the crops, the litter-fall under higher yield classes (YC) being greater quantitatively and having a higher nutrient content. However, breakdown was sufficiently slower on the lower yield-class plots for the accumulation of litter on the forest floor to be greater than under the higher yield-class plots, notwithstanding their higher annual litter-fall (Adams *et al.* 1980).

Litter-fall was also assessed in the Sitka spruce crops subject to nutrient cycle studies (§12.21 above). Litter-fall across six Sitka spruce crops, 25-30 years old, ranged from 2440 to 4450 kg per ha per year (Miller 1986b). Accumulated litter and humus in Sitka spruce in these sites ranged from 22 to 35 tonnes per ha, indicating a turnover time for the annual litter-fall of 8 - 10 years, but reaching 13 years on the coldest site (Miller *et al.* 1996a).

In laboratory trials, samples of litter and humus from Sitka spruce under 10m tall were incubated in order to establish rates of mineralisation of N compared with forest stand requirements. Incubation under laboratory conditions gave higher rates of mineralisation than samples incubated in the forest. Differences in temperature regime were insufficient to account for the slower rate in the forest, implying presence in the forest of inhibitors of decomposition (Williams 1983).

In Denmark, older stands of Sitka spruce have been subject to severe defoliation by spruce aphis *(Elatobium abietinum)*. In species trials on former farmland, heathland and oak woodland, soil water from below the rooting zone was analysed for all major nutrient ions. Litter-fall was also analysed.

During the period 1988-94, at least two severe defoliations occurred on all three sites, linked to heavy infestations by aphis. In years of heavy infestation, N input to the forest floor from litter-fall was 2½ times that from throughfall. In other years, litter-fall and throughfall N totals were fairly similar.

There were insignificant amounts of NH_4^+ in the soil water from any plot. On the former woodland and heathland sites, NO_3^- loss in soil water was less than the NO_3^- in throughfall, amounting to 11 and 2.5% of the annual throughfall + litter-fall N. On the former farm site, however, NO_3^- loss in soil water exceeded throughfall input substantially and amounted to 67% of the annual throughfall + litter-fall N (Pedersen & Bille-Hansen 1995). See also §13.24 (N-saturation).

Litter under other species

Studies of litter from woodlands over a wide area in the Lake District, (Cumbria), showed that most conifer litters when dried and ground had a more acid reaction than most broadleaves. However, in this respect, western red cedar *(Thuja plicata)* litter was more like broadleaf litter than litter from other conifers (Howard & Howard 1990).

Incubation of soil organic matter

Samples of LFH layers from beneath 16 conifer species were incubated with and without starch. For 11 species, mineral nitrogen increased under incubation significantly more in the presence of starch than where it was absent. This nitrogen was almost entirely ammonium nitrogen. Levels of nitrate nitrogen remained low except for litter of *Thuja plicata*, where nitrate levels exceeded ammonium-N levels (Harmer & Alexander 1986).

12.27   Ground flora and cropping

Anderson (1982) concluded that for lowland Britain, by about half rotation age, the ground flora on land that had been planted with conifers had come to resemble that of acid oak woodlands.

At Culbin Forest, five key indicator species were identified as indicating progressively enhanced nitrogen status in the humus layer, both under Corsican pine and Scots pine: *Cladonia* spp (low N status on dry site) < *Agrostis tenuis* < *Rumex acetosella* < *Chamaenerion angustifolium* < *Holcus lanatus* (highest nitrogen status) (Miller *et al.* 1977).

12.3   MYCORRHIZAS, HEATHER CHECK AND NITROGEN

12.31   Mycorrhizas

Mycorrhizas* were first observed in the 1880s as part of investigations in Germany into the natural distribution of truffles. A.B. Frank described the intimate association between plant roots and hyphae of soil fungi surrounding and within root tissues. Subsequent work in several countries showed that such associations were widespread, and were associated with fully healthy plants (*eg* Rayner & Neilson-Jones 1944).

The assumption in the 1930s when Rayner started her work at Wareham, Dorset was that mycorrhizas were beneficial. Work before and since has shown a very widespread symbiotic, *ie* mutually beneficial, relationship between the fungal species present and trees. Fungi take over the absorptive functions of root hairs but draw carbohydrate from trees (Harley *et al.* 1958, Levisohn 1958, Read & Finley 1985, Dighton 1991). Peace (1962) pointed out that strictly, mycorrhizas, because they take carbohydrate from their host plants, were parasitic. However, that view took no account of nutrient returns to the host nor any benefit through reduction of drought stress or protection from pathogens (Harley & Smith 1983).

Dighton (1991) summarised evidence that mycorrhizal fungi are capable of producing enzymes able to secure carbon, N and P from complex organic materials in soil. In practice, mycorrhizas are mostly beneficial, often harmless and never harmful.

* Mycorrhiza = 'fungus root'. Different authors treat the plural form of the word as mycorrhizas, mycorrhiza, mycorrhizae or even mycorrhizaes. Rayner & Neilson-Jones (1944) in their introduction (p.18) gave the derivation of the term. They used *mycorrhizas*, as the plural form; their usage is followed in this text.

The structures formed by mycorrhizal fungi in relation to fine roots fall into two main classes, ectotrophic and endotrophic mycorrhizas.

In plants of many genera, characteristic ectotrophic mycorrhizal rootlets are formed. These are short rootlets about 2.5 mm long, enveloped in a covering mantle of fungal hyphae over the whole root tip. In some species, such visible external root structures (ectomycorrhizas) characteristically occur in clusters; they constitute a high proportion of large parts of the active roots of oaks, beeches, pines and spruces. In other genera, notably *Ericacae* and *Orchidaceae* but also including some tree species, fungal associations within the root ('endomycorrhizas') are formed; these are described as 'vesicular arbuscular' and 'ericoid'. See Fig 7.1 in *Forest Nursery Practice* for illustrations of vesicular arbuscular and ectomycorrhizas (Walker & Wheeler 1994).

On several heathland sites in Britain in the 1920s, attempts were made to establish pine crops by direct sowing. Many of these failed. However, on the heathlands in Dorset best germination and survival was obtained where sowings were adjacent to stumps of Maritime pine, many of which had colonised the heathland from peri-urban plantings around Bournemouth and Poole.

Early Forestry Commission research into the causes of failure of the direct sowings compared the effects of basic slag and bonemeal incorporated into sown patches on heathland near Wareham, Dorset. At the same time, the possibility that ectomycorrhiza-forming fungi might have a critical role led to a series of investigations on Wareham soils under Rayner. She concluded that some inhibitor operated which prevented formation of good mycorrhizal associations. From knowledge that mycorrhiza-forming (m-f) fungi developed well in organic-rich growing media, substantial volumes of various composts were incorporated experimentally into direct sown patches. Seedlings grew extremely well when sown on such patches. While the role of m-f fungi was not clear, it was very evident that potential inhibitory effects could be avoided by adequate provision of composts (Rayner 1936, Rayner & Neilson-Jones 1944).

At that time, the fact that the patches where sowings were made were freed from heather and other vegetation was not considered worthy of remark, nor was any account then taken of the nutrient content of the composts.

Comparisons of bonemeal, basic slag and composts incorporated into sown patches showed that plants on composted plots initially grew faster than those given bonemeal or basic slag. After 10-15 years, however, while early differences in height persisted, the rates of growth of current shoots were similar. The implication was that direct-sown seedlings benefitted from the release of nitrogen in the compost, over and above the response to phosphorus but that the effect was not sustained.

Other work at Wareham showed that:

- the use of composts and other bulky organic supplements was too costly and labour-consuming to be applied to direct sowings in the forest; adequate performance could be obtained using a phosphatic fertilizer;
- by enlarging sown patches into continuous strips, seedlings could be grown of a quality far exceeding the conventional production from forest nurseries at that time.

Seedbeds manured with composts to the specification found successful in direct sowings were set up at Sugar Hill, in Wareham Forest, the site becoming the first heathland nursery and the beginnings of a revolution in nursery plant production. See also §11.13 *Mycorrhizas in forest nurseries*.

12.32   Heather check

While the benefits of cultivation and addition of phosphate resulted in vigorous early growth of young spruce and other crops on many upland heathlands, this was not always sufficient to sustain their long-term growth. From the beginning of the 20th century and before, heather (*Calluna vulgaris*) had been recognised as an associate of poorly growing trees of some species. Scots pine and birches were able to colonise and grow on heathland, especially where the physical local dominance of the heather had been reduced by ploughing. However, on many ploughed and planted sites, heather recolonisation was often associated with slowing or 'checking' growth rates of young planted trees, particularly of Sitka spruce, before sufficient canopy had formed to suppress the heather (Weatherell 1953).

In early experiments where the main emphasis was on effects of forms, amounts and placement of phosphates, it was also noted that removal of *Calluna* was also usually beneficial. At the same time, the manual techniques available (hoe/screef/mulch) were too expensive for general use (Zehetmayr, 1960).

Following work at Allerston Forest, Yorkshire, Leyton (1950) suggested that the primary effect of heather competition was to induce nitrogen deficiency. Investigations into 'check' showed that live *Calluna* (an ericoid mycorrhizal species) was able to generate allelopathic compounds in the soil, inhibiting the activities of ectotrophic mycorrhiza-forming fungi. It was also suggested that there could be rhizosphere effects due to mycorrhizal species or their antagonists, before the full development of mycorrhizal rootlets (Levisohn 1953, 1956).

Additions of a heather mulch to heather-free land improved growth of Lawson cypress, indicating that once dead, heather plant tissue no longer inhibited growth (Leyton 1955). The first trial to compare mulching, nitrogen, trace elements and control of *Calluna* with 2,4-D was laid down in collaboration with Leyton in 1955 (Edwards & Holmes 1956).

Handley (1963) supported Rayner's earlier view that failure of direct sowings was not through lack of available m-f species but through antagonism resulting from a *Calluna*-generated inhibitory factor. Allelopathic compounds could be found in raw humus under *Calluna* but disappeared over time if the *Calluna* was killed or suppressed. Robinson (1972) also supported Handley's views that inhibitors are generated by *Calluna's* own endotrophic mycorrhizas, showing that washings from *Calluna* roots severely inhibited *Amanita muscaria* but had less effect on other soil and root fungi.

The inhibitory effect was found not to be equal across species; Sitka spruce was one of the more 'heather-sensitive species', while birches, Scots pine, juniper, rowan and broom came into the 'more resistant to heather check' group (Handley 1961, Malcolm 1975).

Handley proposed that the current aim in establishing crops should be to use all silviculturally acceptable means to get rid of *Calluna* and to prevent recolonisation as this would lead to further risk of check. Soils of low fertility not colonised with *Calluna* should be treated with great care to prevent *Calluna* invasion. Nitrogen fertilizers should be used to stimulate tree growth.

Mycorrhizas of Sitka spruce growing poorly on basaltic soils have also been examined and classified by type present. While normal ectomycorrhizas predominated where growth was better, non-mycorrhizal 'beaded roots' were common on less vigorous plants; foliage nutrient concentrations indicated P and K deficiency. Heather was not present on these soils (James *et al.* 1978).

More recent work has shown that the mycorrhizas associated with *Calluna*, *Vaccinium* and other ericaceous species are able to take up free amino acids from humus on strongly leached, nutrient poor soil, thereby giving these species a competitive advantage over fungi not able to do this (Abuarghub & Read 1988). Ectomycorrhizas can also mobilise N in soils more acid than pH 5 (Read 1991).

Taking a global view, Read linked ectotrophic mycorrhizas with northern coniferous and deciduous forest, ericoid mycorrhizas with alpine 'heath-tundra' plant associations, and vesicular arbuscular mycorrhizas with plant associations in warmer climatic regions. *Figure 12.4* shows his view of the inter-relationships between species, site factors, humus and mycorrhizas.

12.33   Mycorrhizal symbiosis

Fungi occur in large numbers throughout woodlands and forests. A small number are known pathogens, but most are saprophytes, playing an important role in the forest decomposer cycle. The number of species which can form mycorrhizas with trees is large; more than 50 species have been listed for oak woods (Watling 1974) and over 100 for birch (Watling 1984). Alexander & Watling (1987), in a review of the macrofungi of Sitka spruce in Scotland, noted that the fungal species which form mycorrhizas with Sitka spruce are largely also m-f with birches and pines. Also, of the 107 species of fungi in N. America listed as m-f with Sitka spruce, 58 are on the British check list, but only half of these have actually been recorded in Sitka spruce stands in Britain.

The relationships between mycorrhizas and nursery plant production, and between nitrogen and *Calluna* check in the forest, dominated thinking in the 1950s and 60s. The widely accepted conclusion was that nursery and forest soils in Britain were adequately provided with natural populations of m-f fungi.

Figure 12.4   *Major interactions between soil and plants and mycorrhizas*

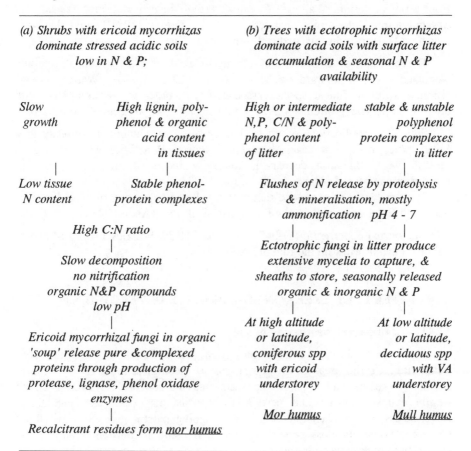

(a) Shrubs with ericoid mycorrhizas dominate stressed acidic soils low in N & P;

(b) Trees with ectotrophic mycorrhizas dominate acid soils with surface litter accumulation & seasonal N & P availability

Slow growth

High lignin, poly-phenol & organic acid content in tissues

High or intermediate N,P, C/N & poly-phenol content of litter

stable & unstable polyphenol protein complexes in litter

Low tissue N content

Stable phenol-protein complexes

Flushes of N release by proteolysis & mineralisation, mostly ammonification   pH 4 - 7

High C:N ratio

Slow decomposition no nitrification organic N&P compounds low pH

Ectotrophic fungi in litter produce extensive mycelia to capture, & sheaths to store, seasonally released organic & inorganic N & P

Ericoid mycorrhizal fungi in organic 'soup' release pure &complexed proteins through production of protease, lignase, phenol oxidase enzymes

At high altitude or latitude, coniferous spp with ericoid understorey

At low altitude or latitude, deciduous spp with VA understorey

Recalcitrant residues form mor humus

Mor humus

Mull humus

Source   Read 1991

At the same time, unnatural sites and techniques *eg* reclaimed mineral workings, partially sterilised soils and micropropagation rooting media, were becoming more commonplace. This led to work studying existing m-f symbionts and the effect of inocula in several nurseries, including young plants of Douglas fir and Sitka spruce (Walker 1981, 1987, Walker & Wheeler 1994, Holden *et al.* 1983, Wilson, 1985a).

Interaction of m-f fungi and added nutrients etc

The use of mycorrhizal fungi to improve nutrient uptake and reduce the need for fertilizers has been investigated (Mason *et al.* 1984). There are indications of benefits to growth and production by inoculation with particular strains of m-f fungi, but not enough yet for these to have found a place in forest production systems (Wilson & Coutts 1985, Walker & Wheeler 1994, Walker & Broome 1995). See also *Calcium and humus in* §10.71.

As part of investigations into the effect of increasing nitrogen deposition from aerial pollution, Wallenda & Kottke (1998) investigated mycorrhizal root tips but found only minor changes. Van der Eerden (1998) noted that with increased N deposition, mycorrhizal infection declined.

The concept of a succession of mycorrhizal species within developing woodland was put forward by Mason *et al.* (1983, 1984). Watling (1984) reported different fungal species associated with older woods compared with those found in young birch woods or woods invading as pioneers of shale heaps.

Following applications of up to 300 kg N/ha as ammonium sulphate to a Sitka spruce plantation near Aberdeen, mycorrhizal activity was affected in several ways; infection decreased, longevity increased and new types of mycorrhizas appeared (Alexander & Fairley 1983).

Concern has been expressed that prolonged acid deposition on woodlands on poorly buffered soils might harm the mycorrhizal flora (Anon 1993a).

The journal *Experientia* (Vol 47, 1991, pp 311-391) carries a series of papers on other current work on the role of mycorrhizas.

12.4   EFFECTS OF TREES ON SOIL PROFILE DEVELOPMENT

12.41   Long-term weathering of soil

While past climatic change has been widely studied and described, weathering, *ie* the break-down of near-surface geological minerals by weather and soil processes, and in particular weathering and soil profile development of freshly exposed glacial and periglacial deposits, has been less publicised. Nevertheless it is a crucial, continuously ongoing natural process. Over long periods of time in most conditions, the more soluble nutrients released by weathering are leached from the site.

Godwin (1975) summarises the slow change in soils in Britain in previous interglacial periods; he points to a succession to podsols and acid soil/plant associations on the less well-buffered soils, a succession which applies to the present post-glacial period as much as previously.

In Scandinavia, the weathering potential of minerals in a limited range of underlying rocks has formed the basis identifying potential susceptibility to acidification by airborne SO_2 and calculation of *Critical loads* (See §14.5) (Nilsson & Grennfelt 1988).

12.42   Changes in soils

Soil types at any time are the resultant of active contemporary processes from the interaction of climate, soil parent material, weathering and vegetation cover.

Anderson (1982, 1983, 1987) found that on recently afforested lowland soils, acidification could be expected in the first 25 years following tree planting. Soils under most broadleaves changed at about half the rate for conifers; under alder, however, soils were acidified as quickly or more quickly than under conifers.

As crops aged, ground flora under a range of species tended to converge to resemble those of acid oakwoods.

Early indications of podsolisation were found under 45-50 year old Norway spruce and European larch in the Forest of Dean, when compared with adjacent acid brown earth soil carrying broadleaved woodland (Grieve 1978).

Cyclical variation in soil nutrient properties was described by Page (1968). In species trials in Kent, Norfolk and Gloucestershire nutrients in litter did not necessarily diminish with time. Nutrient stocks of P and K in litter and the mineral soil increased by varying amounts under 15 species over a 24 year period; N levels decreased slightly on the acid soils in Kent and Gloucestershire but increased slightly on the calcareous sand in Norfolk (Anderson 1986, 1987).

At Balquhidder in central Scotland, a water catchment study was intensively monitored. Profiles from surface water gleys and iron humus podsols that had carried Sitka spruce, Scots pine or European larch for 50 years were analysed for exchangeable cations of Na, K, Ca, Mg and Al, and for forms of P in the upper and lower halves of the top 1m of soil. These were compared with profiles that had been under moorland for the same period and with estimates of the nutrients removed from the soil into the forest crop.

The most acid of the mineral soil profiles was the moorland podsol. On the gley soils, the amount of K in the trees crops was markedly greater than remained in the soil, with indications of soil depletion compared with moorland. Under larch on podsol, some depletion of Mg and Ca was apparent (Anderson *et al.* 1993b).

See also *Birch* in §12.24 above.

12.5   NUTRIENT LOSS IN DRAINAGE WATER

12.51   Effects of trees & cultural operations on stream-water drainage

The effects of forests on drainage water have been studied in parallel with nutrient needs of trees. They are also closely linked to the interaction between trees and pollutant interception.

The nutrient content of water draining from woodland may be affected by:
- the species planted (§12.52);
- drainage and cultivation (§12.53);
- nutrient losses from fertilizers in first-rotation crops (§12.54);
- nutrient losses to drainage water at clear felling (§12.55);
- the scale of interception of water by tree canopies (§14.2);
- the scale, nature and effects of wet and dry deposition of pollutants on surface vegetation (Ch13, §14.3 *et seq.*).

12.52 Effect of species

At Gisburn, (Bowland Forest) Lancashire, water was sampled in 22 year-old oak, alder, Norway spruce and Scots pine. The area is on the edge of a heavily industrialised area and is within 35 km of the sea. Samples were taken fortnightly during 1987 at the soil surface (throughfall), from litter/organic matter (L/F/H) layer and from the A and BC horizons.

Results summarised in *Table 12.4* showed substantial differences in the throughfall both between the two conifer and the two broadleaved species and the concentrations encountered at different depths in the soil.

The effects of tree canopies on interception and concentration of nutrients *etc* are fully discussed in §13.23 and §§13.31-13.33. Some aspects of the figures from Gisburn not discussed there are noted here:

- Sodium and chloride are the dominant ions in the bulk precipitation. This is a characteristic effect of marine influences in deposition found widely throughout Britain. The ions appear to be little affected by canopy or soil chemistry; the changes in their concentration reflect the volume reduction of water as it is taken up from the various horizons of the soil profile. The higher concentrations in throughfall under spruce and pine result from their more efficient capture of depositions.

- While phosphate occurs in solution in much smaller amounts than any of the major plant nutrients, the increase in concentrations between the P content of rainfall and that in throughfall varies markedly between species. Very little appears in the A and BC horizons, implying either immobilisation or active recycling. Under oak and to a lesser extent spruce, the increased levels found in throughfall imply substantial losses from foliage. As these species root heavily in surface litter and soil layers, much of the P is likely to be recycled under their canopy. The increase in P concentrations in the forest floor drainage water under Scots pine may explain the beneficial effect other species show when grown in mixture with pine.

- The pH at any point in the profile reflects the net outcome of soil acidification and buffering processes. The most acidic water is found at the base of the forest floor layer, becoming less acidic with depth as the greater buffering capacity of the soil horizons takes effect.

- The acidifying and nitrogen fixing-effects of alder are apparent. The amount of nitrate at any point in the soil profile under alder is mostly more than twice that of other species at the same horizons.

- Aluminium reaches its maximum concentration in water in the A horizons in contrast to calcium which increases with depth (Brown & Iles 1991).

Similar studies compared throughfall and soil solution chemistry in adjacent stands of Sitka spruce and Japanese larch. Pairs of plots were located in north-east Scotland (Northbrae), northwest England (Dodger Wood) and north Wales (Beddgelert) (Adamson *et al.* 1993). *Table 12.5* shows the

Table 12.4 pH and mean annual concentration of ions in water profiles under oak, alder, Norway spruce & Scots pine at Gisburn

(μ-equivalents per litre**)

Species	Sampling point	pH	PO_4-P	NH_4-N	NO_3-N	K	Mg	Ca	Na	Al	SO_4-S	Cl
	Rain	4.4	2.2	27	24	8	16	26	94	1	71	92
Oak	Throughfall	4.5	13.2	43	25	34	34	55	113	1	113	129
	*Forest floor	4.2	6.7	39	22	34	66	113	130	20	138	131
	A horizon	4.4	1.0	1	21	10	49	132	184	58	202	175
	BC horizon	5.3	0.9	0	8	6	51	296	229	18	238	209
Alder	Throughfall	4.4	2.7	25	20	32	39	69	129	1	114	144
	*Forest floor	3.8	6.1	44	247	66	87	186	162	40	170	190
	A horizon	4.2	1.0	1	265	20	71	240	180	119	197	159
	BC horizon	5.0	0.8	0	174	7	65	344	193	24	207	159
Norway spruce	Throughfall	4.0	8.5	99	76	95	79	130	240	7	328	241
	*Forest floor	4.0	7.1	61	55	74	74	194	217	40	299	316
	A horizon	4.1	0.8	1	19	15	72	193	304	105	356	316
	BC horizon	5.1	0.6	0	24	8	81	506	344	34	356	389
Scots pine	Through-fall	4.0	7.1	99	70	57	60	88	231	14	242	265
	*Forest floor	3.8	14.2	136	127	72	81	108	258	38	278	262
	A horizon	3.9	1.1	3	99	33	79	160	342	167	373	327
	BC horizon	4.6	0.5	0	77	13	95	452	409	8	444	439

Source Brown & Iles 1991   * Below L/F/H layer   ** See Appendix III for conversion factors to mg/litre

Table 12.5   *pH & nutrient concentrations in through-fall & soil solution from H & B horizons under Japanese larch & Sitka spruce on three sites (mg/litre)*

| | | Northbrae | | Dodger Wood | | Beddgelert | |
		J. Larch	Sitka sp	J. Larch	Sitka sp	J. Larch	Sitka sp
pH	Thr fl	4.6	4.8	4.2	4.2	4.4	4.2
	H	4.0	3.6	3.7	3.8	4.2	3.7
	B	4.6	4.4	4.2	4.4	4.6	4.5
PO_4-P	Thr'fl	0.08	0.16	0.01	0.02	0.003	0.007
	H	0.02	0.04	0.005	0.008	0.07	0.1
	B	0.001	0.001	0.000	0.000	0.001	0.000
NH_4-N	Thr'fl	0.6	1.1	0.6	0.7	0.2	0.5
	H	0.3	0.9	0.2	0.8	0.3	0.5
	B	0.01	0.02	0.01	0.04	0.03	0.06
NO_3-N	Thr'fl	0.5	0.6	0.7	0.6	0.4	0.6
	H	0.1	0.2	0.8	0.6	0.3	3.1
	B	0.1	1.1	0.5	1.5	0.2	2.2
K	Thr'fl	2.3	4.0	1.8	2.0	0.9	1.3
	H	3.7	3.2	1.5	3.1	1.8	1.3
	B	0.8	2.0	0.8	1.1	0.7	0.7
Mg	Thr'fl	0.8	0.7	1.4	1.1	1.3	1.9
	H	1.6	1.2	1.7	1.3	1.4	2.2
	B	1.2	1.9	1.7	1.7	1.2	2.1
Ca	Thr'fl	1.2	0.7	1.4	1.1	0.8	1.0
	H	2.0	1.1	1.7	1.1	0.8	1.6
	B	1.5	2.9	0.8	0.9	0.6	1.2
Na	Thr'fl	4.7	5.1	9.0	8.1	9.7	13.8
	H	7.5	8.9	10.3	8.4	10.9	15.1
	B	9.5	15.1	10.7	11.0	11.7	19.3
Al	Thr'fl	0.03	0.05	0.05	0.06	0.02	0.06
	H	0.7	1.6	0.3	0.9	0.2	0.5
	B	0.9	2.0	2.0	2.2	0.5	1.7
SO_4-S	Thr'fl	1.9	2.3	2.7	2.8	1.7	2.4
	H	2.2	3.7	3.8	3.4	1.9	2.7
	B	2.8	3.7	3.4	3.3	2.0	3.0
Cl	Thr'fl	10.2	9.6	18.9	16.3	18.3	27.8
	H	14.2	16.0	21.5	17.3	20.6	30.2
	B	17.9	32.5	24.2	23.7	22.8	37.7

Source   Adamson *et al.* 1993

mean annual concentrations of chemical constituents of throughfall, and soil water collected at the base of the L/F/H layer and the B horizon, positions similar to those used at Gisburn.

From *Tables 12.4* and *12.5*, the major plant nutrients appear to be conserved on site to a greater or lesser degree, depending on exactly where in the profile water enters surface drainage. Under all species, the litter/humus horizon is the most acidic. In the Japanese larch/Sitka spruce comparisons, Sitka spruce plots have a more acidic forest floor layer than Japanese larch. At Gisburn, alder appears to be as acidifying as pine or spruce.

12.53   Nutrient release from drainage and cultivation

A study of drainage from blanket peatland in Rumster Forest in Caithness covered the first four years following planting in 1989 (Miller *et al.* 1996b). Total chemical inputs to the site in precipitation and 'interception', *ie* wet and dry deposition, were dominated by marine-origin sodium and chloride. Over the 4 years of the study, annual rainfall ranged from 740 to 1098 mm, the annual variation being reflected in chemical depositions and the differing proportion of marine- and non-marine-derived materials.

Drainage outflow from the site was between 60 and 70% of the rainfall input. At this time, the planted trees were small and tree canopy interception trivial.

Lodgepole pine on ploughed ground was given P at planting; Sitka spruce was given PK. The chemical losses from the site over the first four years associated with ploughing and fertilizing are given in *Table 12.6*. The losses of P are from the lodgepole pine plot and those for $\bar{K}$ from the Sitka spruce.

The control (unploughed) plot while undisturbed, had an external perimeter drain installed. The NH_4-N in drainage water from the control was over three times greater than the rainfall input and was due to exposed peat on the perimeter drains. The disturbance of ploughing further increased NH_4-N loss. Increased

Table 12.6   *Nutrient losses to drainage water following ploughing & fertilizing on deep peat: Rumster, Caithness*   *(kg/ha/yr)*

Nutrient	Unploughed 1992	Ploughed No PK 1992	Ploughed + PK 1989	1990	1991	1992
NH_4-N	6.3	12.1	No available data			
NO_3-N	0.4	0.7	"			
Ca	5.7	6.4	"			
SO_4-S	10.5	17.0	"			
Total P	0.07	no data	0.2	1.15	0.26	0.67
K	8.5	10.5	9.5	22.0	17.5	15.0

Source   Miller *et al.* 1996b

SO_4-S losses also arose from mineralisation of newly exposed peat surfaces following ploughing.

The losses of P from plots given fertilizer were small in comparison to the amount applied, the total loss over 4 years being 1-2 kg/ha out of 58kg applied. Losses of K, however, were more substantial, amounting over 5 years to approximately 30% of the 108 kg/ha originally applied (Miller *et al.* 1996b).

The nutrient reserves in the 1 metre depth of peat liable to be affected by ploughing and establishment of forest cover were:

- 20 000 kg N/ha
- 500 kg P/ha

- 10 000 kg S/ha
- 500 000 kg C/ha.

12.54   Nutrient losses from fertilizers in first-rotation crops

The use of fertilizers in forests in the UK is described in Chapter 10; rates of use for selected years are given in *Table 10.6*. That table shows that the scale of use in England and Wales is small compared with the uplands of Scotland where, in addition to application at planting, it has been standard practice to use fertilizers 8-20+ years after planting to boost poor rates of growth on nutrient-poor soils.

Forests in upland, high-rainfall areas are also commonly used as catchments for public water supplies. This dual land-use markedly magnifies the potential disturbance that can be caused by sudden increases in the nutrient content of surface waters draining from recently fertilized areas. Such increases can

- affect the rate of build-up of bacteria in sand filters for water supplies,
- supplement nutrients from other land use sources, increasing the risk of eutrophication in dry conditions.

Losses at planting

At Leadburn, near Edinburgh, Scotland, sixteen lysimeters were installed in a section of recently ploughed raised bog. Plots within the area were given standard applications of P and K fertilizer separately or together. For the subsequent three years, run-off was monitored volumetrically and for chemical content (Malcolm & Cuttle 1983a, b).

Loss of K into the drainage water was detected very shortly following application; rates declined over the three year period until they were little more than the controls.

There was a delay of 24 weeks before P was detected. Loss was seasonal, being high in the winter and low in the summer. There was no obvious change from year to year for the period assessed.

N and Ca were also assessed. There was no clear effect of treatments on Ca losses.   Nitrogen showed a seasonal pattern of loss with maxima in September/October.

Annual losses of P and K are shown in *Table 12.7*. Analyses of peat cores showed that after three years, most of the applied P and K remaining on site was in the top 0.3m of peat, *ie* the freely draining fibrous peat. Most of the leachable

K had already gone. If the P continued to be leached at the rate of the first 3 years, reapplication would be needed in 8 years (Malcolm & Cuttle 1983b).

Table 12.7   *Annual loss to drainage water of nutrients where fertilizer had been applied after planting at Leadburn*

Treatment	Losses (kg per ha) in				% of applied nutrient
	Year 1	Year 2	Year 3	Total	
	Potassium losses				
+ K	23.5	12.7	3.9	40.1	39.3%
+ PK	18.6	7.4	1.0	26.9	26.4%
	Phosphorus losses				
+ P	1.61	4.08	2.01	7.70	16.3%
+ PK	0.32	2.21	0.77	3.30	7.0%

Source	Malcolm & Cuttle 1983a

Nutrient loss in thicket-stage crops

Nutrients loss from treatment of an established crop showing deficiencies was assessed in Loch Ard Forest in central Scotland. *Table 12.8* summarises run-off figures from a 20-year-old Sitka spruce plantation following application of N, P and K.

The highest potassium concentrations in run-off were noted three weeks after treatment. Allowing for differences in dates of application, phosphorus losses decreased annually, winter seasonal maxima coinciding with increased stream flows. Nevertheless, P concentrations at $> 100 \mu g/l$ were being recorded at the end of the third year. The maximum nitrogen content in run-off was recorded two weeks after urea application; 63% was as NO_3^-.

Table 12.8   *Annual loss to drainage water of nutrients where fertilizer had been applied to 20-year old Sitka spruce at Loch Ard*

Treatment	Losses (kg per ha) in				% of applied nutrient
	Year 1	Year 2	Year 3	Total	
Loch Ard	*Potassium losses*				
104 kg/ha K	9.1	3.4	trace	12.5	12%
	Phosphorus losses				
47 kg/ha P	1.7	2.5	2.2	6.4	13.7%
	Nitrogen losses				
163 kg /ha N (urea)	3.4	2.3	1.1	6.8	4.2%

Source	Harriman 1978

Stream-water with the phosphorus levels in this study, if flowing into lakes and reservoirs, could give rise to algal blooms. However, where streams remained part of a river system, such enrichment was considered likely to be beneficial (Harriman 1978).

Urea when applied to forests from the air, usually dissolves quickly and enters the soil where it is transformed to ammonium-N and held there. If, however, heavy rain falls either immediately preceding or immediately following fertilizer application, undecomposed urea may enter surface running water and be transformed to ammonium-N there. The EU Water Directive (78/659/EEC) sets a mandatory maximum concentration for ammonium-N of 0.78 mg per litre for the protection of freshwater fisheries. In water under ordinary circumstances, ammonium-N is associated with anaerobic conditions or with fish-farming, and is not significant in forest streams. However, stream water pollution caused by urea occurred twice in large scale forest operations in 1990. The effect was short-lived, the concentration of NH_4-N returning to background levels within 2-4 weeks.

Nisbet (1994a), summarising experience arising from forest fertilizing, considered that the greatest risk of loss occurred in the period after planting. For P as rock phosphate applied from the air, losses were greatest during the first six months, but further loss could be detected for 3-5 years after application. Losses following hand application to the base of individual trees were negligible.

Risks from forest fertilizing can be minimised by careful timing of applications; where more than 50% of a catchment requires treatment, applications should be made over several years. Nisbet & Stonard (1995) recommended that no more than 15% of a catchment should be treated with urea at any one time. See also §13.24.

12.55   Nutrient losses to drainage water at clear felling and restocking

Clear felled land

At Kielder Forest, Northumberland, an experiment was set up in 1980 on land where a previous crop had been clear felled. At clear felling, branches and tree tops (brash) were piled up into long parallel heaps, providing a mat on which harvesting machinery could travel, thereby minimising physical damage to soil. The zone between was clear of brash.

Lysimeters were installed to provide data on leaching losses, based on felling 0 to 5 years prior to first sampling. They were located under the LFH organic matter layer, either under brash heaps or, in areas cleared of brash, under litter.

Leaching losses of NH_4-N, PO_4-P, K, and Ca in drainage water from below the LFH layer exceeded inputs from precipitation. NO_3-N, Mg and Na leaching losses were less than the input from precipitation. Most K was lost during the first two years after felling; thereafter losses were of the same order as inputs from precipitation. Other nutrient losses continued over the period of the study. For most nutrients, losses from under brash swathes were greater than losses from strips that had been cleared of brash, but not for NO₃-N (Titus & Malcolm 1991, 1992).

At Kershope forest, Cumbria, a drainage experiment had been laid out in Sitka spruce planted in 1948. In one block, starting in 1982 and continuing for several years, plots were monitored at weekly intervals for water flow and its nutrient content. In 1983, some plots were clear felled and others left standing.

The total water discharge from one newly felled plot was about double that on the counterpart unfelled plot. Nitrate concentrations peaked in the first year and returned to levels similar to the unfelled controls within four years. K⁺ concentrations rose slowly after felling, peaking two years later; they showed a marked seasonal pattern each year. Aluminium concentrations were slightly lower than the control.

These responses following tree felling were attributed to the cessation both of water interception and evapotranspiration, and also of the scavenging by mature tree crowns of dry deposition Na⁺, Ca⁺⁺, Cl⁻ and SO₄⁻⁻. Potassium is readily leached from plant tissues and its early release was not unexpected. The increased release of nitrogen is attributed to loss from the soil of material that would otherwise have been taken up by the trees.

The soils at Kershope are not particularly nutrient deficient; no fertilizer was applied during the first rotation. At this site, these losses may not therefore have any significant detrimental effect on establishment of the successor crop (Adamson *et al.* 1987, Adamson & Hornung 1990).

At Beddgelert Forest, north Wales, a six-year study of soil water was laid out in Sitka spruce woodland planted in the 1930s. The site was a freely draining iron-podsol on base-poor Ordovician slates and shales and had formerly been unimproved pasture. In the study, the effects of whole-tree harvesting were compared with conventional clear felling where branch and small crown wood was left in the forest.

In the conventionally felled crop area, most of the K (around 100 kg per ha) leached out of the brash within the first year. Of the K passing through the soil profile, <50 kg/ha reached the stream. In the same period, 10 kg per ha P (about a third of the P present) was leached from the brash but was immobilised in the soil and did not reach the stream.

The brash was a net sink for inorganic-N for 3 years after felling. A pulse of nitrate observed in soil and stream water shortly after felling was considered to have originated from death of fine roots, rapid mineralisation and nitrification.

In the whole-tree harvested plots, there were no K and P pulses but a nitrate boost was observed. By the end of the study, above-ground live biomass on these plots was on average about 50% more than on conventionally harvested plots, suggesting that ground vegetation recolonisation was checked (by smothering?) by the presence of brash. Grasses formed a higher percentage of the vegetation cover on whole-tree harvested plots, possibly a response to the higher levels of nitrate available.

Short-term soil reserves of K were estimated to be sufficient to meet the requirements of two further rotations of conifer; long-term reserves would supplement these for up to 30 rotations.

A nutrient budget for P and Ca for a rotation showed net losses, these being greater where whole trees were harvested and likely to be more limiting than K. (Goulding & Stevens 1988, Stevens *et al.* 1989, Fahey *et al.* 1991a, b, Stevens *et al.* 1995). These results support earlier predictions (Miller *et al.* 1980).

In another four plots at Beddgelert, clear felled between July 1983 and September 1984, monitoring of concentration and fluxes of N showed a substantial increase in nitrate after felling. Concentrations peaked about 1 year after felling; fluxes showed a strong seasonal variation with winter maxima, the C horizon flux also peaking in the second year. For the period of the study, loss of nitrate following felling was of the order of 70 kg N/ha/yr, 6 - 10 times more than the nitrate losses before felling. There was no loss of ammonium nitrogen (Stevens & Hornung 1988).

Also at Beddgelert, in 1987/8, small lysimeters were filled with organic matter from the LFH layer. Onto them, spruce seedlings or grass were planted or a layer of fresh brash (composition unspecified) placed. While treatment had no effect on ammonium or organic N, all treatments generated nitrate-N, particularly the brash. One year after planting grass or spruce seedlings, the nitrate-N from these plots was markedly reduced, the effect of grass being greater, probably because it colonised more quickly than tree seedlings (Emmett *et al.* 1991).

These results should be read in conjunction with sections on atmospheric nitrogen and sulphur deposition, and their effects on stream water (§§ 13.24, 14.4, 14.5).

Part IV

Atmospheric & mineral pollutants

CHAPTER 13

Atmospheric Pollutants

'Visibility today is down to 150 m'

13.1   ATMOSPHERIC POLLUTION

13.11   Major polluting gases

Over the last forty years, the effect of deposition of pollutants on forests has become increasingly well known. Initially, the concern was with forest health and the effect of acidic sulphur depositions. The role of sulphur as a nutrient at that time was not considered an issue. Legislation and technical advances over the last twenty years have brought sulphur and soot pollution under control. However, man-made (anthropogenic) nitrogen compounds from combustion of gas and petroleum fuels for energy and transport have increased. They have become important in plant nutrition cycles and as potential pollutants affecting forest and amenity trees, farm crops and wild plants.

This chapter reviews the atmospheric gases which constitute the major sources of pollution affecting trees:

- nitrogen (§13.2);
- sulphur oxides (§13.3);
- ozone (§13.4);
- other air-borne pollutants, (§13.5);
      methane (§13.51);
      fluorine/fluorides (§13.52).

The following three chapters cover:

- Acid rain, forest health and drainage water (§14.1);
- Other mineral pollutants (§15.1);
- Carbon dioxide and global warming (§16.1).

13.12   The national context

Air-borne pollutants have for over 100 years been recognised as threats to plant health. Since the 1950s, however, their sub-continental scale of circulation have increasingly been recognised along with effects measured on a global scale:

   • by deposition of damaging substances on susceptible plants and materials and their subsequent transfer to soils and water;

- on respiratory and other physiological systems (human, animal and plant);
- on the physical properties of the atmosphere, affecting its gas balance and leading to the possibility of climatic change.

Urban pollution and 'smog', 'global warming', 'acid rain' and 'forest decline' have each featured in widespread and prolonged public debate in Europe and America. Concern for human health and the urban environment has provided the impetus for many investigations (*eg* Curtis *et al.* 1996), the findings frequently being relevant to the health of forests and amenity trees.

UK Environmental Review Groups

Following recommendations from a *Royal Commission on Environmental Pollution* and the *House of Commons Select Committee on the Environment*, several advisory groups were set up by the *Department of the Environment* to review and monitor aspects of pollution: -

- Acid Waters Review Group   (AWRG);
- Building Effects Review Group   (BERG);
- Critical Loads Review Group   (CLAG);
- Photo-chemical Oxidants Review Group   (PORG);
- Quality & Urban Air Review Group   (QUARG);
- Review Group on Acid Rain   (RGAR);
- Review Group on Impacts of Atmospheric Nitrogen   (RGIAN);
- Terrestrial Effects Review Group   (TERG).

These groups have each produced one or more reports containing detailed information about pollutant trends, methods of assessment and their implications for land management. Several include important sections on forests.

13.13   Atmospheric gases

Table 13.1 gives figures for the main constituents of the atmosphere and average levels of four pollutants in unpolluted conditions. Carbon dioxide is not conventionally listed as a pollutant; the level in the atmosphere has, however, increased by 25% since 1850 and is significant in relation to global warming. Volatile organic carbons (VOCs) are not included in the table.

Primary pollutants

Primary pollutants in the atmosphere result from combustion of coal, oil *etc*, or the process of combustion.

- Sulphur emissions result from combustion of impurities in fuels, ores *etc.*
- Emissions of nitrogen oxides (NO_x) originate mainly from combustion of fossil fuels in power stations and as vehicle emissions. Nitrogen and oxygen from the atmosphere combine in the high temperature of the combustion processes, even of 'clean' fuels (Cannell & Cape 1991).

Table 13.1   *Composition of the atmosphere at sea level*

*mole fraction**			
Nitrogen	0.78	Oxygen	0.21
Water vapour	0.006 - 0.06	Argon	0.009
parts per million (10⁻⁶)			
Carbon dioxide	350	Neon	18
Helium	5.2	Krypton	1
Methane	1	Nitrous oxide	0.3
parts per billion (10⁻⁹) Unpolluted air			
Nitric oxide	0.01	Nitrogen dioxide	0.01
Ozone	30	Sulphur dioxide	1

Source   Cannell & Cape 1991   * See App'x III (Conversion factors)

• Ammonia as a pollutant is principally generated by livestock husbandry.

• 60 - 80% of nitrous oxide (N_2O) produced each year originates from natural sources and a further 5 - 20% from land husbandry practices. Emissions from industrial and other sources amount to about 10% of the total. Removal is thought to be by photolysis. N_2O has a long residence time in the atmosphere (~ 150 yr), the long-term trend showing a small annual increase. There is no experimental evidence that N_2O is deposited on land or plants in any significant quantity, nor that it is damaging (INDITE 1994).

• Volatile organic carbon (VOC) emissions other than methane, though not shown in *Table 13.1,* are an important source of pollution. In aggregate, their emissions, quantitatively, are of the same order of magnitude as nitrogen oxides.

90% of VOCs arise from industrial processes, the use of solvents and the production and use of petroleum products. They include several hundred different hydrocarbon compounds. The remaining 10% originate from natural sources, forests contributing approx. 3% (PORG 1993). See also §13.51.

Secondary pollutants

Secondary pollutants are formed in the atmosphere by reactions between gases, or in solution in airborne water droplets or on the surface of airborne particles. Much of the chemistry is driven by oxidation in the presence of sunlight, the oxidised products commonly being referred to as 'secondary photochemical pollutants'.

Ozone is a normal component of the atmosphere but its concentration is commonly enhanced because of the presence of other pollutants *eg* VOCs, which are oxidised in the presence of nitrogen oxides to form ozone. Ozone concentrations can reach damaging levels in prolonged periods of bright sunlight.

Particles

Air near the ground contains a diversity of particles, wholly solid, wholly liquid or both liquid and solid. Particle transport and deposition is determined by particle size. Sulphuric acid is formed when SO_2 is oxidised in the gas-phase, rapidly picks up water and forms H_2SO_4, an involatile molecule. Such molecules, if plentiful, combine to form 'condensation nuclei' - very small particles. These may increase in size by further aggregation and, being hygroscopic, by absorbing water when the relative humidity of the air is moderate to high. Particles of sea salt (NaCl) similarly may also absorb water from damp air.

Airborne pollutants may combine. For example, ammonia and sulphuric acid, when both present in a particle, form ammonium sulphate and may be 'co-deposited' in that form. The same applies to ammonia and nitric acid (QUARG 1996). See also §13.36.

Because of such reactions, it is not normally possible, when assessing levels of deposition, to identify the component compounds. Pollutant depositions are described in terms of the principle ions present *eg* sulphate (SO_4^-), chloride (Cl^-), or the key element, *eg* nitrate-N, ammonium-N *etc.* even though they may have been deposited as particles of acids or salts.

The average length of time that particles remain airborne (their mean residence time) depends on their size. Larger particles ($10\mu m$ mean diameter) are likely to be airborne for 10 - 20 hours, travelling 20 - 30 km. In the UK, small particles (0.1 - 1.0 μm diam.) are most likely to be removed in rain; they will have been airborne 10 days on average, and have travelled a few thousand miles. Particles of similar size, not removed in rain but deposited as 'dry deposition' may remain airborne for 100 - 1000 days and travel many thousands of miles (QUARG 1996).

13.14   Damage to trees due to air pollution

Acute damage to trees by smoke and gases emitted from factories and industrial processes has been recognised from before the beginning of the 20th century. Initially, case investigations could link episodes of high emissions to acute local damage. Metal smelting in particular was formerly a significant cause of local damage.

With the increasing scale of industrial processes, specialists in the effects of fume damage in forests have met to report and compare evidence, *eg* in USA in 1951, in Bochum, Germany in 1959, and Essen, Germany in 1970 (Anon. 1971b). While most interests and reports were confined to affairs in individual countries, the Committee on Electric Power of the Economic Commission for Europe (CEP/ECE) began international enquiries into 'damage to forests and parks caused by ashes and sulphurous compounds from thermal power stations'. Many countries reported damage; the United Kingdom, however, responded that such damage was absent because its modern power stations had tall chimneys which efficiently dispersed the products of combustion (FAO/ECE 1970)!!

13.15 Long range transboundary air pollution

In 1967, a newspaper report from Sweden claimed that acidic substances were being deposited in Sweden following long-distance rather than local transport of air pollutants. The term 'acid rain' derives from these beginnings. The main assertions put forward were that in Europe:

- acidic precipitation was a large-scale regional phenomenon with well defined source and sink regions;
- both precipitation and surface waters were becoming more acid;
- long-distance (100-200km) transport of both sulphur- and nitrogen-containing pollutants was occurring across national boundaries (Odén 1976, Brydges & Wilson 1991).

The Swedish initiative was widely recognised as pointing to a problem affecting the whole of Europe and North America. By 1979, an international Convention on Long-Range Transboundary Air Pollution had been adopted. This came into force in 1983 and led to a European Monitoring and Evaluation Programme (EMEP). Commitments have been agreed (protocols) to cut emissions and thereby reduce transboundary fluxes. For SO_2, 1980 levels were to be reduced by 30% by 1993 and by 80% by 2010; for NO_x, a protocol set a target for 1994 of bringing levels back to those in 1987. A Protocol for *Volatile organic carbons* (VOCs) set a target of a 30% reduction of 1988 levels by 1999.

The UK has either achieved the targets set or made a commitment to do so by the date set. The *Review Group on Acid Rain* (RGAR) issued its fourth report in 1997 and is a source of much data relevant to these targets and the interaction of acid rain with land, crops and natural vegetation on agricultural and forest land.

13.2 AIR-BORNE ('FIXED') NITROGEN

13.21 Nitrogen deposition

The current best estimate of the UK atmospheric fixed nitrogen budget for oxides of nitrogen, ammonia and ammonium and nitrate ions is summarised in *Table 13.2*. The figures are for 1993 from the *Report of the Review Group on Acid Rain* (RGAR 1997), supplemented where necessary by figures from a slightly earlier report on *Impacts of Nitrogen Deposition in Terrestrial Ecosystems* (INDITE 1994).

Estimates for 'reduced nitrogen' (ammonia + ammonium+) emission from agriculture and associated activities range from 190 to over 500 kilotonnes (kt) N/yr in the INDITE (1994) report, and from 198 (Pain *et al.* 1998) to 440 kt N/yr for 1993 in the RGAR (1997) report.

The 'official' figure for the UK for 1993 was 225 kt N/yr (RGAR 1997); in an earlier INDITE report for 1988-1992, it was 350 kt N/yr. Such differences reflect differences in modelling methodology as much as real change in emissions.

Table 13.2 shows that for the period 1992-4, the equivalent of about 80% of nitrogen oxides and 12% of reduced nitrogen compounds originating in the UK were carried out of the country on the wind. These figures assume that for all N compounds imported, an equivalent amount was exported.

Table 13.2 *Atmospheric Fixed Nitrogen Emissions and Depositions, 1992-4*
         UK fertilizer usage for the same period.   *(kt N/yr)*

Nitrogen form	Oxidised nitrogen NO_x	HNO₃	NH₃ NH₄⁺ NH_x		Total N
Import from overseas	60			30	90
Emissions (all dry)					
Industry + other	350	(30)*			350
Vehicles	430				430
Agriculture + other			260		260
Emissions + Import	840	(30)*	290		1130
Depositions					
Dry	40	(30)*	110		150
Wet + cloud		110		120	230
Total deposition	40	110 (30)*	110	120	380
Net export		690 (NOₓ)		60 (NHₓ)	750

Total UK fertilizer usage   *(1988-1992)*   1300 - 1500 kt N/yr.

Source   RGAR 1997   *unreliable estimate - not included in totals.
$NO_x = NO + NO_2$. Usually N_2O present is insignificant and can be ignored.

Emission sources and N deposition

Reports from three review groups, INDITE (Nitrogen deposition), CLAG (Critical Loads) and RGAR (Acid rain), each include maps showing the distribution of emissions and wet and dry nitrogen depositions for Great Britain on a 20x20 km grid scale.

NO_2 emissions are shown to originate from urban and industrial areas of Britain.

The NH_3 emissions map is based on estimates of animal numbers and a constant emission factor per animal. It reflects the distribution of the cattle and pig rearing industry in Britain. Emissions and concentrations at ground level are shown as being highest in south Lancashire/Cheshire, the Vale of York, southwest England and East Anglia. Emissions from the more intensive stock-rearing areas may often exceed 50 kg N/ha/yr (RGAR 1997) and have been reported as reaching 100kg/ha/yr (TERG 1993). Cattle are estimated to have accounted for 50% of emissions from UK agriculture, poultry 15%, pigs 12%, sheep 7% and agricultural fertilizers 16%. Of the latter, breakdown of urea used in fertilizer is estimated at 30% from arable and 28% from grassland (*ie* 4.8 and 4.5% of the gross total emissions from agriculture) (Pain *et al.* 1998).

The ammonia estimates have been criticised as being too simplistic; however, while a 'constant emission factor per animal' could be refined to take account of differences in animal weight, season of year, stock-rearing and slurry disposal techniques, such refinements are not yet in practice (Asman *et al.* 1998, Schjorring 1998).

13.22 Wet & Dry Deposition

Wet deposition

Wet deposition is estimated by combining national figures for mean precipitation with data for concentrations of major ions in precipitation collected from 32 sites throughout Britain. The figures take account of the increase of the pollutant content of rain and cloud with altitude.

For ammonium- and nitrate-nitrogen, the areas of greatest wet deposition are the high rainfall areas of Wales, northwest England and south-west Scotland.

The extent to which ammonium and nitrate nitrogen are taken up directly by foliage from depositions is disputed. Wilson & Tiley (1998) reported an experiment using 5-year old Norway spruce trees to which both forms of N were applied as fine rain, using the ^{15}N isotope. They concluded that foliar uptake was unlikely to exceed 5% of the total annual requirement. This is contrary to views expressed by Nihlgard (1985) that ammonia generated from livestock and fertilizer usage would be taken up by foliage sufficiently to stimulate growth at the expense of root development, would increase susceptibility to drought and pests and could be a major contributory factor to 'forest decline'.

Rennenberg *et al.* (1998) described studies on beech and Norway spruce in Bavaria where the Norway spruce canopy is continuously exposed to high loads of N. On this site, there was no root uptake of NO_3^-; root uptake of NH_4^+ occurred mainly in the summer. Estimates of N fluxes show:

'N in bulk precipitation'	12 kg/ha/yr
'N in throughfall'	30 kg/ha/yr
'net canopy uptake (assimilation - emission of N)'	*'a'* (not known) kg/ha/yr
'canopy N interception' (by difference)	$30+a-12 = 18+a$ kg/ha/yr.

Uptake of NO_3^- is thought not to be directly inhibited by NH_4^+ but by accumulation of amino-acids in the roots. One consequence of reduced uptake of soil-N sources has been greater nitrification and denitrification rates. Under beech, there was little leaching but correspondingly greater denitrification; under spruce, NO_3^- was leached into ground-water at concentrations exceeding the current EC Groundwater directive.

While figures for canopy uptake were not given, from the effects reported, the amounts were large enough to affect nitrogen/water balance in soil-water drainage, but not to affect crop health or vigour.

Dry deposition of nitrogen oxides and reduced nitrogen

N_2O, while detectable, is present in such small quantities that it is ignored in emission and deposition calculations (Macdonald *et al.* 1997).

NO_2 uptake measurement is complicated by the presence of NO_2 oxidised from NO released by denitrification within the soil in addition to external airborne sources (Fowler *et al.* 1989). The scale and within-year apportionment of NO_2 flux within forests require further clarification.

Dry-deposited NO_2 is considered to be taken up by plants through stomata while active, *ie* predominantly between April and September. Only small net amounts are believed be taken up by leaf cuticles (CLAG 1997). Deposition rates vary according to vegetation type. The bulk of NO_2 deposition occurs in England south of a line from the Tees to Morecambe Bay, but excluding Devon and Cornwall. Some emission of NO_2 by plants is possible, but the amounts are likely to be small.

NH_3, when dry-deposited, is rapidly absorbed into plants, whether on moorland, heath, forest and unimproved grass vegetation. There are also significant ammonia emissions from arable land and fertilized grass which over a year may cancel out depositions on these crop types. The values of the 'compensation point' *ie* where emissions and depositions are in balance are not established for forest types (Asman *et al.* 1998).

Because of the large quantity of emissions from intensively managed high density livestock production units, local concentrations of emissions in the vicinity of point sources may exceed national and regional averages by a factor of x3 - x4 commonly, and occasionally by x7 or more. Exposure of plants and the rate of ammonia deposition and uptake by plants is correspondingly higher.

Ammonia uptake may also assist SO_2 uptake; see §13.36.

Mapping dry deposition of ammonia

Two reports include maps estimating maximum dry deposition of ammonia to forest in the UK. The earlier map shows highest rates in the range 30 to >45 kg N/ha/yr. These occur in England south of the Lake District, but excluding Devon and Cornwall, and in montane Wales (INDITE 1994).

A later map for nitrogen deposition to forests (CLAG 1997) differs in that lower values are shown for montane Wales and higher values for the lower land to the east and for southwest England. In Scotland, higher values are given for the central Scotland lowland belt and lower values for the Ben Lomond/Ben Nevis area.

The differences between maps reflect the lower weighting in the underlying models to orographic rain, and higher weightings to the areas of emission. Both reports recognise that many of the conclusions and the supporting maps are the result of extrapolating small-scale detailed results to regional scale climatic and soils models. Both emphasise the need for further field work to verify the predictions from the models.

Earlier studies assumed that all ammonia deposited on conifer crowns was rapidly absorbed. In a Dutch study where forest and heathland are close to an agricultural livestock production unit with high ammonia emissions, net deposition on canopies of 15 m tall Douglas fir was estimated at approximately 50 kg N/ha/yr (Duyzer *et al*. 1992).

Ammonia emission has been observed during the day from this Douglas fir; at night, ammonia flux was generally towards the canopy (Wyers & Erisman 1998, Fowler *et al*. 1998a, Van Oss *et al*. 1998)). A study of the diurnal pattern of net flux showed this to be greatest at night (Wyers *et al*. 1992).

Other Dutch work reported depositions to *Calluna* canopies on heathland as 30-45 kg N/ha/yr and 27-33 kg S/ha/yr, of which half was by wet deposition (Bobbink *et al*. 1992).

In Yorkshire, deposition rates of N as ammonia, upwind and downwind of an intensive animal rearing unit were estimated at 30-50 kg N/ha/yr and 100 kg N/ha/yr respectively. Norway spruce stands associated with the highest deposition levels suffered severe needle loss and necrosis (TERG 1993).

In East Anglia, potential broiler chicken waste has been mapped to identify highest concentrations in the context of energy recovery (ETSU 1997e). Earlier, damage by ammonia from broiler house litter dumped in a forest nursery had been noted (Aldhous 1972 p.33).

13.23   Nitrogen depositions on some forest areas

The average annual deposition of N for all the UK for the period 1992-94 was 14 kg/ha, 5-10 times more than prior to the industrial revolution (CLAG 1997).

Accepting the limitations described above in attempting to derive reliable local values from regional averages, *Table 13.3* overleaf, gives regional rates of dry deposition for selected forest areas. The rates are based on values for the 20x20 km squares where a number of forests in Britain are located, based on the 1994 INDITE map. Wet depositions are higher in forests in high rainfall areas; dry depositions increase towards the south. Over the geographical spread of the selected forests, these mostly balance out to give similar total annual depositions. The exception is the most northerly forest listed, Culbin, which lies in a region consistently receiving the lowest amounts of any of the depositions mapped.

Deposition in urban woodland

Evidence is scanty; however, work in the United States showed nitrogen deposition on forest canopies in urban areas equivalent to up to 12 kg NO_2/ha/yr as against 0.1 - 2 kg NO_2/ha/yr in natural forest. Deposition on a large-leaved broadleaved species (*Platanus occidentalis)* was an order of magnitude greater than on a local pine species (*Pinus taeda)* (Hanson *et al*. 1989).

13.24   Nitrate saturation and drainage water from forests

Nitrate levels in water are relevant both to drinking water standards and to the possibility of eutrophication, and are of world-wide interest. In the UK, upland

Table 13.3   *Wet and dry depositions of nitrogen for selected forest areas based on 20 km square mean averages*   *(kg N per ha per yr)*

Forest	Wet deposition as NO_3^-	as NH_4^+	Dry deposition as NO_2	Total N 1989-2*	1992-4*
Culbin	< 2	< 2	1.2 - 2.4	< 10	< 12
Balquhidder	6 - 8	6 - 8	< 1.2	20 - 25	16 - 20
Kielder	6 - 8	6 - 8	1.2 - 2.4	20 - 25	16 - 20
Beddgelert	4 - 6	4 - 6	1.2 - 2.4	10 - 15	16 - 20
Plynlimon	4 - 6	4 - 8	2.4 - 3.6	15 - 30	16 - 24
Grizedale	6 - 8	> 10	2.4 - 3.6	25 - 30	24 - 28
Thetford	4 - 6	4 - 6	4.8 - 6	15 - 20	20 - 24
New Forest	2 - 4	4 - 6	4.8 - 6	15 - 20	20 - 24

Sources   INDITE 1994,   CLAG 1997 (for last column only)   * includes regional estimates for NH_3 and underestimates dry NH_3 deposition to forests

water is preferred for water supplies because of its low nutrient load and relative freedom from urban pollutants; in addition, concern about nitrates is coupled with the possible effects on soils and drainage waters of long-term acidification by nitrate run-off and seepage as a present-day substitute for sulphate acidification.

Nitrogen saturation

Inorganic-nitrate deposition has increased in many parts of Europe over the last 100 years. Several questions arise. One issue is of the capacity of the plants and soil to accommodate the additional deposition. In what sense can soils be said to be 'nitrogen-saturated'? The questions of plant response and displacement of species are discussed later in this section.

There is as yet no widely accepted definition of 'saturated'. Proposals range between:

• when N outputs equal or exceed N inputs (Billett *et al.* 1990, Ågren & Bosatta 1988),
• when there is some trace in outflow from added N (Hultberg *et al.* 1994).

Skeffington & Wilson (1988), reviewing saturation and excess nitrogen deposition, listed 12 'issues for consideration'. These included how to define nitrogen saturation, and whether there are detrimental effects if and when any site becomes 'nitrogen saturated'. They point to the complexity of ecosystems. They link 'saturation' with 'critical load' and offer eight alternative criteria for defining the latter.

In the short term, critical loads may have to be determined empirically in relation to the specific circumstances when a component of an ecosystem can be recognised to be at risk. Critical load levels set in that way would be higher than levels set by the most conservative definition of 'saturation'.

Further studies are needed to link the storage capability of plants and soils to inputs and outputs as the amounts of nitrogen stored on site are often two orders of magnitude greater than annual fluxes.

Nitrogen in drainage water

There have been many studies of water draining from forests and adjoining ground.

Emmett *et al.* (1993) surveyed 25 sites in Wales, recording N-levels in through-fall, in the Bs horizon at 500mm, and in streams. Over a 12 month period in older crops, nitrate-N concentrations in stream-water and Bs horizon varied between 0.40 and 0.80 mg per litre. In younger crops and moorland the range was 0.05 - 0.33 mg per litre.

Table 13.4 summarises results of 7 studies analysing the annual mean concentrations of nitrate-N in streams draining conifer plantations of a range of ages from 'young' to 54 years old.

Table 13.4 *Annual mean nitrate-N concentrations in streams draining forest catchments in the UK*

Site	NO_3-N (mg / l) Moorland	Forest	Forest age	Reference
Beddgelert	0.08	0.63	46 - 54	Stevens & Hornung 1988
Plynlimon	0.21	0.39	21 - 40	Reynolds *et al.* 1989
Llyn Brianne	0.17	0.23	22 - 24	Stoner *et al.* 1984
Scotland	0.01	0.13	various	Harriman *et al.* 1990
Loch Dee	0.10	0.08	8-12	Farley & Werrity 1989
Duchray	0.14	0.14	young	Harriman & Morrison 1982
Loch Chon	0.12	0.14	mature	Harriman & Morrison 1982

Source   INDITE 1994 - Table 6.4

A simple model has been developed for nitrate concentrations in water in two streams draining part of Plynlimon, mid-Wales. The model describes major changes but not short-term episodes (Sloan *et al.* 1994).

A number of other studies also describe trends in nitrogen deposition, nitrogen saturation and nitrate in drainage water.

• In Alltcaileach Forest, north east Scotland, soil organic horizons from 15 sites were sampled in 1949/50 and again in 1987. Six sites were located among crops planted in the late 1880s and nine in crops planted in the 1930/40s. Attributes assessed included pH, organic horizon thickness, %C, %N, and C/N ratio.

The organic horizons of the older stands were more acidic than the younger stands at both assessments, similarly for C in the organic horizons. The mean C% and N% in the organic horizons were less in 1987; however, the horizons themselves were on average twice as thick, so that the total stored C and N in these horizons was much increased. While the C/N ratio was higher in the organic horizons of the older stands, in neither age group was there any significant change over time.

While no data are available about changes in wet and dry deposition over the period of the comparison, the area is considered to be one of 'moderate atmospheric N deposition'. Although atmospheric inputs may exceed the suggested critical load of 5-20kg N/ha/yr, the lack of change in C/N ratio over time was taken to indicate that N saturation was not imminent (Billet *et al.*1990).

• At Beddgelert, north Wales, on a recently felled site, concentrations of nitrate in water from B horizons more than doubled in the first year, but thereafter decreased progressively until, by five years after felling, its concentration was less than before clear felling. The substantial reductions in sulphate and chloride observed in soil-water can be attributed to the removal of the tree crowns and their intercepting surfaces of foliage *etc* (Reynolds *et al.* 1992).

Harrison *et al.* (1995) in a study on Sitka spruce plantations in north Wales showed that very little N leached from the sites in their first 25-30 years of growth but that subsequently, losses increases progressively. A bioassay of the nutrient demands by excised Sitka spruce roots suggested that in older crops on certain soil types, P or K-deficiency might limit growth rates so that available N which otherwise might be taken up is, instead, leached.

Other experiments on the same sites showed that leaching of nitrate to stream-water increased with age, but that it could be reduced by applications of PK fertilizer (Stevens *et al.* 1993, 1994).

Considering saturation in the context of use of fertilizers in forestry plantations and evidence from fertilizer experiments in British forests, Miller & Miller (1987) pointed out that

• nitrogen uptake fluctuates as even-age tree stands develop. Response to fertilizer depends on the age of the stand and its site;
• evidence for excess nitrogen leading to growth disturbance is more appropriately interpreted as indications of other limiting factors coming into

operation. They conclude that fertilizer experiments give no support to the concept that heavy or repeated inputs of nitrogen to forest soils will lead to specific growth abnormalities or site deterioration;

• regarding the levels of nitrate in drainage water, they stressed the need to avoid simplistic approaches to the interaction of fertilizer application to plantations and additional nitrate in stream-water. They pointed out that in natural northern coniferous forest, the quantity of N on site is an order of magnitude less than long-term aggregated deposition, indicating that some loss of nitrogen has to be thought of as part of the normal nutrient cycle.

In Austria, in Norway spruce on south west-facing slopes exposed to the prevailing wind, N deposition in throughfall was 30-70% higher and sulphur deposition 50% higher than on the corresponding north east-facing slope. On the south west slope, nitrogen losses in water draining through the soil were of the same magnitude as atmospheric deposition. On the north east slope, nitrogen inputs could still be stored in soil and vegetation (Katzensteiner *et al.* 1992).

Stark & Hart (1997) found for a range of conifer forest types in the USA, that *in situ* measurements of gross nitrification rates were substantially higher than had previously been recognised from assessment of net soil nitrate levels. These are commonly observed to be low. While this had previously been interpreted as implying low rates of nitrification, Stark & Hart's observation suggested that soil microrganisms could be acting as highly efficient 'sinks' for nitrate-N as it is produced, leaving small 'net' amounts in soil solution.

Other responses to high rates of deposition of N

In Europe, high rates of deposition of atmospheric N have been shown to reduce root uptake of NO_3^-. The mechanism is not clear but differences in concentrations of amino-acids in phloem tissue of roots appear inversely proportional to root uptake. NO_3^- not taken up is either leached or denitrified (Näsholm 1998, Rennenberg *et al.* 1998).

In American studies, it is suggested that ammonium-N is the predominant form of N in forest soils and that conifer species respond well to it. On disturbance such as clear felling, the release of nitrate-N favours nitrophilous pioneers such as aspen; conifer species that would be expected to become established in successional forest in the absence of disturbance, fail on disturbed sites because of the change in form of available nitrogen (Kronzucker *et al.* 1997).

In Holland, deposition of nitrogen is the highest in Europe, two thirds of which is from $(NH_3 + NH_4^+)$. Where there is high N deposition, the species composition of the undergrowth has changed from lichen-dominated to grass-dominated vegetation. Performance of trees in the forest is not well correlated with predictions from laboratory studies of risk of damage due to high concentrations of mobile aluminium (Al^{+++}).

Assumptions of the inter-changeability of forms of nitrogen in relation to critical load norms are questioned. (van der Eerden *et al.* 1998).

Nitrate levels in streams and larger rivers

Statutory water and drainage authorities in Britain run 220 'Harmonised Monitoring' sites. While the sites include lakes and streams over the whole country, most of the river sites are on major drainage systems, often on their lower reaches. Site data include annual mean concentrations of NO_3^-. During 1990, these ranged from 0.05 to 13.6 mg N per litre. A substantial part of this range is an order of magnitude greater than any reported in forest outflows shown in *Table 14.3*. The figures in the table are consistent with the conclusion stated from studies of the N-loading of the Don and the Dee catchments, *ie* 'total N deposition does not significantly contribute to N loadings in either river' (INDITE 1994). Nevertheless upland stream-water is valued for its low nutrient (oligotrophic) quality, requiring inexpensive treatment as a potable water source.

Questions remaining open relate to the significance of 'episodes' or peak concentrations following, for example, rapid snow melt or heavy rain after drought.   However, in the context of the scale of large river catchments, agricultural use of fertilizer and the scale of nitrogen depositions, it is clear that the amount of fertilizers used in forestry (§10.15) is not significant.

13.25   Forest-based simulation of effects of wet & dry deposition of N

Many experiments in Scandinavia, Germany, Holland and the UK have studied the effects of N deposition by applying controlled amounts of N over forests. Some experiments have been combined with application of S. Two projects supported by the European Commission and referred to by the acronyms NITREX and EXMAN have attempted to bring together results from several countries.

At Fetteresso in north east Scotland, nets charged with soluble salts of ammonium or nitrate salts +/- PK salts were hung above 11-year-old Sitka spruce so that over 12 months, N was supplied at approximately 10 kg per ha per month. Other plots were treated with solid fertilizer broadcast over the soil surface monthly. The background wet deposition of N for the forest was low (6 kg N per ha per yr). The N concentrations measured in throughfall below the nets was within the range for polluted areas of Europe.

After a year, foliage N concentrations were approximately 15% higher on all plots receiving nitrogen; there was no difference arising from method of application. Other effects were small. On control plots, soil extractable mineral N was dominated by ammonium; the application of nitrate caused a significant increase in ammonium levels, but addition of ammonium caused only a small increase in nitrate. N applications did not increase the low level of nitrate reductase in foliage but may have had a small effect on roots. Soil water was not assessed (Thomas & Miller 1992). See also *Aphid damage and deposition* below.

At Aber, north Wales, plots in a mature Sitka spruce plantation were sprayed at ground level with solutions of sodium or ammonium nitrate. Either 35 or 75 kg N was applied annually for 2½ years. The intention was to examine the effects of added nitrogen at rates equal or greater than those encountered in throughfall where there is heavy wet and dry deposition.

Nitrate-N in 'below rooting depth' percolation water from the site contained additional nitrate of the same order of magnitude as the nitrate content of the added salts, whether as sodium nitrate or as ammonium nitrate. In contrast, there was no increase in ammonium-N in percolation water from any treatment.

On plots without any added nitrogen, nitrate levels in percolation water were equivalent to losses of 10-20 kg N per ha per yr. Against this level of base load leaching and non-retention of added nitrate-N, the soils at Aber could be said to be 'nitrate-N saturated' but not 'ammonium-N saturated' (Emmett *et al.*1995).

In a subsequent paper, the total ongoing atmospheric nitrogen deposition on this site was estimated at 28 kg N/ha/yr (Emmett & Reynolds 1996). This appreciably exceeded the previously predicted critical load for the site, so that some leaching of nitrate into drainage water was to be expected.

At Grisedale Forest, Cumbria, additional N in the form of ammonium sulphate was applied roughly fortnightly for a year to lysimeters containing the top 17.5 cm of litter and soil. In aggregate 75 kg N /ha/yr was applied. All lysimeter treatments showed a high throughflow of nitrate, raising the question whether the installation of the lysimeters had disturbed the soil profile unduly, setting in train excessive nitrification. On the plots given ammonium sulphate, approximately one third of the ammonium had been leached through the lysimeters and two thirds of the sulphate. There was a corresponding increase in mobile cations, especially aluminium, released from plots given ammonium sulphate (Carnol *et al.* 1997).

In Germany, in an experiment at Göttingen simulating wet deposition, nitrogen was applied by overhead irrigation to young Norway spruce. Evidence was found of foliage uptake of N both as nitrate and ammonium (Eilers *et al.* 1992).

In Norway, 80-year old Norway spruce was treated weekly with ammonium nitrate in water at the equivalent to an annual rate of 30-50kg N/ha. Only nitrate was found in drainage water, only in the winter months and at a rate equivalent to 1% of the year's application. While this was described as nitrogen saturation, this is an extreme use of the term (Hultberg *et al.* 1994).

In another trial in Norway, at Sognal, a small catchment covered with alpine shrub vegetation (*Betula verrucosa, B. nana, Juniperus communis, Salix hastata*), was for nine years treated with equivalent amounts of nitric and sulphuric acids. For nitrogen, the rate was equivalent to 7 kg N/ha/yr. Acids were applied to snowpack in April and at approximately monthly intervals between June and October each year. The area is subject to very low natural pollutant deposition. Over the period of the experiment, 90% of the N was retained on site, a similar proportion to adjacent untreated catchments where the only additional N came from natural atmospheric deposition. There were high peaks in run-off from the treated area. However, these directly followed N applications and were interpreted as evidence that the ecosystem was briefly overloaded by the application. Higher N and S concentrations were found in organic layers in the soil. The increase in stored N on the site due to treatment was not considered significant, being approximately 1% of the total N on site (Wright & Tietema 1995).

Seven sites covering a range of rates of deposition were included in the NITREX project. In three of seven sites, the ground was roofed to reduce deposition; on three others, nitrogen was added and on one site both treatments were applied. Sites included Aber and Sognal mentioned above. At deposition rates below 10 kg N/ha/yr, nearly all the N was retained; between 10 and 25 kg N/ha/yr losses from the sites varied; above 25 kg N/ha/yr, losses of N were substantial (Wright *et al.* 1995).

Aphid damage and simulated deposition

The experiment mentioned above at Feteresso was fortuitously subject to a severe attack of green spruce aphis (*Elatobium abietinum*) in all treatment plots.

Height increment was substantially reduced in proportion to the severity of damage by the aphids, but was unaffected by fertilizer treatment. There was no interaction in height response between aphid severity and fertilizer.

Diameter increment response, however, showed a very clear interaction. On plots receiving no fertilizer, diameter increment in the year after attack was reduced by up to 50% in proportion to the severity of the aphid damage to needles. However, while diameter increment on plots given N or NPKMg was also reduced by aphids, for the trees suffering needle damage similar to the worst on the unfertilized plots, there was only a 25% reduction in increment, indicating that the addition of fertilizer had assisted recovery from attack.

There was no evidence that fertilizer application had increased damage by spruce aphis (Thomas & Miller 1994).

See also §13.35 under *South Wales coalfield*, and §12.26 under *Litter under spruce*, a Danish study of litterfall linked to spruce aphis infection.

13.26   European overview of nitrogen deposition

Dise & Wright (1995) describe a data-base compiled for 'Evaluation of Nitrogen and Sulphur Fluxes' (ENSF). It has records from over 100 catchments and plots throughput central and northwest Europe; from this, 65 sites with data on N inputs and outputs were selected for analysis and report. These included the NITREX series of experiments.

Deposition data were based on throughfall on the questionable assumption that foliage uptake and foliage leaching cancelled each other out. Deposition values for any site reflected the combination of long-range transport of NO_x and more local (regional) concentrations of emission and deposition. Highest depositions were in the Netherlands and northwestern Germany because of high levels of ammonia emission from livestock production areas. Depositions diminished to the south and east and to the far north.

N loss from the site ('output') in the form of run-off or seepage of NH_4^+ and NO_3^- was clearly related to total N deposition. Where it was less than 10 kg N/ha/yr, N in run-off or seepage (soil leaching) was minimal. Where it ranged from 10 - 25 kg N/ha/yr, N losses varied partly according to input, but for a given rate were less when N input contained higher proportions of NH_4^+. At all sites where input exceeded 25 kg N/ha/yr, significant soil leaching occurred.

The relationship between N output and soil properties showed that the pH of seepage water in the B horizon was most relevant. *Figure 13.1* shows the sharp demarcation either side of pH4.2 between insignificant levels of N in soil-water on less acid soils, and substantial loss on the more acid soils.

Output of aluminium in run-off and seepage was highly correlated with the nitrogen content (Dise & Wright 1995).

Figure 13.1 *Rate of N loss from B horizon in relation to soil pH*

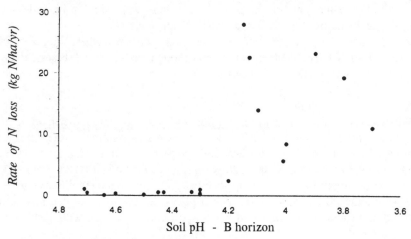

Soil pH - B horizon

Source Dise & Wright 1995

Correlations between input, output and site factors were examined in a more detailed study of NITREX and another group of sites. Plant N content and surface organic matter (L+F+H layer) both reflected N input. Where there was any substantial N content in drainage water, NO_3^- constituted 80 to over 90% of it (Tietema & Beier 1995).

The evidence for nitrogen deposition as a threat to forests in Sweden was comprehensively reviewed by Binkley & Högberg (1997). N deposition was not considered to represent an immediate threat to Swedish forests. In the short term, forest growth rates had increased. Some changes in ground vegetation had occurred, grasses replacing ericaceous species, and the possibility of gradual acidification and nitrate leaching was recognised. On a scale of several decades, losses of diversity might occur in south west coastal areas of Sweden where deposition rates are highest.

While the ratio of aluminium to other cations declined with increasing N fertilization, growth did not deteriorate as had been previously predicted. On the contrary, it improved.

13.27   Denitrification

Loss of nitrogen from woodland sites by denitrification has not been widely studied. However, Dutch & Ineson (1990) have pointed out that if such loss is significant, it is likely to occur when nitrate levels in soil are relatively high.

They describe sequences of core samples taken from an existing Sitka spruce plantation and an adjacent clear-felled area at Kershope Forest, in the north of England. Denitrification loss in the range 3 - 4 kg N/ha/yr was observed from the standing crop. Following felling, there was a substantial increase (up to 40 kg N/ha/yr) for 2-3 years, after which the clear-felled site reverted to the rates of loss similar to the standing trees. These results were similar in scale and nature to that found overseas in a wide range of forest types.

In the studies of N inputs and outputs, losses on sites where inputs of N exceeded 30 kg N/ha/yr, denitrification was thought to account for part of the higher losses observed (Dise & Wright 1995).

13.28   Other studies

Why oxides of nitrogen, particularly NO and NO_2 are toxic rather than being freely usable as a nutrient source has been reviewed in relation to their uptake into foliage and subsequent metabolism and physiology. Phytotoxic effects were considered most likely to arise through generation of nitrite and HNO_2 from NO, and subsequent disturbances of cellular nitrogen metabolism (Wellburn 1990). There were indications that some individual plants can fix NO_x and that selection and breeding programmes could lead to transport corridors lined with such stock (Wellburn 1998).

13.29   Future trends for nitrogen emission and deposition

Papers presented at a symposium in 1997 on atmospheric gases and their effects on plants and reported in *New Phytologist* **139** 1998 summarised current work and thinking. Topics ranged from:

- the huge scale of increase of available atmospheric fixed nitrogen, several authors likening its effect to that of a catastrophic perturbation following major meteorite impacts (Cape 1998, Raven & Yin 1998);
- the interaction of nitrogen deposition and ectomycorrhizas (Wallenda & Kottke 1998, van der Eerden 1998);
- detailed consideration of atmospheric ammonia, (Asman *et al.* 1998, Schjorring 1998);

• commentary on response by plants in N-limited communities. On a local scale these have been compared to semi-natural P-limited communities, (Lee & Caporn 1998), while on a global scale, they were linked with nitrogen depositions and with increased carbon sequestration (Norby 1998), or with ozone and nitrogenous pollutants and natural and semi-natural ecosystems (Bobbink 1998).

Over the next 25 years, emissions of nitrogen oxides are predicted to fall as an increasing proportion of vehicles are fitted with catalytic converters. Between 1989 and 1994, there has been a 20% drop based on 1994 levels; a further 45% fall is predicted by 2010.

Emissions of ammonia are expected to follow the numbers of farm animals. Globally, these are expected to increase in proportion to increases in world population. However, air-borne ammonia is very quickly absorbed, its half-life being approximately 1 day; deposition must be expected to be concentrated on vegetation within 1 km of the sources of emission. Improved agricultural practices may help to keep emissions down (RGAR 1997).

Prospects for UK forests in response to depositions of atmospheric N

If rates of deposition decrease as predicted, there is little to suggest that the tree component of forests, woodlands and amenity trees will suffer to any greater extent than has already been observed, *ie* local damage associated with high short-term levels of ammonium-N. Whether the previously occurring natural associations of shrubs and ground flora will be unaffected is more open to doubt (Lee & Caporn 1998); similarly, whether any novel situations will arise through interactions of plants with N depositions and increasing CO_2 or O_3 cannot be foretold. Disasters cannot be dismissed; it would be prudent to continue to monitor natural and cultivated plant communities for medium and long-term change.

13.3   SULPHUR AND ACID RAIN

13.31   Introduction

Sulphur, like nitrogen is a major plant nutrient. However, it has rarely been considered necessary to take action about its nutrient role; plants secure sulphur needed for growth from soils, from atmospheric depositions or from sulphur included in fertilizers applied for their N, P or K content *eg* superphosphate, sulphate of ammonia, sulphate of potash. The availability of S from fertilizers has, however, diminished with the increasing use of more concentrated compound fertilizers, so that future sulphur requirements of plants must be monitored.

Sulphur also has been the most damaging of the atmospheric pollutants during the last 150 years.

The roles of sulphur are pervasive. In the atmosphere, sulphur compounds exist as gases and in mist or rain droplets, mostly as SO_2, or as SO_4^{-}. They occur in aerosol particles as acid or sulphate salts formed with ammonium or other cations present derived from emissions. H_2S is associated with anaerobic decomposition but, though highly toxic, is insignificant in the atmosphere as a pollutant.

In reports of pollutant deposition, sulphur may be expressed in terms of sulphate, or SO_4^- or S in SO_4^- in precipitation, throughfall, soil water *etc*. In the atmosphere, critical levels of S are in terms of SO_2 or S in SO_2. See Appendix III for conversion factors.

UK atmospheric sulphur balance sheet

Table 13.5 gives figures for emissions and depositions for sulphur for 1986-8 and 1992-4 (RGAR 1997). 'Wet+cloud' deposition accounted for 60% of total UK depositions in 1992-4. Exports of air-borne S pollutants from the UK is 9 - 10 times greater than imports and constitute 78% of UK emissions.

Table 13.5   *Annual UK sulphur balance sheet*   *(kt S/yr)*

Year	Import from Europe	Emissions by Industry	Depositions Dry	Depositions Wet+cloud	Export beyond UK
1986-88	160	1900	240	230	1590
1992-94	160	1600	140	210	1410

Source	RGAR 1997	See also *Table 13.2*

13.32   Air-borne SO_2 and SO_4^-

Air pollution from gaseous compounds of sulphur has a long history. Emissions of SO_2 from industry go back to the beginnings of metal smelting, when metal sulphides were important ores. With the advent of the industrial revolution and the burning of coal, emissions rose through the 19th and 20th centuries, reaching a peak in 1970 both in the UK and in Europe. Damage from SO_2 has been catastrophic around smelters in several parts of the world, notably, Sudbury, Ontario and Mt Lyell, Tasmania. Death or severe damage to forests over areas of several tens of square kilometres have occurred in these places and also in central Europe, centred on the Erzgebirge in Czechoslovakia and East Germany (Sands 1984).

While the pattern of SO_2 air pollution is reasonably well documented for the UK, Germany and other western European countries, emissions from eastern Europe in the period 1950-70 have not been widely discussed but nevertheless account for much of 'forest decline'.

Current concerns

Three main strands of concern about the effects of sulphur compounds on forests and trees have been:

- acute localised damage from traceable pollutant sources (§13.33),
- regional scale damage linked to 'acid rain' and 'forest decline' (§14.4),
- the role of trees in accentuating stream and lake water acidification (§14.5).

13.33   Emissions and depositions in Britain

In the early 1950s, concern about the harm to health and to structures from air pollution led in Britain to the *Clean Air Act 1956*. The first effect of this legislation was to reduce soot and smoke from burning coal. Reduction of SO_2 emissions took longer to take effect. From the 1970s, domestic pressure to clean up the air was reinforced by international obligations under the Convention on Long-range Transboundary Air Pollution (§14.41).

Over the last 25 years, SO_2 emissions in the UK have fallen dramatically:

- 1970   Emission level   3.1 million tonnes S per yr   (100%)
- 1980   2.5   "   "   "   (87%)
- 1990   1.8   "   "   "   (58%)
- 1994   1.4   "   "   "   (45%)

The official UK forecast is that by 2010, UK emissions will be down to 0.6 million tonnes (RGAR 1997).

These figures are important to forestry. The effect of trees in intercepting wet and dry nitrogen depositions has been described in §§13.23 and 14.32. SO_2 and SO_4^- in dry and wet deposition respectively behave similarly to nitrogen compounds.   Any substantial drop in emissions should reduce depositions proportionately and should materially reduce the risk that future depositions of sulphur compounds in forests will exceed the level of critical loads deemed to threaten water supplies from poorly buffered soils (Nisbet 1996a).

Air quality standards for trees

The first attempt to set European air quality standards for SO_2 resulted from activities of the *International Union of Forest Research Organisations (IUFRO)* (Wentzel 1983).   These were overtaken by recom-mendations for agricultural crops, forests, natural vegetation and lichens from the World Health Organisation (WHO) and the United Nations Economic Committee for Europe (UN ECE) at conferences in 1988 and 1992.   *Table 13.6* summarises figures for forests. Annual mean levels for agricultural crops were set at 30 $\mu g/m^3$ (CLAG 1996).

Such standards form the underlying support for international political pressure to reduce industrial emissions.

Table 13.6   *Minimum air quality standards for SO$_2$ and forests*   *(μg/m^3)*

	IUFRO (1978) normal	extreme sites	WHO (1987)	UN ECE (1988)	(1992)	cold sites
Annual mean	50	25	30	20	20	15
24 hour	100	50	100	70	20	15

Source	CLAG 1996

13.34   Local damage to plants by sulphur dioxide

Plant pathologists have been aware of the potential for pollutants to damage trees from the start of the 20th century. Acute injury to plants caused by sulphur dioxide is discussed by Peace (1962) in terms of damage relatable to specific pollution sources. Damage may follow short 'episodes' of unusually high pollutant levels. Injury is most common on younger plants or fresh foliage, damage being worse in misty or humid conditions when foliage is damp.

Damage may also be associated with temperature inversions in still air, when smoke and fumes remain in a low-lying layer.

Symptoms of acute sulphur dioxide damage are bleaching of intervenal leaf regions in broadleaves, and browning and chlorotic banding toward needle tips of conifers (Phillips & Burdekin 1982, Butin 1995). Butin comments that on trees showing the 'comb' crown pattern of Norway spruce (*ie* branchlets that are pendulous), bud damage is particularly prevalent, inferring that some damage to buds may be accentuated by concentration of droplets running down the hanging branchlets. He also notes that European silver fir is particularly susceptible to damage from SO$_2$.

Species differ in susceptibility to damage. Peace rated beech, oak and six conifer genera as 'susceptible' and eleven broadleaved and four conifer genera as 'resistant'. Pines are included in both categories, depending on the species. *Pinus nigra* is listed as resistant; *Pinus sylvestris, Abies* spp., Norway spruce and larches are listed as susceptible. Sitka spruce is not included in either list.

Some fungi, especially those growing superficially on leaves *eg* tar spot of sycamore (*Rhytisma aceroides*), oak mildew (*Microsphaera alphitoides*) and black spot of roses are noted as not occurring in areas of heavy air pollution, sulphur pollutant deposition acting as a fungicide (Peace 1962, Buczacki *et al.* 1981).

13.35   Damage to trees in heavily polluted areas in Britain

Pollution in the Pennines

Some of the earliest developments of the industrial revolution in the United Kingdom occurred in the southern Pennines, a hilly area of sandstones (Millstone grit) and coal measures of Carboniferous age. For more than 150 years, the land

and plants in the region were exposed to the whole range of pollutants from developing coal-based metal-working industries and brickworks. Rainwater collected in 1910 had been noted as 'acid'! In the 1960s, the area was considered a lichen 'desert' (Hawksworth & Rose 1970).

Attempts at afforestation in the late 19th century and first half of the 20th century in the Pennines largely failed, air pollution being recognised during that time as the major local limiting factor.

In the late 1950s, the Forestry Commission (FC) set up a range of species trials, at the same time studying conditions where heavy pollution and exposure were thought to be limiting growth in south Wales.

Trials in 1951 assessed SO_2 using lead dioxide 'candles'. Contrary to the expectation at the time that concentrations would fall with increasing distance from towns, SO_2 in 'truly rural upland sites' frequently had values higher than in suburbs; peak values on several FC sites were similar to those commonly associated with town centre pollution in the mid-1950s (Lines 1984).

The effect of local shelter was compared with full exposure by constructing a fence 50% lath and 50% space and 1.3 m tall, which enclosed a circular area 27 m in diameter. Lead dioxide gauges close to the fence showed 20-30% reduction in SO_2 compared with a 6% reduction in the centre of the enclosed area (Lines 1962). Plants within the shelters were less discoloured than those outside (Edwards *et al.* 1959). Zinc plates were also used to assess pollution (Edwards *et al.* 1960).

Retrospectively estimating mean SO_2 concentrations from sulphation rates of lead dioxide candles is imprecise. However, the best available approximations placed 1950s figures in the range between 75 and 225 μg SO_2/m^3 of air. Volume increment of Sitka spruce on the most heavily polluted sites was 4 yield classes (8 m^3/yr) less than on comparable unpolluted sites. With some exceptions, growth was inversely correlated with mean SO_2 concentrations as estimated by lead dioxide gauges.

By 1978-81, there had been a 52% reduction in SO_2 concentrations following reductions in industrial emissions, the reduction coinciding with marked improvements in growth (Lines 1984, Vann *et al.* 1985).

South Wales Coalfield

Plantations in the south Wales coalfield had been the subject of concern since the 1930s. Spruces, Douglas fir and European larch had performed relatively poorly compared with Corsican pine and Japanese larch. Poor growth had manifested itself through a reduction of height increment of a proportion of the crop, in particular, Douglas fir and Sitka spruce. From about 15 years old, the tops of many trees were losing their spire-like form and bending or flattening, a condition described as 'bent top' (Coutts *et al.* 1985). SO_2 pollution was a prime suspect because of the proximity not only of the coalfields but also metal smelting and steel manufacturing industries.

Studies in the 1960s were impeded by vandalism endemic in the area at that time. However, through use of zinc plates, tatter flags and surveys of the distribution of lichens, SO_2 pollution effects were identified as highest in the immediate hinterland of Newport and Port Talbot. Strong winds were considered to aggravate pollution, their incidence reflecting the deeply dissected local topography. Lichen flora was richer both in the shelter of deep valleys and towards the centre of the larger blocks of forest (Mayhead *et al.* 1974).

Decline persisted and was again evident in crops planted in the 1960s. A multi-disciplinary programme of work between 1984 and 1991 examined the effects of SO_2, nutrition, soil conditions, incidence of insect attack and other factors. Lead dioxide monitors were used to assess SO_2 concentrations. These may have over-estimated concentrations (but not dose) because of underestimates of the windiness of the site. Mean annual values between 1984-7 ranged from 8.0-11.3 nl/l. One monthly peak value of 45 nl/l was observed; other short-term, more severe episodes up to 12 nl/l have occurred in comparable areas (Freer-Smith & Dobson 1995). Nevertheless, the background average annual means recorded are substantially lower than the concentrations associated with good growth found in the Pennines, implying that SO_2 was not the factor limiting growth in south Wales.

Mineral N and readily mineralised N concentrations in peat on Sitka spruce sites in the coalfield were low and the peat was particularly acidic, indicating potential N deficiency. A modest response to fertilizers was obtained in the most recent trials, foliar nutrient levels reverting to marginal/deficient levels after 4 years.

Correlating site factors to basal area increment, the best correlation was with potassium. The next best was with sulphur, followed by nitrogen, suggesting that either sulphur plays a directly or indirectly beneficial role in spruce nutrition on the coalfield rather than being a source of damage, or that its presence is the inadvertent consequence of prolonged acidification of soils by atmospheric sulphur deposition.

Foliar nutrient levels remained at marginal/deficient levels in three out of four years of the trial in spite of heavy applications of fertilizer (Coutts *et al.* 1995).

Cloud-water deposition was not estimated. However, estimates by the Review Group on Acid Rain (RGAR 1997) put annual sulphate deposition over that part of south Wales at 20-30 kg S/yr and placed the area in the highest category for exceedance of critical load, because of existing soil acidity.

SO_2 and aphids

Infestation by Green spruce aphis (*Elatobium*) had been found to be heavier in the presence of SO_2 (Warrington & Whittaker 1990).

Comparing individual 'infested' trees in the coalfield with paired adjacent 'uninfested' trees, basal area increment of the infested trees was reduced by nearly half. However, crop studies showed that overall levels of infestation by spruce aphis were low (Carter 1995) and were insufficient to account for the scale of loss of growth in the region.

Root aphids (*Pachypappa* spp.) were present in large numbers and, in conjunction with *Elatobium* defoliation, were considered to be a possible contributor to the 'forest decline' syndrome in the coalfield. Mild winter temperatures associated with the coalfield's coastal position could have resulted in less winter death among aphid populations than elsewhere (Carter 1995).

A laboratory study in Cardiff of population dynamics of aphids on Scots pine showed that for the needle aphis *Schizolachnus pineti*, the presence of acid mist increased individual and population growth rates. For the large pine aphid, *Cinara pinea*, levels of amino-acids and phenols in host plant nutrition determined population increase. For both species, such variations could be significant in the interactions between air pollution and tree health (Kidd & Lewis 1987, Kidd *et al.* 1987).

It has not been possible to establish the 'pre-industrial revolution' base status of coalfield soils and thus have some indication of the long-term accumulated effect of acidic depositions on soils in the coalfield. Nevertheless, acid deposition conditions in the late 19th century, with the many mines and steelworks in the area must have been severe.

13.36 Sulphur and nitrogen interactions

Co-deposition of NH_3 and SO_2

Where concentrations of SO_2 and NO_x which separately would cause marginal damage are both present, plant damage may be markedly greater than any simple additive effect (Mansfield & Freer-Smith 1981).

Fowler *et al.* (1989) observed that uptake of SO_2 by dry leaf surfaces was lower than would be expected from stomatal conductances, implying an internal resistance to SO_2. However, when surfaces were wet and NH_3 was present, canopy resistance disappeared.

McLeod *et al.* (1990) added SO_2 and O_3 separately and together to Scots pine and Sitka spruce plants established in open-air plots at Liphook, Hampshire. Fumigation with SO_2 was at 1½ and 3 times ambient levels, treatments lasting for 6 - 8 months in each of three growing seasons. Ozone was fumigated at 1½ times ambient levels for the second and third season. Nitrogen was not added as an experimental treatment; the annual mean concentrations in ambient air for the site were quite low at 4 ppb NH_3, 5 ppb NO and 8 ppb NO_2.

Foliage analysis at the end of the second growing season showed that the nitrogen content of Norway spruce needles was 1.7 times higher in plants from the high-SO_2 treatments compared with those exposed to ambient air. Similar results were obtained with Sitka spruce. In quantitative terms, the amounts corresponded to uptake of 7.8 and 12.8 kg N/ha. At this site, the N-content of foliage of both species in the ambient air plots was below 'deficiency' level. Scots pine did not respond in the same way, nor were its needles deficient.

The mechanism considered most likely to have caused this effect is the enhancement by SO_2 of NH_3 deposition to wet surfaces; the solubility of SO_2 or

NH_3 alone is limited by the pH of the solution they form, whereas deposition of both gases together can leave the pH at non-limiting values.

Subsequent studies of throughfall showed the ammonium-N content of throughfall was 1.7 and 1.9 times higher for spruce and pine respectively on the high rate fumigation plots compared to ambient plots, suggesting that some co-deposited NH_3 had been washed off the foliage.

Dishes of deionised water picked up more NH_3 than SO_2 in ambient air. In high SO_2 air, capture of NH_3 was 50% higher.

The authors comment that if this result applied widely, some results attributed to the effects of SO_2 could have been due to co-deposited NH_3.

In a study of the scale of sulphate leaching from a Scots pine canopy, Cape *et al.* (1992) found that sulphur throughput could only be accounted for by co-deposition to canopy surfaces.

Damage due to deposition of very large amounts of NH_4^+ and SO_4^{--} to canopies of Norway spruce adjoining an intensive animal rearing unit in Yorkshire is thought to illustrate co-deposition (Fig. 6.1 in TERG 1993). Deposition rates downwind of the farm are estimated at approximately 100 kg N/ha/yr, and half that amount upwind. On a 'molecular equivalent' basis, the deposition of sulphate corresponds to about 75% of the deposition of ammonium (TERG 1993).

Acute damage, frost susceptibility, leaf loss etc

A number of experiments on young plants under controlled environments indicate that N and S together can increase plants' susceptibility to cold more than either separately.

Red spruce (*Picea rubens*) has been used in trials because, in the north east United States, the species has suffered seriously from pollution-induced apical die-back, loss of foliage and some deaths, related to winter frost injury.

In an attempt to simulate symptoms of damage, small amounts of mist were applied to red spruce seedlings twice weekly from mid-July to December. Mists were composed of equimolar amounts of ammonium sulphate and nitric acid in concentrations such that the pH of the mist ranged from 2.5 to 5.0. Visible injury symptoms appeared 10 weeks after treatments commenced; damage was observed on plants treated with acid mist in the range 2.5-3.0; very slight symptoms were observable on plants treated at pH 3.5. Damage symptoms developed over 3 days; initially, about 40% of affected needles became light brown and 60% were light orange brown orange. Colours darkened over the following 4 weeks. Most seedlings showing any great degree of necrosis were in the orange-brown category. In the following spring, buds of badly damaged plants flushed about 11 days earlier than buds on unaffected plants.

The symptoms were similar to those reported for Norway spruce exposed to acid mist at pH 3. The inputs in the most acidic treatment (pH 2.5) corresponded to applications of 55 and 43 kg/ha of N and S respectively (Leith *et al.* 1989).

Susceptibility to frost

Cape *et al.* (1991) exposed red spruce seedlings in open-top chambers to filtered air and to air + mists containing NH_4^+, SO_4^{--}, NO_3^- and H^+ separately and together. Plants exposed to mists containing NH_4^+ and SO_4^{--} were less hardy than controls.

The susceptibility to winter cold of other red spruce seedlings grown in open-top chambers was also related to exposure to the SO_4^{--} content of acid mists. It was postulated that frost susceptibility could follow build-up of SO_4^{--} within foliage tissues as a result of high external concentrations. In the absence of sufficient assimilate and N, damage to cell membranes and cell proteins led to loss of winter hardiness (Sheppard 1994).

To test resistance to autumn frost, Sitka spruce and red spruce seedlings were raised with NPK fertilizer regimes applied either in May, or in May and August, with and without sulphate. Shoot samples were tested for susceptibility to freezing in November and December of the year of treatment. Differences were small and not significant. There was no evidence to suggest that soil applications of SO_4^{--} had any significant effect on resistance to autumn frost, supporting the view that observed damage is more likely to be due to foliage uptake of sulphate than uptake from the soil (Sheppard 1997).

Foliage of grafts from mature Sitka spruce exposed to acid mist, again in open-top chambers, were less sensitive than seedling foliage. Concentrations of S in foliage were less; reductions in frost hardiness associated with high S in foliage only occurred at temperatures outside the range normally encountered in the winter environment in Britain. For Sitka spruce, the increased risk of winter cold damage due to cloud-water deposition of SO_4^{--} is small (Sheppard *et al.* 1991).

Sitka spruce seedlings exposed to controlled concentrations of SO_2 and NO_2 (30 nl/l) in wind-tunnels prior to freezing showed slight increases in injury to buds rather than needles. Effects occurred only with certain combinations of pollutant dose and cold treatment (Freer-Smith & Mansfield 1987).

Birch *(Betula pendula)* seedlings and cuttings exposed to SO_2 and NO_2 at 40 nl/l shed leaves early in response to exposure to SO_2. Exposure to NO_2 alone increased dry matter production but this did not occur when plants were exposed to SO_2 and NO_2 together. SO_2 gas exchange measurements were significantly correlated with SO_2 flux; there were also indications of small reductions in stomatal conductance and photosynthetic rate in the presence of SO_2 (Freer-Smith 1985).

Earlier trials with six broadleaved species *(Tilia cordata, Malus domestica, Betula pendula, B. pubescens, Populus nigra & Alnus incana)* had shown a slight improvement in growth in response to NO_2 fumigation. By itself, SO_2 had depressed growth of three species but the reduction was markedly greater when NO_2 and SO_2 were applied together (Whitmore & Freer-Smith 1982).

13.37   Other effects

Effect of species

Amounts of throughfall of nutrient and marine depositions (mostly Na+ and Cl-) were studied in Japanese larch and Sitka spruce in Wales and European larch and Sitka spruce at Crathes, northeast Scotland (Reynolds *et al.* 1989).

SO_4^{--} levels were higher in throughfall on all sites, being higher under larch than under Sitka spruce. H+ in throughfall was also higher under larch. On both sites, total amounts of most nutrients other than nitrogen reaching the forest floor in throughfall were greater than in the original precipitation. For the larch at Crathes, the additional depositions were twice those of all other sites, implying that the site was not strictly comparable to the others.

For K+, and to a lesser extent, Mg++ and Ca++, throughfall was enhanced by leaching from the canopy. Also, Sitka spruce on both sites lost small amounts of PO_4^3.

For all species on both sites, nitrogen, whether as NH_4^+ or NO_3^{--}, was absorbed by the canopy. See also §12.51 and *Tables 12.4* and *12.5*.

Search for resistance to pollutants

Species trials set up between 1951 and 1977 in the Pennines showed that Sitka spruce was less tolerant of severe pollution and acid sites than 'North Coastal' origins of *Pinus contorta,* and that *Betula pendula* was the most tolerant broadleaved species (Lines 1984).

In a comparison of species subject to fumigation at high and low levels, growth of spruces was found to be more depressed by low levels of SO_2 than pines (Garsed & Rutter 1982).

In studies of 'decline' in the south Wales coalfield, individual trees were found to be 'susceptible' and others 'resistant' (Coutts *et al.* 1995).

Re-oxidation of reduced sulphur

Atmospheric deposition of S has lessened in eastern North America since 1980. However, in extremely dry years, there have been indications that older deposits of reduced S in wetlands are being re-oxidised, masking any benefit from reduced atmospheric deposition (Dillon & La Zerte 1992).

Canopy leaching of sulphate

In a study of the effect of applying radioactive sulphate to the soil surface under a 37 year old Scots pine plantation, the sulphate was quickly translocated to current year's foliage; however, only 3% of the net throughfall of sulphur at that time could be attributed to canopy leaching. The experiment was not considered to be rigorous enough to dismiss the possibility that higher percentages of soil-sulphate could be leached from the canopy as part of S-cycling in other conditions (Cape *et al.* 1992).

13.4   OZONE (O_3)

13.41   A fluctuating pollutant

Ozone is a natural component of the atmosphere. The largest concentrations and the bulk of the ozone mass are in the stratosphere, formed in clean air by recombination of atomic and molecular oxygen. Some stratospheric ozone is transferred down, but the majority of ground level ozone is formed within the troposphere (PORG 1997).

Ozone is not emitted directly by any industrial or natural process but is created in the troposphere by sunlight-initiated photolysis of NO_2 to O_3 and NO. Independently, NO may be re-oxidised in air by O_2 or O_3 back to NO_2.

In daylight, concentrations of O_3, NO and NO_2 are in fluctuating balance between oxidation and photolysis, depending on intensity and duration of sunlight. Ozone levels are higher in sunny weather in summer, with long daylight hours. There are natural diurnal and seasonal cycles of atmospheric ozone, reflecting the sunniness of the period. In the dark, NO_2 accumulates.

Any polluting process which increases NO_x in the atmosphere will alter the O_3/NO_x/sunlight equilibrium and may raise the ozones levels correspondingly.

The major source of NO_2 results from oxidation of (pollutant) volatile organic compounds (VOCs), involving NO and 'free peroxyradicals', in particular, the hydroxyl (OH) and hydroperoxy (HO_2) radicals which result from photolysis of nitrous acid and aldehydes.

Both types of free radical react with NO to form NO_2, but do not require an external O_2 or O_3 source. Ozone resulting from subsequent photolysis of this NO_2 is a net gain. Reaction pathways involving free radicals in this way account for the larger part of photochemically mediated ozone production in the troposphere.

A second source of NO_x pollution arises from the combustion of fossil fuels. During combustion, even of clean fuels, NO_x are emitted into the boundary layer and consist of approximately 90% NO_2.

In urban areas, the effects of pollution from VOC and fossil fuel combustion are not additive, VOC pollution reactions using up some NO emitted by vehicles.

The Photochemical Oxidants Review Group in their third and fourth reports (PORG 1993, 1997) give comprehensive summaries of current views on the role of ozone. These cover ozone levels, other oxidants, oxidation chemistry, materials most directly involved in oxidative reactions, effects on vegetation, human health and materials, and their relationship with international policy on emission controls. See also Fowler *et al.* (1998b).

Other reactions between ozone, NO_x etc

NO_2 also reacts with the free radical OH to form HNO_3, which is the principal daytime means of removing NO_2 from the atmosphere and an important factor in controlling its concentration.

At night, NO_2 is also slowly converted into NO_3 by reaction with O_3. This reaction takes place during the day as well, but in sunlight the NO_3 rapidly breaks down to NO_2 and O_3 again. In the absence of sunlight, NO_3 can react in the gas phase with NO_2 to form N_2O_5 and then with water vapour to form HNO_3. However, this is a slow reaction; the reaction takes place more rapidly on the surface of cloud-water droplets, the latter accounting for the bulk of nitric acid formation in the atmosphere.

If ammonia is present, it will react with the nitric acid to form ammonium nitrate. In most localities, however, NH_3 is efficiently removed by acidic sulphate aerosols to form ammonium sulphate. Particulate NH_4NO_3 only forms where there is heavy NH_3 pollution and acidic sulphate aerosols have already largely been neutralised (PORG 1997).

A recent report notes substantial additional formation of ozone by reaction with nitrogen oxides from burning tropical grasslands (Pearce 1996b).

Hydrogen peroxide

H_2O_2 can be formed by photochemical processes associated with the breakdown of ozone. It is highly soluble and is considered to be a major oxidant of SO_2 in the atmosphere (Dollard & Davies 1992).

13.42   Ozone distribution

Ground level ozone concentrations throughout Europe are considered to have increased from 10-15 ppb at the beginning of the century, to a mean annual concentration of about 30 ppb in 1990. This increase is observable throughout the troposphere (PORG 1997, Fowler *et al.* 1998b).

The dependency of ozone formation on sunlight level results in a strong diurnal cyclic pattern of concentration and lesser annual cycles, reflecting the seasonal variation in day length.

Other factors affecting ozone distribution include:

• At low windspeeds, dry deposition to ground exceeds downward diffusion so that near-ground concentrations fall in light winds. Above windspeeds of 4 m/second, this effect is small.

• With increasing altitude, the diurnal fluctuation is less. This is due to enhanced windspeeds and turbulent mixing processes associated with hilly ground. The daily peak concentrations on high ground are usually only a little higher than those on adjacent low ground but the night time minima are greater so the daily average is higher.

• Average ozone concentrations in on-shore winds may be of the order 5-7 ppb higher that in air over similar inland sites.

• Ozone in urban areas with heavy vehicle exhaust emissions is lower than in rural areas because of the 'scavenging' effect of NO emission from this source. Even so 'air quality thresholds' for the protection of human health have been exceeded in the 1990s whenever there has been a long period of sunny weather (Wellburn 1998).

• The primary sink for O_3 is dry deposition. It exceeds the combined sinks through reactions in the atmosphere. Ozone is deposited on external surfaces including plant cuticles and is taken up by plants through stomata; deposition takes place throughout the year; atmospheric reactions predominate in daylight in the growing season (Fowler *et al.* 1998b).

• Quantification of ozone deposition is made more complex because of its reactivity with particles, microflora, *etc.* on leaf surfaces and emitted volatile hydrocarbons, usually leaving few traces of the reaction.

13.43 Damage to plants

In the late 1970s and 1980s, ozone was suspected as one of the important potential sources of 'forest decline' in central Europe (Binns 1984). It was postulated that ozone in combination with acid mist accelerated nutrient leaching from foliage, leading to nutrient deficiencies. Evidence to demonstrate any such specific effect of ozone has not been forthcoming. In eastern USA, acute damage associated with 'episodes' of high short-term concentrations of ozone, and reductions in growth under greenhouse or open-top chamber conditions have been frequently reported. However, clearly defined, cause-effect relationships between visible injury and growth losses due to ozone have not yet been validated (Chappelka & Samuelson 1998).

Skärby *et al.*(1998) considered that the effects of ozone were much milder, in respect of growth, than stresses due to drought or nutrient deficiency.

The bulk of damaging ozone uptake is through stomata; when stomata are open to permit transpiration, ozone can enter. Ozone is highly reactive and may react with cell membranes. Damage may include death of tissues or damage to cells and impairment of their function (Roberts *et al.* 1989).

Estimates of ozone uptake can be based on duration of stomatal opening, allowing that ozone may sometimes induce stomatal closure (Fowler *et al.* 1991). There are no residues once ozone has reacted with plant tissue.

Ozone has been long known to damage *Pinus ponderosa* in California and Weymouth pine (*P. strobus)* in eastern United States. There, economic losses due to ozone damage on cereal and other grain crops are estimated at several billion US dollars annually(Sanders *et al.* 1993). In Britain, damage is much less significant. Visible injury to peas and beans was recognised in 1976 following a record hourly mean concentration of 260 ppb (PORG 1993). Damage is also well recognised in Europe (*eg* Godzik 1997).

In Britain, 80-100 ppb is a more typical maximum summer value, such concentrations occasionally persisting for 8 or more days. The duration of peak concentrations of O_3 is usually longest in south east England decreasing to the north. Concentrations in the range 40-60 ppb commonly persist at high altitude sites for longer periods than at lower elevations (Skärby & Sellden 1984, Bell 1996). Ozone concentrations are less around towns, the gas reacting with nitrogen oxides emitted by road vehicles concentrated there.

Damage symptoms on trees

Damage on broadleaved trees appears as irregularly shaped flecks, ranging in colour from white to reddish brown (Phillips & Burdekin 1982). Conifer needles may be banded or mottled or necrotic when severely affected. Within Norway spruce needles fumigated by ozone, most damage occurred in the upper palisade parenchyma, whereas SO_2 induced more general injury to mesophyll cells (Roberts *et al.* 1989, citing Fink 1988).

Analysis of changes in forest condition in Britain showed that the three factors most influencing tree condition were soil moisture deficit (drought), ozone and sulphur deposition. In particular, ozone was linked to beech crown die-back and to reduced Scots pine flowering (Mather *et al.* 1995). Damage to Norway spruce in Germany showed a broad trend to increased damage with higher ozone levels (Schulze & Freer-Smith 1991).

Drought and ozone damage have in common that the weather conditions conducive to high ozone levels *ie* hot sunny weather under calm conditions, are also associated with high moisture stress in trees, and if prolonged, with drought. Lonsdale (1986a), reporting on health of beech in 1985, commented that '1985 was a year when drought was absent and when ozone formation must have been very limited'.

13.44   Open-top growth chamber and glasshouse studies

Open-top growth chambers have been constructed in many places, to compare the effects of different combinations of added pollutants on plant growth or to compare ambient air with air filtered to remove some or all pollutants.

Tests of added ozone, filtered air etc

The Forestry Commission has from the early 1980s studied plant performance in growth chambers.

In 1985 at three sites over three years, increased levels of ozone affected transpiration of clonal stock of Sitka spruce and Scots pine, but only when present in excess of 400 ppb (Willson 1986). In the following year, however, exposure to 500 ppb for 4 hours caused damage to foliage on both species, varying according to stage of growth of the plants. Damage was not correlated to transpiration.

In 1987, in trials of responses to acute damage, dosage rate was increased to 600 ppb for 5 hours, with Norway spruce as an additional species. Differences in within-season weather conditions between 1986 and 1987 causing, for example, differences in ripening of current shoots, was also reflected in susceptibility to ozone (Willson *et al.* 1988).

In another experiment with four different air qualities, Sitka spruce was unaffected by any treatment but Scots pine showed progressive reductions in roots and needle biomass in response to increasing concentrations of ozone (Willson *et al.* 1987). A similar response was found in the following year except that foliage damage was more severe.

Other work reported includes:

- Two-year-old seedlings of Sitka spruce in pots were exposed to ozone at up to 170 nl/l in fumigation chambers at Lancaster during the summer of 1986. No significant effects of ozone on growth were found. However, in tests of autumn frost-hardiness of plants in the following November, shoots of ozone treated plants were significantly more susceptible to frost damage than unfumigated plants (Lucas *et al.* 1988).

- Ozone had no effect either on photosynthesis or stomatal conductance of *Picea abies* one-year-old seedlings growing in drought conditions and exposed to short high concentration episodes of ozone. On Sitka spruce transplants, stomatal conductance was reduced and ozone flux was less on plants left unwatered for 7 or 14 days. Ozone flux to plants reflected its concentration (Dobson *et al.* 1990).

- Norway spruce, which for two summers had been exposed to high ozone concentrations, had at the beginning of the third growing season a much higher transpiration rate, the needles were more easily wetted, and when detached, lost water more quickly than needles from plants raised in filtered (pollutant-free) air. Such effects could reduce affected plants' resistance to drought conditions. No significant effects were found in the amount of wax covering the needle surface (Barnes *et al.* 1990).

- A larger scale open-air fumigation facility was set up at Liphook, Hampshire in 1987. The intention was for longer-term trials to be undertaken than were possible with smaller open-top chambers. Fumigation initially included two levels of SO_2 and one level of O_3 (McLeod *et al.* 1992).

- Ethylenediurea (EDU) is known to protect many plants against injury by ozone. Beech in semi-open-top chambers were injected with EDU prior to exposure to episodes of high concentrations of ozone. While the injection technique appeared to be successful, injection gave no protection compared with untreated plants (Ainsworth & Ashworth 1992).

- In a comparison of autumn leaf fall of four broadleaved species grown in ambient air and filtered air, birch, ash and American ash retained a significantly higher proportion of green leaves when grown in filtered air. Beech was unaffected (Bell 1996).

- In Switzerland, cumulative dosing with ozone corresponded with a 10% reduction of beech root biomass. Seedlings exposed to air polluted with ozone showed more severe symptoms of frost damage following a cold winter with rapid temperature change (Braun & Flückiger 1995).

Ozone and tree health
A four-year study into condition of beech considered 11 descriptors of health on sites spread widely over Britain. While appreciable changes in tree health were noted, these were often linked with heavy mast and drought. Soil

characteristics were considered. However, soil variables had no important effect on crown condition variables.

The possibilities for air pollution damage were reviewed but no direct correlation between crown condition and air-borne acidification could be found. Ozone episode damage was possible, but was also heavily correlated with drought incidence. There were no clear indications of any pollution effect but evidence was insufficient to discriminate between closely correlated possible causes for this type of damage (Innes 1992a).

The continuations of several of these trials and their implications are summarised in *Air pollution and tree health* prepared under the auspices of the UK Government 'Terrestrial Effects Research Group' (TERG 1993).

Non-specific reduction in growth due to atmospheric pollutants
Open-top chambers using filtered air

Studies of plant response to added pollutants have been complemented by studies where comparisons of plant responses to the local 'ambient' air and air which has passed through filters to remove all polluting gases and particles.

Plants in clean air in several of these have grown better than those in ambient 'dirty' air, without there being any clear symptoms of damage. These 'subclinical' reductions imply unspecified physiological effects on plants.

Scots pine, Norway spruce and beech were grown in clean air in open chambers for three years in Hampshire. *Figure 13.2*, opposite, shows the 10 - 25% increase in height observed. Responses were less on two other sites.

The effect of growing in ambient air whether in growth chambers or outside was slight (Lee *et al.* 1990a, b, Durrant *et al.* 1992). Improved growth because of higher temperatures in open chambers has been recorded elsewhere.

Physiological responses were measured on other beech grown at the Hampshire site. Stomatal water loss by the first flush of leaves was reduced in polluted air but slightly increased in late season lammas growth.

Root development of beech was also affected by air quality. The overall root dry weight of beech was greater under a 'clean air' regime. Roots of plants grown in polluted air were longer and thinner than when grown in unfiltered air. There was no further differential effect when plants were not watered and subject to drought stress (Taylor & Davies 1990). Depression of root growth had been identified in a declining Norway spruce stand in Bavaria (Schulze *et al.* 1989).

In 1988, concentrations of ozone reached 85 ppb in the south of England on at least 5 occasions. At Headley, Hampshire, rooted cuttings of 4 clones of Sitka spruce and transplants of beech were raised in filtered and unfiltered air. In unfiltered (*ie* polluted) air, leader growth of two of the spruce clones was reduced, roots of spruce and beech were longer and thinner and beech leaf area was increased. For beech, stomatal water loss initially was reduced in polluted air but for new growth later in the growing season, water loss was greater. In polluted trees, the rate at which photosynthetic substrate regenerated was greater, leading to greater maximum rates of photosynthesis.

These physiological and morphological changes were considered able to affect plants' resistance to drought stress, but had little effect on end of season plant dry weight (Taylor *et al.* 1989a & b).

When well-watered, the overall root dry weight of 'clean air' beech was the greater. There was no further differential effect when plants were not watered and subject to drought stress (Taylor & Davies 1990).

Figure 13.2   *Response of 3-year-old plants to filtered air in growth chambers at Headley, Hampshire*

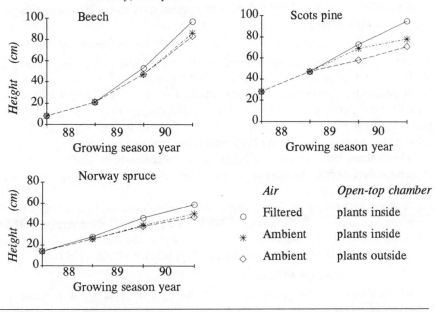

Source   Durrant *et al.* 1992

13.45   Critical levels and Accumulated exposure over threshold (AOT)

In attempting to establish 'critical levels' standards, international studies of ozone have been focused on levels causing damage and have developed the concept of 'accumulated exposure over a threshold' (AOT). The unit accumulated is a 'part per billion hour (ppb.h) over the threshold value'. ('Accumulated day degrees' for assessment of summer heat is a long-standing example of another application of the same threshold value concept.)

Two values are set for AOTs, the threshold average hourly concentration, and the number of accumulated hours over a specified period.

Against a background diurnal cyclic level of 20 - 40 parts per billion by volume (ppb) of O_3, episodes of up to 150 ppb may occur with short-term peaks in excess of 250 ppb. Levels have been recorded and averaged over alternative periods during the day, *eg* during daylight hours, or for 12 hours (0600 - 1800 hrs), or for 7 or 8 hours, *ie* covering the period of the day when concentrations are highest, or for the full 24 hours.

Values for forest trees have been discussed as part of the UN ECE programme on atmospheric pollutants and the environment, at workshops at Bad Harzburg in 1988, Bern in 1993, and most recently, Kuopio in 1996. Current values for forest trees are:

- 40 ppb as the threshold O_3 concentration in the sub-stomatal cavity (AOT40),
- 10000 'AOT40 hours' during daylight in April - September.

For comparison, the threshold quantity applied to wheat is 3000 hours during daylight in May - July.

Ozone levels in Europe

A map of accumulated ppb.h based on current thresholds has been calculated for central and northwestern Europe for 1990-94 (PORG 1997). It shows strong radial gradients, converging towards northern Italy and the northern end of the Adriatic Sea (highest values of threshold exceedance) from points along the northwest European coastline from southern Ireland to northern Norway (lowest values). The map does not cover land outside the region mentioned.

The Italian and French Alps fall in the medium-high zones, while the Erzgebirge and Sudety Mountains on the Polish/Slovak border show low levels of threshold exceedance.

No part of the British Isles shows ozone exceedance on this scale of map. However, on a more detailed map of the United Kingdom, the average AOT40 ppb.h figures for the period 1990-1994 showed almost all the land south of a line from London to Bristol to be at or over 10000 ppb.hours (PORG 1997).

Review of the AOT40 standard

Broadmeadow (1998) pointed out that AOT 40 ('Level I') is a means of mapping relative exposure but does not take into account local environmental conditions. The AOT 40 standard may be too high for Scandinavian conditions. While the existing 'Level I' maps are in general use, they are being supplemented by 'Level II' mapping, based on estimates of impacts in the field for particular species. This is made more difficult because of uncertainties about ozone 'sinks' (Skärby *et al.* 1998).

13.46   Prospects

Regional scale ozone formation, its long range transport, the role of precursors, in particular NO_x, are topics on the agenda of several international bodies (UN ECE, European Union, WHO). One estimate is that ozone pollution will increase by 15-40% over the next 30 years as a consequence of 'global warming', and the attendant increases in ozone precursors (Sanders *et al.* 1993).

A more recent analysis showed no uniform trends for Europe as a whole. Neither study anticipates marked reductions.

If critical levels or AOTs for trees and arable crops at their present levels are not to be exceeded in the long term, the route for reduction in ozone levels and their precursors can only be through international agreement, at least at subcontinental level if not global level.

In a study of the interaction between herbicides and (future) increased ozone levels, crop yields were predicted to fall. However, the response to the interaction between ozone and herbicides was specific to the plant variety; some were more responsive, others less (Sanders *et al.* 1993).

The effect of ozone on wild plants has been neglected; the limited available evidence is reviewed by Davison & Barnes (1998). See also §13.29.

The relationships between trees and ozone remain an area of intensive study.

13.5 OTHER AIR-BORNE POLLUTANTS

13.51 Methane in natural gas and land-fill gas

Methane is a major component of the natural gas associated with oil fields. It is also generated by decaying plant material in land-fill sites (§11.35). Weight for weight, it is second only to hydrogen for its yield of energy when burned (Emsley 1994b). Natural gas over the last 30 years has displaced town gas as a domestic and industrial fuel, a network of gas pipelines covering most of the UK. Methane seeping into and mixing with air in confined spaces is potentially explosive in certain proportions. Legislation to minimise direct venting and increase utilisation or flaring is expected to be implemented under the Environment Protection Act (1990) (UKGovt 1994b).

In 1998, electricity generation utilising land-fill gas accounted for 30% of operating Non-Fossil Fuel Obligation schemes (ETSU 1998a).

Damage to trees

Escapes of natural gas into soil can displace oxygen directly; alternatively, escapes may stimulate rapid multiplication of methane-destroying soil bacteria and consequent depletion of oxygen. Roots close to such escapes are likely to be damaged or killed (Phillips & Burdekin 1982, Strouts & Winter 1994).

Peace (1962) describes analagous damage due to town gas leaking underground. He notes that damage from underground escaped gas was often worst on the side of the tree nearest the leak.

See §16.31 for details of methane in the atmosphere as a greenhouse gas.

13.52 Fluorine (F)

Fluorine compounds such as hydrogen fluoride and silicon tetrafluoride are important. In the United States, they are included among the five most damaging industrial atmospheric pollutants. Damage is mostly found close to brick kilns, oil refineries and aluminium smelters (Peace 1958, 1962, Phillips & Burdekin 1982, Bunce 1989).

Air-borne fluorides are toxic to plants at lower concentrations than many other pollutants, injury being caused at atmospheric concentrations less than 1 ppb (*ca* 0.8 μg F/m³) (Weinstein, cited in Bunce 1989).

Gaseous forms of fluoride enter mainly through stomata.   Some soluble forms may also enter through the cuticle and epidermis.   Fluoride that enters the leaf accumulates at leaf margins and needle tips, concentrations of 50 - 200 ppm in foliage causing necrosis on susceptible plants (Phillips & Burdekin 1982). These authors give a table of relative tolerance of trees to fluorides.   It contains 6 conifer groups and 15 broadleaves.   Larches, pines and Douglas fir are shown as most susceptible.

As well as having toxic effect at high concentrations, exposure to lower concentrations of fluoride can depress growth.

A long-term monitoring study was set up around an aluminium smelter at Kitimat, British Columbia, because it was located well away from other sources of pollution.   Production controls reducing the emission levels by 70-80% were introduced in 1975.

For the period from before the smelter started up in 1954 until 1984, mean annual ring widths of western hemlock (*Tsuga heterophylla*) were analysed.   The mean annual ring index inversely followed emission levels.   Measured evidence of damage on trees enabled the forest to be divided up into inner, outer, surround and unaffected zones.   From aerial photographs, the areas of affected zones were: 1236 ha, 4620 ha and 5624 ha respectively.

Immediately before the reduction of emissions, basal area increment of trees in the inner zone was 40% less than in comparable unaffected growth.   However, once the emissions had been reduced to 25% of their peak levels, growth recovered, indicating that fluoride once taken up into foliage, was not recycled during senescence or by soil uptake (Bunce 1989).

In Scotland, monitoring of fluorine levels around aluminium smelters at Invergordon, Rossshire, and Fort William, Invernessshire commenced in the early 1970s.   In 1977, browning of foliage near Glen Nevis was noted 1.5 km from the Fort William smelter.   Analysis of foliage showed a pattern of high fluorine content, rapidly diminishing with distance from the smelter.   Highest values within a radius of 2 km exceeded 100 ppm, with a maximum of 250 ppm (Neustein 1970, Lines & Jobling 1974, Lines 1978, 1979).

At Invergordon, the fluorine content of spruce foliage was inversely related to the distance from the smelter there.   Changes in lichen populations within a few km of the smelter were also observed (Lines 1976, 1977).

Strouts & Winter (1992) mention fluorine damage in passing, commenting that visible damage is usually localised, near the point of emission.

CHAPTER 14

Acid rain, forest health and drainage water

'Pollution, pollution........ only don't drink the water and don't breath the air'
(Tom Lehrer)

14.1   SOURCES OF CONCERN WITHIN THE UK

14.11   Woodland and upland stream water quality

Silviculture and forest management have become involved in atmospheric environment issues in the UK as a result of:

- the practice of obtaining public water supplies for substantial areas of Britain from upland surface water catchments. In these areas, the quantity and quality of drainage water leaving the forest are of immediate importance to such water supplies. Many water catchments include forest planted as part of 20th century afforestation programmes (Calder & Newson 1979, Binns 1984).

- anxiety about the health of forests, initially in central Europe and affecting silver fir but later extending to include Norway spruce and beech. At first, 'acid rain' and SO_2 were among the prime suspects, along with ozone and drought (Binns *et al.* 1985).

- evidence of reduction or loss of fish stocks in rivers and lakes draining non-calcareous parent geological formations.

The future of woodlands in the uplands of the UK has been questioned on account of their potential 'acidifying' effect of land and drainage water and has received close scrutiny in this context, *eg* in the 1988 report of the *Acid Waters Review Group* (AWRG 1989).

These issues derive from the greater quantities of pollutants intercepted by tree cover compared with grass or heather moorland. The pollutants most affected are the oxides of sulphur, and nitrogen oxides and ammonium nitrogen, (collectively 'fixed nitrogen'). These, under ordinary meteorological conditions reach the surface vegetation, litter or soil surface as gases or aerosol particles in 'dry deposition', or dissolved in rainwater as 'wet deposition'. They are reviewed under:

- standards for stream water - critical loads & critical levels, §§14.12/3;
- water interception, §14.21;

- trees & atmospheric pollutant deposition, §14.3;
- airborne nitrogen, §13.2;
- nitrate saturation and drainage water from forests, §13.24;
- sulphur, §13.3;
- ozone §13.4;
- fluorine §13.52;
- other pollutants, §15.1;
- silvicultural operations and drainage water from forests, §12.5.

14.12   Standards for stream water   -   'Critical loads'

Standards for the protection for surface waters when abstracted for potable supply and for fisheries were set by European Community Water Directives 75/440/EEC and 78/659/EEC, recognising two levels, 'guide' and 'mandatory'. These have subsequently been refined and redefined, introducing the concept of *Critical load*. The definition of 'Critical load' adopted by the United Nations Economic Commission for Europe is:

a quantitative estimate of exposure to one or more pollutants below which significant harmful effects on sensitive elements of the environment do not occur according to our present knowledge (UNECE 1988, INDITE 1994).

A 'low' load rating implies that an area can tolerate only a small additional deposit of the relevant pollutant; a 'high' load rating, correspondingly, implies a more tolerant site.

The exact wording of the definition and interpretation of 'significant' and 'load' have been widely debated, opinions reflecting the background and values of the proponents (Bull 1991, 1992, Skeffington & Wilson 1988, Harriman 1994, CLAG 1997). Proposals include:

- for freshwaters, 'the highest load that will not lead in the long term to decline of natural fish populations';
- 'the highest deposition of acidifying compounds that will not cause chemical changes leading to long-term effects on ecosystem structure and function';
- for forests, 'loads should be such that soils are protected from long-term chemical change due to anthropogenic impact, which cannot be compensated for by natural processes' (Cape 1993);
- also for forests, 'the maximum deposition of sulphur pollutants that will not cause chemical changes in stream waters or lakes, leading to long-term adverse effects on salmonid fish', (Nisbet 1996a) who illustrated the application of this criterion.

Forests and Water Guidelines (4th Edn) (FC WG 1997) review all forest practices affecting water. For new planting proposals within acid sensitive areas, critical load assessments are recommended on a catchment scale. The Guidelines appear to have successfully addressed this and other problems relating to water and forests (Nisbet 1994a).

Standards for mapping are supported through the UN Economic Commission for Europe (UNECE 1988, Nilsson & Grennfelt 1988). Methods for preparing initial 'critical load maps' are reviewed by Langan & Hornung (1992) and outlined in conference papers (Hornung & Skeffington 1993).

Recent maps for critical loads in the UK were published in 1995 and 1997 (CLAG 1995, 1997). While such maps require thorough validation because of the small database and nature of the assumptions on which they are founded, the concept is unlikely to be abandoned.

The UK Government-supported *Critical Loads Advisory Group* in 1995 reviewed progress with alternative models for acidification processes. While their report was primarily focused on acidification by sulphur compounds, nitrogen compounds were noted as of increasing importance (CLAG 1997).

The *UK Forestry Standard* (FC 1998) recognises that it may be necessary for water regulatory authorities to take a strategic view of the cumulative effect of interception by forests of nitrogen and sulphur depositions in individual catchments. Where acidification is of concern and loads of acidifying materials are already high, their agreement to otherwise sound proposals for planting may be withheld.

See also §13.24 - nitrate in drainage of plantations, and §14.52 for details of models for critical loads and their application to forest areas.

14.13   Critical levels for airborne pollutants

For airborne pollutants, *critical levels* are the counterpart of *critical loads* for water-borne pollutants and have been defined as:

the threshold concentrations in the atmosphere above which direct adverse effects on receptors such as plants, ecosystems or material, may occur according to our present knowledge (UNECE 1988, INDITE 1994).

A sub-group of the *Critical Loads Advisory Group* reviewed *'Critical levels of air pollutants for the United Kingdom'* (CLAG 1996) and put data and responses by trees into context with agricultural crops and semi-natural vegetation.

The review gives considerable detail about the current data sources. Attention is given to upland vegetation and forests especially, because of the significance of dry and cloud-water deposition for trees. For SO_2, critical levels are considered seldom to be exceeded in the more remote rural areas. However, winter levels in towns and nearby areas are clearly higher. Special mention is made of particulate sulphate because 'critical levels' for cloud are set for this form of pollutant.

Other pollutants reviewed are ammonia, nitrogen oxides and ozone.

Critical levels of pollutant gases for trees

The earliest approach to acute damage to trees from SO_2 considered high concentration 'episodes' arising from a definable source and having effect under specific adverse conditions within a limited distance of the source. Cape (1993) pointed out, however, that high concentration episodes might also arise from diffuse-source wet deposition to montane forest exposed to high concentrations of pollutants close to cloud base. He proposed an extension of the usage of '*critical level*' to include, for montane forests, the level of exposure to pollutants as wet deposition above which direct damage to forest plants is likely.

The proposed critical level is based on observations of damage to foliage of a wide range of species exposed to acid mist at pH 2.5 - 3.5. The acidity of cloud can be estimated approximately from sulphate ion concentrations; duration of exposure to cloud currently has to be modelled using climatic data.

Application of the method suggests that many montane forests in central Europe are at risk from direct effects of fossil-fuel-derived pollutants in cloud. However, mists in the specified range have seldom been recorded in the UK.

14.2   WATER

14.21   Water interception

At the beginning of the 20th century, the accepted wisdom in relation to surface water used for public consumption was embodied in a 1902 report by the Board of Agriculture. This recommended that public corporations responsible for public water supplies should plant their catchments with trees which would 'retain the rain that falls on the area and thus regulate water supply to prevent floods and water famines, purify the water and yield regular income'. At this time, it was also thought desirable to keep people away from reservoirs, thereby reducing the risk of human-borne diseases.

In 1903, the Liverpool Water Committee responsible for managing Lake Vyrnwy, a reservoir created to supply Liverpool with water, increased the area of woodland around the reservoir so as to '..diminish the risk of infection by drainage from farms...' (Newton & Rivers 1982). In 1948, the report of the 'Gathering Grounds Committee' concluded 'We consider that water undertakers should be encouraged to adopt a policy of afforestation in all gathering grounds where soil and climate are suitable' (Newson 1985).

During the second half of the 20th century, the view that trees were the most appropriate cover for water catchments was challenged in respect of water yield and water quality. These issues have increasingly been recognised as crucial to the future management and expansion of woodlands in Britain. At the same time, modern water purification processes, in particular chlorination, minimise the risk of water-borne disease.

Scale of rainfall interception

In 1956, figures were presented indicating that a Sitka spruce plantation in a Pennines (Lancashire) forest used more water that open moorland (Law 1956, 1958), a view not immediately accepted by some foresters (Chard 1985).

Nevertheless, Law's views have been substantiated by many subsequent studies measuring interception. Typical results include:

- in Norway spruce near Oxford, almost 30% of the annual precipitation did not reach the ground   (Reynolds & Leyton 1963, Leyton & Reynolds 1964, Leyton *et al.* 1965);
- interception ranged between 27 and 38% in four crops between 14 and 29 years old (cited by Anderson & Pyatt 1986);
- at Kielder Forest, Northumberland, interception over three years in 25-year-old Sitka spruce stands averaged 29%; in a 63-year stand 1.5 km away, interception loss averaged 49%. The annual loss varied from the equivalent of 200 to over 500 mm of rainfall (Anderson & Pyatt 1986);
- at Kelty, Perthshire, interception loss was estimated at 25% of the bulk deposition of precipitation outflow (Miller *et al.* 1990a);
- over two sites in Scotland and two in Wales, interception ranged from 30 to 45% (weighted average 35%) (Newson & Calder 1989);
- the amount of precipitation intercepted as snow on branches was substantially more than from any rain shower, in one study, ranging between 40 and 69%. Sitka spruce canopy at any one time was reported to be able to hold more than the equivalent of 20 mm of rain; this could be lost by evaporation at rates of over 0.5 mm /hr (Blackie & Calder 1984, 1985, Calder 1985);
- canopy capacity for broadleaved trees is given as 1mm when in leaf and 0.9mm when leafless (IH/EA 1998);
- interception loss through the year reflects the seasonal pattern of rainfall and may commonly reach a winter peak (Walsh & Walker 1985);
- heather intercepts more water than grassland but direct transpiration rates are lower (Newson & Calder 1989).

The greater interception losses in forests are due to:

- the greater surface area available for wetting,
- the additional advective energy available because of the aerodynamic roughness and consequent air-mixing in woodland canopies, and
- the greater volume of space they occupy (Newson & Calder 1989, Miller *et al.* 1990a).

At the extreme, in those parts of wet upland regions of Britain where 75% of land is afforested, annual evaporation losses may be double those from equivalent areas under grassland; stream-flow typically may be reduced by about 20%.

The proportion of total precipitation intercepted, *ie* the 'interception fraction' or 'interception ratio' is widely used in water balance calculations. *Figure 14.1* shows interception ratios plotted against annual rainfall, for 20 conifer and 17 broadleaved forest sites. Conifer ratios are about double those of broadleaves, with only a little overlap in values. Overall, interception accounts for a higher percentage of rainfall in low rainfall areas than where the annual rainfall is high.

Figure 14.1   *Interception ratios for conifers and broadleaves*

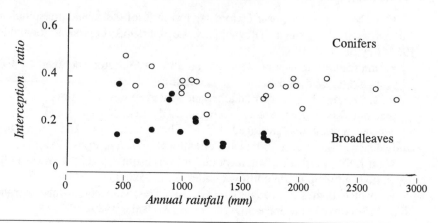

Source   IH/EA 1998

Stem-flow and drip

Interception is taken as meaning 'being retained in the canopy on leaf and stem surfaces'. However, not all water striking woodland canopies is retained; the amount is directly related to total surface area and the physical characteristics of the surface.

Excess water runs off either as drip from foliage and twigs or via the plant stem (stem-flow). Drip is normally considered part of throughfall and measured with rain falling to the ground directly through the canopy. However, because of its brief contact with foliage and stem surfaces, it may have swept off or dissolved any dust or soluble materials previously deposited there. Run-off may also dissolve any soluble material 'excreted' by the foliage and which is described in nutrient balance statements as 'lost from foliage by leaching'.

The proportion of stem-flow to the combined stem-flow+throughfall varied from 14 - 18% in pole-crop Sitka spruce and lodgepole pine to 2% in mature Sitka spruce (Anderson & Pyatt 1986).

14.22   Cloud-water deposition

Woodland canopy not only intercepts water falling as rain or snow. It is also more effective than herbaceous vegetation in intercepting fine droplets in cloud and mist, the intercepted droplets coalescing to the extent that they run off to the ground. While drip from trees on foggy days is a common experience, only in the last two decades have the quantities of water intercepted by woodland canopies has been measured.

Intercepted water has been referred to as *cloud-water deposition, occult precipitation* or *interception deposition*. Here the term *cloud-water* is used.

Quantity of cloud-water intercepted

The amount of cloud-water deposition varies according to the extent to which woodland is likely to be in cloud and fog, and is therefore strongly dependent on altitude and on relative exposure and landform.

Fowler *et al.* (1988a, 1989) estimated that in specified conditions, cloud-water deposition would amount to 0.15 mm/hour or about 10% of typical rainfall rates. The conditions were: 'forest 10-15 m tall, land over 400m, in western Britain, in cloud with wind speeds of 5 - 10 m/second'.

At Glentress Forest, Peeblesshire cloud- and rainwater were collected at 275 and 600 m elevation. Cloud base seldom occurs there below 400 m elevation so that the water collected at 600 m from cloud represented additional deposition. It was estimated that 6% of water in cloud passing through a forest canopy would be deposited there. The water deposited from cloud was the equivalent of an additional 31% of annual rainfall.

At two sites in Perthshire with similar bulk precipitation, interception deposition on the more sheltered and more exposed sites respectively amounted to 20% and 39% of the bulk precipitation (Miller *et al.* 1990a).

Pollutant deposition in cloud-water

The concentration of dissolved impurities in cloud-water is greatest around the average level of cloud base for any locality (Crossley *et al.* 1992).

The effect of topography and the associated orographic rainfall may have a differential effect according to wind direction. In Scotland, about half the observed NW-SE gradient in pollutant deposition results from the combined effect of two trends:

• orographic enhancement of rainfall increasing depositions of relatively polluted rain on high ground in the east from wind on easterly trajectories,

• westerly air streams depositing enhanced quantities of relatively unpolluted rain on high ground in the west (Weston & Fowler 1991).

Not all cloud-water intercepted by crowns becomes part of throughfall. It is subject to the same retention processes as precipitation and may be re-evaporated into the atmosphere (Unsworth 1984).

14.23   Water loss through transpiration

Water transpired by vegetation has to be included in any calculation of water balance in catchments. *Table 14.1* summarises published figures for transpiration from western and central Europe. They demonstrate less variation than might have at first been expected and show the essentially water-conserving nature of the regulation of the transpiration process through decreasing stomatal conductance with increasing specific humidity deficit (= as it gets drier!). For woodland cover, an average value of 333mm has been used in calculations of water loss balances (Roberts 1983b). Transpiration loss in young, vigorously growing 'biomass' coppice appears to be a third as much again as for older trees.

Table 14.1   *Annual water loss by transpiration from woodland & coppice*
                              (mm/yr rainfall equivalent)

Species	No of studies	Transpiration loss Mean	Range	Countries where studied	Age range
Norway spruce	5	320	279-340	Germany UK	20-100
Sitka spruce	1	340	-	UK	26
Scots pine	3	369	327-427	Germany UK	27-43
Ash	2	350	294-407	UK	45-63
Beech	4	327	283-393	Belgium France Germ'y UK	30-90
Oak	8	308	241-342	France Germany UK	18-165
Oak/beech	3	289	239-362	Holland	100
Sw chestnut copp	1	275	-	France	12
Poplar coppice	1	423	-	UK	3
Willow coppice	3	469	445*-481*	Finland Sweden	2

                    *includes soil evaporation

Source   IH/EA 1998

Water loss from understorey vegetation, where present, has also to be allowed for and may account for an increasing proportion of the water loss by transpiration from a site as the specific humidity deficit increases (IH/EA 1998).

14.24   Process modelling

'Process modelling' is the formulation of mathematical models using numerical data to simulate natural processes and predict outcomes from given data inputs. Such models may be developed iteratively, testing the model with successively wider ranges of data and including modifications whenever these improve the fit between predictions and actual outcomes.

One of the first predictive models of interception was based on studies on Corsican pine in Hampshire (Rutter *et al.* 1971, 1975). From that time, models have increasingly been able to predict water flows, incorporating catchment data for rates of water deposition, evaporation, transpiration and drainage in relation to vegetation, soil and climate.

On many sites in Britain and elsewhere, there has been good agreement between process model predictions and field observations.

At Balquhidder in central Scotland, models predicted greater loss from the forested catchment than was observed. However, the initial conclusion in adjacent catchments instrumented by the Institute of Hydrology was that 'water use by the partially forested catchment is lower than that by the unplanted control and also lower than the Penman Evapotranspiration estimate' (Blackie 1987). The site is steeper and more folded than many research sites, indicating the need for more detailed studies of actual precipitation and water loss in relation to landform (Newson and Calder 1989, Calder 1993).

Investigations at Balquhidder and their national and international significance were the sole subject of the 1993 (3) issue of the *Journal of Hydrology* (*eg* Whitehead & Robinson 1993, Johnson & Whitehead 1993).

Detection of airborne pollutants and changes in their concentration involve capture of low quantities of ions in emissions moving in large volumes of air over forests. Both the physics and chemistry of pollutant detection in air in these conditions, and the associated modelling are extremely demanding.

The 4th report of the Review Group on Acid Rain (RGAR 1997) includes a section describing seven modelling systems developed to describe and predict concentration of gases and total depositions. The results using different models vary both within themselves and when compared with measurements from national monitoring networks, models consistently overestimating sulphur depositions and mostly underestimating nitrogen. This is an active and important area of current environmental research in which forests are one of several components of regional land use. The implementation of the concept of 'Critical Loads' and 'Critical Levels' depend on a successful integration of the results of process modelling with forest management. See §14.52.

14.3   TREES AND ATMOSPHERIC POLLUTANT DEPOSITIONS

14.31   Fog

The dominant public perception of air pollution in the 1950s and 1960s was of sulphur- and soot-laden fog. Following the passing of the *Clean Air Act 1956*, mean urban concentrations of smoke in the UK between 1960 and 1993 fell from 150 $\mu g/m^3$ to 11 $\mu g/m^3$, while concentrations of smoke in rural areas fell from approximately 50 $\mu g/m^3$ to 10 $\mu g/m^3$ (CLAG 1996). Sulphur dioxide total emissions rose in the late 1960s to a peak of over 3000 kilotonnes (kt), but by 1994 had fallen by 53% to 1358 kt (RGAR 1997, Harrison 1996).

The smoke reduction was achieved by promotion of smokeless fuels such as coke and coalite and by replacing coal, first by oil and later by natural gas. For sulphur for a period in the 1960s, it was public policy in the UK to construct 'tall chimney stacks' which allowed dispersion of acidic pollutants into airstreams 100 metres or more above the ground. Even at that time, however, concern was voiced that these would be carried towards Scandinavia and come to ground as acidic deposition there (Binkley & Högberg 1997).

While solid particulate emission and to a lesser extent sulphur emissions decreased from the 1960s, discharges to the air of nitrogen compounds from industry, motor vehicles and agriculture increased; by 1970, the contribution of nitrogen compounds to acidic deposition was about a third of the total.

Sulphur oxides (SO_x), and 'fixed' nitrogen compounds (NO_x + NH_y) constitute the dominant present day global air pollutants. Recent emission and deposition budgets are given in *Tables 13.2* and *13.5*. Ozone remains a threat, future trends being dependent on future emission levels of NO_x (§13.4). Methane, although present in air in slightly larger quantities, has some influence on global warming but little on plant health (§13.51). Fluorine, though a notorious pollutant in the 1940s and 50s, is not a current threat (§13.52).

Differential exposure to pollutants

Pollutants are not uniformly distributed. They are initially concentrated round emissions sources, subsequent dispersal depending on each pollutant's 'mean residence time' (the average time taken for the quantity of a pollutant emitted at a specific time to diminish by half) and the course (trajectory) of air flow between emission and the final deposition of the pollutant.

In relation to airborne nitrogen compounds, trees growing in and near to large centres of population (urban forests) are proportionately more at risk from high levels of nitrogen oxides originating from vehicle and industrial emissions; woodland in rural areas where intensive livestock production predominates are at a greater risk of exposure to damaging levels of emissions of ammonia.

Woodlands in uplands are more at risk from increased deposition of all pollutants as a result of higher rainfall, more persistent cloud cover and higher mean windspeed. Increased deposition is usually a concomitant of increased exposure.

14.32   Scale of depositions

Airborne pollutants may reach vegetation as or in:

- *wet deposition* carried in rain, snow or hail:
- *cloud-water* intercepted by foliage and branches;
- *dry deposition* of gases, aerosols or dust particles.

Dry deposition of gases occurs while gas molecules are in contact with plant tissue. The rate of uptake depends on the ability of the tissue to absorb the gas. Uptake may be via leaf cuticle tissue or through stomata; uptake rates differ markedly accordingly.

Dissolved pollutants which are neither immediately absorbed nor washed off become more concentrated on foliage and twig surfaces as the carrier water evaporates (Unsworth 1984).

Tables 14.2 and *14.3* give figures for the quantities of airborne depositions measured in forests and moorland, and the effects of vegetation type, elevation and mode of deposition on the amounts deposited.

Table 14.2 shows, for Kielder forest, both the concentrations of pollutants and rates of annual deposition. Concentrations in cloud-water were 2 - 2½ times that in rainfall. However, the total wet depositions were 4 or more times more than cloud-water depositions except for sulphur, where wet depositions

Table 14.2   Concentrations of atmospheric pollutants in rain & cloudwater; wet & dry depositions to forest and moorland at Kielder, Northumberland (300m elevation: 1500mm annual precipitation)

Pollutant	Pollutant concent'n ($\mu eq/l$)		Pollutant	Annual deposition				
	In rain	In cloud-water		Wet Forest & moorl'd	Cloud-water Forest	moorl'd	Dry (kg/ha) Forest	moorland
Cl^-	64	166	Cl^-	49.8	8.6	1.7	2.0	0.4
SO_4^{--}	46	93	S in SO_4^{--}	13.1	6.5	1.3	3.1*	3.1*
NO_3^-	15	42	N in NO_3^-	3.5	0.9	0.2	4.2*	2.2*
NH_4^+	21	50	N in NH_4^+	4.5	1.0	0.2	9.3*	1.8*
Na^+	58	139	Na^+	28.1	4.6	0.9	-	-
Mg^{++}	11	26	Mg^{++}	3.3	0.4	0.1	-	-
Ca^{++}	10	24	Ca^{++}	6.2	0.7	0.1	-	-

Annual average atmospheric concentrations (parts per billion (10^9) by volume):

HNO_3: 0.3   HCl: 0.1 SO_2: 2.0 NH_3: 3.0 NO_2: 5.0

*Dry deposition as SO_2, NO_2+HNO_3, & NH_3 respectively.

Source   Fowler *et al.* 1989

Table 14.3   Weighted mean annual ion concentrations in bulk & total deposition, canopy throughfall and stem-flow in Norway spruce at Chon forest & Sitka spruce at Kelty forest, Central Scotland (both ~250m elevation)

| | Chon (Norway spruce) | | | | Kelty (Sitka spruce) | | | | Kelty Inputs & outputs | |
	Bulk dep	Total dep	Thro'fll	Stem flow	Bulk dep	Total dep	Thro'fll	Stem flow	Total dep	Stream wtr
	Ion concentration (µeq/l)				Ion concentration (µeq/l)				Kg /ha (1986)	
Cl	130	212	217	312	156	391	272	389	146	144
S in SO_4	56	69	92	136	60	96	164	170	38.6	34.2
P in PO_4	0.4	0.4	0.5	0.7	0.2	0.2	0.7	1.5		
N in NO_3	11	17	19	5	11	24	33	21	6.6	3.0
N in NH_4	16	18	18	11	25	26	29	15	5.7	2.7
Na	112	192	183	238	136	364	266	238	94	89
Mg	19	34	38	51	27	78	76	95	13.8	10.2
Ca	19	22	50	64	17	32	45	62	13.6	12.8
K	8	8	48	106	6	13	32	48	18.8	8.0
Al	1.1	1.6	2.8	3.6	3.7	4.0	4.9	3.8	0.6	4.6
H	22	19	16	17	32	28	56	62	0.8	1.8
Sum cations	197	293	354	485	242	542	504	710		
Sum anions	200	299	328	454	230	513	475	583		
pH	4.6	4.7	4.8	4.8	4.5	4.6	4.7	4.2		4.0
Annual average precipitation (mm)	2132	2568	1317	187	2238	3132	1359	111		

Source Miller, J.D. *et al.* 1990b

totalled twice the cloud-water depositions.

All cloud-water and dry depositions were greater on forest than on moorland.

Dry depositions of both forms of nitrogen accounted for more input to the site than from any other source.

Table 14.3 shows figures of ion concentrations in water falling onto two forests in central Scotland.   For each site, the first two columns show concentrations in pollutants in *Bulk deposition*, in this instance mainly rain, and in *Total deposition ie* 'rain + cloud-water'.   The third and forth columns show separately concentrations at the soil surface in tree throughfall and stem-flow.

For most ions, the higher concentrations in throughfall reflect the effect of rain on foliage washing off deposition from intercepted water that has evaporated, thereby adding to its initial load of pollutants.

After allowing for this concentration, potassium ion concentrations are higher than would be expected, indicating leaching from foliage; ammonium concentrations in throughfall in contrast, are lower, inferring foliar uptake.

The volume of stem-flow is only 8 - 15% of the throughfall; the high ion concentrations reflect the greater opportunity for dissolving surface deposits *en route* to the ground.

The final two columns for Kelty show the total deposition inputs to the catchment and the quantities removed over a year in stream water, expressed in kg per hectare.   There was no substantial gain or loss for chloride, sulphate, sodium or calcium ions.   The ratio of potassium in stream water to the total input is the opposite of the ratio of concentrations in bulk deposition to total deposition, implying an internal cycle of half the mobile K.

Losses of both forms of nitrogen also are less than inputs to the site in bulk deposition.

No allowance was made in the original calculations of inputs for assimilation of nutrients directly by foliage in the crown.   Any N ions taken up directly by foliage are an additional gain to the site.

14.33   Influences of topography and weather on depositions

Local topography and wind direction

Comparison of depositions in rainfall at different elevations at Glentress showed how, to a greater or lesser extent, deposition increases with elevation (Crossley *et al.* 1992).

Table 14.4 shows, for a site at Great Dun Fell, Cumbria, how concentrations of pollutants in rain also varied with elevation and in comparison with cloud-water.   Concentrations of wet deposition each are 2 - 3 times higher at high elevation than in the valley, and in cloud water compared with rainwater.   The combined effect *ie* a comparison of the difference between the concentration of an ion in rain in the valley and in cloud on the summit is given by the product of the two figures shown in the table.   For example, SO_4^{--} ion concentrations in cloud at the summit could be 2.2 x 2.0 (= 4.4) times higher than in rain in the valley.

Table 14.4　　　*Effects of elevation and water source on ratios of wet deposition concentrations, Great Dun Fell, Cumbria*

| | Ratios of concentrations | | | | | Rain |
	SO_4^{--}	NO_3^-	NH_4^{++}	Cl^-	H^+	(amount)
Elevation						
<u>Ratio</u>　Summit:valley	2.2	2.3	3.1	2.9	2.9	
Water source						
<u>Ratio</u>　Cloud:rain	2.0	2.8	2.4	2.6	3.9	2.0

Source　　　　Fowler *et al* 1988

The figures in *Table 14.4* are averages covering substantial variation. The pollutant load of rainfall and cloud depended on the wind direction and route of the air flow before reaching the collectors. Where air is essentially of marine origin, pollutant concentrations are about half those where air had travelled over major sources of pollution such as the English industrial midlands. Also, collections over a three-week period and analysed for 2-3 day periods during that time, showed a twenty-fold range between highest and lowest values. Rainfall is often but not always inversely correlated with concentration.

Effect of rain following a dry spell

Table 14.5 shows how the content of drainage water varies with the preceding weather. Towards the end of a dry period, figures for the chemical load in stream water running at 0.1 of normal flow represented the content of soil drainage water. During this time, dry deposition accumulated on foliage. The first appreciable fall of rain washed off accumulated soluble depositions on foliage; stream-flow increased back to the average, the water became more acidic and the concentration of sulphate ions increased. After the flush of washed off dry deposition passed, the stream reverted to its long-term average chemical load.

Table 14.5　*Chemical composition of an upland stream in the River Forth catchment as a result of drought followed by heavy rain　(mg/litre)*

Date	Estimated proportionate flow	pH	$CaCO_3$	Ca^{++}	Na^+	SO_4^-	NO_3^-	Cl^-
14.9.76	0.1N	6.4	2.8	2.22	3.38	5.86	0.62	5.68
23.9.76	0.5N	5.2	0.25	2.40	3.15	13.34	0.62	3.69
30.9.76	1.0N	4.5	0.0	1.74	3.54	14.69	0.62	4.26
15.4.77	1.0N	5.2	2.0	0.88	2.99	5.23	0.93	5.40

Source　　　Morrison 1987　　　　　　　　N = normal flow rate

14.34   Edge effects etc

While many blocks of woodland are on flat or evenly sloping terrain where
air flow can be studied, much of the landscape of Britain is composed of
hedgerows, small woods and shelter belts, where the edge is of much greater
significance in relation to the unit as a whole than in larger blocks. Similarly, on
broken topography, bluffs and brows of hills will receive much more wind than
more sheltered ground nearby. While these features are recognised as important
in relation to local interception of water and deposition of pollutants, there have
been few attempts to quantify them, even though edge effects such as that of salt
close to the coast are very conspicuous (IH/EA 1998).

Interactions
Interactions and co-deposition of nitrogen and sulphur compounds are
reviewed in §13.36.

Follow up
Fuller details on depositions of airborne pollutants are given in reports on
Acid Deposition in the United Kingdom. These have been published at about 3-
year intervals, the most recent appearing in 1997 (RGAR 1997, CLAG 1997).

Proceedings of a 1992 conference on '*Atmospheric pollutants and forests*' set
British problems into an international context but emphasised the importance of
site factors and regional specificity in assessing interactions (*Forest Ecology &
Management*, vol 51, 1992).

A number of phenomena remain insufficiently investigated. Gibbs & Greig
(1997) in reviewing the causes of oak die-back, for example, point out that inputs
of pollutant nitrogen may be linked with increased frost damage and bark death
both of which are a feature of the recent outbreak of die-back.

14.4   ACID RAIN AND FOREST DECLINE IN EUROPE

14.41   International concern for the health of the forest

Damage and death of trees in heavily industrialised areas of Central Europe
was widespread in the 1950s. In Germany, in the Ruhr, reduced growth and
mortality covered an area estimated at 3000 sq. km; other reports included in a
summary of conference papers describe many cases of pollution around both old
and newly established industrial and chemical complexes (Wenzel in Anon
1971a). Many of the effects observed at that time could be considered local even
though on a large scale, because damage could be related to specific local
pollution sources *eg* new power stations, chemical works *etc*. However, the
recognition of long-range transboundary air pollution and data from the
European Monitoring and Evaluation Programme (EMEP) brought into an
international forum, pollution and the health of trees in central Europe.

When in the late 1970s, extensive crown needle-loss and die-back were
reported, first in European silver fir (*Abies alba*) and then Norway spruce (Binns

& Redfern 1983, Binns 1984) fears were expressed of a broad regional decline in forest health. Damage was most severe in the Erzgebirge and Sudety mountains, the mountainous area between East Germany, Czechoslovakia and Poland - the 'Black Triangle' (Bochenek *et al.* 1997).

The symptoms in the most severely affected areas were at first called 'Waldsterben' - 'forest dying'. This was changed to 'neuartige Waldschäden' - literally, 'new types of forest damage', as there was no consistent set of symptoms involved. In Britain, the term 'forest decline' has been adopted.

Studies on the continent indicated that forest ecosystems may have become more sensitive to changes in climate and other habitat factors. At higher altitudes, continued cation leaching is feared to lead to an accelerating negative feed-back of less growth, lower stand density, lower nutrient uptake, further nutrient leaching, making acid soils even less favourable for tree growth (Schulze & Freer-Smith 1991).

The difference in rotation age of forests in Europe and in Britain is rarely commented on in the context of 'forest decline'; however, it should be noted that many crops in Britain are managed on rotations of 50 - 65 years whereas on the Continent, rotation lengths are commonly twice as long. We do not have in Britain, forests that compare in age and type with those where forest decline has been most widely observed.

Six groups of causes have been proposed to account for 'forest decline':

- bad forest practice (change from mixed broadleaved/conifer forest to Norway spruce monoculture) leading to increased susceptibility to extreme climatic events (frost, drought);
- epidemics attributable to one or more pest or disease, independent of any atmospheric pollutants;
- direct effect of locally generated toxic levels of pollutants acting on above ground parts of plants, both acutely and chronically;
- cumulative long-term deposition of atmospheric pollutants, present in concentrations below the critical level for acute damage, such effects being most obviously reflected in changes in soil acidity and nutrient status, and in ground flora;
- indirect effects brought about by deficiencies or toxicities resulting from the effects of pollutants on soil and plant biochemical processes;
- complex interactions between soil, climate and habitat, in which pollutants may or may not play a part (Schulze & Freer-Smith 1991).

The first two groups of hypotheses are not supported by observations of the distribution and symptoms of 'decline'.

The third hypothesis can be accepted for a number of specific, sub-regional cases of damage to trees. However, where these give rise to 'acute' symptoms that can be linked to specific pollution sources and pollutants, they fall outside the range of conditions thought of as 'forest decline'.

Effects of deposition on soils and changes in ground flora have certainly occurred; however, decline of trees is not consistently observed.

The fifth hypothesis could apply to soils where magnesium deficiency has occurred (§14.43); sub-acute chronic effects have been identified in experiments on small plants (§13.44) but are more difficult to recognise in the forest. Reports of improved growth following reductions in ambient SO_2 levels could be interpreted as indicating a reduction of chronic depression of growth rates.

The last hypothesis, while vague and unspecific, reflects the reality that 'forest decline' has never been a single factor phenomenon. Trees have limited ways in which they can respond to changing environments. It should be expected that under changing environmental conditions, different balances of factors may cause decline (or enhancement) of forests.

Last *et al.* (1986) related pollution damage in the more central parts of western Europe to three 'pollution climates'. While this could be a starting point for multi-factor 'integrating' approach, it has not been developed. However, if extended geographically, summaries such as in *Table 13.3* which show the different scale and proportions of N depositions within Britain could embody that concept.

14.42   Annual surveys of forest health

Decline implies adverse change. To establish what changes could be identified, annual surveys of forest health (subsequently renamed 'Surveys of forest condition') were set up in several European countries. Trees were to be assessed primarily on the condition of their crowns, in terms of foliage density, foliage colour and twig and branch habit. Site characteristics would be assessed for each stand surveyed.

The Commission of European Communities 'Forest and Silviculture Division' issued a booklet containing 120 photographs showing classes of tree health as an initial guide towards standardisation of survey criteria (CEC 1985).

In 1985, an *International Cooperative Programme on the Assessment and Monitoring of Air Pollution Effects on Forests* (ICP Forests) was launched. In Great Britain, between 1985 and 1988, national survey procedures were progressively modified so that assessment methods following recommendations under the ICP Forests programme also conformed with European Community regulations (EC 1986, 1987, 1989, Innes 1990 a & b, Innes & Boswell 1989, IPCC 1990, 1992).

Forest condition surveys in Britain

The first British survey took place in 1984. The country was divided into six zones from which survey areas were selected, taking account of:

- rainfall - above and below 1000mm annual average rainfall,
- sulphur deposition - above and below 20 kg S/ha/yr acid deposition as shown on modelled maps. This was included because of its past

association with acute damage even though it was known that depositions throughout western Europe had been declining since the mid-1970s.

Zones were sampled by elevation - above & below 244m (800ft). Crops examined were initially selected to fall in the 30-45 year age band. These were big enough to show symptoms, if affected, and not likely to come within a felling programme in the short term, so that they could be re-examined at a later date. In subsequent surveys, the age range of crops was extended and results for spruces and Scots pine tabulated as under or over 50 years old (Binns 1985, Innes & Boswell 1987).

Ozone as a pollutant had been more prevalent in Germany than in Britain (Binns *et al.* 1985). Initially, it was excluded as a site characteristic because of difficulties in mapping and because of the considerable fluctuations which characterise ozone levels near the ground. However, it was introduced later.

Assessments were extended to include nutrient status through foliage analysis, and also the extent of top-dying (Binns *et al.* 1986).

Methods and locations were modified whenever necessary to keep them in line with the ICP and European Community schemes. Some changes were to the crop characteristics assessed, others related to the sampling location, method and disposition of samples and intensity of study at sample points.

Results of surveys have been reported annually (Binns *et al.* 1985, 1986, Londsale 1986a, b, Innes 1987a, b, c, d, 1990a, b, 1992b, Innes & Boswell 1987, 1988, 1989, 1990, 1991a, b, Innes 1992a, Redfern *et al.* 1994, 1995, 1996, 1997, 1998, 1999, Gibbs 1994c, Mather *et al.* 1995, Strouts 1995b, 1996, Gregory *et al.* 1996, 1997,).

For the UK, the dominant feature affecting 'Forest condition' appears to have been climatic stresses associated with the frequency and intensity of drought and evapotranspiration. Other factors affecting crown condition included severity of exposure to strong winds and damaging storms, and the incidence of late spring frost and seed cropping. For Sitka spruce, incidence of green spruce aphis contributed to reductions in crown density, crops being slow to recover (Redfern *et al.* 1999). For beech, a marked deterioration of crown condition of beech in 1995 was associated with heavy mast production (Redfern *et al.* 1996).

In a study of health of beech, based on measurements of annual shoot increment of leading shoots, Lonsdale & Hickman (1988) found that over the period 1955-1985, there was a general reduction of shoot length in the trees studied. Superimposed on those reductions was clear evidence of further short-term reductions associated with summer drought. Gradients of environmental N deposition and concentration of atmospheric S did not appear correlated to the pattern of shoot growth.

In respect of the general condition of trees throughout Britain, the role of pollution is probably small. There was no evidence that forests in the UK have been declining directly because of sulphur or nitrogen deposition due to air pollution (Innes 1992b, Mather *et al.* 1995).

Acid rain and mycorrhizas

Concern had been expressed that continuing acid deposition would reduce the effectiveness of mycorrhizas and thereby contribute to forest decline.

In one investigation, root-tip mycorrhizas were examined from three-year-old Scots pine grown from seed on humo-ferric podzol monoliths in lysimeters. For two years before sowing and for the period of the trial, the monoliths were treated, either with simulated rain or rain to which sulphuric acid had been added to bring it to pH 3.

Root tips with six mycorrhizal types were found. For three corralloid types, the number of root tips in the top 6 cm was reduced by 30-50% on the acid-treated plots; below this depth, differences were not significant. The number of non-mycorrhizal root tips was less than 1%.

Non-mycorrhizal roots were more prevalent on the acid plots. The concentration of aluminium ions in the soil solution from acid treated plots was approximately 8 times the concentration from the plain water soils. This was considered to be the most likely explanation for the differences in numbers of the different mycorrhizal types (Dighton & Skeffington 1987).

14.43   Forest decline in Europe

Magnesium deficiency in Germany

In the UK surveys of tree health, yellowing of Norway spruce foliage has seldom been recorded. On the Continent, however, it had been one of the major indications of 'decline'. In Germany, careful analysis found five types of Norway spruce forest 'decline'. Type 1, associated with yellowing of foliage of trees at higher elevations in the Black Forest and Bavarian forest has been identified as due to magnesium deficiency associated with low available soil magnesium. *Figure 14.2* overleaf illustrates the causes and relationships between deposition, species, and foliage deficiency symptoms (Roberts *et al.* 1989). This cause of yellowing had previously been suspected from the evidence of foliage analysis (Binns & Redfern 1983).

Analyses of the occurrence of yellowing in trees assessed in Britain had not found any strong relationship with Mg deficiency (Mather *et al.*1995). The maritime position of Britain and the presence of magnesium ions of marine origin in atmospheric depositions appears to have been sufficient to minimise the occurrence of this type of decline in the UK.

Other symptoms

Other types of 'decline' recognised in Germany are:

- thinning of tree crowns at medium-high altitudes;
- needle reddening of older stands in the foothills of the Bavarian Alps;
- needle-yellowing on trees at higher altitudes in calcareous Alps
- foliage thinning of trees in coastal areas.

Figure 14.2 *Causes & relationships in Type 1 forest decline & Mg deficiency*

Causes	Effects			
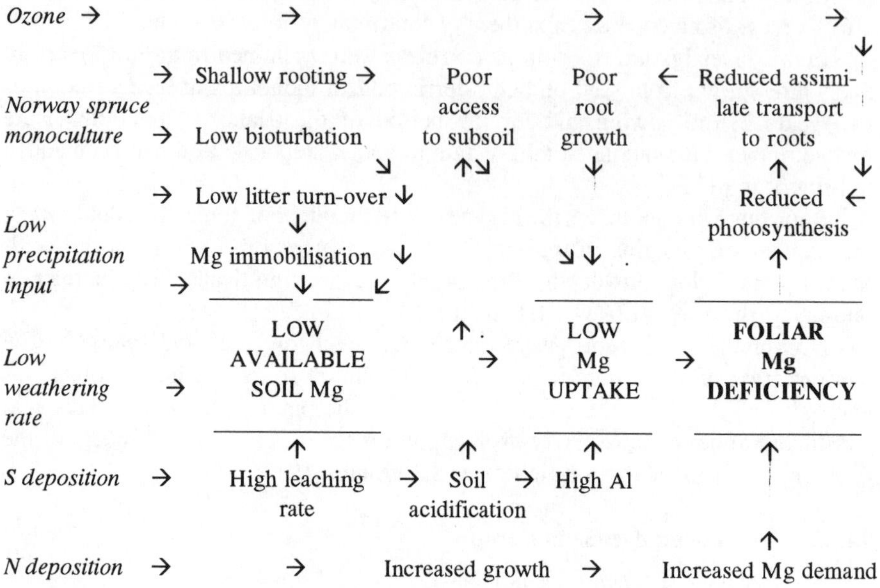 Ozone →	→		→	→ ↓
	→ Shallow rooting →	Poor access	Poor root	← Reduced assimi-
Norway spruce				late transport
monoculture →	Low bioturbation	to subsoil	growth	to roots
	↘	↑↘	↓	↑ ↓
	→ Low litter turn-over ↓			Reduced ←
Low	↓			photosynthesis
precipitation	Mg immobilisation ↓		↘↓	↑
input →	↓ ↙			
	LOW	↑	LOW	**FOLIAR**
Low	AVAILABLE	→	Mg	**Mg**
weathering →	SOIL Mg		UPTAKE	**DEFICIENCY**
rate				
	↑	↑	↑	↑
S deposition →	High leaching	→ Soil →	High Al	
	rate	acidification		
				↑
N deposition →	→	Increased growth →		Increased Mg demand

Source	Roberts *et al.* 1989

In West Germany, damage classes by forest districts were compared with measurements of combined deposition of S and N pollutants and with summer concentrations of ozone. Increasing damage was associated with increasing deposition. In a multiple regression, acidic deposition and ozone accounted for about 50% of the variation.

In the longer term, in Germany (and Scandinavia), forest soils have become more acidic over recent decades. Consequences include:

- reduction in base saturation of soils;
- increased mobility of aluminium ions, (§14.44).

Deposited ammonium may also interfere with cation uptake (Schulze & Freer-Smith 1991).

Following the political division of Czechoslovakia, assessment of forest health in Slovakia showed that over the 10 years between 1985 and 1994, a slight reduction in the level of defoliation in a wide range of species was detectable. At the same time, however, the proportion of 'salvage cutting' in the country's total annual cut had risen from 29.7% to 59.6%, the amount of the annual salvage cut ranging during this period from 1.7 to 2.9 million m³ (Oszláni 1997). Over the period, SO_2 emissions from Slovak industry had diminished substantially and NO_x emissions slightly.

A survey of forest health in Poland disclosed highest values for SO_2 deposition in southeast Poland. These were associated with higher sulphur concentrations in spruce needles than in needles from trees growing in less polluted mountains to the east (Grodzinska & S-Lukaszewska 1997).

High needle sulphur had previously been described around pollutant sources in Austria and Yugoslavia (Donaubauer & Stefan 1971, Kerin 1971).

In Hungary, although there has been extensive die-back, especially of oak, the main cause is thought to be lack of water, possibly with pollution as a predisposing factor (Szepesi 1997).

Forest decline and improved rates of growth

There have been a number of reports that, in spite of reductions in crown density characteristic of 'forest decline', forest current annual increment has increased above what had been expected from forest site quality classifications (Kenk & Fischer 1988, Spiecker *et al.*1996, Binkley & Högberg 1997).

One hypothesis is that, with the diminution of SO_x and possible increase in fixed nitrogen and CO_2, trees are responding to the added nitrogen. Another is that trees are responding to a reduction of litter removal for farming rather than any effect of atmospheric N or reduction in S. See also §16.29.

Infra-red photography to record forest condition

A programme to develop colour infra-red (CIR) photography as a means of determining forest condition started in about 1990. In the Republic of Ireland, Stanley *et al.* (1996) found the technique useful for identifying nutrient deficiency and incidence of 'top-dying' of Norway spruce. Their conclusion was that, organised with links to geographic information systems, CIR photography could provide a valuable means of obtaining information on the distribution of visible symptoms of disease and damage quickly and less expensively than by ground survey.

14.44 Soil aluminium

Aluminium and tree health

Root death due to high concentrations of labile Al in the soil was put forward in 1980 as a possible cause of 'forest decline' of beech in Germany (Ulrich *et al.* 1980). This followed a lengthy period of observations on changes in loess soils under beech. As these gradually became more acidic, the soil aluminium increased and the beech progressively looked less healthy.

This hypothesis has been taken seriously by many investigators. However, while there has been clear evidence of acidity at levels which could mobilise Al and many studies have shown increased Al in soil profiles under spruce (*eg* Reynolds *et al.* 1992), overt symptoms of ill-health are not reported.

Aluminium in stream water and in soil profiles

Soil acidification caused by water-borne SO_4^{--} is commonly linked to mobilisation of aluminium ions in the soil. The increase in the NO_3^- in water draining from clear felled sites on acid soils can also be accompanied by increased inorganic aluminium and more acidic outflow.

Ormerod *et al.* (1989) reported a study carried out by the Welsh Water Authority in the mid 1980s, where weekly stream water samples were taken for a year from 117 sites in Wales and analysed for 'filterable' aluminium. Sites were grouped by 'water hardness' and the amount of forest cover in the catchment supplying each sampling point was assessed. For each of the hardness groups, stream water aluminium concentration increased and waters became more acidic with increasing forest cover. Sulphate values were not given in this study.

In a 10-year study of stream water in streams feeding Loch Dee, Galloway, effects of Sitka spruce plantations in the thicket stage (age 6 - 17) on stream water chemistry were small. Increased scavenging due to trees on this site was predicted to be small. Over the period of the study, there were small declines in aluminium, sulphate and nitrate concentrations in stream water (Nisbet *et al.* 1995a).

At Gisburn Forest (*Table 12.4*), substantial differences in pH, Al, sulphate and nitrate were recorded in water taken from the surface of the forest floor and in the A and BC horizons under pine, spruce, oak and alder.

Soil solution in the A horizon was markedly more acidic and contained more available aluminium that in the B horizon. At this site, the pH of the soil solution ranged between 5.0 and 5.3 in the B horizon soil under 3 species, but was between 3.8 to 4.5 in throughfall, forest floor and A horizon water. The pathway of drainage from such profiles is clearly critical to the water quality of drainage water from the site.

At this site, the nitrate content of water in the B horizon under the alder was over 8 times greater than under any other species

At Plynlimon, mid-Wales, in soil water from the B horizon of a podsol where trees had been clear-felled, nitrate concentrations in the first two years following felling rose from being insignificant to levels that were more than twice those of sulphate, whereas sulphate and chloride levels decreased by approximately 30%. Aluminium concentrations increased by 60% by the end of the second year.

There was no similar change in concentrations in water from the B horizon of gley soil. Increased nitrate concentrations were not detected in the two years following felling; concentrations of sulphate, chloride and aluminium decreased.

At another site at Plynlimon in stream water from felled areas, nitrate levels increased and chloride and sulphate levels decreased slightly, while aluminium concentrations were not affected (Reynolds *et al.* 1992).

At Kershope, north west England, in ditch water over a four year period, nitrate levels increased three-fold in the year following felling and then slowly decreased; sulphate and chloride levels decreased progressively from the second year. Aluminium concentrations also decreased from the second year.

At Aber, north Wales, total aluminium in soil water from the B horizon was strongly correlated with nitrate levels. The latter rapidly increased with clear felling of the forest crop (Emmett *et al.* 1995). See also §13.24.

Calcium + magnesium/Aluminium ratio

(Ca+Mg):Al ion ratios have been proposed as a yardstick to assess the nutrient status of soils in relation to potential decline, on the hypothesis that over a certain range, Ca and Mg concentrations are linearly related to uptake by plants, and that in acidified soils, the ratio between (Ca+Mg):Al may be critical at a value of 1:1.

In six oak and six Sitka spruce plots which had been assessed annually for forest condition, the (Ca+Mg):Al ratio was not correlated with any of the indicators of decline (Freer-Smith & Read 1995).

Hunter & McDonald (1992) drew attention to studies showing elevated levels of aluminium in drainage from heather and bracken moorland where no trees had been planted, and stressed the need for more thorough examination of the natural processes of acidification and Al mobilisation, to back up recent empirical observations. In particular, more attention needs to be given to the function of Al/organic matter complexes in acid soils, in the context of processes which have given rise to soil organic matter 'pans', increased atmospheric nitrogen depositions and the role of mycorrhizal fungi as a protection against Al.

Accepting that monomeric Al+++ is potentially toxic to plants, there is evidence of some root reduction associated with increased Al levels. However, Sitka spruce, Douglas fir and western hemlock are reported as tolerant to Al in solution. Al-organic matter complexes appear to have little effect on root structure and vitality. An overall summary of factors affecting plant health is given in *Air Pollution and Tree Health in the United Kingdom*. This report concludes that there is little concrete evidence of pollutant-induced changes in soil chemistry adversely affecting tree health in the UK (TERG 1993).

14.5   ACIDIFICATION & SURFACE WATERS

14.51   The UK water resource

The UK water resource includes about 1500 river systems with 200 000 km of water courses and some 5000 lakes and reservoirs. Below ground there are major aquifers, ranging from the chalk, geologically older sands and gravels (Devonian, Bunter and Keuper sandstones, Greensands *etc.*) to recent river gravels.

Water from upland reservoirs and lakes, fed by gravity direct to urban and industrial users, requires filtration and disinfection but often little else. The cost (mid 1980s) of processing a megalitre (one million litres) was approx £5 compared with £20 - £50 for water extracted from aquifers or lowland rivers.

The proportion of water used for potable water supplies in relation to measured run-off varies from 1.4% for Scotland, 2.4% in Wales, 5.4% for the west of England to over 20% for southeast England (Hornung & Adamson 1991).

Demand for potable water has been growing at about 1% per annum, most growth being in England and Wales. There is limited scope for increase in the south and east of England, natural flow in dry period in rivers is getting close to the allowable minimum and scope for abstraction from aquifers is limited.

The value of recreational uses of surface waters is not readily identifiable. The value of angling in terms of licences and rentals, manufacture and purchase of tackle, associated tourist spending *etc.* in total is believed to amount to several hundred million pounds annually. Two thirds of Scottish streams in aggregate are designated as 'salmonid' (salmon and trout) waters but less than 0.5% as 'cyprinid' waters (coarse fishing waters). Any substantial deterioration in stream quality leading to loss of fisheries would have a serious economic and social knock-on effect. The most recent threats to salmon, however, appear more to be linked to disease, louse infestation and overfishing by off-shore commercial fisheries than anything to do with land use (Gammon 1998).

In Scotland, out of 173 major lochs surveyed for the impact of human activity, 82% were classed as unaffected, 13% were nutrient-enriched and 5% have become acidified.

While nutrient enrichment can follow the application of fertilizers in upland catchments, its scale in forestry is small in relation to agriculture. Acidification is the more relevant problem.

Hypotheses put forward to explain the observed acidification of many surface waters have included:

- afforestation,
- acidic deposition,
- natural soil processes,
- heathland regeneration.

All have to be reviewed in the context of

- the chronological evidence for when acidification started, §14.52,
- the characteristics of catchments most likely to be affected, § 14.53,
- the possibilities of alleviating the effects of acidification § 14.61.

14.52   Acidification, the diatom chronology and critical loads

Changes in the acidity of freshwater lakes affect many of their diatom microfauna; their characteristic skeletal remains accumulate at the bottom of lakes in deposition layers which can be dated. The range and abundance of species can be used to plot changes with time which reflect the conditions of the water in which they lived. Where stratified sediments from the lake can be recovered, analysis of strata enable dates to be identified.

Any change in the composition of the range of diatoms present may be significant, not only for the diatom assemblage itself, but also as a marker relevant to other species. For a lake to carry a stock of fish, the water acidity should not exceed pH 5.3 - 5.5; diatoms can indicate when this threshold has been crossed.

Diatom deposits in acidified lakes in south west Scotland show acidification to have been in progress since the middle of the 19th century. This pre-dates any major afforestation, is too rapid to be attributable to natural soil evolution, and cannot be linked to farming practice or increasing cover by *Calluna* heathland (Batterbee *et al.* 1985, 1995).

In many other lakes and lochs also, the start of acidification can be traced back to that time. Its cause is attributed to the impact of the burgeoning industrial revolution and depositions originating from the uncontrolled emission of sulphurous gases at the time.

For other organisms, the critical pH has not been studied and may be quite different. At the same time, if conditions suitable for fish are recreated, any locally extinct organisms that were exclusively part of the fishes food chain, may become re-established. For example, because of reduced sulphur deposition, lake water in the Round Loch of Glenhead, Galloway, became slightly less acidic (0.2 pH units). This change was accompanied by a recovery of diatom flora (Allott *et al.* 1992)

Critical load models for deposition

The studies on diatoms and the establishment of a reference point of direct relevance to water resource managers have been applied to predict future trends for critical loads for sulphur deposition, and to a lesser extent for nitrogen deposition.

To determine critical loads based on acidity as shown by change in diatom distribution, 40 lakes, 30 of which were in Scotland or the north of England, were sampled. The ratio of stream water calcium to total S deposition provided the basis for segregating acidic from non-acidic waters, the discriminating ratio being 94:1::Ca:S. Critical loads were calculated on the basis of the amount of S deposition required to reach the 94:1 ratio (Batterbee *et al.* 1995).

Based on this work, change in diatom fauna indicating a change to acid conditions unsuited to fish has been taken as the first baseline for determining critical loads for freshwater bodies (Freer-Smith 1995b).

The frequency of episodes when the water was on the acid side of the threshold, and the associated Ca and Al critical levels during those periods, has also been used (Harriman & Christie 1995).

Other models have been described that:
- allow for changes in rates of acidic deposition over time (Ferrier *et al.* 1993, Jenkins 1995, Ormerod 1995);
- may be better able to predict future critical loads either assuming deposition continuing at present levels (Waters & Jenkins 1992) or anticipating predicted decreases;
- introduce data from other biological indicators (Juggins *et al.* 1995);
- propose means of predicting calcium and hydrogen ion contents of streams based on chemical parameters of stream water and the chemical characteristics of the upper and lower soil horizons (Billet & Cresser 1992).

The existence of these different models and approaches illustrates how the science and art of modelling is still in a developmental state, needing refinement and field verification before any one model can be taken as more than an indicative approximation.

Nevertheless, in practice, where scale of new forest planting or restocking is a current issue, the threshold pH 5.3 associated with diatom change and the presence or absence of fish, has been taken as a current working yardstick for assessing the effects of forests on stream water in respect of fish and the appropriate critical load (Nisbet 1996a).

14.53   Susceptibility of stream water to acidification

Geology, weathering and soil drainage water

The acidity of water draining from permeable soils may be reduced wherever weathering soil minerals release calcium and other mineral cations.   For freshwaters in northern Europe, estimates of *Acid neutralising capacity* (ANC) based on weatherable subsoil minerals have been used for classifying soils for their critical load (Henriksen *et al.* 1986), based on the principle that for any given pre-determined water condition, excess base cation production in a catchment should be equal to or greater than acidic anion input.   Harriman and Christie (1995) describe some practical problems in its application.

Water hardness has been used as a alternative discriminant.   Values of > 15 mg, 10-15 mg and < 10 mg $CaCO_3$ per litre total hardness were used to differentiate between waters in relation to effects of forest cover and extent of mobilisation of aluminium in Welsh soils (Ormerod *et al.* 1989).

Streams and lakes most susceptible to further acid deposition are those where acidic, poorly buffered soils are prevalent and where any acidity in soil drainage water has not been neutralised while passing through the soil.   They are commonly associated with non-calcareous underlying geological formations and with soils of low % base saturation.   A map showing distribution of soils classified on the basis of % base saturation is given in Figure 5.8 in *Air Pollution & Tree health in the UK* (TERG 1993).

Acidification and sulphur

Preceding sections gave details of studies, some on nutrients and fertilizer usage, others on plant nutrition or nutrient cycles and some relating to effects of pollutants. Few give a complete picture of movement of the more important nutrients and pollutants, even for one site.

Table 14.3 (p.394) is one of the more complete tabulations and shows for Kelty Forest, that total deposition of SO_4^- is closely balanced by its outflow in stream water. At the same time, it is accompanied by increased levels of aluminium and hydrogen ions in acidified stream water.

This may be taken as typical of the effect of sulphate deposition. Most sulphate deposition passes through the woodland canopy and the soil and enters drainage water; it is neither converted nor stored to any extent. Unless there is a substantial reserve of calcium carbonate-rich material in the lower horizons of the soil, where there are high rates of acidic sulphate deposition, drainage water will be acidic. Where there are susceptible waters therefore, the prime interest is to avert or at least minimise further acidification.

The long term prospect is of a diminishing problem; SO_2 emission levels have fallen since 1970 and are predicted to continue to be reduced. This trend is of great importance for forest and stream water management.

Recent decreases in acidic depositions in North America and parts of Europe are recognised as having initiated a reversal of acidification and some recovery of affected ecosystems. However, increased nitrogen depositions may be on such a scale as to offset such recovery (Wright & Hauhs 1991).

Acid formation following drying out of organic deposits and oxidation of captured reduced S has also been reported (§13.37).

Acidification and nitrogen

Increasingly NO_3^- is being recognised as a potential additional contributor to future stream water acidification. Maps estimating deposition of NO_x and NH_x ($=NH_3 + NH_4^+$) have been published and updated frequently over the last 15 years (*eg* Derwent 1986, INDITE 1994, RGAR 1997).

Nitrogen deposition and its movement through plants and soils is discussed in §13.2.

The evidence is reasonably consistent that NH_x deposition is seldom directly linked to NO_3^- in drainage.

For the highest deposition rates of NO_x on acid soils, deposition is clearly reflected in increased stream water NO_3^-; at lower rates, effects are at best indirect.

The amount of nitrate in drainage from forests can be influenced by forest management, especially harvesting operations when nitrate loss from the area felled can rise sharply and persist for 3-4 years.

Preliminary 'critical load' maps for nitrogen have been published; these specify values in the range <2.8 to >28 kg N/ha/yr, most woodland areas being placed in the band 7 - 28 kg N/ha/yr (INDITE 1994).

14.6   EFFECTS OF FOREST PRACTICE

14.61   Off-setting the effects of acidification

Liming in acid catchments

The use of ground limestone or chalk to 'sweeten' the soil is of very long standing. The practice has usually been applied to reduce soil acidity for the benefit of one or a group of crop plants.

Several sites on non-calcareous parent material were selected in the mid-1980s in Wales and Scotland to study the effectiveness of liming to off-set acidification. The inter-relationships with water quality, geology and land use were examined, testing alternative patterns of distribution of lime (Carnell 1986, Nisbet 1989, 1993). Further monitoring was continuing; however, interim conclusions were:

 • whole-catchment liming intended to improve reservoir water quality, while it reduced surface water acidity in the short term, could adversely affect acid-loving moorland plants and was too expensive. Part had no effect on water quality; the effective part was that applied to areas forming the main source of run-off in periods of high flows. Treatment of just these areas with powdered limestone has given promising results, remaining effective after 8 years.
 • liming bank-side strips is not effective; too little of the catchment water passes through the treated ground;
 • treatment of tree-covered areas requires helicopters and pelletised material in order to penetrate the canopy and minimise drift;
 • how best to apply lime to flowing waters in a way that can alleviate acidity at times of high flow remains under investigation (Nisbet 1996a).

In a general revue of the effects of nitrogen deposition in Swedish forests, Binkley & Högberg (1997) report varying results from the effects of adding lime to growing forests. Especially in the more northern countries, growth rates were commonly depressed by 5 - 20%. In Bavaria, Germany, a growth response was obtained. While some desired effects on soil water were obtained, the lime in many cases also materially affected organic matter mineralisation.

Olen (1991) reviewed the effects of liming acidic surface waters and responses by fish, mostly in the United States but also elsewhere.

Restricting plantation areas

Recent practice has been to assess current and future rates of SO_4 deposition for the forested area of a catchment and to compare this with the acid deposition that would maintain stream water on the less acid side of pH 5.3. One recent case-study involved an acidified catchment where for the period up to 1990, non-marine sulphur deposition was 45-50% above the critical load level. It was forecast that if SO_2 emissions continue to drop according to national predictions, and if stream water quality responds immediately and quantitatively, the S-deposition would fall below the critical load level by between 2001 and 2005.

The effect of the forest cover should result in no more than a two year delay in meeting the critical load (Nisbet *et al.* 1995b).

Many new planting proposals in parts of Wales and southwest Scotland are scrutinised for their potential long-term effect on stream water quality (Nisbet 1996a), a practice also encouraged in the *UK Forestry Standard* (FC 1998).

14.62 Forest operations and water quality

Several aspects of land and forest management affecting water quality are discussed in other sections or chapters: -

- nutrient loss in drainage water in relation to forest operations §12.5;
- critical load standards for stream water §14.12;
- scale of water interception by forests §§14.21, 14.22;
- deposition of pollutants, scale, effects of elevation, increased interception by forest canopies §14.32 *et seq.*;
- pesticides in woodlands §4.29;
- soil erosion §14.63;
- nitrate in drainage water from forests, including felling areas §13.24;
      (for figures for sulphate see tables of other pollutants or nutrients).
- use of salt for de-icing §15.2;
- casual pollution from accidental spillage *etc.* §15.3.

14.63 Sediment and soil erosion

Forest operations, through

- initial cultivation,
- drainage,
- creation of access roads,
- excessive wheel loadings and damage to soil by harvesting equipment

may expose soil and increase sediment load (Blackie & Calder 1985, Moffat 1988, 1989, Soutar 1989a & b, Worrell & Hampson 1997). Past failure to appreciate the vulnerability of soils, and poor design of drainage systems and harvesting track networks has led to excessive scouring of drainage channels and wash-off from forest roads.

Increased sedimentation due to forest ploughing, and to road making and use have equally been criticised (SEPA 1996).

Ground vegetation on the banks of minor streams in the forest may also be suppressed by closed forest canopy, reducing its protective value and increasing the risk of erosion (Mills 1980).

Comparisons of sediment load have been included in some water catchment studies. At Balquhidder, in a comparison of undisturbed moorland and areas with clear-felling and ploughing for new planting, suspended sediments in stream water were 595% and 121% of their former levels. The increase linked to clear-felling was associated with heavy use of forest roads (Johnson 1993).

There are no data comparing forests in Britain with arable farming. Globally, however, major losses of soil are associated with farm crop production on soils formerly covered by natural woodland and grassland (Fu 1989).

Soil erosion because of its small scale and localised occurrence within the forest has little effect on the nutrient status of stream water. Detrimental effects are found where sediments choke fish spawning areas and where accumulated silt reduces the capacity of reservoirs and other impounded waters.

Recommendations to minimise erosion by design of land drainage, maintenance of streamside vegetation cover, construction of sumps in drains on erodible soil *etc.*, are currently available *Forests and Water Guidelines 4th Edn* (FCWG 1997) and *Forests and Soil Conservation Guidelines* (FCSG 1998). See also §12.41, Mills (1980) and Pyatt & Low (1986).

CHAPTER 15

Mineral & other pollutants

15.1   MINERAL POLLUTANTS

15.11   Introduction

Mineral pollutants affecting trees fall into three groups:

- those derived from naturally occurring minerals within the soil and which may be found *in situ*, or may have been redistributed through mineral extraction or industrial processing.   These largely consist of materials containing heavy metals; see §15.12;
- particulate matter deposited as radio-active dust; see §15.14;
- salts deposited under storm conditions as sea spray, or intentionally deposited to disperse snow and ice from roads; see §15.2.

For pollution from forest operations, other than use of pesticides, see §15.3.

15.12   Heavy Metals

A limited number of metals or their ores are hazardous. Soils containing enhanced amounts of such elements may constitute an environmental hazard through entering animal and human food chains, or through physical contact or through toxicity to plants.

Most of the potentially toxic elements (PTEs) are metals of high atomic weight and are commonly referred to as 'heavy metals'. They occur in soils:

- as a direct result of metal ore mining, extraction and processing;
- indirectly through discarding of waste products, surpluses *etc.*;
- through relocating accumulated wastes, *eg* canal and river dredgings.

In urban areas, lead, cadmium, zinc and copper in particular are found at higher concentrations than in rural soils as a result of burning fossil fuel, vehicle emissions and industrial emissions.

Peace (1962) commented that soils contaminated with heavy metals are often associated with stunted growth of agricultural crops but that while similar toxicities certainly occur in forests, little information is available either on their frequency or the specific symptoms they produce in trees.   His observation that they are not widespread or important in rural parts of the UK reflects industrial history.   Metal-rich areas drew people to them and became heavily populated industrial areas. As the mineral resources were exhausted, sites were abandoned, often leaving a legacy of hazardous toxic residues.

Because of the potential risks to food supplies and a desire to foster re-use of abandoned (brownfield) sites for industry, recreation or housing, there have been many national or more localised surveys of concentrations of potentially toxic elements in soils (*eg* Welsh Office 1975, Davies & Roberts 1978).

Guidance has also been given on use of potentially contaminated materials on agricultural land (*eg* Chumbley 1971, ICRCL 1990), and on assessment of contaminated land with potential for redevelopment (ICRCL 1987). Results of surveys into the distribution of PTEs and their uptake by vegetables and cereals are briefly summarised by Bridges (1989).

Threshold values for heavy metals considered an environmental risk if present in excessive quantities are summarised in *Table 15.1*.

Limit values which should not be exceeded in soil when using sewage sludge in forestry are given in *Table 11.6* in §11.23.

In addition to total content of a heavy metal in a soil or a waste, their 'availability' or solubility in its environmental setting has to be determined. This may depend both on the chemical form in which it exists, particularly for chromium and mercury, and the physical form of its occurrence. Metals in slags, for example, are generally not as available as in fresh mining wastes or tailings.

Table 15.1   *Heavy metals - soil contamination thresholds   (mg/kg dry matter)*

Heavy metal *(Potentially toxic element)*	*Background levels*	*ICRCL thresholds for contamination*		
		uncontaminated soil	*garden soil*	*public open space*
Primary hazard to human health				
Arsenic (As)	0.1-4.0		10	40
Cadmium (Cd)	0.01-2.0	3	3	15
Chromium (Cr)	5.0-1500 (total)		600	1000
Lead (Pb)	2.0-300	300	500	2000
Mercury (Hg)	0.01-0.5		1	20
Selenium (Se)			3	6
Primary hazard to plant health				
Boron (B)				3
Copper (Cu)	2.0-250	240		130
Nickel (Ni)	2.0-750			70
Zinc (Zn)	1.0-900	1000		300

Sources   Bridges 1989, ICRCL 1987, 1990, Palmer 1995, Bradshaw *et al.* 1995

Iron is not considered a potentially toxic element even though some ironstone wastes and iron-bearing colliery wastes are hostile to plant life. This is due to iron pyrites (FeS_2) which, when exposed to air and moisture, oxidises to form soluble ferrous salts and then ferric compounds. On spoil heaps, the surface material becomes extremely acid; water from abandoned pyrites-containing coal mines causes severe problems for public water supplies and rivers because of the acidity of the sulphur acids formed and the unsightliness of the hydrated iron oxides carried by water coming out of the ground. See §11.34 *Ironstone*.

Sites contaminated with heavy metals

These include:

* mining and quarrying spoil, and sites where ores have been separated from the associated non-ore-bearing rocks. Particularly high concentrations may be found when ore has been crushed and mineral-rich particles separated by water and gravity, leaving local high concentrations in 'tailings',
* manufacturing sites where metals have been processed,
* land within range of fall-out from airborne emissions of metalliferous dusts and gases, deposits often being heavier on the downwind side of the prevailing wind,
* sites where worked out and waste materials have been dumped,
* farmland treated with sewage sludge (§11.21).

Natural outcrops of metalliferous ores are too local to be significant for tree growth, and, if present in any quantity, are likely to have been worked.

Identification and treatment of contamination

Bare patches of discoloured soil, poor growth *etc.* may indicate concentrations of toxic materials. However, in most sites, the exact nature and distribution of potential contaminants is uncertain, detailed survey being necessary to identify the extent and concentration of toxic materials.

Risk of exposure to potentially toxic elements through food chains, contact on recreation areas, leaching into potential drinking water sources *etc.* may be reduced by:

* physical removal as special controlled waste to a site licensed to receive it (§11.35);
* encapsulation with sufficient soil that vegetation can grow satisfactorily without breaking the surface seal over the potentially toxic material (§11.35);
* burying toxic material *in situ* under a sufficient depth of soil to enable plants to grow;
* tree planting or natural vegetation colonisation without any pre-treatment of the site.

Damage to trees directly attributable to heavy metals

In considering the causes of poor growth of Sitka spruce on the South Wales coalfield, soil solution was extracted and used in tests of grass seedling growth.

Several heavy metals were shown to be present, and there were indications of better growth with lower concentrations of metals, in particular copper. However, it was concluded that there was little to suggest that heavy metal toxicity is involved in the decline of Sitka spruce (Skeffington & Graham 1995, Burton & Morgan 1981).

An investigation of poor growth over metalliferous mining wastes in Cornwall showed that where the added soil cover was shallow, copper diffused upward into the soil cover. Sycamore roots did not penetrate the underlying waste. Nevertheless, added organic matter and suitable choice of pioneer species resulted in adequate or good growth. The presence of copper was not limiting (Arbutat *et al.* 1995).

Studies in south Wales on the distribution of lead found evidence that trees were redistributing it, concentrations under the canopy being markedly less than outside the canopy (Bridges 1989).

Models for the processes involved are being sought in studies of the impact of potentially toxic elements on soil quality (MLURI 1997).

Zinc

Peace (1962) describes zinc damage by drip from galvanised netting used in cages as protection against bird damage to seedlings. Death of seedlings has also been observed where rolls of galvanised netting have remained over seedbeds for several months.

Mycorrhizas

The presence of mycorrhizas associated with *Amanita muscaria* improved growth in the presence of zinc, indicating that the mycorrhizas were sequestering the heavy metal (Wilkins 1997). However, in studies on a range of heavy metal wastes in the lower Swansea valley, south Wales, rhizobial microflora were hard to find (Bridges 1989).

Cambell & Jones (1996) point out that different mechanisms may be involved for each metal, *eg* tolerance of zinc does not induce tolerance of copper, and *vice versa*. They noted that *Theleophora terrestris* was particularly resistant to metals and had improved survival of Scots pine when exposed to zinc.

Other toxic materials in soils

Toxic residues from manufacture of organic chemicals are found on many industrial sites, especially those connected with coal. The same treatments are available for these as for heavy metal residues.

15.13   Bio-remediation using trees

'Bio-remediation' covers the use of living organisms to treat toxic materials directly or indirectly. To exploit such processes, an understanding of the movement of heavy metals within trees and soils and their interaction is crucial. However, while some information is available, there are many gaps (Lepp 1996).

Direct bio-remediation applies to pesticides or their residues that can be broken down by one or other micro-organisms, *eg* pentachlorphenol-based wood preservatives around sawmills. While other micro-organisms can be utilised to assist in extracting minerals, their use to detoxify metalliferous mining residues is not yet adequately developed.

Indirect bio-remediation (*synonym* phyto-remediation) involves using tolerant plants to concentrate toxic chemicals. By their repeated cutting and removal, the concentration of particular toxic chemicals may be reduced significantly.

A third remedial role is the use of trees on contaminated sites as a biological 'shield' or 'cap' to reduce access to sites.

Bio-concentration of heavy metals in trees

In a study on derelict ground at the site of the former Lanarkshire Steelworks (Salt *et al.* 1995), Scots pine, grey alder and silver birch were planted on:

- ridges (bunds) of contaminated material scraped from the surface of the site. These were well-drained and uncompacted;
- *in situ* clay-loam subsoil, intended as the uncontaminated contrast. However, the site was flat with poor natural drainage and compacted.

In the contaminated soil, concentrations of copper (Cu), nickel (Ni) and zinc (Zn) were respectively 16, 4 and 6 times higher than in the uncontaminated clay subsoil. The pH of all sites was between 7.6 and 7.9.

Trees grew better on the bund sites, grey alder and silver birch performing well. Scots pine, while healthy, grew at about a third of the rate of the two broadleaved species. On the clay-loam former subsoil, tree performance reflected the depth of drained soil.

The differences in the concentrations of Cu, Ni and Zn in trees on the two site types was small and did not reflect the higher heavy metal content of the bund soils. The slightly alkaline soils of the site were thought to have blocked uptake.

There was a notable difference in the heavy metal content between species as shown in *Table 15.2*. However, the transfer of metal to tree biomass was less than 10% of the metal contained in the top 10 cm of soil; it was concluded that harvesting of timber was unlikely to lead to any marked reduction in ground contamination, making phyto-remediation unfeasible at this site.

Table 15.2   *Heavy metal content of birch, alder and pine on contaminated soil*

| | *Range of heavy metal content* | | *mg/kg dry matter* |
	Alder	*Birch*	*Pine*
Metal			
Zinc	63-89	390-396	56
Copper	11-13	7.5-7.7	6.0
Nickel	3.3-5.1	1.8-4.3	2.2
Source	Salt *et al.* 1995		

Willows for phyto-remediation

Many varieties of willow have been developed for biomass. Because of their high rates of growth, recent studies have been extended to include an assessment of their potential for phyto-remediation. In a laboratory study, 23 clones of willow species or hybrids were screened for their resistance to harmful effects of copper, cadmium and zinc added to the nutrient solution. For one or more of the test metals, ten clones showed less than 30% depression of root growth compared with untreated plants.

While Cu was least tolerated, further experiments showed that plant tolerance could be increased by initial exposure to short-term pretreatment exposure to a low level of toxic metal, and that tolerance increased if the concentration of the contaminant started at a relatively low level and increased steadily (Punshon *et al.* 1996).

Woodland shielding - tolerance of heavy metals

Recommendations to grow trees densely over contaminated sites are based on the observation from natural colonisation and trials. Many plant species can grow adequately on contaminated sites (Dickinson 1996), the dense thicket of young woodland initially deterring access. As canopy develops, the effects of litter fall and the network of tree roots will bind and stabilise the soil so that the risks to visitors to the wood from contaminated soil is minimised. Whether there will be a complete representation of native species that would occur on uncontaminated sites is open to doubt.

Some plants have developed a tolerance to specific metals in specific contaminated locations (Dickinson 1996). For example, tolerances appear to have developed in grasses within the 200 year timespan of industrial mining. Other studies of grasses around electricity pylons showed that some species could tolerate the associated zinc pollution while others could not (Wilkins 1997).

Birch seedlings which were progeny from trees growing near a zinc mine showed a greatly variable metal tolerance when grown in culture (Wilkins 1997).

Other laboratory studies with willows and sycamore showed little evidence of genotypic selection among trees but both species showed considerable resilience in withstanding metal toxicity.

Food chains

On any contaminated soil, vascular plants are but one component of the food chains supporting many vertebrates. Cases of poisoning through the toxic effects of mercury (Hg), lead (Pb) and selenium (Se) are recorded.

There is also appreciable risk to birds of build up of cadmium (Cd) through eating earthworms with a high Cd body load. There is no indication whether such a risk in forests differs from that on grassland (Furness 1996).

Cd and Pb similarly can build up in small woodland mammals, Cd accumulating mainly in the liver and kidneys and Pb mainly in bone. The risk to predators is related to the ratio of such small mammals caught on contaminated land to those in the rest of the predator's hunting range (Shore 1996).

Metal hyper-accumulation

Some herbaceous species appear able to accumulate extremely high concentrations of metals; through harvesting and removing off-site the herbaceous species, significant reductions in heavy metal content of the sites seem possible. At present, no timber-producing tree species falls into this category. The nearest appear to be some willow varieties grown on short rotation as coppice.

Short-rotation willow coppice offered promise as a means of cleaning up soils contaminated with cadmium (Dickinson 1996). If the coppice is subsequently burned for energy, cadmium residues in the ash would need to be safely disposed of. See also McGregor *et al.* 1996.

15.14   Radio-activity (Air-borne pollution)

Chernobyl

The explosion at the Ukrainian nuclear reactor in April 1986 generated a plume of radio-active dust which left fall-out over a wide swathe of northern and north-west Europe. In Great Britain, deposition occurred in highland, central and southwest Scotland, in Cumbria and parts of Wales.

The dust included several radio-active elements, some with a short life and others that were more persistent. The element most widely studied was caesium; this has two radio-active isotopes, Cs^{134} and Cs^{137}, with half-lives of approximately 2 yrs and 30 yrs respectively. It had initially been expected that, being allied to potassium in behaviour and chemical properties, caesium would either be leached out of the soil or would become attached to a clay mineral and be immobilised. Levels were expected to drop relatively quickly.

Monitoring of the fall-out showed, however, that initial concentrations of $Cs^{134/137}$ were sufficiently high to prohibit sale of livestock feeding in the most heavily affected areas. A series of Food Protection (Emergency Prohibitions) (Radio-activity in sheep) Orders were made or amended to this end (Edwards 1998); some were still in force in early 1999.

In Scotland, the Scottish Office sampled farmed and wild animals, game and fish. Most samples were from farmed sheep, live sheep being monitored as well as fresh meat samples from slaughterhouses. Summary figures were published regularly in Scottish Office press releases between 1987 and 1995.

The concentrations of radioactive materials did slowly drop and the number of farms subject to restriction has been slowly reduced. However, it is clear that caesium has been captured and circulated in the nutrient cycle to a far greater extent than was initially expected. In particular, litter layers and soil fungi retained more Cs than expected (Nylén 1996).

A study of soil properties on the mobility of Cs in soils in Cumbria showed that it was possible to classify soils' potential to immobilise Cs according to the type and amount of clay minerals present. Podsols and peaty soils were least able to immobilise Cs, while brown earths and calcareous soils were best. Farms with restrictions on marketing were mostly on soils least capable of immobilising Cs (Livens & Loveland 1988).

Venison, grouse and fish were included in the Scottish sampling for Cs. Sufficiently high radioactivity levels occurred for the question to be raised informally as to whether any person or animal frequently undertaking deer stalking was at risk if regularly eating high-caesium venison or offal.

Edwards (1998) reviewed a study comparing the 'critical loads' for soils in Europe in respect of milk supplies from the cattle grazing on them. Critical loads of Cs[137] for peat soils were 6 times less than for sandy soils and about 32 times less than for clays. These figures reflect the tightness with which the soil minerals bind Cs[137]. On peat soils, while Cs is retained, it is also more readily recycled than when bound to clay minerals.

15.2   SALINITY

15.21   Introduction

Ions of sodium (Na+) and chloride (Cl-) commonly appear as the most plentiful mineral contaminants in rain and cloud-water in Britain (see *Tables 13.3-5*) and are carried through the soil profile (*Tables 12.5, 14.3*). Both elements are considered to be essential nutrients at trace element level; however, air-borne sources almost universally meet these requirements.

Excessive salt in soil can prevent or damage tree growth. While naturally occurring saline soils exist in many parts of the world, in Britain they are restricted to coastal salt marshes. Local high concentrations of salt sufficient to damage trees occur naturally, or as a result of use by man. However, 300 mm of rain is considered sufficient to remove all chloride ions from the top 0.6 m depth of soil (Dobson 1991a), so that local temporary salt concentrations are washed away within a few months.

Prolonged exposure of soils to saline water results in the displacement of calcium and to a lesser extent potassium ions. Lack of potassium may show up discolorations typical of K deficiency; on clayey soils, displacement of Ca by Na can lead to loss of soil structure through deflocculation.

Damage by excessive salt is a recurring hazard, whether natural or man-made, because of its association with recurring strong winds or harsh winter climate. However, the location of damage is limited geographically to coasts, urban streets and paths, and the main road transport network.

15.22   Natural sources of damage

Salt damage by sea spray

Strong winds over rough seas pick up droplets of sea water. If air-borne for any length of time, the water may evaporate, creating aerosol size droplets that can be carried long distances. Effects of wind and salt 'cutting' trees and other vegetation are readily seen in exposed coastal localities and sand dune systems

(Edwards & Holmes 1968). While damage from salt-laden winds occurs commonly within 5-6 kilometres of the coast, injury has occurred up to 20 km inland, being most severe on exposed ground and often related to particular gales (*eg* Edlin 1957).

Salt-spray injury on foliage of broadleaved trees appears as withering, or marginal and interveinal browning. On conifers, damage appears as browning of needle tips and death of mature needles on the sides of plants most heavily exposed.

The holm oak (*Quercus ilex*) and the shrubs *Tamarix*, *Escallonia* and *Olearia* have been found most resistant to sea spray. Beech is particularly susceptible (Peace 1962).

In contrast to damage following use of salt in winter on roads, damage by marine salt spray can occur at any time of year. Recently flushed foliage is marked more susceptible to salt spray damage than mature or dormant foliage (Dobson 1991b, c).

Salt damage by sea-water flooding

Severe flooding by sea water is a relatively rare event in the UK. A notable and widespread inundation occurred in Britain and Holland in January 1953, damaging or killing many trees. The trees were dormant at the time. However, roots were submerged for several days and quantities of salt remained in the soil after the floods receded. Severely affected trees came into leaf but the new growth died as did subsequent flushes. Of the more susceptible species or where salt concentration was high, trees progressively died back and often died. Oak, some willows and white poplars were relatively resistant; beech, sycamore, ash and black poplar were particularly susceptible. Conifers were also highly susceptible.

Rehabilitation of flooded areas depended on leaching out of salt by rainfall or irrigation with fresh water (Peace 1962).

Where soil structure had been damaged by loss of calcium, calcium as gypsum (calcium sulphate) was applied to restore soil structure.

15.23   Salt as de-icing material

The use of chemicals to disperse ice on public highways goes back to the 1940s at least, with calcium chloride reported in North America as causing leaf scorch, die-back and tree death. In the 1960s in Britain and Germany, the risk to trees from use of salt+grit on roads was recognised; deaths of elms, lime and horse chestnut were attributed to its use (Peace 1962, Dobson 1991d).

Between 1960 and 1990, the amount of salt, by itself or in mixture with grit, applied annually to roads in mainland Britain, rose from 200 kilotonnes (kt) to 2000 kt, with peaks of 2500 kt in icy winters. The scale of usage in Britain has been substantially higher than in many European countries. While this may be partly due to the repeated fluctuations above and below freezing point in many winters, part is also due to excessive rates of application. A study in the late

1980s showed that up to 60% of salt purchased was wasted through inadequate storage and inaccurate spreading (Audit Commission 1988).

With increasing use of salt have come reports, world-wide, of increasing damage to roadside trees. *De-icing salt damage to trees and shrubs* (Dobson 1991a) contains summaries of literature reports on 332 woody species, and data on foliage concentrations of sodium (47 species) and chloride (130 species) in relation to occurrence of symptoms of damage.

Damage to foliage is related to levels of chloride, foliage showing symptoms of scorch where the concentration of Cl⁻ in leaf tissue exceeds 1%.

In Britain, significant crown die-back of London plane (*Platanus acerifolia*) in towns coincided with years of high salt usage (Gibbs & Burdekin 1983).

Causes of damage

The two main causes of damage follow from:

- dissolved salt percolating into soil and being taken up through plant roots; damage due to uptake from soil occurs predominantly in towns;
- wind-borne spray droplets accumulating on foliage, branches and stems and causing damage *in situ*. Spray damage occurs on salted roads with fast-moving traffic and is most noticeable on motorways and trunk roads. Such deposits subsequently dissolve and enter the soil; spray damage is nevertheless distinguishable from soil-water-borne damage.

Percolating saline water

Damage commonly occurs where relatively concentrated salt-laden water has been able to percolate around rooting zones of trees, *eg*:

- when salt or salted grit has been stored or dumped on open ground prior to spreading;
- where snow clearance by snow-plough, scraper blade or blower has dumped large volumes of salt-treated snow on roadside verges;
- where salt-laden run-off water accumulates near the base of trees.

Raised (water shedding) planting positions are less affected than where plants are in slight hollows (water receiving positions) (Bradshaw *et al.* 1995). Roots may pick up saline water from run-off drains, at considerable distances from roads.

Roadside spray damage

Damage is commonly seen where roads have been salted and moving traffic creates a fine spray. Salt-laden water is then deposited on stems and foliage if present. Its persistence is entirely dependent on the immediately following weather conditions. Heavy rain will remove much of the deposit in run-off to surface drains. Continuing dry, frosty conditions will worsen damage by allowing sprays to dry and accrete. Deposition and damage are more acute and more widely distributed where traffic is fast-moving.

- Worst damage occurs within 5m of treated road or path, reducing to minimal levels at 30m.
- The distribution of damage is influenced by gravity and wind, being worse down hill and on the most exposed trees. Deposition by turbulent wind induced by fast-moving traffic is often more important than natural winds, so that damage is often worse facing on-coming traffic;
- Where exposed to spray, damage is greatest on lowest branches.

15.24   Symptoms of salt damage

For a full description and illustrations of damage, see Dobson (1991a) especially plates 4-17, Bradshaw *et al.* (1995) plates 12.14 & 17.7, and Strouts & Winter (1994) figures 74 &75.

Damage symptoms differ according to whether they are due to soil uptake or spray.

Percolating water and root uptake
Broadleaved trees, while dormant, may fail to flush. Alternatively, buds may open but the leaves wither and die while still small - 'post-flushing die-back' (Gibbs & Burdekin 1983, Gibbs & Palmer 1994). Other leaves may show marginal browning within a few weeks of flushing.

Damage to stems appearing as tapering areas of dead bark stretching from the roots to several feet up the stem have been associated with salt damage, though no direct cause of death has been identified (Dobson 1991a, plates 10 & 12).

On roadside trees, small annual growth rings in the woody stem may indicate poor growth following salt damage. Long-term effects show up as decreased radial growth of stems.

Conifers have been less widely studied but failure to flush and needle browning in the summer are characteristic symptoms of salt damage.

High salt concentration in soil water may reduce the effect of natural osmotic processes in water uptake by roots, inducing additional water stress or aggravating drought.

Salt spray
The distribution of damage on a tree is significant in relation to salt spray damage. Foliage taking the heaviest impact of spray shows the worst damage. Damage from road salt spray is usually worst nearest ground level.

Broadleaved trees may show death of dormant buds and die-back of twigs or late leaf emergence. While death of trees is not common, damage may recur in any year salt is applied.

Conifers show damage on first year needles, turning yellow then brown. In Scots pine, salt crystals have been found in the intercellular spaces of needles at the boundary between healthy and necrotic tissues.

Salt-tolerance
Trees may be able to tolerate salt by avoidance and/or physiological tolerance. Roots of some plants appear more able to limit uptake of salt ions than

others.   Physical characteristics *eg* resinous buds, well-shielded buds, more waxy foliage, thicker cuticle, provide some protection against surface deposits.

Identifying the relative tolerance of species to salt damage is hampered by limited comparisons.   *Table 15.3* lists 19 tree and shrub species in order of tolerance based on Dobson (1991a).   The most tolerant species is *Ulmus glabra*, the most sensitive species is sycamore.   Each species is more sensitive than the one above it on the list; however, the author points out that differences of one or two places on this list are not significant.

Table 15.3   *Salt-tolerance of a range of tree species*

Tolerant	Intermediate	Sensitive
Most tolerant		
Ulmus glabra	*Picea pungens*	*Alnus incana*
Robinia pseudoacacia	*Fraxinus excelsior*	*Sorbus aucuparia*
Quercus robur	*Crataegus monogyna*	*Fagus sylvatica*
Salix alba	*Acer campestre*	*Carpinus betulus*
Gleditsia triacanthos	*Picea abies*	*Acer pseudoplatanus*
Alnus glutinosa	*Pseudotsuga menziesii*	
Eleagnus angustifolia	*Aesculus hippocastanum*	**Most sensitive**

Source   Dobson 1991a

Anticipating effects of de-icing salt

Recent best practice includes:
  • ensuring that de-icing salt is kept in a dry store until required;
  • ensuring that road equipment can apply correct amounts evenly over treated road surfaces;
  • maintaining close liaison with weather forecasters so that salt is spread when necessary and not otherwise;
  • where there are bins for use by the public on pathways, mixing salt with sand at between 1:4 and 1:10, to reduce salt usage (Gibbs 1998).

In a review of recent local authority practices, Pinchin (1999) notes also that some local authorities are applying gypsum at $2kg/m^2$ when back-filling holes dug for amenity plantings in areas likely to be subject to de-icing salt.   Prompt and thorough irrigation is also recommended following heavy use of salt, especially if a period of dry weather has followed the snow and ice (Gibbs 1998).

15.25   Alternatives to salt

A number of chloride and sulphate salts have been tested as alternatives (Dobson 1991a).   Calcium chloride ($CaCl_2$) in particular is used in parts of North

America and Europe. Less is needed on a weight-for-weight basis; however, $CaCl_2$ is more expensive and, being hygroscropic, is more difficult to store.

Urea is currently used on a small scale in Britain on some bridges and city centres, where it is applied by spraying.

15.3 OTHER POLLUTANT SOURCES

15.31 Forest machinery

In a summary of pollution incidents in Scotland between January 1991 and May 1994, out of 9500 incidents reported to the Scottish 'River Purification Boards', only 52 related to forestry. Of these, 36 related to harvesting, ground preparation or maintenance and road construction, 13 were listed as oil spills, 2 attributed to use of fertilizer and 1 to herbicide use (Virtue 1995).

The use of heavy vehicles and attachments in the forest gives rise to local pollution risks from two sources, physical and chemical.

Physical disturbance

Physical sources arise through the disturbance and mobilisation of silty material from the ground or from forest road surfaces. If such material is carried in rainfall run-off into streams, it may be deposited on river beds threatening fish-spawning areas, or in reservoirs threatening water quality.

In a study of erosion of forest roads at Balquhidder, north Scotland, the surface-water run-off from an unused road was greater than from a road in current use by timber lorries. However, the sediment load from the used road was twice that from the unused road, rising to five times as much shortly after regrading. After eight months, sediment load from the regraded road had fallen to a little above the level before regrading.

Erosion associated with heavy use of roads started with the loosening of the surface and rutting. The material eroded was 75-85% 'very fine' ($<355\mu$ in diameter). Regrading removed ruts but fine sediment at the road surface was subsequently washed out until a protective layer of coarser gravel on the road surface had developed (Johnson & Bronsdon 1995).

Chemical risk

· Chemical risk arises through accidental rupture of hydraulic or sprayline pipes, spillage, production of oily wastes from on-site vehicle maintenance and loss of fuel through overturning. Measures to minimise the effects of all these should be part of contingency planning to meet health and safety requirements and should be based on use of absorbent materials, containment of spillage and safe removal of wastes (Murgatroyd 1995, Anon 1996g).

Bio-degradable lubricants based on rapeseed oil produced to ISO standards have been accepted for hydraulic systems. They have also come into widespread use for 'total loss' systems, eg chain/chain bar lubrication of saws used when timber harvesting (Davies 1995).

15.32   Forest fire fighting

Fire retardants as additives to water applied to fires and herbicides to controlling vegetation on fire-rides have been tested but have not come into general practice (Holmes & Fourt 1960, Connell & Cousins 1969).

Foams applied close to fire fronts by air or from the ground have found a role (Ingoldby & Smith 1982, Aldhous & Scott 1993), and are not considered to be an environmental risk.

15.33   Civil engineering materials

Construction and maintenance of bridges and structures require care to avoid pollution *eg* if using Zn/Cr or Ca/Pb based paints on steel out of doors (Granfield 1972).

Construction of concrete structures near streams requires care to avoid pollution from cement dust and disturbance that may generate silt in the watercourse (Freedman 1995).

15.34   Wet storage of logs

A violent storm in October 1987 uprooted trees containing over 4 million cubic metres (m^3) of timber in the south of England. A trial of a method of timber storage not previously used in Britain was consequently set up.

70 thousand m^3 of Scots and Corsican pine were stored under water-sprinklers at Thetford forest, Norfolk/Suffolk borders. A full description of the trial is given in Forestry Commission Bulletin 117 *Water storage of timber; experience in Britain* (Webber & Gibbs 1996). Comments on control of 'blue stain' fungal infection are give in §7.32.

Water used for sprinkling was held in a flooded gravel pit adjacent to the storage area. There were no specific mineral pollutants emanating from the logs in the store; leaching of nutrients was considered insignificant.

Sugars, tannins and other soluble organic matter coloured the drainage water appreciably in the first two years of the trial, the effect diminishing with time.

The 'biological oxygen demand' (BOD) fluctuated markedly for the period of the trial, again diminishing somewhat over the period of the trial. However, it was considered that BOD fluctuations in the gravel pit were associated with algal growth and the eutrophic quality of the local ground waters rather than leachates from the logs (Nisbet 1996b).

15.4   POLLUTION AND NON-WOODLAND TREES

The interaction of non-woodland trees and pollutants can be viewed from two perspectives:
- how vulnerable are non-woodland trees compared with those in woodland;
- is there any potentially beneficial role which urban and peri-urban trees can play in reducing aerial pollutant loads?

15.41   Vulnerability of non-woodland trees to pollutants

The most conspicuous effect of pollutants on non-woodland trees currently is the effect of de-icing salt, discussed above. Here, it is the proximity of street trees to the pollution source which renders them susceptible rather than any biological factor.

In the 1950s, similarly, trees closest to major sources of SO_2 often showed damage; these were frequently 'amenity' trees in urban or industrial settings. Such damage is now infrequent.

Interception of particulate matter and gaseous pollutants

From their position, specimen trees, avenues and shelter belts are likely to intercept particulate deposition to a greater extent than woodland trees. Town trees are also being exposed to increasing concentrations of traffic-induced nitrogen oxides pollution.

Offsetting these adverse factors, trees in towns will have benefitted from reductions in soot and SO_2 over the last 20 years, following the introduction of 'clean air' legislation. They are also exposed to lower concentrations of ozone than in nearby rural areas because of its interaction with vehicle emissions.

Conclusions about the role of woodlands on soils contaminated by heavy metals apply equally to non-woodland plantings on such sites.

15.42   Beneficial effects of urban trees in reducing pollution loads

The climate of cities differs according to attribute, compared with surrounding countryside. Solar radiation is less but temperatures are higher. Average annual windspeed is less; air pollution is, however, 5-15 times more (Deelstra 1994).

Reviewing whether air pollution is a growing or declining threat to arboriculture, Good (1991) concluded that the atmosphere in British towns and cities is more conducive to healthy growth of trees than formerly.

Provision of more vegetated areas within towns primarily to act as air filters has been strongly recommended (Madders & Lawrence 1982). Deelstra (1994) quotes figures showing that dust above urban woodland could be 80% less than in the adjoining built environment. Conifers, while they filter dust more efficiently than broadleaved trees are generally less tolerant of the pollutants which accumulate on their leaves.

Published data for uptake of pollutants by 12 broadleaved species and 6 conifer species is summarised by Freer-Smith *et al.* (1997). They concluded that the greater the stand leaf area, the greater the uptake of pollutants.

In a study of the predicted effects of a community forest near Nottingham on O_3, NO_x and SO_2, a cover of equal proportions of pine and oak was expected to bring about a 1-2% decrease in the ambient air pollutant concentration. It was also pointed out that NO_x and SO_2 pollution is worst during the winter so that the scale of the reductions is influenced markedly by the leaf area of any planted trees and whether they are deciduous or evergreen. 4-7% reductions in pollutant concentrations would be brought about by use of evergreen species with a high leaf area, such as Norway spruce, more than double the effect of the oak and pine plantings proposed (Freer-Smith & Broadmeadow 1996).

15.43   Die-back of hedgerow ash

Ash trees form an important component of the agricultural landscape in many parts of lowland Britain. Crown die-back has been observed for many years.

Die-back of older hedgerow ash was linked with an inability of trees to tolerate major variations in soil water content associated with heavy soils (Peace 1962). A reconnaissance survey of trees in a broad band of countryside between York and Aylesbury (Buckinghamshire) confirmed its prevalence but failed to identify apparent causes of ash decline, though damage to superficial roots by ploughing was suspected (Pawsey 1983, 1987).

In a subsequent, more detailed study of ash growing as non-woodland trees, die-back was most strongly associated with arable soils. There were indications that die-back was worse in drier areas and that pollutants exacerbated the effect.

Disturbance by cultivation of the surface 60 cm of soil, a zone normally intensively rooted by ash was thought to account for the association of die-back with arable cropping (Hull & Gibbs 1991).

CHAPTER 16

Carbon dioxide & climatic change

... but baby, its hot outside

16.1   INTRODUCTION

This chapter reviews:

- carbon, carbon dioxide and climatic change (§16.2);
- other gases contributing to the 'greenhouse effect'. (§16.3):
      methane (§16.31);
      nitrous oxide (§16.32);
- renewable energy (§16.4).

International cooperation and agreements about reductions of emissions are briefly described in §16.5.

16.2   CARBON, CARBON DIOXIDE AND CLIMATIC CHANGE

16.21   Global distribution of organic carbon in vegetation and soils

Carbon is at the heart of the energy storage processes that have evolved on earth.   Energy is a non-recoverable resource; however, live plants through photosynthesis provide the mechanism for capture of incident solar energy. Plants store it as reduced organic carbon compounds from which energy is available for release later through oxidation.   Fossilised organically combined carbon (fossil fuels) resulted from situations where oxidation of carbon in deposits of organic materials in lakes, seas and bogs was prevented by the exclusion of free atmospheric oxygen by water.

Table 16.1 overleaf sets out global estimates of the distribution of carbon in vegetation by types, in soils and in peat.   It also shows estimates of annual fixation rates.   The wide range of values in several of the estimates reflects the scantiness of data and difficulties of estimation.

No account is given here of the scale of inorganic carbon (carbonates) in soils and rocks, other than to note that the carbon in the biosphere is less than 0.1% of global total C.   The bulk is in limestones and other geological formations.   The marine shellfish carbonate cycle is substantial, dead shells forming the basis for massive calcareous deposits in several geological periods. Nevertheless, that cycle has always been dependent on organic carbon cycles.

The total amount of carbon in vegetation is at least as great as the total amount in CO_2 in the atmosphere.

Table 16.1   *Global distribution of carbon in living vegetation, in soils and in non-living peats; annual carbon fixation rate etc.   (Gt*)*

Carbon in forest vegetation and soils

Low latitude forests	550-670 Gt	Mid-latitude forests	210-250 Gt
High latitude forests	740-880 Gt		

Total global   1500 - 1800 Gt

Carbon in live vegetation

Forest, woodl'd & savannah	400-550 Gt	Grasslands	20-50 Gt
Croplands	17-30 Gt	Wetlands	5-15 Gt
Tundra & arid lands	0-40 Gt	Oceans	5 (1-45) Gt

Total global   450 - 700 Gt

Soil organic carbon

Forest soils & associated peat lands	700-1500 Gt
Peats and wetlands	110-1100 Gt

Total global   1100 - 2200 Gt

Carbon in CO_2 in the atmosphere   Total global   700 Gt

Carbon in the animal kingdom   Total global   1 - 2 Gt

Soil organic carbon in top metre of undisturbed soil

Latitude 0-10°	10-15 kg/m³	Subtrop'l thorn w'dl'd	5.4 kg/m³
Latitude 61-65°	20-24 kg/m³	Boreal rain forest	32 kg/m³

Annual carbon fixation rate

Forest, woodland & savannah	50 Gt	Grasslands & scrub	17 Gt
Croplands	25 Gt	Wetlands	8 Gt

Sources   Harrison *et al.* 1995, Holdgate 1995, Freer-Smith 1990,
Scurlock & Hall 1991 *1 Gt = 1 Gigatonne = 1 billion tonnes = 10^9 t

16.22   Greenhouse gases and climatic change

The warming effect of gases in the atmosphere has occurred for as long as they have been present. Their combined effect is to maintain earth's annual mean surface temperature at 15°C. Without such absorption, the corresponding earth surface temperature would be 33°C lower, at -18°C (Cannel & Cape 1991). Without such warming, life in its present form on earth would not have developed (Rowntree 1993). This 'greenhouse effect' is totally natural, based on the physical and optical properties of gases.

The term *Global warming* has commonly been used to encapsulate the effects of anthropogenically caused *additional* emissions of 'greenhouse gases' on global climate. However, there is no clear agreement on the quantitative relationship between increases in greenhouse gases and possible changes in climate. It is most unlikely that warming will be uniformly distributed. Consequently, the more non-committal term 'Climatic change' has been substituted for 'global warming' in many documents.

16.23　　　Atmospheric 'greenhouse' gases

The atmospheric gases that bring about the greenhouse effect currently make up less than 1% of the atmosphere. The gases are transparent to sunlight and incoming short-wave solar radiation outside the visible spectrum; however, they absorb particular wavelengths of the long-wave infra-red radiation emitted by the earth. The lower part of the earth's atmosphere, the troposphere, is thereby warmed and re-radiates infra-red energy, acting as a blanket and reflector. The result is that the surface of the earth is warmer than if there were no absorption (Gribbin & Gribbin 1996).

Water vapour has the biggest effect as a greenhouse gas. However, this is off-set by the cooling effect of thick cloud reflecting sunlight back to space. For thin clouds, the greenhouse effect dominates up to a certain thickness, then the reflecting (cooling) effect takes over (Emsley 1994a, Rind 1995).

Other greenhouse gases are listed in *Table 16.2* below, with figures for their relative warming potency and relative contribution to atmospheric warming. Their absorptive capacities differ by up to nearly 4 orders of magnitude.

The right-hand column of *Table 16.2* shows the forecast increase in contribution to global warming from each gas listed, on the assumption that further increases in warming will be proportional to present effects.

All the greenhouse gases sooner or later are either absorbed organically or inorganically, or they break down. Increases in concentrations observed over time indicate that annual rates of absorption and breakdown together are currently

Table 16.2　*Relative contribution of greenhouse gases to global warming in 1990 and 2020 (excluding water vapour)*

	Relative contribution in 1990	Relative warming potency**	% Predicted increase by 2020 over 1990	Relative contribution in 2020
	%			%
Carbon dioxide	47	1	+35%	37
Ozone*	11	n.a	+147%	15
CFCs	20	6000	+130%	27
Nitrous oxide	6	200	+144%	9
Methane	16	90	+22%	12
			Combined　+70%	
Total	100%			100%

Source　Cannell & Cape 1991　　n.a = data not available　　* See also §13.4
** temperature increase/molecule relative to the increase/molecule for CO_2

less than the rate of formation/emission. Concentrations are likely to increase without any change in rate of emissions, until rate of gain equals rate of loss.

There is clear evidence that increases in many of the more significant 'greenhouse' gases are due to man's activities over the last 130 years.

16.24   Carbon dioxide increase

CO_2 is, after water vapour, the biggest component of the greenhouse gases.

Table 16.3 gives a summary annual global balance sheet for atmospheric CO_2 in 1990 from Jarvis & Dewar (1993). The table shows that emission of CO_2 to the atmosphere due to fossil fuel combustion constitutes about two thirds of the annual total CO_2 emissions due to man's activities. Of this total, an amount equivalent to over half remains in the air. This annual net increase corresponds to a 0.5% annual increase in the concentration of atmospheric CO_2.

Table 16.3   *Annual global atmospheric CO_2 balance sheet in 1990 (as Gt* C)*

Sources (emitted into atmosphere)	Gt/yr	Sinks (absorbed from atmosphere)	Gt/yr
Anthropogenic			
Fossil fuel combustion	5.7	Oceans	1.0
Tropical deforestation	2.1	Temperate and boreal	
CO and CH_4 from burning		forests	1.8
vegetation & soil changes	0.7	Tropical forest & grasslands	2.5
Total anthropogenic sources	8.5	*Total anthropogenic sinks*	5.3
Net gain from anthropogenic sources		3.2 Gt/yr	
Natural sources		*Natural sinks*	
eg respiration and decomposition,		*eg photosynthesis*	
Oceans	90		93
Terrestrial biomass	63		120
Soils & litter	60		nil
Carbon storage pools (Gt)**			
Atmosphere	725	Terrestrial & freshwater biomass	560
Soils & litter	1300-1400	Oceans	37 000
Fossil fuels	5000-10 000	Sea-bed frozen methane	10 000-20 000

Sources   Tinker 1993, Jarvis & Dewar 1993, Cannell & Cape 1991, Thorpe 1998, Clennell 1998.

*   1Gt = 1 Gigatonne = 10^9 t = 1 thousand million tonnes

** Differences in values between those above and those in *Table 13.1* arise because of differing assumptions made by the authors of these figures.

Among the many natural sinks, the boreal forests, old growth tropical forest and southern oceans are each considered to be important sinks for CO_2. However, there is considerable uncertainty as to the size of each of these sinks.

A reduction in atmospheric oxygen has also been noted, in proportion to the additional CO_2 accumulated (Gribbin 1992).

The scale of emissions of CO_2 is small in comparison with the total amounts of carbon ('carbon pools') in the global circulation. Nevertheless, the concentration of atmospheric carbon is currently clearly increasing and showing no sign of slowing. By implication, firstly, a new equilibrium level where sources and sinks are balanced cannot yet be predicted, and secondly, there are no large unsatisfied and rapidly activatable sinks for increased emissions.

Carbon in geological deposits

The amount of carbon in carbonate rocks is greater again by many orders of magnitude that all the carbon in the active biological cycle; however, magnesium and calcium carbonates are unlikely to be involved in the short term in the global biological carbon balance, except possibly through oceanic carbonate/bicarbonate buffering. Some geologically recent carbon and carbonate deposits become deeply buried and inaccessible as a result of on-going tectonic subduction processes; however, these processes have little influence on the annual atmospheric CO_2 balance.

Two 'deposition layer' types of combined carbon are potential contributors to the long-term biological carbon balance sheet:

- dead organic matter accumulating in peats and marshes (§16.28), and
- methane combining with freezing water to form seabed deposits of frozen methane hydrate (Clennell 1998) (§16.31).

Changes in CO_2 concentration

Past concentrations of CO_2 can be determined by sampling dissolved gases in datable ice cores. For the period between the 11th and late 18th centuries, such sampling showed concentrations to have been reasonably steady at 270 to 290 ppm. 270-280 ppm is the current best estimate of the 'natural sources' level.

Concentrations started to increase noticeably in the late 18th century, since then rising steadily, being 315 ppm in 1957 and 360 ppm in 1999.

Estimates of further increases depend on assumptions of future growth and the effect of attempts to reduce emissions. Cannell & Cape (1991) put the figure in a range between 380 and 650 by the year 2060; Tinker (1993) proposed 600ppm by the same date, while in several IPCC papers and elsewhere, the effects on temperature are considered assuming a doubling of 1990s CO_2 levels 60 - 100 years hence.

While the bulk of recent increases can reasonably be attributed to combustion of fuel and industrial activity, the earliest increases coincide with mid-18th century emigration, early widespread destruction of forests and the ploughing of extensive areas of organically rich grassland soils (Jarvis 1989a). However, by

1980-1990, deforestation was responsible only for between 12 and 17% of the total CO_2 emission (Tinker 1993). Holdgate (1995) stresses that deforestation can still be a significant source of greenhouse gases.

A small observable annual cyclic fluctuation in CO_2 concentration is attributed to heavier demands for CO_2 during the growing season of vegetation in the northern hemisphere (Scurlock & Hall 1991, Gribbin & Gribbin 1996).

16.25   Potential climatic changes and greenhouse gases

Temperature changes

Figure 16.1 shows changes in mean annual global temperature over the period 1860 - 1997. The curve is not smooth. Some perturbations can be attributed to natural events like major volcanic eruptions and others to identifiable short-term activities of mankind. However, from such data, the United Nations Intergovernmental Panel on Climatic Change (UN IPCC) concluded in 1995 that 'the balance of evidence suggests there is a discernible human influence on global climate'.

The temperature rise shown in *Figure 16.1* of 0.6-0.7°C over 130 years has been widely used as the baseline for long-term predictions, assuming long-term increases in the concentration of CO_2. Future mean annual land temperature increases have been estimated to be in the range 3-6°C in winter and 2-4°C in summer (Tinker 1993, Harrison *et al.* 1995, Gribbin & Gribbin 1996, DETR/MO 1997).

Figure 16.1   *Global mean surface temperature changes 1860-1997*

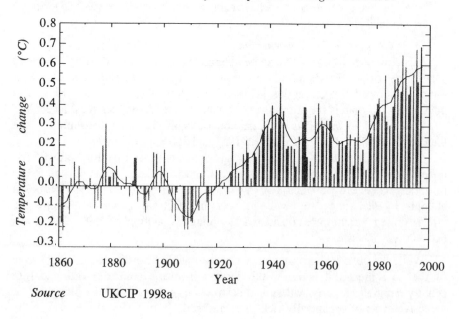

Source   UKCIP 1998a

Parry & Rosenzweig (1993) pointed out that, in practice, CO_2 increase cannot be taken in isolation; it has to be linked with other gases shown in *Table 16.2*. The greenhouse warming effect of the 'doubling of CO_2 concentrations' forecast for 2060 could be brought forward by 20-30 years through the supplementary effects of additional methane, nitrous oxide and chlorofluorocarbons.

Cooling through interception of solar energy

Over shorter periods, warming has been less than predicted from changes in greenhouse gas concentrations because of interception and dissipation of solar energy by aerosol and dust particles.

- The heavy loads of aerosol sulphate and particulate carbon from mid-20th century coal-burning intercepted appreciable amounts of radiation so that there were no observable changes in global temperature between 1940 and 1970. Their removal as a result of 'clean air' legislation has been accompanied by an increase in temperature.
- Volcanic eruptions periodically throw immense amounts of dust into the stratosphere. The loss of energy at ground level is evident in lower mean temperatures lasting for 2 - 3 years. In 1991, Mt Pinatubo in the Phillipines erupted; the predicted 2-year period of slightly lower temperatures occurred but is now past (Gribbin & Gribbin 1996).

Changes in precipitation

Changes in amounts and distribution of precipitation are predicted to result from increases in mean temperature. Differences in the UK are likely to be less than 0.5mm per day, but greater changes are predicted for other parts of the world (DETR/MO 1997). Soil moisture evaporation is predicted to increase by about 0.35 mm per day (Tinker 1993).

Sea level change

Changes in sea level have occurred as a continuing response to major climatic changes, in particular, glaciation. While sea levels have fallen or risen according to the volume of water accumulated as ice on ice caps, changes in the weight of ice caps have caused the supporting land to sink or rise, offsetting the effect of the ice cap (isostatic reaction) on sea levels. The speed of these effects is quite different; at the end of the last glacial period, melting was relatively rapid and sea levels rose sharply. The isostatic uplift of land in response to the removal of the weight of ice was slow and is still continuing.

It was at first assumed that climatic warming would lead to rapid melting of polar ice-caps and that consequent rises in sea level would inundate many low-lying and heavily populated parts of the earth (Boorman *et al.* 1989, Kleiner 1994). More recent views predict heavier snowfall at the poles and smaller increases in mean sea level. Most of any recent rise in sea levels can be attributed to thermal expansion due to slight increases in sea temperature (Tinker 1993, Gribbin & Gribbin 1996, UKCIP 1998b).

Modelling climatic change

Climatic 'general circulation models' (GCMs) currently used for predicting temperature changes in the atmosphere can track trends during the 20th century closely. These include changes in types of combustibles, effects of variation in aerosol particulate interception following volcanic activity, variable solar energy output *etc.* However, none of the variations examined has improved predictive models based on greenhouse gases (Rowntree 1993, Matthews 1994).

A European Union international study of the *Long-term effects of climatic change on European forests* (LTEEF) is investigating the fluxes of water and carbon between the vegetation and the atmosphere, aiming to identify strategies to minimise risk of decline or damage to forest and to establish complete carbon budgets (IGBP 1998).

Smaller scale modelling the effects of increasing levels of CO_2 requires the development GCM data at higher temporal and spatial resolution than has previously been available (Barrow & Semenov 1995). Friend *et al.* (*1998*) review modelling at leaf, tree and forest scale.

Table 16.4 gives a summary of predicted changes used as the basis for assessments of the impact of climatic change (DETR/MO 1997).

Table 16.4   *Assumptions for assessments of impacts of climatic change*

	Present day	2020s	2050s	2080s
CO_2 concentration (ppm)	365	441	565	731
Temperature increase (°C)	0	1.2	2.1	3.2
(including sulphate aerosol)	0	1.0	1.6	2.6
Precipitation increase (%)	0	1.6	2.9	4.5
Sea-level rise (cm)	0	10	26	44
Population (millions)	5266	8121	9759	10672

Source   DETR/MO 1997

16.26   Effects of climatic change on crop production

The former continuum of tree cover from boreal scrub to warm-temperature mixed high forest during periods of climatic change clearly indicates that forest cover has been mobile, forest types successfully migrating during periods of climatic change. However, the speed of present-day changes is quicker than most past natural climatic changes and may bring unforeseeable consequences.

Tree species well established in the UK should thrive if adequate moisture accompanies warmer summer temperatures. Increased summer warmth near the limits to tree growth could allow tree growth at elevations now above the tree limit. A 1.5°C mean increase in annual temperature would displace the limit of tolerance of species some 225 km towards the poles (Holdgate 1995). However, if a rise in temperature changed current relationships with other climatic factors, *eg* winter minimum temperatures and seasonal distribution of rainfall, more detail predictions would be required before meaningful predictions could be made of expected species responses.

Not all effects would necessarily be beneficial. For example, top-dying of Norway spruce in Britain, associated with mild winters, could become more widespread if there were a general increase in mean winter temperatures.

Increased spruce budworm activity in Alaska has been attributed to insects benefitting immediately from recent higher summer temperatures (Mulvaney 1998). Such short-term ecological opportunism may occur again.

16.27 Carbon pools and increases in carbon storage in forests

From the late 1980s, the distribution of carbon in UK forests and the potential for increasing rates of carbon fixing and storage by woodland management have been studied intensively. *Forestry Vol. 68 (4) 1995* contains important papers from a conference on *Greenhouse gas balance in forestry*.

In *Sustainable forestry - the UK programme*, the UK government recognised the role of forests as stores and sinks of carbon (UK Govt 1994b), implicitly endorsing this work.

Table 16.5 (overleaf) shows the distribution of carbon in UK forests. It reflects the history of woodlands and, in particular, the relative juvenility of much of the coniferous woodland. Carbon-rich mature old natural forest scarcely exists in the UK; woodlands in Britain are largely of managed plantation origin.

In a broader review, Cannel & Dewar (1995) estimated the annual rate of accumulation of carbon in forest plantations in Britain at 2.5 million tonnes (Mt). They used Sitka spruce (Yield class 14) as a conifer representative of plantings over the last 70 years; beech (Yield class 6) and poplar (YC 12) were added in models for the early part of the 21st century.

The current annual rate of accumulation of carbon in UK forest will rise in the short term because of the high proportion of young plantations.

To maintain the highest rate of accumulation in the long term requires additional annual planting of 25-30000 hectares of fast-growing conifer and 10000 ha of poplar. If slower-growing species are planted, larger areas proportionate to their rates of growth will be required to secure the equivalent carbon-accumulation.

The annual aggregate accumulation of carbon, 2.5 Mt C, corresponds to 4% of the current C reservoir in plantation trees. It offsets 1.5% of the C emitted yearly by burning fossil fuels in the UK (Cannell & Dewar 1995, Cannell 1999).

Figures for carbon storage in Northern Ireland are given in *Table 16.6*. They show that N. Ireland forests account for about 7% of current carbon accumulating in UK forests. To maintain this rate requires an annual afforestation programme of 1500-2000 ha/ann (Cannell *et al.*1996).

Table 16.5   *Distribution of carbon in forests etc. in the UK*

Carbon in vegetation (all types)   <u>*Total UK*</u> *110 (90 - 130) Mt*C*

Carbon in forest vegetation
Broadleaves   53 - 58 Mt   Conifers   34 Mt
                                        <u>*Total UK*</u>   *90 - 92 Mt*

Carbon in soil (incl. plant litter)
In peat and lowland fens   4500 Mt   In other soils   5000 Mt
In peat excl. lowl'd fens   3000 Mt
                                        <u>*Total UK*</u>   *9500 (7500 - 22000) Mt*

Annual accumulation of carbon in woodlands planted between 1921 & 1996
In existing trees   1.86 Mt C/yr   In litter   0.56 Mt C/yr
In wood products   0.31 Mt C
                                        <u>*Total UK*</u>   *2.79 Mt/yr*

C in forest vegetation per ha
Broadleaves per ha   62 tC**   Conifers per ha   21 tC
** Higher than conifer because older
                                        <u>*UK Mean*</u>   *30 tC in 1991 < 65 tC by 2050*

C in soils per ha
In soils under broadl'ves   410 tC/ha   Under conifers   1060 tC /ha

---

Sources   Matthews 1991, UK Govt 1994b, Harrison *et al* 1995, Cannell *et al.* 1993, Cannell & Dewar 1995, Cannell & Milne 1995, Milne *et al.,* 1998.
*   1Mt = 1Megatonne = 10[6] t = 1million tonnes

Table 16.6   *Distribution of carbon in forests in Northern Ireland*   *(Mt*)*

Total forest area (1993)
Conifers   65 100 ha   Broadleaves   13200 ha   <u>*Total N.Ire*</u> 78 300 ha

Carbon in forest vegetation
Conifers   3-4 Mt   Broadleaves   0.8 Mt   <u>*Total N.Ire*</u>   4 -5 Mt

Annual accumulation of carbon
Conifers 0.15-0.2 Mt   Broadleaves   0.025 Mt   <u>*Total N.Ire*</u>   0.175-0.225 Mt

---

Source   Cannell *et al.* 1996   * 1Mt = 1Megatonne = 1million tonnes = 10[6] t

Management of plantations to maximise carbon accumulation

In one of the first studies on greenhouse effects, Thompson & Matthews (1989) developed a model representing carbon stored in managed plantations.

Besides considering silvicultural options, they reviewed the uses of products from felled timber, arguing that carbon is just as much 'stored' in wood products as in a standing tree. A 'half-life' can be assigned to each wood product based on estimates of appropriate 'periods in use'. They assigned a 'half-life' of 1 year to pulpwood, 18 years to particle board and 70 years for construction and engineering uses, with other products within that range.

Matthews (1995) pointed out that not only may sustainable harvested wood act as a direct substitute for fossil fuel where used as a heat source, it also has an indirect effect when, for example, using energy-economical laminated wood beams instead of energy-expensive reinforced concrete beams.

Table 16.7 shows the effect of including or excluding carbon in manufactured wood and wood products, when evaluating silvicultural options for C fixation and storage rates over a 50 year period.

Table 16.7   *Change in annual carbon-sequestrating potential of UK forests*
                +/- benefit of substituting wood for energy-consuming products

Management option	Change over 50 year period No substitution	With direct & indirect substitution
Shorten rotation (-20 yrs)	-1.2 MtC/yr	-1.3 MtC/yr
Lengthen rotation (+20 yrs)	+0.6 "	-0.8 "
Harvest whole trees	0	0
Utilise unmanaged forests	-0.3 "	+0.1 "
Improve timber quality	0	-0.1 "
Stop all felling & harvesting	+5.4 "	-1.5 "
Increase forest area by 20% over 50 years		
with conifers	+0.7 "	+1.1 "
with broadleaves	+0.5 "	+0.8 "

Source   Matthews 1995

The House of Commons Energy Committee (EnCom 1992) when discussing the case for development of renewable energy recognised that any comparison of costs of energy from fossil fuels with costs from renewable sources must clearly state what account has been taken of external costs of pollution and environmental damage arising from mining or drilling for fossil fuels. Omission of such costs has a very large effect on the costs which renewables have to match.

16.28   Organic carbon in peats, cultivated soils and plantation soils

Soil organic matter accumulates under undisturbed vegetation to an equilibrium level. Peats and peaty surfaced soils contain 75% of the soil organic carbon in Britain, 86% of these soils being in Scotland. In a study of correlations between soil organic carbon and environmental and geographical variables, the highest positive correlation was with latitude; the highest negative correlation was with mean monthly soil moisture deficit (Harrison *et al.* 1995a).

Table 16.1 shows that globally, the amount of carbon in soils exceeds that in vegetation. *Table 16.4* shows that in the United Kingdom, with its high rainfall and cool climate in many parts of the country, humus and peat together contain in the order of a hundred times more total organic carbon than is in the surface vegetation. The difference is more marked in the UK because of the comparatively small area of forest land and the extensive areas of peats.

Undrained and drained peat

The effect of undrained peats on greenhouse gases is considered to be 'neutral'. Sufficient methane is emitted from peats in the course of anaerobic decomposition of peat to fully offset carbon sequestration by live bog vegetation.

The equilibrium is disturbed by any operation affecting soil aeration. Soil cultivation and/or drainage improve aeration so that oxidation rates rise, reducing the organic matter equilibrium level.

Cannell & Milne (1995) describe four processes which determine whether drained peats will become net sources or sinks of greenhouse gases.

• Methane production will continue from lower layers of peat. However, where there is a surface aerated layer of peat 20 cm or more thick through which the methane has to pass, almost all will be oxidised. In most situations, aeration resulting from drainage will result in the rate of surface methane emission falling to near zero.

• Oxygenation as a result of drainage can increase the rate of organic matter breakdown 50-fold, so that drained peats become a source of CO_2.

• Microbial decomposition of peat causes some mineralisation of nitrogen which stimulates growth and CO_2 removal but also leads to release of nitrous oxide (N_2O) and other greenhouse gases. In respect of global warming, these effects cancel each other out.

• Trees growing on peat will accumulate C. The amount of litter from planted trees falling on the peat surface is thought to compensate for the C that would have been accumulated by the undrained peat vegetation. Carbon accumulated by the trees as wood is additional.

Future rates of methane emission are likely to increase markedly if the rise in global temperature at higher latitudes is of the order of 2-4°C and there is no change in the water table. A drier climate at high latitudes would reduce methane emissions and vice versa (Fowler *et al.* 1995).

Peat as a fuel and a horticultural resource

While peat is no longer a major source of fuel for farmers on peat lands in western Europe, peat has been worked for fuel for large scale electricity generation in Ireland for much of the latter part of the 20th century. *Sphagnum* and fen peats also remain widely used by the horticultural industry for rooting cuttings and production of plants in pots. While outside the scope of this text, these industrial uses are significant in the overall formulation of national and international policies for management of peatlands, both as habitats to be conserved and for the impact of their disturbance and use on greenhouse gases.

Organic matter in regularly cultivated soils

Annual cultivation is not a normal part of woodland management, except where fire-breaks are maintained by cultivation, and in forest nurseries.

In Scotland at Teindland, Morayshire, the organic matter in soil fell from 17% to 7% where conversion of a woodland to nursery ground was followed by 22 years regular cultivation for nursery production. Most of this loss occurred in the first 10-12 years. On the assumption that a 1% loss of organic matter over the top 20 cm of soil corresponds approximately to a loss of 12 tonnes carbon/ha, the mean annual loss due to regular cultivation was about 5.5 tC/ha. Annual applications of raw hopwaste equivalent to about 3.3 tC/ha prevented net loss and increased soil organic matter by about 1.1 tC/ha/yr.

In the two English nurseries referred to in *Table 11.2*, the soils had previously been regularly cultivated. Effects of a further 15 years nursery cultivations on soil organic matter were small - a loss of 0.1% on one site and a gain of 0.2% in the other over that period. These figures correspond to annual losses or gains of 0.08 and 0.16 tC/ha respectively. Where bulky organic matter equivalent to 3.8 tC/ha had been applied annually, soil organic matter had risen by 0.3 and 1.2%, equivalent to annual gains of about 0.24 and 0.96 tC/ha.

Plantations and carbon storage

Plantations on peat

For plantations on peat, a peat loss rate equivalent to 0.5 tC/ha could be sustained for 6-7 rotations of Sitka spruce Yield class 12 before CO_2 emissions from peat oxidation exceeded accumulation by the tree crop. However, a loss rate of 3 tC/ha could only be compensated for by the first rotation. Estimates of losses suggest that for commercial forests, 3 tC/ha may be representative (Cannell *et al.* 1993, Cannell & Milne 1995, Fowler *et al.* 1995, Cannell 1999).

In Norway, 1.2 and 2.7 tC/ha were lost annually from the top 40 cm of peat respectively in spruce and in pine stands.

In northern Scotland, in an intensively drained experimental plot at Rumster, losses of 6-9 tC/ha were recorded (Harrison *et al.* 1995a).

Plantations on mineral soils

For aerated mineral soils carrying crops on a regular rotation, organic matter fluctuates cyclically, reflecting the age of the crop.

Mineralisation of litter and soil organic matter following clear felling is described in §13.24 in the context of nitrate release into drainage water.

Accumulation recommences at about the time of canopy closure (Harrison *et al.* 1995, Figure 4).

For land being reclaimed, sand dunes and other sites where soil has been lost or where soil profiles are not yet fully developed, soil organic carbon can be expected to accumulate to the point where the soil organic matter pool is in equilibrium with site, species and silviculture.

Soil organic matter can also be expected to increase where soils previously under arable cultivation are planted with trees to create woodland.

Conventional forest management rotations as practised in Britain do not achieve maximum carbon sequestration for the site and crop. *Figure 16.2* shows the relationship between the mean annual increment (MAI) and the amount of carbon stored in the forest. Maximum MAI is reached while forests are still accumulating carbon. A rotation of Maximum Carbon Accumulation (MCA) would be about twice the length of a maximum MAI rotation. At MCA age, the total C sequestered is about 50% more than at the time of the maximum MAI rotation age (Thompson & Matthews 1989, Cannell & Milne, 1995). In the UK, the 'financial rotation' on which many felling programmes have been based may be 5 years less than the maximum MAI rotation. Plantations on a no-thin, premature-clear-fell regime because of threatened instability may be cut even earlier.

Figure 16.2   *Relationship between age of maximum MAI & C accumulation*

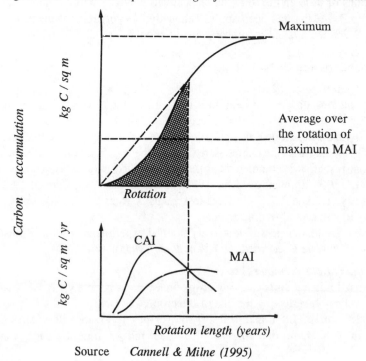

Source   *Cannell & Milne (1995)*

Figure 16.3 shows the predicted long-term effect on carbon accumulation when carbon cycling in a Corsican pine crop on a conventional rotation is combined with provision for accumulation in wood products.

Accumulation values are shown for four successive rotations and their aggregate value. A line showing mean accumulated C values is shown for the period from the end of the first rotation. It slowly rises, reflecting the long life of a proportion of timber products.

Figure 16.3 *Carbon accumulated in a managed pine forest & its derived wood products*

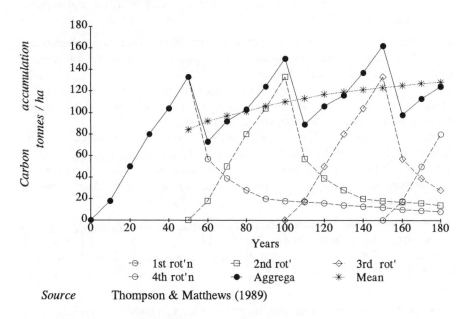

-⊖- 1st rot'n	-⊟- 2nd rot'	-◇- 3rd rot'
-⊖- 4th rot'n	-●- Aggrega	-*- Mean

Source Thompson & Matthews (1989)

Values for fixing carbon

Once it has been established, nationally or internationally, that carbon dioxide levels have to be reduced and that costs will be incurred to do so, values can be assigned to achieving this through accumulating C in the form of wood.

In *Assessing the returns to the economy and society from investment in forestry* (Pearce 1991), the value per unit carbon fixed was equated to either:

• the cost of damage which would have occurred if the C had not been fixed *ie* the value of avoidance, or

• the least cost of the best alternative means of achieving the same effect, *ie* the cost of substitution.

The value proposed in America in 1989 following the 'damage avoidance' approach was £8/tC. Figures for the 'alternative method' approach were approximately double. Pearce in his study used the more conservative figure to calculate 'Net Present Values' (NPVs) for eight forest types, with and without values for carbon fixation. He concluded that present values for fixing carbon, using a 6% discount rate, ranged between £142 and £254 per hectare.

16.29   Growth responses to higher concentrations of atmospheric CO_2

Increasing the concentration of atmospheric CO_2 around plant foliage has a direct effect on plant photosynthesis and growth. This has given rise to more detailed studies on interactions of growth rates with other environmental factors.

In one of the early studies, conifer seedlings raised in CO_2-enriched greenhouses for 16 weeks after sowing showed 20 - 28% increase in dry matter production (Canham 1976).

Later, Jarvis (1989b) reviewed approximately 30 experiments on conifers and 30 on broadleaves, where plants had been exposed to double the atmospheric concentration of CO_2. In all cases, the rate of growth of dry matter increased by 20-120%. In most experiments, the leaf mass increased, whether in number, surface area or thickness. Root mass also increased, often linked to an increase in the root/shoot ratio, especially where nutrient supply was limited.

Evidence suggests that increases in CO_2 concentration compensate for shading of leaves in the canopy. However, at that time, the net effect of increased leaf mass and leaf area on water use, water stress, associated increased transpiration and interception of rainfall, were beyond the capability of process models to incorporate.

In other recent experiments, beech, wild cherry, Sitka spruce and pedunculate oak each produced a greater dry mass under a 250 ppm CO_2 supplement. For most species, transpiration was slightly reduced (Kersteins *et al.* 1995).

Assessments of growth in many European forests in the 1990s showed there to be consistently greater increases in growth rate than had been predicted from management tables. Spiecker *et al.* (1996) attributed this effect to increased CO_2 levels, claiming that 'forest decline' was a localised short-term occurrence associated with excess pollution in the post-war period and not a permanent loss of fertility.

In a recent study to examine the basic physiological responses to increased CO_2, trees in growth chambers were exposed to doubled concentrations of CO_2 with and without extra ozone and water. Added CO_2 stimulated dry matter production of oak by 40 - 75%; effects on growth of ash and Scots pine were smaller but also positive. Ozone reduced dry weight production of oak and ash, the effect being mitigated by irrigation (Broadmeadow *et al.* 1996).

Tree growth models based on tree physiological processes, site climatic factors and their interaction, are being developed by the Forestry Commission. Changes in tree response to increased CO_2 may necessitate adjustment to management table models of tree growth and yield (Broadmeadow *et al.* 1996).

Site studies showed 20-40% increases in growth rates in conifer plantations in northern Britain. Process models to simulate these increases tested the effects of increased N, CO_2 and higher temperatures. The temperature effect was less than anticipated but N and CO_2 effects were predicted to continue into the 21st century (Cannell *et al.* 1998).

16.3   OTHER GREENHOUSE GASES

16.31   Methane as a greenhouse gas

Methane is present in the atmosphere at a concentration of about 1 part per million (ppm) compared with 350 ppm for carbon dioxide. Nevertheless, it is an important greenhouse gas and a commercial energy source. The concentration in the environment has increased by 250% since the beginning of the industrial revolution.

The particular circumstances under which methane acts as a pollutant, causing damage to trees are described in §13.51.

Table 16.8 summarises estimated sources and sinks of methane and should be read in conjunction with *Tables 13.1* and *16.2*. *Table 16.8* shows that wetlands produce about 75% of naturally occurring methane. §16.28 above describes the gas emissions of undrained and drained peat in Britain.

Table 16.8   *Sources and sinks of atmospheric methane*   *(Tg CH_4/yr)*

Sources			Sinks		
Natural					
Wetlands	115	(100-200)	Atmospheric		
Termites	20	(10-50)	removal	470	(420-520)
Ocean	10	(5-25)	Removal		
Freshwater	5	(1-25)	by soils	30	(15-45)
CH_4 hydrate	5	(0-5)			
Total natural sources	155		*Total sinks*	500	
Anthropogenic					
Rice paddies	60	(20-150)			
Animal husbandry	105	(85-130)			
Coal, gas & petrol ind'try	100	(70-120)			
Urban sewage & landfill	55	(20-70)			
Biomass burning	40	(20-80)			
Other	17				
Total anthropogenic sources	360		*Annual atmospheric increase*	32	(28-37)

Source   IPCC, 1992*   1Tg = 1 Teragramme = 10^{12}g = 1million tonnes

The figures in *Table 16.8* are global and include very substantial areas of coastal and tropical wetlands where methane emission is higher than from UK peats. The table also shows that while anthropogenic sources of methane exceed natural sources, they are not in any way dependent on woodland practice.

Marine deposits

Methane at temperatures below 25°C and under high pressure, can form ice-like crystals of methane hydrate, where the methane molecule is contained in a 'cage' of water molecules (a 'clathrate' lattice). The global reserves of such methane are enormous but are located at 500m+ depth in oceans and at present are too dispersed for economic recovery (Clennell 1998, Pendick 1998).

16.32   Nitrous oxide as a greenhouse gas

Nitrous oxide is the commonest of the oxides of nitrogen occurring in the atmosphere; its concentration is about one third that of methane. *Table 16.9* gives estimates of its sources and sinks.

N_2O is a potent greenhouse gas accounting for 6% of the total heating effect. It is decomposed by photolysis in the stratosphere (INDITE 1994).

Concentrations of N_2O have increased from a pre-industrial level of 285 ppb to the present 307 ppb. Although natural emissions from soil as a result of microbial action in nitrification and denitrification dominate over anthropogenic production (Bouwman 1990), its greenhouse gas properties ensure that industrial emissions remain a target for reduction.

There is no evidence that N_2O is deposited onto the land (INDITE 1994); it seems to have no material role in plant health or growth.

16.4   RENEWABLE ENERGY

16.41   Alternatives to fossil fuels

Until the late 1960s, there were few voices expressing concern about increasing use of fossil fuels. However, in 1970, the Middle-east producers sought substantial increases of crude oil and gas, provoking a politico-economic crisis. Since then oil and gas prices have continued to fluctuate. While the cost of exploring for and developing reserves in ever deeper water or more remote parts of the land mass continually increases, the conflicting pressures of over-production and need for hard currencies has placed many individual countries in difficulty and has been at the base of continuing international political tension and low oil prices.

Following the 1970 oil crisis, international programmes were initiated to seek renewable sources of energy. In Europe, the International Energy Agency (IEA) was formed under the auspices of OECD to promote cooperation in reducing excessive dependence on oil (Mitchell 1991). The European Commission also introduced several programmes *eg* ALTENER I and II ALTernative ENERgy), JOULE, LEBEN, ARBRE, THERMIE,

Table 16.9   *Sources and sinks of nitrous oxide*   *(Tg*N/yr)*

	Global sources		*Global sinks*
Natural			
Oceans	1.4-2.6	Removal by soils	(no data given -
Tropical soils			not significant)
Wet forests	2.2-3.7	Photolysis in stratosphere	7-13
Dry savannas	0.5-2.0		
Temperate soils			
Forests	0.05-2.0		
Grasslands	(no data given)		
Anthropogenic			
Cultivated soils	0.03-3.0		
Biomass burning	0.2-1.0		
Stationary combustion	0.1-0.3		
Mobile sources	0.2-0.6		
Adipic acid production	0.4-0.6		
Nitric acid production	0.1-0.3		
Total emissions	*5.2-16.1+?*	*Total sinks*	*7 - 13*
	Atmospheric increase		*3-4.5*

	UK Sources	*(Gg*N/yr)*
	(No estimates of sinks)	
Fertilized land	22-36	
Livestock waste	12	
Forest & woodland	5	
Semi-natural land	6	
Road transport	3	
Industry	49	
Other fuel combustion	2	*total emissions 100-115*

Source   INDITE 1994   *1Tg = 1000 Gg;   1Gg = 1000 tonnes

to encourage inter-state cooperation in developing new renewable energy uses. The target for 1996 was a 6% contribution from renewable resources to total European energy consumption; the aim is to double this by 2010. Within that target, a three-fold increase is sought for biomass.

Energy cropping has also been encouraged against a background of agricultural surpluses through operation of the Common Market Agricultural Policy (CAP). Potentially surplus land on the scale of 1 million hectares in 2000, rising to 5 million ha by 2010 is predicted, much of it well suited to energy crops of willow or poplar (Carter 1990a).

Nuclear power currently provides a substantial proportion of electricity generating capacity world-wide, and had been considered as the principal long-term alternative to fossil fuels. However, nationally and internationally, public support for such long-term reliance has diminished through concern about repetition of accidents such as at Chernobyl and Four Mile Island, and the difficulties of finding acceptable sites for radio-active wastes and residues arising from decommissioning.

The House of Commons Energy Committee, in their 4th Report, Session 1991-92, (EnCom 1992) strongly supported all aspects of renewable energy, but in their forecast for 2025, saw forestry as contributing materially as a supplier of sources of renewable heat rather than electricity. They commented also that while in the 1970s the impetus to develop renewable resources arose out of concern about oil supplies, in the 1990s, the concern was chiefly to reduce further global warming, using carbon in trees as a recyclable resource.

Energy and carbon accumulation

Dry weight yields of established short rotation willow coppices cut on a 2-3 year cycle are expected to average 10-12 tonnes dry matter per hectare per year, corresponding to 4-5t C/ha/yr (Tabbush & Parfitt 1996). Current experiments may show this figure to be conservative (Armstrong 1997, 1999).

Creation of new areas of fast-growing woodland on mineral soils is, on a unit area basis, a most effective way of realising a net reduction of carbon dioxide (Cannell 1988). However, the high population density in the UK and the scale of energy use is such that covering every potentially suitable hectare would not be sufficient to offset the country's current excess CO_2 emission.

On a larger scale, Cannell & Cape (1991) point out that globally, the total area of fast-growing plantation forest would have to increase five- or six-fold to offset global increases in CO_2; alternatively, the net annual growth rate of all the world's forests would have to be increased by 50% over the next 100 years to have a similar effect.

If targets for reduction in net global CO_2 emission are to be met, additional tree planting has to be one element of a substantial portfolio of programmes including both direct actions to conserve energy and also the use of renewable energy sources.

16.42   Development of renewable energy sources in the UK

From the mid-1970s, the *Energy Technology Support Unit*, (ETSU) acting for the Department of Energy and its successor, the Department of Trade and Industry, sponsored research over the whole range of potential renewable energy sources. The scale of support for forestry is reflected in the substantial number of published reports (ETSU 1998f).

Potential renewable energy sources include:

- wind energy;   • hydroelectric power;

- tidal & off-shore wave energy;   • solar cells (photovoltaics);
- solar direct heat;
- geothermal heat from dry rocks or hot springs;
- fermentable organic materials: *ie* landfill gas, sewage, straw, wood;
- 'biofuels'.   Besides willow and poplar energy crops and forest
harvesting residues, this category includes municipal wastes and agricultural
wastes *eg* straw, broiler house litter.

By 1989, Department of Energy expenditure on Renewable Energy R & D over the preceding eleven years exceeded £135 million, biofuels accounting for between 2 and 9% annually. In a review of prospects at that time, combustion of dry wastes and anaerobic digestion (landfill gas) were categorised as 'economically attractive'. Energy forestry was 'promising but uncertain' (DEngy 1989a, b, c).

Much work on coppice and forest residues systems and costs has been channelled through the 'Wood supply research group' at Aberdeen University (Ford-Robertson *et al.* 1991).

Over the same period, the Northern Ireland Department of Agriculture Horticulture and Plant Breeding Station embarked on an extensive programme, seeking alternative crops for 200 kha of marginal agricultural land. Their initial trials, starting in 1973, identified willows as the most promising potential crop, with yields of 12-15 t/ha dry matter (Dawson 1988).

16.43   Wood as an energy source

Wood as fuelwood and as charcoal was the principal source of energy, world-wide, until the advent of the industrial revolution. Fuelwood remains a dominant wood use in many tropical countries; it still accounts for nearly half the global consumption of wood (*Table 1.4*).

Firewood continues as a significant source of fuel in those parts of Europe with substantial forests. In Britain, while most households no longer burn wood for domestic heating or cooking, in the 1990s 50K tonnes of charcoal were used each year, mostly for barbecues. Of this 95% was imported (Hemsley 1998).

Contemporary uses of wood for heating in the UK range from the open fire to sophisticated and instrumented wood-burning systems. These may be grouped into:

- Heat-only systems, §16.44;
- Electricity generation, §16.45;
- Biofuels, §16.46.

Waste heat from systems involving electricity generation may be utilised in 'Combined heat and power' (CHP) schemes, for heating hot water, drying *etc.* thereby reducing unit costs and increasing % energy recovery.

Specially planted energy crops

As part of the programme to develop uses of wood for renewable energy, the component stages of potential high-yielding production systems have been studied, including:

- trials of willow and poplar clones to determine rates of mean annual dry matter production by site type (Stott & Parfitt 1986, ETSU 1989b, 1998e, Potter 1989, Dawson 1992, Potter & Tabbush 1991, Armstrong, 1997). By the mid-1990s, 17 clones of poplar hybrids and 24 clones of willow hybrids had been recommended for short rotation energy coppice (Tabbush & Parfitt 1996, 1999, Armstrong 1999). Many of the willow clones under trial are selections developed in Sweden for energy cropping;

- comparison of fast-growing tree species and non-woody plants *eg Miscanthus* (DTI 1993a);

- cultural systems - plant type, stocking density, spatial arrangement on the ground, nutrient regime, weed competition (Aldhous 1991, Ford-Robertson *et al.* 1991, ETSU 1996a);

- risks from pests and diseases. Rusts on willow have been sufficiently serious that single clone plantings are not recommended (Dawson 1988, Tabbush & Parfitt 1996). Suitable clones have been grouped by susceptibility; polyclonal plantings are recommended, ensuring that clones of the same susceptibility group are not planted in adjacent rows (DANI 1993b, DTI 1994a, ETSU 1995a, McCracken & Dawson 1996). Willows are recommended not to be planted in close proximity to larches because of the latter's role as alternate host for rusts (ETSU 1998c). In 1999, carefully selected mixtures of clones were recommended, so as to dilute the distribution on the ground of clone groups which had been found to be susceptible to newly developing stains of rusts (Tabbush & Parfitt 1999);

- harvesting systems and technology of mechanical harvesting of coppice willow (Dawson 1988, Deboys 1994, FC TDB 1995, ETSU 1989a, 1997c, 1998d, Anderson 1999); storage of cut or chipped material (ETSU 1997b, d);

- minimisation of loss of dry matter during storage. Coarse chips with natural ventilation sufficient to avoid heating in the chip pile is desirable (ETSU 1997b);

- combustion equipment (DTI 1994b, ETSU 1998b);

- environmental impact.

A study on the hydrological effects of short rotation coppice showed that water loss by transpiration of short rotation coppice of poplar clones was about 35% higher than for ash or beech. The high rate was attributed to high stomatal conductance linked to absence of response to low air humidity and delayed response to soil water deficits (ETSU 1996b).

This higher water demand could become a threat to ground-water recharge in low rainfall areas; on that account, large-scale short rotation coppice was to be preferred in the wetter parts of the country.

The low nitrogen input associated with short rotation coppice could make it favoured in 'Nitrate Sensitive Areas'; use of sewage sludge in conjunction with short rotation coppice is feasible but should be closely monitored (ETSU 1996b).

Short rotation coppice for energy production (ETSU 1996a) summarises conclusions from this work.

Unmanaged or under-managed woodlands and woodland residues

Reviews have been made of the potential of

- abandoned coppice;
- small woodlands (Humphries 1983);
- unharvested wood in managed crops;
- potential of lop and top residues following felling. A preliminary trial showed that the moisture content of residues baled and stacked at the forest rideside for 12 months fell to 16-19% (wet weight basis), and offered net economies in transport costs on that account (ETSU 1997d). It is necessary, once the bales have dried, to prevent rewetting;
- potential of wastes at other stages in the wood chain (Martindale 1986).

Wood fuel from forestry & arboriculture (ETSU 1998h) sets out good practice guidelines for the development of a sustainable energy production industry using small roundwood and woodland residues.

Economics of operations

Wood has not been an important fuel in the UK over the last 150 years because of limited supply and availability of alternative fuels which are cheaper, more efficient and more convenient. Analyses of the economic feasibility of energy crops demonstrate their sensitivity to productivity, operational costs and grant aid (Mitchell 1991b, Ford-Robertson *et al.* 1991).

The cost of haulage is frequently a major component in wood supply of low-value products. Energy crop systems that have a heavy concentration of material within a short-haul radius are advantageously placed.

16.44   Small roundwood & wood chips as a heat source

The market for logs for open fires and wood-burning stoves in the UK has never quite died out but, in the 1950s and 60s, it was at a low ebb. However, contemporaneously with the 1970 oil crisis, Dutch elm disease was killing trees in large numbers throughout southern and central Britain, leading to a glut of dead and dying wood. Much of the timber-dimension material was felled and sawn. Nevertheless, there was plentiful surplus branchwood and smaller stemwood. Substantial areas of unmanaged or under-managed broadleaved woodland in southern England also contain much potential energy fuel as logs or chips. Further supplies came forward following the severe gales in the south of England in 1987.

These events stimulated and sustained a resurgence of small-scale use of wood for domestic heating, encouragement of local marketing, and efficient use of fuelwood in small log form (Keighley 1985, ETSU 1996a, c).

The advent of mobile wood chippers opened up a further range of opportunities for small dimension branchwood previously unutilised; at the same time improvements in the technology of harvesting and comminution were sought to reduce costs (ETSU 1990b, Alexander & Chadwick 1991).

Equipment currently available ranges from manually fed log-burning stoves to fully automatic wood-chip boilers. Chipped wood wastes have been evaluated in a number of heating trials in institutional or industrial premises. An early trial at a distillery demonstrated how the practical difficulty of forecasting heat requirement combined with a substantial fall in oil prices increased the costs of wood chip heating beyond the cost of alternatives (ETSU 1989b). Nevertheless, there has been a steady flow of reports of successful installation of wood burners for small-scale rural heating needs, usually where heat could be utilised close to the point of wood production (*eg* DTI 1993b, Hague 1993, ETSU 1996c, 1998c).

Larger wood-burning units may be used to raise steam to drive turbines linked to electricity generators. Virtually all the existing wood-fire driven electricity generators in the world are steam turbine systems (as are fossil-fuel driven generators).

While figures are not readily available, all large sawmills, chipboard and pulpwood-using factories burn their unsalable waste for process heat. Reviewing future potential markets for British timber, Jaako Poyry (1998) forecast the opportunity for a large combined bleached softwood Kraft pulp and bio-energy mill. This is envisaged as producing 500-600 Ktonnes of pulp and 40-50 MW of power from its waste products and from forest and wood residues available in the vicinity (FCNR 1998a).

16.45   Electricity generation

As part of the 1989 Electricity Act, the UK government introduced the *Non-Fossil Fuel Obligation* (NFFO). In Scotland, a similar requirement is referred to as *the Scottish Renewables Obligation* (SRO). These obligations required public electricity supply companies to secure specified amounts of new generating capacity from non-fossil sources. It was accepted that initially, electricity would be purchased at a price appreciably above the 'base/pool' price. However, it was also expected that prices would 'converge' as experience was gained (ETSU 1998g).

Under the NFFO, all potential suppliers are invited to tender to supply electricity from non-fossil fuels at a unit price per kWhour of electricity, for whatever amount they believe they can produce. Once a NFFO contract has been signed, *ie* a market at a specified price for a quantity of electricity output guaranteed, the generating facility has to be constructed and commissioned. This has taken anything from 2 - 6+ years.

By 1998, there had been five rounds of NFFOs, the first being in 1990. A scheme at Eggborough, Yorkshire for 8 Megawatts (MW) Declared Net Capacity (DNC) won a NFFO3 contract under an ARBRE programme. For an area approximately 40 miles in radius round the site, the Forestry Commission introduced a Locational Supplement for Short Rotation Coppice to supply the projected power station. The supplement could be applied for during a three year period commencing in August 1998 (FA 1998).

In the 4th NFFO, 7 contracts were won for energy by gasification or pyrolysis of biomass and forestry wood waste. Five were in the north of England and two in south Wales. Their total capacity, 67 MW DNC, constituted about 8% of the capacity accepted in NFFO4.

The results of a fifth tranche were announced in September 1998 but included no schemes based on forestry waste.

While NFFO bids are reviewed against the 'pool price', the marginal cost of supply at times of peak demand may be 2 - 3 times the pool price. The wood residue NFFO4 bids accepted were at about twice the pool price. Short rotation coppice supplies are available most readily from the autumn through to late spring and could be offered to generators able to supply at times of peak load and so justify higher prices - a seasonally based 'price convergence'. However, that would depend on the annual charge for generating equipment not being cripplingly higher because of shorter running hours than if the equipment were in use all the year round.

In mid-1998, in aggregate, over 200 NFFO schemes with an annual capacity (DNC) of 578 MW were running. Of these, combustion of landfill gas, incineration of municipal waste, and wind generation accounted for 170, 154 and 133 MW DNC respectively (ETSU 1998a).

Gasification
If wood is heated with a restricted air supply, combustible gases are given off. These, cleaned to remove tars, can be used to drive an engine coupled to a generator. The principle and technology have been long practised with gas from coal and, on a lesser scale, from charcoal. However, in mid-1998, while prototypes were under trial, electricity generation derived from wood gasification had not been fully demonstrated commercially (ETSU 1998h).

NFFO schemes using wood biomass energy crops for gasification
Up to the middle of 1998, energy from wood biomass has barely registered as a potential electricity producer.

Northern Ireland pioneered medium scale, short rotation coppice for energy. Willow coppice was grown on a three-year cycle and a combined heat and power (CHP) gasifying plant installed at the agricultural college at Eniskillen (DANI 1993a). After an initial period of monitoring, a NFFO contract was secured. This was the first NFFO wood-based scheme using willow coppice to be commissioned in the UK and was the first farm-based CHP unit anywhere in Europe. Electricity is produced by a 100 kW DNC generator together with 150 kW heat output. The latter is used for space heating, hot water for estate houses, grain drying and chip drying (ETSU 1998b).

16.46   Biofuels

In reviews of the potential for alternative energy sources from wood, biofuels have always been listed (eg Dawson 1992). Two types of process are possible:

- pyrolysis, or wood distillation;
- high pressure thermal liquifaction at lower temperatures than are involved in pyrolysis, followed by fermentation and/or distillation.

In pyrolysis, wood is heated completely without air and degrades into combustible gases, charcoal and 'bio-oil liquor', the last of these being a product that can be easily stored and transported (ETSU 1998h).

In the 1940s-80s, wood distillation plants operated, *eg* in the Forest of Dean and Nottinghamshire (Reynolds 1953, Hart 1967), but have since closed. They produced charcoal and a range of distillates, including methanol (hence its older name 'wood alcohol'). This can be used directly as a petrol extender or as a feedstock for methyl tertiary butyl ether - another petrol extender.

Fuels from wood bio-oil are not cheap to produce. While oil-prices are at their 1996-1998 level, their commercial prospects are poor unless a premium can be afforded for clean-burning fuel.

One NFFO scheme involving pyrolysis is under development in south Wales.

16.47   Waste paper as an energy source

It has been conventional wisdom that recycling of paper is a desirable means of utilising a waste product and reducing the call on pulpwood from forests. In the UK, one of the largest recycling mills at Aylesford, Kent recycles 450 K tonnes annually, the output constituting about 5% of the newsprint used in Europe. It takes 14 days to complete the cycle from mill to printing press, newsagent, reader, recycling bin and back to the mill.

This operation however has hidden environmental costs, *eg* fuel used to convey paper to collecting points. The recycling process itself, particularly the de-inking and cleaning are energy-hungry. A case has been made that in gross energy conservation terms, rather than recycle as pulp, it would be preferable to recover the contained energy by burning waste paper in local energy-producing incineration schemes.

For small roundwood, continuing outlets to pulpmills are vital if forests are to be economically sustainable (Pearce 1997b).

16.5   PERSPECTIVES ON CLIMATIC CHANGE AND FORESTS

16.51   International collaboration

International cooperation, established in the 1960s by joint actions on acid rain and forest decline, continued in the 1980s. Concern about climatic change led in 1988 to an international joint initiative by the United Nations Environment Programme and the World Meteorological Organisation. A conference was called which led to the formation of the *Intergovernmental Panel on Climate Change* (IPCC). Its role has been to assess research and policy options on climatic change and report on risks of global warming.

Action by the IPCC led to the *Framework Convention on Climatic Change* (FCCC) agreed at the UNCED conference in Rio de Janeiro in 1992 and aimed at 'stabilization of greenhouse gases in the atmosphere at a level that will prevent dangerous anthropogenic interference with the climatic system'. The UK ratified its support for the Convention in December 1993; by 1998, it had been ratified by 174 countries.

Under the convention, the UK and other signatories undertook to seek to prevent further increases in greenhouse gases, and by 2000 to restore levels to what they were in 1990. The principle of quantitative cuts in emissions of CO_2 was embodied in a Protocol agreed at Kyoto in 1997 as a first step towards the reduction of emissions from fossil fuel combustion and from some terrestrial ecosystems. In 1998, within the European Union, a burden-sharing agreement was reached, the UK undertaking to obtain a reduction of 12½% from 1990 CO_2 levels by 2010. Application of the Kyoto principles was discussed in a larger forum at Buenos Aires in 1998 but with less agreement (UK Govt 1994b, DETR/MO 1997, IGBP 1998, UKCIP 1998a, b, Hulme & Jenkins 1998).

UK government programmes to restrict increases in some greenhouse gases were introduced under the FCCC and have been summarised in *Climatic change - the UK programme* (UK Govt 1994b). Studies on climate prediction and associated subjects were to be co-ordinated at the Meteorological Office Hadley Centre set up in 1990 at Bracknell. UK expenditure in 1992/3 on climate research and supporting data collection totalled close to £200 million. Part of this programme was to restore, protect and enhance carbon sinks, particularly forests (UK Govt 1994a, 1994b).

Details of programmes to increase efficiency in use of energy in the UK were set out in *Climate change - Progress report on carbon dioxide emissions* (DEnv 1995a). A consultation paper on *UK Climatic Change Programme* indicated that about half of the reduction could come from increased use of *Combined heat and power* technology, and almost all the rest from increase use of renewables (DETR 1998).

Fuller details of programmes are given in *Climatic change and its impact - a global perspective* (DETR/MO 1997) and *Climatic Change Scenarios for the United Kingdom: Technical Report No 1, Summary Report* and *Scientific Report* (respectively UKCIP 1998b, Hulme & Jenkins 1998).

16.52   Recent technical and scientific reviews

Mean global temperature is predicted to continue to show evidence of slow increase to the year 2000 and beyond (Jones *et al.* 1998).

European Forest & Global Change (Jarvis 1998c) contains 11 reviews covering the scientific aspects of studies of the effects of increasing CO_2, including methodology, photosynthesis, respiration and growth, water use, interactions with temperature, nutrients and process modelling. The question of the scale, location and plant science of the terrestrial sinks for carbon are discussed, but left unanswered pending further research.

Forest Ecosystems, Forest Management and the Global Carbon Cycle (Apps & Price 1996) with 34 papers from an advanced research workshop covers similar ground but includes also forest products and the carbon cycle, and socio-economic issues.

Boreal forest and global change (Apps *et al.* 1995), a collection of 45 papers, constitutes a broadly based reference text on effects of change for the whole circum-polar boreal forest region.

A slightly earlier conference discussed the implications for global climate change on crop protection (Atkinson 1993a, b). While some shifts in the geographical location of production of particular crops were anticipated, there was not enough data to come to any firm conclusions on the interaction of climatic change with crops and pests.

16.53   Prospects

While nations are still struggling to come to terms with the actions necessary to check greenhouse gas emissions, it is premature to anticipate the outcomes. Nevertheless, the evidence of the need for action continues to accumulate.

There is a body of opinion that pressure to economise in energy use and CO_2 generation is best supported by a carbon tax. Cheaper domestic heating costs through improved insulation may, for example, only encourage setting a higher living room temperature (Pearce 1998b). In Sweden, a carbon tax has been used to favour use of wood-burning heating systems. In the UK, tax duties on petrol and diesel fuels have been raised above cost-of-living inflation rates as an incentive to fuel economy.

However, it is wrong to think that the problems are not global. By 1970, 9 million sq km of forest had been cleared for agriculture. At that time, tropical forest destruction in preparation for agriculture was running at 0.113 million sq km/yr and in the 1980s averaged 0.16 million sq km/yr. Since then, annual rates of clearance have accelerated, such clearance being principally for agriculture, not timber production (Holdgate 1995). In those parts of the world where population growth is greatest, pressure to maintain such rates of clearance is likely to be high.

Part V

Our future

CHAPTER 17

The woodland environment for the 21st Century

'In sustainable forests, trees may be used but the forests and their people remain'

17.1   HOLISTIC LIVING ON A SHRINKING GLOBE

17.11   Ethics, Economy and Environment

The global context in which 21st century forestry in the UK will operate is likely to feature:

- a world population increasing substantially over the next 30-50 years and requiring food, housing, wood and wood products;
- greenhouse gases increasing substantially over the next 30-50 years, as a direct result of increased population and increased industrialisation, and probably causing climatic change, and possibly rapid climatic warming;
- unequal distribution of food, wealth and knowledge, the three often being associated together;
- political factions seeking power, feuding and fighting, with genocide locally being practised as a means of domination, or revenge for past wrongs;
- a trend of large international commercial businesses to enlarge, becoming less amenable to national control, and at the same time being apparently indifferent to the effect of their actions on the economy and society of the countries in which they operate;
- public distrust of the role of chemicals in the countryside provoked by practices in agriculture and on industrial land, inside and outside the UK;
- single-subject pressure groups often stimulated by what are seen to be anti-social activities of commercial or state bodies.

None of these situations applies uniformly to all countries; in the UK for example, population is, if anything decreasing, and activities of pressure groups are well developed.

Sustainability

Since the Rio Conference, there has been intensive and widespread debate about sustainability. The debate has been pursued over a whole range of topics. Many are unrelated to forestry, *eg* transport policy, housing and urban capacity (Southwood 1997). Some have considered sustainability from an ethical and philosophical viewpoint, *eg* consensual participation by interested parties (stakeholders) (Hurka 1996). Others, *eg* on land use options or paper production (IIED 1996), have a direct link to sustainable productive woodland,

The 11th World Forestry Congress debated *Forestry for sustainable development; towards the 21st century*. As well as 38 commissioned papers, over 1200 voluntary papers were presented! The first keynote speaker (Harcharik 1997) emphasised the point made in many studies, that sustainable development has to integrate three interests, environment, economics and society. Difficulties arise from differences in the ethics and values attributed by stakeholders to outcomes for each of these interests.

Another paper explored 'pluralism' *ie* means of accommodating multiple interests (Anderson *et al.* 1998).

A report from the Macauley Land Use Research Institute comments:

'Sustainable land use systems provide a wide range of goods and services. Some are easy to value (*eg* food and fibre). Others are not marketed (*eg* conservation, amenity, recreation, environment) and are difficult to value in money terms. Nevertheless, all have costs and benefits in social, economic and environmental terms. In analysing land use options, the problem is how to express and value the trade-offs being made between these different dimensions.

Two other effects are also important; ... firstly, costs may not be felt in the same place as the benefits - *ie* a geographical effect. Secondly, costs and benefits are experienced by different groups in society in different ways, so that there is no single value either for 'cost' or for 'benefit' - *ie* there is a societal effect. As a consequence, the term 'sustainable' has no absolute definition; its use reflects a spectrum of beliefs in our society, ranging from the technocentric view of man winning over nature, to the ecocentric view of man within nature. Scientific results must be interpreted in the context of these different viewpoints.' (MLURI 1997b).

The same could be said for discussions between stakeholders seeking consensus. Indeed, until stakeholders fully recognise such differences in ethical values and find a basis for discussing the technical and managerial issues in the context of differences in values, the prospects for consensus are slim. Any agreements reached are likely to involve compromise, leaving all parties feeling that reluctantly they have had to give more ground than they wished, but that the agreements reached justified the sacrifice.

Bio-diversity in the forest

Maintaining bio-diversity was the second basic principle of the Rio and Helsinki agreements. Apart from issues of restoring native woodland under habitat action plans, the thrust of thinking has been to favour maintenance of a diverse vegetation overall within the forest, as the base of food chains for many micro-organisms.

This fails to give adequate recognition to the natural stages of growth through which forests pass, and the changing ecological and economic pressures that have been accommodated to get to the present. The much admired 'old growth' forests of today are the outcome of past responses to natural disturbing processes such as fire, devastating windblow and grazing/browsing pressure from vertebrates.

Kimmins (1999) stresses the need for a full understanding of the natural cycles in forests, the frequency and type of natural disturbance as stimuli for regeneration, and the temporal and spatial variability that results, if attempts to maintain bio-diversity are not at risk of being nullified by the effect of natural processes that should have been foreseen.

In the UK, problems of sustaining forest and maintaining bio-diversity are particularly acute because:

- a relatively small number of tree and shrub species constitute the native woodland flora;
- most native woodland has long been converted to farm land. Much of what woodland remains has been subject to intensive 'creaming' of the best stems and the most favoured species;
- forestry and horticulture are dominated by introduced species, chosen for their greater productivity and, in landscape and amenity plantings, their wider range of colour, texture and form;
- the major native timber-producing species are relatively low-yielding. Conifers introduced from North America commonly grow 3 - 4 times more rapidly than oak and beech and produce harvestable crops in 45 - 60 years compared with 120 - 160 for beech and oak.

Sustainability in forestry has been associated in most of the world with sustaining the natural ecosystem (Innes 1993). Plans are in hand to manage and extend the relics of British natural woodland types where these exist. Listings of semi-natural ancient woodlands are a starting point (*eg* Hodge *et al.* 1998).

For most of managed plantation forestry in Britain, however, the most appropriate target is to increase woodland bio-diversity, allowing natural processes to mould new forest types and accepting that the many introductions, welcome and unwelcome, are here to stay.

Economic sustainability of non-market benefits

Pearce (1991) identified ten potential non-market benefits of forestry in the UK, and discussed sustainability in terms of financial capital. He noted the recognised need to 'keep capital intact' and not to live off capital. Applied to forestry, different results follow according to how 'constant capital stock' is defined, and how far its valuation goes beyond land values and growing stock to take account of non-market benefits.

Not all non-market attributes are universally appreciated. Public access is viewed as a benefit by many but nevertheless resisted by some. Opposition as well as support has to be included if particular non-market benefits are not to be over-valued (Macmillan & Duff 1998).

17.12   Spiritual values

Fifteen million dead!
Fifteen million dead what?
'Trees' I said.
Like great elephants they lie
And die,
Their tangled roots exposed to sky,
And I
Cry.

Hurricane 1987 Jane (1987)

While seldom explicitly debated, the sense of spirituality is part of human experience. Sacred groves and holy trees can be found world-wide; in Britain while yew is particularly associated with churchyards (Chean & Brueton 1994), throughout the country, memorial trees and groves continue to be planted.

That forests can offer stillness and awe, spiritual comfort and sadness, must not be forgotten in the context of consensus and sustainability.

Kimmins (1999), in his analysis of the requirements of sustainable forest management, started from a view of *Respect for nature*. He pointed out that the common usage of 'respect' includes both to 'esteem or honour' and 'to be aware of, so as to avoid harming'. 'Esteem and honour' recognises spiritual and emotional values; 'awareness' requires detailed understanding of the structure and behaviour of the forest over time.

Such recognition and understanding is essential if foresters are to manage sustainable forests so that the spiritual values are also not neglected.

17.2   FORESTRY IN THE UK IN THE 21ST CENTURY

17.21   The UK Forestry Standard

The pattern for forestry in the United Kingdom at the start of the 21st century is set in the *UK Forestry Standard: the Government's approach to sustainable forestry* (FC FS 1998a). This embraces woodlands in Great Britain and Northern Ireland, large or small, whatever their objects of management. It sets standards for forest practice in an international context and is the culmination of thinking developed in *Our Common Future*, the 1993 report of the World Commission of Environment & Development, and the 1992 UN Conference on Environment & Development in Rio de Janeiro.

Prospects for pesticides and fertilizers

The *Forestry Standard* envisages prudent use of pesticides and fertilizers within current approved practices.

Usage of pesticides in forests in Britain is so small that foresters are likely to remain dependent on compounds for which there is an agricultural market sufficient to carry the main development and marketing costs.

Future acceptance will depend on openness of practice and strict attention to operational details so that mishaps are rare.

Dialogue on the ethics of pesticide use

There remains public unease about use of pesticides. The informed dialogues that formed part of the debate when formulating the UK Forestry Standard on overall forestry practice should be taken as a significant precedent. Such dialogues should be maintained as required at local and national levels to ensure that principles underlying the roles of chemicals in forests are better understood and concerns fully explored.

As a corollary, all parties in such dialogues should be alert to recognise and minimise use of emotively loaded language. Use of 'politically correct' language was a response to sexually or ethnically offensive usages. That readjustment may at times have been overdone; however, nuances of language must not be overlooked. Views underlain by alternative ethics or values relating to the use of pesticides need to be expressed as objectively and dispassionately as possible.

17.22   Forest Certification

During the 1980s and 90s, overt 'green' pressure was applied against trade in tropical hardwoods, to conserve tropical rainforest against exploitative logging. That pressure continued with occasional threats against specific retailers handling tropical woods (TWP 1998a). The campaign subsequently expanded to become a demand that all timber harvested and marketed should come from sustainably managed sources. Third world countries commented that 'green' groups from wealthy countries should not seek to impose standards on them that were not being applied in the richer countries.

Certification proposals are in train in the UK and Europe and are certain to be widely adopted (FBT 1997, 1998a, b, TWP 1998b, Jeffree 1997, FC NR 1999).

Attempts have been made to impose additional restrictions on uses of chemicals as part of the certification process. These have been resisted, both as being beyond the competence of the bodies such as the Forest Stewardship Council, and being adequately covered by the *UK Forestry Standard* and the *Audit protocol* (FC FS 1998a, b), and by the over-arching framework of national and EU regulations on pesticide and fertilizer use.

17.3   VIGILANCE (AGAIN)

It is not possible to understate the need for vigilance over the whole range of woodland and tree management activities. The preceding chapters have attempted to set the background to present best practices in relation to chemicals. None is so entrenched that some improvement is not possible. Nor are external circumstances - climate, availability of materials, the pressures of society - sufficiently stable that adjustments to practice will not be required.

Vigilance is required:

- at national strategic level to avert introduction of additional pest species or varieties, and any untoward effects of climatic change;
- locally, by monitoring any change such as build up of insects or to detect and respond to falls in standards because of complacency.

Vigilance also underlies sustainability.   Definition and application of appropriate cultural practices requires well-informed managers, familiar with:

- the function of trees and forests and the full range of associated values;
- the interaction of forest culture and protection with the other demands on the land and environment;
- the national and international contexts in which forestry operates.

17.31   Prevention is better than cure

Britain has suffered over the last 20 years from the results of an insufficiently developed knowledge of the biology of Dutch elm disease. Recognition of the more aggressively pathogenic strain of Dutch elm disease and its biology was the result of meticulous study (§7.31). However, it only came after most of the elms had died.

The economic ill-effects of the introduction of the great spruce bark beetle *Dendroctonus micans* (§6.53) have been marginally less severe, but only because there happened to be an effective biological control system available more or less 'off the shelf'.

Gibbs & Wainhouse (1986) describe spread of pests and pathogens throughout the Northern hemisphere and record many instances of devastating losses through introduced pests. Novel forms of disease are still emerging, *eg* new strains of *Phytophthora* and unexplained die-back of pedunculate oak (Brasier 1999, Gibbs 1999). Their concerns remain valid.

Current provisions for control of pests and avoidance of introductions are summarised in § 5.3. To sustain trees and forests against pests and to avoid repetition of past disasters requires:

- legislation giving effective powers to intercept and destroy novel pests and recognised potential pests not yet naturalised in the country, wherever found;
- effective internationally approved plant health inspections for all stock passing in international trade;
- a widespread network of monitors for invading mobile insects, mammals *etc.* which have evaded detection at entry;
- a back-up system of local treatment before invaders become established;
- a communication network, capable of informing all interested parties of the appearance of a novel pest;

- an adequate pool of skilled pathologists, entomologists and silviculturists, able to prescribe appropriate action.

In addition to these more established disciplines, skills are required in plant physiology, physical and biochemical sciences, so as to identify, measure and model atmospheric pollutants and their reaction with trees and woodlands.

17.32   Technological advance

Genetic engineering

Advances in genetic engineering have been applied to organisms ranging from bacteria to vascular plants and insects. Some are designed to synthesise complex chemicals, others to facilitate insect and weed control in agricultural crops. Developments in British agriculture and the range of uses being tested are described in a government-sponsored review (Pierpoint & Shrewry (1996) . Potential uses for genetically engineered forestry products are also being explored (Tickell 1999).

Because of the scale of study of genes and the breadth of spread of organisms studied, the role, potential and risks of genetic engineering is likely to be controversial. The danger that good and bad projects will be uncritically lumped together and condemned is analagous to some more glib attitudes to pesticides. There are undoubtedly serious risks but also serious benefits; what is crucial is well-informed public debate.

Information technology and remote sensing

The speed and power of data processing and the increased ability to view land from satellites is bound to have some benefits for forest management.

Earlier warnings of disease, mineral deficiency and, on a shorter time scale, outbreaks of fire may be secured through better surveillance.

Process modelling

Models of forest processes of growth and response to changes in the environment are likely to be widely developed and are likely to lead, or example, to more sophisticated use of fertilizers and the possibility of lower rates of application.

In the context of forest meteorology, Monteith (1989) commented:

'Models are the best way of using physics to sort out ... the complexities of interactions between forests and the atmospheric boundary layer. Success in modelling depends on identifying what corners can safely be cut without violating the laws of physics or the principles of biology, so that the final product is no more complicated than it needs to be to fit independent observations, or guide decisions for management.'

Recent advances in modelling in relation to atmospheric sulphur and nitrogen and other topics, have been mentioned in earlier chapters. Rapid advances can be expected as other applications are found and existing models improved.

17.4   THE CHALLENGE

In the coming half century, global humanity will face a greater ethical challenge than ever in the past. The challenge is to improve the quality of life of the billions in today's population who suffer poverty and great social insecurity, and to maintain those standards sustainably for those likely to be living in 2050 (Holdgate 1995).

In his *Introduction to World Forestry,* Westoby (1989) wrote:

'It is time to dispel the notion that forestry is about tending woodyards. ... Foresters can do more than this. ... They have a key responsibility in providing society with understanding of the opportunities and constraints upon natural resource use. It can be difficult for lay men and women to grasp the trade-offs involved in resource use decisions, particularly if these are wrapped up in jargon, but none of the issues is so complex as to be beyond the understanding of average citizens, if professionals take the trouble to express themselves in comprehensible terms. Those who speak in jargon either have something to hide or do not themselves understand.'

17.41   The prudence principle

With present atmospheric, climatic and population trends, strenuous efforts will be required to secure the long-term well-being of mankind and the global environment Land has to be managed to sustain people, to sustain the environment and maintain its productive capacity.

For all land managers who may use fertilizers and pesticides, or may have to react to chemical pollutants that affect trees or woods, there are four necessities:

* *Integrity* on the part of all those who control, earn their living or benefit from use of chemicals, to use them only where necessary, using the minimum quantities required and to do so following best current practice;
* *Ethical awareness* both of one's own values and of those held by others with an interest in woodland management or the benefits of woodland;
* *Continuing vigilance and monitoring* for unexpected effects of any sort, whether in woods, pests, people or the environment, recording, and reporting faithfully anything unexpected;
* *Determination* to secure for land management, a sustainable legal, administrative and economic framework seen to be effective, impartial and equitable between economic, social and environmental interests.

Such an approach might be called the 'prudence principle'. Within this framework, pesticides, fertilizers and other chemicals should be used to maximise benefits and minimise adverse effects in forests, small woods and trees.

Prudent use of available resources to achieve economic, social and environmental sustainability offers a practical way forward.

'In sustainable forests, trees may be used but the forests and people remain'.

Appendices

Bibliography

Index

Appendix I *Tables 1a-d and Table 2*

References to woodland pesticides in literature

Appendix I, Tables 1a to 1d have been compiled in the course of scrutinising literature, in particular, research and development references to pesticides for forestry in the UK. The tables list active ingredients/active substances, not commercial products.

Each single or double letter entry on these tables refers to a published report in which the pesticide indicated is mentioned. References are set out by decade.

Appendix I Table 2 provides a key to the one or two letter references. These are arranged in alphabetical order; against each is the name of the author(s) and year of publication. The full reference can be found under author and year in the main bibliography at the end of the book.

The tables in Appendix I also illustrate how interest has moved between pesticides. For example, *Appendix Table 1a* shows DDT to have been of active interest in the 1950s to 1970s. Diflubenzuron and permethrin were subjects of report in the 1980s and 1990s. Lindane has been reported over the whole of the period.

Appendix I, Table 1a *References in literature to **insecticides** of possible use in forests, farm or amenity woodlands & trees, or tree nurseries*

Active substance	Decade in which publication referring to insecticide appeared				
	1950s & before	1960s	1970s	1980s	1990s
Aldicarb	-	-	-	ef	-
Aldrin	e	-	-	ef	-
Amitraz	-	-	-	-	mk
Carbaryl	-	hw	-	u,ef	-
Carbon disulphide	d,et,ff	-	-	-	-
Carbosulfan	-	-	-	-	bx,cc,mi
Chlorpyrifos	-	-	j,fq,fr,ft,fu,gk,gl,ht	k,n,u,ef	bx,ed,nb
Clofentezine	-	-	-	-	mk
Cypermethrin	-	-	-	u,ef, ga,gv,iv	ed,mk,na
DDT)	e,f,du,et,fg,fh,fi, fm,gn,go,gp,hd	h,bb,be, es,gl,hw	a,eh,gk,gl, hm,hs,ht,lt	-	nb
Deltamethrin	-	-	-	ef	-
Demeton-S-methyl	-	-	-	ef	-
Diazinon	-	-	bk,hw	n,u,ef	-
Dicofol	-	-	a	n,u,ep	ed,mk
Dieldrin	e,fm,gp	-			
Diflubenzuron))	-	-	-	k,u,am,ch,ck, ef,gc,gh,gi,gz	q,bq,ed,mk,na
Dimethoate	-	-	bk	u,ef	nb
DNOC winter wash	bc, gp	-	-	-	-
Fenitrothion	-	hy	ah,bk,hm,hs,lt	n,u,ch,ep,fx,fy,gb,gh,gi	bq,ed,mk
Fenpropathrin	-	-	-	ef	mk
Fenvalerate	-	-	-	-	mk
Horticultural oils	-	-	-	-	mk
Lead arsenate	d	-	-	-	-
Lime sulphur	bc	-	-	-	-
Lindane) =γHCH)	e,az,et,fh, fm,gn,gp	h,bj,dq, hw,lo	a,bb,bl,eh,fq, fs,fu,gk,gl,gx	k,n,u,cf,cg, ef,ep,iv,jz	bq,bx,bz,ca,cx,ed, mk,mu,mv,mw,mx
Malathion	d	h,bb,hw,hx,ln	a,bk	n,u,ak,ef	nb
Methoxychlor	-	gl	fn,fq,gk,ht	-	-
Nicotine	d,er,gy	ba	-	u,ef	-
Oxydemeton-methyl	-	-	-	ef	-
Paraffin emulsion	d	ba	-	-	-
Paris green	d,e	-	-	-	-
Permethrin))	-	-	-	cf,cg,ef,ga, gf,gv,iv	p,bq,bx,by,bz, ca,dk,ia,mk,na
Phorate	-	-	fw	-	-
Phosalone	-	-	lt	-	-
Pirimicarb	-	-	-	u	ed,mk,nb
Rotenone (Derris)	d	-	-	-	-
Soaps (horticultural)	-	-	-	-	mk,nb
Tar Oils winter wash	d,bc	-	bk,fa	n,u,ef,ey,ez	ed
Tetrachlorvinphos	-	-	ag,gi,gj,hi,ht,lt	-	-
Tetradifon	-	-	-	ed,ef	mk
Trichlorfon	-	-	-	ef	na

See *Appendix I, Table 2* for the key to the one- or two-letter references in the table above.

Appendix I,   Table 1b   *References in literature to **fungicides** of possible use in forests, farm or amenity woodlands & trees, or tree nurseries*

Active substance	1950s & before	1960s	1970s	1980s	1990s
Decade in which publication referring to fungicide appeared					
Benodanil	-	-	-	ef	-
Benomyl	-	-	hg	n,u,y,ef,ei,ep	dp,dz,ed,ey,mz
Borax/borates	fd	az,dq,if	fa	y	dl,dm,dn,ek,el,ho
Bordeaux mixture	ha	h,av	a,eh	y,ef,eq,hr,hu	-
Bupirimate	-	-	-	ef	-
Captafol	-	-	-	y	ed
Captan	-	-	a,dc,eh,lw	n,u,y,ef	dp,ey
Carbendazim	-	-	fl	y,dt,ef,hq	
Cheshunt compound	-	-	-	ef	-
Chlorothalonil	-	-	-	ef	-
Copper ammonium carbonate			-	y,ef,eq	-
Creosote (stump treat't)	dy,fc	h,bh,bi,dq,dr	az	y	
Cresylic acid (tar acids)	-	-	-	hu	
Cycloheximide	ls	ay,gu	a,eh,ez	n,y,ep	-
Dichlofluanid	-	-	-	ef	-
Dinocap	-	-	-	n,u,y,ef,ep	dp,ey
Dodemorph	-	-	-	ef	-
Etridiazole	-	-	-	y,ef,eq	-
Fenarimol	-	-	-	ef	-
Fenpropimorph	-	-	-	-	ee
Fosetyl-aluminium	-	-	-	ef,eq	ed
Furalaxyl	-	-	-	ef	-
Imazalil	-	-	-	dt,ef	
Iprodione	-	-	-	ef	-
Mancozeb	-	-	-	ef	ed,mn
Maneb	-	-	-	n,u,y,ef,ep,eq	dp,ed,ey
Metalaxyl	-	-	-	ef	-
Octhilinone	-	-	-	y	ed
Oxycarboxin	-	-	-	ef	-
Penconazole	-	-	-	ef,mo	-
Prochloraz	-	-	-	u,ef	ey
Propamocarb hydrochloride	-	-	-	ef	
Propiconazole	-	-	-	dt,ef	-
Pyrazophos	-	-	-	ef	-
Quinomethionate	-	-	-	y,ef	ed
Sodium nitrite	fd	h	az,fa,fb	y,ml	-
Sulphur/lime sulphur	bf,gy	bd,du	a,eh	n,u,y,ef,ep	dp,ed,ey
Thiabendazole	-	-	fl,hg	y,aa,ab,al,dt	cw,hf
Thiophanate-methyl	-	-	-	y,ef,hp	-
Thiram	-	kt	a,dc,eh	n,u,y,ef	dp,ey
Tolclofos-methyl	-	-	-	ef	-
Triadimefon	-	-	-	u,hq	
Tribromophenol	-	-	gx	-	-
Triforine	-	-	-	ef	-
Urea	-	h	az,eh,fa	n,u,y,ep,hu	bq,ca,dm,dp,el
Vincozolin	-	-	-	ef	-
Zineb	-	aw	a	n,u,y,ef,ep,eq	dp,ey

See *Appendix I, Table 2* for the key to the one- or two-letter references in the table above.

Appendix I, Table 1c   *References in literature to **herbicides** of possible use in forests, farm or amenity woodlands & trees, or tree nurseries*

Active substance	Decade in which publication referring to pesticide appeared				
	1950s & before	1960s	1970s	1980s	1990s
2,4,5-T)))	bg,ew,hb, ih,jc,ls	b,g,h,ic, if,ij,ja, lu,ki,ma	j,z,ae,af, eb,ev,in, iz,jb,jo	k,dd,ds,ey,mf	-
2,4-D/2,4,5-T	ih	-	cn,eb	ds	-
2,4-D))	bg,ew,hb, ih,jc	b,g,h,dv, ij,lt,ma	j,z,ae,af, ar,eb,ev	k,l,dd,,ey,ji,mf	o,w,aj,bp bu,cm,dj
2,4-D + Dicamba + triclopyr	-	-	-	jt	m,o,w,bt,bu,cm,dj,en
Alloxydim-sodium	-	-	-	bs,ke	db,en,eo
Amitrole	ew,hb,ih,jc	c,v,if,ij	a,jd	-	bu,ce,dj,md,mh
Amm. sulphamate))	bg,hb	b,g,h,v,if	j,z,ae,cn,eb	k,l,ac,dd,ds,jk,mm	o,w,bp,bu ca,ct,dj
Asulam))	-	-	j,z,af,cn,eb, iz,jb,jp,iq,me	k,l,dd	o,q,w,bp,bt,bu,ca cm,dj,dx,hk,md,ll,lz
Atrazine))	-	ij,jl	j,ae,af,eb,ik, il,iz,jd,jp,kd	l,bn,bs,dd,ke	m,o,w,bp,bu,ca,ce cm,db,dj,hk,hv,nf
Atrazine + cyanazine	-	-	iz	-	lz
Atrazine + dalapon	-	-	-	l,bn,dd,jf,jg	o,w,bp,ca
Cadocylic acid	-	-	jn,jo	-	-
Chlorbufam + c'dazon	-	-	-	bs,ke	db,hv
Chlorpropham	-	v	-	-	-
Chlorpropham + fenuron	-	v	-	-	-
Chlorthal-dimethyl	-	-	-	bs,ix,ka	bu,db,dj,hv
Chlorthiamid))	-	b,g,v,co, cq,ja,kh	j,z,ae,af, cn,eb,iz,jo	k,ey	ec,eo
Clorpyralid))	-	-	-	bs,ke	o,bt,bu,ce,cj,cm,db,di dj,dp,eo,hk,md,mg,ne,nf
Clorpyralid + cyanazine	-	-	-	ke-	dx
Cyanazine	-	-	-	bs	m,bu,cm,db,hk,hv,md,nf
Cyanazine + atrazine	-	-	-	l	m,o,w,bp,bu,ce,cj,dj
Cycloxidime	-	-	-	-	bu,ce,dj,nf
Cyprazine	-	-	-	jf,jg,jh	-
Dalapon	ew,ih,jc	b,c,g,v,ie,ij	a,j,z,ae,cn,eb	k	-
Dalapon + dichlobenil	-	-	af,cn	iq,jf,jg	o,w,bu,cm
Dicamba	-	b,g,v,cp,kh,ki	ae	iq,jr,jt	o,w,bu,cm,dj
Dichlobenil	-	v,ij,kl	j,z,ae,eb,iz	k,ey	w,en,lz,mg,nf
Diflufenican	-	-	-	-	nf
Diphenamid	-	-	kd	bs,ke,ko	m,w,bu,db,dj,eo,hv,nd
Diquat	ih	b,v,if,ij,lu	cn	-	-
Diuron	mb	-	-	-	en,eo
Fluazifop-P-butyl))	-	-	-	-	o,bu,ce,cj,cm,db dj,dp,en,hk,md,ne
Fluroxypyr	-	-	-	is	-
Fosamine-ammonium	-	-	hs,jb	k,dd,hh,jh,mf	ca,cm,en
Glufosinate ammonium	-	-	-	hh,ji	o,bu,ce,cm,db,dj,hk,hv,nf
Glyphosate))))	-	-	z,hs,jb,je	k,l,ac,bm, bn,bo,bs, cy,dd,hh, ig,mp,nc	o,v,aj,ap,bp,bt,bu,ca ce,cm,ct,cu,cv,db,dj dj,do,dp,dx,en,hk,hv,is jz,lk,ll,md,mg,mh,ne,nf

Appendix I, Table 1c (continued)   *Herbicides referred to in literature*

Active substance	1950s & before	1960s	1970s	1980s	1990s
Hexazinone))	-	-	jb	k,l,bn,dd,iq,it,jb, ji,jk,jq,ka,kb,ke	w,bp,ca,cm,cu
Imazapyr)))	-	-	-	it,js,la	m,o,x,bt,bu,bv ca,cm,cu,da,dj dw,ec,hz,ks,jg,lz
Isoxaben)))	-	-	-	bs,ka	m,o,v,bu,ce cm,db,dj,dx,hk ll,md,mg,nd
Lenacil	-	-	-	-	p,bu,ce,db,dj,dx,en,nf
Light fuel oils	lw	-	-	-	-
Vaporising oil	ad,kf,kg	c,g,h,v	a,j	-	-
White spirit	ad,kg	c,h,v	a,j	-	-
MCPA	hb,kd	ma	jd	-	en
Mecaprop	-	-	-	-	en
Metamitron	-	-	-	iy,ka	bu,ce,db,dj,dx,hv,nf
Metazachlor))	-	-	-	bs,ke	bu,ce,ci,cj,cm,db dx,hk,hv,ll,md,ne,nf
Napropamide))	-	-	-	bs,cz,ix,iy,kd	bu,ce,db,dj,dx eo,hv,nd,nf
Nitrofen	-	-	kd	-	-
Oryzalin	-	-	-	bs,iy,ke	db,dx,eo,hv
Oxadiazon	-	-	-	bs,cz,ke	db,en,eo,hv,mg,nd,nf
Oxyfluorfen					nd
Paraquat))	-	b,g,v,cq,ij,ja ki,kl,kr,mr	a,j,z,ae af,cn,eb	k,bm,bn,bs, dd,ji,ke	v,w,bu,cm,hv, db,dj,en,eo,ne
Paraquat+diquat	v	-	k,ey	o,bu,cm,dj	-
Pendimethalin))	-	-	-	-	o,bu,ce,cj,cm,db,dj,dp dx,hk,hvll,md,ne,nf
Picloram	-	g,v,ja,kh,ki	ev,jn,jo	-	-
Propachlor	-	-	-	-	en
Propaquizafop	-	-	-	-	o,bu,ce,cm,dj,hk
Propyzamide)))	-	-	j,z,af,eb, eu,iz	k,l,bm,bn, bs,dd,ke	o,v,w,bp,bu,ca,ce cm,cv,db,dj,en,hk hv,ll,md,mg,nf
Simazine))	ih,km	c,g,h,v bw,id,ij,kn	a,j	k,bs,ey ig,ke	w,ce,db,en,hv md,nf
Sodium arsenite	hc,jc	if	fk	-	-
Sodium chlorate	lw	c,if,hj	a	bs,ds,ey,ke	w,hv
Sodium pentachlorphenate	ls	-	-	-	-
Sulphuric acid	gy,lw	-	-	-	-
TCA (Trichloracetic acid)	ex,jc	ma	a	-	-
Terbuthylazine+atrazine	-	-	-	l,cr,iw	w,x ,bp,dx,jf,jg de,di
Thifensulfuron-methyl	-	-	-	-	de,di
Tordon	-	kl	-	-	-
Tribenuron-methyl	-	-	-	-	de,di
Triclopyr))	-	-	jb,je	l,ac,bo,dd,jf,jg jh,jt,ju,jv,mf	w,bp,bt,bu,ca,cm dj,en,jz,ks,md

See *Appendix I, Table 2* for the key to the one- or two-letter references in the table above.

Appendix I, Table 1d   *References in literature to 'other pesticides' of possible use in forests, farm or amenity woodlands & trees, or tree nurseries*

Active substance	1950s & before	1960s	1970s	1980s	1990s
Biological agents					
Bacillus thuringensis	-	-	r	ah,ch,ef	na
Heterorhabditis spp.(nematode)	-	-	-	-	ef
NPV	-	-	-	u,ch,ck,ep,gd,ge,gg,gh,gi	bq,ea
Phlebopsis (Peniophora) gigantea	-	fe	eh,fb	n,u,y,ao,ep, ey,ml	bq,ca,dm,dy, el,ms,mt
Rhizophagus grandis	-	-	-	ep	(See §6.53)
Trichoderma viride	-	-	-	aa,an,ef	bq,dp,dy,ed
Animal poisons & repellants					
Aluminum phosphide	-	-	ai	n,cd,cl,ep	bq,ca,cb
Arbinol	ls	-	-	-	-
Bone oil	ls	-	-	-	-
Endrin	fj,ls	-	-	-	-
Hydrogen cyanide	et	-	-	-	-
Sodium cyanide	-	as	ai	n,ep	bq,ca,cb
Strychnine	-	-	a	ep	t
Sulphonated cod liver oil	-	-	-	-	mp
Tetramine	ls	-	-	-	-
Toxaphene	fj,ls	-	-	-	-
Warfarin	-	ax,kt	ai,at	n	t,bq,ca,cs,la,lb,kv,kw
Ziram	-	fo,kt	ai,aq,ku	-	bq,br,ed,ky,mq
Soil and/or timber fumigants					
Carbon disulphide	-	lm	-	-	-
Chloropicrin	-	i,v	-	ei	-
Dazomet	-	-	a,dg,eh	k,n,u,ef,ep,eq,ey	p,w,y,bs,db,df,dp
Dichloropropene	-	-	-	-	em
Formaldehyde	dh	h,i,v	a,eh	n,ef,ep,eq,ey	-
Methyl bromide+chloropicrin	-	-	ex	-	ap,ef,ei,eq,w,y,bs,db,em
Metham sodium	-	-	hg	eq	-
Molluscicides					
Metaldehyde	-	-	-	ef	-
Moss/algicides					
Methiocarb	-	-	-	ef	-
Wood preservatives & Miscellaneous					
Bitumen emul'n (wound sealant)	-	-	-	ef	-
Borax	-	-	-	-	lj
Borax + PCP	-	au,dq,dr	le	lf,lg	-
Copper chrome arsenate	-	-	-	-	lh
Creosote	lb	lc,ld	le	lf,lg	am,lj
Fluor chrome arsenate	lb	lc,ld	le	-	-
Lead paint (wound sealant)	-	-	-	ef	-
Mercury salts	-	dr	-	-	-
Methylene-bis-thiocyanate	-	-	-	eg	-
PCP	-	dq,dr,if,lo	-	-	-
Sodium pentaborate	ls	-	-	-	-

Appendix I, Table 1d (continued)   *'Other pesticides' referred to in literature*

Active substance	Decade in which publication referring to pesticide appeared				
	1950s & before	1960s	1970s	1980s	1990s
Growth Inhibitors					
Daminozide	-	-	-	hh	-
Dikegulac	-	-	-	hh	-
1-naphthalenic acid	-	-	-	hh	-
Maleic hydrazide	-	b,v	-	hh,ir,ks,kt	en
Mefluidide	-	-	-	hh,ks	en
Paclobutrazol	-	-	-	hh,ks	en

See *Appendix I, Table 2* for the key to the one- or two-letter references in the table above.

Appendix I   *Table 2*

This table provides the keys for *Appendix I, Tables 1a - d*.

For each of the one- or two-letter references in the preceding tables, the table following gives the name of the author(s) and year of publication. The full reference can then be found using 'author and year', in the main bibliography at the end of the book.

Key	Author(s)	Year	Key	Author(s)	Year
a	Aldhous	1972a	aa	Brasier & Webber	1987
b	Aldhous	1969	ab	Grieg & Coxwell,	1983
c	Aldhous	1962	ac	Tabbush & Williamson	1987
d	Chrystal	1937	ad	Anon	1958
e	Aldhous	1959	ae	Wittering	1974
f	Crooke	1959	af	Crowther	1976
g	Aldhous	1967a	ag	Brown & Barbour	1978
h	Aldhous	1968a	ah	Stoakley	1979
i	Benzian	1965	ai	Pepper	1976
j	MAFF	1978	aj	Taylor & Tabbush	1990
k	MAFF	1984	ak	Lonsdale & Wainhouse	1987
l	Williamson & Lane	1989	al	Greig	1985a
m	Palmer	1993	am	Spencer	1999
n	Blatchford	1983	an	Mercer et al	1983
o	Willoughby & Dewar	1995	ao	Greig	1985b
p	Aldhous & Mason	1994	ap	Garnett & Williamson	1990
q	MAFF	1996	aq	Pepper	1978
r	Holden & Bevan	1978	ar	Everard	1974
s	Davies	1995	as	Rogers	1965
t	Springthorpe & Myhill	1983	at	Rowe	1973
u	Carter & Gibbs	1989	au	Holtam	1966
v	Fryer & Evans	1968	av	Pawsey	1964b
w	Hance & Holly	1990	aw	Pawsey	1964a.
x	Anon.	1988	ax	Rogers Bramwell	1958
y	Phillips & Burdekin	1982	ay	Anon.	1967
z	Brown	1975	az	Anon.	1970a

Appendix I Table 2 contd.   *Key to 'Author/years' for entries in Appendix I Table 1*

Key	Author(s)	Year	Key	Author(s)	Year
ba	Crooke	1960	ca	Willoughby	1995
bb	Davies (J.M.)	1968	cb	Pepper	1995
bc	Anon.	1956a	cc	Ellis	1995
bd	Phillips	1963	cd	Pepper	1982
be	Bevan	1966	ce	Willoughby & Clay	1996
bf	Anon.	1956b	cf	Stoakley & Heritage	1988
bg	Holmes	1957	cg	Stoakley & Heritage	1989
bh	Low & Gladman	1960	ch	Leather	1986
bi	Small	1960	ci	Williamson & Mason	1990a
bj	Bevan	1962	cj	Williamson	1991b
bk	Parker	1974	ck	Sterling	1985
bl	Carter	1975	cl	Pepper	1987b
bm	Davies	1987e	cm	Edwards et al	1994
bn	Wilson	1985	cn	Aldhous	1970
bo	Sale	1985	co	Aldhous	1967b
bp	Williamson	1991a	cp	Aldhous	1965b
bq	Evans & Stoakley	1990	cq	Aldhous	1965c
br	Hibberd	1988	cr	Nelson & Williamson	1989
bs	Williamson & Mason	1990c	cs	Pepper	1990
bt	Willoughy	1996a	ct	Staples & Nelson	1990
bu	Willoughy	1996b	cu	Nelson	1990
bv	Edwards et al	1993	cv	Tracey & Nelson	1991
bw	Aldhous	1961	cw	Greig	1990
bx	Heritage	1996a	cx	Stoakley & Heritage	1989
by	Stoakley & Heritage	1990a	cy	Lane	1984
bz	Stoakley & Heritage	1990b	cz	Sale & Mason	1986
da	Clay et al	1992	ea	Anon	1992d
db	Williamson et al	1993	eb	F&HGT	1976
dc	Gordon et al	1976	ec	Anon	1992e
dd	Sale et al	1986	ed	Strouts & Winter	1994
de	Lawrie & Clay	1994a	ee	Brasier	1987
df	Moffat	1994	ef	Brooks et al	1989
dg	Low	1974	eg	ACP/Ann Rpt year	Year
dh	Edwards	1952	eh	Blatchford	1978
di	Lawrie & Clay	1994b	ei	Pierce & Gibbs	1981
dj	Willoughby	1996c	ej	Garrett	1982
dk	Scott, Wyatt & Lane	1990	ek	Pratt & Lloyd	1996
dl	Butin	1995	el	Pratt et al	1998
dm	Pratt	1996b	em	Cooper	1993
dn	Pratt	1996a	en	Carter	1990
do	Anon.	1994	eo	Clay et al	1990
dp	Strouts	1990	ep	Hibberd	1986
dq	FPRL	1968	eq	Strouts	1981
dr	FPRL	1971	er	Anon	1948
ds	Wilson	1981	es	Anon (Pine weevil)	1960
dt	Burdekin	1983	et	Anon (Megastigmus)	1954
du	Peace	1952	eu	Mayhead & McCavish	1975
dv	Aldhous	1967	ev	McCavish & Smith	1978
dw	Palmer	1994	ew	Anon (Rep For Res)	1959
dx	Nelson et al	1991	ex	King	1977
dy	Gibbs, Grieg & Risbeth	1996	ey	Strouts et al	1994
dz	Harmer	1991	ez	Burdekin & Phillips	1982

Appendix I Table 2 (cont)   *Key to 'Author/years' for entries in Appendix I Table 1*

Key	Author(s)	Year	Key	Author(s)	Year
fa	Phillips & Greig	1970	ga	Stoakley, King & Heritage	1983
fb	Greig & Redfern	1974	gb	Stoakley	1984
fc	Risbeth	1959a	gc	Stoakley et al	1985
fd	Risbeth	1959b	gd	Evans, Stoakley & Heritag	1985
fe	Risbeth	1963	ge	Stoakley & Heritage	1986
ff	Findlay	1959	gf	Stoakley & Heritage	1987
fg	Bevan	1954	gg	Stoakley & Patterson	1987
fh	Crooke	1954a	gh	Stoakley	1987
fi	Crooke & Bevan	1958	gi	Stoakley	1988
fj	Holmes, Brown & Ure	1958	gj	Brown & Barbour	1978
fk	Greig & Low	1975	gk	King & Scott	1975
fl	Gibbs & Dickinson	1975	gl	Bevan	1964
fm	Bevan, Davies & Brown	1957	gm	Balfour & Kirkland	1962
fn	King & Scott	1975a	gn	Crooke	1954b
fo	Rowe	1966	go	Crooke	1955a
fp	Bevan & Davies	1971	gp	Crooke	1953
fq	Scott & King	1973	gq	Bevan & Davies	1970
fr	Stoakley	1976	gr	Holmes	1962
fs	Carter	1973	gs	Scott & King	1974
ft	Davies & King	1973	gt	Crooke	1955b
fu	Stoakley	1975	gu	Pawsey	1964c
fw	King & Heritage	1979	gv	Stoakley et al	1984
fv	Stoakley	1977b	gw	Stoakley	1985
fx	Stoakley	1979	gx	Savory et al	1970
fy	Stoakley	1981	gy	Gray	1953
fz	Stoakley	1982	gz	Stoakley	1984a
ha	Peace	1952b	ia	Heritage & Johnson	1997
hb	Holmes	1952	ib	Greenwood & Halstead	1997
hc	Holmes	1957a	ic	Dannatt & Wittering	1967
hd	Peace	1954	id	Aldhous	1966
he	Bevan & Davies	1972	ie	Wood & Nimmo	1962
hf	Grieg	1978	if	Holmes	1961
hg	Burdekin & Gibbs	1974	ig	Salter & Darke	1988
hh	Evans (J.)	1987	ih	Wood & Holmes	1959
hi	Bevan & Brown	1978	ii	Wood & Holmes	1958
hj	Zehetmayr	1960	ij	Wood, Holmes & Fraser	1961
hk	Willoughby	1996e	ik	Brown	1970
hl	Burdekin & Phillips	1977	il	Brown	1972
hm	Scott	1972	im	Mackenzie et al	1976
hn	Gibbs, Liese & Pinon	1984	in	Brown & Mackenzie	1974
ho	Pratt	1997	io	Rogers	1974
hp	Mercer	1984	ip	Brown & Thompson	1974
hq	Clifford & Gendle	1987	iq	Sale, Malone & Cooper	1986
hr	Alford & Locke	1989	ir	Evans	1986
hs	McCavish & Smith	1977	is	Gibbs	1984
ht	Scott & Walker	1975	it	Williamson et al.	1987
hu	Buczacki & Harris	1981	iu	Sale et al	1986
hv	Williamson & Morgan	1994	iv	Tabbush & Heritage	1988
hw	Bevan	1966	iw	Williamson & Tabbush	1988
hx	Bevan	1968	ix	Williamson & Mason	1988
hy	Bevan	1967	iy	Mason & Williamson	1987
hz	Edwards & Morgan	1997	iz	McCavish & Smith	1976

Appendix I Table 2 (cont)   *Key to 'Author/years' for entries in Appendix I Table 1*

Key	Author(s)	Year	Key	Author(s)	Year
ja	Aldhous & Atterson	1967	ka	Ogilvie	1984
jb	McCavish	1978	kb	Wolfendon	1988
jc	Holmes	1956b	kc	Williamson et al.	1990
jd	Brown & Mackenzie	1972	kd	Biggin	1979
je	McCavish	1979	ke	Williamson	1989
jf	McCavish	1981	kf	Edwards & Holmes	1949
jg	Tabbush	1982	kg	Edwards & Holmes	1950
jh	Sale et al.	1982	kh	Aldhous	1966
ji	Tabbush	1984	ki	Aldhous & Hendrie	1966
jj	Tabbush	1985	kj	Aldhous	1964a
jk	Tabbush	1986	kk	Aldhous	1964b
jl	Aldhous et al	1963	kl	Aldhous	1964c
jm	Brown & Mackenzie	1969	km	Faulkner & Aldhous	1959
jn	Brown & Mackenzie	1970	kn	Aldhous & Atterson	1960
jo	Brown & Mackenzie	1971	ko	Mason & Williamson	1988
jp	Brown & Mackenzie	1973	kp	Ross	1988
jq	Biggin & McIntosh	1981	kq	Mason	1988
jr	Palmer	1988	kr	Holmes	1962
js	Winfield	1988	ks	Clay & Dixon	1996a
jt	Palmer et al.	1988	kt	Rowe	1967
ju	Valkova	1988	ku	Rowe	1977
jv	deAth	1988	kv	Pepper & Stocks	1993
jw	Darrall	1988	kw	Pepper	1989
jx	Baker	1996	kx	Pepper & Hodge	1996
jy	Page	1996	ky	Pepper et al	1996
jz	Tabbush	1985a	kz	Rowe	1967
la	Christensen	1988	ma	Salisbury	1964
lb	Richards	1957	mb	Holmes	1956a
lc	Richards	1961	mc	Scott	1972
ld	Aaron	1962	md	ETSU	1995b
le	Clarke & Boswell	1976	me	Rogers	1974
lf	Aaron & Oakley	1985	mf	Sargent	1984
lg	Aaron	1983	mg	Willoughby	1998
lh	Palfreyman et al	1995	mh	Willoughby	1996e
li	Eason	1996	mi	Heritage et al	1997a
lj	Dixon & Lloyd	1997	mj	Heritage et al	1997b
lk	Willoughby	1997b	mk	Carter & Winter	1998
ll	Wall	1998	ml	Pratt & Greig	1988
lm	Anon	1948	mm	Risbeth	1987
ln	Stoakley	1967	mn	Rose & Gregory	1997
lo	Rawlinson	1968	mo	Rose	1989
lp	Peace	1936	mp	Turner & Tabbush	1984
lq	Holmes & Barnsley	1953	mq	Pepper et al	1995
lr	Binns & Aldhous	1966	mr	Neustein	1966
ls	Wood & Holmes	1957	ms	Pratt et al	1999
lt	Winter & Scott	1977	mt	MAFF Eval	1998
lu	Semple	1964	mu	MAFF Eval	1992b
lv	Wood	1974	mv	MAFF Eval	1992c
lw	Denne & Atkinson	1973	mw	MAFF Eval	1996
lx	Morris & Whipp	1998	mx	MAFF Eval	1997
ly	Gurnell & Pepper	1998	my	MAFF Eval	1990
lz	Palmer	1998	mz	MAFF Eval	1992a
na	TAT	1999	nd	O'Carroll & O'Reilly	1997
nb	Shaw & Fielding	1998	ne	Willoughby & MacDonald	1999
nc	Sale et al	1983	nf	Willoughby & Clay	1999

Appendix II - *Pesticide tables.*

Appendix II, Tables 1a - 1d give background details of the type of use, toxicity and evaluations for most of the pesticides listed in the *Appendix I* tables. In both appendices, the tables refer to *'Active ingredients/substances'*, not *'Approved products'*.

Recommendations for selecting currently products containing the active substance required at any time are given in *Appendix II, Table 2.*

Appendix II, Tables 1a - 1d

Entries in roman type refer to active ingredients/substances which are included in the 1999 list of products approved for use in the United Kingdom.

Entries *in italic* refer to refer to active ingredients/substances *not* on the 1999 list.

Column 1 in the tables lists active ingredients/substances.

Col 2   the relevant *Fields of use, crop type or situation* are indicated:
F = Forest and woodland;   M = Farm woodlands
N = Nursery production for forestry and ornamental trees and shrubs
O = Hardy ornamental trees & shrubs;   V = Vertebrate control
W = Wood preservatives

Col 3   *Year first used*: an approximate date when the pesticide was first used commercially.

Col 4   Whether products containing active ingredient listed were approved under the Control of Pesticides Regulations and listed in the 1999 MAFF/HSE Reference Book 500 *Pesticides 1999.*

Col 5   If data on the active ingredient have been published in *FAO residue reviews*, the year of publication, based on information in Tomlin (1997).

Col 6   Toxicity ratings for the active ingredient issued by the World Health Organisation and United States Environmental Protection Agency. See §4.23.

Col 7   Published data on minimum acceptable residues in food.   See §4.28.

Col 8   If a review or evaluation has been carried out for the Advisory Committee on Pesticides and a report is available, the year of publication.

Caution in use

Many products contain mixtures of pesticides together with adjuvants and other chemicals in the formulation which may modify properties when used. The current *Product label* is *always* the definitive source of precautions to be taken for all approved applications.

Appendix II, Table 1a   *Insecticides: Field of use, toxicity classifications, etc.*

Active ingredient or substance	Fields of use, crop type or situation	Year first used	In '99 MAFF bklt	In FAO r'due reviews	WHO tox'ty rating	EPA tox'ty rating	Accep' min residue level ppm	ACP review/ eval'n date
Aldicarb	O	1965	yes	1992	Ia	I	-	1994
Aldrin	*NO*	*1949*	*no*	*-*	*-*	*-*	*-*	*-*
Carbaryl	NO	1957	yes	1973	II	I/II/III	-	1998
Carbon disulphide	*N*	*1854*	*no*	*-*	*-*	*-*	*-*	*-*
Carbofuran	O	1970s	yes	1983	Ib	I/II		
Carbosulfan	N	1979	yes	1986	II	I/II	0.01	-
Chlorpyrifos	FNO	1965	yes	1982	II	II	0.01	-
Cypermethrin	FNO	1975	yes	1981	II	II	0.05	1984/93
DDT	*FN*	*1939*	*no*	*1984*	*II*	*II*	*0.02*	*-*
Deltamethrin	O	1974	yes	1982	II	II	0.01	1993
Demeton-S-methyl	O	1957	yes	1989	Ib	I	0.0003	1993
Diazinon	NO	1953	yes	1993	II	II/III	0.002	1991/2
Dicofol	N	1956	yes	1992	III	II/III	0.002	1998
Dieldrin	*N*	*1949*	*no*	*-*	*-*	*-*	*-*	*-*
Diflubenzuron	FO	1972	yes	1985	table 5	III	0.002	-
Dimethoate	FNO	1951	yes	1987	II	II	0.01	1990/93
DNOC winter wash	NO	1892	no	1965	Ib	I	-	-
Fenitrothion	I	1960	yes	1988	II	II	0.005	
Fenpropathrin	O	1981	yes	1993	II	II	0.03	1989
Lead arsenate	*N*	*-*	*no*	*-*	*-*	*-*	*-*	*-*
Lime sulphur	*N*	*19th C*	*no*	*-*	*-*	*I*	*-*	*-*
Lindane/γHCH	FNOW	1942	yes	1989	II	II	0.008	1992/4 1996/7/9
Malathion	FNO	1952	yes	1967	III	III	0.02	1994/7
Methoxychlor	*F*	*1950s*	*no*	*1978*	*III*	*-*	*-*	*-*
Nicotine	NO	19th C	yes	1967	Ib	I	0.02	
Oxydemeton-methyl	*O*	*1950s*	*no*	*1989*	*Ib*	*I*	*0.0003*	*1993*
Paraffin emulsion	*F*	*-*	*no*	*-*	*-*	*-*	*-*	*-*
Paris green	*N*	*-*	*no*	*-*	*-*	*-*	*-*	*-*
Permethrin	FNO	1973	yes	1987	II	II/III	0.05	1993
Phorate	N	1970s	yes	1995	Ia	I	0.01	1994
Pirimicarb	FN	1969	yes	1982	II	II	0.02	1995
Rotenone	O	<1895	yes	-	II	III/I	-	-
Tar Oils winter wash	NO	-	yes	-	-	-	-	-
Tetrachlorvinphos	*F*	*1966*	*no*	*-*	*table 5*	*III*	*-*	*-*
Tetradifon	O	1955	yes	1986	table 5	III	-	-
Trichlorfon	O	1957	yes	1978	II	II	0.01	-

Appendix II, Table 1b — *Fungicides: Field of use, toxicity classifications, etc.*

Active ingredient or substance	Field of use crop type or situation	Year first used	In '99 MAFF bklt	In FAO r'due reviews	WHO tox'ty rating	EPA tox'ty rating	Accep' min residue level ppm	ACP review/ eval'n date
Benodanil	O	1973	no	-	-	-	-	-
Benomyl	NO	1968	yes	1983	II	IV	0.2	1992/4
Borax/borates	FW	pre 1900	yes	1994	table 5	III	-	-
Bordeaux mixture	NO	1885	yes	-	-	-	-	-
Bupirimate	O	1975	yes	-	table 5	III	-	-
Captafol	O	1962	no	1985	Ia	IV	-	-
Captan	NO	1952	yes	1990	table 5	IV	0.1	1998
Carbendazim	O	1973	yes	1985	III	IV	0.01	1992
Cheshunt compound	O	-	no	-	-	-	-	-
Chlorothalonil	O	1964	yes	1992	table 5	IV	0.003	1994
Copper amm. carbonate	O		yes	-	-	-	-	-
Creosote	FOW		yes	n/a	-	-	n/a	-
Cresylic acid/Tar acid	OW	1949	(yes)	n/a	-	-	n/a	-
Cycloheximide	N	1946	no	-	n.i	n.i	n.i	-
Dichlofluanid	O	1964	yes	1983	table 5	III	0.3	-
Dinocap	NO	1965	no	1989	III	III	0.001	1990
Dodemorph	O	1967	yes	-	-	-	-	-
Etridiazole	O		yes	-	III	III	0.025	-
Fenarimol	O	1977	yes	-	table 5	III	-	1984
Fenpropimorph	F	1983	yes	-	-	-	0.003	1984
Fosetyl-aluminium	O	1977	yes	-	table 5	III	0.3	1984
Furalaxyl	O	1977	yes	-	III	-	0.0009	-
Imazalil	O	1973	yes	1991	II	II	0.03	1997
Iprodione	NO	1974	yes	1992	table 5	IV	0.2	1990
Mancozeb	O	1961	yes	1967	table 5	IV	0.03	-
Maneb	NO	by 1955	yes	-	table 5	IV	0.03	1997
Metalaxyl	O	1977	yes	1982	III	-	0.03	-
Octhilinone	O	1960	yes	-	-	-	-	-
Oxycarboxin	O	1966	yes	-	table 5	IV	0.15	-
Penconazole	O	1981	yes	1992	III	-	0.03	1984
Prochloraz	FO	1977	yes	1983	III	III	0.01	-
Propamocarb hyd'chl'de	O	1986	yes	1986	table 5	IV	0.1	-
Propiconazole	O	1979	yes	1987	III	-	0.04	1993
Pyrazophos	O	1969	yes	1992	II	II	0.0004	-
Quinomethionate	O	1960	yes	1987	table 5	III	0.006	-
Sodium nitrite	F	-	no	-	-	-	-	-
Sulphur	NO	-	yes	-	-	-	-	-
Thiabendazole	O	1964	yes	1993	table 5	IV	0.1	1992
Thiophanate-methyl	O	1970	yes	1977	table 5	IV	0.08	1992
Thiram	NO		yes	1992	III	III	0.5	1997
Tolclofos-methyl	O	-	yes	-	table 5	III	-	1993
Triadimefon	N	1973	yes	1985	III	III	0.03	-
Tribromophenol	W	1960s	no	-	-	-	-	-
Triforine	O	1969	yes	1978	table 5	IV	-	-
Urea	F		yes	-	-	-	-	-
Vincozolin	O	1975	yes	1988	table 5	IV	0.07	1991/95
Zineb	NO		yes	1993	table 5	IV	-	-

Appendix II, Table 1c   **Herbicides:** *Field of use, toxicity classifications, etc.*

Active ingredient or substance	Fields of use, crop type or situation	Year first used	In '99 MAFF bklt	In FAO r'due reviews	WHO tox'ty rating	EPA tox'ty rating	Accep' min residue level ppm	ACP review/ eval'n date
2,4,5-T	F	1944	no	-	-	-	-	-
2,4-D	F	1942	yes	1984	II	II	0.3	1993
Alloxydim-sodium	NO	1980	yes	-	table 5	III	-	-
Amitrole	N	1960	yes	1993	table 5	IV	0.0005	1984
Ammonium sulphamate	F	1940s	yes	-	-	III	-	-
Asulam	F	1968	yes	1975	table 5	IV	-	-
Atrazine	FN	1958	yes	-	table 5	III	-	1992/93
Chlorbufam*	N	1960s	yes	-	table 5	-	-	-
Chlorpropham	O	1964	yes	1965	table 5	III	-	1993
Chlorthal-dimethyl	N	1960	yes	-	table 5	IV	-	-
Chlorthiamid	*FO*	*1964*	*no*	-	*III*	*III*	-	-
Clopyralid	FMNO	1979	yes	-	table 5	IV	0.15	-
Cyanazine	MN	1971	yes	-	II	II	-	-
Cyanazine + atrazine	*FM*	-	*no*	-	-	-	-	-
Cycloxidime	F	1985	yes	1992	-	IV	0.07	1990
Dalapon*	FNO	1957	no	1974	table 5	II	-	-
Dicamba	F	1961	yes	-	table 5	III	-	-
Dichlobenil	FO	1965	yes	-	table 5	III	0.025	-
Diphenamid	N	1972	yes	-	III	III	-	-
Diquat	FNO	1961	yes	1993	II	II	0.002	-
Diuron	O	1951	yes	-	table 5	III	0.002	-
Fluazifop-P-butyl	FMNO	1981	yes	-	II	III	-	1984/8
Fosamine-ammonium	F	1979	yes	-	table 5	IV	-	-
Glufosinate ammonium	FNO	1984	yes	1991	III	III	0.02	1990/1
Glyphosate	FNO	1974	yes	1986	table 5	IV	0.3	-
Hexazinone	*F*	*1975*	*no*	-	*III*	*II*	-	-
Imazapyr	F	1985	yes	-	table 5	IV	-	1984
Isoxaben	FN	1984	yes	-	table 5	IV	0.056	-
Lenacil	NO	1965	yes	-	-	IV	-	-
Light fuel oil								
Tractor vaporising oil	N	1949	no	-	-	-	-	-
Light mineral oil - white spirit	N	1949	no	-	-	-	-	-
MCPA	O	1945	yes	-	III	III	-	-
Mecaprop	O	1953	yes	-	III	III	-	1996
Metamitron	N	1977	yes	-	III	III	0.13	-
Metazachlor	MN	1982	yes	-	-	-	0.036	-
Napropamide	NO	1983	yes	-	table 5	III	-	-
Oryzalin	*NO*	*1973*	*no*	-	*table 5*	*IV*	*0.15*	-
Oxadiazon	NO	1977	yes	-	table 5	IV	-	-
Paraquat	FNO	1962	yes	1986	II	II	0.004	-
Pendimethalin	F	1974	yes	-	III	III	-	-
Pentanochlor	FMN	1979	yes	1994	table 5	IV	0.08	-
Picloram	F	1967	yes	-	table 5	IV	-	-
Propachlor	O	1964	yes	-	III	III	-	-

App. II, Table 1c *contd*   **Herbicides:** *Field of use, toxicity classifications, etc.*

Active ingredient or substance	Field of use crop type or situation	Year first used	In '99 MAFF bklt	In FAO r'due reviews	WHO tox'ty rating	EPA tox'ty rating	Accep' min residue level ppm	ACP review/ eval'n date
Propaquizafop	FM	1987	yes	-	III	-	-	1993
Propyzamide	FMNO	1971	yes	1994	table 5	IV	0.08	1996
Simazine	NO	1956	yes	-	table 5	IV	0.005	1992/3
Sodium chlorate	N	1910	yes	-	III	III	-	-
TCA (Trichloracetic acid)	N	1950	yes	-	table 5	III	-	-
Terbuthylazine*	FNO	1966	no	-	III	III	0.0035	-
Thifensulfuron-methyl	N	1985	yes	-	III	IV	100*	1991/4
Tribenuron-methyl	N	1985	yes	-	III	III	0.008	1992
Triclopyr	FO	1982	yes	-	III	III	0.03	-

*available only in mixtures

Appendix II, Table 1d   *Other pesticides: Field of use, toxicity classification etc*

Active ingredient or substance	Field of use, crop type or situation	Year first used	In '99 MAFF bklt	In FAO r'due reviews	WHO tox'ty rating	EPA tox'ty rating	Accep' min residue level ppm	ACP review/ eval'n date
Biological agents								
Bacillus thuringiensis	F	1938	yes	n/a	-	-	n/a	-
Heterorhabditis spp.(nematode)	FO	*		n/a	n/a	n/a	n/a	-
Neodiprion sertifer NPV	F		no	n/a	-	-	n/a	-
Phlebiopsis gigantea	FO		yes	n/a		-	n/a	1998
Rhizophagus grandis	F	*		n/a	n/a	n/a	n/a	-
Trichoderma viride	FO		no	n/a	-	-	n/a	-

* Not within scope of the *Control of Pesticides Regulations*

Soil and/or timber fumigants								
Chloropicrin	NO	1908	yes	1965	-	II	-	-
Dazomet	N	After 1897	yes	-	III	III	-	-
Dichloropropene /dichloropropane**	O	1956	yes	-	-	II	n.a	-
Formaldehyde	NO	1888	***	1985	-	-	-	-
Meth bromide(+chl'pcr'n)	NO	1923	yes	1967	-	II	1.0	1992
Metam sodium	O	1956	yes	-	II	II	-	-

** only sold as a mixture   *** Commodity chemical - see §3.43

Molluscicides								
Metaldehyde	O	1936	yes	-	III	III/II	-	-
Moss/algicides								
Methiocarb	NO	1962	yes	1987	Ib	I	0.001	-
Animal poisons & repellants								
Aluminum phosphide	V	-	yes	1967	Ib	I	-	-
Endrin	V	*1950s*	no	-	-	V	-	-
Sodium cyanide	V	-	yes	-	Ib	I	0.05	1992/4/5
Strychnine	V	*Long known*	no	-	*Ib*	-	-	*1998*
Sulphonated cod liver oil	V	-	yes	-	-	-	-	-
Toxaphene	V	-	no	-	-	-	-	-
Warfarin	V	1944	yes	-	Ib	I	-	-
Zinc phosphide	V		yes					
Ziram	V	bfr1964	yes	1980	III	III	-	1997
Wood preservatives & Miscellaneous								
Bitumen emul'n (wound sealant)	O	-	yes	-	-	-	-	-
Boron compounds	W	-	yes#	-	-	-	-	-
Lead paint (wound sealant)	O	-	no	-	-	-	-	-
Mercury salts	W	*1891*	no	-	*Ib*	*I*	-	-
Methylene-bis-thiocyanate	W	-	yes##	-	-	-	-	-
PCP	W	1936	yes	-	Ib	II	-	1994

\# a component of many mixtures - see HSE section of *Pesticides 1999*   \#\# in HSE section

Growth inhibitors								
Maleic hydrazide	FO	1949	yes	1984	table 5	III	0.19	1984
Mefluidide	O	1974	yes	-	-	-	-	-
Paclobutrazol	O	1982	yes	1988	III	IV	0.1	-

Appendix II Table 2

Choice of Approved Products

To find products with a current approval for a task in mind, current editions of the following publications should be consulted:

- MAFF Booklet 500 *Pesticides 1999*
- *UK Pesticide Guide 1999* (Whitehead, 1999).

The choice may range from several to 'no choice' (ie only one approved product).

Both texts are revised annually. They include list addresses of manufacturers, from whom current brochures and product safety sheets can be requested. While old data sheets might still be current, marketing companies, the products they are marketing and the legal conditions of use may change at any time. It is essential to check that *all* relevant up-to-date information is available to whoever has to prescribe, control and apply any pesticide.

Appendix II Table 2, overleaf, gives a brief check-list of sources of pesticide product information that may be required in order to control a pest.

No news may be good news

When considering choice of pesticides, if there is no relevant warning and no precautions are prescribed either on a current label for a product or in the texts listed above, any foreseen hazard can be treated as not requiring more precautions when using that product than would apply to any pesticide application.

Appendix II Table 2   *Check list of sources of information about approved products*

Information sought	Comment	Source
Active ingredient/ substance	Seek current published recommend- ations in technical literature	Current Forestry Commission & Arboricultural Information & Advisory Service publications lists
Marketing company for selected product	Names given in *Pesticides 1999* & *UK Pesticide Guide*	MAFF B'klet 500: Stationery Office Whitehead (1999)
Product registered number	Refers to latest approval; earlier numbers may be obsolete.	Listed in *Booklet 500* and *UK Pesticides Guide*
Product toxicity ratings	Include toxicity, irritancy, sensi- tisation, risks to skin & eyes;	Current ratings summarised in *UK Pesticides Guide*; also see §4.22.
Content of organophos- phorus/carbamate	Identifies whether to restrict use & need for medical surveillance.	Noted in *UK Pesticides Guide*
Respiratory hazards & Occupational exposure limits	More general background information rather than of immediate need.	See §4.24 & current issue of HSE *Occupational Exposure Limits (Booklet EH40/year)*
Environmental safety	Risks to bees, fish, wildlife.	Noted in *UK Pesticides Guide*
PSPS approvals	'Label' & 'Off-label' approvals for forest nursery, forest, ornamentals and farm forestry.	Outlined in *Booklet 500*; fuller details in *UK Pesticides Guide*

Before work commences with any product, a current label from its container should be checked for more detailed information relating to these headings and other terms of its current approval.

Appendix III

Glossary

I   Abbreviations/Acronyms

ACP	Advisory Committee on Pesticides
AOT40	Accumulated Exposure Over Threshold 40 (Ozone standard)
BAA	British Agrochemicals Association
BCPC	British Crop Protection Council
CAI	Current Annual Increment, *ie* the mean recent rate of growth of timber in a woodland; conventionally taken as the last 5 years and expressed in m³/yr. See MAI.
CFCs	Chlorofluorocarbons
CHP	Combined heat and power
DMCA	*Dendroctonus micans* Control Area
DMPZ	*Dendroctonus micans* Protection Zone
DNC	Declared Net Capacity (electricity generator rating)
FA	Forestry Authority (part of the Forestry Commission)
FC	Forestry Commission
FCCC	Framework Convention on Climatic Change
FEPA	Food and Environment Protection Act
FAO	Food and Agricultural Organisation (of the United Nations)
HSC	Health and Safety Commission
HSE	Health and Safety Executive
IPCC	International Panel on Climatic Change
IUPAC	International Union of Pure & Applied Chemistry
LERAP	Local environmental risk assessment for pesticides
LFG	Landfill gas
MAI	Maximum Annual Increment, *ie* the maximum rate of growth of a woodland in m³/yr. A rotation of MAI is the number of years necessary for a woodland to achieve its maximum mean rate of growth averaged over its age at the time. See CAI
NFFO	Non Fossil Fuel Obligation
NO_x	Oxides of Nitrogen (usually $NO + NO_2$)
PSPS	Pesticides Safety Precautions Scheme
SRC	Short rotation coppice
SO_x	Oxides of sulphur
SRO	Scottish Renewables Obligation
VMD	Volume Median Diameter
YC	Yield Class; *ie* the predicted rate of growth of a woodland at the time of Maximum Annual Increment; expressed in '2 m³' classes, *eg* YC 12, YC10

II   Chemical & other conversion factors

Concentrations in stream-flow or precipitation
1 ppm =   1 mg/kg = approx 1mg/litre;   1 g/m² =   10 kg/ha

(for Tables 14.2, 14.5, conversion factors for microequivalents μeq)
Per litre Calcium   1 μeq =   0.020 mg Ca
              Chloride   1 μeq =   0.035 mg Cl
              Ammonium   1 μeq =   0.014 mg N   =   0.018 mg NH_4^+
              Magnesium   1 μeq =   0.012 mg Mg (= millegrams magnesium!)
              Sodium   1 μeq =   0.023 mg Na
              Nitrate   1 μeq =   0.014 mg N   =   0.062 mg NO_3^-
              Phosphate   1 μeq =   0.010 mg P   =   0.032 mg PO_4^{--}
              Sulphate   1 μeq =   0.016 mg S   =   048 mg SO_4^{--}

Orders of magnitude

10^{-9}	10^{-6}	10^{-3}	10^0	10^3	10^6	10^9	10^{12}
	μ	milli	1	Kilo	Mega	Giga	Tera
ppb	ppm						
				thousand	million	billion	

III   Terms relating to atmospheric gases

Mole fraction (Table 13.1): the proportion of gases by number of molecules.

Gas concentrations given in *parts per million* (ppm) or *billion* (ppb) reflect the relative proportions of molecules of each of the gases present.

Stratosphere: Layer of atmospheric gases extending to about 50 km above the earth's surface, from the *troposphere* to the ionosphere.

Troposphere: Layer of atmospheric gas extending from the boundary with the earth's surface to the stratosphere. The thickness of the troposphere varies between 6 - 10 km; its temperature decreases with increasing height.

Boundary layer: Layer of the atmosphere in contact with the ground. Movement of air in the layer is directly affected by surface features. The depth of the boundary layer varies continually in response to surface temperature, the pattern of winds and other events (*eg* volcanic eruptions).

IV Terms of pesticide use

Contact: a pesticide applied to pest/plants and affecting the target treated has *contact* action. Compare with *'systemic'* and *'translocated'*.

Differential uptake: one of the factors accounting for selectivity between species in response to a given product. It may be the result of differences in foliage surface, age, physiological condition *etc*.

Directed treatment: a major means of achieving selective control of weeds, avoiding or minimising exposure of crop species to pesticide. Application pattern may depend on available equipment. Hand controlled equipment *ie* knapsack sprayers, spot guns, weed wipers *etc*. can readily be applied to spots or patches around the planting site, with or without a tree guard to ensure the tree is not treated. Tractor-mounted sprayers usually treat a continuous band along the line of planting.

Pre-emergence sprays: control susceptible weeds that germinate rapidly before more slowly germinating crop seedlings have emerged above the soil surface.

Post-emergence sprays: are intended to control recently germinated susceptible weeds. They rely for selectivity on crop-plant tissues being either inherently resistant or having acquired resistance with age.

Residual action: pesticides applied to a surface have *residual* action if they persist for sufficient time to control pests which subsequently encounter the pesticide by contact or feeding. Residual pesticides may be washed into the topmost layer of the soil and there be taken up by weed roots or other pests.

Selectivity: *Selective treatments* are those intended to kill undesired pest or crop organisms and leave desired organisms unharmed.

Soil-applied: pesticides applied to the soil surface. Pesticides applied to the soil surface have *soil residual* action if intended to be taken up by superficial roots of plants or germinating seedlings in the treated zone.

Spectrum of control: where the range of species controlled by a pesticide is large, the pesticide may be designated *broad spectrum*. The terms *'limited'*, *'narrow spectrum'* or *'specific'* may be applied to pesticides controlling few pests.

Susceptibility: describes the effectiveness of a pesticide on its target. A pest or crop species treated in particular conditions is *susceptible* if the treatment has a clear controlling effect and *resistant* if the control is insignificant or absent.

Systemic: a fungicide or insecticide has a *systemic* effect if it is translocated within the target plant species and controls pests which did not come directly into contact with the application. A herbicide has a systemic effect if it is translocated within the plant to take effect, *eg* to roots or rhizomes.

Translocation: a pesticide is *translocated* if after entering a plant, it moves away from its entry point to take effect. Compare with *contact* and *systemic*.

Timing of applications may seek to exploit small differences in age of plant, speed of germination or stage of growth. The aim is commonly to treat weeds shortly after fresh growth has emerged and whilst their tissues are still soft and readily penetrated by herbicide.

Bibliography

Introduction

The journals and publications listed below are given their full titles, except (mostly) for Forestry Commission publications for which the following abbreviations have been used:

For. Comm. = *Forestry Commission;* otherwise *For.* = *Forest.*

Ann. = *Annual,*	*Bull.* = *Bulletin,*	*Conf.* = *Conference,*
Devel. = *Development,*	*Handbk* = *Handbook,*	*Leaf* = *Leaflet,*
Occ. = *Occasional,*	*Pap.* = *Paper,*	*Proc.* = *Proceedings,*
Rec. = *Record,*	*Rep.* = *Report,*	*Res.* = *Research,*
R.. = *Royal,*	*Soc.* = *Society,*	*Tech.* = *Technical.*

For any surname, 'single author' references are listed first, followed by 'two author' references and then by 'multiple author' references.

Single author references are listed in alphabetical order of the surname of the author and year of publication.

References by two authors are listed by the first author surname, within that group, by second author surname and, if necessary by year.

References by three or more authors are referred to in the text as Other, A.N. *et al.* In the bibliography, they are listed by first author and year.

Aaron, J.R. 1962 Interim report on field trials of treated and untreated round fencing timber. *For. Comm. Rep. For. Res.* **1962** 133-136 HMSO, London.

Aaron, J.R. 1970 Utilisation of bark. *For. Comm. Res. & Dev. Pap.* **32** 22 pp Forestry Commission, London.

Aaron, J.R. 1976 Conifer bark: its properties and uses. *For. Comm. Forest Rec.* **110** 31 pp HMSO, London.

Aaron, J.R. 1978 Timber utilisation - spruce power transmission poles. *For. Comm. Rep. For. Res.* **1978** 52 HMSO, London.

Aaron, J.R. 1979 Harvesting & marketing - telegraph & power transmission poles. *For. Comm. Rep. For. Res.* **1979** 49-51 HMSO, London.

Aaron, J.R. 1982 Wood utilisation. *For. Comm. Rep. For. Res.* **1982** 44 HMSO, London.

Aaron, J.R. 1983 Wood utilisation. *For. Comm. Rep. For. Res.* **1983** 52 HMSO, London.

Aaron, J.R. & Oakley, J.S. 1985 The production of poles for electricity supply and tele-communications. *For. Comm. Forest Rec.* **128** 1-12 HMSO, London.

ABMAC 1963 *Code of Practice for the safe use of agricultural pesticides in Great Britain.* Association of British Manufacturers of Agricultural Chemicals, London.

ABMAC 1968 Pesticides 'A Code of Conduct' (In) *Some safety aspects of pesticides in the countryside.* Proc. Conf. 'Countryside in 1970', London, 20 Nov., 1967. (Eds.) N.W. Moore & W.P. Evans. ABMAC/Wildlife c'ttee, Alembic House, London, SE1.

Abramovitz, J.N. 1998 Forest decline continues. (In) *Vital Signs 1998-1999.* (Eds.) L.R. Brown *et al* 124-125 Worldwatch Institute, c/o Earthscan Publications, Pentonville Road, London.

Abuarghub, S.M. & Read, D.J. 1988 The distribution of nitrogen in soil of a typical upland Callunetum with special reference to 'free' amino-acids. *New Phytologist* **108** 425-431.

Ackers, C.P. 1947 *Practical British forestry.* 394pp Oxford University Press.

Acland, F.D. (Chairman) 1917 *Report of the Forestry sub-committee of the Reconstruction Committee.* Report to Parliament **Cd 8881** HMSO, London.

ACP (Advisory Committee on Pesticides). *(Yearly) Annual Report.* Ministry of Agriculture, Fisheries & Food; Health & Safety Executive, HMSO, London.

Adams, S.N. & Dickson, D.A. 1973 Some short-term effects of lime and fertilizer on a Sitka spruce plantation. *Forestry* **46** (1) 31-37.

Adams, S.N., Jack, W.H. & Dickson, D.A. 1970 The growth of Sitka spruce on poorly drained ground in Northern Ireland. *Forestry* **43** (2) 125-133.

Adams, S.N., Cooper, J.E., Dickson, D.A., Dickson, E.L. & Seaby, D.A. 1978 Some effects of lime and fertilizer on a Sitka spruce plantation. *Forestry* **51** (1) 57-65.

Adams, S.N., Dickson, E.L. & Quinn, C. 1980 Amount and nutrient content of litter fall under Sitka spruce on poorly drained soils. *Forestry* **53** (1) 65-70.

Adamson, J.K. & Hornung, M. 1990 The effect of clear-felling a Sitka spruce (*Picea sitchensis*) plantation on solute concentration in drainage water. *Journal of Hydrology* **116** 287-297.

Adamson, J.K., Hornung, M., Pyatt, D.G. & Anderson, A.R. 1987 Changes in solute chemistry of drainage waters following clearfelling of a Sitka spruce plantation. *Forestry* **60** (2) 165-177.

Adamson, J.K., Hornung, M., Kennedy, V.H., Norris, D.A., Paterson, I.S. & Stevens, P.A. 1993 Soil solution chemistry & throughfall under stands of Japanese larch and Sitka spruce in three locations in Britain *Forestry* **66** (1) 51-68.

Agarwal, A. 1992 *Statement on global environmental democracy.* Centre for Science & the Environment. F-6 Kailash Colony, New Delhi 110048 India

Ågren G.I. & Bosatta, E. 1988 Nitrogen saturation of terrestrial ecosystems. *Environmental Pollution* **54** 185-197.

Ainsworth, N. & Ashmore, M.R. 1992 Assessment of ozone effects on beech by injection of a protectant chemical. *Forest Ecology & Management* **51** 129-136.

Aldhous, J.R. 1959 Control of cutworm in forest nurseries. *Forestry* **32** (2) 155-165.

Aldhous, J.R. 1961a Experiments in handweeding conifer seedbeds in forest nurseries. *Weed Research* **1** (1) 59-67.

Aldhous, J.R. 1961b Simazine - a weedkiller for forest nurseries. *For. Comm. Rep. For. Res.* **1961** 154-165 HMSO, London.

Aldhous, J.R. 1962a Weed control in forest nurseries. *For. Comm. Res. Br. Pap.* **24** Forestry Commission, London (Departmental use).

Aldhous, J.R. 1962b A survey of Dunemann seedbeds in Great Britain. *Quarterly Journal of Forestry* **56** (3) 185-196.

Aldhous, J.R. 1964a Paraquat as a pre-emergence spray for nursery seedbeds. *Proc. 7th British Weed Control Conf. Brighton* 256-266 British Crop Protection Council.

Aldhous, J.R. 1964b Bracken control using dicamba. *Proc. 7th British Weed Control Conf., Brighton* 896-898 British Crop Protection Council.

Aldhous, J.R. 1964c Effect of paraquat, 2,6-dichlorothiobenzamide, and tordon on species planted in the forest. *Proc. 7th British Weed Control Conf., Brighton.* 267-275 British Crop Protection Council.

Aldhous, J.R. 1964d Nursery investigations; nutrition; fertilizer damage to transplants. *For. Comm. Rep. For. Res.* **1964** 16 HMSO, London.

Aldhous, J.R. 1965a Chemical control of weeds in the forest. 1st edn *For. Comm. Leaflet* **51** HMSO, London.

Aldhous, J.R. 1965b Bracken control using dicamba. *For. Comm. Rep. For. Res.* **1965** 141-149 HMSO, London.

Aldhous, J.R. 1965c Effect of paraquat, 2,6-dichlorothiobenzamide & 4-amino-3,5,6-trichlor-picolinic acid on species planted in the forest. *For. Comm. Rep. For. Res.* **1965** 150-153 HMSO, London.

Aldhous, J.R. 1966a Simazine residues in two forest nursery soils. *For. Comm. Res. & Dev. Pap.* **31** 1-9 Forestry Commission, London.

Aldhous, J.R. 1966b Bracken control with dicamba, picloram and chlorthiamid. Proc. 8th British Weed Control Conf. Brighton. 150-159 British Crop Protection Council.

Aldhous, J.R. 1967a Review of Weed control in forestry in Great Britain. *For. Comm. Res. & Dev. Pap.* **40** Forestry Commission, London (Departmental use).

Aldhous, J.R. 1967b Progress report on chlorthiamid ('Prefix') in forestry; 1962-66. *For. Comm. Res. & Dev. Pap.* **49** 1-6 Forestry Commission, London (Departmental use)

Aldhous, J.R. 1967c 2,4-D residues in water following spraying in a Scottish forest. *Weed Research* **7** (3) 239-241.

Aldhous, J.R. 1968a Countryside in 1970 - Report on Pesticides Conference. *Quarterly Journal of Forestry* **62** (3) 225-230.

Aldhous, J.R. 1968b Maintenance of fertility in forest nurseries. *For. Comm. Res. & Dev. Pap.* **68** 24pp Forestry Commission, London (Departmental use).

Aldhous, J.R. 1969a Aircraft and British forestry. *Quarterly Journal of Forestry* **63** (2) 105-113.

Aldhous, J.R. 1969b Chemical control of weeds in the forest. 2nd edn. *For. Comm. Leaflet* **51** 49pp HMSO, London.

Aldhous, J.R. 1969c Aerial spraying - notes for landowners. *Quarterly Journal of Forestry* **63** (2) 152-155.

Aldhous, J.R. 1970 UK: trends in forest weed control. *Span* **13** 21-23 Shell Chemicals Ltd.

Aldhous, J.R. 1972a Nursery Practice. *For. Comm. Bull.* **43** HMSO, London.

Aldhous, J.R. 1972b Silvicultural techniques & problems with special reference to timber production. (In) Lowland forestry & wildlife conservation. *Nature Conservancy Symposium* **6** 65-79 Nature Conservancy, Monkswood.

Aldhous, J.R. 1972c Trials of formalin & chloropicrin 1951-56. (In) Nursery Practice. *For. Comm. Bull.* **43** 157-161 HMSO, London.

Aldhous, J.R. 1972d Observations on the response of annual meadow grass, *Poa annua*, and other common weeds to soil pH. *For. Comm. Bull.* **43** 153-156 HMSO, London.

Aldhous, J.R. 1983 Fuelwood from woodlands in southern England. (In) Proc. Conf. *Growing wood for fuel* Paper 6 National Agricultural Centre, Stoneleigh, Warwick.

Aldhous, J.R. 1988 Background to second draft Code of 'Practice for the use of pesticides in forestry.' (Subsequently published as 'Forestry Commission Occasional Paper 21') *Aspects of Applied Biology* **16** 5-7 Association of Applied Biologists.

Aldhous, J.R. 1991 *Short rotation coppice prospects in Great Britain.* Report to the Forestry Commission 13 pp Forestry Commission, Edinburgh.

Aldhous, J.R. 1994 Nursery policy & planning; legislation & the nursery manager. (In) *For. Comm. Bull.* **111** 1-12; 223-246 HMSO, London

Aldhous, J.R. 1995 Land rehabilitation & tree growth on former industrial sites in the Central Belt of Scotland. (In) *Heavy metals & trees* Proc. conf. Glasgow 1995 (Ed.) I. Glimmerveen. 196-206 Institute of Chartered Foresters, Edinburgh.

Aldhous, J.R. 1997 British forestry: 70 years of achievement. *Forestry* **70** (4) 283-291.

Aldhous, J.R. & Atterson, J. 1967 Weed control in the forest. *For. Comm. Rep. For. Res.* **1967** 70-73 HMSO, London.

Aldhous, J.R. & Hendrie, R. 1966 Control of *Rhododendron ponticum* in forest plantations. Proc. 8th British Weed Control Conf. Brighton. 160-166 British Crop Protection Council

Aldhous, J.R. & Low, A.J. 1974 The potential of western hemlock, western red cedar, grand fir and noble fir in Britain. *For. Comm. Bull.* **49** 105 pp HMSO, London.

Aldhous, J.R. & Mason, W.L. 1994 Forestry nursery practice. *For. Comm. Bull.* **111** 268 pp HMSO, London.

Aldhous, J.R. & Scott, A.H.A. 1993 Forest fire protection in the UK; 1950-1990. *Commonwealth Forestry Review* **72** (1) 39-47.

Aldhous, J.R., Atterson, J., Brown, R.M., Low, A.J. & Stoakley, J.T. 1967 Nursery investigations: trials of new fertilizers. *For. Comm. Rep. For. Res.* **1967** 30-35.

Aldhous, J.R., Brown, R.M. & Atterson, J. 1968a Weed control in the forest. *For. Comm. Rep. For. Res.* **1968** 75-82 HMSO, London.

Aldhous, J.R., Atterson, J., Low, A.J. & Brown, R.M. 1968b Nursery investigations. *For. Comm. Rep. For. Res.* **1968** 28-36 HMSO, London.

Alexander, A.C. & Chadwick, D.J. 1991 Harvesting & yield of forest residues. (In) *Wood for energy - implication for harvesting, utilisation & marketing.* Proc. Discussion meeting (Ed.) J.R. Aldhous 69-87 Institute of Chartered Foresters, Edinburgh.

Alexander, I. & Fairley, R.I. 1983 Effects of N fertilization on populations of fine roots & mycorrhizas in spruce humus. *Plant & Soil* **71** 49-53.

Alexander, I. & Watling, R. 1987 Macrofungi of Sitka spruce in Scotland. *Proc. R. Soc. Edinburgh* **93B** 112-115.

Alexander, J.H. 1954 The forestry nursery trade in Scotland. *Scottish Forestry* **8** (2) 75-84.

Alford, D.V. 1995 *A colour atlas of pests of ornamental trees, shrubs & flowers.* Manson publishing, London.

Alford, D.V & Locke, T. 1989 Pests & diseases of fruit & hops. (In) *Pest & Disease Handbook.* (Eds.) N. Scopes & L. Stables 3rd Edn 323-403 British Crop Protection Council.

Allen, S.E. 1964 Chemical aspects of heather burning. *Journal of Applied Ecology* **1** 347-67.

Allot, T.E.H., Harriman, R. & Battarbee, R.W. 1992 Reversibility of lake acidification at the Round Loch of Galloway, Scotland. *Environmental Pollution* **77** 219-225.

Anderson, M.A. 1982 Effects of trees species on vegetation & nutrient supply in lowland Britain. *For. Comm. Res. Inf. Note* **75/82/SSS** 4pp Forestry Commission, Forest Research Station, Farnham.

Anderson, M.A. 1983 Effects of tree species on vegetation & nutrient supply in lowland Britain. *Arboriculture Research Note* **44/83/SSS** Arboricultural Advisory & Information Service, Farnham, Surrey.

Anderson, M.A. 1985 The impact of whole-tree harvesting in British forests. *Quarterly Journal of Forestry* **79** (1) 33-39.

Anderson, M.A. 1986a Conserving soil fertility (In) *Forestry's social & environmental responsibilities* Proc. discussion meeting. (Ed.) R.J. Davies 65-78 Institute of Chartered Foresters, Edinburgh.

Anderson, M.A. 1986b The effects of trees on sites. *For. Comm. Rep. For. Res.* **1986** 22-23 HMSO, London.

Anderson, M.A. 1987 The effects of forest plantations on some lowland soils. *Forestry* **60** (1) 69-86.

Anderson, T.A. 1991 Influence of stem flow and throughfall from common oak on soil chemistry and vegetation patterns. *Canadian Journal of Forest Research* **21** (6) 917-924.

Anderson, T.A. 1992 Impact of legislation in Scotland. (In) *Proc. conf.: Sewage sludge disposal, Glasgow, March 1992* 3/1-3/17 Institute of Water & Environment Management, London.

Anderson, A.R. 1997 Impacts of conifer plantations on blanket bogs and prospects of restoration. (In) *Restoration of temperate wetlands.* (Eds) Wheeler, B.D. *et al.* 533-548 Wiley, Chichester.

Anderson, G. 1999 Fiberpac system gives bioenergy fuel a boost. *SkogForst News* **2**(1999) 3.

Anderson, I. & Nowak, R. 1997 Australia's giant lab. *New Scientist* **22 Feb. 1997** 4pp.

Anderson, A.R. & Pyatt, D.G. 1986 Interception of precipitation by pole stage Sitka spruce and lodgepole pine, and mature Sitka spruce at Kielder Forest Northumberland. *Forestry* **59** (1) 29-38.

Anderson, A.R., Pyatt, D.G. & Stannard, J.P. 1990 Effects of clearfelling a Sitka spruce stand on the water balance of a peaty gley soil at Kershope forest, Cumbria. *Forestry* **63** (1) 51-69.

Anderson, A.R., Pyatt, D.G., Sayers, J.M. *et al.* 1993a Volume and mass budgets of blanket peats in the north of Scotland. *Suo* **43** (4-5) 195-198.

Anderson, H.A., Miller, J.D., Gauld, J.H., Hepburn, A. & Stewart, M. 1993b Some effects of 50 years of afforestation on soils in Kirkton Glen, Balquhidder. *Journal of Hydrology* **145** 439-451.

Anderson, A.R., Pyatt, D.G. & White, I.M.S. 1995 Forestry & peatlands. (In) *Conserving peatlands* (Eds) L. Parkyn, R.E. Stoneman & H.A.P Ingram 234-244 CAB International, Wallingford, Oxon.

Anderson, J., Clément, J & Crowder, L.V. 1998 Accommodating conflicting interests in forestry - concepts emerging from pluralism. *Unasylva* **49** (194) 3-10 FAO.

Anon 1920 Pine weevils *For. Comm. Leaflet* **1** HMSO, London.

Anon 1921 *Adelges cooleyi* A pest on Sitka spruce and Douglas fir. *For. Comm. Leaflet* **2** HMSO, London.

Anon 1933, 1937 Forestry Practice (1st & 2nd editions). *For. Comm. Bull.* **14** HMSO, London.

Anon 1936 Studies on the Pine shoot moth. *For. Comm. Bull.* **16** 36 pp HMSO, London.

Anon 1946 The black pine beetles (*Hylastes ater*) and other closely allied beetles. *For. Comm. Leaflet* **4** 10pp HMSO, London.

Anon 1948a Chafer beetles. *For. Comm. Leaflet* **17** 10 pp HMSO, London.

Anon 1948b *Adelges cooleyi*. *For. Comm. Leaflet* **2** (revised) 4pp HMSO, London.

Anon 1951 Forestry Practice 3rd Edn. *For. Comm. Bull.* **14** HMSO, London.

Anon 1952 *Pissodes* weevils. *For. Comm. Leaflet* **29** HMSO, London.

Anon 1954a *Megastigmus* flies attacking conifer seed. *For. Comm. Leaflet* **8** 10pp HMSO, London.

Anon 1954b Pine looper moth. *For. Comm. Leaflet* **32** HMSO, London.

Anon 1955 Pine sawflies. *For. Comm. Leaflet* **35** HMSO, London.

Anon 1956a Felted beech coccus. *For. Comm. Leaflet* **15** 8pp HMSO, London.

Anon 1956b Oak mildew. *For. Comm. Leaflet* **38** 6pp HMSO, London.

Anon 1956c Two leaf-cast diseases of Douglas fir. *For. Comm. Leaflet* **18** (revised 1961) 6pp HMSO, London.

Anon 1958a Honey fungus (see also 1967 edn). *For. Comm. Leaflet* **6** 5pp HMSO, London.

Anon 1958b *Weed control handbook*. 1st edn. Blackwell, Oxford.

Anon 1958c Elm disease, *Ceratostomella ulmi*. *For. Comm. Leaflet* **19** 7pp HMSO, London.

Anon 1960 The large pine weevil. *For. Comm. Leaflet* **1** (revised) 10pp HMSO, London.

Anon 1961 Megastigmus flies attacking conifer seed. *For. Comm. Leaflet* **8** 10pp HMSO, London.

Anon 1962a The grey squirrel. *For. Comm. Leaflet* **31** 3rd edn 18pp HMSO, London.

Anon 1962b Watermark disease of willow. *For. Comm. Leaflet* **20** 5pp HMSO, London.

Anon 1967a *Review of the present safety arrangements for the use of toxic chemicals in agriculture and food storage.* Department of Education & Science 72pp HMSO, London.

Anon 1967b *Keithia* disease of Western Red cedar, *Thuja plicata*. *For. Comm. Leaflet* **43** 7pp HMSO, London.

Anon 1968a Bluestain prevention in the UK. *Forest Products Research Lab Technical note* **2** 4pp.

Anon 1968b Pesticides: a code of conduct. *The countryside in 1970* Proc. conf. London, 20.11.67. (Eds) N.W. Moore & W.P. Evans. 12-20 Joint ABMAC/Wildlife C'ttee, Alembic House, London.

Anon 1970 *Fomes annosus* (see also 1974 edn). *For. Comm. Leaflet* **5** 1-10 HMSO, London.

Anon 1971a Fume damage to forests. *For. Comm. Res. & Dev. Pap.* **82** 50pp Forestry Commission, London (Departmental use).

Anon 1971b Dutch elm Disease. *For. Comm. Leaflet* **19** 11pp HMSO, London.

Anon 1972 *Limits to growth.* Club of Rome.

Anon 1974 *Non-agricultural uses of pesticides in GB.* Central unit on environmental pollution. HMSO, London.

Anon 1980 *Draft proposals for 'Poisonous Substances in Agriculture Regulations'.* HSE consultative document, Health & Safety Executive.

Anon 1988 The practice of weed control & vegetation management in forestry, amenity & conservation areas. *Aspects of Applied Biology* **16** 422pp Association of Applied Biologists.

Anon 1989 Provisional code of practice for the use of pesticides in forestry, 1989. *For. Comm. Occasional Pap.* **21** Forestry Commission, Edinburgh.

Anon 1990 The use of pesticides in forestry. (In) *Proc. FAO/ECE/ILO seminar on forest technology, management and training.* 1-340 pp Forestry Commission, Edinburgh.

Anon 1992a Forestry impact on upland water quality. *Institute of Hydrology Reports* **119 (July 1992)** Institute of Hydrology, Wallingford.

Anon 1992b Hawks and herons return to England. *New Scientist* **26.12.92** 5.

Anon 1992d Moth damage. *Forestry & British Timber* **Aug. 1992** 2.

Anon 1992e Controlling difficult weeds. *Forestry & British Timber* **Feb. 1992** 34.

Anon 1993a *Air pollution & tree health in the United Kingdom.* Department of the Environment 88 pp HMSO, London.

Anon 1993b Nematode timber discovered in Canadian shipments to UK. *Timber Trades Journal* **12.6.93** 3.

Anon 1994 (& yearly) *Annual report of the committees on toxicity, mutagenicity, carcinogenicity of chemicals in food.* MAFF Pesticides Safety Directorate 1994 68pp.

Anon 1994a Ready-to-use comes top. *Forestry & British Timber* **Feb. 1994** 42.

Anon 1994b Menzi takes a walk on the wild side. *Forestry & British Timber* **Feb. 1994** 40-42.

Anon 1994c New weapon in bracken battle. *Forestry & British Timber* **Nov. 1994** 10.

Anon 1995a *Biodiversity: the UK Steering Group report. Vol 2: Action plans.* 91-92 HMSO, London.

Anon 1995b Root growth regulator waits for approval in UK. *Horticulture Week* **June 8 1995** 5.

Anon 1995c How low can we go? *New Scientist* **2.12.95** 3.

Anon 1996a Changes to pesticide regulations. *ENDS Report* **256** 37 MAFF, London.

Anon 1996b Biodegradable spray to wipe out UK weeds. *Horticulture Week* **Dec. 12 1996.**

Anon 1996c 'Euro-pallet' threat to spruce. *Forestry & British Timber* **May 1995** 7.

Anon 1996d Carbon tax leads to burial at sea. *New Scientist* **3.8.96** 11.

Anon 1996e What the world needs now. *New Scientist* **15.6.96** 3.

Anon 1996f What price progress? *New Scientist* **7.9.96** 3.

Anon 1996g Machine cleaning on site. *Forestry & British Timber* **April 1996** 24-27.

Anon 1997a Foam to fight forest fires? *Forestry & British Timber* **Nov. 1997** 3.

Anon 1997b Spruce beetles evade capture. *Forestry & British Timber* **Nov. 1997** 4.

Anon 1997c *Ips typographus* update *ICF News* **4/97** 13 Institute of Chartered Foresters, Edinburgh.

Anon 1998 Bark beetle hunt goes on. *Timber Grower* **147** (Summer) 22.

Anon 1999 *Occupational exposure limits 1999.* HSE Books (Revised annually) **EH40/99** Health & Safety Executive.

Apps, M.J. & Price, D.T. (Eds) 1996 Forest ecosystems, forest management & the global carbon cycle. *Springer NATO Advanced Science Institutes Series, Series 1 Global Environmental Change* **40** 454pp Springer Verlag.

Apps, M.J., Price, D. & Wisniewski, J 1995 *Boreal Forest and global change.* 548pp Kluwer Academic Publishers.

Arbutat, P.H.W., Edwards, R.P. & Wilkins, C. 1995 Improving the establishment of trees on metalliferous mine wastes. *For. Comm. Rep. For. Res.* **1995** 13-14.

Armstrong, A. 1997 The UK network of experiments on site/yield relationships for short rotation coppice. *For. Comm. Res. Inf. Note* **294** 6pp Forestry Commission, Edinburgh.

Armstrong, A. 1999 Establishment of short rotation coppice. *For. Comm. Practice Note* **FCPN 7** Forestry Commission, Edinburgh.

Arnold, G. 1992 Soil acidification as caused by the nitrogen uptake pattern of Scots pine (*P. sylvestris*). *Plant & Soil* **142** 41-51.

Arnold, M. 1991 The long term global demand for & supply of wood. *For. Comm. Occasional Pap.* **36** 27pp Forestry Commission, Edinburgh.

Asman, W.A.H., Sutton, M.A. & Schjorring, J.K. 1998 Ammonia: emission, atmospheric transport & deposition. *New Phytologist* **139** 27-48.

Atkinson, D. 1993a Global climatic change - implications for crop protection. *Proc. BCPC Symposium.* **Monograph 56** 95-102 British Crop Protection Council, Farnham, Surrey.

Atkinson, D. (Ed.) 1993b Global climatic change - implications for crop protection. *Proc. BCPC Symposium.* **Monograph 56** 102pp British Crop Protection Council, Farnham, Surrey.

Atterson, J. 1964 Silvicultural investigations: use of lime. *For. Comm. Rep. For. Res.* **1964** 23 HMSO, London.

Atterson, J. 1965 Silvicultural investigations: forest nutrition in Scotland & N. England. *For. Comm. Rep. For. Res.* **1965** 21-25 HMSO, London.

Atterson, J. 1966a Nursery investigations: slow acting inorganic fertilizers. *For. Comm. Rep. For. Res.* **1966** 21-23 HMSO, London.

Atterson, J. 1966b Manuring of young crops on peat soils. *For. Comm. Rep. For. Res.* **1966** 32 HMSO, London.

Atterson, J. & Binns, W.O. 1968 Crop nutrition; manuring of pole-stage crops. *For. Comm. Rep. For. Res.* **1968** 45-54 HMSO, London.

Atterson, J. & Binns, W.O. 1975 Peat nutrients & tree requirements in Forestry Commission plantations. Proc. NERC symposium, Edinburgh, 1968 ITE, Edinburgh.

Atterson, J. & Davies, E.J.M. 1967 Fertilizers - their use and methods of application in British forestry. *Scottish Forestry* **21** (4) 222-228.

Audit Commission 1988 *Improving highways management - a management handbook.* HMSO, London.

Austarå, O. 1987 Defoliation by *Neodiprion sertifer* - effect on growth & economic consequences. (In) Population biology & control of Pine beauty moth *For. Comm. Bull.* **67** 57-60 HMSO, London.

AWRG 1989 *Acidity in United Kingdom Fresh Waters.* UK Acid Waters Review Group 2nd Report HMSO, London.

BAA (British Agrochemicals Association) 1987 *Why use pesticides? Pesticides in perspective.* 4pp British Agrochemicals Association, Peterborough.

BAA 1992 *Pesticides and water quality.* Booklet 11pp British Agrochemicals Association, Peterborough.

BAA 1993 *Suspected pesticide poisoning - guidance for doctors.* 2pp British Agrochemicals Association, Peterborough.

BAA 1994a *Crop protection: the way forward.* Booklet 19pp British Agrochemicals Association, Peterborough.

BAA 1994b Container rinsing. *BAA Leaflet* **U3** British Agrochemicals Association, Peterborough.

BAA 1994c Glove hygiene. *BAA Leaflet* **U4** British Agrochemicals Association, Peterborough.

BAA 1994d Pesticide disposal. *BAA Leaflet* **U5** British Agrochemicals Association, Peterborough.

BAA 1994e Think water. Keep it clean. *BAA Leaflet* **A6** 8pp British Agrochemicals Association, Peterborough.

BAA 1996 Good agrochemical storage. Checklist for annual safety and storage practice audit. *Leaflet* 2pp British Agrochemicals Association, Peterborough.

BAA 1998a *Annual review & handbook* 1998 British Agrochemicals Association, Peterborough, Cambridgeshire.

BAA 1998b *Guide to the selection and use of amenity pesticides.* British Agrochemicals Association, Peterborough, Cambridgeshire.

Baake, A. Saethe, T. & Kvamme, T. 1983 Mass trapping of the Spruce bark beetle *Ips typographus.* Pheromone & trap technology. *Meddelelser, Norsk Institut for Skogforsking* **38** 2-35 HMSO, London.

Baker, R.M. 1996 The future of the invasive shrub, Sea buckthorn *Hippophaë rhamnoides* on the west coast of Britain. (In) Vegetation management in forestry, amenity & conservation areas. *Aspects of applied biology* **44** 461-468 Association of Applied Biologists, Warwick.

Balfour, R.M. & Kirkland, R.C. 1962 Effect of creosote on populations of *Trypodendron* breeding in stumps. *For. Comm. Rep. For. Res.* **1962** 163-165 HMSO, London.

Barbour, D.A. 1978 *Bupalus piniaria For. Comm. Rep. For. Res.* **1978** 38 HMSO, London.

Barbour, D.A. 1985 Pattern of population fluctuation in the Pine looper moth (*Bupalus piniaria*) in Britain. *For. Comm. Res. & Dev. Pap.* **135** 8-20 HMSO, London.

Barbour, D.A. 1987a Pine beauty moth population dynamics. (In) Population biology and control of the Pine beauty moth. *For. Comm. Bull.* **67** 7-13 HMSO, London.

Barbour, D.A. 1987b Monitoring Pine beauty moth by means of pheromone traps; the effect of moth dispersal. (In) *For. Comm. Bull.* **67** 49-56 HMSO, London.

Barbour, D.A. 1987c, d Pine looper moth *Bupalus piniaria*; Pheromone trapping for exotic beetles. *For. Comm. Rep. For. Res.* **1987** 53 HMSO, London.

Barclay, G. & Foster, S. 1996 Burrowing threat to our archaeological heritage. *Farming & Conservation.* Oct. 1996 19-22.

Barker, R. 1996 CAP reform; agriculture versus environment. *Farming & Conservation* **2** (4) 30-31.

Barnett, D.W. 1987 Pine beauty outbreaks - associations with soil type, host nutrient status & tree vigour. (In) *For. Comm. Bull.* **67** 14-20 HMSO, London.

Barnes, J.D., Eamus, D., Davison, A.W., Ro-Poulsen, H. & Mortensen, L. 1990 Persistent effects of ozone on needle water loss and wettability in Norway spruce. *Environmental Pollution* **63** 345-363

Barrow, E.M. & Semenov, M.A. 1995 Climate change scenarios with high spatial and temporal resolution for agricultural applications. *Forestry* **68** (4) 349-360.

Battarbee, R.W., Flower, R.J., Stevenson, R.B. & Rippey, B. 1985 Lake acidity in Galloway: a paleoecological test of competing hypotheses. *Nature.* **314** 350-352.

Battarbee, R.W., Ormerod, S.J., Juggins, S. & Kreiser, A.M. 1995 Estimating the base critical load: the diatom model. (In) *Critical loads for acid deposition for United Kingdom freshwaters* 3-4 Dept. of the Environment (Critical Loads Advisory Group)

Bayes, C.D., Davies, J.M. & Taylor, C.M.A. 1987 Sewage as a forest fertilizer; experiences to date. *Journal of the Institute of Water Pollution Control* **86** 158-171.

Bayes, C.D., Taylor, C.M.A. & Moffat, A.J. 1991 Sewage sludge utilisation in forestry; the UK research programme. (In) *Alternative uses for sewage sludge* Proceedings, Water Research Council Conference, York, 1987. (Ed.) J.E. Hall Pergamon Press.

BCPC 1994a *Hand-operated sprayers handbook.* 1989 edn., rev. 1994 44pp British Crop Protection Council.

BCPC 1994b *Boom sprayers handbook.* 1991 edn., rev. 1994 60pp British Crop Protection Council.

BCPC 1996 *Using pesticides; a complete guide to safe and effective spraying.* British Crop Protection Council/ATB-Landbase

Bell, E.A. 1986 Plants as sources of novel pest and disease control agents. *Proc. 1986 British Crop Protection Conference* 661-667 British Crop Protection Council, Farnham, Surrey.

Bell, J.N.B. 1996 Ozone effects on trees. *Arboriculture Research & Information Note.* 132/96/EXT 4 pp Arboricultural Advisory & Information Service, Farnham, Surrey.

Bending, N.A.D. & Moffat, A.J. 1997 *Tree establishment on landfill sites. Research & updated guidance.* 53 pp HMSO, London (for the Dept. of Environment, Transport & the Regions).

Bennett, K.D. 1994 Post-glacial pinewoods. (In) *Our pinewood heritage* (Ed.) J.R. Aldhous 1994 23-39 Forestry Commission, RSPB & Scottish Natural Heritage joint publication.

Benson, J.F. & Willis, K.G. 1991 The demand for forests for recreation. (In) Forestry expansion: a study of technical, economic and ecological factors. *For. Comm. Occasional Pap.* **39** 25pp Forestry Commission, Edinburgh.

Bentham, G. & Hooker, J.D. 1924 (7th edn) *Handbook of the British flora.* 606pp Reeve, Ashford, Kent.

Benzian, B. 1965 Experiments on nutrition problems in forest nurseries. *For. Comm. Bull.* **37** HMSO, London.

Benzian, B. 1966 Risk of damage from certain fertilizer salts to transplants of Norway spruce. (In) Physiology in forestry. *Forestry* **1966 Supplement** 65-69.

Benzian, B. 1972 Cumulative dressings of potassium metaphosphate and soluble PK fertilizers ... in two English nurseries. *Plant & Soil* **36** 243-245.

Benzian, B. & Freeman, S.C.R. 1967 Effect of late-season top dressing of N & K to conifer seedlings in the nursery on their survival & growth in the British Isles. *For. Comm. Rep. For. Res.* **1967** 135-139 HMSO, London.

Benzian, B. & Freeman, S.C.R. 1968 Nutrition experiments in forest nurseries. *For. Comm. Rep. For. Res.* **1968** 140-142 HMSO, London.

Benzian, B. & Smith, H.A. 1973 Nutrient concentrations of healthy seedlings and transplants of *Picea sitchensis* grown in English forest nurseries. *Forestry* **46** 55-69.

Benzian, B. & Warren, R.G. 1956a Copper deficiency in Sitka spruce seedlings. *Nature* **178** 864-5

Benzian, B. & Warren, R.G. 1956b Nutritional problems in forest nurseries: needle tip-burn. *For. Comm. Rep. For. Res.* **1956** 80-81 HMSO, London.

Benzian, B., Bolton, J. & Mattingly, G.E.G. 1967 Nutrition experiments in forest nurseries. *For. Comm. Rep. For. Res.* **1967** 133-134 HMSO, London.

Benzian, B., Bolton, J. & Mattingly, G.E.G. 1969 Soluble and slow release PK fertilisers for seedlings and transplants of *Picea sitchensis*. *Plant & Soil* **31** 238-256

Benzian, B., Freeman, S.C.R. & Patterson, H.D. 1972a Comparisons of crop rotations, and of fertilizers with composts, in 15-year experiments with Sitka spruce. *For. Comm. Rep. For. Res.* **1972** 139-142 HMSO, London.

Benzian, B., Freeman, S.C.R. & Patterson, H.D. 1972b Comparisons of crop rotations, and of fertilizer with compost, in long-term experiments with Sitka spruce in two English nurseries. *Forestry* **45** (2) 145-176.

Benzian, B., Bolton, J. & Coulter, J.K. 1974 Fertilizer effects on the growth and composition of SS and NS nursery transplants and on the composition of a podzol. *For. Comm. Res. & Dev. Pap.* **109** 1-14 Forestry Commission, London.

Bevan, D. 1954 The status of the Pine looper (*Bupalus piniarius*) in Britain in 1953. *For. Comm. Rep. For. Res.* **1954** 159-163 HMSO, London.

Bevan, D. 1962 Pine shoot beetles (see also 1972 edn). *For. Comm. Leaflet* **3** 8pp HMSO, London.

Bevan, D. 1964 Forest entomology. *For. Comm. Rep. For. Res.* **1964** 61-62 HMSO, London.

Bevan, D. 1966a Forest entomology. *For. Comm. Rep. For. Res.* **1966** 75-76 HMSO, London.

Bevan, D. 1966b Pine looper moth. *For. Comm. Leaflet* **32** 12pp HMSO, London.

Bevan, D. 1966c Douglas fir seed wasp, *Megastigmus spermatrophus*. *For. Comm. Rep. For. Res.* **1966** 75 HMSO, London.

Bevan, D. 1967 Douglas fir seed wasp, *Megastigmus spermatrophus*. *For. Comm. Rep. For. Res.* **1967** 103 HMSO, London.

Bevan, D. 1968 Douglas fir seed wasp, *Megastigmus spermatrophus*; control of cutworm. *For. Comm. Rep. For. Res.* **1968** 112-113 HMSO, London.

Bevan, D. 1982 Insect pests - what kind of control? *Arboriculture Research Note* **11/82/ENT** Arboricultural Advisory & Information Service, Farnham, Surrey.

Bevan, D. 1983 Forest entomology, *Dendroctonus micans*. *For. Comm. Rep. For. Res.* **1983** 37 HMSO, London.

Bevan, D. 1987 Forest insects. *For. Comm. Handbk.* **1** 153pp HMSO, London.

Bevan, D. & Brown, R.M. 1960 Pine looper moth *Bupalus piniarius* in Rendlesham & Sherwood forests. *For. Comm. Rep. For. Res.* **1960** 172-179 HMSO, London.

Bevan, D. & Brown, R.M. 1978 Pine looper moth. *For. Comm. Leaflet* **119** 13pp HMSO, London.

Bevan, D. & Davies, J.M. 1970 Forest entomology. *For. Comm. Rep. For. Res.* **1970** 120-124 HMSO, London.

Bevan, D. & Davies, J.M. 1971 Pine looper moth *Bupalus piniaria*. *For. Comm. Rep. For. Res.* **1971** 85-86 HMSO, London.

Bevan, D. & Davies, J.M. 1972 Dutch elm disease. *For. Comm. Rep. For. Res.* **1972** 102 HMSO, London.

Bevan, D. & King, C.J. 1983 *Dendroctonus micans* - a new pest of spruce in the UK. *Commonwealth Forestry Review* **62** (1) 41-51.

Bevan. D. & Paramonov, A. 1956 Fecundity of *Bupalus piniaria* in Britain. *For. Comm. Rep. For. Res.* **1956** 155-162 HMSO, London.

Bevan, D., Davies, J.M. & Brown, R.M. 1957 Forest entomology. *For. Comm. Rep. For. Res.* **1957** 68-75 HMSO, London.

Biggin, P. 1979 Herbicides for use in forest nurseries. *Scottish Forestry* **33** (1) 9-14.

Biggin, P. 1981 Production of planting stock. *For. Comm. Rep. For. Res.* **1981** 18-19 HMSO, London.

Biggin, P. & McCavish, W.J. 1980 Glyphosate - a herbicide for nursery and forest use. *For. Comm. Res. Inf. Note* **50** Forestry Commission, Forest Research Station, Farnham.

Biggin, P. & McIntosh, R. 1981 Heather control. *For. Comm. Rep. For. Res.* **1981** 22 HMSO, London.

Billany, D., Carter, C.J., Winter, T.G. & Gould, I.D. 1985 *Olescampe monticola* redescribed, with notes on its biology as parasite of *Cephalcia lariciphila*. *Bulletin of Entomological Research* **75** 267-274.

Billany, D.J. 1977 Web-spinning sawfly, *Cephalcia lariciphila*. *For. Comm. Rep. For. Res.* **1977** 36 HMSO.

Billany, D.J. 1978 *Gilpinia hercyniae*, a pest of spruce. *For. Comm. Forest Rec.* **117** 11pp HMSO, London.

Billany, D.J. & Brown, R.M. 1980 The web-spinning sawfly *Cephalcia lariciphila* - a new pest of larch in England and Wales. *Forestry* **53** (1) 71-81.

Billett, M.F., & Cresser, M.S. 1992 Predicting stream water quality using catchment and stream water characteristics. *Environmental Pollution* **77** 263-268.

Billett, M.F., Fitzpatrick, E.A. & Cresser, M.S. 1990 Changes in carbon & nitrogen status of forest soil organic horizons between 1949/50 and 1987. *Environmental Pollution* **66** 67-79.

Billett, M.F., Fitzpatrick, E.A. & Cresser, M.S. 1993 Long term changes in the nutrient pools of forest organic horizons between 1949/50 and 1987, Alltcailleach forest, Scotland. *Applied Geochemistry* Supplement 2 179-183.

Binkley, D. & Högberg, P. 1997 Does atmospheric deposition of nitrogen threaten Swedish forests? *Forest Ecology & Management* **92** 119-152.

Binns, W.O. 1960 Forest soils research in Scotland. *For. Comm. Rep. For. Res.* **1960** 93-4 HMSO, London.

Binns, W.O. 1964 Silvicultural investigations: potassium deficiencies on peat. *For. Comm. Rep. For. Res.* **1964** 22 HMSO, London.

Binns, W.O. 1965 Silvicultural investigations: potassium deficiencies on peat. *For. Comm. Rep. For. Res.* **1965** 27-28 HMSO, London.

Binns, W.O. 1966 Current fertilizer research in the Forestry Commission. (In) Physiology in forestry *Forestry* **Supplement 1966** 60-64

Binns, W.O. 1975a Whole-tree utilisation: consequences for soil & environment: experience and opinion in Britain. Proc. Konf. SK2, Elmia 75, Jönköping, Sweden. 18-25.

Binns, W.O. 1975b Fertilizers in the forest: a guide to materials. *For. Comm. Leaflet* **63** 1-14 HMSO, London.

Binns, W.O. 1980 Trees and water. *Arboricultural Leaflet* **6** 1-20pp HMSO, London. (Prepared for the Dept of the Environment by the Forestry Commission)

Binns, W.O. 1983 Reclamation of Mineral workings to forestry - treatment of surface workings. *For. Comm. Res. & Dev. Pap.* **132** 9-16 Forestry Commission, Edinburgh.

Binns, W.O. 1984 Acid rain and forestry. *For. Comm. Res. & Dev. Pap.* **134** 1-18 Forestry Commission, Edinburgh.

Binns, W.O. 1985 Forest health (air pollution survey). *For. Comm. Rep. For. Res.* **1985** 24 HMSO, London.

Binns, W.O. 1986 Forestry & fresh waters: problems & remedies. Proc. Conf. *Effects of land use on fresh waters: agriculture, forestry, mineral exploitation, urbanisation etc.* (Eds) J.F. & L.G. Solbé 364-377 Ellis Horwood, New York.

Binns, W.O. & Aldhous, J.R. 1965 Silvicultural investigations: Intensive manuring & weedkillers applied at planting time. *For. Comm. Rep. For. Res.* **1965** 28 HMSO, London.

Binns, W.O. & Aldhous, J.R. 1966 Nitrogen top-dressing on checked crops. *For. Comm. Rep. For. Res.* **1966** 33 HMSO, London.

Binns, W.O. & Atterson, J. 1967 Nutrition of forest crops; manuring of young crops on peat soils; manuring of pole-crops. *For. Comm. Rep. For. Res.* **1967** 48-53 HMSO, London.

Binns, W.O. & Coates, A.E. 1965 Silvicultural investigations; manuring of pole stage crops. *For. Comm. Rep. For. Res.* **1965** 25-26 HMSO, London.

Binns, W.O. & Fourt, D.F. 1981 Surface workings & trees. (In) Research for Practical Arboriculture. Proc. Seminar at Preston, 1980. *For. Comm. Occasional Pap.* **10** 60-75 Forestry Commission, Farnham.

Binns, W.O. & Fourt, D.F. 1984 Reclamation of surface workings for trees. II. Nitrogen nutrition. *Arboriculture Research Note* **38/84/SSS** Arboricultural Advisory & Information Service, Farnham, Surrey.

Binns, W.O. & Grayson, A.J. 1967 Fertilization of established crops: prospects in Britain. *Scottish Forestry* **21** (2) 81-98.

Binns, W.O. & Keay, J. 1962 Research on Scottish forest & nursery soils. *For. Comm. Rep. For. Res.* **1962** 88-89 HMSO, London.

Binns, W.O. & Mackenzie, J. 1969 Nutrition of forest crops. *For. Comm. Rep. For. Res.* **1969** 63-74 HMSO, London.

Binns, W.O. & Redfern, D.B. 1983 Acid decline and forest decline in West Germany. *For. Comm. Res. & Dev. Pap.* **131** Forestry Commission, Farnham.

Binns, W.O., Mackenzie, J. & Everard, J.E. 1970 Nutrition of forest crops. *For. Comm. Rep. For. Res.* **1970** 67-80 HMSO, London.

Binns, W.O., Everard, J.E. & Mackenzie, J.M. 1971 Nutrition of forest crops; distribution patterns of fertilizers applied from the air. *For. Comm. Rep. For. Res.* **1971** 52 HMSO, London.

Binns, W.O., Mackenzie, J. & Everard, J.E. 1972 Nutrition of forest crops. *For. Comm. Rep. For. Res.* **1972** 47-53 HMSO, London.

Binns, W.O., Everard, J.E., Fourt, D.F., Hinson, W.H. & Mackenzie, J.M. 1973 Nutrition of forest crops. *For. Comm. Rep. For. Res.* **1973** 50-59 HMSO, London.

Binns, W.O., Mayhead, G.J. & MacKenzie, J.M. 1980 Nutrient deficiencies of conifers in British forests. *For. Comm. Leaflet* **76** 24pp HMSO, London.

Binns, W.O., Redfern, D.B., Boswell, R. & Betts, A. 1985 Forest health and air pollution: 1984 survey. *For. Comm. Res. & Dev. Pap.* **142** 18pp Forestry Commission, Edinburgh.

Binns, W.O., Redfern, D.B., Boswell, R. & Betts, A. 1986 Forest health and air pollution: 1985 survey. *For. Comm. Res. & Dev. Pap.* **147** 16pp Forestry Commission, Edinburgh.

Binns, W.O., Redfern, D.B., & Reynolds, K 1987 Forest Decline; a view from Britain. *Effects of atmospheric pollution on forests wetlands & agricultural ecosystems.* 69-81 Springer Verlag.

Binns, W.O., Insley, H. & Gardiner, J.B.H. 1989 Nutrition of broadleaved amenity trees. *Arboriculture Research Note* **50/89/ARB** Forestry Commission, Farnham.

Blackburn, P. & Brown, I.R. 1988 Some effects of exposure and frost on selected birch progenies. *Forestry* **60** (1) 57-67.

Blackie, J.R. 1987 The Balquhidder catchments: the first four years. *Transactions of the Royal Society of Edinburgh.* **78** 227-239.

Blackie, J.R. & Calder, I.R. 1984 Effects of afforestation on water resources *For. Comm. Rep. For. Res.* **1984** 63 HMSO, London.

Blackie, J.R. & Calder, I.R. 1985 Effects of afforestation on water resources *For. Comm. Rep. For. Res.* **1985** 60 HMSO, London.

Blatchford, N. (Ed.) 1978 Forestry Practice 9th Edn. *For. Comm. Bull.* **14** HMSO, London.

Blatchford, N. (Ed.) 1983 Use of chemicals (other than herbicides) in forest and nursery. *For. Comm. Booklet* **52** Forestry Commission, Edinburgh.

Bletchly, J.D. & White, M.G. 1962 Significance and control of attack by the ambrosia beetle, *Trypodendron lineatum (Oliv.).* in Argyll forests. *Forestry* **35** (2) 139-163.

Blight, M.M., Wadhams, L.J., Wenham, M.J. & King, C.J. 1979 Field attraction of *Scolytus scolytus* to the enantiomers of 4-methyl-3-heptanol, the major component of the aggregation pheromone. *Forestry* **52** (1) 83-90.

Blight, M.M., King, C.J., Wadhams, L.J. & Wenham, M.J. 1980 Studies on chemically mediated behaviour in the large elm bark beetle *Scolytus scolytus* F. *For. Comm. Res. & Dev. Pap.* **129** 34pp Forestry Commission, Edinburgh.

Blumenthal, E.M., Fusco, R.A. & Reardon, R.C. 1979 Augmentative release of two established parasite species to suppress ... gypsy moth. *Journal of Economic Entomology* **72** 281-288.

Bobbink, R. 1998 Impacts of tropospheric ozone and airborne nitrogenous pollutants on natural and semi-natural ecosystems *New Phytologist* **139** 161-168.

Bobbink, R., Heil, G.W. & Raessen, M.B. 1992 Atmospheric deposition & canopy exchange processes in heathland ecosystems. *Environmental Pollution* **75** 29-37.

Bochenek, Z., Ciolkosz, A. & Iracka, M. 1997 Deterioration of forests in the Sudety Mountains, detected on satellite images. *Environmental Pollution.* **98** (3) 375-379.

Body, Sir R. (Chairman) 1987 *Effects of pesticides on human health.* House of Commons Agriculture Committee. 2nd special report. H. of C **371** (I-III) HMSO, London.

Boggie, R. & Miller, H.G. 1976 Growth of *Pinus contorta* on different water-table levels on deep blanket peat. *Forestry* **49** (2) 123-129.

Bolton, J. & Coulter, J.K. 1965 Distribution of fertilizer residues in a forest nursery manuring experiment at Wareham, Dorset. *For. Comm. Rep. For. Res.* **1965** 90-92 HMSO, London.

Bolton, T 1996 Thistle control - a prickly problem. *Farming & Conservation* **2** (4) 32-34.

Bonneau, M. 1995 *Fertilisation des forêts dans les pays tempérés.* 367 pp ENGREF, Nancy.

Bonner, J. 1996 Red or dead? *New Scientist* **20.1.96** 29-31.

Boorman, L.A., Goss-Custard, J.D. & McGorty, S. 1989 *Climatic change, rising sea level & the British coast line.* Institute of Terrestrial Ecology. 24 pp HMSO, London.

Booth, T.C. 1990 Alternative techniques and methods of protection. *Proc. Joint FAO/ECE/ILO C'ttee seminar on forest technology, management & training, Sparsholt.* 106-115 Forestry Commission, Edinburgh.

Booth, D.C. 1997 Integrated pest management: gypsy moth. (In) Arboricultural Practice Present & Future (Ed.) J. Claridge. *Research for Amenity Trees* **6** 61-68 HMSO, London.

Bosman, A.J.W. & Hoekstra, F. 1990 Watch out for those toxicants: towards an environmentally sound forest management. *Proc. FAO/ECE/ILO seminar on forest technology, management and training.* 138-144 Forestry Commission, Edinburgh.

Bould, C., Hewitt, E.J. & Needham, P. 1983 *Diagnosis of mineral disorders in plants; principles.* Vol. 1 170pp HMSO, London.

Bouwman, A.F. 1990 *Soils and the greenhouse effect.* John Wiley, Chichester.

Boylan, H. 1988 Establishing plantings on the urban fringe. *Aspects of applied biology* **16** 351-354.

Bradbury, A. 1996 Beetle-infested dunnage could trigger Czech embargo. *Timber Trades Journal* **6.7.96** 2.

Bradshaw, A.D. 1981 Growing trees in difficult environments. (In) Research for Practical Arboriculture. Proc. Seminar at Preston, 1980 *For. Comm. Occasional Pap.* **10** 93-106 Forestry Commission, Edinburgh.

Bradshaw, A., Hunt, B. & Walmsley, T. 1995 *Trees in the urban landscape - principles and practice.* 220pp E & FN Spon, London.

Brambell, F.W. Rogers 1958 Voles & field mice. *For. Comm. Leaflet* **44** 12pp HMSO, London.

Bramryd. T. 1980 Sewage fertilization in pine forests; ecological effects on soil and vegetation. *Biochemistry of ancient and modern environments.* (Eds.) Trudinger, Walter & Ralph Australian Academy of Science, Canberra.

Brasier, C.M. 1979 Dual origin of recent Dutch elm disease outbreaks in Europe. *Nature* **218** 78-79.

Brasier, C.M. 1983 The future of Dutch elm disease in Europe. (In) Research on Dutch Elm disease in Europe. *For. Comm. Bull.* **60** 96-104 HMSO, London.

Brasier, C.M. 1996 New horizons in Dutch elm disease control. *For. Comm. Rep. For. Res.* **1996** 20-28 HMSO, London.

Brasier, C.M. 1997 Controlling the Dutch invader. *Tree News* **Spring 1997** 8-11 Tree Council, London.

Brasier, C.M. 1999 *Phytophthora* pathogens: their rising profile in Europe. *For. Comm. Information Note* **30** 6pp Forestry Commission, Edinburgh.

Brasier, C.M. & Gibbs, J.N. 1973 Origin of the Dutch elm disease epidemic. *Nature* **242** 607-609.

Brasier, C.M. & Mehrotra, M.D. 1995 *Ophiostoma nova-ulmi* sp. nov., a new species of Dutch elm disease fungus endemic in the Himalayas. *Mycological Research* **99** 205-215.

Brasier, C.M. & Webber, J.F. 1987 Recent advances in Dutch Elm Research in (Ed.) D. Patch 'Advances in practical arboriculture'. *For. Comm. Bull.* **65** 166-179 HMSO, London.

Braun, S & Flückiger, W. 1995 Effects of ambient ozone on seedlings of *Fagus sylvatica* L & *Picea abies* (L) Karst. *New Phytologist* **129** 33-44.

Bravery, T. 1992 Water-stored timber. *For. Comm. Rep. For. Res.* **1992** 57 HMSO, London.

Brazier, J.D. 1977 The effect of forest practices on quality of the harvested crop. *Forestry* **50** (1) 49-66.

Bridges, E.M. 1989 Toxic metals in amenity soil. *Soil Use & Management* **5** (3) 91-100.

Britt, C.P. & Smith, J.M. 1996 Non-triazine herbicides with residual activity for weed control in newly planted farm woodlands. *Aspects of applied biology* **44** 81-88.

Britt, C., Buckland, M., Ryan, M., Whiteman, A. & Wilson, A. 1996 Economic surveys of farm woodland establishment. *For. Comm. Res. Inf. Note* **276** 4 pp Forestry Commission, Farnham.

Brixey, J. 1997 The potential for biological control to reduce *Hylobius* damage. *For. Comm. Res. Inf. Note* **273** Forestry Commission, Forest Research Station, Farnham.

Broadmeadow, M. 1998 Ozone & forest trees. *New Phytologist* **139** 123-125.

Broadmeadow, M., Ludlow, T., & Randle, T. 1996 Modelling the effects of global change on European forests. *For. Comm. Rep. For. Res.* **1996** 34-37.

Brookes, P.C., Wigston, D.L. & Bourne, W.F. 1980 Dependence of *Quercus robur* & *Q. petraea* seedlings on cotyledon potassium, magnesium, calcium & phosphorus during the first year of growth. *Forestry* **53** (2) 167-178.

Brooks, A.V., Halstead, A.J., Smith, P.M. & Evans, E.J. 1989 Pests & diseases of outdoor ornamentals, etc. (In) *Pest & Disease Handbook (3rd Edn)*. Ch 13 513-601 British Crop Protection Council.

Brown, R.M. 1969 Herbicides in forestry. *Journal of Food & Science in Agriculture* **20** 509-512.

Brown, R.M. 1970 Atrazine and ametryne for grass weed control in British forestry. *Proc. 10th British Weed Control Conf.* 1970 718-726 British Crop Protection Council.

Brown, R.M. 1972 Production and use of planting stock. *For. Comm. Rep. For. Res.* **1972** 27 HMSO, London.

Brown, R.M. 1973 Pine looper moth, Spruce sawfly. *For. Comm. Rep. For. Res.* **1973** 105 HMSO, London.

Brown, R.M. 1975 Chemical control of weeds in the forest. *For. Comm. Booklet* **40** 66pp HMSO, London.

Brown, R.M. & Barbour, D.A. 1978 Chemical control - pine looper moth. *For. Comm. Rep. For. Res.* **1978** 38 HMSO, London.

Brown, J.M.B. & Bevan, D. 1966 The great spruce bark beetle, *Dendroctonus micans* in NW Europe. *For. Comm. Bull.* **38** 1-41 HMSO, London.

Brown, R.M. & Billany, D. 1973 Web-spinning larch sawfly - *Cephalcia alpina*. *For. Comm. Rep. For. Res.* **1973** 107 HMSO, London.

Brown, A.H.F. & Iles, M.A. 1991 Water chemistry profiles under four tree species at Gisburn, NW England. *Forestry* **64** (2) 169-187.

Brown, L. & Kane, H. 1994 *Full house: reassessing the Earth's population carrying capacity.* Worldwatch Institute, Norton.

Brown, R.M. & Low, A.J. 1972 Production and use of planting stock: Japanese paper pot seedlings. *For. Comm. Rep. For. Res.* **1972** 43-46 HMSO, London.

Brown, R.M. & Mackenzie, J.M. 1969 Forest weed control. *For. Comm. Rep. For. Res.* **1969** 74-83 HMSO, London.

Brown, R.M. & Mackenzie, J.M. 1970 Forest weed control. *For. Comm. Rep. For. Res.* **1970** 81-85 HMSO, London.

Brown, R.M. & Mackenzie, J.M. 1971 Forest weed control. *For. Comm. Rep. For. Res.* **1971** 54-57 HMSO, London.

Brown, R.M. & Mackenzie, J.M. 1972 Forest weed control. *For. Comm. Rep. For. Res.* **1972** 43-46 HMSO, London.

Brown, R.M. & Mackenzie, J.M. 1973 Woody weed control. *For. Comm. Rep. For. Res.* **1973** 48 HMSO, London.

Brown, R.M. & Thomson, J.H. 1974 Trials of ULV applications of herbicides in British forestry. *BCPC Monograph* **13** British Crop Protection Council, Farnham, Surrey.

Brown, R.M. & Winter, T.G. 1981 Forest insects imported from Canada. *For. Comm. Rep. For. Res.* **1981** 39-40 HMSO, London.

Brown, A.H.F., Carlisle, A. & White, E.J. 1966 Some aspects of nutrition of Scots pine on peat. *Forestry* **1966** (Supplement) 78-87.

Brydges, T.G. & Wilson, R.B. 1991 Acid rain since 1985 - times are changing. *Proc. R. Soc. Edinburgh* **97B** 1-16.

BSI 1983 British Standard glossary for terms for equipment for crop protection. *BSI 6355: 1983* British Standards Institution.

BSI 1989 Building structures for agriculture; Code of practice for design and construction of chemical stores. *British Standard* **BS 5502** Pt 81 British Standards Institute.

Buczacki, S. & Harris, K. 1981 *Collins Guide to the pests, diseases & disorders of garden plants.* 1981 512 pp Collins, London.

Bull, K.R. 1991 The critical loads/levels approach to gaseous pollutant emission control. *Environmental Pollution* **69** 107-123.

Bull, K.R. 1992 An introduction to critical loads. *Environmental Pollution* **77** 173-176.

Bunce, H.W.F. 1989 The continuing effects of aluminium smelter emissions on coniferous forest growth. *Forestry* **62** (3) 223-231.

Burdekin, D.A. 1972 Study of losses in Scots pine caused by *Fomes annosus. Forestry* **45** (2) 189-196.

Burdekin, D.A. 1980 Tree health insurance policies. British Association for the Advancement of Science (Forestry Section) Sept. 1980 5 pp.

Burdekin, D.A. 1980 Trees at risk, a balanced view of tree diseases. *Quarterly Journal of Forestry* **74** (3) 177-179.

Burdekin, D.A. 1983 Research into Dutch elm disease in Europe. *For. Comm. Bull.* **60** 1-114 HMSO, London.

Burdekin, D.A. 1986 European plant health requirements for pests. *EPPO Bulletin.* **16** 509-512.

Burdekin, D.A. & Gibbs, J.N. 1972/4 The control of Dutch elm disease. *For. Comm. Leaflet* **54** (1st & 2nd edns) HMSO, London.

Burdekin, D.A. & Phillips, D.H. 1970 Chemical control of *Didymascella thujina* on W. red cedar in forest nurseries. *Annals of Applied Biology.* **67** 131-136.

Burdekin, D.A. & Phillips, D.H. 1977 Some important foreign diseases of broadleaved trees. *For. Comm. Forest Rec.* **111** 11pp HMSO, London.

Burdekin, D.A. & Rushworth, K.D. 1988 Breeding elms resistant to Dutch elm disease. *Arboriculture Research Note* **2/88/PATH** Arboricultural Advisory & Information Service, Farnham, Surrey.

Burge, M.N., Lawton, J.H. & Taylor, J.A. 1988 The prospects of biological control of bracken in Britain. *Aspects of Applied Biology* **16** 299-309 Association of Applied Biologists.

Burgess, R. 1998 Progress report on *Ips typographus ICF News* 4/98.

Burnett, S. 1996 Risk analysis - a key issue for independent developers. *New Review* **28** 10-12 Department of Trade & Industry.

Burton, K.W. & Morgan, E. 1981 Effect of heavy metals on South Wales forests. *For. Comm. Rep. For. Res.* **1981** 55-56.

Burton, K.W., Morgan, E. & Roig A. 1983 The influence of heavy metals upon the growth of Sitka spruce in South Wales. *Plant & Soil* **73** 327-336.

Busvine, J.R. 1989 DDT - Fifty years for good or ill. *Pesticide Outlook* **1** (1) 4-8.

Butin, H. 1995 *Tree diseases & disorders.* Oxford University Press.

BWPDA 1995 *Safe design & operation of timber treatment plants. Code of Practice* British Wood Preserving & Damp Proofing Association.

Cain, R.B. & Head, I.M. 1991 Enhanced degradation of pesticides: its biochemical & molecular biological basis. (In) Pesticides in soils & water: current perspectives *Proc. BCPC Conf. Monograph* **47** 23-40 British Crop Protection Council, Farnham, Surrey.

Calder, I.R. 1985 What are the limits to forest evaporation? *Journal of Hydrology* **82** 179-184.

Calder, I.R. 1990 *Evaporation in the uplands.* J. Wiley, Chichester.

Calder, I.R. 1993 The Balquhidder catchment water balance and process experiment results in context - what do they reveal? *Journal of Hydrology* **145** 467-477.

Calder, I.R. & Newson, M.D. 1979 Land-use & upland water resources in Britain - a strategic look. *American Water Resources Association Bulletin* **15** 1628-1639.

Calderwood, R. 1996 Plant health & the single market. *ICF News* **3/96** 7-8 Institute of Chartered Foresters, Edinburgh.

Campbell, B. 1958 Crested tit. *For. Comm. Leaflet* **41** HMSO, London.

Campbell, B. 1964 Birds & woodlands. *For. Comm. Leaflet* **47** (Revised) HMSO, London.

Campbell, B. 1965 Crossbills. *For. Comm. Leaflet* **36** (Revised) HMSO, London.

Campbell, B. 1967 The Dean nest-box study. *Forestry* **1967 Supplement** 13-14.

Campbell, C.D. & Jones, D. 1996 The effect of heavy metals on forest soil processes and mycorrhizal functioning. (In) *Heavy metals & trees* Proc. conf. Glasgow 1995 (Ed.) I. Glimmerveen. 33-39 Institute of Chartered Foresters, Edinburgh.

Canham, A.E. 1976 Effect of environmental factors on tree seedling growth. *For. Comm. Rep. For. Res.* **1976** 55-56 HMSO, London.

Canham, A.E. & McCavish, W.J. 1981 Effects of CO_2, day-length & nutrition on the growth of young forest tree plants. *Forestry* **54** (2) 169-182.

Cannell, M.G.R. 1982 Short rotation coppice. (In) *Broadleaves in Britain* Proc. symposium, Loughborough (Eds.) D.C. Malcolm, J. Evans & P.N. Edwards. 150-160 Institute of Chartered Foresters, Edinburgh.

Cannell, M.G.R. 1988 Harnessing solar energy. *For. Comm. Occasional Pap.* **16** 16-24 Forestry Commission, Edinburgh.

Cannell, M.G.R. 1995 Growing trees to sequester carbon in the UK: answers to some common questions. *Forestry* **72** (3) 237-247.

Cannell, M. & Cape, J. 1991 International environmental impacts: acid rain & greenhouse effects. *For. Comm. Occasional Pap.* **35** 1-28 Forestry Commission, Edinburgh.

Cannell, M.G.R. & Dewar, R.C. 1995 The carbon sink provided by plantation forests and their products in Britain. *Forestry* **68** (1) 35-48.

Cannel, M.G.R. & Hooper, M.D. 1990 *The greenhouse effect and terrestrial ecosystems of the UK.* 56 pp Institute of Terrestrial Ecology, Huntingdon.

Cannell, M.G.R., & Milne, R. 1995 Carbon pools and sequestration in forest ecosystems in Britain. *Forestry* **68** (4) 361-378.

Cannell, M.G.R., Sheppard, L.J., Smith, R.I. & Murray, M.B. 1985 Autumn frost damage on young *Picea sitchensis*: 2 Shoot frost hardening and probability of frost damage in Scotland. *Forestry* **58** (2) 145-166.

Cannell, M.G.R., Murray, M.B. & Sheppard, L.J. 1987 Frost hardiness of red alder provenances in Britain. *Forestry* **60** (1) 57-67.

Cannell, M.G.R., Grace, J. & Booth, A. 1989 Possible impacts of climatic warming on trees & forests in the United Kingdom: a review. *Forestry* **62** (4) 337-364.

Cannell, M.G.R., Tabbush, P.M. *et al.* 1990 Sitka spruce & Douglas fir seedlings in the nursery and in cold storage: root growth potential, carbohydrate, dormancy, frost hardiness & mitotic index. *Forestry* **63** (1) 9-27.

Cannell, M.G.R., Dewar, R.C. & Pyatt, D.G. 1993 Conifer plantations on drained peatlands in Britain: a net gain or loss of carbon? *Forestry* **66** (4) 353-369.

Cannell, M.G.R., Cruikshank, M.M. & Mobbs, D.C. 1996 Carbon storage and sequestration in the forests of Northern Ireland. *Forestry* **69** (2) 155-165.

Cannell, M.G.R., Thornley, J.H.M., Mobbs, D.C. & Friend, A.D. 1998 UK conifer forest may be growing faster in response to increased N deposition, atmospheric CO_2 and temperature. *Forestry* **71** (4) 277-296.

Cape, J.N. 1993 Direct damage to vegetation caused by acid rain & polluted cloud: definition of critical levels for forest trees. *Environmental Pollution* **82** 167-180.

Cape, J.N. 1998 Uptake & fate of gaseous pollutants in leaves. *New Phytologist* **139** 221-223.

Cape, J.N., Leith, I.D., Fowler, D. *et al.* 1991 Sulphate & ammonium in mist impair the frost hardening of red spruce seedlings. *New Phytologist* **118** 119-126.

Cape, J.N., Sheppard, L.J., Fowler, D. *et al.* 1992 Contribution of canopy leaching to sulphate deposition in a Scots pine forest. *Environmental Pollution* **75** 229-236.

Carey, M.L. 1980 Whole-tree harvesting in Sitka spruce. Possibilities and implications. *Irish Forestry* **37** 48-63.

Carey, M.L., McCarthy, R.G. & Miller, H.G. 1988 More on nursing mixtures. *Irish Forestry* **45** 7-20.

Carlyle, J.C. 1986 Nitrogen cycling in forested ecosystems. *Forestry Abstracts* **47** (5) 307-336.

Carnell, R. 1986 Impact of forestry on water resources. *For. Comm. Rep. For. Res.* **1986** 24-26 HMSO, London.

Carnol, M., Ineson P. & Dickinson, A.L. 1997 Soil solution nitrogen & cations influenced by $(NH_4)_2SO_4$ deposition in a coniferous forest. *Environmental Pollution* **97** (1) 1-10

Carson, R. 1963 *Silent Spring*. Hamish Hamilton, London.

Carter, A.D. & Heather, A.I.J. 1996 Atrazine & simazine levels in selected untreated water sources. (In) Vegetation management in forestry, amenity & conservation areas. *Aspects of applied biology* **44** 165-170 Association of Applied Biologists, Warwick.

Carter, C.I. & Winter, T.M. 1998 Christmas tree pests. *For. Comm. Field Bk* **17** The Stationery Office, London.

Carter, C.I. 1971 Conifer woolly aphids in Britain. *For. Comm. Bull.* **42** HMSO, London.

Carter, C.I. 1972 Winter temperatures and survival of Green spruce aphis. *For. Comm. Forest Rec.* **84** HMSO, London.

Carter, C.I. 1974 Pineapple galls caused by *Adelges abietis*. *For. Comm. Rep. For. Res.* **1974** 39 HMSO, London.

Carter, C.I. 1975 Towards integrated control of tree aphids. *For. Comm. Forest Rec.* **104** 17pp HMSO, London.

Carter, C.I. 1977 Impact of green spruce aphis on growth. *For. Comm. Res. & Dev. Pap.* **116** Forestry Commission, Edinburgh.

Carter, C.I. 1995 The green spruce aphid, spruce root aphids and tree growth at Afan Forest. *For. Comm. Technical Pap.* **9** 95-104 Forestry Commission, Edinburgh.

Carter, C.I. & Anderson, M.A. 1987 Enhancement of lowland forest ride sides & roadsides to benefit wild plants & butterflies *For. Comm. Res. Inf. Note* **126** 6 pp Forestry Commission, Farnham.

Carter, C.I. & Gibbs, J.N. 1989 Pests & diseases of forest crops. (In) *Pest & Disease Handbook* (3rd Edn). 619-634 British Crop Protection Council.

Carter, C.I. & Maslen, N.R. 1982 Conifer lachnids. *For. Comm. Bull.* **58** 73pp HMSO, London.

Carter, M.M. 1990a Positive use of set-aside. *REView (Renewable Energy view)* **12** 5-7 Department of Energy

Carter, E.S. 1990b Weed control in amenity areas & other non-agricultural land. (In) *Weed Control Handbook: Principles* (Eds.) R.J. Hance & K. Holly. 8th edn 431-456 Blackwell Scientific Publications, Oxford.

Catt, J.A. 1985 Natural soil acidity. *Soil Use & Management* **1** (1) 8-10.

Cayford, J.T. 1988 Control of heather with glyphosate in young forest plantations; effect on habitat use by black grouse in summer. *Aspects of Applied Biology* **16** 355-362

CEC 1980 Council directive of 15 July 1980 relating to the quality of water intended for human consumption. *Official J. of the European Communities* **L229** 11-29 Commission of the European Communities.

CEC 1999 Proposal for a Council Directive on the marketing of forest reproductive material. *Documents* COM(1999) 188 final **EN 03 10 15 02** 48 pp Commission of the European Communities

CEC (Commission of the European Communities) 1985 Diagnosis and classification of new types of damage affecting forests. *Allgemeine Forst Zeitschrift* **Special Ed'n. EEC** 20pp.

Chadwick, A.H., Hodge, S.J. & Ratcliffe, P.R. 1997 Foxes & forestry. *For. Comm. Technical Pap.* **23** Forestry Commission, Farnham.

Chadwick, D.J. 1992a Rest allowances and protective clothing - heat stress. *Technical Development Branch Report* **7/92.**

Chadwick, D.J. 1992b Heat stress in protective clothing. *Technical Development Branch Information Note* **7/92.**

Chadwick, D.J. 1992c Health and Safety in forestry. *Forestry & British Timber* **July 1992** 22.

Chandra Nigram, P. 1990 Use of chemical insecticides against spruce budworm in Eastern Canada & status of *Bacillus thuringiensis*. *Proc. FAO/ECE/ILO seminar on forest technology, management and training* 116-132 Forestry Commission, Edinburgh.

Chapman, S.C. 1967 Nutrient budget for a dry heath ecosystem in the south of England. *Journal of Ecology.* **58** 445-452.

Chappelka, A.H. & Freer-Smith, P.H. 1994 Predisposition of trees by air pollution to low temperature and moisture stress. *Environmental Pollution* **87** (1) 105-117.

Chappelka, A.H. & Samuelson, L.J. 1998 Ambient ozone effects on forest trees of the eastern United States - a review. *New Phytologist* **139** 91-108.

Chard, J.S.R. 1964 The roe deer. *For. Comm. Leaflet* **45** 26 pp HMSO, London.

Chard, J.S.R. 1985 Forestry & water in the uplands of Britain. *Quarterly Journal of Forestry* **89** (3) 209-210.

Charters, H. 1961 Fires in state forests, 1929-1956. *For. Comm. Forest Rec.* **45** HMSO, London.

Chinery, M. 1973 *A field guide to the insects of Britain and Northern Europe.* 352 pp Collins, London.

Christensen, P. 1988 Danish results with a new herbicide, imazapyr in forestry. The practice of weed control & vegetation management in forestry amenity & conservation areas. *Aspects of Applied Biology* **16** Association of Applied Biologists.

Chrystal, R. Neil, 1937 *Insects of the British Woodlands* (Reprinted 1944) F. Warne, London.

Chumbley, C.G. 1971 Permissible levels of toxic metals in sewage sludge used on agricultural land. *ADAS Advisory Paper* **10** MAFF, London.

CLAG (Critical Loads Advisory Group) 1995 *Critical loads for acid deposition for United Kingdom freshwaters.* 80 pp Institute of Terrestrial Ecology, Edinburgh.

CLAG (Critical Loads Advisory Group) 1996 *Critical levels for air pollutants for the United Kingdom.* 100pp Institute of Terrestrial Ecology, Edinburgh.

CLAG (Critical Loads Advisory Group) 1997 *Deposition fluxes of acidifying compounds in the United Kingdom. A compilation of the current deposition maps & methods used for Critical Loads exceedance assessment in the UK.* 45 pp Institute of Terrestrial Ecology, Edinburgh.

Claridge, M.F. & Evans, H.F. 1990 Species-area relationships: relevance to pest problems of British trees. (In) *Population dynamics of forest insects* (Eds.) A.D. Watt, S.D. Leather, et al. 59-69 Intercept Ltd., Andover.

Clarke, J.C. & Boswell, R.C. 1976 Tests on round timber fence posts. *For. Comm. Forest Rec.* **108** 44 pp HMSO, London.

Clay, D. 1992 Herbicide evaluation on forest weeds and crops. *For. Comm. Rep. For. Res.* **1992** 55 HMSO, London.

Clay, D.V. & Dixon, F.L. 1993 Factors affecting the toxicity of imazapyr to *Rhododendron ponticum*. Proc. conf.: *Crop Protection in Northern Britain* 295-300.

Clay, D.V. & Dixon, F.L. 1996a Investigations of control methods for *Clematis vitalba*. (In) Vegetation management in forestry, amenity & conservation areas. *Aspects of applied biology* **44** 313-318 Association of Applied Biologists, Warwick.

Clay, D.V. & Dixon, F.L. 1996b Weed management in short-rotation energy coppice. Selectivity of foliar acting herbicides. (In) Vegetation management in forestry, amenity & conservation areas. *Aspects of applied biology* **44** 109-116 Association of Applied Biologists, Warwick.

Clay, D.V., Lawrie, J. & Richardson, W.G. 1988 New herbicides for forestry seed beds: pot experiments to evaluate crop tolerance. *Aspects of Applied Biology.* **16** 223-230.

Clay, D.V., Lawson, H.M. & Greenfield, A.J. 1990 Weed control in fruit & other perennial crops. (In) *Weed Control Handbook: Principles* (Eds.) R.J. Hance & K. Holly. 8th edn 367-386 Blackwell Scientific Publications, Oxford.

Clay, D.V., Goodall, J.S. & Nelson, D.G. 1992a Evaluation of imazapyr on *Rhododendron ponticum*. Vegetation management in forestry, amenity & conservation areas. *Aspects of Applied Biology* **29** 287-294.

Clay, D.V., Goodall, J.S. & Williamson, D.R. 1992b Evaluation of post-emergence herbicides for forestry seedbeds. *Aspects of applied biology* **29** 139-147.

Clay, D.V., Dixon, F.L. & Morgan, J.L. 1996 Effect of post-spraying rainfall on the response of forest tree seedlings to herbicides. *Aspects of applied biology* **44** 39-46.

Clennell, B. 1998 Ice-cold heat - the fuel of the future. Natural gas hydrates beneath the sea. British Association Annual Festival of Science. Cardiff Univ. Sept 1998 3+4 NERC Briefing Notes 7th September 1998. NERC Swindon.

Clifford, D.R.& Gendle, P. 1987 Treatment of fresh wound parasites and treatment of cankers. *For. Comm. Bull.* **65** 145-148 HMSO, London.

Coghlan, A. 1994a Will the scorpion gene run wild? *New Scientist* **25.6.94** 14-15.

Coghlan, A. 1994b Scorpion gene virus on trial in Oxford. *New Scientist* **3.12.94** 11.

Coghlan, A. 1995 Moths safe from scorpion virus. *New Scientist* **25.11.95** 8.

Coghlan, A. 1996a Doomsday has been postponed. *New Scientist (Two issues)* **17.2.96** (p8) & **5.10.96**, (p8).

Coghlan, A. 1996b Nations clash over genetic protocol *New Scientist* **3.8.96** 5.

Coghlan, A. 1996c Andes flower is champion pest killer. *New Scientist* **20.1.96** 7.

Coghlan, A. 1996d Europe halts march of supermaize. *New Scientist* **4.5.96** 7.

Cohen, P. 1996 Brazil acts on the incredible shrinking rainforest. *New Scientist* **3.8.96** 4.

Collins, S. 1987 American nematodes can't cut the ice. *Entopath News* **May 1987** 14-16 Forestry Commission Research Station, Farnham, Surrey (Departmental publication).

Connell, C.A. 1967 Current problems facing fire research. *For. Comm. Res. & Dev. Pap.* **38** Forestry Commission, Edinburgh. (Departmental publication).

Connell, C.A. & Cousins, D.A. 1969 Practical developments in the use of chemicals for forest fire control. *Forestry* **42** (2) 119-132.

Cook, J. (Chairman) 1964a *Review of persistent organochlorine pesticides.* Report by the Advisory Committee on Poisonous Substances used in Agriculture and Food Storage. (24-314) HMSO, London.

Cook, J. (Chairman) 1964b *Review of persistent organochlorine pesticides.* Supplementary report by the Advisory Committee on Poisonous Substances used in Agriculture and Food Storage. (24-314-2) HMSO, London.

Cook, J.W.(Chairman) 1967 *Review of the present safety arrangements for the use of toxic chemicals in agriculture and food storage.* Report by the Advisory Committee on Pesticides and Other Toxic Chemicals (SO 27-402) HMSO, London.

Cooper, J.I. 1978 Virus & virus-like diseases of trees. *Arboricultural Leaflet* **4** 1-12 HMSO, London. (Prepared for Dept of the Environment by the Inst. of Terrestrial Ecology.)

Cooper, J.I. 1993 *Virus diseases of trees & shrubs* 2nd Edn 205 pp Chapman & Hall

Cooper, M.R. & Johnson, A.W. 1984 Poisonous plants in Britain & their effects on man. *Ministry of Agriculture Fisheries & Food Reference Book* **161** 305pp HMSO, London.

Copping, L.G., Hewitt, H.G. & Rowe, R.R. 1990 Evaluation of a new herbicide. (In) *Weed Control Handbook: Principles* (Eds.) R.J. Hance & K. Holly. 8th edn 261-300 Blackwell Scientific Publications, Oxford.

Cory, J., 1996 Biocides: do they present a risk? *Farming & Conservation* **2** (4) 28-29.

Costigan, P.A., Bradshaw, A.D. & Gemmel, R.P. 1982 The reclamation of acidic colliery spoil. III Problems associated with the use of high rates of limestone. *Journal of Applied Ecology* **19** 193-201.

Costigan, P.A., Bradshaw, A.D. & Gemmell, R.P. 1981 Reclamation of acidic colliery spoil. I. Acid production potential. *Journal of Applied Ecology* **18** 865-878.

Cousens, J.E. 1988 Report of a 12-yr study of the litter fall and productivity in a stand of mature Scots pine. *Forestry* **61** (4) 255-266.

Cousins, J. 1996 Amenity on farm: asset, income or encumbrance? *Farming & Conservation* **2** (4) 5-7.

Coutts, M.P. 1983 Development of the structural root system in Sitka spruce. *Forestry* **56** (1) 1-16.

Coutts, M.P. (Ed.) 1995 Decline in Sitka spruce on the South Wales coalfield. *For. Comm. Technical Pap.* **9** 121 pp Forestry Commission, Edinburgh.

Coutts, M.P., Low, A,J., Pyatt, D.G., Binns, W.O. & Carter, C.I. 1985 Reduced growth and bent top of Sitka spruce. *For. Comm. Rep. For. Res.* **1985** 35 HMSO, London.

Coutts, M.P., Winter, J.A. & Ashenden, T.W. 1995 Atmospheric pollution in forests of the South Wales coalfield. *For. Comm. Technical Pap.* **9** 5-9 Forestry Commission, Edinburgh.

Craig, J.B. & Miller, H.G. 1966 Research on forest soils & nutrition. *For. Comm. Rep. For. Res.* **1966** 94-98 HMSO, London.

Cramb, A. 1993a A sickness which stems from healthy practice. *Scotsman* 18.3.93.

Cramb, A. 1993b Missing links in the safety chain. *Scotsman* 17.7.93 6.

Crick, H.Q.P. & Spray, C.J. 1987 The impact of fenitrothion on forest bird population. (In) *For. Comm. Bull.* **67** 76-86 HMSO, London.

Crooke, M. 1952 Forest Entomology - *Sirex* parasites for New Zealand. *For. Comm. Rep. For. Res.* **1952** 88 HMSO, London.

Crooke, M. 1953 Forest Entomology - Insecticidal control of *Hylobius abietis*. *For. Comm. Rep. For. Res.* **1953** 73 HMSO, London.

Crooke, M. 1954a Forest insects in gale-damaged woodlands of NE Scotland. *For. Comm. Rep. For. Res.* **1954** 163-169 HMSO, London.

Crooke, M. 1954b Forest Entomology. *For. Comm. Rep. For. Res.* **1954** 33 HMSO, London.

Crooke, M. 1955a Forest Entomology. - the Pine looper moth, *Bupalus piniarius For. Comm. Rep. For. Res.* **1955** 57-59 HMSO, London.

Crooke, M. 1955b Forest Entomology - Insecticidal control of Pine weevil. *For. Comm. Rep. For. Res.* **1955** 60 HMSO, London.

Crooke, M. 1959 Insecticidal control of Pine looper in Great Britain. *Forestry* **32** (2) 166-196.

Crooke, M. 1960 *Adelges cooleyi*, an insect pest of Douglas fir and Sitka spruce. *For. Comm. Leaflet* **2** (revision) 8pp HMSO, London.

Crooke, M. & Bevan, D. 1956 Forest Entomology - the Pine weevil, *Hylobius abietis. For. Comm. Rep. For. Res.* **1956** 70-71 HMSO, London.

Crooke, M. & Bevan, D. 1957 Notes on the first appearance of *Ips cembrae*. *Forestry* **30** (1) 21-28.

Crooke, M. & Bevan, D. 1958 Forest Entomology - Control operation at Tentsmuir. *For. Comm. Rep. For. Res.* **1958** 73 HMSO, London.

Crooke, M. & Bevan, D. 1959 Forest Entomology. *For. Comm. Rep. For. Res.* **1959** 70-74 HMSO, London.

Crooke, M. & Kirkland, R.C. 1956 The gale of 1953: an appraisal of its influence on forest pest populations in pine areas. *Scottish Forestry* **10** 135-145.

Crossley, A., Wilson, D.B. & Milne, R. 1992 Pollution in the upland environment. *Environmental Pollution.* **75** 81-88.

Crowther, R.E. 1976 Guidelines to forest weed control. *For. Comm. Leaflet* **66** 8pp HMSO, London.

Crowther, R.E. & Benzian, B. 1951 Sub-committee on Nutrition Problems in Forest Nurseries - summary report on 1950 experiments. Soil acidification in seedbeds *For. Comm. Rep. For. Res.* **1951** 117-118.

Culleton, N., Murphy, W.E. & McLoughlin, A. 1996 The use of fertilizers in the establishment phase of common ash. *Irish Forestry* **53** (1&2) 28-35.

Currie, F.A., Elgy, D.,& Petty, S.J. 1977 Starling roost dispersal from woodlands. *Journal of Applied Ecology* **6** 403-410.

Curtis, C.J., Reed, J.M., Battarbee, R.W. & Harrison, R.M. (Eds.) 1996 *Urban air pollution & public health.* Proc. conf. at University College, London, Sept. 1994 96 pp Ensis Publishing, 26, Bedford Way, London.

DANI 1993a/b Unnumbered leaflet containing 2 papers:
Paper 1 Dispersed heat and power production: the Enniskillen gasifier project 5 pp.
Paper 2 The effects of clonal mixtures. 4 pp Dept. of Agriculture for Northern Ireland.

DANI 1998 Forest operations. *Forest Service Annual Report* **1998** 5-12.

Dannatt, N. & Wittering, W.O. 1967 Work study in silvicultural operations with particular reference to weeding. *For. Comm. Res. & Dev. Pap.* **58** 23pp Forestry Commission, London. (Printed for departmental use.)

Darlington, A. 1974 The galls on oak. (In) *The British Oak* Proc. BSBI conf. (Eds.) M.G. Morris & F.H. Perring 298-311 E.W. Classey, for the Botanical Society of the British Isles.

Darrall, N.M. 1984 Vegetation control near overhead power lines. *Aspects of Applied Biology* **5** 161-181.

Darrall, N.M. 1988 Woody vegetation control under power lines: results of herbicide trials. (In) Practice of weed control and vegetation management in forestry, amenity and conservation areas. *Aspects of Applied Biology.* **16** 151-162 Association of Applied Biologists, Warwick.

Davidson, A.M. & Adams, W. 1973 Grey squirrel and tree damage. *Quarterly Journal of Forestry* **72** (1) 16-26.

Davidson, G. 1992 Icy prospects for a warmer world. *New Scientist* **8 Aug 1992** 23-26.

Davies, E.J.M. 1967 Aerial applications of fertilizers at Kilmory Forest. *Scottish Forestry* **21** 99-104.

Davies, J.M. 1968 Adelgids attacking spruce and other conifers. *For. Comm. Leaflet* **7** 1-12 HMSO, London.

Davies, R.J. 1985 The importance of weed control and the use of tree shelters for establishing broadleaved trees on grass-dominated sites in England. *Forestry* **58** (2) 167-180.

Davies, R.J. 1987a Black polythene mulches to aid tree establishment. *Arboriculture Research Note* **71/87/ARB** Forestry Commission, Farnham, Surrey.

Davies, R.J. 1987b Weed competition & broadleaved tree establishment. (In) Advances in practical arboriculture. Ed. D. Patch. *For. Comm. Bull.* **65** 91-99 HMSO, London.

Davies, R.J. 1987c Sheet mulches: suitable materials and how to use them. *Arboriculture Research Note* **72/87/ARB** Forestry Commission, Farnham, Surrey.

Davies, R.J. 1987d Comparison of survival & growth of plants.+/- weed control. *Arboriculture Research Note* **67/87/SILS** Arboricultural Advisory & Information Service, Farnham, Surrey.

Davies, R.J. 1987e Trees & weeds - weed control for successful establishment. *For. Comm. Handbk.* **2** 36pp HMSO, London.

Davies, R.J. 1987f Do soil ameliorants help tree establishment? *Arboriculture Research Note* **69/87/SILS** Arboricultural Advisory & Information Service, Farnham, Surrey.

Davies, R.J. 1987g Fertilizing broadleaved landscape trees. (In) D. Patch (Ed.) Advances in practical arboriculture. *For. Comm. Bull.* **65** 107-114.

Davies, R.J. 1988a Sheet mulching as an aid to broadleaved tree establishment I Effectiveness of synthetic sheets. *Forestry* **61** (2) 89-105.

Davies, R.J. 1988b Sheet mulching as an aid to broadleaved tree establishment II Comparison of black polythene mulch and herbicide treated spot. *Forestry* **61** (2) 107-124.

Davies, R.J. 1988c Alginure root dip and tree establishment. *Arboriculture Research Note* **75/88/ARB** Arboricultural Advisory & Information Service, Farnham, Surrey.

Davies, R. 1995 Biofuels & biodegradable lubricants. *Prevention of pollution during forest engineering operations.* Proc. FAO/ECE/ILO seminar on forest technology, management and training. 13 pp Institute of Agricultural Engineers.

Davies, R.J. & Colderick, S.M. 1986 Landscape tree establishment. *For. Comm. Rep. For. Res.* **1986** 11-12 HMSO, London.

Davies, R.J. & Gardiner, J.B.H. 1985 The effects of weed competition on broadleaved tree establishment. *For. Comm. Res. Inf. Note* **98/85/SILS** Forestry Commission, Forest Research Station, Farnham.

Davies, R.J. & Gardiner, J.H.B. 1989 The effects of weed competition on tree establishment. *Arboriculture Research Note* **59/89/ARB** Arboricultural Advisory & Information Service, Farnham, Surrey.

Davies, J.M. & King, C.J. 1977 Pine shoot beetles. *For. Comm. Leaflet* **3** HMSO, London.

Davies, R.J. & Pepper, H.W. 1989 Influence of small plastic guards, tree shelters and weed control on damage to young broadleaved trees by field voles. *Journal of Environmental Management* **28** 117-125.

Davies, R.J. & Pepper, H.W. 1990 Protecting trees from field voles. *Arboriculture Research Note* **74/90:74/93** DOE Arboricultural Advisory & Information Service.

Davies, B.E. & Roberts, L.J. 1978 The distribution of heavy metal contaminated soils in northeast Clwyd, Wales. *Water, Air & Soil Pollution* **9** 507-518.

Davies, R.J. & Tabbush, P.M. 1987 The need to weed. *For. Comm. Rep. For. Res.* **1987** 19 HMSO, London.

Davies, H.L., Taylor, C.M.A. & Allen, M.W. 1988 Farm forestry. *For. Comm. Rep. For. Res.* **1988** 13.

Davison, A.W. & Barnes, J.D. 1998 Effects of ozone on wild plants. *New Phytologist* **139** 135-151.

Dawson, W.M. 1988 Production of biomass from short-rotation coppice willow in Northern Ireland. (Ed.) A. Ferm Proc. IEA, Task II meeting and workshops, Oulu, Finland 1987. *Bulletin of the Finnish Forest Research Institute* **304** 91-100.

Dawson, W.M. 1991 Short rotation coppice willow: the northern Ireland experience. (In) *Wood for energy - implication for harvesting, utilisation & marketing.* Proc. Discussion meeting (Ed.) J.R. Aldhous 235-247 Institute of Chartered Foresters, Edinburgh.

Dawson, W.M. 1992 Aspects of development of short-rotation coppice willow for biomass in Northern Ireland. *Proc. R. Soc. Edinburgh* **98 B** 193-205.

Day, M. 1996 Secrecy destroys faith in drug safety. *New Scientist* **28.9.96** 4.

Day, S. 1993 A shot in the arm for plants. *New Scientist* **9.1.93.**

Day, W.R. 1924 Watermark disease of the Cricket bat willow. *Oxford Forest Memoirs.* **3** 30pp.

Dayton, L. 1992 Self-dipping sheep will poison parasites. *New Scientist* **4.4.92** 19.

de Groot, P. 1992 The future goes up in flames. *New Scientist* **12 Sept. 92** 39.

de Silva, W. 1996 Long dry spells, outlook gloomy. *New Scientist* **12.10.96** 9.

De'Ath, M.R. 1988 Triclopyr - a review of its forestry and industrial weed control uses. (In) Practice of weed control and vegetation management in forestry, amenity and conservation areas. *Aspects of Applied Biology.* **16** 183-188 Association of Applied Biologists, Warwick.

Deboys, R. 1994 No myth - machines can harvest coppice. *Forestry & British Timber* **Mar. 1994** 21-27.

Deelstra, T. 1994 Trees, people and the urban environment. Scottish Forestry **48** (3) 163-176.

Delatour, C. Weissenberg, K. von & Dimitri, L. 1998 Host resistance. (In) *Heterobasidion annosum, biology, ecology, impact & control* (Eds.) S. Woodward *et al.* 143-166 CAB International, Wallingford, Oxon. England.

Delfosse, E.S., Leppla, N.C. & Oraze, M.J. 1996 Developing & supporting a national biological control programme. Biological introductions. *Proc. BCPC Symposium* **67** 115-132 British Crop Protection Council.

DEngy 1989a,b,c Dept of Energy. a) Renewable energy R&D programme. b) Planning for the future. c) Renewable energy in Britain. *IREN* **1(2)**, **3 & 15** Department of Energy, Harwell, Oxfordshire.

DEngy 1990 Harvesting & chipping of wood fuels. *REView (Renewable Energy view)* **11** Spring 20 Department of Energy, Harwell, Oxfordshire.

DEnv 1974 *Report on non-agricultural uses of pesticides in the United Kingdom.* Dept. of the Environment **158/74** 27.11.74 HMSO, London.

DEnv 1986 Land filling wastes. *Waste Management Paper* **26** HMSO, London.

DEnv 1989b *Digest of environmental protection and water statistics.* Dept. of the Environment HMSO, London.

DEnv 1991a The new law on waste; the duty of care. *Leaflet* **91 EP 0298** Department of the Environment.

DEnv 1991c *Potential effects of climatic change in the United Kingdom.* UK Climatic change impacts review group, Dept of Env. 108 pp HMSO, London.

DEnv 1991c *Survey of derelict land in England, 1988.* Dept. of the Environment HMSO, London.

DEnv 1992 *Code of practice for the agricultural use of sewage.* Dept. of the Environment Revised 1996 12 pp HMSO, London.

DEnv 1994 *Householders' response to radon risk.* Dept. of the Environment 38 pp HMSO, London.

DEnv 1995a *Climatic change. Progress report on carbon dioxide emissions.* Dept. of the Environment **ENVI J08-5041 TC** 48 pp Central Office of Information, London.

DEnv 1995b *This common inheritance, 1995.* Dept. of the Environment London.

DEnv 1995c *A guide to risk assessment and risk management for environmental protection.* Dept. of the Environment 98 pp HMSO, London.

DEnv 1995d *Expert panel on air quality standards - sulphur dioxide.* Dept. of the Environment HMSO, London.

DEnv 1995e *Climate change - the UK programme: progress report on carbon dioxide emissions.* 48pp Central Office of Information London.

DEnv 1996 *Waste management: the duty of care. A code of practice.* (Revised edn.) 54pp HMSO, London.

Denne, M.P. & Atkinson, L.D. 1973 Phytotoxic effect of captan on growth of conifer seedlings *Forestry* **46** (1) 49-53.

Derwent, R.G. 1986 The nitrogen budget for the UK and NW Europe. *ETSU Report* **37** AERE, Harwell.

DETR 1998 *UK Climatic Change Programme - Consultation Paper* 60pp DETR (Dept. of Environment, Transport & Regions)

DETR/MO 1997 *Climatic change & its impacts: a global perspective.* 16pp Dept. of Environment, Transport & Regions & Meteorological Office.

Dickinson, N.M. 1996 Metal resistance in trees (In) *Heavy metals & trees* Proc. conf. Glasgow 1995 (Ed.) I. Glimmerveen. 93-104 Institute of Chartered Foresters, Edinburgh.

Dickson, D.A. 1971 The effect of form, rate and position of phosphatic fertilizers on growth and nutrient uptake of Sitka spruce on deep peat. *Forestry* **44** (1) 17-26.

Dickson, D.A. 1977 Nutrition of Sitka spruce on peat - problems and speculations. *Irish Forestry* **34** 31-39.

Dickson, D.A. & Savill, P.S. 1974 Early growth of *Picea sitchensis* (Bong.) Carr. on deep oligotrophic peat in N. Ireland. *Forestry* **47** (1) 57-88.

Dighton, J. 1991 Acquisition of nutrients from organic resources by mycorrhizal autotrophic plants. *Experientia* **47** 362-331.

Dighton, J. & Skeffington, R.A. 1987 Effects of artificial acid precipitation on the mycorrhizas of Scots pine seedlings. *New Phytologist* **107** 191-202.

Dillon P.J. & La Zerte, B.D. 1992 Response of the Plastic Lake catchment to reduced sulphur deposition. *Environmental Pollution* **77** 211-217.

Dimbleby, G.W. 1952 Soil regeneration on the North Yorkshire moors. *Ecology* **40** 331-341.

Dise, N.B. & Wright, R.F. 1995 Nitrogen leaching from European forests in relation to nitrogen deposition. *Forest Ecology & Management* **71** 133-142.

Dixon, A.F.G. 1971a The role of aphids in wood formation. I. Sycamore aphid and the growth of sycamore *Journal of Applied Ecology* **8** (1) 165-179.

Dixon, A.F.G. 1971b The role of aphids in wood formation. II. Lime tree aphid and the growth of lime. *Journal of Applied Ecology* **8** (2) 393-399.

Dixon, F.L. & Clay, D.V. 1996 Control of heather (*Calluna vulgaris*) as a weed in forest nursery seed-beds. (In) Vegetation management in forestry, amenity & conservation areas. *Aspects of applied biology* **44** 69-74 Association of Applied Biologists, Warwick.

Dobson, M.C. 1991a De-icing salt damage to trees and shrubs. *For. Comm. Bull.* **101** 76pp HMSO, London.

Dobson, M.C. 1991b Diagnosis of de-icing salt damage to trees. *Arboriculture Research Note* **96/91/PATH** Arboricultural Advisory & Information Service, Farnham, Surrey.

Dobson, M.C. 1991c Tolerance of trees and shrubs to de-icing salt. *Arboriculture Research Note* **99/91/PATH** Arboricultural Advisory & Information Service, Farnham, Surrey.

Dobson, M.C. 1991d Prevention & amelioration of de-icing salt damage to trees. *Arboriculture Research Note* **100/91/PATH** Arboricultural Advisory & Information Service, Farnham, Surrey.

Dobson, M.C. & Moffat. A.J. 1993 *The potential for woodland establishment on landfill sites.* 88 pp HMSO, London.

Dobson, M.C., Taylor, G. & Freer-Smith, P. 1990 The control of ozone uptake by *Picea abies* & *P. sitchensis* during drought & interacting effects on shoot water relations. *New Phytologist* **116** 465-474.

Dollard, G.J. & Davies, T.J. 1992 Observations of H_2O_2 and PAN in a rural atmosphere. *Environmental Pollution* **75** 45-52.

Donaubauer, E. & Stefan, K. 1971 Proof of increase in pollution effects after construction of a thermal power station in a fume-damage area. *For. Comm. Res. & Devel. Pap.* **82** 12.

Dowding, P. 1970 Colonisation of newly bared pine sapwood surfaces by staining fungi. *Transactions of the British mycological Society* **55** 399-412.

Drury, S. 1995 Woodland wrecker. *Horticulture Week* **June 1, 1995** 20-21.

DTde 1968 Census of production - Formulated pesticides & disinfectants. *Department of Trade* **40** HMSO, London.

DTde 1970 Census of production - Formulated pesticides & disinfectants. *Department of Trade* **C 40** HMSO, London.

DTde 1971 Census of production - Formulated pesticides & disinfectants. *Department of Trade* **PA 279.4** HMSO, London.

DTde 1972 Census of production - Formulated pesticides & disinfectants. *Department of Trade* **PA 279.4** HMSO, London.

DTI 1993a Crops for energy. *REView (Renewable Energy view)* **21** 12-13 Dept. of Trade & Industry

DTI 1993b Specialised wood burning units. *Wood fuel now* **3** 1p Department of Trade & Industry

DTI 1994a Short rotation coppice. Plantation design: combating diseases & pests. *Agriculture & Forestry Fact Sheet* **SRC 14** 2 pp ETSU, Harwell, Oxfordshire. (for Dept. of Trade & Industry).

DTI 1994b Biomass conversion gasification of wood for small-scale CHP. *Agriculture & Forestry Fact Sheet* **CON 1** 2 pp ETSU, Harwell, Oxfordshire. (for Dept. of Trade & Industry).

DTI 1994c Storage & drying of wood chips. *Agriculture & Forestry Fact Sheet* **CON 3** 2 pp ETSU, Harwell, Oxfordshire. (for Dept of Trade & Industry).

DTI 1994d New & renewable energy. Future prospects for the UK. *Department of Trade & Industry Energy Paper* **62** HMSO, London.

DTI 1995a Energy projections for the UK. *Department of Trade & Industry Energy Paper* **65** 162 pp HMSO, London.

DTI 1995b *Digest of United Kingdom energy statistics, 1995.* Department of Trade & Industry 200 pp HMSO, London.

DTI 1995c NFFO-3/SRO-1 results announced. *REView (Renewable Energy view)* **Supplement** ETSU, Harwell, Oxfordshire. (for Dept of Trade & Industry).

DTI 1996 Spore production in stored wood chips & straw. *Agriculture & Forestry Fact Sheet* **CON 9** 2 pp ETSU, Harwell, Oxfordshire. (for Dept of Trade & Industry).

Duffy, M.J. 1991 The characterisation of herbicide persistence. (In) Pesticides in soils & water: current perspectives *Proc. BCPC Conf. Monograph* **47** 85-92 British Crop Protection Council, Farnham, Surrey.

Durrant, D.W.H. 1988 Site studies (South): Air pollution - Biomonitors. *For. Comm. Rep. For. Res.* **1988** 27-28 HMSO, London.

Durrant, D.W.H., Waddell, D.A., Benham, S.E. & Houston, T.J. 1992 Air quality and tree growth in open top chambers. *For. Comm. Res. Inf. Note* **208** Forestry Commission, Farnham, Surrey.

Durrant, D.W.H., Lee, H., Barton, C. & Jarvis, P. 1993 Long-term carbon enrichment experiment examining the interactions with nutrition of Sitka spruce. *For. Comm. Res. Inf. Note* **238** Forest Research Station, Farnham.

Durrant, P. 1996 Meeting in paradise. (Report of meeting of International research group on wood preservation.) *Timber Trades Journal* **8 June, 1996.**

Dutch, J.C. & Ineson, P. 1990 Denitrification of an upland forest site. *Forestry* **63** (4) 363-377.

Dutch, J.C., Taylor, C.M.A. & Worrell, R. 1990 Potassium fertilizers: effects of rates, types and time of application on height of Sitka spruce. *For. Comm. Res. Inf. Note* **188** Forest Research Station, Farnham.

Dutton, J.C.F. 1993a Grey squirrel control in Britain Pt 1. *Forestry & British Timber* **Sept. 1993** 30-35.

Dutton, J.C.F. 1993b Grey squirrel control in Britain Pt 2. *Forestry & British Timber* **Oct. 1993** 31-35.

Duyzer, J H., Verhagen, H.L.M. & Weststrate, J.H. 1992 Measurement of dry deposition flux of NH_3 on to coniferous forest. *Environmental Pollution* **75** 3-13.

Eason, I. 1996 ENs set a new agenda. *Timber Trades Journal* **22 June 1966** 14.

EC 1977 Council Directive of 21 December 1976 on protective measures against the introduction into Member States of harmful organisms of plants or plant products. *Official J. of the European Communities* **L26** 20-54.

EC 1986 Council Directive of 11 Nov. 1986 on the protection of Community forests against atmospheric pollution. No. 3528/86 *Official J. of the European Communities* **L326** 2.

EC 1987 Directive of 10 June, 1987 on requirements for inventories of forest condition. No. 1696/87 *Official J. of the European Communities* **L121** 1-22.

EC 1989 Directive of 4th Oct. 1989 on forest condition; assessment standards No. 2995/89 *Official J. of the European Communities* **L287** 11.

EC 1992a Directive of 3 November establishing obligations to which producers and importers ... details of their registration. *Official J. of the European Communities* **L334** 38.

EC 1992b Commission Directive of November 1992. Plants, plant products ... which must be subject to Plant Health Inspection. *Official J. of the European Communities* **L352** 1-5.

EC 1992c Commission Directive of 3 December 1992 establishing plant passports. *HSE Leaflet* **L4** 22.

Edlin, H.L. 1957 Saltburn following a summer gale in south-east England *Quarterly Journal of Forestry* **37** 46-50.

Edwards, M.V. 1952 The effects of partial soil sterilisation with formalin on the raising of Sitka spruce and other conifer seedlings. *For. Comm. Forest Rec.* **16** HMSO, London.

Edwards, M.V. 1958 Use of triple superphosphate for forest manuring. *For. Comm. Rep. For. Res.* **1958** 117-130 HMSO, London.

Edwards, M.V. 1959 Effects & amounts of basic slag and mineral phosphate on JL on blanket bog. *For. Comm. Rep. For. Res.* **1959** 116-125 HMSO, London.

Edwards, P.N. 1981 Yield models for forest management. *For. Comm. Booklet* **48** 32pp+tables HMSO, London.

Edwards, R. 1998 Take the low road. *New Scientist* **21.11.1998** 12.

Edwards, M.V. & Holmes, G.D. 1949 Experimental work in the nursery. *For. Comm. Rep. For. Res.* **1949** 35-43 HMSO, London.

Edwards, M.V. & Holmes, G.D. 1950 Experimental work in the nursery. *For. Comm. Rep. For. Res.* **1950** 12-24 HMSO, London.

Edwards, M.V. & Holmes, G.D. 1956 Problems in afforestation. *For. Comm. Rep. For. Res.* **1956** 36-39 HMSO, London.

Edwards, R.S. & Holmes, G.D. 1968 Studies of airborne salt deposition in some North Wales forests. *Forestry* **41** (2) 155-174.

Edwards, C. & Morgan, J. 1996 Control of *Rhododendron ponticum* by stump application of herbicides following mechanical clearance. (In) *Proc. conf.: Crop Protection in Northern Britain* (Ed.) T.D. Heilbronn **1996** 213-218 SCRI, Dundee.

Edwards, C. & Morgan, J. 1997 Cut-stump applications with imazapyr. *For. Comm. Res. Inf. Note* **293** 4pp Forestry Commission, Edinburgh.

Edwards, M.V. & Stewart, G.G. 1958 Silvicultural investigations in the forest; Scotland & N. England: manuring in the forest. *For. Comm. Rep. For. Res.* **1958** 46-51 HMSO, London.

Edwards, M.V., Stewart, G.G. & Henman, D.W. 1959 Silvicultural investigations in the forest; Scotland & N. England: manuring in the forest. *For. Comm. Rep. For. Res.* **1959** 42-46 HMSO, London.

Edwards, M.V., Stewart, G.G., Lines, R. & Henman, D.W. 1960 Silvicultural investigations in the forest; Scotland & N. England: afforestation problems; manuring in the forest. *For. Comm. Rep. For. Res.* **1960** 38-43 HMSO, London.

Edwards, C., Tracy, D. & Morgan, J.L. 1993 *Rhododendron* control by imazapyr. *For. Comm. Res. Inf. Note* **233** 5pp Forest Research Station, Farnham.

Edwards, C., Morgan, J. & Willoughby, I. 1994 Approved herbicides for use in forestry. *For. Comm. Res. Inf. Note* **246** Forest Research Station, Farnham.

Eilers, G., Brumme, R. & Matzner, E. 1992 Above-ground N-uptake from wet deposition by Norway spruce. *Forest Ecology & Management* **51** 239-249.

Ellis, G. 1995 New strategy in weevil control. *Forestry & British Timber* **Oct. 1995** 42.

Emmett, B.A. & Reynolds, B. 1996 Nitrogen critical loads for spruce plantations in Wales: is there too much nitrogen? *Forestry* **69** (3) 206-214.

Emmett, B.A., Anderson, J.M. & Hornung, M. 1991 The controls on dissolved nitrogen losses following two intensities of harvesting in Sitka spruce forest. *Forest Ecology & Management* **41** 65-79.

Emmett, B.A., Reynolds, B., Stevens P.A. *et al.* 1993 Nitrate leaching from afforested Welsh catchments - interaction between stand age & nitrogen deposition. *Ambio* **22** (6) 386-394.

Emmett, B.A., Brittain, S.A. *et al.* 1995 Nitrogen additions ($NaNO_3$ and $NHNO_3$) at Aber Forest, Wales; response of throughfall and soil water chemistry. *Forest Ecology & Management* **71** 45-59.

Emsley, J. 1994a Cool reception for warming predictions. *New Scientist* **8.10.94** 19.

Emsley, J. 1994b Energy & fuels. *New Scientist* **15.1.94.**

EnCom 1992 Renewable energy. House of Commons, Session 1991-92 *Energy Committee* **4th report** Vol 1 42pp HMSO, London

English, C. 1990 Synthetic pyrethroids, medical hazards & monitoring. *Proc. FAO/ECE/ILO seminar on forest technology, management and training.* 152-153 Forestry Commission, Edinburgh.

Entwistle, P.F. & Evans, H.F. 1984 Analysis of the natural spread of insect virus diseases in relation to pest control. *Proc. 17th International Entomology Conference, Hamburg* 752.

Entwistle, P.F. & Evans, H.F. 1985 Viral control. A review. (In) *Comprehensive insect physiology, biochemistry and pharmacology.* (Eds) G.A. Kerkut, & L.I. Gilbert Pergamon Press, Oxford.

Entwistle, P.F. & Evans, H.F. 1987 Trials on the control of *Panolis flammea* with a nuclear polyhedrosis virus. (In) *For. Comm. Bull.* **67** 61-75 HMSO, London.

Entwistle, P.F., Evans, H.F., Harrap, K.A. & Robertson, J.S. 1985 Control of European pine sawfly with its nuclear polyhedrosis virus in Scotland. *For. Comm. Res. & Dev. Pap.* **135** 36-46 Forestry Commission, Edinburgh.

Entwistle, P.F., Adams, P.H.W. & Evans, H.F. 1977 Epizootology of NPV in European spruce sawfly: status of birds as dispersal agents *For. Comm. Leaflet* **7** 12pp HMSO, London.

EPA 1990 Environment Protection Act. Act of Parliament Chapter 43 1990 HMSO, London.

ETSU 1989a Monitoring a commercial demonstration of harvesting and combustion of forestry wastes. *ETSU Contractor Report* **B 1187-P1** ETSU, Harwell, for Dept. of Energy.

ETSU 1989b Establishment & monitoring of large-scale trials of short rotation forestry for energy. *ETSU Contractor Report* **B 1171** ETSU, Harwell, for Department of Trade & Industry.

ETSU 1990a Forestry waste firing of industrial boilers. *ETSU Contractor Report* **B 1178** 86 pp Energy Technology Support Unit, Harwell, for Dept. of Energy.

ETSU 1990b Wood fuel supply strategies. *ETSU Contractor Report* **B 1176-P1** 122 pp Energy Technology Support Unit, Harwell, for Dept. of Energy.

ETSU 1994a Wood logs as a domestic fuel. *Agriculture & Forestry Fact Sheet* **CON 8** 2 pp ETSU, Harwell, for Department of Trade & Industry.

ETSU 1994b Chemical weed control. *Agriculture & Forestry Fact Sheet Short Rotation Coppice* **4** ETSU, Harwell, for Department of Trade & Industry.

ETSU 1994c *An assessment of renewable energy for the UK*. Department of Trade & Industry 300 pp HMSO, London.

ETSU 1995a Epidemiology, population dynamics and management of rust diseases in willow energy plantations. *ETSU Contractor Report* **B/W6/00124/REP** Harwell Laboratories, Oxford.

ETSU 1995b Herbicides ... tried on SRC plantations. *Agriculture & Forestry Fact Sheet Short Rotation Coppice* **5** ETSU, Harwell, for Dept. of Trade & Industry.

ETSU 1996a *Short rotation coppice for energy production. The development of an economically & environmentally sustainable industry. Good practice guidelines.* 46pp ETSU, Harwell, Oxfordshire.

ETSU 1996b Hydrological effects of short rotation coppice. *ETSU Contractor Report* **B/W5/00275/REP** 201 pp ETSU, Harwell, for Dept. of Trade & Industry.

ETSU 1996c Small-scale industrial & domestic heating. *ETSU Extended Renewable energy case study* **14** 6 pp ETSU, Harwell, for Dept. of Trade & Industry.

ETSU 1997a Evaluation of methodology for managing existing broadleaved and coniferous woodlands for timber & energy. *ETSU Contractor Report* **B 1156** 223pp ETSU, Harwell, for Dept. of Energy.

ETSU 1997b Storage & drying of short rotation coppice. *ETSU Contractor Report* **B/W2/00391/REP** 146pp ETSU, Harwell, for Dept. of Trade & Industry.

ETSU 1997c Mechanised harvesting. *Agriculture & Forestry Fact Sheet* **CON 8** 2 pp ETSU, Harwell, for Dept. of Trade & Industry.

ETSU 1997d UK Industry wood fuel baling storage trial. *ETSU Contractor Report* **B/W2/00548/07/REP** ETSU, Harwell, for Dept. of Trade & Industry.

ETSU 1997e Putting renewables on the map. *New REView* **32** 11 ETSU, Harwell, for Dept. of Trade & Industry.

ETSU 1998a NFFO News - Update. *New REView* **37** 8 ETSU, Harwell, for Dept. of Trade & Industry.

ETSU 1998b Wood-fuelled gasifier opens in Londonderry.; School harness wood for heat. (In) *New REView* **35** 3 (2 paras) ETSU, Harwell, for Dept. of Trade & Industry.

ETSU 1998c Population dynamics of *Melampsora* rusts & disease control in renewable energy plantations. *ETSU Contractor Report* **B/W6/00508** 61 pp ETSU, Harwell, for Dept. of Trade & Industry.

ETSU 1998d Short rotation coppice harvesting: evaluation of Salix Maskiner Bender III. *ETSU Contractor Report* **B/W2/00548/19/REP** ETSU, Harwell, for Dept. of Trade & Industry.

ETSU 1998e Establishment and monitoring of large-scale trials of short rotation coppice. *ETSU Contractor Report* **B/W2/00514/REP** Energy Technology Support Unit, Harwell, for Dept. of Energy.

ETSU 1998f *Renewable energy. General literature lists.* ETSU, Harwell, Didcot Oxfordshire OX11 0RA

ETSU 1998g The renewables obligations, NFFO, SRO & NI-NFFO policy. *NFFO Fact sheet* **1** 2pp ETSU, Harwell, for Dept. of Trade & Industry.

ETSU 1998h *Wood fuel from forestry & arboriculture. The development of a sustainable energy production industry. Good practice guidelines.* 55pp ETSU, Harwell, Oxfordshire.

Evans, J. 1984 Silviculture of broadleaved woodland. *For. Comm. Bull.* **62** 232pp HMSO, London.

Evans, J. 1986a Nutrition experiments in broadleaved stands: I pole stage oak and ash. *Quarterly Journal of Forestry* **80** (2) 85-94.

Evans, J. 1986b Nutrition experiments in broadleaved stands: II sweet chestnut and stored coppice. *Quarterly Journal of Forestry* **80** (2) 95-104.

Evans, J. 1986c Epicormic branches on oak. *For. Comm. Rep. For. Res.* **1986** 11 HMSO, London.

Evans, J. 1987 The control of epicormic branches. (In) Advances in Arboriculture (Ed. D. Patch) *For. Comm. Bull.* **65** 115-120 HMSO, London.

Evans, J. 1993b Silviculture of broadleaved woodland. *For. Comm. Bull.* **62** 232 pp HMSO, London.

Evans, H.F. 1995 Entomology overview. *For. Comm. Rep. For. Res.* **1995** 3 HMSO, London.

Evans, H.F. 1996 Entomological threats to British forestry. *ICF News* **3/96** 5-7 Institute of Chartered Foresters, Edinburgh.

Evans, H.F. 1997a Microbial insecticides: novelty or necessity? *Symposium proceedings* **68** 302 pp British Crop Protection Council, Farnham, Surrey.

Evans, H.F. 1997b The role of insecticides in forest pest management. (In) H.F. Evans (Ed.) *Symposium proceedings* **68** 29-40 British Crop Protection Council, Farnham, Surrey.

Evans, H.F. 1997c The present position of forest entomology in Great Britain. *Forestry* **70** (4) 327-336.

Evans, C.C. & Allen, S.E. 1971 Nutrient losses in smoke produced during heather burning. *Oikos* **22** 149-154.

Evans, H.F. & Fielding, N.J. 1994 Integrated management of *Dendroctonus micans* in the UK. *Forest Ecology & Management* **65** (1) 17-30.

Evans, H.F. & Fielding, N.J. 1996 Restoring the natural balance: biological control of *Dendroctonus micans* in Great Britain. Biological introductions *Proc. BCPC Symposium* **67** 47-57 British Crop Protection Council, Farnham, Surrey.

Evans, H.F. & King, C.J. 1988 *Dendroctonus micans* - guidelines for forest managers. *For. Comm. Res. Inf. Note* **128** 8pp Forestry Commission, Forest Research Station, Farnham.

Evans, H.F. & King, C.J. 1989a Great spruce bark beetle, *Dendroctonus micans*. *For. Comm. Rep. For. Res.* **1989** HMSO, London.

Evans, H.F. & King, C.J. 1989b Biological control of *Dendroctonus micans*: British experience of rearing and release of *Rhizophagus grandis*. (In) *Potential for biological control of Dendroctonus and Ips bark beetles.* (Eds.) D.L. Kuhlevy & M.C. Miller 109-128.

Evans, H.F. & Stoakley, J.T. 1990 The use of insecticides, fungicides and vertebrate control products in British forestry. *Proc. FAO/ECE/ILO seminar on forest technology, management and training.* 54-62 Forestry Commission, Edinburgh.

Evans, H.F., King, C.J. & Wainhouse, D. 1984 *Dendroctonus micans* in the UK - two years experience in survey & control. Proceedings EEC Seminar 1984 20-32 EC, Brussels.

Evans, H.F., King, C.J., Fielding, N.J. & Martin, A.F. 1985a Great spruce bark beetle, *Dendroctonus micans*. *For. Comm. Rep. For. Res.* **1985** 37 HMSO, London.

Evans, H.F., Stoakley, J.T. & Heritage, S.G. 1985b European pine sawfly, *Neodiprion sertifer*. *For. Comm. Rep. For. Res.* **1985** 39-40 Forestry Commission, Edinburgh.

Evans, H.F., Stoakley, J.T., Leather, S.R. & Watt, A.D. 1991 Development of an integrated approach to control of pine beauty moth in Scotland. *Forest Ecology & Management* **39** 19-28 Elsevier, Amsterdam.

Everard, J.E. 1973 Foliar analysis, sampling methods, interpretation and application of results. *Quarterly Journal of Forestry* **67** (1) 51-66.

Everard, J.E. 1974 Fertilizers in the establishment of conifers in Wales and southern England. *For. Comm. Booklet* **41** 1-46 HMSO, London.

Everard, J.E. 1990 The goal of normality. (In) *Silvicultural systems*. Proceedings of Discussion Meeting 1990 111-114 Institute of Chartered Foresters, Edinburgh.

Everard, J.E. & Mackenzie, J.M. 1974 Nutrition of forest crops - trace elements. *For. Comm. Rep. For. Res.* **1974** 21-22.

F&HGT 1976 Guide to the use of herbicides in the forest. *Forestry & Home Grown Timber* Feb/Mar. **1976** Separate page reprint.

FA (Forestry Authority) 1998 *Challenge scheme for short-rotation energy crops in Yorkshire.* Forestry Commission, Edinburgh.

Fahey, T.J., Hill, M.O., Stevens, P.A., Hornung, M. & Rowland, P. 1991a Nutrient accumulation in vegetation following conventional & whole-tree harvesting of Sitka spruce ... in North Wales. *Forestry* **64** (3) 271-288.

Fahey, T.J., Stevens, P.A., Hornung, M. & Rowland, P. 1991b Decomposition and nutrient release from logging residue ... of Sitka spruce in North Wales. *Forestry* **64** (3) 289-303.

Fairhurst, C.P. & Atkins, P.M. 1987 Dutch elm disease: the vectors. (In) *For. Comm. Bull.* **65** 160-165 HMSO, London.

FAO 1967 Wood; world trends and prospects. *Basic study* **16** Food & Agriculture Organisation, Rome.

FAO 1970 *Control of pesticides.* 152 pp FAO, Rome.

FAO 1988 Forest products; world outlook projections. *Forestry Paper* **84** FAO, Rome.

FAO 1989 *FAO yearbook: forest products, 1976-87.* Food & Agriculture Organisation, Rome.

FAO 1992 Financial analysis in agricultural project preparation. *FAO Investment Centre Technical Paper* **8** 222 pp Food & Agriculture Organisation, Rome.

FAO 1994 Prevention of water pollution by agriculture and related activities. *FAO Water report* **1** 357 pp Food & Agriculture Organisation, Rome.

FAO/ECE 1970 Damage to forests caused by ashes & sulphurous compounds in smoke from thermal power stations. *For. Comm. Res. & Dev. Pap.* **82** 47-49 Forestry Commission, London.

Farley, D.A. & Werrity, A. 1989 Hydrochemical budgets for the Loch Dee catchment in SW Scotland 1981-85. *Journal of Hydrology* **109** 351-368.

Farmer, R.A., Alexander, A. & Acton, M. 1985 Aerial fertilizing - monitoring spread is vital operation. *Forestry & British Timber* **April 1985** 15-17.

Farr, W.A., Smith, H.A. & Benzian, B. 1977 Nutrient concentrations in naturally regenerating seedlings of *Picea sitchensis* in South-east Alaska. *Forestry* **50** (2) 103-112.

FASTCo 1990 Dipping plants in insecticide. *Safety Guide* **101** Forestry & Arboriculture Safety & Training Council, Edinburgh.

FASTCo 1994a Pre-plant spraying container-grown seedlings. *Safety Guide* **102** Forestry & Arboriculture Safety & Training Council, Edinburgh.

FASTCo 1994b Application of pesticides by hand-held applicators. *Safety Guide* **202** Forestry & Arboriculture Safety & Training Council, Edinburgh.

FASTCo 1994c Hand-tool weeding & cleaning. *Safety Guide* **201** Forestry & Arboriculture Safety & Training Council, Edinburgh.

Faulkner, R.F. 1956 Greencrops compared with hopwaste as heathland nursery treatments. *For. Comm. Res. Br. Pap.* **19** 10 pp Departmental Paper, Forestry Commission, London.

Faulkner, R.F. 1957 Scottish experiments comparing chloropicrin with formalin as a partial sterilizer for conifer beds. *For. Comm. Rep. For. Res.* **1957** 159-170 HMSO, London.

Faulkner, R.F. & Holmes, G.D. 1953 Experimental work in nurseries. *For. Comm. Rep. For. Res.* **1953** 17-31 HMSO, London.

Faulkner, R.F. & MacDonald, A. 1954 Inchnacardoch long-term soil fertility trial. *For. Comm. Res. Br. Pap.* **15** 15pp Departmental Paper, Forestry Commission, London.

Fawell, J.K. 1991 Pesticide residues in water - imaginary threat or imminent disaster. (In) Pesticides in soils & water: current perspectives *Proc. BCPC Conf. Monograph* **47** 205-208 British Crop Protection Council, Farnham, Surrey.

FBT 1992 Moth damage. *Forestry & British Timber* **Aug. 1992**

FBT 1996 Wales fights killer weevil. *Forestry & British Timber* **May 1996** 6.

FBT 1997a Public (and private) enemy No 1. *Forestry & British Timber* **April 1997** 12-15.

FBT 1997b FE 'will get certified'. *Forestry & British Timber* **Feb. 1997** 2.

FBT 1998a Standard raised to much applause. *Forestry & British Timber* **Mar. 1998** 2-3.

FBT 1998b UK *Woodland Assurance Scheme* goes ahead. *Forestry & British Timber* **Nov 1998** 3.

FC 1989a Provisional Code of practice for the use of pesticides in forestry. *Forestry Commission Occasional Paper* **21** 52pp Forestry Commission, Edinburgh.

FC AR Annual *Annual report of Forestry Commissioners.* HMSO, London.

FC AR 1979 Worker protection. *For. Comm. Ann. Rep.* 23.

FC EA 1997 *Phytophthora disease of alder.* 6pp Environment Agency, Bristol.

FC FF 1996 Forestry Commission Facts & Figures (issued annually). Forestry Commission, Edinburgh.

FC FS 1998a *The UK Forestry standard. The government's approach to sustainable forestry.* 74 pp Forestry Commission, Edinburgh.

FC FS 1998b *UK Forestry Standard - Audit protocol.* (draft) Forestry Commission, Edinburgh.

FC NR 1996a Early warning of moth attacks. *News Release* **304** 3 pp Forestry Commission, Edinburgh.

FC NR 1996b Great spruce bark beetle discovered in Kent. *News Release* **496** 3 pp Forestry Commission, Edinburgh.

FC NR 1998a Forest growth points to new employment opportunities in Scotland & N. England. *News Release* **1581** 3 pp Forestry Commission, Edinburgh.

FC NR 1998b Commission acts to combat new beetle threat. *News Release* **1709** 2 pp Forestry Commission, Edinburgh.

FC NR 1999 Tony Blair welcomes world-first for UK forestry. *News Release* **2048** 4pp Forestry Commission, Edinburgh.

FC PH 1986 *Dendroctonus micans* - the great spruce bark beetle. General control strategy. *Plant Health Information Sheet* **10** Forestry Commission, Edinburgh.

FC PH 1989 *Dendroctonus micans* - the great spruce bark beetle. General control strategy. *Plant Health Information Sheet* **13** Forestry Commission, Edinburgh.

FC PH 1992 Plant Health and the Single Market Explanatory Booklet. *Plant Health Booklet* **7** 15pp Forestry Commission, Edinburgh.

FC PH 1993a Single market registration & passporting. *Plant Health Newsletter* **4** July 1993 Forestry Authority, 231, Corstorphine Road, Edinburgh.

FC PH 1993b Mill certificates; plant passports; *Dendroctonus. Plant Health Newsletter* **5** Oct. 1993 Forestry Authority, 231, Corstorphine Road, Edinburgh.

FC PH 1994a Industry (Mill) Certificates, Dunnage, Bark, etc. *Plant Health Newsletter* **6** Mar. 1994 Forestry Commission, Edinburgh.

FC PH 1994b Christmas trees, *Ips*, Inspection of EC Goods, Gypsy moth. *Plant Health Newsletter* **7** Nov. 1994 Forestry Commission, Edinburgh.

FC PH 1994c Plant Health and the Common Market. *Plant Health Leaflet* **8** 1994 (Under revision) Forestry Commission, Edinburgh.

FC PH 1995 Eight-toothed spruce bark beetle (*Ips typographus*); Gypsy moth (*Lymantria dispar*); guidance on plant health controls. *Plant Health Newsletter* **8** Aug. 1995 Forestry Commission, Edinburgh.

FC PH 1996 *Dendroctonus micans* controls. Gypsy moth update. *Plant Health Newsletter* **9** Aug. 1996 Forestry Commission, Edinburgh.

FC PH 1998a *Ips typographus*; *Dendroctonus micans*. *Plant Health Newsletter* **10** Jan. 1998
Forestry Commission, Edinburgh.

FC PH 1998b *Ips typographus*; dunnage. *Plant Health Newsletter* **11** July 1998 Forestry
Commission, Edinburgh.

FC PH 1998c *Dendroctonus micans* - a guide for forest managers on control techniques. *For.
Comm. Plant Health Leaflet* **9** 12 pp Forestry Commission, Edinburgh.

FC PH 1999 *Ips typographus*: Asian longhorn beetle; termites. *For. Comm. Plant Health
Newsletter* **12** 4 pp Forestry Commission, Edinburgh.

FC SG 1998 *Forests & soil conservation guidelines*. Issued jointly by the Forestry Authority and
Department of Agriculture for Northern Ireland Forest Service. 24 pp Forestry Commission,
Edinburgh.

FC TDB 1995 Second field trial of short rotation coppice harvesters. *Technical Development
Branch Report* **1/95** Forestry Commission, Edinburgh.

FC WG 1997 *Forests & Water Guidelines*. 4th Edn. 32 pp HMSO, London.

FC WS 1987 A trial planting with Teno Collar. *FC Work Study, Southwest Region Report* **No 4**
Forestry Commission, Edinburgh.

FCms (Forestry Commissioners) 1943 *Post-war forest policy*. Cmd 6447 HMSO, London.

Fell, N & Liss, P. 1993 Can algae cool the planet? *New Scientist* **21.8.93** 34.

Ferrier, R.C., Whitehead, P.G. & Miller, J.D. 1993 Potential impacts of afforestation and
climate change on the stream water chemistry of the Monachyle catchment. *Journal of
Hydrology* **145** 453-466.

FICGB 1998 *The forestry industry handbook*. 72 pp Forest Industry Council of Great Britain,
Stirling Business Centre, Stirling FK8 2DZ

Fielding, N.J. 1991 The pine wood nematode. *Entopath News* **97 (Oct 1991)** 15-16 Forestry
Commission Research Station, Farnham, Surrey.

Fielding, N.J. 1992 *Rhizophagus grandis* as a means of biological control against *Dendroctonus
micans* in Britain. *For. Comm. Res. Inf. Note* **224** 3pp Forestry Commission, Forest
Research Station, Farnham.

Fielding, N.J. & Evans, H.F. 1996 Pine wood nematode *Bursaphelenchus xylophilus* ... : an
assessment of the current position. *Forestry* **69** (1) 35-46.

Fielding, N.J. & Waters, A. 1994 Great European spruce bark beetle - *Dendroctonus micans*.
For. Comm. Rep. For. Res. **1994** 5-6 HMSO, London

Fielding, N.J., Evans, H.F., Williams, B.J. & Evans, B. 1991 Distribution and spread of the
great European spruce bark beetle, *Dendroctonus micans* in Britain: 1982-1989. *Forestry* **64**
(4) 345-358.

Fielding, N.J., Evans, B., Burgess, R.L. & Evans, H.F. 1994 Protected zone surveys in GB for
Ips spp & *Dendroctonus micans*. *For. Comm. Res. Inf. Note* **253** Forestry Commission,
Forest Research Station, Farnham.

Finch, R.R. 1997 Gypsy moth: a threat to Britain's trees. (In) Arboricultural practice: present &
future (Ed. J. Claridge). *Research for Amenity Trees* **6** 57 - 60 HMSO.

Findlay, W.P.K. 1959 Sap stain of timber. *Forestry Abstracts* **20** 167-174.

Fisher, G.G., Garnett, R. & De'Ath, M. 1990 Industrial weed control. (In) *Weed Control
Handbook: Principles* (Eds.) R.J. Hance, & K. Holly. 8th edn 491-500 BCPC/Blackwell,
Oxford.

Fisher, G.G., Clark, L. & Ramsey, P.M. 1991 Pesticides in a chalk catchment: inputs and
aquatic residues. (In) Pesticides in soils & water: current perspectives *Proc. BCPC Conf.
Monograph* **47** 193-199 British Crop Protection Council, Farnham, Surrey.

FOE (Friends of the earth) 1987 *Briefing sheet: pesticides*. 4pp

Ford, E.D. 1982 High productivity in a polestage spruce stand and its relation to canopy structure.
Forestry **55** (1) 1-18.

Ford-Robertson, J.B., Watters, M.P. & Mitchell, C.P. 1991 Short rotation coppice willows for
energy. (In) *Wood for energy - implication for harvesting, utilisation & marketing*. Proc.
Discussion meeting (Ed.) J.R. Aldhous 218-234 Institute of Chartered Foresters, Edinburgh.

Foster, S.S.D. & Chilton, P.J. 1991 Pesticides in groundwater: some preliminary observations on behaviour & transport. (In) Pesticides in soils & water: current perspectives *Proc. BCPC Conf. Monograph* **47** 203-204 British Crop Protection Council, Farnham, Surrey.

Foulkes, D.M. 1989 Pesticide regulation - perceptions and reality. *Pesticide Outlook* **1** (1) 19-22.

Foulkes, D.M. 1990 Symposium overview. Future changes in pesticide registration in the EC. *Proc. BCPC Conf. Monograph* **44** 79-85 British Crop Protection Council, Farnham, Surrey.

Fourt, D.F. 1979 Winged tines & gravel pits. *Saga Bulletin* **11** 4, 12 HMSO, London.

Fourt, D.F. 1980 Reclamation methods. *For. Comm. Rep. For. Res.* **1980** 24 HMSO, London.

Fourt, D.F. 1984 Preparation of mine spoil for tree colonisation or planting. (Cited in): *For. Comm. Rep. For. Res.* **1984** 74 HMSO, London.

Fourt, D.F. 1985 Reclamation. *For. Comm. Rep. For. Res.* **1985** 21 HMSO, London.

Fourt, D.F. & Best, N. 1983 Reclamation: sand and gravel. *For. Comm. Rep. For. Res.* **1983** 19 HMSO, London.

Fowler, D., Cape, J.N., Leith, I.D., Choularton, T.W., Gay, M.J. & Jones, A. 1988 The influence of altitude on rainfall composition at Great Dun Fell. *Atmospheric Environment* **22** (7) 1355-1362.

Fowler, D., Cape, J.N. & Unsworth, M.H. 1989 Deposition of atmospheric pollutants on forests. *Phil. Trans. R. Soc. Lond.* B **324** 247-265.

Fowler, D., Duyzer, J.H. & Baldocchi, D.D. 1991 Inputs of trace gases, particles and cloud droplets to terrestrial surfaces. *Proc. R. Soc. Edinburgh* **97B** 35-59.

Fowler, D., Hargreaves, K.J., Macdonald, J.A. & Gardiner, B. 1995 Methane & CO_2 exchange over peatland and the effects of afforestation. *Forestry* **68** (4) 327-334.

Fowler, D., Flechard, C.R., Sutton, M.A. & Storeton-West, R.L. 1998a Long-term measurements of the land-atmosphere exchange of ammonia over moorland. *Atmospheric Environment* **32** 453-459.

Fowler, D., Flechard, C., Skiba, U., Coyle, M. & Cape, J.N. 1998b The atmospheric budget of oxidised N and its role in ozone formation and deposition. *New Phytologist* **139** 11-23.

FPRL 1968 Blue-stain prevention in the United Kingdom. *Technical Note* **2** 6pp Forest Products Research Laboratory, Princes Risborough.

FPRL 1971 Sap-stain in timber its cause, recognition & prevention. *Technical Note* **50** 11pp Forest Products Research Laboratory, Princes Risborough.

Fraser, G.K. 1933 Studies of certain Scottish moorlands in relation to tree growth. *For. Comm. Bull.* **15** 112pp HMSO, London.

Frazer, A.C. (Chairman) 1964 *Report of research committee on toxic chemicals.* Agricultural Research Council. HMSO, London.

Frazer, A.C. (Chairman) 1965 *Supplementary report of research committee on toxic chemicals.* Agricultural Research Council. HMSO, London.

Frazer, A.C. (Chairman) 1970 *Third report of research committee on toxic chemicals.* Agricultural Research Council. HMSO, London.

Freedman, G. 1995 Potential pollution from structures. Paper to Conf. 'Prevention of pollution during forest engineering operations.' Proc. Joint FAO/ECE/ILO Committee seminar on forest technology, management and training. 31.8.95 10 pp Institute of Agric'l Engineers.

Freer-Smith, P.H. 1985 Influence of SO_2 and NO_2 on the growth, development and gas exchange of *Betula pendula*. *New Phytologist* **99** 417-430.

Freer-Smith, P.H. 1990 Climatic change - the contribution of forestry to response strategies. *For. Comm. Res. Inf. Note* **189** 3pp.

Freer-Smith, P.H. (Ed.) 1992 Acidic deposition, its nature and impacts: atmospheric pollution and forests. *Forest Ecology & Management* **51** (1-3) 1-249.**Freer-Smith, P.H.** 1995a Environmental research *For. Comm. Rep. For. Res.* **1995** 9 HMSO, London.

Freer-Smith, P.H. 1995a Environmental research *For. Comm. Rep. For. Res.* **1995** 9 HMSO, London.

Freer-Smith, P.H. 1995b Critical loads in forestry. (In) *Acid rain & its impact: the critical loads debate* (Ed.) A.R. Battarbee 69-71 Ensis Publishing, London.

Freer-Smith, P.H. & Benham, S. 1994 Critical loads of air pollutants for direct effects on forests and woodlands. (In) *Mapping critical levels/loads* (Eds.) M.R. Ashmore & R.B. Wilson 94-109 Creative Press, Reading.

Freer-Smith, P.H. & Broadmeadow, M.S.J. 1996 The improvement of urban air-quality by trees. *Arboriculture Research & Information Note.* **135/ERB/96** 4 Arboricultural Advisory & Information Service, Alice Holt Lodge, Farnham, Surrey.

Freer-Smith, P.H. & Dobson, M.C. 1995 Assessment of the role of physiological response to sulphur dioxide in Sitka spruce. *For. Comm. Technical Pap.* **9** 111-121 Forestry Commission, Edinburgh.

Freer-Smith, P.H. & Mansfield, T.A. 1987 The combined effect of low temperature and $SO_2 + NO_2$ on growth and water relations of *Picea sitchensis. New Phytologist* **106** 237-250.

Freer-Smith, P.H. & Read, D.B. 1995 The relationship between crown condition and soil solution chemistry in oak and Sitka spruce in England and Wales. *Forest Ecology & Management* **79** 185-196.

Freer-Smith, P.H. & Taylor, G. 1992 Comparative evaluation of the effects of gaseous pollutants, acidic deposition and mineral deficiencies. *Agriculture, ecosystems & the environment* **42** 321-332.

Freer-Smith, P.H., Broadmeadow, M.S.J. & Jackson, S. 1997 The uptake of pollutants by trees: benefits to air quality. (In) Trees for shelter (Eds). H. Palmer et al. *For. Comm. Technical Pap.* **21** 16-22.

Friend, A., Kellomäki, S. & Kruijt, B. 1998 Modelling leaf, tree & forest responses to increasing atmospheric CO_2 and temperature. (In) *European Forests & Global Change* (Ed.) P.G. Jarvis. 293-335 Cambridge University Press.

Fryer, J.D. & Evans, S.A. 1968a Evolution of methods of weed control; safeguards for the user. (In) *Weed Control Handbook: Principles* (Eds.) J.D. Fryer & S.A. Evans 5th edn 25-43, 325-341 Blackwell Scientific Publications, Oxford.

Fryer, J.D. & Evans, S.A. 1968b *Weed Control Handbook - II Recommendations.* 1-324 Blackwell Scientific Publications, Oxford.

Fryer, D. & Stevens, C. 1996 Control freaks. *Horticulture Week.* **Oct. 31 1996** 16-21.

FSC 1981 Tractor-mounted weeding machines. *Forestry Safety Guide* **FSC 2** Forestry Safety Council, Edinburgh.

Fu, Bojie, 1989 Soil erosion and its control in the loess plateau of China. *Soil Use & Management* **5** (2) 76-81.

Furness, R.W. 1996 Transfers of heavy metals from contaminated land to top predators: implications for birds of growing trees for soil amelioration. (In) *Heavy metals & trees* Proc. conf. Glasgow 1995 (Ed.) I. Glimmerveen. 107-121 Institute of Chartered Foresters, Edinburgh.

Galley, R.A.E. 1968 Development of a new pesticide. *Some aspects of pesticides in the Countryside.* (Eds. N.W. Moore & W.P. Evans) 12-20 Association of British Manufacturers of Agricultural Chemicals, London.

Gammon, C. 1998 The noble salmon, wild, mysterious, prized above all other fish, and dying out in rivers all over Europe. *Independent* **31.7.98** Review p1.

Gardiner, A.S. 1968 The reputation of birch for soil improvement. *For. Comm. Res. & Dev. Pap.* **67** Forestry Commission, London (Departmental use)

Garnett, R.P. & Williamson, D.R. 1992 Susceptibility of trees to over-the-top applications of glyphosate during the winter dormant season. Vegetation management in forestry, amenity & conservation *Aspects of Applied Biology* **29** 131-135 Association of Applied Biologists.

Garrett, C.M.E. 1982 Bacterial canker of cherry & plum. *MAFF Leaflet* **592** (Amended, 1982) 6 pp HMSO, London.

Garrett, S.D. 1953 Rhizomorph behaviour in *Armillaria mellea*: I Factors controlling rhizomorph initiation in pure culture. *Annals of Botany* **17** (NS) 63-79.

Garrett, S.D. 1956 Rhizomorph behaviour in *Armillaria mellea*: II Logistics of infection. *Annals of Botany* **20** (NS) 193-200.

Garsed, S.G. & Rutter, A.J. 1982 Relative performance of conifer populations to various tests for sensitivity to SO$_2$ and implications for selecting trees for planting in polluted areas. *New Phytologist* **1982** 349-367.

Gemmel, R.P. 1981 Reclamation of acidic colliery spoil. Use of lime wastes. *Journal of Applied Ecology* **18** 879-887.

Gessel, S.P. & Turner, J. 1976 Litter production in western Washington Douglas fir stands. *Forestry* **49** (1) 63-72.

Gibbs, J.N. 1971 Dutch elm disease. *For. Comm. Leaflet* **19 (Revised)** 11 pp HMSO, London.

Gibbs, J.N. 1974 Biology of Dutch elm disease. *For. Comm. Forest Rec.* **94** 1-9 HMSO, London.

Gibbs, J.N. 1979 Dutch elm disease: survey 1978. *For. Comm. Rep. For. Res.* **1979** 30-31 HMSO, London.

Gibbs, J.N. 1984 Oak wilt. *For. Comm. Forest Rec.* **126** 1-8 HMSO, London.

Gibbs, J.N. 1994a Pathology. *For. Comm. Rep. For. Res.* **1994** 24 HMSO, London.

Gibbs, J.N. 1994b De-icing salt damage to trees: the current position. *Arboriculture Research Note* **119/94/PATH** Arboricultural Advisory & Information Service, Farnham, Surrey.

Gibbs, J.N. 1994c The health of non-woodland trees in England, 1993. *Arboriculture Research Note* **121/94/PATH** Arboricultural Advisory & Information Service, Farnham, Surrey.

Gibbs, J.N. 1994d *Phytophthora* root disease of common alder. *For. Comm. Res. Inf. Note* **258** 4pp Forestry Commission, Forest Research Station, Farnham.

Gibbs, J.N. 1997 Fifty years of sooty bark disease of sycamore. *Quarterly Journal of Forestry* **91** (3) 215-221.

Gibbs, J.N. 1998 Will there be de-icing salt damage this winter? *Tree Damage Alert* **52** 1p Arboricultural Advisory & Information Service, Farnham, Surrey.

Gibbs, J.N. 1999 Die-back of pedunculate oak. *For. Comm. Information Note* **FCIN 22** 6pp Forestry Commission, Edinburgh.

Gibbs, J.N. & Burdekin, D.A. 1983 De-icing salt and crown damage to London plane. *Arboricultural Journal* **6** 227-237.

Gibbs, J.N. & Dickinson, J. 1975 Fungicidal injection for the control of Dutch elm disease. *Forestry* **48** (2) 165-176.

Gibbs, J.N. & Evans, H. 1996 Disease and pest problems. *For. Comm. Rep. For. Res.* **1996** 15-19 HMSO, London.

Gibbs, J.N. & French, D.W.J. 1980 Transmission of oak wilt. *USDA Forest Service Research Paper* **NC-185.**

Gibbs, J.N. & Greig, B.J.W. 1997 Biotic and abiotic factors affecting the dying back of pedunculate oak *Q. robur*. *Forestry* **70** (4) 399-406.

Gibbs, J.N. & Howell, R.S. 1972 Dutch elm disease survey, 1971. *For. Comm. Forest Rec.* **82** 34pp HMSO, London.

Gibbs, J.N. & Howell, R.S. 1974 Dutch elm disease survey, 1972-73. *For. Comm. Forest Rec.* **100** 34pp HMSO, London.

Gibbs, J.N. & Inman, A 1991 The pine shoot beetle as a vector of blue stain fungus to windblown pine. *Forestry* **64** (3) 239-249.

Gibbs, J.N. & Lonsdale, D. 1996 Phytophthora disease of alder: the situation in 1995. *For. Comm. Res. Inf. Note* **277** Forestry Commission, Forest Research Station, Farnham.

Gibbs, J.N. & Palmer, J.N. 1994 A survey of damage to roadside trees in London caused by application of de-icing salt during the 1990/91 winter. *Arboricultural Journal* **18** (3) 321-343.

Gibbs, J.N. & Wainhouse, D. 1986 The spread of pathogens in the northern hemisphere. *Forestry* **59** (2) 142-153.

Gibbs, J.N., Brasier, C.M. & Burdekin, D.A. 1973 Dutch elm disease. *For. Comm. Rep. For. Res.* **1973** 94-97.

Gibbs, J.N., Burdekin, D.A. & Brasier, C.M. 1977 Dutch elm disease. *For. Comm. Forest Rec.* **115** 10pp HMSO, London.

Gibbs, J.N., Liese, W. & Pinon, J. 1984 Oak wilt for Europe? *Outlook for Agriculture* **13** 203-207.

Gibbs, J.N., Brasier, C.M. & Webber, J.F. 1994 Dutch Elm Disease in Britain. *For. Comm. Res. Inf. Note* **252** Forestry Commission, Forest Research Station, Farnham.

Gibbs, J.N., Greig, B.W. & Risbeth, J. 1996 Tree diseases of Thetford Forest and their influence on its ecology & management. *For. Comm. Technical Pap.* **13** 26-32 Arboricultural Advisory & Information Service, Alice Holt Lodge, Farnham, Surrey.

Gill, R.G.S. 1983 Comparison of production costs and genetic benefits of transplants and rooted cuttings. *Forestry* **56** (2) 61-74.

Gill, R.M.A. 1992a Review of damage by mammals in N. temperate forests. 1 Deer. *Forestry* **65** (2) 145-169.

Gill, R.M.A. 1992b Review of damage by mammals in N. temperate forests. 2 Small mammals. *Forestry* **65** (3) 281-308.

Gill, R.M.A. 1992c Review of damage by mammals in N. temperate forests. 3 Impact on trees and forests. *Forestry* **65** (4) 363-388.

Gillander, A.T. 1912 *Forest entomology.* William Blackwood & Sons, Edinburgh & London.

Gimingham, C.H. 1972 *Ecology of heathlands.* Chapman & Hall, London.

Godwin, H. 1975 *History of the British flora.* 540 pp Cambridge University Press

Godson, T. 1996 Defining the situations where pesticides may be used. *Aspects of applied biology* **44** 143-148.

Godzik, B 1997 Ground level ozone concentrations in the Krakov region, Southern Poland. *Environmental Pollution* **98** (3) 273-280.

Good, J.E.G. 1991 Air pollution and tree health in relation to arboriculture. (In) Research for practical arboriculture. *For. Comm. Bull.* **97** 107-119.

Goodey, J.B. 1965 The relationship between the nematode *Hoplolaimus uniformis* and Sitka spruce. (In) Experiments on nutrition problems in Forest Nurseries *For. Comm. Bull.* **37** 210-211 HMSO, London.

Gordon, A.G., Salt, G.A. & Brown, R.M. 1976 Effect of pre-sowing moist-chilling on seedbed emergence of Sitka spruce infected by *Geniculodendron pyrifome*, Salt. *Forestry* **49** (2) 143-151.

Goulding, K.W.T. & Stevens, P.A. 1988 Potassium reserves in a forested, acid upland soil & the effect of clear-felling v. whole-tree harvesting. *Soil Use & Management* **4** (2) 45-51.

Goulding, K.W.T., Bailey, N.J. *et al.* 1997 Nitrogen deposition & its contribution to N-cycling and associated soil processes. *New Phytologist* **139** 49-58.

Gowers, E. 1949 *Health and welfare of employed persons not covered by Factories Act or Mines & Quarries Act.* Cmd. Pap 7664 HMSO, London,.

Grace, J. 1995 The role of undisturbed forest as carbon sinks. *Science* **270** 778-780 ECTF, Edinburgh.

Gradwell, G. 1974 The effect of defoliators on tree growth. (In) *The British Oak* Proc. BSBI conf. (Eds.) M.G. Morris & F.H. Perring 182-193 E.W. Classey, for the Botanical Society of the British Isles.

Granfield, E.F. 1972 Protection of small steel structures from corrosion. *For. Comm. Forest Rec.* **81** 82pp HMSO, London.

Gray, W.G. 1953 Nursery notes on broadleaved trees. *For. Comm. Res. Br. Pap.* **10** 40pp Departmental Circulation, Forestry Commission, London.

Grayson, A.J. 1989a Carbon dioxide, global warming and forestry. *For. Comm. Res. Inf. Note* **146** 1-3 Forestry Commission, Forest Research Station, Farnham.

Grayson, A.J. (Ed.) 1989b The 1987 storm: impacts and responses. *For. Comm. Bull.* **87** HMSO, London.

Green, R.G. & Wood, R.F. 1956 Manuring of seedlings directly planted in the forest. *For. Comm. Rep. For. Res.* **1956** 132-138.

Greenwood, P. & Halstead, A. 1997 *The Royal Horticultural Society: Pests & diseases - the complete guide.* 1-224 Dorling Kindersley, London.

Gregoire, J-C. & Pasteels, J.M. (Eds) 1985 *Biological control of bark beetles.* Proc. seminar, CEC & Université Libre, Brussels. Commission of the European Communities, Brussels.

Gregory, S.C. 1983 *Armillaria. For. Comm. Rep. For. Res.* **1983** 33-34 HMSO, London.

Gregory, S.C. 1989 *Armillaria* species in Northern Britain. *Plant Pathology* **38** 93-97.

Gregory, S.C., Rose, D.R., Gibbs, J.N. & Winter, T.G. 1996 Health of non-woodland trees in England in 1996. *Arboriculture Research & Information Note.* **136/96.**

Gregory, S.C., MacAskill, G.A., Rose, D.R., Gibbs, J.N. & Tilbury, C.A. 1997 Health of non-woodland trees in England in 1997. *Arboriculture Research & Information Note.* **140/97.**

Greig, B.J.W. 1967 Honey fungus. *For. Comm. Leaflet* **6** 10 pp HMSO, London.

Greig, B.J.W. 1976a Biological control of *Fomes annosus* by *Peniophora gigantea. European Journal of Forest Pathology* **6** (2) 65-71.

Greig, B.J.W. 1976b Inoculation of pine stumps by chainsaw felling. *European Journal of Forest Pathology* **6** (5) 286-290.

Greig, B.J.W. 1978 Chemical, biological & silvicultural control of *Fomes annosus. Proc. 5th Int. Conf on problems in Root and butt rot in conifers. Kassel 1978* IUFRO working party S2.o6.01 75-84.

Greig, B.J.W. 1984 Management of East England pine plantation affected by *Heterobasidion annosus* root rot. *European Journal of Forest Pathology* **14** 393-397.

Greig, B.J.W. 1985a Ceratotect - a fungicide treatment for Dutch elm disease. *Arboriculture Research Note* **61/85/PATH** Forestry Commission, Edinburgh.

Greig, B.J.W. 1985b Management of east England pine plantations affected by *Heterobasidion annosum* root rot. *European Journal of Forest Pathology* **14** (6) 41-48.

Greig, B.J.W. 1986 Further experiments with thiabendazole for the control of Dutch elm disease. *Arboricultural Journal* **10** 191-201.

Greig, B.J.W. 1990 Ceratotect - a fungicide treatment for Dutch elm disease. *Arboriculture Research Note* **61/90/PATH** Forestry Commission, Farnham.

Greig, B.J.W. 1992 Twentieth anniversary of the Dutch Elm Disease Control Orders. *Arboricultural Journal* **16** (3) 253-254.

Greig, B.J.W. 1996 Fungicide treatments for the control of Dutch elm disease. *Arboriculture Research Note* **61/96/PATH** 4 pp.

Greig, B.J.W. & Coxwell, R.A.G. 1983 Experiments with thiabendazole for the control of Dutch elm Disease. *Arboricultural Journal* **7** 119-126.

Greig, B.J.W. & Low, J.D. 1975 An experiment to control *Fomes annosus* in second rotation pine crops. *Forestry* **48** (2) 147-163.

Greig, B.J.W. & Pratt, J.E. 1973 Death & decay caused by *Fomes annosus. For. Comm. Rep. For. Res.* **1973** 94.

Greig, B.J.W. & Redfern, D.B. 1974 *Fomes annosus. For. Comm. Leaflet* **5** HMSO, London.

Greig, B.J.W. & Strouts, R.G. 1983 Honey fungus. *For. Comm. Arboricultural Leaflet* **2** HMSO, London.

Greig, B.J.W., Gregory, SC. & Strouts, R.G. 1991 Honey fungus. *For. Comm. Bull.* **100** 1-11 HMSO, London.

Gribben, J. 1992 Decline & fall of atmospheric oxygen. *New Scientist* **12 Sept. 1992** 16.

Gribbin, J. & M. 1996 The greenhouse effect. *New Scientist* **6.7.96** (Inset) 4pp.

Grieve, I.C. 1978 Some effects of plantation conifers on freely drained lowland soil. *Forestry* **51** (1) 21-28.

Griffin, E, Carey, M.L. & McCarthy, R.G. 1984 Treatment of checked Sitka spruce crops in Republic of Ireland. (In) Weed control & vegetation management in forests & amenity areas *Aspects of applied biology.* **5** 211-222 Association of Applied Biologists, Warwick.

Gritten, R. 1994 Menzi takes a walk on the wild side. *Forestry & British Timber* **Feb. 1994** 40-41.

Grodzinska, K., & Szarek-Lukaszewska, G. 1997 Polish mountain forests, past, present & future. *Environmental Pollution* **98** (3) 369-374.

Guillebaud, W.H. 1937 Green manuring in forest nurseries. *For. Comm. Journal* **16** Departmental publication.

Gurnell, J. & Pepper, H.W. 1988 Perspectives in the management of red & grey squirrels. Proc Conf. *Wildlife management in forests.* (Ed.) J. Jardine. 92-109 Institute of Chartered Foresters, Edinburgh.

Gurnell, J. & Pepper, H.W. 1991 Conserving red squirrel. *For. Comm. Res. Inf. Note* **205** Forestry Commission, Farnham.

Gurnell, J. & Pepper, H.W. 1998 Grey squirrel damage to broadleaf woodland in the New Forest. *Quarterly Journal of Forestry* **92** (2) 117-124.

Hagner, S. 1966 Timber production by forest fertilization. *Proc. Fertilizer Society* **94** Fertilizer Society of London.

Hague, B. 1993 A growing issue. *REView (Renewable Energy view)* **20** 3-5 Department of Trade & Industry.

Hamilton, G.J. & Christie, J.M. 1971 Forest management tables. *For. Comm. Booklet* **34** HMSO, London.

Hance, R.J. 1989 Accuracy and precision in pesticide analysis. *Pesticide Outlook* **1** (1) 23-26.

Hance, R.J. & Holly, K. (Eds.) 1990 *Weed control handbook: principles.* 8th edn. Blackwell, Oxford.

Hand, S.C., Ellis, N.W. & Stoakley, J.T. 1987 Development of a pheromone monitoring system for monitoring the winter moth *Operophthera brumata. Crop Protection* **6** (3) 191-196.

Handley, W.R.C. 1954 Mull & mor formation in relation to forest soils. *For. Comm. Bull.* **23** 115 pp HMSO, London.

Handley, W.R.C. 1961 Further evidence for the importance of residual leaf litter complexes in litter decomposition and supply of nitrogen for plant growth. *Plant & Soil* **1961** 37-73.

Handley, W.R.C. 1963 Mycorrhizal associations and *Calluna* heathland afforestation. *For. Comm. Bull.* **36** HMSO, London.

Handley, J.F. & Perry, D. 1998 Woodland expansion on damaged land - reviewing the potential. *Quarterly Journal of Forestry* **92** (4) 297-306.

Hands, R.G. 1985 Wood utilisation; transmission/telecommunication poles. *For. Comm. Rep. For. Res.* **1985** 45 HMSO, London.

Hansen, E.A., Dawson, D.H. & Tolsted. D.N. 1980 Irrigation of intensively cultured plantations with paper mill effluent. *Tappi* **63** (1) 139-143.

Hansen, E.M. 1985 Forest pathogens of NW North America and their potential for damage in Britain. *For. Comm. Forest Rec.* **129** 14pp HMSO, London.

Hanson, H.S. 1949 Entomology. *For. Comm. Rep. For. Res.* **1949** 18-26 HMSO, London.

Hanson, H.S. 1950 Forest Entomology. *For. Comm. Rep. For. Res.* **1950** 83-91 HMSO, London.

Hanson, H.S. 1951 *Megastigmus* seedflies. *For. Comm. Res. Br. Pap.* **2** 6 pp Departmental Circulation, Forestry Commission, London.

Hanson, P.J., Rott, K., Taylor, G.E. *et al.* 1989 NO_2 deposition to elements representative of forest landscape. *Atmospheric Environment* **23** (8) 1783-1794.

Harcharik, D.A. 1997 The future of world forestry: sustainable forest management. *Unasylva* **48** (190/191) 4-8 FAO

Harding, J. & Adam, G. 1994 Respacing of Sitka natural regeneration. *Forestry & British Timber* Sept. **1994** 22-25.

Harley, J.L. & Smith, S.E. 1983 *Mycorrhizal symbiosis.* Academic Press, London.

Harley, J.L., McCready, C.C. & Brierley, J.K. 1958 Uptake of phosphorus by excised mycorrhizal roots of beech; VIII Translocation of phosphorus. *New Phytologist* **57** 353.

Harmer, R. 1991 Vegetative propagation of oak from coppice shoots. *For. Comm. Res. Inf. Note* **198/91** 1-3 Forestry Commission, Forest Research Station, Farnham.

Harmer, R. & Alexander, I. 1986 Effect of starch amendment on nitrogen mineralisation from the forest floor beneath a range of conifers. *Forestry* **59** (1) 39-46.

Harriman, R & Morrison, B.R.S. 1981 Forestry, fisheries and acid rain in Scotland. *Scottish Forestry* **35** 89-95.

Harriman, R. 1978 Nutrient leaching from fertilized forest watersheds in Scotland. *Journal of Applied Ecology* **15** 933-942.

Harriman, R. 1994 The acidification issue - the role of forestry & the freshwater critical loads approach. (In) *Forests & Water.* Proc. discussion meeting. (Ed.) I.R. Brown March 1994 104-111 Institute of Chartered Foresters, Edinburgh.

Harriman, R. & Christie, A.E.G. 1995 Estimating critical loads for biota: the steady-state water chemistry model. (In) *Critical loads for acid deposition for United Kingdom freshwaters* 7-8 Dept. of the Environment (Critical Loads Advisory Group)

Harriman, R. & Morrison, B.R.S. 1982 Ecology of streams draining forested and non-forested catchments in an area of central Scotland subject to acid precipitation. *Hydrobiologia* **88** 251-263.

Harriman, R., Ferrier, R.C., Jenkins, A. & Miller, A.D. 1990 Long- and short-term budgets in Scottish catchments. (In) *The surface water acidification programme.* (Ed.) B.J. Mason. 31-43 Cambridge University Press.

Harris, E.H.M. 1996 The European white elm (*Ulmus laevis*) in Britain. *Quarterly Journal of Forestry* **90** (2) 121-125.

Harris, E.H.M. 1997 Grey squirrel control. *Quarterly Journal of Forestry* **91** (4) 337-8.

Harris, J.G. 1997 Microbial insecticides - an industry perspective. (In) H.F. Evans (Ed.) *Symposium proceedings* **68** 41-50 British Crop Protection Council, Farnham, Surrey.

Harris, E. & Harris, J. 1991 *Wildlife conservation in managed woodlands & forests.* 358 pp Blackwell, Oxford.

Harrison, R.M. 1996 Urban air pollution and public health. (In) *Urban air pollution & public health.* (Eds.) C.J. Curtis, J.M. Reed et al. 9-14 Ensis Publishing, London, for University College, London.

Harrison, A.F., Miles, J. & Howard, D.M. 1988 Phosphorus uptake by birch from various depths in the soil. *Forestry* **61** (4) 349-358.

Harrison, A.F., Howard, P.J.A., et al 1995a Carbon storage in forest soils. *Forestry* **68** (4) 335-348.

Harrison, A.F., Stevens, P.A. et al 1995b Critical load of nitrogen for Sitka spruce forests on stagnopodsols in Wales; role of nutrient limitations. *Forest Ecology & Management* **76** 139-148.

Hart, C.E. 1967 *Practical forestry for the land agent & surveyor.* 217pp Estates Gazette Ltd, London.

Hart, C.E. 1995 *The Forest of Dean - New History 1550-1818.* 330pp Alan Sutton Publishing Co., Stroud, Gloucestershire.

Hawksworth, D.L. & Rose, F. 1970 Qualitative scale for estimating sulphur dioxide air pollution in England and Wales using epiphytic lichens. *Nature* **227** 145-8.

Hecht, J. 1994 Clouds hold key to Global warming theory. *New Scientist* **22 Jan. 1994** 16.

Hemsley, S. 1998 A coppice industry. *Timber & Wood Products* **1/8 Aug 1998** 12-13.

Hendry, A. 1992 Pine beauty is on the up. *Entopath News* **98** 6-7 Forestry Commission Research Station, Farnham, Surrey.

Hendry, G.A.F. 1993 Climatic change in perspective - vegetation responses to global climate changes of the past. Global climatic change - implications for crop protection. *Proc. BCPC Symposium.* **Monograph 56** 57-70 British Crop Protection Council, Farnham, Surrey.

Henningsen, J. 1990 European Commission objectives and the Single European Act. Future changes in pesticide registration in the EC *Proc. BCPC Conf.* **Monograph 44** 1-10 British Crop Protection Council, Farnham, Surrey.

Henriksen, A. Dickson, W. & Brakke, D.F. 1986 Estimation of critical loads to surface waters (In) (Ed. J. Nilsson) *Critical loads for sulphur & nitrogen.* 87-120 Nordic Council of Ministers, Copenhagen.

Heritage, S.G. 1989 Monitoring the efficacy of the Electrodyn Conveyor sprayer during 1988. Departmental report. 2pp Forestry Commission Research Station, Farnham, Surrey.

Heritage, S.G. 1996 Protecting plants from damage by the large pine weevil and black pine beetles. *For. Comm. Res. Inf. Note* **268** 8pp Forestry Commission, Forest Research Station, Farnham.

Heritage, S.G. 1997a Protecting plants from weevil damage by dipping or spraying using aqueous insecticides. *For. Comm. Res. Inf. Note* **270** Forestry Commission, Forest Research Station, Farnham.

Heritage, S. 1997b Pine beauty moth: its biology, monitoring & control. *For. Comm. Res. Inf. Note* **290** 7pp Forestry Commission, Farnham, Surrey

Heritage, S.G. & Johnson, D. 1997 Use of post-planting sprays to improve protection of plants from damage by *Hylobius abietis*. *For. Comm. Res. Inf. Note* **272** 12pp Forestry Commission, Edinburgh

Heritage, S.G. & Stoakley, J.T. 1988 Entomology - the Pine weevil and black pine beetles. *For. Comm. Rep. For. Res.* **1988** 46 HMSO, London.

Heritage, S.G., Collins, S. & Evans, H.F. 1989 A survey of damage by *Hylobius abietis* and *Hylastes* spp in Britain. (In) *Insects affecting reforestation: biology and damage* (Eds.) R.I. Alfaro & S.G. Glover 36-42 Forestry Canada, Victoria, British Columbia.

Heritage, S.G., Collins, S.A., Jennings, T. & Watt, I. 1990 Nematodes against *Hylobius*. *For. Comm. Rep. For. Res.* **1990** 58-59.

Heritage, S., et al 1993 Control of restocking pests: biological control against *Hylobius abietis*. *For. Comm. Rep. For. Res.* **1993** 3.

Heritage, S.G., Hendry, A., Moore, R. *et al*. 1994a Control of Pine beauty moth in Scotland. *For. Comm. Rep. For. Res.* **1994** 6-7 HMSO, London.

Heritage, S.G., Hendry, A., Moore, R., Johnson, D. & Evans, H. 1994b Entomology: Pine beauty moth *Panolis flammea*. *For. Comm. Rep. For. Res.* **1994** 6-7 HMSO, London.

Heritage, S.G., Moore, R., Brixey, J & Henry, C. 1995 Possibilities for the use of biological agents against *Hylobius abietis*. *For. Comm. Rep. For. Res.* **1995** 5-8 HMSO, London.

Heritage, S.G., Johnson, D. & Jennings, T. 1997a The use of Marshall suSCon granules to protect plants from *Hylobius* damage. *For. Comm. Res. Inf. Note* **269** Forestry Commission, Forest Research Station, Farnham.

Heritage, S.G., Johnson, D. & Jennings, T. 1997b The use of the Electrodyn sprayer conveyor to protect plants from *Hylobius* damage. *For. Comm. Res. Inf. Note* **271** Forestry Commission, Forest Research Station, Farnham.

Hewitt, E.J. 1966 A physiological approach to the study of tree nutrition. *Forestry* **1966 Supplement** 49-59.

Hewson, A. 1996 On top of the pots. *Horticulture Week* **28 Nov. 1996.**

Heybroek, H.M. 1983 Resistant elms for Europe. (In) *For. Comm. Bull.* **60** 108-113 HMSO, London.

Hibberd, B.G. (Ed.) 1986 Forestry practice. 10th Edn. *For. Comm. Bull.* **14** HMSO, London.

Hibberd, B. 1988 Farm woodland practice. *For. Comm. Handbk.* **3** 106pp HMSO, London.

Hibberd, B.G. (Ed.) 1989 Urban forestry practice. *For. Comm. Handbk.* **5** 150pp HMSO, London.

Hinson, W.H. & Reynolds, E.R.C. 1958 Cation adsorption & forest fertilisation. *Chemistry & Industry* **1958** (7) 194-196

Hodge, S.J. 1990 Organic soil amendments for tree establishment. *Arboriculture Research Note* **86/90/ARB** Arboricultural Advisory & Information Service, Farnham, Surrey.

Hodge, S.J. 1991a Improving growth of established amenity trees: site physical condition. *Arboriculture Research Note* **102/91/ARB** Arboricultural Advisory & Information Service, Farnham, Surrey.

Hodge, S.J. 1991b Improving growth of established amenity trees: fertilizer & weed control. *Arboriculture Research Note* **103/91/ARB** Arboricultural Advisory & Information Service, Farnham, Surrey.

Hodge, S.J. 1993 Nutrient injection into trees. *Arboriculture Research Note* **112/93/ARB** Arboricultural Advisory & Information Service, Farnham, Surrey.

Hodge, S.J. 1995 Creating & managing woodlands around towns. *For. Comm. Handbook* **11** 176pp HMSO, London.

Hodge, S. & Pepper, H.W. 1998 The prevention of mammal damage to trees in woodland. *Forestry Authority Practice Note* **FCPN 3** 12 pp Forestry Commission, Edinburgh.

Hodge, S.J., Patterson, G. & McIntosh, R. 1998 The approach to the British Forestry Commission to the conservation of forest diversity. *Scottish Forestry* **52** (1) 30-36.

Hodgkiss, D.L. 1990 Recovery from wearing protective clothing and equipment: a scientific review. *Proc. FAO/ECE/ILO seminar on forest technology, management and training.* 233-239 Forestry Commission, Edinburgh.

Holden, A.V. & Bevan, D. (Eds.) 1978 Control of Pine beauty moth by fenitrothion in Scotland. *For. Comm. Occasional Pap.* **4** Forestry Commission, Edinburgh.

Holden, A.V. & Bevan, D. (Eds.) 1981 Aerial application of insecticide against Pine beauty moth. *For. Comm. Occasional Pap.* **11** Forestry Commission, Edinburgh.

Holden, J.M., Thomas, G.W. & Jackson, R.M. 1983 Effect of mycorrhizal fungi on the growth of Sitka spruce seedlings in different soils. *Plant & Soil* **71** 313-317.

Holdenrieder, O. & Greig, B.J.W. 1998 Biological methods of control. (In) *Heterobasidion annosum, biology, ecology, impact & control* (Eds.) S. Woodward *et al.* 235-258 CAB International.

Holdgate, M. 1995 Greenhouse gas balance in forestry. *Forestry* **68** (4) 297-302.

Hollingsworth, M.K. & Mason, W.L. 1989 Provisional regimes for growing containerised Douglas fir and Sitka spruce. *For. Comm. Res. Inf. Note* **141** 4 pp.

Hollis, G. 1990 Status of proposed registration directive and views of member states - the UK view. *Future changes in pesticide registration in the EC Proc. BCPC Conf.* **Monograph 44** 19-21 British Crop Protection Council, Farnham, Surrey.

Hollis, J.M. 1991 Mapping vulnerability of aquifers & surface waters to pesticide contamination at national/regional scale. (In) *Pesticides in soils & water: current perspectives Proc. BCPC Conf. Monograph* **47** 165-174 British Crop Protection Council, Farnham, Surrey.

Holmes, G.D. 1952 Experiments on the chemical control of hazel. *For. Comm. Res. Br. Pap.* **9** 15 pp Departmental Circulation, Forestry Commission, London.

Holmes, G.D. 1956a Experiments on the chemical control of *Rhododendron ponticum. Proc. British Weed Control Conference* **1956.**

Holmes, G.D. 1956b Chemical weed control, bark peeling & animal repellents. *For. Comm. Rep. For. Res.* **1956** 47-48 HMSO, London.

Holmes, G.D. 1957a Notes on Forest use of chemicals for control of unwanted trees and woody growth. *For. Comm. Res. Br. Pap.* **21** 15pp Departmental Circulation, Forestry Commission, London.

Holmes, G.D. 1957b Experiments on the chemical control of *Rhododendron ponticum. For. Comm. Forest Rec.* **34** 1-8 HMSO, London.

Holmes, G.D. 1961 Preliminary trials of chemicals for debarking hardwood pulpwood. *For. Comm. Rep. For. Res.* **1961** 184-196 HMSO, London.

Holmes, G.D. 1962a Trials of 2,4,5-T for removal of epicormic shoots on hardwoods. *For. Comm. Rep. For. Res.* **1962** 156-162 HMSO, London.

Holmes, G.D. 1962b Fire control. *For. Comm. Rep. For. Res.* **1962** 43 HMSO, London.

Holmes, G.D. & Barnsley, G.E. 1953 Control of heather with 2,4-D. *Proc. British Weed Control Conference* **1953** 289-296.

Holmes, G.D. & Cousins, D.A. 1960 Applications of fertilizers to checked plantations. *Forestry* **33** (1) 54-73.

Holmes, G.D. & Fourt, D. 1960 The use of herbicides for controlling vegetation in forest fire breaks and uncropped land. *For. Comm. Rep. For. Res.* **1960** 119-137 HMSO, London.

Holmes, G.D. & Ivens, G.W. 1952 Chemical control of weeds in nursery seedbeds. *For. Comm. Forest Rec.* **13** HMSO, London.

Holmes, G.D., Brown, R.M. & Ure, R.M. 1958 Preliminary experiments on the use of toxaphene and endrin (against) short-tailed voles. *For. Comm. Rep. For. Res.* **1958** 148-156 HMSO, London.

Holtam, B.W. 1966 Blue stain. *For. Comm. Leaflet* **53** HMSO, London.

Hornung, M. 1985 Acidification of soils by trees and forests. *Soil Use & Management* **1** (1) 24-28.

Hornung, M. & Adamson, J. 1991 Impacts of forestry expansion on water quality & quantity. *For. Comm. Occasional Pap.* **42** 1-38 Forestry Commission, Edinburgh.

Hornung, M. & Skeffington, R.A. (Eds.) 1993 Critical loads: concepts & applications. (Proceedings of symposium) *Institute of Terrestrial Ecology Symposium* **28** HMSO, London.

Howard P.J.A. & Howard, D.M. 1990 Titratable acids and bases in tree and shrub leaf litters. *Forestry* **63** (2) 177-196.

HSC 1988 Safety representatives & safety committees. *Approved Code of Practice* HMSO, London.

HSC 1991a The safe use of pesticides for non-agricultural purposes. *Approved Code of Practice* **L9** Health & Safety Executive

HSC 1991b First aid at work. *Approved Code of Practice* **COP 42** HMSO, London.

HSC 1992a Management of Health & Safety at Work. *Approved Code of Practice.* **L21** 25pp Health & Safety Executive.

HSC 1992b Guide to the Health & Safety at Work Act, 1974. *HSE guide* **L1 5th edn.** 1-32+6 Health & Safety Executive.

HSC 1995a Safe use of pesticides for non-agricultural purposes. *Approved Code of Practice under COSHH Regulations, 1994.* **L9(Rev)** 46pp Health & Safety Executive.

HSC 1995b (revised 1997) General COSSH ACOP & Carcinogens ACOP & Biological agents ACOP. *Approved Codes of Practice under COSHH Regulations, 1994.* **L5** 74pp Health & Safety Executive.

HSC 1996a *Information approved for the classification, packaging and labelling of dangerous substances for supply & conveyance by road.* **3rd Edn** Health & Safety Commission, London.

HSC 1996b Workplace health, safety and welfare. Workplace (H.S & W) regulations, 1992. *Approved Code of Practice and guidance.* **L24** 6th imp. 1-51 Health & Safety Executive.

HSC 1996c Proposals for amendment to COSHH Regs. 1994 and Approved Code of Practice (General ACOP) *Consultative Document.* **CD 103** 1-13 HMSO, London.

HSC 1996d COSHH in fumigation operations. *Approved Code of Practice under COSHH Regulations, 1994.* **L86** Health & Safety Executive.

HSC 1997 The management of Health & Safety at Work Regulations, Updated. *HSC* **MISC079** Health & Safety Executive.

HSC 1998a *Information approved for the classification and labelling of dangerous substances for supply and conveyance by road.* Authorised and approved list. **4th edn** HMSO, London.

HSC 1998b Safe use of work equipment. *Approved Code of Practice and guidance.* **L22** Health & Safety Executive.

HSE 1988 Storage of approved pesticides: guidance for farmers & other professional users. *Guidance Note* **CS19** Health & Safety Executive

HSE 1989a Hazard & risk explained. *Leaflet* **IND(G)67(L)** Health & Safety Executive

HSE 1989b Poisoning by pesticides - First aid. *HSC* **MS(B)7** Health & Safety Executive

HSE 1992 Training in the use of pesticides. *HSE Agricultural safety leaflet* **AS 25 (rev)** Health & Safety Executive.

HSE 1994a Preventing asthma at work: how to control respiratory sensitisers. *HSE Leaflet* **L 55** Health & Safety Executive

HSE 1994b *Review of Health & Safety regulations.* 1-12 Health & Safety Executive.

HSE 1994c Health & safety guide for gamekeepers *HSE guide* **C300:IND(G)177L** 1-12 Health & Safety Executive.

HSE 1994d 7 steps to successful substitution of hazardous substances. *Booklet* **HS(G) 110** Health & Safety Executive.

HSE 1995a Staying healthy: a guide for workers in farming, forestry & horticulture. *Leaflet* **C250** 24 pp Health & Safety Executive

HSE 1995b Safety data sheets for substances & preparations dangerous for supply. *HSE guidance on Reg 6 of the CHIP regulations, 1994* **L 62** Health & Safety Executive.

HSE 1995c Agricultural pesticides. *HSE Agricultural safety leaflet* **AS 27 (rev)** Health & Safety Executive.

HSE 1995d *Health risk management: a practical guide for managers in small and medium sized enterprises.* HS(G) 137 49pp Health & Safety Executive.

HSE 1995e Farm forestry operations. *Agricultural safety leaflet* **AS 15** 12 pp Health & Safety Executive.

HSE 1995f Personal Protective Equipment at Work Regulations, 1992. *HSE guidance on the regulations; (5th impression with revisions).* **L25** 48pp Health & Safety Executive.

HSE 1995g *Essentials of health & safety at work.* Health & Safety Executive.

HSE 1995h Biological monitoring of workers exposed to organo-phosphorus pesticides. *HSE guidance note* **MS 17 (2nd rev)** 4pp Health & Safety Executive.

HSE 1995i *Health surveillance under COSHH: Guide for employers.* Health & Safety Executive

HSE 1996a Everyone's guide to RIDDOR (Reporting of injuries, diseases and dangerous occurrences regulations 1995) *HSE leaflet* **567** Health & Safety Executive.

HSE 1996b Guide to the Genetically Modified Organisms (Contained Use) Reg's 1992, amended 1996. *HSE Leaflet* **L29** 56pp Health & Safety Executive.

HSE 1996c A short guide to the Personal Protective Equipment at Work Regulations 1992. *Guidance Note* **IND(G)1754(l)** (Rev. 3/96) Health & Safety Executive

HSE 1996d Proposals for amendment to COSHH Regs. 1994 and Approved Code of Practice (General ACOP) *Consultative Document.* **CD 103** 1-13 HMSO, London.

HSE 1997a Guidance on storing pesticides for farmers & other professional users. *HSE information sheet* **AIS 16** 4pp Health & Safety Executive.

HSE 1997b A step-by-step guide to COSHH assessment. *Guidance Note* **HS(G)97** Health & Safety Executive

HSE 1997c Gassing of rabbits & vertebrate pests. *Leaflet* **AIS22** Health & Safety Executive.

HSE 1998a *Occupational exposure limits, 1998* Issued annually. **EH40/98** 61pp Health & Safety Executive.

HSE 1998b HSE seeks public evidence for its review of organophosphate and carbamate pesticides. *HSE News Release* **E232:98 15.10.98** Health & Safety Executive.

HSE 1999a Reporting incidents of exposure to pesticides and veterinary medicines. *Leaflet* **INDG 141** (rev.) Health & Safety Executive.

HSE 1999b Managing health & safety in forestry *Leaflet* **INDG 294** Health & Safety Executive.

HSE 1999c COSHH in forestry *HSE Agricultural safety leaflet.* **AS30 (rev)** 12pp Health & Safety Executive.

HSE 1999d RIDDOR explained. *Leaflet* **HSE 31 (rev.1)** 12pp Health & Safety Executive.

Hudson, G.H. 1990 Status of the proposed directive. *Future changes in pesticide registration in the EC Proc. BCPC Conf.* **Monograph 44** 13-17 British Crop Protection Council, Farnham, Surrey.

Hull, S.K. & Gibbs, J.N. 1991 Ash dieback a survey of non-woodland trees. *For. Comm. Bull.* **93** 32pp HMSO, London.

Hultberg, H., Dise, N.B., Wright, R.F., Anderson, I. & Nystrom, U. 1994 Nitrogen saturation induced during winter by experimental NH_4NO_3 addition to a forested catchment. *Environmental Pollution* **86** 145-147.

Humphries, S. 1983 Managing small woods for fuel. *Proc. Conference 'Growing wood for fuel'* **Paper 7** 5 pp National Agricultural Centre, Stoneleigh, Warwick.

Hunter, C. & McDonald, A. 1992 Plantation forestry & aluminium in upland streams. *Quarterly Journal of Forestry* **86** (2) 114-115.

Hunter, I. 1997 Europe's forests are growing faster - but why? *EFI (European Forest Institute News)* **5** (1) 3, 16.

Hurka, T 1998 Sustainable development: what do we owe to future generations. *Unasylva* **187** 38-43.

Hussey, N.W. & Scopes, N. 1985 *Biological pest control - the glasshouse experience.* Blandford.

ICF 1996 *The UK Forestry Accord.* Institute of Chartered Foresters, 7A St. Colme St., Edinburgh EH3 6AA.

ICF 1998 *Report of the UK Forestry Accord workshops.* 48 pp UK Forestry Accord, 3 St Colme Road, Dalgety Bay, Fife KY11 9LH.

ICRCL 1987 Guidance on the assessment and redevelopment of contaminated land. *Interdepartmental committee on the redevelopment of contaminated land.* **Guidance Note 59/83 2nd edn.** Dept of the Environment, London.

ICRCL 1990 Notes on the restoration and aftercare of metalliferous mining sites for pasture & grazing. *Interdepartmental committee on the redevelopment of contaminated land.* **Guidance Note 70/90** Dept of the Environment, London.

IGBP 1998 The terrestrial carbon cycle implications for the Kyoto protocol. *Science* **280** 29 May 98.

IH/EA (Inst. of Hydrology & Environment Agency) 1998 Broadleaf woodlands. Implications for water quality & health. *R & D Publication* **No 5** Environment Agency, Bristol.

IIED (International Institute for Environment & Development) 1996 *Towards a sustainable paper cycle.* World Business Council for Sustainable Development, Poole, Dorset.

INDITE 1994 *Impacts of Nitrogen Deposition in Terrestrial Ecosystems.* Report of the UK Review Group on Impacts of Atmospheric Nitrogen 110pp Dept. of the Environment, London.

Ingoldby, M.J.R. & Smith, R.O. 1982 Forest fire fighting with foam. *For. Comm. Leaflet* **80** 19pp HMSO, London.

Innes, J.L. 1987a Interpretation of international forest health data. (In) *Acid rain: scientific and technical advances.* (Eds) R. Perry, R.M. Harrison, J.N.B. Bell & J.N. Lester 633-640 Selper Ltd., London.

Innes, J.L. 1987b Air pollution and forestry. *For. Comm. Bull.* **70** 40pp HMSO, London.

Innes, J.L. 1987c Acid rain, pollution and trees. *Quarterly Journal of Forestry* **81** 191-193.

Innes, J.L. 1987d Surveys of tree health: 1987. *For. Comm. Res. Inf. Note* **117/87/SSS** 2pp Forestry Commission, Forest Research Station, Farnham.

Innes, J.L. 1988 Acid rain and other forms of air pollution - recent developments. *European Environmental Review* **1** (5) 38-45.

Innes, J.L. 1990a Assessment of tree condition. *For. Comm. Field Bk* **12** HMSO, London.

Innes, J.L. 1990b Surveys of tree condition: 1990. *Arboriculture Research Note* **83/90/PATH** 2pp Forestry Commission, Farnham.

Innes, J.L. 1992a Observations on the condition of beech (*Fagus sylvatica* L) in Britain in 1990. *Forestry* **65** (1) 35 - 60.

Innes, J.L. 1992b Forest condition and air pollution in the UK. *Forest Ecology & Management* **51** (1-3) 17-27.

Innes, J.L. & Boswell, R.C. 1987 Forest health surveys: 1987. Part 1: results. *For. Comm. Bull.* **74** HMSO, London.

Innes, J.L. & Boswell, R.C. 1991a Monitoring of forest condition in Great Britain 1990. *For. Comm. Bull.* **98** HMSO, London.

Innes, J.L. & Boswell, R.C. 1991b Monitoring of forest condition in Great Britain 1991. *For. Comm. Res. Inf. Note* **209** Forestry Commission, Farnham, Surrey.

Innes, J.L. & Boswell, R.C. 1988 Forest health surveys 1987. Pt 2 Analysis & interpretation. *For. Comm. Bull.* **79** 52 pp HMSO, London.

Innes, J.L. & Boswell, R.C. 1989 Monitoring of forest condition in Great Britain, 1988. *For. Comm. Bull.* **88** HMSO, London.

Innes, J.L. & Boswell, R.C. 1990 Monitoring of forest condition in Great Britain, 1989. *For. Comm. Bull.* **94** HMSO, London.

IPCC 1990 *Climatic change: the IPCC Scientific assessment.* (Eds) J.T. Houghton, G.J. Jenkins & J.J. Ephraums Cambridge University Press, Cambridge.

IPCC 1992 *Climatic change: Supplementary report to IPCC Scientific assessment.* (Eds) J.T. Houghton, B.A. Callander, & S.K. Varney Cambridge University Press, Cambridge.

Izat, R. 1996 Pest watch. *Horticulture week* **20.6.96** 31-33.

Jackson, J.E. 1994 The edible or fat dormouse in Britain. *Quarterly Journal of Forestry* **88** (2) 119-125.

James, H., Court, N.N., McLeod, D.A. & Parsons, J.W. 1978 Relationship between growth of Sitka spruce, soil factors & mycorrhizal activity on basaltic soils in W Scotland. *Forestry* **51** (2) 105-119.

Jane, Romola, 1987 *Impressions of the hurricane, October 16th,1987.* Privately printed All Saints Centre, Lewes Sussex.

Jarvis, P.G. 1989a Production efficiency of coniferous forest in the UK. (In) *Physiological processes limiting crop growth.* (Ed.) J.P. Cooper 81-107 Butterworth, London.

Jarvis, P.G. 1989b Atmospheric carbon dioxide and forests. *Phil. Trans. R. Soc. Lond.* **B 324** 369-392.

Jarvis, P.G. (Ed). 1998 *European forests & global change.* 380pp Cambridge University Press.

Jarvis, P.G. & Dewar, R.C. 1993 Forests in the global carbon balance: from stand to region. (In) *Scaling physiological processes* (Eds.) J.R. Ehleringer & C.B. Field 191-221 Academic Press Inc., San Diego.

Jeffree, M. 1997 Swedish forestry certification poised to get FSC approval. *Timber Trades Journal.* **1.11.97** 4.

Jenkins, A. 1995 Dynamic modelling of chemical responses to future changes in acid deposition and land-use. (In) *Critical loads for acid deposition for United Kingdom freshwaters* 35-45 Dept. of the Environment (Critical Loads Advisory Group)

Jinks, R.L. 1994 Container production of tree seedlings. *For. Comm. Bull.* **111** 122-134 HMSO, London.

JNCC 1996 *UK strategy for red squirrel conservation.* JNCC, Peterborough.

Jobling, J. 1960 Establishment methods for poplars. *For. Comm. Forest Rec.* **43** HMSO, London.

Jobling, J. 1981 Reworked spoil & trees. (In) Research for Practical Arboriculture. Proc. Seminar at Preston, 1980 *For. Comm. Occasional Pap.* **10** 76-92 Forestry Commission, Farnham, Surrey.

Jobling, J. 1983 Reclamation of Mineral workings to forestry - treatment of deep-mined colliery spoil. *For. Comm. Res. & Dev. Pap.* **132** 17-19 Forestry Commission, Edinburgh.

Jobling, J. 1987 Reclamation : colliery spoil. (In) Advances in practical arboriculture. (Ed) D. Patch. *For. Comm. Bull.* **65** 42-51 HMSO, London.

Jobling, J. 1990 Poplars for wood production and amenity. *For. Comm. Bull.* **92** 1-84 HMSO, London.

Jobling, J. & Carnell, R. 1985 Tree planting on colliery spoil. *For. Comm. Res. & Dev. Pap.* **136** 1-7 Forestry Commission, Edinburgh.

Jobling, J. & Stevens, F.R.W. 1980 Establishment and management of trees on reclaimed colliery spoil. *For. Comm. Occasional Pap.* **7** Forestry Commission, Edinburgh.

Johansson, M-B. 1995 The chemical composition of needle and leaf litter from Scots pine, Norway spruce and white birch in Scandinavian forests. *Forestry* **68** (1) 49-62.

Johnen, B.G. 1990 An industry view of data harmonisation. *Future changes in pesticide registration in the EC Proc.* BCPC Conf. **Monograph 44** 79-85 British Crop Protection Council, Farnham, Surrey.

Johnson, R.C. 1993 Effects of forestry on suspended solids & bedload yields in the Balquhidder catchments. *Journal of Hydrology* **145** 403-417.

Johnson, R.C. & Bronsdon, R.K. 1995 The erosion of forest roads: experimental results. *Prevention of pollution during forest engineering operations.* Proc. FAO/ECE/ILO seminar on forest technology, management and training. 14 pp Institute of Agricultural Engineers.

Johnson, R.C. & Whitehead, P.C. 1993 An introduction to the research in the Balquhidder experimental catchments. *Journal of Hydrology* **145** 231-238.

Jones, E.W. 1959 *Quercus* L. Biological flora of the British Isles. *Journal of Ecology* **47** 169-222.

Jones, O.T. 1987 The use of pheromone traps in monitoring Pine beauty moth populations. (In) *For. Comm. Bull.* **67** 46-48 HMSO, London.

Jones, P.D., Briffa, K.R., Barnett, T.P. & Tett, S.F.B. 1998 High-resolution palaeological records for the last millenium. *The Holocene* **8** (4) 455-471.

Juggins, S., Ormerod, S.J. & Harriman, R. 1995 Relating critical loads to aquatic biota. (In) *Critical loads for acid deposition for United Kingdom freshwaters* 9-12 Dept. of the Environment (Critical Loads Advisory Group)

Jukes, M.R. 1984 The Knopper gall. *Arboriculture Research Note* **55/84/ENT** DoE Arboricultural Advisory & Information Service, Forestry Commission, Farnham.

Karjalainen, R., Ernst, D. & Woodward, S. 1998 Molecular biology of host defence in *Heterobasidion annosum*. (In) Heterobasidion annosum *Biology, impact & control* (Eds S.Woodward *et al.*) CAB International.

Katzensteiner, K., Glatzel, G. & Kazda, M. 1992 Nitrogen induced nutritional imbalances - a contributing factor to Norway spruce decline in the Bohemian Forest. *Forest Ecology & Management* **51** 29-42.

Kaufmann, P.R., Herlihy, A.T. & Baker, L.A. 1992 Sources of lake acidity & streams of the United States. *Environmental Pollution* **77** 115-122.

Keay, J. 1964 Nutrient deficiencies in conifers. *Scottish Forestry* **18** (1) 22-29.

Keighley, G.D. 1985 *Wood as fuel.* (leaflet, not in any series) 4 pp Forestry Commission, Alice Holt Lodge, Farnham, Surrey.

Kenk, G. & Fischer, H. 1988 Effects from nitrogen fertilization in the forests of Germany. *Environmental Pollution* **54** 199-218.

Kenward, R.E. 1982 Bark stripping by grey squirrels - some recent research. *Quarterly Journal of Forestry* **76** (2) 108-121.

Kenward, R.E. & Dutton, J.C.F. 1996 Damage by grey squirrels. II The value of prediction. *Quarterly Journal of Forestry* **90** (3) 211-218.

Kenward, R.E., Parish, T. Holm, J. & Harris, E.H.M. 1988a Grey squirrel bark-stripping. I The roles of tree quality, squirrel learning & food abundance. *Quarterly Journal of Forestry* **82** (1) 9-20.

Kenward, R.E., Parish, T. & Doyle, F.I.B. 1988b Grey squirrel bark-stripping. II Management of woodland habitats. *Quarterly Journal of Forestry* **82** (2) 87-94.

Kenward, R.E., Dutton, J.C.F., Parish, T, et al. 1996 Damage by grey squirrels I Bark stripping correlates & treatment. *Quarterly Journal of Forestry* **90** (2) 135-142.

Kerin, D. 1971 Fume damage to forests by the lead & steel industry in the fore-alps region of Jugoslavia. (In) *For. Comm. Res. & Devel. Pap.* **82** 9.

Kerr, G. 1995 Silviculture of ash is southern England. *Forestry* **68** (1) 63-70.

Kerstiens, G., Townend, J., Heath, J. & Mansfield, T.A. 1995 Effects of water & nutrient availability on physiological response to elevated CO_2. *Forestry* **68** (4) 303-315.

Kidd, N.A.C. & Lewis, G.B. 1987 Influence of *P. sylvestris* chemistry on population dynamics of the Large pine aphid. *For. Comm. Rep. For. Res.* **1987** 77 HMSO, London.

Kidd, N.A.C., Lewis, G.B. & Thomas, M.B. 1987 Acid mist can increase damage to conifers. *For. Comm. Rep. For. Res.* **1987** 77 HMSO, London.

Kimmins, J.P.H. 1999 *Respect for nature - the essential foundation for sustainable forest management in Canada & British Columbia.* (Text of open lecture) 8 pp Canadian High Commission, London.

King, C.J. 1977 Chemical control: control of beetles in logs. *For. Comm. Rep. For. Res.* **1977** 37 HMSO, London.

King, C.J. 1987 *Rhizophagus grandis* as a means of biological control of *Dendroctonus micans* in Britain. *For. Comm. Res. Inf. Note* **124** Forestry Commission, Forest Research Station, Farnham.

King, C.J. & Fielding, N.J. 1989 *Dendroctonus micans* in Britain - its biology and control. *For. Comm. Bull.* **85** 1-8 HMSO, London.

King, C.J. & Heritage, S.G. 1979 Chemical control - *Hylobius abietis*. *For. Comm. Rep. For. Res.* **1979** 37 HMSO, London.

King, C.J. & Scott, T.M. 1975a Testing dosage rates of methoxychlor applied by helicopter for control of Dutch elm disease. *Forestry* **48** (2) 147-163.

King, C.J. & Scott, T.M. 1975b Control of Large pine weevil & bark beetles of the genus *Hylastes*. *Forestry* **48** (1) 87-97.

King, C.J., Evans H.F., Martin, A.F. & Fielding, J. 1984 Biological control - Great spruce bark beetle, *Dendroctonus micans*. *For. Comm. Rep. For. Res.* **1984** 40 HMSO, London.

King, C.J., Fielding, N.J. & O'Keefe, T. 1987 Biological control of *Dendroctonus micans. For. Comm. Rep. For. Res.* **1987** 49 HMSO, London.

Kleiner, K. 1994 Climate change threatens southern Asia. *New Scientist* **27.8.94** 6

Koch, A.S. & Matzner, E. 1993 Heterogeneity of soil and soil solution chemistry under Norway spruce & beech as influenced by distance from stem base. *Plant & Soil* **151** 227-237.

Korhonen, K., Lipponen, K., Bendz, M. *et al.* 1994 Control of *H. annosum* by stump treatment with 'Rotstop', a new commercial formulation of *Phlebopsis gigantea.* (In) *Proc. 8th IUFRO conf. on root and butt rots, Aug. 1993* (Eds.) M. Johansson & J. Stenlid 675-685.

Kronzucker, H.J. & Siddiqi, M.Y. 1997 Conifer root discrimination against soil nitrate and ecology of forest succession. *Nature* **385** 59-61.

Labous, P. & Willis, S. 1997 Spin out spins off. *Horticulture Week* **March 6 1997** 20-21.

Laidlaw, W.B.R. 1947 On the appearance of the bark beetle *Ips typographus* in Britain on imported timber. *Forestry* **20** (1) 52-56.

Landers, A.J. 1989 Injection closed system sprayers - a review. *Pesticide Outlook* **1** (1) 27-30.

Lane, P.B. 1983 Adaptation of ULVA forest tractor-mounted CDA to apply Medium volume gamma-HCH to plantation trees. *Forestry Commission Work Study Eastern Region Team Report 90 addendum 1* **Oct. 1983** 4pp Forestry Commission Departmental Use.

Lane, P.B. 1984 Chemical weeding: direct applicators. *For. Comm. Res. Inf. Note* **78/84/WS** Forestry Commission, Forest Research Station, Farnham.

Lane, P.B. 1990a Dribble bar for applying herbicides around sensitive species. *Eastern Region Work Study Team Report* **37/90** Forestry Commission, Edinburgh.

Lane, P.B. 1990b Chemical weeding - hand-held direct applicators. *Arboriculture Research Note* **53/89/ARB** Forestry Commission, Farnham.

Langan, S.J. & Hornung, M. 1992 An application and review of the critical load concept to the soils of northern England. *Environmental Pollution* **77** 205-210.

Last, F.T., Cape, J.N. & Fowler, D. 1986 Acid rain or 'pollution climate'? *Span* **29** 2-4.

Latham, P. 1994 Modern conservation practices and the image of the competent forester. *Quarterly Journal of Forestry* **89** (3) 217-223 Royal Forestry Society of England, Wales & N. Ireland.

Law, F. 1956 The effect of afforestation upon the yield of water catchment areas. *Journal of the British Waterworks Association* **38** 489-494.

Law, F. 1958 Measurement of rainfall, interception and evaporation loss in a plantation of Sitka spruce trees. Paper to 11th Gen Assembly Int. Assoc. Hydrology, Toronto, 1957 397-411.

Lawrie, J. & Clay, D.V. 1994a Tolerance of broadleaved and coniferous forestry species to herbicides with potential for bracken control. *Forestry* **67** (3) 237-244.

Lawrie, J. & Clay, D.V. 1994b Tolerance of 2-year-old forestry trees to five herbicides. *Forestry* **67** (4) 287-295.

Lawrie, J., Clay, D.V. & Nelson, D.G. 1992 Evaluation of herbicides, herbicide mixtures and adjuvants for the control of naturally regenerated Sitka spruce. (In) Vegetation management in forestry, amenity & conservation areas. *Aspects of applied biology* **29** 317-324 Association of Applied Biologists, Warwick.

Le Tacon, F. & Oswald, H. 1977 Influence de la fertilization minérale sur la fructification de la hêtre. *Annales Scientifiques Forestières* **34** (2)

Leather, S.R. 1986 *Panolis flammea* - a threat to Scottish forestry. *Antenna* **10** 167-170.

Leather, S.R. 1987 Lodgepole pine provenance & the Pine beauty moth. *For. Comm. Bull.* **67** 27-30 HMSO, London.

Leather, S.R. 1992 Forest management practice to minimise the impact of the Pine beauty moth. *For. Comm. Res. Inf. Note* **217** 3pp Forestry Commission, Forest Research Station, Farnham.

Leather, S.R. & Knight, J.D. 1997 Pines pheromones & parasites: a modelling approach to the integrated control of the Pine beauty moth. *Scottish Forestry* **51** (2) 76-83.

Leather, S.R., Stoakley, J.T. & Evans, H.F.(Eds) 1987a Population biology and control of the Pine beauty moth. *For. Comm. Bull.* **67** HMSO, London.

Leather, S.R., Watt, A.D. & Forrest, G.I., 1987b Insect-induced chemical changes in young Lodgepole pine: effect of previous defoliation. *Ecological Entomology* **12** 275-281.

Leather, S.R., Docherty, M., Walsh, P. & Aegerter, J. 1993 Population ecology of *Panolis flammea. For. Comm. Rep. For. Res.* **1993** 1-2.

Lee, J.A. & Caporn, S.J.M. 1998 Ecological effects of atmospheric reactive N deposition on semi-natural terrestrial ecosystems. *New Phytologist* **139** 127-134.

Lee, K, & Gibbs, J.N. 1996 Investigation of the influence of harvesting practice on development of blue-stain in Corsican pine logs. *Forestry* **69** (2)

Lee, L.S.J, Wilson, A., Benham, S.E., Durrant, D.W.H., Houston, T. & Waddell, D.A. 1990a Effect of air quality on tree growth. *For. Comm. Res. Inf. Note* **182** 4 pp.

Lee, L.S.J, Wilson, A., Benham, S.E., Durrant, D.W.H., Houston, T. & Waddell, D.A. 1990b Air quality on timing of tree shoot development. *For. Comm. Res. Inf. Note* **183** 6pp.

Lee-Harwood, B. 1991 Pesticides in water - an environmentalist's perspective. (In) Pesticides in soils & water: current perspectives *Proc. BCPC Conf. Monograph* **47** 209-210 British Crop Protection Council, Farnham, Surrey.

Leith, I.D., Murray, M.R., Sheppard, L.J. *et al.* 1989 Visible foliar injury of red spruce seedlings subject to simulated acid mist. *New Phytologist* **113** 313-320.

Lepp, N.W. 1996 Uptake, mobility & loci of concentration of heavy metals in trees. (In) *Heavy metals & trees* Proc. conf. Glasgow 1995 (Ed.) I. Glimmerveen. 68-92 Institute of Chartered Foresters, Edinburgh.

Levisohn, I. 1953 Growth response of tree seedlings to mycorrhizal mycelia in the absence of a mycorrhizal association. *Nature* **172** 316-317 London.

Levisohn, I. 1956 Growth stimulation of forest tree seedlings by the activity of free-living mycorrhizal mycelia. *Forestry* **29** (1) 53-59.

Levisohn, I. 1958 Effects of mycorrhiza on tree growth. *Soils & Fertilizers* **21** 73-82 HMSO, London.

Levisohn, I. 1965 Mycorrhizal investigations. (In) Experiments on nutrition problems in forest nurseries. *For. Comm. Bull.* **37** Vol 1 228-235 HMSO, London.

Leyton, L. 1950 Nutrient uptake of conifers. *For. Comm. Rep. For. Res.* **1950** 118-119.

Leyton, L. 1955 The influence of heather mulching on growth and nutrient status of Lawson cypress. *Forestry* **28** (2) 147-151.

Leyton, L. 1957 The relationship between growth and mineral composition of foliage of Japanese larch: II Evidence from manuring trials. *Plant & Soil* **9** 31.

Leyton, L. 1958 Mineral relationships of forest trees. *Forestry Abstracts* **9** 399-408.

Leyton, L. & Armson, K.A. 1955 Mineral composition of foliage in relation to growth of Scots pine. *Forest Science* **1** 210-218.

Leyton, L. & Reynolds. E.C.R. 1964 Hydrological relations of forest stands. *For. Comm. Rep. For. Res.* **1964** 116-120 HMSO, London.

Leyton, L. & Weatherell, J. 1959 Coniferous litter amendments and the growth of Sitka spruce. *Forestry* **32** (1) 7-13.

Leyton, L., Reynolds, E.C.R. & Thompson, F.B. 1965 Water relations of trees & forests. *For. Comm. Rep. For. Res.* **1965** 119-123 HMSO, London.

Liese, W. & Peeke, R.D. 1984 Experiences with wet storage of logs. *Dansk Skovforeningens Tidsskrift* **69** 73-91.

Lim, M.T. & Cousens, J.M. 1986a Internal transfer of nutrients in a Scots pine stand: I Biomass components, current growth and nutrient content. *Forestry* **59** (1) 1-16.

Lim, M.T. & Cousens, J.M. 1986b The internal transfer of nutrients in a Scots pine stand: II Patterns of transfer and effects of nitrogen availability. *Forestry* **59** (1) 17-27.

Lines, R. 1957 (In) Exotic forest trees in Great Britain *For. Comm. Bull.* **30** 57-66 HMSO, London.

Lines, R. 1962 Experiments with artificial shelters (In) Silvicultural investigations. *For. Comm. Rep. For. Res.* **1962** 40-41 HMSO, London.

Lines, R. 1976 Air pollution. *For. Comm. Rep. For. Res.* **1976** 17 HMSO, London.

Lines, R. 1977 Polluted sites. *For. Comm. Rep. For. Res.* **1977** 18-19 HMSO, London.

Lines, R. 1978 Polluted sites. *For. Comm. Rep. For. Res.* **1978** 17-18 HMSO, London.

Lines, R. 1979 Air pollution effects on trees. *For. Comm. Rep. For. Res.* **1979** 17-18 HMSO, London.

Lines, R. 1984 Species & seed origin trials in the industrial pennines. *Quarterly Journal of Forestry* **78** 9-23.

Lines, R. & Jobling, J. 1974 Air pollution *For. Comm. Rep. For. Res.* **1974** 15 HMSO, London.

Lisansky, S. 1997 Microbial pesticides. (In) *Microbial insecticides: novelty or necessity?* (Chaired) H.F. Evans *Symposium proceedings* **68** 3-10 British Crop Protection Council, Farnham, Surrey.

Livens, F.R. & Loveland, P.J. 1988 Influence of soil properties on the environmental mobility of caesium in Cumbria. *Soil Use & Management* **4** (3) 69-75.

Lloyd, H.G. & Hewson, R. 1986 The fox. *For. Comm. Forest Rec.* **131** 20 pp HMSO, London.

Lloyd, H.G. & Taylor, K.D. 1960 Grey squirrel research. *For. Comm. Rep. For. Res.* **1960** 71 HMSO, London.

Locke, G.M.L. 1987 Census of woodlands & trees 1979-82. *For. Comm. Bull.* **63** 124 pp HMSO, London.

Lockhart, J.A.R., Samuel, A. & Greaves, M.P. 1990 The evolution of weed control in British agriculture. (In) *Weed Control Handbook: Principles* (Eds.) R.J. Hance, & K. Holly. 8th edn 43-74 Blackwells Scientific Publications, Oxford.

Longhurst, C. & Billany, D.J. 1982 Biological control *Rhyaconia buoliana. For. Comm. Rep. For. Res.* **1982** 30 HMSO, London.

Lonsdale, D. 1980 *Nectria* infection of beech bark: variations in disease in relation to predisposing factors. *Annales des Sciences Forestières* **37** 307-317.

Lonsdale, D. 1983a A definition of the best pruning position. *Arboriculture Research Note* **48/83/PATH** Arboricultural Advisory & Information Service, Farnham, Surrey.

Lonsdale, D. 1983b Decay in amenity trees. *For. Comm. Rep. For. Res.* **1983** 34 HMSO, London.

Lonsdale, D. 1984a Decay in amenity trees. *For. Comm. Rep. For. Res.* **1984** 36-7 HMSO, London.

Lonsdale, D. 1984b The external signs of decay in trees. *Arboricultural Leaflet (2nd edition)* **1** HMSO, London.

Lonsdale, D. 1986a Beech health study, 1985. *For. Comm. Res. & Dev. Pap.* **146** Forestry Commission, Edinburgh.

Lonsdale, D. 1986b Beech health study, 1986. *For. Comm. Res. & Dev. Pap.* **149** Forestry Commission, Edinburgh.

Lonsdale, D. 1986c Decay in amenity trees. *For. Comm. Rep. For. Res.* **1986** 39-40 HMSO, London.

Lonsdale, D. 1987 Prospects for long-term protection against decay in trees. (In) Advances in practical arboriculture. *For. Comm. Bull.* **65** 149-155 HMSO, London.

Lonsdale, D. 1992 Treatment of tree wounds. *Arboriculture Research Note* **109/92/PATH** Arboricultural Advisory & Information Service, Farnham, Surrey.

Lonsdale, D. 1993a Comparison of 'target' pruning versus flush cuts & stub pruning. *Arboriculture Research Note* **116/93/PATH** Arboricultural Advisory & Information Service, Farnham, Surrey.

Lonsdale, D. 1993b Choosing time of year to prune trees. *Arboriculture Research Note* **117/93/PATH** Arboricultural Advisory & Information Service, Farnham, Surrey.

Lonsdale, D. & Hickman, I.T. 1988 Beech health study. *For. Comm. Rep. For. Res.* **1988** 42-44 HMSO, London.

Lonsdale, D. & Pratt, J.E. 1981 Some aspects of growth of young beech trees and incidence of beech bark disease on chalk soils. *Forestry* **54** (2) 183-195.

Lonsdale, D. & Tabbush, P.M. 1998 Poplar rust & its recent impact in Great Britain. *For. Comm. Information Note* **FCIN 7** 4pp.

Lonsdale, D. & Wainhouse, D. 1987 Beech bark disease. *For. Comm. Bull.* **69** 1-15 HMSO, London.

Low, A.J. 1971 Tubed seeding research & development. *Forestry* **44** (1) 27-41.

Low, A.J. 1974 The use of dazomet for partial sterilisation of forest nursery soils. *Forestry* **47** (1) 31-43.

Low, A.J. 1975 Production & use of tubed seedlings. *For. Comm. Bull.* **53** 46 pp HMSO, London.

Low, A.J. & Brown, R.M. 1972 Production and use of ball-rooted planting stock in Sweden and Finland. *For. Comm. Res. & Dev. Pap.* **87** 25 pp Departmental Paper, Forestry Commission, London.

Low, A.J. & Brown, R.M. 1974 Production and use of planting stock; Seedbed herbicides. *For. Comm. Rep. For. Res.* **1974** 12.

Low, J.D. & Gladman, R.D. 1960 *Fomes annosus* in Great Britain: an assessment of the situation in 1959. *For. Comm. Forest Rec.* **41** 1-22 HMSO, London.

Low, A.J. & Sharpe, A.L. 1973 The long-term effects of organic and inorganic fertilizer regimes at Teindland nursery. *Scottish Forestry* **27** 287-295.

Lowday, J.E. & Marrs, R.H. 1992 Control of bracken and the restoration of heathlands. 1. Control of bracken. *Journal of Applied Ecology* **29** 195-203.

Lucas, P.W., Cottam, D.A., Sheppard, L.J. & Francis, B.J. 1988 Growth responses and delayed winter hardening in Sitka spruce following summer exposure to ozone. *New Phytologist* **108** 495-504.

Lund-Hoie, K. 1988 New trends in chemical vegetation control in Norwegian forestry. (In) The practice of weed control & vegetation management in forestry, amenity & conservation areas. ... *Aspects of Applied Biology* **16** 169-176 Association of Applied Biologists.

Macdonald, J., Wood, R.F., Edwards, M.V. & Aldhous, J.R. 1957 Exotic forest trees in Great Britain. *For. Comm. Bull.* **30** 168 pp HMSO, London.

Macdonald, J.A., Skiba, U., Sheppard, L.J. *et al.* 1997 Effect of N deposition & seasonal variability on methane oxidation & nitrous oxide emission rates in upland spruce and moorland. *Atmospheric Environment* **31** (22) 3693-3706.

Macdonald, J.A.B. 1936 The effect of introducing pine species among checked Sitka spruce on a dry *Calluna*-clad slope. *Scottish Forestry Journal.* **50** (2) 83-86.

Macdonald, J.A.B. & Macdonald, A. 1952 The effect of interplanting with pine on the emergence of Sitka spruce from check on heather land. *Scottish Forestry* **6** (3) 72-81.

MacGillivray, S. 1998 Woodland expansion on damaged land - the Scottish experience. *Quarterly Journal of Forestry* **92** (4) 307-310.

Mackenzie, J.M. 1972 Early effects of types, rates and methods of application of phosphate rock on peatland. Proc. 4th International Peat Congress, Finland **3** 531-546.

Mackenzie, J.M. 1974a Fertilizer/herbicide trials on Sitka spruce in east Scotland. *Scottish Forestry* **28** 211-221.

Mackenzie, J.M. 1974b Weed problems & control in forestry. *BCPC Monograph* **9** 48-55.

MacKenzie, D. 1994a Where has all the carbon gone? *New Scientist* **8.1.94** 30-33.

MacKenzie, D. 1994b Will tomorrow's children starve? *New Scientist* **3.9.94** 24-29.

MacKenzie, D. 1995 Killing crops with cleanliness. *New Scientist* **23 Sept 1995** 4 (Discussion of findings of A. Auclair, Science & Policy Associates, Washington, DC

MacKenzie, D. 1997 Poison in the air. *New Scientist* **18 Oct. 1997.**

Mackenzie, J.M., Thompson, J.H. & Wallis, K.E. 1976 Control of heather by 2,4-D. *For. Comm. Leaflet* **64** 19pp HMSO, London.

Mackie-Dawson, H.A., Millard, P. & Proe, M.F. 1995 The effect of nitrogen supply on root growth & development in sycamore and Sitka spruce trees. *Forestry* **68** (2) 107-114.

Macmillan, D.C. & Duff, E.I. 1998 Estimating the non-market costs and benefits of native woodland restoration using the contingent valuation method. *Forestry* **71** (3) 247-259.

Madders, M. & Lawrence, M.I.G. 1982 The role of woodland in air pollution control. *Quarterly Journal of Forestry* **76** (4) 256-261.

MAFF (year) Chemical compounds used in agriculture and food storage in GB. User & consumer safety. Advice of Govt depts. *MAFF Recommendation Sheets* Identified by name of active ingredient MAFF, London.

MAFF 1966 *Pesticides safety precautions scheme agreed between governments and industry.* MAFF, London.

MAFF 1967 (revised 1985) *Code of practice for use of herbicides in water-courses and lakes.* MAFF 5pp Ministry of Agriculture, Fisheries and Food.

MAFF 1975a Safe use of poisonous chemicals on the farm. *MAFF Leaflet* **APS/1** 44pp Ministry of Agriculture, Fisheries and Food.

MAFF 1975b Code of practice for the disposal of unwanted pesticides and containers on farms and holdings. (Duplicated booklet) 19 pp Ministry of Agriculture, Fisheries and Food.

MAFF 1975c *Code of practice for ground spraying.* MAFF 4pp Ministry of Agriculture, Fisheries and Food. Also earlier editions *eg* 1962.

MAFF 1980 Guidelines for the disposal of unwanted pesticides and containers on farms and holdings. *Booklet* (Revised 1980) **2198** 1-14 Ministry of Agriculture, Fisheries and Food.

MAFF 1983 Guidelines for applying crop protection chemicals: ground sprayers for agriculture. *Booklet* (Revised 1983) **2272** 1-32 Ministry of Agriculture, Fisheries and Food.

MAFF 1987 Off-label use of pesticides. *Press Release* **330/87** 8 pp Ministry of Agriculture, Fisheries and Food.

MAFF 1989 Consents under Food & Environment Protection Act & Control of Pesticides Regulations. *London Gazette* **20.1.85** 783-785.

MAFF 1990 *Code of practice for suppliers of pesticides to agriculture, horticulture and forestry.* **PB0091** 1-75 Ministry of Agriculture, Fisheries and Food.

MAFF 1992a Dioxins in food. *MAAF Food Surveillance Paper* **31** 79 pp HMSO, London.

MAFF 1992b Veterinary residues in animal products. *MAAF Food Surveillance Paper* **33** 82 pp HMSO, London.

MAFF 1992c Rpt of working party on pesticide residues. *MAAF Food Surveillance Paper* **34** 120 pp HMSO, London.

MAFF 1998a *Code of practice for suppliers of pesticides to agriculture, horticulture and forestry.* *2nd Edn.* **PB3529** Ministry of Agriculture, Fisheries and Food.

MAFF 1998b *Code of good agricultural practice for the protection of water.* Ministry of Agriculture, Fisheries and Food; Welsh Office Agricultural Dept.

MAFF 1999 *Local Environment Risk Assessment for Pesticides (LERAP), a practical guide.* MAFF, London.

MAFF ACAS (year) Agricultural Chemicals Approval Scheme. Approved products for farmers and growers (published annually). *MAFF Reference Book.* **380** (1984) HMSO, London.

MAFF Eval 1990 Evaluation on HOE 39866 (Glufosinate-ammonium). *Evaluation of fully approved or provisionally approved products* **33** MAFF Pesticides Safety Directorate

MAFF Eval 1992a Evaluation on Benomyl. *Evaluation of fully approved or provisionally approved products* **57** MAFF Pesticides Safety Directorate

MAFF Eval 1992b Evaluation on gamma-HCH (Lindane 1). *Evaluation of fully approved or provisionally approved products* **47** MAFF Pesticides Safety Directorate

MAFF Eval 1992c Evaluation on gamma-HCH (Lindane 2). *Evaluation of fully approved or provisionally approved products* **64** MAFF Pesticides Safety Directorate

MAFF Eval 1993a Evaluation on 2,4-Dichlorophenoxy acetic acid and its salts and esters. *Evaluation of fully approved or provisionally approved products* **68** 186+20pp MAFF, London.

MAFF Eval 1993b Evaluation on simazine (2). *Evaluation of fully approved or provisionally approved products* **72** 182pp MAFF, London.

MAFF Eval 1993c Evaluation on atrazine (2). *Evaluation of fully approved or provisionally approved products* **71** 203pp MAFF, London.

MAFF Eval 1996 Evaluation on gamma-HCH (Lindane 3). *Evaluation of fully approved or provisionally approved products* **151** MAFF Pesticides Safety Directorate

MAFF Eval 1997 Evaluation on Lindane - reproductive toxicity effects in dogs. *Evaluation of fully approved or provisionally approved products* **164** MAFF Pesticides Safety Directorate

MAFF Eval 1998 Evaluation on: *Phlebopsis gigantea. Evaluation of fully approved or provisionally approved products* **173** 60pp MAFF Pesticides Safety Directorate

MAFF Mon (several issues per year) *The pesticide monitor.* MAFF HMSO, London.

MAFF PReg (monthly) *The pesticide register* MAFF HMSO, London. (Replaced by *Pst Monitor*.)

MAFF PRes 1995 Annual report of the working party on pesticide residues, 1994 Supplement to the Pesticides Register, 1995 166 pp HMSO, London.

MAFF Pst Year Pesticides (Year of Issue) *Reference Book* **500** HMSO, London.

MAFF Reg 1995 Revised long-term arrangements for extension of use. *Pesticides Register.* **1 Jan. 1995.**

MAFF/HSC 1990 Code of practice for the safe use of pesticides on farms and holdings.(1st Edn) *Joint Code of Practice (MAFF, HSC & DoEnv)* 75pp HMSO, London.

MAFF/HSC 1998 Code of practice for the safe use of pesticides on farms and holdings.(2nd Edn.) *Joint Code of Practice (MAFF, HSC & DoEnv)* **PB3528** HMSO, London.

Makepeace, R. 1996 Sticking up for adjuvants. *Farming & Conservation* **2** (4) 26-27.

Malcolm, D.C. 1972 The effect of repeated urea applications on properties of drained peat. Proc. 4th International Peat Congress, Finland **3** 451-460 International Peat Society.

Malcolm, D.C. 1975 The influence of heather on silvicultural practice - an appraisal. *Scottish Forestry* **29** 14-24.

Malcolm, D.C. 1997 The silviculture of conifers in Great Britain. *Forestry* **70** (4) 293-307.

Malcolm, D.C. & Cuttle, S.P. 1983a The application of fertilizers to drained peat. 1 Nutrient losses in drainage. *Forestry* **56** (2) 155-174.

Malcolm, D.C. & Cuttle, S.P. 1983b The application of fertilizers to drained peat. 2 Uptake by vegetation and residual distribution in peat. *Forestry* **56** (2) 175-183.

Malcolm, D.C. & Freezaillah, B.C.Y. 1973 *Interaction of climate & P fertilization on Sitka spruce seedlings.* FAO/IUFRO International Symposium of forest fertilization. FOR:FAO/IUFRO /F/73/32 October 1973 8pp.

Malcolm, D.C. & Freezaillah, B.C.Y. 1975 Early frost damage on Sitka spruce seedlings & the influence of P nutrition. *Forestry* **48** (2) 139-146.

Malcolm, D.C. & Titus, B.D. 1983 Decomposing litter as a source of nutrients for second rotation Sitka spruce on peaty gley soils. (In) *USDA Forest Service Gen. Tech. Rpt : Forest site and continuous productivity* **PNW-163** 138-145.

Malcolm, D.C., Hooker, J.D. & Wheeler, C.T. 1985 *Frankia* symbiosis as a source of nitrogen in a mixed *Alnus - Picea* plantation in Scotland. *Proc. R. Soc. Edinburgh* **85B** 263-282.

Mansfield, T.A. & Freer-Smith, P.H. 1981 Effects of urban pollution on plant growth. *Biological Review* **56** 343-368.

Marchington, J. 1992 Herbicides aid habitat creation. *Forestry & British Timber* **Feb. 1992** 36-39.

Martindale, L. 1986 Potential for wood as a fuel in the UK. *For. Comm. Occasional Pap.* **16** 26-35 British Assoc. for the Advancement of Science; Forestry Section Proceedings

Mason, W.L. 1986 Control of simazine-resistant weeds in forest nurseries. *Forestry & British Timber* **Feb. 1986** 27-28.

Mason, W.L. 1994 Production of bare-root seedlings & transplants. Production of undercut stock. (in 'Nursery Practice') *For. Comm. Bull.* **111** 84-103, 112-120 HMSO, London.

Mason, W.L. & Gill, J.G.S. 1986 Vegetative propagation of conifers as a means of intensifying wood production in Britain. *Forestry* **59** (2) 155-172.

Mason, W.L. & Jinks, R.L. 1990 Use of containers as a means of raising tree seedlings. *For. Comm. Res. Inf. Note* **179** 2 pp.

Mason, W.L. & Jinks, R.L. 1994 Vegetative propagation. (In) Forest Nursery Practice (Eds) J.R. Aldhous & W.L. Mason *For. Comm. Bull.* **111** 135-147 HMSO, London.

Mason, W.L. & Williamson, D.R. 1987 Nursery herbicides. *For. Comm. Rep. For. Res.* **1987** 18 HMSO, London.

Mason, W.L. & Williamson, D.R. 1988 Recent research into weed control on seedbeds in forest nurseries. *Aspects of applied biology* **16** 231-238.

Mason, W.L. & Williamson, D.R. 1992 Alternatives to triazines for weed control in forest nurseries. *Aspects of applied biology* **29** 149-155.

Mason, P.A., Wilson, J., Last, F.T. & Walker, C. 1983 The concept of succession in relation to the spread of sheathing mycorrhizal fungi on inoculated tree seedlings growing in unsterile soils. *Plant & Soil* **71** 247-256.

Mason, P.A., Wilson, J. & Last, F.T. 1984 Mycorrhizal fungi of *Betula* spp: factors affecting their occurrence. *Proc. R. Soc. Edinburgh* **85B** 141-151.

Mather, R.A., Freer-Smith, P.H. & Savill, P.S. 1995 Analysis of the changes in forest conditions in Britain, 1989 - 1992. *For. Comm. Bull.* **116** 52pp HMSO, London.

Mathur, S.B. & Mandanhar, H.K. 1994 Quarantine for seed. *FAO Plant production & protection paper* **119** 296 pp Food & Agriculture Organisation, Rome.

Matthews, R.W. 1991 Biomass production & carbon storage by British forests. (In) *Wood for energy - implication for harvesting, utilisation & marketing.* (Ed.) J.R. Aldhous 162-176 Institute of Chartered Foresters, Edinburgh.

Matthews, R.W. 1994 The rise and rise of global warming. *New Scientist* 26Nov. 1994 6.

Matthews, R.W. 1995 Modelling impacts of forest policy on carbon sequestration. *For. Comm. Rep. For. Res.* **1995** 22-23 HMSO, London.

Matthews, G. 1999 Catching the drift. *Horticulture Week - Landscape & amenity buyer* **April 1999** 26-27.

Matthews, R.W., Robinson, R.L., Abbott, S.R. & Fearis, N. 1994 *Modelling carbon and energy budgets of wood fuel coppice systems. ETSU Contractor's report* B/W5/00337/REP. ETSU, Harwell, Oxon.

Mattingly, G.E.G. 1965 Residual value of cumulative dressings of superphosphate, rock phosphate and basic slag on a sandy soil at Wareham, Dorset. *For. Comm. Rep. For. Res.* **1965** 93-96 HMSO, London.

Mayhead, G.J. 1976 Forest fertilizing in Great Britain. *Proc. Fertilizer Society.* **158** Fertilizer Society of London.

Mayhead, G.J. & McCavish, W.J. 1975 Forest weed control of grass using propyzamide. *For. Comm. Res. Inf. Note* **9/75/SILS** 4pp Forestry Commission, Forest Research Station, Farnham.

Mayhead, G.J., Broad, K. & Marsh, P. 1974 Tree growth on the South Wales coalfield. *For. Comm. Res. & Dev. Pap.* **108** 29pp Forestry Commission, London.

Mayle, B.A., Pepper, H.W., Ratcliffe, P.A., Rowe, J.J. & Tee, L.A. 1984 Management of deer, squirrels & other animals. *For. Comm. Rep. For. Res.* **1984** 42 HMSO, London.

Mayle, B. 1999 Managing deer in the countryside. *For. Comm. Practice Note* **FCPN 6** 12pp.

MCA 1996 *Towards safe medicines.* Revised edn. 92pp. Medical Control Agency, London.

McCary, J. 1989 CCWA's land treatment programme - a model of forest utilisation. Timber Harvesting **June 1989** 31-32 (USA periodical)

McCavish, W.J. 1978 Forest weed control. *For. Comm. Rep. For. Res.* **1978** 11-12 HMSO, London.

McCavish, W.J. 1979 Forest weed control. *For. Comm. Rep. For. Res.* **1979** 11-12 HMSO, London.

McCavish, W.J. 1980 Forest weed control. *For. Comm. Rep. For. Res.* **1980** 11-12 HMSO, London.

McCavish, W.J. 1981 Seedbed and transplant herbicides; Forest weed control. *For. Comm. Rep. For. Res.* **1981** 11, 14-15 HMSO, London.

McCavish, W.J. 1990 Forestry Commission Great Britain. Weeding and Cleaning Survey, 1989. *Proc. Joint FAO/ECE/ILO Committee seminar on forest technology, management & training, Sparsholt.* 70-101 Forestry Commission, Edinburgh.

McCavish, W.J. & Smith, F.S. 1976 Forest weed control. *For. Comm. Rep. For. Res.* **1976** 11-12 HMSO, London.

McCavish, W.J. & Smith, F.S. 1977 Forest weed control. *For. Comm. Rep. For. Res.* **1977** 12-13 HMSO, London.

McCavish, W.J. & Smith F.S. 1978 Forest weed control - Removal of unwanted trees using stem incision techniques. *For. Comm. Res. Inf. Note* **38/78/SILS** 4 pp Forestry Commission, Forest Research Station, Farnham.

McCracken, A.R. & Dawson, W.M. 1996 Growing short rotation coppice in clonal mixtures as a method of reducing the impact of rust disease caused by *Melampsora* sp. *European Journal of Forest Pathology* **1996.**

McGregor, S.D., Duncan, H.J., Pulford, I.D. & Wheeler, C.T. 1996 Uptake of heavy metals from contaminated soil by trees. (In) *Heavy metals & trees* Proc. conf. Glasgow 1995 (Ed.) I. Glimmerveen. 171-176 Institute of Chartered Foresters, Edinburgh.

McIntosh, R. 1978 Response of Sitka spruce to remedial fertilization in Galloway. *Scottish Forestry* **32** 271-282.

McIntosh, R. 1979 Nutrition, establishment phase. *For. Comm. Rep. For. Res.* **1979** 19 HMSO, London.

McIntosh, R. 1980 Effect of weed control and fertilization on the growth and nutrient status of Sitka spruce on some upland soils. *Weed control in Forestry (Conference proceedings - Nottingham)* 55-63 Association of Applied Biologists.

McIntosh, R. 1981 Fertilizer treatment of Sitka spruce in the establishment phase in upland Britain. *Scottish Forestry* **35** (1) 3-13.

McIntosh, R. 1982 Effects of different forms & rates of nitrogen fertilizer on growth of Lodgepole pine. *Forestry* **55** (1) 61-68.

McIntosh, R. 1983a Effect of fertilizer & herbicide applications on Sitka spruce on mineral soils with a dominant grass/herb vegetation. *For. Comm. Res. Inf. Note* **76/83/SILN** 5pp Forestry Commission, Forest Research Station, Farnham.

McIntosh, R. 1983b Nitrogen deficiency in establishment phase Sitka spruce in upland Britain. *Scottish Forestry* **37** 185-193.

McIntosh, R. 1984a Fertilizer experiments in established conifer stands. *For. Comm. Forest Rec.* **127** 1-24 HMSO, London.

McIntosh, R. 1984b Phosphate fertilizers in upland forestry - types, application rates & placement methods. *For. Comm. Res. Inf. Note* **89/84/SILN** 4 pp Forestry Commission, Forest Research Station, Farnham.

McIntosh, R. 1984c Silviculture (North): nutrition. *For. Comm. Rep. For. Res.* **1984** 17-18 HMSO, London.

McIntosh, R. & Tabbush, P. 1981 Silviculture (North): nutrition. *For. Comm. Rep. For. Res.* **1981** 22 HMSO, London.

McKay, H.M. 1998 Root electrolyte leakage & root growth potential as indicators of spruce & larch establishment. *Silva Fennica* **32** (3) 241-252.

McKay, H.M., Aldhous, J.R. & Mason, W.L. 1994 Lifting, storage, handling & despatch (In) Forest Nursery Practice (Eds) J.R. Aldhous & W.L. Mason *For. Comm. Bull.* **111** 198-222 HMSO, London.

McKay, H.M., Thorn, C. & Kempton, E. 1998 Frost and insecticide damage to Corsican pine seedlings between lifting and planting. *Quarterly Journal of Forestry* **92** (1) 55-62.

McKinlay, R. & Atkinson, D. 1995 Integrated crop protection: towards sustainability? *British Crop Protection Council; proceedings* **63** 488 pp British Crop Protection Council.

McLeod, A.R., Holland, M.R., Shaw, P.J.A. *et al.* 1990 Enhancement of nitrogen deposition to forest trees exposed to SO_2 *Nature* **347** 277-279 London.

McLeod, A.R., Shaw, P.J.A. & Holland, M.R. 1992 Liphook forest fumigation project: studies of sulphur dioxide & ozone on coniferous trees. *Forest Ecology & Management* **50** 121-127.

McNeill, J.D., Hollingsworth, M.K., Mason, W.L., Sheppard, L.J. & Wheeler, C.T. 1989 Inoculation of *Alnus rubra* seedlings to improve seedling growth and forest performance. *For. Comm. Res. Inf. Note* **144** 5pp Forestry Commission, Farnham, Surrey.

Mellanby, K. 1992 *The DDT story.* 113pp British Crop Protection Council.

Mercer, P.C. 1984a Callus growth and the effect of a wound dressing. *Annals of Applied Biology* **103** 527-540.

Mercer, P.C. 1984b The effect of bark stripping by grey squirrels. *Forestry* **57** (2) 199-203.

Mercer, P.C. & Kirk, S.A. 1984 Biological treatments for the control of decay in tree wounds II Field tests. *Annals of Applied Biology* **104** 221-229.

Mercer, P.C. Kirk, S.A. Gendle, P. & Clifford, D.R.J. 1983 Chemical treatments for control of decay in pruning wounds. *Annals of Applied Biology* **102** 435-453.

Meredith, D.S. 1959 The infection of pine stumps by *Fomes annosus* and other fungi. *Annals of Botany* **23** (NS) 455-476.

Meredith, D.S. 1960 Further observations on fungi inhabiting pine stumps. *Annals of Botany* **24** (NS) 63-78.

Meulemans, M. & Parmentier, C. 1983 Studies on *Ceratocystis ulmi* in Belgium. *For. Comm. Bull.* **60** 114 pp HMSO, London.

Middleton, A.D. 1931 *The grey squirrel.* Sidgwick & Jackson.

Miles, J. 1981 *Effect of birch on moorlands.* Institute of Terrestrial Ecology, Cambridge.

Miles, J. 1985 Pedogenic effects of different species & vegetation types and implications for succession. *Journal of Soil Science* **36** 571-584.

Miles, J. 1986 What are the effects of trees on soils? *Proc. ITE Symposium* (Ed.) Jenkins. **17** Institute of Terrestrial Ecology, Huntingdon.

Miles, J. & Young, W.F. 1980 Effects on heathland & moorland soils in Scotland & N. England of birch. *Bulletin of Ecology* **11** 233-242.

Miller, A.D.S. 1951 Comparative costs of converting hazel coppice to beech high forest. *For. Comm. Res. Br. Pap.* **1** 6 pp Forestry Commission, London. (Departmental paper).

Miller, A.D.S. 1955 Derelict woodland investigations. *For. Comm. Rep. For. Res.* **1955** 37-39 HMSO, London.

Miller, H.G. 1966 Current research into nitrogen nutrition of Corsican pine. *Forestry* **Supplement 1966** 70-77 Oxford University Press.

Miller, H.G. 1978 Forest soils and tree nutrition. *For. Comm. Rep. For. Res.* **1978** 53-54 HMSO, London.

Miller, H.G. 1979a The nutrient budgets of even aged forests. (In) *The ecology of even aged plantations.* Proc. IUFRO meeting, Sept 1978 (Eds) E.D. Ford, D.C. Malcolm & J. Atterson. 221-256 ITE, NERC, Cambridge.

Miller, H.G. 1979b Forest soils. *For. Comm. Rep. For. Res.* **1979** 52-53 HMSO, London.

Miller, H.G. 1980 Throughfall, stemflow, crown leaching & wet deposition. (In) *Methods for studying acid precipitation in forest ecosystems* (Eds.) I.A. Nicholson et al. 29-33 ITE, NERC, Cambridge.

Miller, H.G. 1981a Forest fertilization: some guiding concepts. *Forestry* **54** (2) 157-167.

Miller, H.G. 1981b Nutrition and forest soils. *For. Comm. Rep. For. Res.* **1981** 54-55 HMSO, London.

Miller, H.G. 1983 Studies of proton flux in forests and heaths in Scotland. (In) *Effects of accumulation of air pollutants in forest ecosystems.* (Eds.) Ulrich & D. Reidel 183-193 Reidel, Dortrecht

Miller, H.G. 1984a Maintenance & improvement of forest productivity; an overview. *Proc. IUFRO symposium on forest site & continuous productivity* USDA Forest Service: PNW-13 280-285.

Miller, H.G. 1984b Deposition-plant-soil interactions. *Phil. Trans. R. Soc. Lond. B* **305** 30-35.

Miller, H.G. 1984c Dynamics of nutrient cycling in plantations. (In) *Nutrition of plantation forests* (Eds.) G.D. Bowen & E.K.S. Namibar 53-78 Academic Press, London.

Miller, H.G. 1984d The nutrition of hardwoods. (In) *National Hardwoods programme. Report of 5th meeting* Commonwealth Forestry Institute, Oxford.

Miller, H.G. 1984e Nutrient cycles in birchwoods. *Proc. R. Soc. Edinburgh* **85b** 83-96.

Miller, H.G. 1984f Water in forests. *Scottish Forestry* **38** 165-181.

Miller, H.G. 1986 Carbon x nutrient interactions - the limits to productivity. *Tree Physiology* **2** 373-385.

Miller, H.G. 1989 *Forests and acidification.* Proc. Symposium 52-71 Scottish Development Dept. Edinburgh.

Miller, H.G. 1991 British forestry in 1990. *For. Comm. Occasional Pap.* **34** 1-23 Forestry Commission, Edinburgh.

Miller, H.G. & Cooper, J.M. 1973 Changes in amount and distribution of stem growth in pole-stage Corsican pine following N fertilizer. *Forestry* **46** (2) 157-190.

Miller, H.G. & Mackenzie, R.C. 1964 Research on Scottish forest and nursery soils; nitrogen nutrition. *For. Comm. Rep. For. Res.* **1964** 83-86 HMSO, London.

Miller, H.G. & Miller, J.D. 1976a Effect of nitrogen supply on net primary production in Corsican pine. *Journal of Applied Ecology* **13** 249-256.

Miller, H.G. & Miller, J.D. 1976b Analysis of needle fall as means of assessing nitrogen status in pine. *Forestry* **49** (1) 57-61.

Miller, H.G. & Miller, J.D. 1987 Nutrition requirements of Sitka spruce. *Proc. R. Soc. Edinburgh* **93** (1/2) 75-84.

Miller, H.G. & Miller, J.D. 1988 Responses to heavy N applications in fertilizer experiments in British forests. *Environmental Pollution* **54** 219-231.

Miller, H.G. & Miller, J.D. 1991 Energy forestry; the nutrition equation. (In) *Wood for Energy: Implications for harvesting, utilisation & marketing.* Proc. Discussion Meeting (Ed.) J.R. Aldhous 137-147 Institute of Chartered Foresters, Edinburgh.

Miller, H.G. & Proe, M.F. 1986 Nutrient flow modelling. (In) *Computers in forestry* Proc. Conference, Edinburgh, 1984 (Eds.) W.L. Mason, & R. Muetzelfeldt, 228-237 Institute of Chartered Foresters, Edinburgh.

Miller, H.G. & Williams, B.L. 1968 Research on forest and tree nutrition. *For. Comm. Rep. For. Res.* **1968** 143-147 HMSO, London.

Miller, H.G. & Williams, B.L. 1969 Research on forest soils and tree nutrition. *For. Comm. Rep. For. Res.* **1969** 142-146 HMSO, London.

Miller, H.G. & Williams, B.L. 1972 Research on forest soils and tree nutrition. *For. Comm. Rep. For. Res.* **1972** 143-147 HMSO, London.

Miller, H.G. & Williams, B.L. 1973 Research on forest soils and tree nutrition. *For. Comm. Rep. For. Res.* **1973** 145-147 HMSO, London.

Miller, H.G. & Williams, B.L. 1974 Research on forest soils and tree nutrition. *For. Comm. Rep. For. Res.* **1974** 60-61 HMSO, London.

Miller, H.G. & Williams, B.L. 1976 Forest soils and tree nutrition. *For. Comm. Rep. For. Res.* **1976** 53-55 HMSO, London.

Miller, H.G. & Williams, B.L. 1977 Nutrition and forest soils *For. Comm. Rep. For. Res.* **1977** 51-52 HMSO, London.

Miller, H.G., Cooper, J.M. & Miller, J.D. 1976a Effect of nitrogen supply on nutrients in litter fall and crown leaching in a stand of Corsican pine. *Journal of Applied Ecology* **13** 233-248.

Miller, H.G., Miller, J.D. & Pauline, O.J.L. 1976b Effect of nutrient supply on nutrient uptake in Corsican pine. *Journal of Applied Ecology* **13** 955-966.

Miller, H.G., Williams, B.L., Millar, C.S. & Warin, T.R. 1977 Vegetation & humus nitrogen indicators of N-status in established sand dune forest. *Forestry* **50** (2) 93-101.

Miller, H.G., Cooper, J.M., Miller, J.D. & Pauline, O.J.L. 1979 Nutrient cycles in pine & their adaptation to poor soils. *Canadian Journal of Forest Research.* **9** 19-26.

Miller, H.G., Miller, J.D. & Cooper, J.M. 1980 Biomass & nutrient accumulation at different growth rates in thinned plantations of Corsican pine. *Forestry* **53** (1) 23-39.

Miller, H.G., Alexander, C., Cooper, J.M., Keenleyside, J., McKay, H.M., Miller, J.D. & Williams, B.L. 1986 *Maintenance and enhancement of forest productivity through manipulation of the nitrogen cycle.* Report on Contract BOS-093 UK 300 pp Macaulay Institute for Soil Research, Aberdeen.

Miller, J.D., Anderson, H.A., Ferrier, R.C. & Walker, T.A.B. 1990a Comparison of the hydrological budgets & detailed hydrological responses in two forested catchments. *Forestry* **63** (3) 251-269.

Miller, J.D., Anderson, H.A., Ferrier, R.C. & Walker, T.A.B. 1990b Biochemical fluxes & their effects on stream acidity in two forested catchments in central Scotland. *Forestry* **63** (4) 311-331

Miller, H.G., Cooper, J.M. & Miller, J.D. 1992 Response of pole-stage Sitka spruce to applications of fertilizer N, P and K in upland Britain. *Forestry* **65** (1) 15-33.

Miller, J.D., Cooper, J.M. & Miller, H.G. 1993 A comparison of above-ground component weights and element amounts in four forest species at Kirkton Glen. *Journal of Hydrology* **145** 419-438.

Miller, J.D., Cooper, J.M. & Miller, H.G. 1996a Amounts & nutrient weight in litterfall, and their annual cycles from a series of experiments on pole-stage Sitka spruce. *Forestry* **69** (4) 289-302.

Miller, J.D., Anderson, H.A., Ray & Anderson, A.R. 1996b Impact of initial forestry practices on drainage waters from blanket peatlands. *Forestry* **69** (3) 193-203.

Mills, D.H. 1980 Management of forest streams. *For. Comm. Leaflet* **78** 20 pp HMSO, London.

Millward Forestry 1998 ATV sprayers on test. *Forestry & British Timber* **April 1998** 30.

Milne, R., Brown, T.A.W. & Murray, T.D. 1998 Effect of geographical variation of planting rate on the uptake of carbon by new forests of GB. *Forestry* **71** (4) 297-309.

Mitchell, A.F. 1974 *Field guide to the trees of Britain and Northern Europe.* Collins, London.

Mitchell, C.P. 1991 Energy coppice. Proc. conf.: *Energy cropping on Farms.* 5 March 1991. Paper 4 11pp ETSU, Harwell/RASE, National Agricultural Centre, Stoneleigh, Warwick.

Mitchell, J.D. 1998 Population growth continues. (In) *Vital Signs 1998-1999.* (Eds.) L.R. Brown et al. 102-103 Worldwatch Institute, c/o Earthscan Publications, Pentonville Road, London.

Mitchell, C.P., Cooper, R.J., Ball, J., Gunneberg, B. & Swift, J.J. 1994 Private woodland survey. *For. Comm. Technical Pap.* **5** 50pp Forestry Commission, Edinburgh.

MLURI (Macauley Land Use Research Institute) 1997 Soil quality, contaminated land and waste utilisation. Geographic & resource analysis. Long-term change *Annual Report* 1997 28-37; 3-11; 82-87.

Moffat, A.J. 1987 The use of sewage sludge in forestry, an assessment of the potential in England and Wales. *Water Research Report* **PRU 1707-M** Water Research Centre & Forestry Commission.

Moffat, A.J. 1988 Forestry & soil erosion in Britain - a review. *Soil Use & Management* **4** (2) 41-44.

Moffat, A.J. 1989 Forestry & soil erosion in Britain - a reply. *Soil Use & Management* **5** (2) 199.

Moffat, A.J. 1994a Using waste materials in the forest; implications of research. *For. Comm. Res. Inf. Note* **256** 4pp Forestry Commission, Farnham, Surrey.

Moffat, A.J. 1994b Nursery sterilisation and inoculation regimes for alder production. *Forestry* **67** (4) 313-327.

Moffat, A.J. 1994c Sewage sludge as a fertilizer in amenity and reclamation plantings. *Arboriculture Research Note* **76/94/SSS** Arboricultural Advisory & Information Service, Farnham, Surrey.

Moffat, A.J. 1994d Forestry & soil erosion in Britain - a review. *Soil Management* **7** 145-151.

Moffat, A.J. & Bird, D. 1989 Potential for using sewage sludge in forestry in England and Wales. *Forestry* **62** (1) 1-17.

Moffat, A.J. & Boswell, R.C. 1990 Effect of tree species and species mixtures on soil properties at Gisburn Forest Yorks. *Soil Use & Management* **6** (1) 46-51.

Moffat, A.J. & McNeil, J. 1994 Reclaiming disturbed land for forestry. *For. Comm. Bull.* **110** 103pp HMSO, London.

Moffat, A.J. & Roberts, C.J. 1987 Effects of nitrogen-fixing plants. *For. Comm. Rep. For. Res.* **1987** 28 HMSO, London.

Moffat, A.J. & Roberts, C.J. 1989a Experimental tree planting on china clay spoils in Cornwall. *Quarterly Journal of Forestry* **83** (3) 149-156.

Moffat, A.J. & Roberts, C.J. 1989b Use of large scale ridge & furrow landforms in forestry reclamation of mineral workings. *Forestry* **62** (3) 233-248.

Moffat, A.J., Roberts, C.J. & McNeil, J.D. 1989 Use of N-fixing plants in forest reclamation. *For. Comm. Res. Inf. Note* **158** 3pp+App. Forestry Commission, Forest Research Station, Farnham.

Monteith, J.L. 1989 Reflection and review. *Phil. Trans. R. Soc. Lond. B* **324** 433-436.

Moore, N.W. 1968 The value of pesticides for conservation & ecology. (In) *Some safety aspects of pesticides in the countryside* Proc. Conf. Countryside in 1970, Nov. 1967. (Eds.) N.W. Moore, & W.P. Evans, 104-108 Joint ABMAC/Wildlife C'tee, Alembic House, London.

Moore, N.W. & Evans, W.P. (Eds) 1968 *Some safety aspects of pesticides in the countryside.* Proc. Conf. Countryside in 1970, Nov., 1967 124pp Joint Association of British Manufacturers of Agricultural Chemicals/Wildlife C'tee, Alembic House, London.

Moore, R. 1997 Scale of restocking and cost of *Hylastes* control. *Forestry Commission Report on Forest Research* 1997.

Moore, T.R. & Knowles, R. 1989 Influence of water table levels on methane and carbon dioxide emissions from peat soils. *Canadian Journal of Soil Science* **69** 33-38.

Morgan, J. 1997 Taking the spin out of roots. *Forestry & British Timber* **March 1997** 26-28.

Morgan, J.L. 1994 Fertilization regimes to produce different size classes of cell-grown birch. *For. Comm. Res. Inf. Note* **247** 1-3 Forestry Commission, Forest Research Station, Farnham.

Morris, M.G. 1974 Oak as a habitat for insect life. (In) *The British Oak* Proc. BSBI conf. (Eds.) M.G. Morris & F.H. Perring 274-297 E.W. Classey for the Botanical Society of the British Isles.

Morris, J. & Temple, R.K. 1998 Nest-tubes: a potential new method for controlling edible plant dormouse in plantations. *Quarterly Journal of Forestry* **92** (3) 201-205.

Morris, J. & Whipp, P. 1998 Grey squirrel control. Annual variations in uptake of warfarin from hoppers in a small Chiltern woodland. *Quarterly Journal of Forestry* **92** (3) 206-208.

Morris, R.A.C., Royle, D.J., Whellan, M.J. & Arnold, G.M. 1994 Biocontrol of willow rusts with the hyperparasite *Sphaerellopsis filum.* Proc. Brighton Crop Protection Conf., - Pests & Diseases **1994** 1121-1126 British Crop Protection Council.

Morris, P.A., Temple, R.K. & Jackson, J.E. 1997 Studies of the edible dormouse in British woodlands - some preliminary results. *Quarterly Journal of Forestry* **91** (4) 321-326.

Morrison, B.R.S. 1987 The effects of forest management on the biology of watercourses. Proc. Discussion meeting *Wildlife management in forests.* (Ed) D.C. Jardine. 34-42 Institute of Chartered Foresters, Edinburgh.

Mortimer, A.M. 1990 The biology of weeds. (In) *Weed Control Handbook: Principles* (Eds.) R.J. Hance, & K. Holly. 8th edn 1-42 Blackwell, Oxford.

Morton, O. 1998 The storm in the machine. *New Scientist* **31 Jan.1998** 22-27.

Mountford, E.P. 1997 A decade of grey squirrel bark-stripping damage to beech in Lady Park wood, UK. *Forestry* **70** (1) 17-29.

MTB 1998 Is certification affordable by emerging economies? *Malaysian Timber Bulletin* **4** (4) 9.

Mulvaney, K. 1998 Alaska is heating up - that's good news for caterpillars & bad news for trees. *New Scientist* **18 July 1998** 12.

Munroe, J.W. 1914 A braconid parasite of the pine weevil. *Hylobius abietis Annals of Applied Biology* **1** 170-176.

Munroe, J.W. 1926 British bark-beetles. *For. Comm. Bull.* **8** HMSO, London.

Murgatroyd, I.R. 1993 Menzi flail mounted on a tracked excavator. *For. Comm. Technical Development Br. Rpt.* **23/93** Forestry Commission, Edinburgh.

Murgatroyd, I.R. 1996 Motor-manual & mechanised *Rhododendron* clearance. *Technical Development Branch: Technical Note,* **2/96** Forestry Commission, TDB, Smithton, Inverness.

Murgatroyd, I.R. 1995 Machine cleaning on site. *For. Comm. Technical Development Br. Technical Note* 14/95 Forestry Commission Technical Devel't Br., Ae, Dumfries.

Murray, J.S. 1955 Rusts of British forest trees. *For. Comm. Booklet* **4** HMSO, London.

Murray, J.S. & Young, C.W.T. 1961 Group dying of conifers. *For. Comm. Forest Rec.* **46** 20pp HMSO, London.

Murray, J.B., Cannell, M.G.R. & Sheppard, L.J. 1986 Frost hardiness of *Nothofagus procera* & *N. obliqua* in Britain. *Forestry* **59** (2) 209-222.

Namibar, E.K.S & Fife, D.N. 1991 Nutrient retranslocation in temperate conifers. *Tree Physiology* **9** 185-207.

Näsholm, T. 1998 Qualitative & quantitative changes in plant nitrogen acquisition induced by anthropogenic nitrogen deposition. *New Phytologist* **139** 87-90.

NCB 1967 Plant growth on pit heaps - a literature survey. Technical Intelligence Branch Report, Research & Development Dept. (Coal Processing & Combustion) **No. 42** 24 pp + Appendices National Coal Board.

Nelson, D.G. 1989 Specification for Sitka spruce restocking. *For. Comm. Rep. For. Res.* **1989** 22-23 HMSO, London.

Nelson, D.G. 1990 Chemical control of Sitka spruce natural regeneration. *For. Comm. Res. Inf. Note* **187** Forestry Commission, Forest Research Station, Farnham.

Nelson, D.G. 1991 Management of Sitka spruce natural regeneration. *For. Comm. Res. Inf. Note* **204** Forestry Commission, Farnham, Surrey.

Nelson, R.G. & Williamson, D.R. 1989 Gardaprim-A liquid: experimental results on grass weed control and crop tolerances. *For. Comm. Res. Inf. Note* **152** 1-4 Forestry Commission, Forest Research Station, Farnham.

Nelson, D.G., Williamson, D.L. Mason, W.L. & Clay, D. 1991 Herbicide update. *Forestry & British Timber* Feb. **1991** 24-27.

Neustein, S.A.N. 1964 Deer deterrents. *For. Comm. Rep. For. Res.* **1964** 38 HMSO, London.

Neustein, S.A.N. 1966 Weed control before planting. *For. Comm. Rep. For. Res.* **1966** 47 HMSO, London.

Neustein, S.A.N. 1967 Slash disposal to aid regeneration. *For. Comm. Res. & Dev. Pap.* **1967** 39 Forestry Commission, Edinburgh.

Neustein, S.A.N. 1970 Choice of species: Atmospheric pollution. *For. Comm. Rep. For. Res.* **1970** 55-56 HMSO, London.

Neustein, S.A.N. & Jobling, J. 1969 Choice of species: Industrial waste sites. *For. Comm. Rep. For. Res.* **1969** 50-51 HMSO, London.

Newson, M. 1985 Forestry & water in the uplands of Britain; the background of hydrological research and option for harmonious land use. *Quarterly Journal of Forestry* **79** 113-120.

Newson, M.D. & Calder, I.R. 1989 Forestry & water resources: problems of prediction on a regional scale. *Phil. Trans. R. Soc. Lond. B* **324** 283-298.

Newton, J.P. & Rivers, M.J. 1982 Lake Vyrnwy Estate, an example of multiple use of rural land. *Quarterly Journal of Forestry* **76** 92-102.

Nicholls, P.H. 1988 Factors influencing the entry of pesticides into soil. *Pesticide Science* **22** 123-137.

Nicholson, I.A., Cape, N. et al. 1980 Effects of a Scots pine canopy on the chemical composition and deposition pattern of precipitation. (In) *Proc. Sandefjord Conf.* (Eds.) Drablós & Tollan

Niemelä, T & Korhonen, K. 1998 Taxonomy of the genus *Heterobasidion*. (In) *Heterobasidion annosum, biology, ecology, impact & control* (Eds.) S. Woodward et al. 27-33 CAB International.

Nigam, P.C. 1990 Use of chemical insecticides against spruce budworm ... in E. Canada. *Proc. Joint FAO/ECE/ILO Committee seminar on forest technology, management & training, Sparsholt.* 116-132 Forestry Commission, Edinburgh.

Nihlgard, B. 1985 Ammonium hypothesis: an additional explanation for dieback in Europe. *Ambio* **14** 2-8.

Niles, J. 1994 Ethics in forestry. *Quarterly Journal of Forestry* **89** (3) 213-216

Nilsson, J. & Grennfelt, P. (Eds) 1988 Critical loads for sulphur & nitrogen. *Miljorapport 1988* **15** Nordic Council of Ministers.

Nilsson, S.I., Miller, H.G. & Miller, J.D. 1982 Forest growth as a possible cause of soil and water acidification, an examination of the concept. *Oikos* **39** 40-49.

Nimmo, M. 1952 The 1945 broom & pine nursing experiments at Coldharbour, Wareham forest, Dorset. *For. Comm. Rep. For. Res.* **1952** 48-49 HMSO, London.

Nimmo, M. 1957 Grey squirrel research. *For. Comm. Rep. For. Res.* **1957** 61-63 HMSO, London.

Nimmo, M. & Weatherell, J. 1961 Experiences with leguminous nurses in forestry. *For. Comm. Rep. For. Res.* **1961** 126-147 HMSO, London.

Nisbet, T.R. 1989 Liming to alleviate surface water acidity. *For. Comm. Res. Inf. Note* **148** 1-3 Forestry Commission, Forest Research Station, Farnham.

Nisbet, T.R. 1990 Forest and surface water acidification. *For. Comm. Bull.* **86** 1-8 HMSO, London.

Nisbet, T.R. 1993 Effects of pelletised limestone on drainage water acidity in a forest catchment in mid-Wales. *Journal of Hydrology* **145** 521-539.

Nisbet, T.R. 1994a Forests and Water Guidelines - how well do they tackle the water issues? (In) *Forests & Water.* Proc. discussion meeting. (Ed.) I.R. Brown March 1994 104-111 Institute of Chartered Foresters, Edinburgh.

Nisbet, T.R. 1994b Liming in forestry. (In) *Agricultural lime and the environment* 28-29 Agricultural Lime Producers' Council.

Nisbet, T.R. 1996a Application of the freshwater critical loads approach to forestry. *For. Comm. Rep. For. Res.* **1996** 43-50.

Nisbet, T.R. 1996b Impact of the Lyndford wet store on water quality. (In) Water storage of timber; experience in Britain. (Eds) J. Webber & J. Gibbs *For. Comm. Bull.* **117** 11-16 HMSO, London.

Nisbet, T.R. & Stonard, J. 1995 The effect of aerial applications of urea fertilizer on stream water quality. *For. Comm. Res. Inf. Note* **266** 4pp Forestry Commission, Forest Research Station, Farnham.

Nisbet, T.R., Fowler, D. & Smith, R.I. 1995a Investigation of the impact of afforestation on stream-water chemistry in the Loch Dee catchment, SW Scotland. *Environmental Pollution* **90** 111-120.

Nisbet, T.R., Fowler, D. & Smith, R.I. 1995b Use of the critical loads approach to quantify the impact of afforestation on surface water acidification. (In) *Acid rain & its impact: the critical loads debate* (Ed.) A.R. Battarbee 116-118 Ensis Publications, London.

Nixon, C.J., Rogers, D.G. & Nelson, D.G. 1992 Protection of trees in silvipastoral systems. *For. Comm. Res. Inf. Note* **219** 3pp Forestry Commission, Forest Research Station, Farnham.

Norby. R.J. 1998 Nitrogen deposition: a component of global change analysis. *New Phytologist* **139** 189-200.

Norden, U. 1994a Influence of tree species on acidification and mineral pools in deciduous forests of S. Sweden. *Water, Air & Soil Pollution* **76** 363-381.

Norden, U. 1994b Influence of broad-leaved species on pH and Organic matter content of forest top soils in Scania, S. Sweden. *Scandinavian Journal of Forest Research* **9** 1-8.

NPTC 1994 *Schedule of standards - certificate of competence in the use of pesticides.* National Proficiency Tests Council 96 pp NPTC, NAC, Kenilworth, Warwickshire CV8 2LG

Nylén, T. 1996 *Uptake, turnover & transport of radiocaesium in boreal forest ecosystems.* FOA R96-00242-4.3 41pp Swedish University of Agricultural Sciences, Uppsala, Sweden

O'Carroll, J, & O'Reilly, C. 1997 Effects of selected pre-emergence herbicides on germination *etc* of Sitka spruce. *Irish Forestry* **54** (1) 2-10.

O'Carroll, N. 1978 The nursing of Sitka spruce: 1 Japanese larch. *Irish Forestry* **35** (1) 60-65.

O'Carroll, N. 1995 The nature of forestry. *Irish Forestry* **52** (1/2) 70-74.

Odén, S. 1976 The acidity problem - an outline of concepts Proc. 1st International Symposium on *Acid Precipitation and forest ecosystems.* 1-35 USDA Forest Service General Technical Report NE-23.

Ogilvie, J.F. 1984 Control of unwanted conifers - chemical thinning. (In) Weed control & vegetation management in forests & amenity areas *Aspects of applied biology.* **5** 195-209 Association of Applied Biologists., Warwick.

Olen, H. 1991 *Liming acidic surface waters.* ISBN 0 87371 243 9 Lewis Publishers Inc., Michigan.

Örke, E-C., Dehne, H-W., Schönbeck, F. & Weber, A. 1994 *Crop production & Crop protection; estimated losses in major food & cash crops.* 808pp Elsevier Science BV, Netherlands.

Ormerod, S.J. 1995 Modelling biological responses, present and future. (In) *Critical loads for acid deposition for United Kingdom freshwaters* 47-57 Dept. of the Environment (Critical Loads Advisory Group)

Ormerod, S.J., Donald, A.P. & Brown, S.J. 1989 The influence of plantation forestry on the pH and aluminium concentration of upland Welsh streams: a re-examination. *Environmental Pollution* 62 (1) 47-62.

Oswald, H. 1982 Sylviculture of oak & beech high forests in France. (In) *Broadleaves in Britain* Proc. symposium, Loughborough (Eds.) D.C. Malcolm, J. Evans & P.N. Edwards. 31-39 Institute of Chartered Foresters, Edinburgh.

Oszlanyi, J. 1997 Forest health & environmental pollution in Slovakia. *Environmental Pollution* 98 (3) 389-392.

Ovington, J.D. 1950 The afforestation of Culbin Sands. *Journal of Ecology* 38 303-19.

Ovington, J.D. 1956 The composition of tree leaves. *Forestry* 29 (1) 22-28.

Ovington, J.D. 1962 Quantitative ecology and the woodland ecosystem concept. *Advances in ecological research* 1 103-192.

Ovington, J.D. & MacRae, C. 1960 Growth of seedlings of Q. petraea. *Journal of Ecology.* 48 549-555.

Ovington, J.D. & Madgewick, H.A.I. 1957 Afforestation & soil reaction. *Journal of Soil Science* 8 141-149.

Ovington, J.D. & Madgewick, H.A.I. 1959 Growth & composition of natural stands of birch; 2 uptake of natural nutrients. *Plant & Soil* 10 389-400.

Owen, T.H. 1954 Observations on the monthly litter-fall and nutrient content of Sitka spruce litter. *Forestry* 27 (1) 7-14.

Page, G. 1968 Some effects of conifer crops on soil properties. *Commonwealth Forestry Review* 47 (1) 52-62.

Page, L. 1993 Forest establishment on opencast restoration sites; some observations at Maesgwyn. *Quarterly Journal of Forestry* 87 (4) 284-285.

Pain, S. 1999 Fiendish fungus. *New Scientist* 15.5.99 7.

Pain, B.F., van der Weerden, T.J., Chambers, B.J., Phillips, V.R. & Jarvis, S.C. 1998 A new inventory for ammonia emissions from UK agriculture. *Atmospheric Environment* 32 (3) 309-313.

Palfreyman, J.W., Smith, G.M. & Sinclair, D.C.R. 1995 Disposal of preservative treated timber: research into methods of decommissioning timber treated with metal based preservatives. (In) *Heavy metals & trees* Proc. conf. Glasgow 1995 (Ed.) I. Glimmerveen. 150-159 Institute of Chartered Foresters, Edinburgh.

Palmar, C.E. 1958a The capercailzie. *For. Comm. Leaflet* 37 HMSO, London.

Palmar, C.E. 1958b Woodpeckers in woodlands. *For. Comm. Leaflet* 43 HMSO, London.

Palmar, C.E. 1962 Titmice in woodlands. *For. Comm. Leaflet* 46 HMSO, London.

Palmer, C. 1988 Bracken control in new forestry plantations using dicamba as a concentrate ribbon. *Aspects of applied biology* 16 289-297 Association of Applied Biologists., Warwick.

Palmer, C. 1993 Choosing the herbicide for the job. *Timber Grower* Winter, 1993 31-32.

Palmer, C. 1994 Arsenal 50F. *Forestry & British Timber* Feb. 1994 36-38.

Palmer, J. 1995 Heavy metals in derelict & contaminated land. (In) *Heavy metals & trees* Proc. conf. Glasgow 1995 (Ed.) I. Glimmerveen. 10-17 Inst. of Chartered Foresters, Edinburgh.

Palmer, C. 1998 Welcome back casuron, all is forgiven; atrazine update; scrub control with arsenal. *Forestry & British Timber* Feb. 1998 36-41.

Palmer, C., Reid, D.F. & Godding, S.J. 1988 A review of forestry trials with a formulation of trichlorpyr, dicamba & 2,4-D. *Aspects of applied biology* 16 207-214 AAB, Warwick.

Parker, J. 1974 Beech bark disease. *For. Comm. Forest Rec.* 96 1-16 HMSO, London.

Parr, T.W. 1988 Long-term effects of grass growth retardants with particular reference to the ecology and management of vegetation on roadside verges. (In) The practice of weed control

& vegetation management in forestry, amenity & conservation areas. *Aspects of applied biology* 16 35-45 Association of Applied Biologists, Warwick.

Parry, M. & Rosenzweig, C. 1993 The potential effects of climate change on world food supply. (In) Global climatic change - implications for crop protection *Proc. BCPC Symposium.* **Monograph 56** 33-56 British Crop Protection Council, Farnham, Surrey.

Parsons, D. & Evans, J. 1977 Forest fire protection in the Neath District of South Wales. *Quarterly Journal of Forestry* **71** 186-198.

Patch, D. & Denyer, A. 1992 Blight to trees caused by vegetation control machinery. *Arboriculture Research Note* **107-92-ARB** Arboricultural Advisory & Information Service, Farnham.

Patch, D., Coutts, M.P. & Evans, J. 1984 Control of epicormic shoots on amenity trees. *Arboriculture Research Note* **54/84/SILS** DOE Arboricultural Advisory & Information Service, Farnham.

Patch, D., Coutts, M.P. & Evans, J. 1989 Control of epicormic shoots on amenity trees. *Arboriculture Research Note* **54/89/SILS** Arboricultural Advisory & Information Service, Farnham, Surrey.

Patrick, K.N. 1991 Watermark disease of cricket bat willow: guidelines for growers. *For. Comm. Res. Inf. Note* **197** Forestry Commission, Forest Research Station, Farnham.

Pawsey, R.G. 1960 An investigation into *Keithia* disease of *Thuya plicata*. *Forestry* **33** (2) 174-186.

Pawsey, R.G. 1963 Rotational sowing of *Thuya plicata* to avoid infection by *Keithia thujina*. *Quarterly Journal of Forestry* **56** 206-209.

Pawsey, R.G. 1964a Needle-cast of pine (*Lophodermium pinastri*). *For. Comm. Leaflet* **48** 1-6 HMSO, London.

Pawsey, R.G. 1964b Grey mould in forest nurseries. *For. Comm. Leaflet* **50** 1-7 HMSO, London.

Pawsey, R.G. 1964c Cycloheximide fungicide trials against *Didymascella thujina* on Western red cedar. *For. Comm. Rep. For. Res.* **1964** 141-150 HMSO, London.

Pawsey, R.G. 1983 Ash die-back survey, summer 1983. *Commonwealth Forestry Institute Occasional Paper* **24** CFI, Oxford.

Pawsey, R.G. 1987 Ash decline. (In) Advances in practical arboriculture (Ed) D. Patch *For. Comm. Bull.* **65** 156-159.

Pawsey, R.G. & Rahman, M.A. 1976 Chemical control of infection by honey fungus, *Armillaria mellea*. *Arboricultural Journal* **2** 468-479 HMSO, London.

Peace, T.R. 1936 *Meria laricis*, the leaf cast disease of larch. *Forestry* **10** (1) 79-82.

Peace, T.R. 1938 Butt rot of conifers in Great Britain. *Quarterly Journal of Forestry* **32** 81-104.

Peace, T.R. 1952a Poplars. *For. Comm. Bull.* **19** 1-50 HMSO, London.

Peace, T.R. 1952b The defoliation of pines with particular reference to *Lophodermium pinastri*. *For. Comm. Res. Br. Pap.* **6** 8pp Departmental Circulation, Forestry Commission, London.

Peace, T.R. 1954 Experiments on spraying with DDT to prevent feeding of *Scolytus* beetles on elm and consequent infection with *Ceratostomella ulmi*. *Annals of Applied Biology* **41** 155-164.

Peace, T.R. 1955 Sooty bark disease of sycamore - a disease in eclipse. *Quarterly Journal of Forestry* **49** 197-204.

Peace, T.R. 1958a A single case of fume damage. *Quarterly Journal of Forestry* **52** 41-45.

Peace, T.R. 1958b Raising *Thuya* in isolated nurseries to avoid infection by *Keithia thujina*. *Scottish Forestry* **12** 7-10.

Peace, T.R. 1960 Status and development of Elm disease in Britain. *For. Comm. Bull.* **33** HMSO, London.

Peace, T.R. 1962 *Pathology of trees and shrubs with special reference to Britain.* 754pp Oxford University Press.

Pearce, D. 1991 Assessing the returns to the economy & to society from investments in forestry. *For. Comm. Occasional Pap.* **47** 1-44 Forestry Commission, Edinburgh.

Pearce, F. 1992 No southern comfort at Rio. *New Scientist* **16.5.92** 36-41.

Pearce, F. 1996a Greedy patenting could starve poor of biotech promise. *New Scientist* **16.11.96** 6.

Pearce, F. 1996b Tropical smogs rival big city smoke. *New Scientist* **18.5.1996** 4.

Pearce, F. 1997 Burn me. *New Scientist* **22.11.1997** 31-34.

Pearce, F. 1998a Population bombshell. *New Scientist* **11.7.98** Inside Science (Inset) 4pp.

Pearce, F. 1998b Is solar activity responsible for changes in climate? *New Scientist* **11.7.1998** 45-48.

Pearce, F. 1999 Chill in the air. *New Scientist* **22.11** 1997 28-32.

Pedersen L.B. & Bille-Hansen, J. 1995 Effects of nitrogen load to the forest floor in Sitka spruce stands; deposition and spruce aphis infestations. *Water, Air & Soil Pollution* **85.**

Pendrick, D. 1998 Methane hydrate - tomorrow's fuel. *New Scientist* 1998 39-41

Penn, R.G. 1979 The state control of medicines: the first 3000 years. *British Journal of Clinical Pharmacology* **8** 293-305.

Pepper, H.W. 1976 Rabbit management in woodlands. *For. Comm. Leaflet* **67** 15pp HMSO, London.

Pepper, H.W. 1978 Chemical repellants. *For. Comm. Leaflet* **73** HMSO, London.

Pepper, H.W. 1982 Rabbit control - phostoxin. *Arboriculture Research Note* **43/82/WILD** Forestry Commission, Edinburgh.

Pepper, H.W. 1987 Rabbit control - phostoxin. *Arboriculture Research Note* **43/87/WLD** Forestry Commission, Edinburgh.

Pepper, H.W. 1988 Grey squirrels. *For. Comm. Rep. For. Res.* **1988** 51-52 HMSO, London.

Pepper, H.W. 1989 Hopper modification for grey squirrel control. *For. Comm. Res. Inf. Note* **153** 1-4 Forestry Commission, Forest Research Station, Farnham.

Pepper, H.W. 1990a Grey squirrel damage control with warfarin. *For. Comm. Res. Inf. Note* **180** Forestry Commission, Forest Research Station, Farnham.

Pepper, H.W. 1990b Grey squirrels & the law. *For. Comm. Res. Inf. Note* **191** 1-2 Forestry Commission, Forest Research Station, Farnham.

Pepper, H.W. 1992a Forest fencing. *For. Comm. Bull.* **102** 42 pp HMSO, London.

Pepper, H.W. 1992b Survey of squirrels, 1992. *Timber Grower* **July 1992** Timber Growers UK, Edinburgh.

Pepper, H.W. 1993 Red squirrel supplementary food hopper. *For. Comm. Res. Inf. Note* **235** Forestry Commission, Forest Research Station, Farnham.

Pepper, H.W. 1995a Mixing squirrel bait. *Quarterly Journal of Forestry* **89** (2) 138.

Pepper, H.W. 1995b How effective is your rabbit control? *Forestry & British Timber* **Feb. 1995** 41.

Pepper, H.W. 1995c Squirrel management in woodlands. *For. Comm. Rep. For. Res.* **1995** 59 HMSO, London.

Pepper, H.W. 1997a Mammals & birds associated with trees: the damage they cause. (In) Arboricultural practice, present & future (Ed. J. Claridge) *Research for amenity trees* **6** 87-92 Dept of Environment, Transport and the Regions. (formerly Dept of Environment)

Pepper, H.W. 1997b Grey squirrels. *Quarterly Journal of Forestry* **91** (4) 338-9.

Pepper, H. 1998 The prevention of rabbit damage to trees in woodland. *Forestry Authority Practice Note* **FCPN 2** 6 pp Forestry Commission, Edinburgh.

Pepper, H. 1999 Recommendations for fallow, roe & muntjac deer fencing; new proposals for temporary fencing. *For. Comm. Practice Note* **FCPN 9** 6 pp Forestry Commission, Edinburgh.

Pepper, H. & Currie, F. 1998 Controlling grey squirrel damage to woodlands. *For. Comm. Practice Note* **FCPN 4** 4 pp Forestry Commission, Farnham, Surrey.

Pepper, H.W. & Hodge, S. 1996 Squirrel population & habitat management. *For. Comm. Rep. For. Res.* **1996** 40-42 HMSO, London.

Pepper, H. & Patterson, G. 1998 Red squirrel conservation. *For. Comm. Practice Note* **FCPN 5** 8 pp Forestry Commission, Farnham, Surrey.

Pepper, H.W. & Stocker, D. 1993 Grey squirrel control using modified hoppers, *For. Comm. Res. Inf. Note* **232** Forestry Commission, Forest Research Station, Farnham.

552 *Pesticides, pollutants, fertilizers & trees*

Pepper, H.W. & Tee, L.A. 1972 Forest fencing. *For. Comm. Forest Rec.* **80** 50pp HMSO, London.

Pepper, H.W. & Tee, L.A. 1984 Wildlife management - chemical & mechanical repellants. *For. Comm. Rep. For. Res.* **1984** 43 HMSO, London.

Pepper, H.W. & Tee, L.A. 1987 Forest fencing. *For. Comm. Leaflet* **87** 42pp HMSO, London.

Pepper, H.W., Rowe, J.J. & Tee, L.A. 1984 Individual tree protection. *For. Comm. Arboricultural Leaflet* **10** HMSO, London.

Pepper, H.W., Davies, R.J. & Mayle, B.A. 1986 Wildlife and conservation - voles. *For. Comm. Rep. For. Res.* **1986** 46 HMSO, London.

Pepper, H.W., Chadwick, A.H. & Butt, R. 1991 Electric fencing against deer. *For. Comm. Res. Inf. Note* **206** 4pp Forestry Commission, Farnham, Surrey.

Pepper, H.W., Dagnall, J. & Gurnell, J. 1994 Grey squirrel damage prediction: index trapping. *For. Comm. Rep. For. Res.* **1994** 50-51 HMSO, London.

Pepper, H.W., Neil, D & Hemmings, J. 1996 Application of chemical repellent Aaprotect to prevent winter browsing. *For Comm. Res Inf. Note* **289** 4pp Forestry Commission, Farnham, Surrey.

Perlin, J. 1989 *A forest journey.* 445pp Harvard University Press.

Perry, H. 1999 Tree-eating beetles from China spark off an alert. *Horticulture Week* **Jan. 21 1999** 14.

Petty, S.J. 1983 Wildlife management - Bird studies. *For. Comm. Rep. For. Res.* **1983** 41 HMSO, London.

Petty, S.J. 1987 Wildlife & Conservation - Birds. *For. Comm. Rep. For. Res.* **1987** 55-56 HMSO, London.

Phillips, D.H. 1963 Leaf cast of larch. *For. Comm. Leaflet* **21** 4pp HMSO, London.

Phillips, D.H. 1978 The EEC plant health directive and British forestry. *For. Comm. Forest Rec.* **116** HMSO, London.

Phillips, D.H. 1980a Plant health legislation & trees. *Arboricultural Journal* **4** (2) 152-157.

Phillips, D.H. 1980b International plant health: controls, conflicts, problems & co-operation: a European experience. *For. Comm. Res. & Dev. Pap.* **125** 1-20 Forestry Commission, Edinburgh.

Phillips, D.H. & Bevan, D. 1967 Forestry quarantine and its biological background. *For. Comm. Forest Rec.* **63** 12pp HMSO, London.

Phillips, D.H. & Burdekin, D. 1982 *Diseases of forest and ornamental trees.* 435pp. Macmillan, London.

Phillips, D.H. & Greig, B.J.W. 1970 Some chemicals to prevent stump colonisation by *Fomes annosus.* *Annals of Applied Biology* **66** 441-452.

Pierce, G.D. & Gibbs, J.N. 1981 *Verticillium* wilt. *Arboricultural Leaflet* **9** 9pp HMSO, London. (Prepared for DoEnv by the Forestry Commission.)

Pierpont, W.S. & Shewry, P.R. (Eds.) 1996 *Genetic engineering of crop plants for resistance to pests and diseases.* 104pp British Crop Protection Council, Farnham, Surrey.

Pinchin, A. 1999a A shot in the bark. *Horticulture Week* **March 11 1999** 18-21.

Pinchin, A. 1999b De-icing with death. *Horticulture Week* **Jan. 28 1999** 26-30

Pinchin, R.D. 1951 Plantations on opencast ironstone mining areas in the Midlands. *For. Comm. Rep. For. Res.* **1951** 31-34 HMSO, London.

Pinder, P.S. & Hayes, A.J. 1986 An outbreak of Vapourer moth on Sitka spruce in C. Scotland. *Forestry* **59** (1) 97-105.

Pitman, R. & Webber, J. 1998 Bracken as a peat alternative. *Forestry Authority Information Note* 6pp Forestry Commission, Edinburgh.

PORG 1993 *Ozone in the United Kingdom 1993.* 3rd report of the Photochemical Oxidants Review Group. 170pp Department of the Environment, London.

PORG 1997 *Ozone in the United Kingdom.* 4th report of the Photochemical Oxidants Review Group. 234pp Department of the Environment, London.

Potter, C.J. 1988 An evaluation of weed control techniques for tree establishment. (In) The practice of weed control & vegetation management in forestry, amenity & conservation areas. *Aspects of applied biology* **16** 337-346 Association of Applied Biologists.

Potter, C.J. 1989 *Coppiced trees as energy crops.* Final report on contract E/5A/CON/1078/ 174/063 98pp Forestry Commission, Alice Holt Lodge, Farnham, Surrey.

Potter, M.J. 1991 Tree shelters. *For. Comm. Handbk.* **7** 26-27 HMSO, London.

Potter, C.J. & Tabbush, P.M. 1991 Cultivation of poplar as an energy crop. (In) *Wood for Energy: Implications for harvesting, utilisation & marketing.* Proc. Discussion Meeting (Ed.) J.R. Aldhous 206-234 Institute of Chartered Foresters, Edinburgh.

Pratt, J.E. 1974 Report on survey into timber losses in western hemlock and grand fir as a result of infection by *Fomes annosus.* (In) The potential of western hemlock, western red cedar, grand fir & noble fir in Britain. *For. Comm. Bull.* **49** 98-100 HMSO, London.

Pratt, J.E. 1979a *Fomes annosus* butt-rot of Sitka spruce: I Observations on the development of butt-rot in individual trees and stands. *Forestry* **52** (1) 11-29.

Pratt, J.E. 1979b *Fomes annosus* butt-rot of Sitka spruce: II Loss of strength of wood in various categories of rot. *Forestry* **52** (1) 31-45.

Pratt, J.E. 1979c *Fomes annosus* butt-rot of Sitka spruce: III Losses in yield and value of timber in diseased trees and stands. *Forestry* **52** (2) 113-127.

Pratt, J.E. 1996a Borates for stump treatment; a literature review. *For. Comm. Technical Pap.* **15** 1-19 Forestry Commission, Edinburgh.

Pratt, J.E. 1996b *Fomes* stump treatment - an update. *For Comm. Res Inf. Note* **287** 3pp Forestry Commission, Forest Research Station, Farnham.

Pratt, J.E. 1997 *Fomes* stump treatment - an update. *Quarterly Journal of Forestry* **91** (1) 63-65.

Pratt, J.E. 1998 Economic appraisal of the benefits of control treatments. (In) *Heterobasidion annosum, biology, ecology, impact & control* (Eds.) S. Woodward et al. 315-332 CAB International, Wallingford, Oxon. England.

Pratt, J. 1999 PG suspension for the control of *Fomes* root rot of pine. *For. Comm. Information Note* FCIN **18** 4pp.

Pratt, J.E. & Greig, B.J.W. 1988 *Heterobasidion annosum*: development of butt rot following thinnings in two young first rotation stands of Norway spruce. *Forestry* **61** (4) 339-347.

Pratt, J.E. & Lloyd, J.D. 1996 Use of disodium octaborate tetrahydrate to control conifer butt rot caused by *Heterobasidion annosum. Proc. conf.: Crop protection in northern Britain, 1996* 207-212 SCRI, Dundee.

Pratt, J.E., Redfern, D.B. & Burnand, A.C. 1989 Modelling the spread of *Heterobasidion annosum* in Sitka spruce plantations in Britain. (In) *Proc. 7th Int Conf. on root & butt rots, Canada* 308-319 Forestry Canada, Victoria, B.C.

Pratt, J.E., Johansson, M. & Hüttermann, A. 1998a Chemical control of *Heterobasidion annosum* (In) *Heterobasidion annosum, biology, ecology, impact & control* (Eds.) S. Woodward et al. 259-282 CAB International.

Pratt, J.E., Shaw III, C.G. & Vollbrecht, G. 1998b Modelling disease development in forest stands. (In) *Heterobasidion annosum, biology, ecology, impact & control* (Eds.) S. Woodward et al. 213-234 CAB International, Wallingford, Oxon. England.

Pratt, J.E., Gibbs, J.N. & Webber, J.F. 1999 Registration of *Phlebiopsis gigantea* as a forest biocontrol agent in the UK: recent experience. *Biocontrol Science & Technology* **9** 113-118.

Preece, T.F. 1977 Watermark disease of the cricket bat willow (3rd edn). *For. Comm. Leaflet* **20** 1-9 HMSO, London.

Proe, M.F. 1994 Plant Nutrition. (In) Forest Nursery Practice *For. Comm. Bull.* **111** 39-65 HMSO, London.

Proe, M.F. & Dutch, J.C. 1994 Impact of whole tree harvesting on second rotation growth of Sitka spruce: the first 10 years. *Forest Ecology & Management* **66** (1-3) 39-54.

Proe, M.F., Dutch, J.C., Miller, H.G. & Sutherland, J. 1992 Long-term partitioning of biomass and nitrogen following application of nitrogen fertilizer to Corsican pine. *Canadian Journal of Forest Research.* **22** (1) 82-87.

Proe, M.F., Cameron, J.D., Dutch, J. & Christodoulou, X.C. 1996 Effect of whole-tree harvesting on growth of second rotation Sitka spruce. *Forestry* **69** (4) 389-401.

Pryor, S.N. 1985 The silviculture of wild cherry or gean (*Prunus avium* L.). *Quarterly Journal of Forestry* **79** (2) 95-109.

Pryor, S.N. 1988 The silviculture and yield of wild cherry. *For. Comm. Bull.* **75** 23pp HMSO, London.

Punshon, T., Dickinson, N.M. & Lepp, N.W. 1996 Potential of Salix clones for bioremediating metal polluted soil. (In) *Heavy metals & trees* Proc. conf. Glasgow 1995 (Ed.) I. Glimmerveen. 93-104 Institute of Chartered Foresters, Edinburgh.

Putwain, P.D., Evans, B.E. & Kerry, S. 1988 Early establishment of amenity woodland on roadsides by direct seeding. (In) The practice of weed control & vegetation management in forestry, amenity & conservation areas. *Aspects of applied biology* **16** 63-72 Association of Applied Biologists.

Pyatt, D.G. 1970 Soil groups of upland forests. *For. Comm. Forest Rec.* **71** 51pp HMSO, London.

Pyatt, D.G. 1982 Soil classification. *For Comm. Res Inf. Note* **68/82/SSN** Forestry Commission, Edinburgh.

Pyatt, D.G. 1984 Effect of afforestation on quality of water run off. *For Comm. Res Inf. Note* **83/84/SSN** 4pp Forestry Commission, Forest Research Station, Farnham.

Pyatt, D.G. 1994 Forestry nursery soils. (In) Forest Nursery Practice (Eds) J.R. Aldhous & W.L. Mason *For. Comm. Bull.* **111** 25-33 HMSO, London.

Pyatt, D.G. 1995 An ecological site classification for forestry in Great Britain. *For Comm. Res Inf. Note* **260** Forestry Commission, Edinburgh.

Pyatt, D.G. & Anderson, A.R. 1986 Increased streamflow after clear felling a spruce plantation. (In) *Effects of land use on fresh waters* (Eds.) J.F.& L.G. Solbé 538-540 Water Research Centre.

Pyatt, D.G. & Suarez, J.C. 1997 An ecological site classification for forestry in Great Britain (with special reference to Grampian Scotland). *For. Comm. Technical Pap.* **20** Forestry Commission, Edinburgh.

Pyatt, D.G., Craven, M.M. & Williams, B.L. 1979 Peat classification for forestry in Great Britain. Proc. International Symposium on classification of peat & peatlands **46** 351-366 International Peat Society.

Pyatt, D.G., Harrison, D. & Ford, A.S. 1969 Guide to the site types in forests of North & mid-Wales. *For. Comm. Forest Rec.* **69** 35pp HMSO, London. (See also 2nd edn, 1982)

QUARG (Quality of Urban Air Review Group) 1996 *Airborne particulate matter in the United Kingdom.* 176pp Department of the Environment, London.

Rackham, O. 1980 *Ancient woodland, its history, vegetation and uses in England.* Edward Arnold, London.

Rackham, O 1986 *The History of the Countryside.* Ch. 11, Elms. 232-247 J.M. Dent, London.

Ramsay, W.J.H. 1986 Bulk soil handling for quarry restoration. *Soil Use & Management* **2** 30-39.

Ranger, J & Nys, C. 1996 Biomass & nutrient content of extensively and intensively managed coppice stands. *Forestry* **69** (2) 91-110

Ratcliffe, P.R. 1985 Glades for deer control in upland forests. *For. Comm. Leaflet* **86** HMSO, London.

Raven, J.A. & Yin, Z-H. 1998 The past, present and future of nitrogenous compounds in the atmosphere, and their interactions with plants. *New Phytologist* **139** 205-219.

Rawlinson, A.S. 1968 Log storage and prevention of degrade during storage. *For. Comm. Res. & Dev. Pap.* **70** 1-20 Forestry Commission, London (Departmental use)

Ray, D. & Anderson, A.R. 1990 Soil regimes on mounds of gleyed soil. *For Comm. Res Inf. Note* **168** 4pp Forestry Commission, Forest Research Station, Farnham.

Rayner, M.C. 1936 Mycorrhizal habit in forestry. *Forestry* **10** (2) 1-22.

Rayner, M.C. & Neilson-Jones, W. 1944 *Problems in tree nutrition.* 184pp Faber & Faber Ltd, London.

RCEP 1996 *Sustainable use of soil.* Royal Commission on Environmental Pollution; 19th report. **Cm 3165** HMSO, London.

Read, D.J. 1984 Interaction between ericaceous plants and their competitors with special reference to soil toxicity. (In) Weed control & vegetation management in forests & amenity areas *Aspects of applied biology.* **5** 195-209 Association of Applied Biologists., Warwick.

Read, D.J. 1991 Mycorrhizas in ecosystems. *Experientia* **47** 376-391.

Read, D.J. & Finlay, R.D. 1987 Mycorrhizas and inter-plant transfer of nutrients. (In) *Proc. 6th North American Conference on mycorrhizae* (Ed.) R. Molina 319-320 Forest Research Laboratory, Corvallis, Oregon.

Redfern, D.B. 1978 Infection by *Armillaria mellea* and some factors affecting host resistance and the severity of the disease. *Forestry* **51** (2) 121-136.

Redfern, D.B. 1982 Infection of *Picea sitchensis* and *Pinus contorta* stumps by basidiospores of *Heterobasidion annosum.* *European Journal of Forest Pathology* **12** 11-25.

Redfern, D.B. 1993 The effect of wood moisture on infection of Sitka spruce stumps by basidiospores of *Heterobasidion annosum.* *European Journal of Forest Pathology* **23** 218-235.

Redfern, D.B. & Cannell, M.G.R. 1982 Needle damage in Sitka spruce caused by early autumn frosts. *Forestry* **55** (1) 39-45.

Redfern, D.B. & Low, J.D. 1972 Pathology (Advisory service, Northern Research Station). *For. Comm. Rep. For. Res.* **1972** 97.

Redfern, D.B. & Stenlid, J. 1998 Spore dispersal and infection. (In) *Heterobasidion annosum, biology, ecology, impact & control* (Eds.) S. Woodward et al. 105-141 CAB International, Wallingford, Oxon. England.

Redfern, D.B. & Ward, D. 1998 The UK and Ireland. (In) *Heterobasidion annosum, biology, ecology, impact & control* (Eds.) S. Woodward et al. 347-354 CAB International, Wallingford, Oxon. England.

Redfern, D.B., Stoakley, J.T. & Minter, D.W. 1987 Dieback and death of larch caused by *Ceratocystis laricicola* following attack by *Ips cembrae.* *Plant Pathology* **36** 467-480

Redfern, D.B., Boswell, R.C. & Proudfoot, J.C. 1994a Forest condition, 1993. *For Comm. Res Inf. Note* **251** 4pp Forestry Commission, Forest Research Station, Farnham.

Redfern, D.B., Pratt, J. & Whiteman, A. 1994b Stump treatment against *Fomes.* Comparison of costs and benefits. *For Comm. Res Inf. Note* **248** Forestry Commission, Forest Research Station, Farnham.

Redfern, D.B., Boswell, R.C. & Proudfoot, J.C. 1995 Forest condition, 1994. *For Comm. Res Inf. Note* **262** 4pp Forestry Commission, Forest Research Station, Farnham.

Redfern, D.B., Boswell, R.C. & Proudfoot, J.C. 1996 Forest condition, 1995. *For Comm. Res Inf. Note* **282** 4pp Forestry Commission, Forest Research Station, Farnham.

Redfern, D.B., Boswell, R.C. & Proudfoot, J.C. 1997a Forest condition, 1996. *For Comm. Res Inf. Note* **291** 4pp Forestry Commission, Forest Research Station, Farnham.

Redfern, D.B., Gregory, S.C. & MacAskill, G.A. 1997b Inoculum concentration and colonization of *Picea sitchensis* stumps by basidiospores of *Heterobasidion annosum.* *Scandinavian Journal of Forest Research* **12** 41-49.

Redfern, D.B., Boswell, R.C. & Proudfoot, J.C. 1998 Forest condition, 1997. *For. Comm. Information Note* FCIN 4 6pp Forestry Authority, Edinburgh.

Redfern, D.B., Boswell, R.C. & Proudfoot, J.C. 1999 Forest condition, 1998. *For. Comm. Information Note* FCIN 19 6pp Forestry Authority, Edinburgh.

Rennenberg, H., Kreutzer, K., Papen, H. & Weber, P. 1998 Consequences of high loads of nitrogen for spruce and beech forests. *New Phytologist* **139** 71-86.

Rennie, P.J. 1954 Physical chemistry of forest soils. *For. Comm. Rep. For. Res.* **1954** 62-63 HMSO, London.

Rennie, P.J. 1955 The uptake of nutrients by mature forest growth. *Plant & Soil* **7** 49-95.

Reynolds, B., Cape, J.N. & Paterson, I. S. 1989 A comparison of elemental fluxes in throughfall beneath larch and Sitka spruce at two contrasting sites in the United Kingdom. *Forestry* **62** (1) 29-39.

Reynolds, B., Stevens, P.A., Adamson, J.K., Hughes, S. & Roberts, J.D. 1992 Effects of clearfelling on stream and soil water aluminium chemistry in three UK forests. *Environmental Pollution* **77** 157-165.

Reynolds, E.R.C. & Leyton, L. 1963 Measurement & significance of throughfall in forest stands. (In) *The water relations of plants* Blackwell Scientific Publications, Oxford.

Reynolds, L. 1953 The manufacture of wood charcoal in Great Britain. *For. Comm. Forest Rec.* **19** HMSO, London.

RGAR 1997 *Acid deposition in the United Kingdom* 4th Report of the UK Review Group on Acid Rain. Prepared for Department of the Environment, Transport & the Regions. 174pp AEA Technology, Culham, Oxfordshire.

Richards, E.G. 1957 Utilisation development *For. Comm. Rep. For. Res.* **1957** 83 HMSO, London.

Richards, E.G. 1961 Utilisation development *For. Comm. Rep. For. Res.* **1961** 79 HMSO, London.

Richardson, J.A. 1993 Long-term experiments on tree growth on colliery & limestone waste. *Quarterly Journal of Forestry* **87** (3) 195-202.

Ridgeway, R.L., Inscoe, M. & Arn, H. (Eds.) 1992 Insect pheromones & other behaviour-modifying chemicals. *BCPC Monograph* **51** 135 British Crop Protection Council.

Riley, D. 1991 Using soil residue data to assess the environmental safety of pesticides. Proc. BCPC symposium, Warwick **Monograph 47** 11-20 British Crop Protection Council, Farnham, Surrey.

Rind, D. 1995 Drying out the tropics. *New Scientist* **6.6.95** 36-40

Risbeth, J. 1951a Observations on the biology of *Fomes annosus* with particular reference to East Anglian plantations. II. Spore production, stump infection and saprophytic activity in stumps. *Annals of Botany* **15** (NS) 1-21.

Risbeth, J. 1951b Observations on the biology of *Fomes annosus* with particular reference to East Anglian plantations. III. Natural and experimental infection of pine and some factors affecting the severity of the disease. *Annals of Botany* **15** (NS) 221-246.

Risbeth, J. 1952 Control of *Fomes annosus*. *Forestry* **25** (1) 41-50

Risbeth, J. 1957 Some further observations on *Fomes annosus* Fr. *Forestry* **30** (1) 69-89.

Risbeth, J. 1959a Stump protection against *Fomes annosus* I Treatments with creosote. *Annals of Applied Biology* **47** 519-528.

Risbeth, J. 1959b Stump protection against *Fomes annosus* II Treatment with substances other than creosote. *Annals of Applied Biology* **47** 529-541.

Risbeth, J. 1963 Stump protection against *Fomes annosus* III Inoculation with *Peniophera gigantea* *Annals of Applied Biology* **52** (1) 63-77.

Risbeth, J. 1976 Chemical treatment and inoculation of hardwood stumps for control of *Armillaria mellea*. *Annals of Applied Biology* **42** 1131-1139.

Risbeth, J. 1987 Honey fungus. (In) Practical advances in arboriculture. *For. Comm. Bull.* **65** 181-185 HMSO, London.

Roberts, G. 1983a Effects of different land uses and changes of land use on water resources in upland Britain. *Man & the bio-sphere workshop, Project V, Land-use impacts on aquatic systems.* 193-216

Roberts, J. 1983b Forest transpiration: a conservative hydrological process? *Journal of Hydrology* **66** 133-141.

Roberts, T.M., Skeffington, R.A. & Blank, L.W. 1989 Cause of 'Type 1' spruce decline in Europe. *Forestry* **62** (3) 179-222.

Robinson, R.K. 1973 The production by roots of *Calluna vulgaris* of a factor inhibitory to some mycorrhizal fungi. *Journal of Ecology* **60** (1) 219-224.

Roche, L. 1988 Forestry & famine - arguments against growth without development. *For. Comm. Occasional Pap.* **16** 54-60

Roe, M. & Tee, L.A. 1980 Electric and mesh (fencing). *Forestry & British Timber* **May 1980** 24-27.

Rogers Brambell, G.W. 1958 Voles and field mice. *For. Comm. Leaflet* **44** 11pp HMSO, London.

Rogers, E.V. 1965 Rabbit control in woodlands. *For. Comm. Booklet* **14** HMSO, London.

Rogers, E.V. 1974 Selection & development of equipment for ULV herbicide spraying in forestry. (In) *BCPC Monograph* **13** BCPC, Farnham, Surrey.

Rogers, E.V. 1975 Ultra low volume herbicide spraying. *For. Comm. Leaflet* **62** 20pp HMSO, London.

Rose, D.R. 1989a *Marssonina* canker and leaf spot of weeping willows. *Arboriculture Research Note* **78/89/PATH** Arboricultural Advisory & Information Service, Farnham, Surrey.

Rose, D.R. 1989b Willow scab and canker. *Arboriculture Research Note* **79/89/PATH** Arboricultural Advisory & Information Service, Farnham, Surrey.

Rose, D.R. & Gregory, S.J. 1997 Spotting cherries. *Tree Damage Alert* **43** 1page Produced by the Forestry Commission under the auspices of the Department of the Environment

Ross, P.G. 1988 Herbicide use in British tree nurseries: problems & perspectives. *Aspects of applied biology* **16** 245-248 Association of Applied Biologists., Warwick.

Rotherham, I.D. & Read, D.J. 1988 Aspects of the competitive ecology of *Rhododendron* with reference to its competitive & invasive properties. *Aspects of applied biology* **16** 327-335 Association of Applied Biologists., Warwick.

Rowe, J.J. 1964 Field & bank voles. *For. Comm. Rep. For. Res.* **1964** 63 HMSO, London.

Rowe, J.J. 1965 Voles & mice. *For. Comm. Rep. For. Res.* **1965** 67 HMSO, London.

Rowe, J.J. 1966 Mammals & birds. *For. Comm. Rep. For. Res.* **1966** 77-78 HMSO, London.

Rowe, J.J. 1967a Mammals & birds. *For. Comm. Rep. For. Res.* **1967** 106-108 HMSO, London.

Rowe, J.J. 1967b The grey squirrel and its control. *For. Comm. Res. & Dev. Pap.* **61** 7pp Forestry Commission, London (Departmental use)

Rowe, J.J. 1968 Mammals & birds. *For. Comm. Rep. For. Res.* **1968** 116-117 HMSO, London.

Rowe, J.J. 1969 Mammals & birds. *For. Comm. Rep. For. Res.* **1969** 114-116 HMSO, London.

Rowe, J.J. 1970 Mammals & birds. *For. Comm. Rep. For. Res.* **1970** 125-126 HMSO, London.

Rowe, J.J. 1971 Mammals & birds. *For. Comm. Rep. For. Res.* **1971** 91-92 HMSO, London.

Rowe, J.J. 1973 Grey squirrel control. *For. Comm. Leaflet* **56** 1-17 HMSO, London.

Rowe, J.J. 1974 Wildlife management. *For. Comm. Rep. For. Res.* **1974** 43 HMSO, London.

Rowe, J.J. 1977 Wildlife management - chemical & mechanical repellants. *For. Comm. Rep. For. Res.* **1977** 38-39 HMSO, London.

Rowe, J.J. 1980 Grey squirrel control. *For. Comm. Leaflet* **56** 2nd edn 18pp HMSO, London.

Rowe, J.J. 1982 Wildlife management. *For. Comm. Rep. For. Res.* **1982** 33-34 HMSO, London.

Rowe, J.J. 1984 Grey squirrel bark stripping damage to broadleaved trees in southern England. *Quarterly Journal of Forestry* **78** (4) 231-236.

Rowe, J.J. & Gill, R.M.A. 1985 The susceptibility of tree species to damage by grey squirrels in England and Wales. *Quarterly Journal of Forestry* **79** (3) 183-190

Rowe, J.J., Pepper, H.W., Ratcliffe, P.R. & Mayle, B.A. 1985 Management of squirrels and other mammals. *For. Comm. Rep. For. Res.* **1985** 41 HMSO, London.

Rowntree, P.R. 1993 Climatic models - changes in environmental conditions. (In) Global climatic change - implications for crop protection *Proc. BCPC Symposium.* **Monograph 56** 13-32 British Crop Protection Council, Farnham, Surrey.

Royle, D.J., Hunter, T. & Pei, M.H. 1996 Evaluation of the biology and importance of diseases and pests in willow energy plantations. *ETSU Contractor Report* **ETSU B 1258** ETSU Harwell, Oxford.

Rutter, A.J., Kershaw, K.A., Robins, P.C. & Morton, A.J. 1971 A predictive model of rainfall interception in forests. *Agricultural Meteorology* **9** 367-384.

Rutter, A.J., Morton, A.J. & Robins, P.C. 1975 A predictive model of rainfall interception in forests. II: comparisons in some coniferous & hardwood stands *Journal of Applied Ecology* **12** 367-380.

Sale, J.S.P. 1985 Forest weed control. *For. Comm. Rep. For. Res.* **1985** 9 HMSO, London.

Sale, J.S.P. & Mason, W.L. 1985 Nursery herbicides. *For. Comm. Rep. For. Res.* **1985** 7 HMSO, London.

Sale, J.S.P., Elgy, D., Risby, P. & Shanks, C. 1982 Forest weed control. *For. Comm. Rep. For. Res.* **1982** 9 HMSO, London.

Sale, J.S.P., Tabbush, P.M. & Lane, P.B. 1983a/1986c The use of herbicides in the forest. *For. Comm. Booklet* **51** 1st & 2nd edns. Forestry Commission, Edinburgh.

Sale, J.S.P., Baker, K.F., Barwick, P.R., Elgy, D. & Shanks, C.W. 1983b Forest weed control. *For. Comm. Rep. For. Res.* **1983** 9 HMSO, London.

Sale, J.S.P., Malone, S. & Cooper, T. 1986a Forest weed control - bracken. *For. Comm. Rep. For. Res.* **1986** 10 HMSO, London.

Sale, J.S.P., Baker, K.F. & Rogers, D. 1986b Forest weed control - *Rhododendron*. *For. Comm. Rep. For. Res.* **1986** 10 HMSO, London.

Salisbury, E.J. 1964 *Weeds & aliens*. 1964 (2nd edn) 384pp Collins, London (New Naturalist series).

Salt, G.A. 1967 Pathology experiments on Sitka Spruce seedlings. *For. Comm. Rep. For. Res.* **1967** 141-146.

Salt, C.A., Hipkin, J.A. & Davidson. B. 1995 Phytoremediation - a feasible option at Lanarkshire steelworks? (In) *Heavy metals & trees* Proc. conf. Glasgow 1995 (Ed.) I. Glimmerveen. 51-60 Institute of Chartered Foresters, Edinburgh.

Salter, B.R. & Darke, R.F. 1988 Use of herbicides in the establishment of amenity plantings in Milton Keynes. *Aspects of applied biology* **16** 347-354 Association of Applied Biologists., Warwick.

Sandeman, H., Wellburn, A.R. & Heath, R.L. 1996 Forest decline & ozone - a comparison of controlled chamber & field experiments. *Ecological Studies* **127** 398pp Springer-Verlag, Berlin.

Sanders, G.E., Dixon, J. & Cobb, A.H. 1993 Will increasing ozone pollution associated with global change alter crop tolerance to herbicides? (In) Global climatic change - implications for crop protection *Proc. BCPC Symposium.* **Monograph 56** 83-94 British Crop Protection Council, Farnham, Surrey.

Sanders, H.G. (Chairman) 1961 *Toxic chemicals in agriculture & food storage.* Report of Research study group on toxic chemicals. 24-190-3 HMSO, London.

Sands, R. 1984 Environmental aspects of plantation management. (In) *Nutrition of plantation forests* (Eds.) G.D. Bowen & E.K.S. Namibar 413-438 Academic Press, London.

Sargent, C. 1984 *Britain's railway vegetation.* 34pp Institute of Terrestrial Ecology, Hills Road, Cambridge..

Satchell, J.E. 1980 Soil and vegetation changes in experimental birch plots on a *Calluna* podzol. *Soil Biology & Biochemistry* **12** 303-310

Satchell, J.E. 1989 *The history of woodlands in Cumbria.* (Ed.) J.K. Adamson *Proc. ITE symposium.* **25** 1-11 HMSO, London.

Savory, J.G., Pawsey, R.G. & Lawrence, J.S. 1965 Prevention of blue-stain in unpeeled Scots pine logs. *Forestry* **38** (1) 59-81.

Savory, J.G., Nash-Wortham, J., Phillips, D.H. & Stewart, D.H. 1970 Control of blue-stain in unbarked logs by a fungicide & an insecticide. *Forestry* **43** (2) 161-174.

Schaible, R. 1998 Effects of fertilizer applications on growth & nutrition of pole-size Sitka spruce on oligotrophic peat. *Forestry* **71** (4) 59-81.

Schjorring, J.K. 1998 Atmospheric ammonia & impacts on nitrogen deposition: uncertainties and challenges. *New Phytologist* **139** 59-60

Schulze, E-D, Oren, R. & Lange, O.L. 1989 Processes leading to forest decline. *Ecological studies* **77** 459-468.

Schulze, E-D. & Freer-Smith, P.H. 1991 Evaluation of forest decline based on field observations focussed on Norway spruce. *Proc. R. Soc. Edinburgh* **97B** 155-168.

Schumacher, 1973 *Small is beautiful.*

Scopes, N. & Stables, L. 1989 *Pest and Diseases Handbook* 732pp British Crop Protection Council.

Scott, T.M. 1972 The pine shoot moth and related species. *For. Comm. Forest Rec.* **83** 15pp HMSO, London.

Scott, T.M. & King, C.J. 1973 Control of *Hylobius abietis* & *Hylastes* spp. *For. Comm. Rep. For. Res.* **1973** 108 HMSO, London.

Scott, T.M. & King, C.J. 1974 The large pine weevil and black pine beetles. *For. Comm. Leaflet* **58** 12pp HMSO, London.

Scott, T.M. & Walker, C. 1975 Experiments with insecticides for the control of Dutch elm disease. *For. Comm. Forest Rec.* **105** 21pp HMSO, London.

Scott, J.C., Wyatt, G. & Lane, P. 1990 The 'Electrodyne' sprayer conveyor. *FC Work Study, Northern Region Report + User manual* **109** 34pp+Apps Forestry Commission, Edinburgh.

Scurlock, J. & Hall, D. 1991 The carbon cycle. *New Scientist* **2.11.91** (inset) 1-4.

Semple, R.M.G. 1964 Further trials of chemicals for debarking hardwood pulpwood. *For. Comm. Rep. For. Res.* **1964** 195-199 HMSO, London.

SEPA 1996 *State of the environment report 1996.* 102pp Scottish Environment Protection Agency, Stirling.

SEPA 1998 The groundwater regulations 1998. *SEPA Guidance Note* **1** 8pp Scottish Environment Protection Agency, Stirling.

SEPA 1999 The groundwater regulations 1998. *SEPA Guidance Note* **3** 8pp Scottish Environment Protection Agency, Stirling.

Shall, F. 1999 Stop the rot. *Horticulture Week* **April 29 1999** 22-27.

Sheppard, L.J. 1994 Causal mechanisms by which sulphur, nitrate & acidity influence frost hardiness in red spruce, review & hypothesis. *New Phytologist* **127** 69-82.

Sheppard, L.J. 1997 Soil dressing with sulphur: does it reduce the frost-hardiness in spruce seedlings? *Environmental & Experimental Botany* **37** 137-146.

Sheppard, L.J. & Cannell, M.G.R. 1985 Performance and frost hardiness of *Picea sitchensis* x *P. glauca* hybrids in Scotland. *Forestry* **58** (1) 67-74.

Sheppard, L.J., Leith, I.D. & Cape, J.N. 1991 Effects of acid mist on frost hardiness of Sitka spruce: 1 Frost hardiness & nutrient concentrations. *Environmental Pollution* **85** 229-238.

Sheppard, L.J., Leith, I.D. & Cape, J.N. 1997 Effects of acid mist on mature grafts of Sitka spruce. Pt I Frost hardiness & nutrient concentration. *Environmental & Experimental Botany* **85** 229-238.

Shi, J.L. & Brasier, C.M. 1986 Experiments on the control of Dutch elm disease by injection of *Pseudomonas* species. *European Journal of Forest Pathology* **16** 280-292.

Shore, R.F. 1996 Accumulation and significance of cadmium and lead in woodland mammals. (In) *Heavy metals & trees* Proc. conf. Glasgow 1995 (Ed.) I. Glimmerveen. 124-149 Institute of Chartered Foresters, Edinburgh.

Shorten, M. 1957 Damage caused by squirrels in Forestry Commission areas, 1954-6. *Forestry* **30** (2) 151-162.

Shorten, M. 1962 Squirrels, their biology and control. *MAFF Bulletin* **184** 44pp.

SI 1952/1929 1952 Destructive Insects and Pests, Great Britain; Importation of Forest Trees (Prohibition) Order *Statutory Instrument* **SI 1952/1929** HMSO, London.

SI 1959/428 1959 Agriculture (Safeguarding of Workplaces) Regulations. *Statutory Instrument* **SI 1959/428** HMSO, London.

SI 1961/656 1961 Landing of Unbarked Coniferous Timber Order. *Statutory Instrument* **SI 1961/656** HMSO, London.

SI 1966/1063 1966 Agricultural Employment: Safety, Health and Welfare; Agriculture (Poisonous substances) Regulations, 1966. *Statutory Instrument* **SI 1966/1063** HMSO, London.

SI 1971/1708 1971 The Dutch Elm Disease (Local Authorities) Order 1974. *Statutory Instrument* **SI 1971/1708** (revoked by SI 1972/1937) HMSO, London.

SI 1973/744, 1973 The Grey Squirrel (Warfarin) Order, 1973. *Statutory Instrument* **SI 1973/744** (under Wild Life & Countryside Act) HMSO London

SI 1974/1 1974 Importation of Forest Trees (Prohibition) Order. *Statutory Instrument* **SI 1974/1** HMSO, London.

SI 1974/2 1974 Importation of Wood (Prohibition) (Great Britain) Order. *Statutory Instrument* **SI 1974/2** HMSO, London.

SI 1974/767 1974 The Dutch Elm Disease (Restriction on Movement of Elms) Order 1974. *Statutory Instrument* **SI 1974/767** HMSO, London.

SI 1974/768 1974 Watermark (Local Authorities) Order 1974. *Statutory Instrument* **SI 1974/768** HMSO, London.

SI 1975/282 1975 Health & Safety (Agriculture) (Poisonous substances) Regulations. *Statutory Instrument* **SI 1975/282** HMSO, London.

SI 1975/1904 1975 Dutch Elm Disease (Restriction of Movement of Elms) Order. *Statutory Instrument* **SI 1975/1904** HMSO, London.

SI 1975/3004 1975 Health and Safety (Agriculture) (Poisonous substances) Regulations 1975. *Statutory Instrument* **SI 1975/3004** HMSO, London.

SI 1977/500 1977 Safety Representatives & Safety Committees Regulations. *Statutory Instrument* **SI 1977/500** HMSO, London.

SI 1980/449 1980 Import & Export of Trees, Wood & Bark (Health) (Great Britain) Order; 1980. *Statutory Instrument* **SI 1980/449** HMSO, London.

SI 1980/450 1980 Tree Pests (Great Britain) Order. *Statutory Instrument* **SI 1980/450** HMSO, London.

SI 1981/917 1981 Health & Safety (First-aid) Regulations, 1981. *Statutory Instrument* **SI 1981/917** HMSO, London.

SI 1982/217 1982 Poisons List Order, 1982. *Statutory Instrument* **SI 1982/217** (amended by SI 1992/2292) HMSO, London.

SI 1982/218 1982 Poisons Rules 1982. *Statutory Instrument* **SI 1982/218** HMSO, London.

SI 1982/1457 1982 Restriction on Movement of Spruce Wood Order 1982. *Statutory Instrument* **SI 1982/1457** HMSO, London.

SI 1983/1457 1983 Restriction on the Movement of Spruce Wood Order 1983. *Statutory Instrument* **SI 1983/1457** HMSO, London.

SI 1986/1510 1986 Control of Pesticides Regulations. *Statutory Instrument* **SI 1986/1510** HMSO, London.

SI 1987/1758 1987 Plant Health (Great Britain) Order. *Statutory Instrument* **SI 1987/1758** HMSO, London.

SI 1988/1657 1988 Control of Substances Hazardous to Health Regulations 1988. *Statutory Instrument* **SI 1988/1657** HMSO, London.

SI 1989/823 1989 Plant Health (Forestry) (Great Britain) Order. *Statutory Instrument* **SI 1989/823** HMSO, London.

SI 1989/1263 1989 Sludge (Use in Agriculture) Regulations 1989. *Statutory Instrument* **SI 1989/1263** HMSO, London.

SI 1989/1790 1989 Noise at Work Regulations. *Statutory Instrument* **SI 1989/1790** HMSO, London.

SI 1991/2839 1991 Environmental Protection Act (Duty of Care) Regulations. *Statutory Instrument* **SI 1989/2839** HMSO, London.

SI 1992/44 1992 The Watermark Disease (Local Authorities) Order 1992. *Statutory Instrument* **SI 1992/44** HMSO, London.

SI 1992/2051 1992 Management of Health & Safety at Work Regulations. *Statutory Instrument* **SI 1992/2051** HMSO, London.

SI 1992/2932 1992 Provision and Use of Work Equipment Regulations, 1992. *Statutory Instrument* **SI 1992/2932** HMSO, London.

SI 1992/2966 1992 Personal Protective Equipment at Work Regulations 1992. *Statutory Instrument* **SI 1992/2966** HMSO, London.

SI 1992/3004 1992 Workplace (Health, Safety & Welfare) Regulations. *Statutory Instrument* **SI 1992/3004** HMSO, London.

SI 1992/3139 1992 Personal Protective Equipment (EC Directive) Regulations. *Statutory Instrument* **SI 1992/3139** HMSO, London.

SI 1993/1282 1993 Treatment of Spruce Bark Order. *Statutory Instrument* **SI 1993/1282** HMSO, London.

SI 1993/1283 1993 Plant Health (Forestry) (Great Britain) Order. *Statutory Instrument* **SI 1993/1283** HMSO, London.

SI 1993/1320 1993 Plant Health (Great Britain) Order. *Statutory Instrument* **SI 1993/1320** HMSO, London.

SI 1993/2938 1993 Medicines (Veterinary Medicinal Products) (Application for Product Licence) Regulations. *Statutory Instrument* **SI 1993/2398** HMSO, London.

SI 1994/1985 1994 The Pesticide (Maximum Residue Levels in Crops, Food and Feeding Stuffs) Regulations 1994 *Statutory Instrument* **SI 1994/1985** 42pp HMSO, London.

SI 1994/2157 1994 Medicines (Veterinary Medicinal Products) (Application for Product Licence) (Amendment) Regulations. *Statutory Instrument* **SI 1994/2157** HMSO, London.

SI 1994/3142 1994 Marketing Authorisations for Veterinary Medicinal Products Regulations. *Statutory Instrument* **SI 1994/3142** HMSO, London.

SI 1994/3246 1994 Control of Substances Hazardous to Health Regulations 1994. *Statutory Instrument* **SI 1994/3246** HMSO, London.

SI 1994/3247 1994 Chemicals (Hazard Information and Packaging) Regulations. *Statutory Instrument* **SI 1994/3247** HMSO, London.

SI 1995/3163 1995 Reporting of Injuries, Diseases and Dangerous Occurrences Regulations,1995. *Statutory Instrument* **SI 1995/3163** (previously SI1985/2023) HMSO, London.

SI 1995/887 1995 Plant Protection Products Regulations. *Statutory Instrument* **SI 1995/887** HMSO, London.

SI 1996/1092 1996 Chemicals (Hazard Information and Packaging)(Amendment) Regulations. *Statutory Instrument* **SI 1996/1092** HMSO, London.

SI 1996/1940 1996 Plant Protection Products (Amendment) Regulations. *Statutory Instrument* **SI 1996/1940** HMSO, London.

SI 1996/3138 1996 Control of Substances Hazardous to Health (Amendment) Regulations. 1996. *Statutory Instrument* **SI 1996/3138** HMSO, London.

SI 1997/11 1997 Control of Substances Hazardous to Health (Amendment) Regulations 1997. *Statutory Instrument* **SI 1997/11** HMSO, London.

SI 1997/188 1997 Control of Pesticides (Amendments) Regulations 1997. *Statutory Instrument* **SI 1997/188** HMSO, London.

SI 1997/189 1997 Plant Protection Products (Basic Conditions) Regulations. *Statutory Instrument* **SI 1997/189** HMSO, London.

SI 1998/2746 1998 Groundwater Regulations. *Statutory Instrument* **SI 1998/2746** HMSO, London.

SI 1998/3109 1998 Plant Health (Forestry) (Great Britain) (Amendment) (No 2) Order. *Statutory Instrument* **SI 1998/3109** HMSO, London.

Sinclair, G. 1832 *Useful & ornamental planting.* Baldwin & Craddock, London.

Skärby, L. & Sellden, G. 1984 Effects of ozone on crops & forests. *Ambio* **13** (2) 68-72.

Skärby, L., Ro-Poulsen, H., Wellburn, F.A.M. & Sheppard, L.J. 1998 Impacts of ozone on forests: a European perspective. *New Phytologist* **139** 109-122.

Skeffington, R.A. & Graham, A.M. 1995 The relationship between metal content of the soil solution and tree growth at Afan forest in South Wales. (In) Decline in Sitka spruce in the South Wales coalfield. (Ed.) M.P. Coutts *For. Comm. Technical paper* **9** 65-73 Forestry Commission, Edinburgh.

Skeffington, R.A. & Wilson, E.J. 1988 Excess nitrogen deposition: issues for consideration. *Environmental Pollution* **54** 159-184, 285-295.

Slater, J.M. (Ed.) 1991 *Fifty years of the National Food Survey 1940-1990.* Proc. Symposium, London, Dec. 1990. 114pp HMSO, London.

Sloan, W.T., Jenkins, A. & Eatherall, A. 1994 A simple model of stream nitrate concentrations in forested and deforested catchments in Mid-Wales. *Journal of Hydrology* **158** 61-78.

Small, D. 1960 A creosoting plant at Thetford Chase. *For. Comm. Forest Rec.* **44** 27pp HMSO, London.

Smart, A. & Andrews, J. 1985 *Birds & broadleaves handbook.* 128pp Royal Society for the Protection of Birds, Sandy, Bedford.

Smith, C.J., Wellwood, R.W. &Elliott, G.K. 1977 Effects of nitrogen fertilizer and current climate on wood properties of Corsican pine. *Forestry* **50** (2) 117-138.

SOAEFD 1997 *Prevention of environmental pollution from agricultural activity. Code of good practice* (2nd Edn - 1st Edn publ 1992) Scottish Office Agriculture, Environment and Fisheries Dept., Edinburgh.

Soutar, R.G. 1989a Afforestation & sediment yields in British fresh waters. *Soil Use & Management* **5** (2) 82-85.

Soutar, R.G. 1989b Afforestation, soil erosion & sediment yields in British fresh waters. *Soil Use & Management* **5** (4) 200

Southwood, R (Co-chairman) 1997 *UK Round Table on Sustainable Development Second Annual Report.* 52pp Department of the Environment, London.

Spark, D. 1996 The Med slides towards disaster. *Independent* **28.10.96** 20

Speight, M.R. & Wainhouse, D. 1989 *Ecology & management of forest insects.* 384pp Oxford Scientific Publications

Spencer, S. 1996 The regeneration game. *Timber Trades Journal* **2.11.96** 10-12.

Spencer, S. 1999 Pole position. *Timber & Wood Products International* **6.3.99** 42-44.

Spiecker, H., Mielikainen, K., Kohl, M. & Skovsgaard, J.P. (Eds) 1996 Growth trends in European forests. *European Forest Research Institute Research Report* **5** 372pp Springer-Verlag.

Springthorpe, G. and Myhill, N.G. 1985/1993 *Wildlife rangers handbook.* 1st/2nd edns. Forestry Commission, Edinburgh.

Stables, S. & Nelson, D.G. 1990 *Rhododendron ponticum* control. *For Comm. Res Inf. Note* **186** Forestry Commission, Forest Research Station, Farnham.

Stace, C. 1997 *New Flora of the British Isles.* 2nd edn. 1130pp Cambridge University Press.

Stanley, B., Dunne, S. & Keane, M. 1996 Forest condition assessment & other applications of colour infrared (CIR) aerial photography in Ireland. *Irish Forestry* **53** (1&2) 19-27.

Stark, J.M. & Hart, S.C. 1997 High rates of nitrification and nitrate turnover in undisturbed coniferous forests. *Nature* **385** 61-64 London.

Steele, R.C. 1975 Wildlife conservation in woodlands. *For. Comm. Booklet* **29** HMSO, London.

Stenlid, J & Redfern, D.B. 1998 Spread within tree and stand. (In) *Heterobasidion annosum, biology, ecology, impact & control* (Eds.) S. Woodward et al. 125-142 CAB International.

Sterling, P.H. 1985/1994/1999 The Brown-tail moth. *Arboriculture Research Note* **57/93/EXT** **(and revisions)** 2pp.

Steven, H.M. 1928 Nursery investigations. *For. Comm. Bull.* **11** HMSO, London.

Stevens P.A., Harrison, A.R., Jones, H.E., Williams, T.G. & Hughes, S. 1993 Nitrate leaching from a Sitka spruce plantation & the effect of fertilization with phosphorus and potassium. *Forest Ecology & Management* **58** 233-247.

Stevens P.A., Norris D.A., Sparks, T.H. & Hodgson, A.L. 1994 Impacts of atmospheric N inputs on throughfall, soil & streamwater interactions of different age forest and moorland catchments in Wales. *Water, Air & Soil Pollution* **73** 297-317.

Stevens, P.A., & Hornung, M. 1988 Nitrate leaching from a felled Sitka spruce plantation in Beddgelert Forest, N. Wales. *Soil Use & Management* **4** 3-9.

Stevens, P.A., Adamson, J.K. , Anderson, M.A. & Hornung, M 1988 Effects of clearfelling on surface water quality & site nutrient status. *Ecological Society Special Publication* **7** Blackwell Scientific Publications, Oxford.

Stevens, P.A., Hornung, M. & Hughes, S. 1989 Solute concentrations, fluxes and major nutrient cycles in a mature Sitka spruce plantation in Beddgelert forest, North Wales. *Forest Ecology & Management* **27** 1-20.

Stevens, P.A., Norris, D.A. *et al.* 1995 Nutrient loss after clearfelling in Beddgelert forest; comparison of the effects of conventional and whole-tree harvest on soil water chemistry. *Forestry* **68** (2) 115-131.

Stewart, G.G. & Neustein, S.A. 1961 Protection of replanted groups within conifer forest against roe deer. *For. Comm. Rep. For. Res.* **1961** 170-174 HMSO, London.

Stewart, H.T.L., Allender, E. & Kube, P. 1986 Irrigation of tree plantations with recycled water. *Australian Forestry* **49** (2) 81-88.

Stewart, W.D. & Bond, G. 1961 The effect of nitrogen on fixation of elemental nitrogen in *Alnus* and *Myrica*. *Plant & Soil* **14** 347-359.

Stirling Maxwell, Sir J. 1907 The planting of high moorland. *Transactions of the Royal Scottish Arboricultural Society* **20** 1-7.

Stirling Maxwell, Sir J. 1910 Belgian system of planting on turfs. *Transactions of the Royal Scottish Arboricultural Society* **23** 153-157.

Stirling Maxwell, Sir J. 1925 On the use of manures in peat planting. *Transactions of the Royal Scottish Arboricultural Society* **39** 103-109.

Stoakley, J.T. 1962 Chemical control of coppice and scrub canopy at Bernwood forest and Whaddon Chase. *Quarterly Journal of Forestry* **56** 276-292.

Stoakley, J.T. 1965 The period of oviposition by the Douglas fir seed wasp. *For. Comm. Rep. For. Res.* **1965** 185-189 HMSO, London.

Stoakley, J.T. 1967 Oviposition period of the Douglas fir seed wasp. *For. Comm. Res. & Dev. Pap.* **43** 4pp HMSO, London.

Stoakley, J.T. 1973 Laboratory and field tests of insecticides against Douglas fir seed wasp (*Megastigmus spermatrophus*). *Plant Pathology* **22** 79-87.

Stoakley, J.T. 1975 Chemical control - the larch bark beetle, *Ips cembrae*. *For. Comm. Rep. For. Res.* **1975** 39 HMSO, London.

Stoakley, J.T. 1976 The larch bark beetle, *Ips cembrae*. *For. Comm. Rep. For. Res.* **1976** 38 HMSO, London.

Stoakley, J.T. 1977a A severe outbreak of Pine beauty moth on Lodgepole pine in Sutherland. *Scottish Forestry* **31** (2) 113-125.

Stoakley, J.T. 1977b Population studies - Pine beauty moth, larch bark beetle. *For. Comm. Rep. For. Res.* **1977** 34 HMSO, London.

Stoakley, J.T. 1978 Pine beauty moth - its distribution, life cycle and importance as a pest in Scottish forests. *For. Comm. Occasional Pap.* **4** 7-12 Forestry Commission, Edinburgh.

Stoakley, J.T. 1979a Chemical control - Pine beauty moth. *For. Comm. Rep. For. Res.* **1979** 36-37 HMSO, London.

Stoakley, J.T. 1979b Pine beauty moth. *For. Comm. Forest Rec.* **120** HMSO, London.

Stoakley, J.T. 1981a Chemical control - Pine beauty moth. *For. Comm. Rep. For. Res.* **1981** 41 HMSO, London.

Stoakley, J.T. 1981b Control of pine beauty moth, *Panolis flammea* by aerial application of fenitrothion. *For. Comm. Occasional Pap.* **11** Forestry Commission, Edinburgh.

Stoakley, J.T. 1982 Chemical control - Pine beauty moth, *Panolis flammea*. *For. Comm. Rep. For. Res.* **1982** 31 HMSO, London.

Stoakley, J.T. 1984a The use of Dimilin against Pine looper moth, *Bupalus piniaria* outbreaks in Scotland. *Proc. Dimilin Forestry Seminar,* Dupham B.V., Weesp, Holland.

Stoakley, J.T. 1984b Chemical control - winter moth *Operophthera brumata*. *For. Comm. Rep. For. Res.* **1984** 39, 40 HMSO, London.

Stoakley, J.T. 1985a Entomology - Pine beauty moth. *For. Comm. Rep. For. Res.* **1985** 37 HMSO, London.

Stoakley, J.T. 1985b The Pine beauty problem in Scotland. (In) *For. Comm. Res. & Dev. Pap.* **135** 75-76 Forestry Commission, Edinburgh.

Stoakley, J.T. 1987a The Pine beauty moth, *Panolis flammea*. *For. Comm. Rep. For. Res.* **1987** 50 HMSO, London.

Stoakley, J.T. 1987b Pine beauty control, past, present & future. (In) Population biology & control of the pine beauty moth. *For. Comm. Bull.* **67** 87-90 HMSO, London.

Stoakley, J.T. 1988 The Pine Beauty moth *Panolis flammea*. *For. Comm. Rep. For. Res.* **1988** 45 HMSO, London.

Stoakley, J.T. & Evans, H.F. 1986 The Pine Beauty moth *Panolis flammea*. *For. Comm. Rep. For. Res.* **1986** 41-42 HMSO, London.

Stoakley, J.T. & Heritage, S.G. 1986 The European sawfly, *Neodiprion sertifer*. *For. Comm. Rep. For. Res.* **1986** 43 HMSO, London.

Stoakley, J.T. & Heritage, S.G. 1987 Pine weevil, *Hylobius abietis* and black pine beetles, *Hylastes* spp. *For. Comm. Rep. For. Res.* **1987** 52 HMSO, London.

Stoakley, J.T. & Heritage, S.G. 1988a The pine weevil *Hylobius* and black pine beetles *Hylastes* spp - physical barriers. *For. Comm. Rep. For. Res.* **1988** 46 HMSO, London.

Stoakley, J.T. & Heritage, S.G. 1988b Approved methods for insecticidal protection of young trees against *Hylobius abietis* & *Hylastes* spp. *For Comm. Res Inf. Note* **140** 3pp+App. Forestry Commission, Forest Research Station, Farnham.

Stoakley, J.T. & Heritage, S.G. 1989a Application leaflet on the use of 'Gammacol' and 'Lindane flowable' for pre-planting treatment of planting stock against *Hylobius* & *Hylastes* spp. *For Comm. Res Inf. Note* **147** Forestry Commission, Forest Research Station, Farnham.

Stoakley, J.T. & Heritage, S.G. 1989b Application leaflet on the use of 'Permit' & 'Permasect 25 EC' for post-planting treatment against *Hylobius* & *Hylastes* spp. attacking young trees. *For Comm. Res Inf. Note* **151** 3pp+App. Forestry Commission, Forest Research Station, Farnham.

Stoakley, J.T. & Heritage, S.G. 1990a Application leaflet on the use of 'Permit' & 'Permasect 25 EC' for pre-planting treatment of young trees against *Hylobius* & *Hylastes* spp. *For Comm. Res Inf. Note* **177** 6pp Forestry Commission, Forest Research Station, Farnham.

Stoakley, J.T. & Heritage, S.G. 1990b Approved methods for insecticidal protection of young trees against *Hylobius abietis* & *Hylastes* spp. *For Comm. Res Inf. Note* **185** 3pp+App. Forestry Commission, Forest Research Station, Farnham.

Stoakley, J.T. & Longhurst, C. 1982 Biological control - Pine beauty moth *Panolis flammea*. *For. Comm. Rep. For. Res.* **1982** 31 HMSO, London.

Stoakley, J.T. & Patterson, J.C.G. 1987 European pine sawfly, *Neodiprion sertifer*. *For. Comm. Rep. For. Res.* **1987** 52 HMSO, London.

Stoakley, J.T. & Stickland, R.E. 1968 Applying sprays to tall trees by ground methods. *For. Comm. Res & Dev. Pap.* **73** 12pp Forestry Commission, London (Departmental use).

Stoakley, J.T., Jones, O.T. & Lisk, J.C. 1983a Biological control. The pine beauty moth, *Panolis flammea*. *For. Comm. Rep. For. Res.* **1983** 39 HMSO.

Stoakley, J.T., King, C.J. & Heritage, S.G. 1983b Chemical control - the pine weevil *Hylobius abietis*. *For. Comm. Rep. For. Res.* **1983** 40 HMSO, London.

Stoakley, J.T., Heritage, S.G. & Martin, A.F. 1984 The Pine weevil and Black pine beetles. *For. Comm. Rep. For. Res.* **1984** 40 HMSO, London.

Stoakley, J.T., Heritage, S.G. & Martin, A.F. 1985a Pine weevil & black pine beetles. *For. Comm. Rep. For. Res.* **1985** 40 HMSO.

Stoakley, J.T., Heritage, S.G., Evans, H.F. & Barbour, D.A. 1985b Pine looper moth *Bupalus piniaria*. *For. Comm. Rep. For. Res.* **1985** 38-39 HMSO, London.

Stoakley, J.T. , Baake, A., Renwick, J.A.A. & Vite, J.P. 1987 The aggregation pheromone system of the Larch bark beetle, *Ips cembrae*. *Zeitschrift für angewandte Entomologie* **86 (2)** 174-177 Forestry Commission, Edinburgh.

Stockfors, J., Linder, S. & Aronsson, A. 1997 Effects of long-term fertilisation & growth on micronutrient status in Norway spruce trees. *Studia Forestalia Suecica* **202** 15 pp.

Stoner, J.H., Gee, A.S., & Wade, K. 1984 The effects of acidification on the ecology of streams in the upper Tywi catchment in west Wales. *Environmental Pollution (A)* **35** 152-157.

Stott, K. & Parfitt, R. 1986 Willows on farms for amenity & profit. *For. Comm. Occasional Pap.* **16** 36-37.

Straw, N. & Fielding, N. 1997 The impact of aphids on Sitka spruce. *For. Comm. Rep. For. Res.* **1997** 40-44 HMSO.

Straw, N. & Fielding, N. 1998 Phytotoxicity of insecticides to control aphids on Sitka spruce. *For. Comm. Information Note* **FCIN 5** 4pp.

Straw, N., Fielding, N. & Green, G. 1994 Pine shoot moth on *Pinus muricata. For. Comm. Rep. For. Res.* **1994** 8-9 HMSO.

Strouts, R.G. 1981 *Phytophthora* diseases of trees and shrubs. *Arboricultural Leaflet* **8** 16pp HMSO, London. (Prepared for DoEnv by the Forestry Commission.)

Strouts, R.G. 1995a *Phytophthora* root disease. *Arboriculture Research Note* **58/95/PATH** Arboricultural Advisory & Information Service, Farnham, Surrey.

Strouts, R.G. 1995b Health of non-woodland trees in England, 1994. *Arboriculture Research Note* **129/95/PATH** Arboricultural Advisory & Information Service, Farnham, Surrey.

Strouts, R.G. 1996 Health of non-woodland amenity trees in England, 1995. *Arboriculture Research & Information Note.* **133/96** Forestry Commission, Farnham, Surrey.

Strouts, R.G. 1990 Control of Pesticides Regulations: 'Chapter IV' and still more to come? *Entopath News* **94** Apl 1990 11-16 Forestry Commission Research Station, Farnham, Surrey (Departmental circulation).

Strouts, R.G. & Patch, D. 1994 Fireblight of ornamental trees and shrubs. *Arboriculture Research Note* **118/94/PATH** Arboricultural Advisory & Information Service, Farnham, Surrey.

Strouts, R.G. & Winter, T.G. 1994 Diagnosis of ill-health in trees. *Research for amenity trees* **2** 380pp HMSO, London.

Strouts, R.G., Gregory, S.C. & Heritage, S.G. 1994 Protection against climatic damage, fungal diseases, insect & animal pests. (In) Forest Nursery Practice *For. Comm. Bull.* **111** 181-197 HMSO, London.

Summerscales, J. 1990 Responsibility for securing adequate health and safety standards by legislators, manufacturers, suppliers, employers and employees. *Proc. Joint FAO/ECE/ILO Committee seminar on forest technology, management & training, Sparsholt.* 169-177 Forestry Commission, Edinburgh.

Sutton, M.A., Moncrieff, J.B. & Fowler, D. 1992 Deposition of atmospheric ammonia to moorlands. *Environmental Pollution* **75** 15-24.

Syme, R.F. 1988 Herbicide use in large British tree nurseries: problems & perspectives. *Aspects of applied biology* **16** 249-255 Association of Applied Biologists, Warwick.

Szepesi, A. 1997 Forest health status in Hungary. *Environmental Pollution* **98** (3) 393-398.

Tabbush, P.M. 1982 Herbicides for grass weeding in the forest. *Forestry & British Timber* **Jan. 1982** 20-22.

Tabbush, P.M. 1984a Forest weed control. *For. Comm. Rep. For. Res.* **1984** 16-17 HMSO, London.

Tabbush, P.M. 1984b Effects of different levels of grass weeding on the establishment of Sitka spruce. (In) *Proc. conf.: Crop Protection in Northern Britain* **1984** 339-346 SCRI Invergowrie.

Tabbush, P.M. 1985a Phytotoxic effects of gamma-HCH and bare rooted conifer transplants. *Scottish Forestry* **39** (3) 167-172.

Tabbush, P.M. 1985b Forest weed control. *For. Comm. Rep. For. Res.* **1985** 16 HMSO, London.

Tabbush, P.M. 1986 Forest weed control. *For. Comm. Rep. For. Res.* **1986** 18-19 HMSO, London.

Tabbush, P.M. 1987a Chemical thinning. *For. Comm. Rep. For. Res.* **1987** 23 HMSO, London.

Tabbush, P.M. 1987b The use of mixture B to enhance the effect of glyphosate herbicide on *Rhododendron ponticum* and coarse grasses. *For Comm. Res Inf. Note* **110/87/SILN** Forestry Commission, Forest Research Station, Farnham.

Tabbush, P.M. 1988 Silvicultural principles for upland restocking. *For. Comm. Bull.* **76** 22pp HMSO, London.

Tabbush, P.M. 1992a Herbicides. *For. Comm. Rep. For. Res.* **1992** 4 HMSO, London.

Tabbush, P.M. 1992b Approved poplar clones. *For Comm. Res Inf. Note* **265** 1-3 Forestry Commission, Forest Research Station, Farnham.

Tabbush, P.M. 1994 Poplar and willow clones for short rotation coppice. (In) *Coppice:- looking beyond set-aside* Proc. symposium, R. Agric Centre, Stoneleigh, Warwicks. R. Agricultural Society of England (Ed.) J.K. Adamson

Tabbush, P.M. 1999 Reply on poplars. *Quarterly Journal of Forestry* **93** (1) 57.

Tabbush, P.M. & Heritage, S.G. 1987 Insecticide phytotoxicity. *For. Comm. Rep. For. Res.* **1987** 26 HMSO, London.

Tabbush, P.M. & Heritage, S.G. 1988 Effect of insecticidal treatment and storage on root growth potential (RGP). *For. Comm. Rep. For. Res.* **1988** 24-25 HMSO, London.

Tabbush, P.M. & Lonsdale, D. 1999 Approved poplar varieties. *For. Comm. Information Note* FCIN 21 4pp Forestry Commission, Edinburgh.

Tabbush, P.M. & Parfitt, R. 1996 Poplar and willow clones for short rotation coppice. *For Comm. Res Inf. Note* **278** 5pp Forestry Commission, Forest Research Station, Farnham.

Tabbush, P.M. & Parfitt, R. 1999 Poplar & willow varieties for short rotation coppice. *For. Comm. Information Note* FCIN 17 4pp Forestry Commission, Edinburgh.

Tabbush, P.M. & Ray, D. 1988 Restocking project: mound temperature experiments. *For. Comm. Rep. For. Res.* **1988** 22 HMSO, London.

Tabbush, P.M. & Sale, J.S.P. 1984 Experiments on the chemical control of *Rhododendron ponticum* L. *Aspects of Applied Biology* **5** 231-242 Association of Applied Biologists.

Tabbush, P.M. & Williamson. D.R. 1987 *Rhododendron* as a forest weed. *For. Comm. Bull.* **73** 7pp HMSO, London.

Tamm, C.O. 1955 Seasonal variation in the nutrient content of conifer needles. *Medd. St. Skogsforskn. Inst. Stockholm.* 45(5) **9** 34pp.

TAT 1999 Revision to Appendix to *Arboricultural Research Note* **567/94/EXT**. Tree Advice Trust, Alice Holt Lodge, Farnham, Surrey.

Taylor, G.G.M 1970 Ploughing practice in the Forestry Commission. *For. Comm. Forest Rec.* **73** 44pp HMSO, London.

Taylor, C.M.A. 1985 The return of nursing mixtures. *Forestry & British Timber* May **1985** 18-19.

Taylor, C.M.A. 1986a Forest fertilization in Great Britain. *Proc. Fertilizer Society* **251** 1-23 Fertilizer Society of London.

Taylor, J.A. 1986b The bracken problem; a local hazard and a global issue. (In) *Bracken: ecology, land-use and control technology* (Eds.) R.T. Smith, & J.A. Taylor 21-42 Parthenon Press, Carnforth.

Taylor, C.M.A. 1987a The use of sewage sludge in forestry, an assessment of the potential in Scotland. *Water Research Report.* **PRU 1707-M** Water Research Centre & Forestry Commission.

Taylor, C.M.A. 1987b The effects of nitrogen fertilizer at different rates and times of application on growth of Sitka spruce in upland Britain. *Forestry* **60** (1) 87-99.

Taylor, C.M.A. 1988 Silviculture North: nutrition. *For. Comm. Rep. For. Res.* **1988** 17-18 HMSO, London.

Taylor, C.M.A. 1990a Survey of fertilizer prescriptions in Scotland. *Scottish Forestry* **44** (1) 3-9.

Taylor, C.M.A. 1990b The nutrition of Sitka spruce on upland restock sites. *For Comm. Res Inf. Note* **164** 4pp Forestry Commission, Forest Research Station, Farnham.

Taylor, C.M.A. 1991 Forest fertilization in Britain. *For. Comm. Bull.* **95** 45pp HMSO, London.

Taylor, G. & Davies, W.J. 1990 Root growth of *Fagus sylvatica*; impact of air quality & drought in southern Britain. *New Phytologist* **116** 457-464.

Taylor, C.M.A. & Moffat, A.J. 1991 Potential for using sewage sludge in forestry in Great Britain. (In) *Alternative uses for sewage sludge.* Proceedings, Water Research Council Conference, York, 1987. (Ed.) J.E. Hall Pergamon Press.

Taylor, C.M.A. & Tabbush, P.M. 1990 Nitrogen deficiency in Sitka spruce plantations. *For. Comm. Bull.* **89** 20pp HMSO, London.

Taylor, C.M.A. & Worrell, R. 1991 Influence of site factors on the response of Sitka spruce to fertilizer at planting in upland Britain. *Forestry* **64** (1) 13-27.

Taylor, G., Dobson, M.C., Freer-Smith, P.H. & Davies, W. 1989a Tree physiology and air pollution in southern Britain. *For. Comm. Technical Information Note* **1/94** Forestry Commission Technical Development Branch, Ae, Dumfries.

Taylor, G., Dobson, M.C., Freer-Smith, P.H. & Davies, W. 1989b Tree physiology & air pollution in Southern Britain. *For Comm. Res Inf. Note* **145** 3pp Forestry Commission, Farnham, Surrey.

TDB (Technical Development Branch) 1994 Biodegradable chain oil. *For. Comm. Technical Development Br. Inf. Note* 3.12.94 11pp Forestry Commission Technical Devel't Br., Ae, Dumfries.

TDB (Technical Development Branch) 1997 *Index of reports, technical notes and information notes, 1990 to April 1997.* Forestry Commission Technical Development, Ae Village, Dumfries, DG1 1QB.

Teasdale, J.B. 1990a The safe use of lindane in forestry. *Proc. Joint FAO/ECE/ILO C'ttee seminar on forest technology, management & training, Sparsholt.* 185-196 Forestry Commission, Edinburgh.

Teasdale, J.B. 1990b The safe use of permethrin in the pre-planting treatment of trees for restocking. *Proc. Joint FAO/ECE/ILO C'ttee seminar on forest technology, management & training, Sparsholt.* 197-199 Forestry Commission, Edinburgh.

Tee, L.A. 1984 Salt licks as a method of preventing grey squirrel bark-stripping damage. *Quarterly Journal of Forestry* **78** (3) 188-190

Tee, L.A. & Petty, S.J. 1973 Survey of losses of first year conifer seedlings in Forestry Commission nurseries. *For. Comm. Res. & Dev. Pap.* **103** 54pp Forestry Commission, Edinburgh.

Tee, L.A. & Rowe, J.J. 1985 Appraisal of revenue loss from conifer thinnings due to Grey squirrel bark stripping damage. *Quarterly Journal of Forestry* **79** (1) 27-28.

TERG 1988 *The effects of acid deposition on the terrestrial environment in the UK.* Terrestrial Effects Review Group 1st report 120pp HMSO, London (Department of the Environment).

TERG 1993 *Air pollution & tree health in the United Kingdom.* 88pp HMSO, London (Department of the Environment).

Thomas, B. 1990 The GIFAP view of European registration of pesticides. (In) *Future changes in pesticide registration within the EC. Proc. BCPC symposium, Reading* **Monograph 44** 43-48 British Crop Protection Council, Farnham, Surrey.

Thomas, D. & Middleton, N.J. 1994 *Desertification: exploding the myth.* John Wiley & Sons.

Thomas, R.C. & Miller, H.G. 1992 Impact of foliar and soil applications on nitrogen on the nutritional status of a young Sitka spruce plantation. *Forest Ecology & Management* **51** 217-225.

Thomas, R.C. & Miller, H.G. 1994 Interaction of green spruce aphis and fertilizer application on the growth of Sitka spruce. *Forestry* **67** (4) 329-342.

Thompson, W.R. 1930 The problems of biological control. *Biology* **1** (3) 96-103.

Thompson, D.A. 1972 Post-establishment treatment of Sitka spruce. *Scottish Forestry* **26** (4) 288-291.

Thompson, D.A. 1978 Forest ploughs. *For. Comm. Leaflet* **70** 20pp HMSO, London.

Thompson, D.A. 1984 Ploughing forest soils. *For. Comm. Leaflet* **71** 24pp HMSO, London.

Thompson, J. 1996 Forestry Commission goes on alert for Baltic infestation. *Timber Trades Journal* 6.4.96 5.

Thompson, D.A. & Gibbs, J.N. 1990 Water-stored timber after 12 months. *For Comm. Res Inf. Note* **178** Forestry Commission, Forest Research Station, Farnham.

Thompson, D.A. & Matthews, R.W. 1989 The storage of carbon in trees and timber. *For Comm. Res Inf. Note* **160** 5pp Forestry Commission, Forest Research Station, Farnham.

Thomson, J.H. & Neustein, S.A. 1973 An experiment in intensive cultivation of an upland heath. *Scottish Forestry* **27** 211-221

Thorpe, N. 1998 Seabed holds 'double' world reserves of fuel reserves in frozen gas. *Scotsman* 8.9.98 2.

Thorpe, W.R. 1930 Observations on the parasites of the Pine shoot moth *(Rhyaconia (Evetria) buoliana) Bulletin of Entomological Research.* **XXI** (3) 387-412.

Tickell, O. 1999a Devastating diversity. *New Scientist* **21.8.99** 17.

Tickell, O. 1999 GM Trees Flowerless forests *Forestry & British Timber* **Sept 1999** 4, 36-39.

Tietema, A. & Beier, C. 1995 A correlative evaluation of nitrogen cycling in forest ecosystems of EC projects NITREX and EXMAN. *Forest Ecology & Management* **71** 143-151.

Tilney-Bassett, H.A.E. 1988 Forestry in the region of the Chilterns. *Forestry* **61** (3) 267-286.

Tinker, P.B. 1993 Climatic change & its implications (In) *Global climatic change - its implications for crop protection. Proc. BCPC Symposium* **Monograph 56** 3-12 British Crop Protection Council, Farnham, Surrey.

Tinsley, J. & Hutcheon, A.A. 1965 Chemical changes in forest litter, 4th report. *For. Comm. Rep. For. Res.* **1965** 17 HMSO, London.

Tittensor, A.M. 1975 Red squirrel. *For. Comm. Forest Rec.* **101** 36pp HMSO, London.

Tittensor, A.M. & Lloyd, H.G. 1983 Rabbits. *For. Comm. Forest Rec.* **125** 20pp HMSO, London.

Titus, B.D. & Malcolm, D.C. 1991 Nutrient changes in peaty gley soils after clear-felling Sitka spruce stands. *Forestry* **64** (3) 251-270

Titus, B.D. & Malcolm, D.C. 1992 Nutrient leaching from the litter layer after clear-felling Sitka spruce on peaty gley soils. *Forestry* **65** (4) 389-416.

Titus, B.D.& Malcolm, D.C. 1987 The effect of fertilization on litter decomposition in clear-felled spruce stands. *Plant & Soil* **100** 297-322.

Tomlin, C.D.S. 1997 *The pesticide manual.* 11th edn. 1341pp British Crop Protection Publications, Farnham, Surrey.

Tooby, T.E. 1990a The likely impact on national regulatory authorities. (In) *Future changes in pesticide registration within the EC. Proc. BCPC symposium, Reading* **Monograph 44** 111-115 British Crop Protection Council, Farnham, Surrey.

Tooby, T.E. 1990b Regulation of pesticides in forestry operations. *Proc. Joint FAO/ECE/ILO C'ttee seminar on forest technology, management & training, Sparsholt.* 145-151 Forestry Commission, Edinburgh.

Tooby, T.E. & Marsden, P.K. 1991 Interpretation of environmental fate & behavioural data for regulatory purposes. *Proc. BCPC symposium, Warwick* **Monograph 47** 3-10 British Crop Protection Council, Farnham, Surrey.

Tracy, D.R. & Nelson, D.G. 1991 Methods of grass control in the uplands. *For Comm. Res Inf. Note* 203 Forestry Commission, Forest Research Station, Farnham.

Troup, R.S. 1952 *Silvicultural systems.* 2nd edition, (Ed.) E.W. Jones 216pp Oxford University Press.

TTJ 1992 EC and Canada team up against nematode. *Timber Trades Journal* **2.5.92** 5.

TTJ 1997 FC acts to stop ugly result of 'Beauty' caterpillars. *Timber Trades Journal* **24.5.97** 3.

Tuley, G. 1984 Trees in shelters do need to be weeded. *Aspects of Applied Biology* **5** 315-318.

Turner, D.J. 1977 The safety of the herbicides 2,4-D and 2,4,5-T. *For. Comm. Bull.* **57** 1-56 HMSO, London.

Turner, D.J. & Tabbush, P.M. 1984 Studies with alternative glyphosate formulations. *BCPC Monograph* **28** 135-145 British Crop Protection Council, Farnham, Surrey.

Turner, D.J., Richardson, W.G. & Tabbush, P. 1988 The use of additives to improve the activity of herbicides for bracken and heather control. *Aspects of applied biology* **16** 271-280 Association of Applied Biologists.

Turner, J.G., Davis, J.M.L. & Guven. K. 1992 Watermark disease of tree willows. *Proc. R. Soc. Edinburgh* **98B** 105-117.

Turner, P. (Chairman) 1982 Guidelines on Testing Chemicals for Toxicity. *Report of Committee on Toxicity of Chemicals in Food* **DHSS Report 27** HMSO, London.

Turner, S. 1996 Employees to be consulted under new Health and Safety regulations. Timber Trades Journal **13.7.96** p34.

TWP 1998a FoE protests at Meyer's mahogany policy. *Timber & Wood Products* **24 Oct 1998** p5.

TWP 1998b FSC endorses the UK forestry certification standard. *Timber & Wood Products* **17**
   Oct. **1998** p4
UK Govt 1990 Government White Paper, *'Our Common Inheritance'*. **Cmd. 1200** HMSO.
UK Govt 1994a *Sustainable forestry; the UK programme.* **Cm 2429** 32pp HMSO, London.
UK Govt 1994b *Climatic change; the UK programme.* **Cm 2427** 80pp HMSO, London.
UK Govt 1994c *Sustainable development; the UK strategy* **Cm 2426** 268pp HMSO, London.
UK Govt 1994d *Biodiversity; the UK strategy.* **Cm 2428** 32pp HMSO, London.
UKCIP (UK Climatic Impacts Programme) 1998a *Climatic Change: Impacts in the UK Agenda
   for assessment & action.* 18pp Dept. of Environment, Transport & the Regions, London.
UKCIP (UK Climatic Impacts Programme) 1998b *Climatic Change: Scenarios for the UK.* 16pp
   Climatic Research Unit, Univ. of E. Anglia, Norwich,
Ulrich, B., Mayer, B. & Khanna, P.K. 1980 Chemical changes due to acid precipitation in a
   loess-derived soil in central Europe. *Soil Science* **130** 193-199.
UN (United Nations) 1992 Only one world: our own to make and keep. United Nations 367pp.
UN 1996 *An Urbanising World.* UN Centre for Human Settlement.
UNECE 1988 *Critical Loads Workshop Report.* United Nations Economic Commission for
   Europe. Bad Harzburg, FRG, March 1988.
UNPF 1994 *State of world population.* United Nations Population Fund, 1994 Report UN.
Unsworth, M.H. 1984 Evaporation from forests in cloud enhances the effects of acid deposition.
   Nature **312** 262-264 London.
Valinger, E & Lindqvist, L. 1992 The influence of thinning & nitrogen fertilization on the
   frequency of snow & wind induced stand damage. *Scottish Forestry* **46** (4) 311-320
Valkova, O 1988 Weeding effects of triclopyr on trees & shrubs. (In) The practice of weed
   control & vegetation management in forestry, amenity & conservation areas. *Aspects of
   applied biology* **16** 193-200 Association of Applied Biologists.
van dem Bussche, G.H. 1993 Storage of timber under permanent irrigation. *Suid-Afrikaans
   Bosboutydeskrif* **164** 59-64.
van der Eerden, L., de Vries, W. & van Dobben, H. 1998 Effects of ammonia deposition on
   forests in the Netherlands. *Atmospheric Environment* **32** (3) 525-532.
van der Eerden. L 1998 Nitrogen on microbial & global scales. *New Phytologist* **139** 201-204.
van Oss, R., Duyzer, J. & Wyers, P. 1998 Influence of gas to particle conversion on
   measurement of ammonia exchange over forest. *Atmospheric Environment* **32** 465-471.
van Voorenburg, F. & van Veen, H.J. 1992 Treatment & disposal of municipal sludge in the
   Netherlands. (In) *Proc. conf.: Sewage sludge disposal, Glasgow, March 1992* 7/1-7/13
   Institute of Water & Environment Management, London.
Vann, A.R., Brown, L., Chew, E., Denison-Smith, G. & Miller, E. 1988 Early performance of
   four species of *Alnus* on derelict and in the industrial Pennines. *Quarterly Journal of Forestry*
   82 (3) 165-170.
Vann, A.R., Makepeace, A.E. & Shergill, Z. 1985 Improvement of radial increment growth of
   Pinus sylvestris in the industrial pennines. *Quarterly Journal of Forestry* **79** (3) 195-203.
Varley, G.C. & Gradwell, G.R. 1962 Effect of partial defoliation by caterpillars on timber
   production of oak trees in England. *Proc. 11th International Congress of Entomology* **2** 211-
   214 Forestry Commission, Edinburgh.
Virtue, A.W. 1995 Forestry engineering operations and water quality. *Proc. Joint FAO/ECE/ILO
   Committee seminar on forest technology, management and training.* Prevention of pollution
   during forest engineering operations. 6pp Institute of Agricultural Engineers.
Visoso, M.R. 1957 Grey squirrels. *For. Comm. Rep. For. Res.* **1957** 133-136 HMSO, London.
Visoso, M.R. 1967 Squirrel populations and their control. *Forestry* **1967 Supplement** 15-21.
VMD 1995a An introduction to the 'Marketing authorisations for Veterinary Medicinal Products
   Regulations, 1994. *VMD Guidance Note* **Amelia 2** Veterinary Medicines Directorate,
   Addlestone, Surrey.
VMD 1995b How veterinary medicines may be available in the United Kingdom. *VMD Guidance
   Note* **Amelia 7** Veterinary Medicines Directorate, Addlestone, Surrey.

Waage, J.K. 1996 Introduction to the symposium. *Biological control introductions; opportunities for improved crop production.* **67** 3-11 BCPC, Farnham, Surrey.

Wainhouse, D. 1979 Dispersal of beech scale in relation to the development of beech bark disease. *Mitteilungen der Entomologische Gesellschaft, Basel* **52** 181-183.

Wainhouse, D. 1987 Forests. (In) *Integrated pest management* (Eds.) A.J. Burns, T.H. Coaker, & P.C. Jepson. 361-401 Academic Press.

Wainhouse, D. 1994 The Horse chestnut scale, a pest of town trees. *Arboriculture Research & Information Note.* **121/94/PATH** Arboricultural Advisory & Information Service, Farnham, Surrey.

Wainhouse, D. & Beech-Garwood, P.A. 1994 Growth and survival of *Dendroctonus* larvae on six species of conifer. *Journal of Applied Entomology* **117** (4) 393-399.

Wainhouse, D., Beech-Garwood, P.A. *et al.* 1993 Field response of predator *Rhizophagus grandis* to prey frass and synthetic attractants. *Journal of Chemical Ecology* **18** (10) 1693-1705.

Wainhouse, D., Evans, H.F. & Winter, H.G. 1998 *Ips typographus in Great Britain. Draft for consultation.* 23pp Plant Health Service, Forestry Commission, Edinburgh.

Waldegrave, W. 1987 Sustaining the environment in a developing world. *NERC Annual Lecture given at the Royal Society, April 1, 1987.* Natural Environment Research Council.

Walker, A. (Ed.) 1991 Pesticides in soils & water: current perspectives. Symposium proceedings. *Monograph* **47** 234pp British Crop Protection Council, Farnham, Surrey.

Walker, C. 1973 Interception of N. American bark beetle imported from Canada into Britain in logs of rock elm. *Plant Pathology* **22** (3) 147.

Walker, C. 1981 Mycorrhizae. *For. Comm. Rep. For. Res.* **1981** 32 HMSO, London.

Walker, C. 1987 Sitka spruce mycorrhizas. *Proc. R. Soc. Edinburgh* **93B** 117-129.

Walker, C. & Broome, A. 1995 Mycorrhiza research. *For. Comm. Rep. For. Res.* **1995** 24-25 HMSO, London.

Walker, C. & Wheeler, C.T. 1994 Mycorrhizas, actinorhizas & rhizobia. (In) Forest Nursery Practice *For. Comm. Bull.* **111** 104-111 HMSO, London.

Wallace, T 1943 *Diagnosis of mineral deficiencies in plants by visual symptoms.* 1st edn. HMSO, London.

Wallenda, T. & Kottke, I. 1998 Nitrogen deposition & ectotrophic mycorrhizas. *New Phytologist* **139** 169-187.

Walsh , P.D. & Walker, S. 1986 Seasonal variation in water losses from Law's forest lysimeter at Stocks Reservoir. (In) *Effects of land use on fresh waters.* (Eds.) J.F.& L.G. Solbé 541-545 Water Research Centre.

Walsh, P. 1991 Predators keep Pine beauty at bay. *Entopath News* **96** 9-10 Forestry Commission Research Station, Farnham, Surrey (Departmental circulation).

Wapshere, A.J., Delfosse, E.S. & Cullen, J.M. 1989 Recent developments in biological control of weeds. *Crop Protection* **8** (Aug) 227-250

Warren, R.G. & Benzian, B. 1959 High levels of phosphorus and die-back in yellow lupins. *Nature* **184** 1588 London.

Warrington, S. & Whittaker, J.B. 1990 Interactions between Sitka spruce, the green spruce aphid, sulphur dioxide pollution and drought. *Environmental Pollution* **65** 363-370

Waters, D. & Jenkins, A. 1992 Impacts of afforestation on water quality trends in two catchments in mid-Wales. *Environmental Pollution* **77** 167-172.

Watling, R. 1974 Macrofungi in the oakwoods of Britain. (In) *The British Oak* Proc. BSBI conf. (Eds.) M.G. Morris & F.H. Perring 222-234 E.W. Classey, Farringdon, Berks. for the Botanical Society of the British Isles.

Watling, R. 1984 Macrofungi of birchwoods. *Proc. R. Soc. Edinburgh* **85B** 129-140

Watt, T.A., Kirby, K.J. & Savill, P.S. 1972 Effect of herbicides on woodland plant communities. (In) Lowland forestry & wildlife conservation. *Nature Conservancy Symposium* **6** Nature Conservancy.

Watts, J.W. 1968 Economic, social & legal implications of use. (In) *Some safety aspects of pesticides in the countryside* Proc. Conf. Countryside in 1970, Nov., 1967. (Eds.) N.W. Moore, & W.P. Evans, 45-50 Joint ABMAC/Wildlife C'tee, Alembic House, London.

WCED (World Commission on Environment & Development) 1993 *Our common future.* WCED, Geneva.

Weatherell, J. 1953 The checking of forest trees by heather. *Forestry* **26** (1) 37-40

Weatherell, J. 1957 The use of nurse species in afforestation of upland heaths. *Quarterly Journal of Forestry* **51** 298-304.

Webb, J. 1996 Kenya's little miracle. *New Scientist* **2.12.95** 16-17.

Webb, P.J. 1973 An alternative to chemical stump protection against *Fomes annosus* on pine in state & private forestry. *Scottish Forestry* **27** 24-25.

Webb, N. 1986 *Heathlands.* New Naturalist Library 223 pp. Collins, London.

Webber, J. & Gee, C. 1994 Wood chips as a mulch or soil amendment. *Arboriculture Research & Information Note* **123/94/FP** Arboricultural Advisory & Information Service, Farnham, Surrey.

Webber, J. & Gee, C. 1996 Composts from woody wastes. *Arboricultural Practice Notes* **APN 2** 4pp Arboricultural Advisory & Information Service, Farnham, Surrey.

Webber, J. & Gibbs, J.M. (Eds) 1996 Water storage of timber - experience in Britain. *For. Comm. Bull.* **117** 1-48 HMSO, London.

Webber, J.F. 1981 A natural biological control of Dutch elm disease. *Nature* **292** 449-451.

Wellburn, A.R. 1990 Why are atmospheric oxides of nitrogen usually phytotoxic and not alternative fertilizers? *New Phytologist* **115** 395-421.

Wellburn, A.R. 1998 Atmospheric nitrogenous compounds and ozone - in NO_x. *New Phytologist* **139** 5-9.

Welsh Office 1975 Survey of toxic metals in soils in Wales. (see Bridges 1989 - page 91).

Wentzel, K.F. 1983 IUFRO studies on maximal emission standards to protect forests. (In) *Effects of accumulation of air pollutants in forest ecosystems.* (Eds.) Ulrich & D. Reidel Dordrecht.

Westoby, J.C. 1987 Forestry and underdevelopment revisited. *The purpose of forests.* Ch 15 304-318 Basil Blackwell.

Westoby, J.C. 1989 *Introduction to world forestry.* 228pp Basil Blackwell.

Weston, K. & Fowler, D. 1991 Importance of orography in spatial pattern of rainfall acidity in Scotland. *Atmospheric Environment* **25A** (8) 1517-1522.

Wheeler, C.T., Hollingsworth, M.K., Hooker, J.E. *et al.* 1991 The effect of inoculation with .. *Frankia* ... on nodulation and growth of *Alnus* ... seedlings in forest nurseries. *Forest Ecology & Management* **43** 153-166.

White, B. 1998 The Dublin Bay project. Paper to conference 'Beneficial approach to sewage sludge treatment & management'. Edinburgh Oct 1998 Unpublished.

Whitehead, P.C. & Robinson, M. 1993 Experimental basin studies - international and historical perspectives of forest impacts. *Journal of Hydrology* **145** 217-230

Whitehead, R. (Ed.) annual (1996-9) *The UK Pesticide Guide. (Revised annually)* 628pp CAB International & BCPC, Bracknell, Berks.

Whiteman, A. 1988 Appraising fertilizer applications - the case of the Moine schists. *Scottish Forestry* **42** (4) 255-275.

Whiteman, A. 1991 UK demand for & supply of wood & wood products. *For. Comm. Occasional Pap.* **37** 1-17 Forestry Commission, Edinburgh.

Whiteman, A. & Sinclair, 1992 (Values of fixed carbon for proposed UK forests - Mercia et al).

Whitmore, M.E. & Freer-Smith P.H. 1982 Growth effects of SO_2 and/or NO_2 on woody plants and grasses during spring and summer. *Nature* **300** 55-56

WHO (World Health Organisation) 1970 *Control of pesticides.* 152pp World Health Organisation, Geneva.

WHO (World Health Organisation) 1991 Safe use of pesticides. *WHO Technical Report* **813** 28pp World Health Organisation, Geneva.

Wilkins, D.A. 1997 Potential for tree growth on sites contaminated with heavy metals. (In) Proc. conf. on Arboricultural practice present & future. (Ed.) J. Claridge *Research for amenity trees* **6** 125-130 Dept. of the Environment, Transport & the Regions.

Williams, B.L. 1972 Nitrogen mineralisation and organic matter decomposition in Scots pine humus. *Forestry* **45** (2) 177-188.

Williams, B.L. 1974 Effect of water table on mineralisation of peat. *Forestry* **47** (2) 195-202.

Williams, B.L. 1983 Nitrogen transformation and decomposition in litter & humus from beneath closed-canopy Sitka spruce. *Forestry* **56** (1) 17-32.

Williams, B.L., Cooper, J.M. & Pyatt, D.G. 1979 Some effects of afforestation with lodgepole pine on rates of nitrogen mineralisation in peat. *Forestry* **52** (2) 151-160

Williamson, D.R. 1990 The use of herbicides in UK forestry. *Proc. Joint FAO/ECE/ILO Committee seminar on forest technology, management & training, Sparsholt.* 36-43 Forestry Commission, Edinburgh.

Williamson, D.R. 1991 Herbicides for farm woodlands and short rotation coppice. *For. Comm. Res Inf. Note* **201** 7pp Forestry Commission, Farnham.

Williamson, D.R. 1992 Vegetation management on bare land - farm woodland sites. *For. Comm. Rep. For. Res.* **1992** 5 HMSO, London.

Williamson, D.R. & Ferris-Kaan, R. 1990 Herbicides in forestry: herbicides can also provide conservation benefits. *Timber Grower* **Spring 1990** 30-32 HMSO, London.

Williamson, D.R. & Lane, P.B. 1989 Use of herbicides in the forest. *For. Comm. Field Bk* **8** 151pp HMSO, London.

Williamson, D.R. & Mason, W.L. 1988 Nursery herbicides. *For. Comm. Rep. For. Res.* **1988** 12 HMSO, London.

Williamson, D.R. & Mason, W.L. 1989 Forest nursery herbicides. *For. Comm. Occasional Pap.* **22** 6pp Forestry Commission, Edinburgh.

Williamson, D.R. & Mason, W.L. 1990a Butisan 'S': weed control in forest nursery transplant lines. *For Comm. Res Inf. Note* **171** 4pp Forestry Commission, Forest Research Station, Farnham.

Williamson, D.R. & Mason, W.L. 1990b Weed control in forest nurseries and forests. (In) *Weed Control Handbook: Principles* (Eds.) R.J. Hance, & K. Holly. 8th edn 457-471 Blackwell, Oxford.

Williamson, D.R. & Mason, W.L. 1990c Nursery herbicides. *For. Comm. Rep. For. Res.* **1990** 9 HMSO, London.

Williamson, D.R. & Morgan, J.L. 1994 Nursery weed control. (In) Forest Nursery Practice *For. Comm. Bull.* **111** 167-180 HMSO, London.

Williamson, D.R. & Tabbush, P.M. 1988 Forest weed control. *For. Comm. Rep. For. Res.* **1988** 12 HMSO, London.

Williamson, D.R., Elgy, D. & Shanks, C. 1987 Forest weed control - bracken. *For. Comm. Rep. For. Res.* **1987** 15 HMSO, London.

Williamson, D.R., Mason, W.L. & Clay, D.V. 1990 Repeat low dose herbicide application for weed control in forest nursery seedbeds. (In) *Proc. conf.: Crop Protection in Northern Britain* **1990** 187-192 SCRI, Dundee.

Williamson, D.R., Mason, W.L., Morgan, J.L. & Clay, D.V. 1993 Forest nursery herbicides. *For. Comm. Technical Pap.* **3** 11pp Forestry Commission, Edinburgh.

Willis, A.J. 1988 Effects of growth retardant and selective herbicide on road-side verges at Bibury...over a 30 year period. *Aspects of applied biology* **16** 1-422 Association of Applied Biologists.

Willis, A.J. 1990 Ecological consequences of weed control systems. (In) *Weed Control Handbook Principles* (Eds.) R.J. Hance, & K. Holly. 8th edn 501-520 BCPC, Farnham, Surrey.

Willoughby, I. 1995 Herbicides in forests and farm woodlands. *Forestry & British Timber* **Feb. 1995** 23-25.

Willoughby, I. 1996a Noxious weeds. *For. Comm. Res. Inf. Note* **274** 8pp Forestry Commission, Forest Research Station, Farnham.

Willoughby, I. 1996b Herbicide update (Spring, 1996). *For. Comm. Res. Inf. Note* **279** Forestry Commission, Forest Research Station, Farnham.

Willoughby, I. 1996c Herbicide update. *Forestry & British Timber* **Feb. 1996** 26-32.

Willoughby, I. 1996d Weed control when establishing new farm woodlands by direct seeding. *For. Comm. Res. Inf. Note* **286** 6pp Forestry Commission, Farnham.

Willoughby, I. 1996e Dormant season application of broad spectrum herbicides in forestry. *Aspects of applied biology* **44** 55-62 Association of Applied Biologists, Warwick.

Willoughby, I. 1997a Weeding young trees - avoiding trouble. *Farming & Conservation* **3** (1) 27-30.

Willoughby, I. 1997b Glyphosate rain fastness. *Quarterly Journal of Forestry* **91** (3) 203-210.

Willoughby, I. 1998 Herbicides for sward control among broadleaved amenity trees. *Arboriculture Research Note* **27 98 SILS** 5pp (Supersedes previous revisions)

Willoughby, I. & Clay, D 1996 Herbicides for farm woodlands and short rotation coppice. *For. Comm. Field Bk* **14** 60pp HMSO, London.

Willoughby, I. & Clay, D 1999 Herbicide update. *For. Comm. Tech. Pap.* **28** 49 pp Forestry Commission, Edinburgh.

Willoughby, I. & Dewar, J. 1995 Use of herbicides in the forest. *For. Comm. Field Bk* **8** (4th edn) 318pp HMSO, London.

Willoughby, I. & McDonald, H.G. 1999 Vegetation management in farm forestry: a comparison of inter-row management. *Forestry* **72** (2) 109-121.

Willoughby, I., Kerr, G., Jinks, R. & Gosling, P. 1996 Establishing new woodlands by direct sowing. *For Comm. Res Inf. Note* **285** Forestry Commission, Farnham, Surrey.

Wilson, A. (Chairman) 1964 *Further review of persistent organochlorine pesticides.* Report by the Advisory Committee on Poisonous Substances used in Agriculture and Food Storage. **(24-314-1)** HMSO, London.

Wilson, G.S. 1968 Agricultural pesticides: training and education of the user. (In) *Some safety aspects of pesticides in the countryside* Proc. Conf. Countryside in 1970, London, 1967. (Eds.) N.W. Moore, & W.P. Evans, 40-44 Joint ABMAC/Wildlife committee, Alembic House, London, SE1.

Wilson, K. 1985a A guide to the reclamation of mineral workings for forestry. *For. Comm. Res. & Dev. Pap.* **141** 56pp Forestry Commission, Edinburgh.

Wilson, K. 1985b Removal of tree stumps. *Arboricultural Leaflet* **7** 16pp HMSO, London. (Prepared for DoEnv by the Forestry Commission.)

Willson, A. 1986 Air pollution. *For. Comm. Rep. For. Res.* **1986** 27 HMSO, London.

Wilson, K. 1987 Reclamation of mineral workings to forestry. (In) Advances in practical arboriculture. (Ed) D. Patch. *For. Comm. Bull.* **65** 38-44 HMSO, London.

Wilson, K & Coutts, M.P. 1985 Exploiting tree-crop symbiont specificity. (In) *Attributes of trees as crop plants* (Eds.) M.G.R. Cannell & J.E. Jackson 359-379 Institute of Terrestrial Ecology. Edinburgh.

Wilson, K. & Pyatt, D.G. 1984 An experiment in intensive cultivation of an upland heath. *Forestry* **57** (2) 117-141.

Wilson, E.J. & Tiley, C. 1998 Foliar uptake of wet-deposited N by Norway spruce, using ^{15}N. *Atmospheric Environment* **32** (2) 513-8.

Willson, A., Waddell, D.A. & Durrant, D.W.H. 1987 Air pollution. *For. Comm. Rep. For. Res.* **1987** 30-31 HMSO, London.

Willson, A., Waddell, D.A. & Lee, S.E. 1988 Air pollution. *For. Comm. Rep. For. Res.* **1988** 27 HMSO, London.

Wilson, S. McG., Pyatt, D.G., Malcolm, D.C. & Connolly, T. 1998 Ecological site classification: soil nutrient regime in British woodlands. *Scottish Forestry* **52** (2) 86-92.

Winfield, R.J. 1988 Imazapyr for the control of bracken. (In) The practice of weed control & vegetation management in forestry, amenity & conservation areas. *Aspects of applied biology* **16** 281-288 Association of Applied Biologists, Warwick.

Winfield, R.J. & Bannister, C.J. 1988 Imazapyr for broad-spectrum weed control in forestry. *Aspects of applied biology* **16** 79-88 Association of Applied Biologists.

Winter, T.G. 1983 Catalogue of phytophagous insects and mites on trees in Great Britain. *For. Comm. Booklet* **53** HMSO, London.

Winter, T.G. 1989 Cypress and juniper aphids. *Arboriculture Research Note* **89/89/ENT** Arboricultural Advisory & Information Service, Farnham, Surrey.

Winter, T.G. 1991 Pine shoot beetles and ball-rooted semi-mature pines. *Arboriculture Research Note* **101/91/ENT** Arboricultural Advisory & Information Service, Farnham, Surrey.

Winter, T.G. 1995 Gypsy moth - larvae found. *Tree Damage Alert* **15** 2pp Arboricultural Advisory & Information Service, Farnham, Surrey.

Winter, T.G. & Burdekin, D.A. 1987 The poem and the pest. *Quarterly Journal of Forestry* **81** 234-238.

Winter, T. & Carter, C. 1998 Christmas tree pests. *For. Comm. Field Book* **17**

Winter, T.G. & Evans, H.F. 1994 The Asian strain of Gypsy moth, *Lymantria dispar*: a significant threat to trees. *Arboriculture Research Note* **124/94/ENT** Arboricultural Advisory & Information Service, Farnham, Surrey.

Winter, T. & Hendry, A. 1993 Moving south. *Entopath News* **99** 4 Forestry Commission Research Station, Farnham, Surrey (Departmental circulation).

Winter, T.G. & Scott, T.M. 1977 Chemical control of pine shoot moth *Rhyaconia buoliana* in seed orchards in Britain. *Forestry* **50** (2) 161-164.

Winter, T., Brown, R.M., Carter, C.I. *et al.* 1984 Entomology, advisory services. *For. Comm. Rep. For. Res.* **1984** 40 HMSO, London.

Wittering, W.O. 1974 Weeding in the forest. *For. Comm. Bull.* **48** 1-168 HMSO, London.

Wolstenholme, R. & Dutch, J. 1995 Amelioration of soil with sewage sludge. (In) *Heavy metals & trees* Proc. conf. Glasgow 1995 (Ed.) I. Glimmerveen. 30-32 Institute of Chartered Foresters, Edinburgh.

Wolstenholme, R., Dutch, J., Moffat, A.J., Bayes, C.D. & Taylor, C.M.A. 1992 A manual for the good practice for the use of sewage sludge in forestry. *For. Comm. Bull.* **107** 1-20 HMSO, London.

Wong, W.C., Nash, T.H. & Preece, T.F. 1974 A field survey of watermark disease of cricket bat willow & observations on the disease. *Plant Pathology* **23** 25-9.

Wood, R.F. 1957 Climate of Great Britain. (In) Exotic forest trees in Great Britain. (Eds.) J. Macdonald *et al*. *For. Comm. Bull.* **30** 6-20 HMSO, London.

Wood, R.F. 1965 Historical notes (In) Experiments on nutrition problems in forest nurseries. *For. Comm. Bull.* **37** 1-5 HMSO, London.

Wood, R.F. 1968 Pesticides in British forestry. (In) *Some safety aspects of pesticides in the countryside* Proc. Conf. 'Countryside in 1970', London, 20 Nov., 1967. (Eds.) N.W. Moore, & W.P. Evans, 63-70 Joint ABMAC/Wildlife Committee, Alembic House, London.

Wood, R.F. 1974 Fifty years of forest research. *For. Comm. Bull.* **50** 134pp HMSO, London.

Wood, R.F. & Holmes, G.D. 1957 Afforestation problems, stand improvement, chemical weed control; chemical bark peeling. *For. Comm. Rep. For. Res.* **1957** 36-44 HMSO, London.

Wood, R.F. & Holmes, G.D. 1958 Afforestation problems, stand improvement, chemical weed control, chemical fire retardants. *For. Comm. Rep. For. Res.* **1958** 38-45 HMSO, London:

Wood, R.F. & Holmes, G.D. 1959 Silviculture in the south: nutrition of pole stage crops, weed control. *For. Comm. Rep. For. Res.* **1959** 32-42 HMSO, London.

Wood, R.F. & Nimmo, M. 1952 Direct sowing experiments at Wareham forest, Dorset. *For. Comm. Res. Br. Pap.* **5** 32pp Forestry Commission, London. (Departmental paper)

Wood, R.F. & Nimmo, M. 1962 Chalk downland afforestation. *For. Comm. Bull.* **34** 45pp HMSO, London.

Wood, R.F. & Thirgood, J.V. 1955 Tree planting on colliery spoil tips. *For. Comm. Res. Br. Pap.* **17** 18pp Forestry Commission, London. (Departmental paper)

Wood, R.F. & Zehetmayr, J.W.L. 1955 Problems of afforestation. Manuring in the forest. *For. Comm. Rep. For. Res.* **1955** 34 HMSO, London.

Wood, R.F., Holmes, G.D. & Fraser, A.I. 1960 Derelict woodlands, manuring of trees at planting, improvement of checked plantations, pole stage manuring. *For. Comm. Rep. For. Res.* **1960** 31-32 HMSO, London.

Wood, R.F., Holmes, G.D. & Fraser, A.I. 1961 Weed control, protection against animals, fire control. *For. Comm. Rep. For. Res.* **1961** 30-32, 32-33 HMSO, London.

Wood, R.F., Miller, A.D.S. & Nimmo, M. 1967 Experiments on the rehabilitation of derelict woodlands *For. Comm. Res. & Dev. Pap.* **51** 51pp Forestry Commission, London. (Departmental paper)

Woodward, S., Stenlid, J., Karjalainen, R. & Hüttermann, A. (Eds) 1998 *Heterobasidion annosum, biology, ecology, impact & control* 585pp CAB International.

Woolfenden, D. 1988 Chemical respacing at Eskdalemuir. (In) The practice of weed control & vegetation management in forestry, amenity & conservation areas. *Aspects of applied biology* **16** 319-325 Association of Applied Biologists, Warwick.

Worrell, R. 1987 Predicting the productivity of Sitka spruce on upland sites in Northern Britain. *For. Comm. Bull.* **72** 12pp HMSO, London.

Worrell, R. & Hampson, A. 1997 The influence of some forest operations on the sustainable management of forest soils - a review. *Forestry* **70** (1) 61-85.

Worrell, R. & Malcolm, D.C. 1990 Productivity of Sitka spruce in Northern Britain. 2. Prediction from site factors. *Forestry* **63** (2) 119-128.

Wright R.F. & Tietema, A. 1995 Ecosystem response to 9 year of added nitrogen at Sogndal, Norway. *Forest Ecology & Management* **71** 133-142.

Wright, B. 1996 Forestry in Northern Ireland. *ICF News* **3/96** 1-4 Institute of Chartered Foresters, Edinburgh.

Wright, I.A. & Milne, J.A. 1996 Aversion of red and roe deer to denatonium benzoate in the diet. *Forestry* **69** (1) 1-4.

Wright, R.F. & Hauhs, M. 1991 Reversibility of acidification: soil & surface waters. *Proc. R. Soc. Edinburgh* **97B** 169-191.

Wright, R.S. et al. 1995 NITREX: responses of coniferous forest ecosystems to experimentally changed deposition of nitrogen. *Forest Ecology & Management* **71** 163-169.

Wright, T.W. 1958 Effects of tree growth on the soil. *For. Comm. Rep. For. Res.* **1958** 105-106 HMSO, London.

Wright, T.W. & Will, G.M. 1958 Nutrient content of Scots & Corsican pines on sand dunes. *Forestry* **31** (1) 13-25.

Wyers, G.P. & Erisman, J.W. 1998 Ammonia exchange over coniferous forest. *Atmospheric Environment* **32** 441-451.

Wyers, G.P., Vermeulen, A.T. & Slanina, J. 1992 Measurement of dry deposition of ammonia on a forest. *Environmental Pollution* **75** 25-28.

Young, C.W.T. 1977 External signs of decay in trees. *Arboricultural Leaflet* **1** 8pp HMSO, London. (Prepared for Department of the Environment by the Forestry Commission.)

Young, C.W.T. 1978 Sooty bark disease of sycamore. *Arboricultural Leaflet* **3** 8pp HMSO, London. (Prepared for Department of the Environment by the Forestry Commission.)

Young, M.R. & Armstrong, G. 1994 The effect of age, stand structure etc on insect communities in native pinewoods. (In) *Our pinewood heritage* (Ed.) J.R. Aldhous 1994 206-221 Jointly Forestry Commission, RSPB & Scottish Natural Heritage.

Zehetmayr, J.W.L. 1951 Afforestation work in the north. *For. Comm. Rep. For. Res.* **1951** 33-46 HMSO, London.

Zehetmayr, J.W.L. 1952 Relief of checked Sitka spruce. *For. Comm. Rep. For. Res.* **1952** 30 HMSO, London.

Zehetmayr, J.W.L 1954 Experiments in tree planting on peat. *For. Comm. Bull.* **22** 110pp HMSO, London.

Zehetmayr, J.W.L. 1960 Afforestation of upland heaths. *For. Comm. Bull.* **32** 145pp HMSO, London.

Zuckermann, S. (Chairman) 1951 *Toxic chemicals in agriculture.* Report of Working Party on Precautionary Measures in Agriculture. (24-190) HMSO, London.

Zuckermann, S. (Chairman) 1953 *Toxic chemicals in agriculture; residues in food.* Report of Working Party on Precautionary Measures in Agriculture. (24-190-0-53) HMSO, London.

Zuckermann, S. (Chairman) 1955 *Toxic chemicals in agriculture - risks to wildlife.* Report of Working Party on Precautionary Measures in Agriculture. (24-190-2) HMSO, London.

Index

Abies spp: scorch	292
Accessibility of data	53
Accumulated exposure: ozone	379
Acid neutralising capacity	408
Acid rain	363, 383 *et seq*
Acid rain & mycorrhizas	401
Acidification of nursery soils	284
Acidification of rivers	406
Acidification by S & NO$_x$	409
ACP	55
reviews	38, 57, 479
Active ingredient	19 *footnote*
Active substance	19 *footnote*
Adelges cooleyi	142
Adhesives for pesticides	26
Adjuvants	26, 27, 69
Advice: legislation	47
Advisory C'ttee on Pesticides	55
Aerial application:	
B. thuringensis	108
DDT	107
diflubenzuron	107, 108, 112
fenitrothion	109, 112
fertilizers	246, 248
herbicides	34, 204, 206, 208
insecticides	105, 106
NPV	112
pattern of use	34
tetrachlorvinphos	107
Agenda 21	4
Agricultural Chemicals Approvals Scheme	
(ACAS)	37
Agricultural policy:	
food & forestry	244
Agricultural use: definition	44
Agrotis spp	136, 137
Air quality standards - SO$_2$	365, 366
Alder species: as N fixers	262
Allelopathy	92
Alternatives to fossil fuels	445 *et seq*
Aluminium: as trace element	253
Aluminium (soil) & tree health	403
Aluminium phosphide	226
Aluminium smelters	382
Ambrosia beetles	18
Amenity plant establishment	280
Amenity trees	427

Ammonia	
sources & dry deposition	350, 352
Ammonium nitrate	261
Ammonium phosphates	221
Ammonium sulphamate	197, 208, 209,
Ammonium sulphate	261
Animal repellent	225
Anoplopora glabripennis	81, 134
Anthracnose of willow	166
AOT40	379
Aphids: on nursery stock	138
Aphids	98, 136, 141, 400
Aphids & SO$_2$	368
Approvals: costs	28
for pesticide products	55
Approvals system	27, 28
Approved codes of practice	48
farms & holdings	20, 48
guidance notes	50
health & safety at work	49
non-agricultural purposes	48
suppliers of pesticides	49
ARBRE	452
Area sprayed:	
confinement to intended area	44
Armillaria	155
Arsenic	47, 302, 414
Ash: response to fertilizer	277
Ash die-back	427
Asian longhorn beetle	81, 134
Asulam	36, 203, 210
aerial application	34
Atmospheric gases	346, 347
Atmospheric pollution	345 *et seq*
Atrazine	199, 201
Autumn frosts	293
Bacillus thuringiensis:	
on *Panolis flammea*	109
Bacterial canker of cherry	166
Bacterial canker of poplar	167
Baculoviruses	94, 131
Bad working practice	73
Bank vole	237
Banned pesticides	36
Barium	47, 253, 271
Bark	298
Bark (phloem) thickness	229, 230

Bark as mulch   189
Bark beetles   98, 117 *et seq*
Bark stripping   226
Basic slag   255
Beaded roots   330
Beech bark disease   163
Beech scale   138
Beech snap   164
Bees: safety   69, 204
Belgian turf system   256
'Bent top'   367
Bio-concentration   417
Bio-degradable lubricants   425
Bio-diversity Convention   4
Bio-diversity in the forest   460
Bio-energy & industry   451
Bio-fuels   453
Bio-remediation   416
Biological agents   91
Biological control: 89, 91, 106, 113, 114, 124
  spontaneous   132
  weeds   190
Biological oxygen demand   426
Birch: nutrient response   277
Birch rust   178
Birds in forest nurseries: netting   238
Black cherry aphid   136
Black pine beetles   116
Blowlamp: weed control   215
Blue stain   161
Bone meal   255
Bootlaces (honey fungus)   155
Boron: 47, 253, 414
  as wood preservative   239
Botrytis   175
Bracken: as compost ingredient   299
  as weed   205
Bracon hylobii   118
Bramble   209
Brash mats   188
Brash removal   278
Broadleaves:
  N responses in nursery   295
  nutrient cycle   323
  response to fertilizer   277
  stump removal/honey fungus   156
  as weed   209
Broom: as N fixer   261
  as a weed   209
Brown-field sites: reclamation   8, 414
Brown-tail moth   115
Buffer zones   65
  pupal surveys   101, 106
Bupalus piniaria: control   105, 106
  pheromones   88

Bursephelenchus xylophilus   83, 141
Cadmium   47, 302, 414, 418
Caesium   419
Cage trapping - squirrels   232, 233
Calcareous seedbed grits   284
Calcium   252, 253, 269, 273, 284
Calcium chloride:
  alternative to road salt   424
Calcium+magnesium:aluminium ratios   405
Calluna   203, 241
Calluna in seedbeds   217
Calluna-check   263
Canker of larch   168
  of poplar   167
  of willow   167
Cannock forest: *Bupalus piniaria*   107
Canopy capacity (water retention)   387
Carbamates: health monitoring   74
Carbon accumulation 436, 437, 438, 441, 442
Carbon dioxide   see CO$_2$
Carbon in biosphere   430, 438
Carbon sequestering   437, 438, 441, 442
Carbonate (geological)   433
Carbosulfan granules   120
Carrier   26
Cement dust   426
Cephalcia lariciphila   101, 132
Ceratocystis, Ceratostomella spp
                            125, 145, 156, 164
Certificate of competence:
  users   45
  storage for supply   45
CFCs   431
Chalk downland   249
Charcoal   9, 449, 453
Check of spruce   203, 263, 327
Chemical Abstracts names   21
Chemical debarking   220
Chemical inhibitors of Fomes rot   150
Chemical names of pesticides   21
Chemical thinning   211
Chernobyl   419
Chestnut blight   43
China clay   311
Chipped arboricultural waste   298
Chlorpyrifos: against *Hylobius*   119
Chlorthal-dimethyl   217
Chlorthiamid   201
Chondrostereum purpureum   165, 166
Christmas tree pests   134
Chromium   47, 253, 302, 414
CIPAC   56
Classes of pest'de active substances   24, 25
Clean air & growth   378, 379
Clear felling & nutrient loss   340

Clematis vitalba 209
Climatic change 429 *et seq*, 454
Climatic Change Convention 4, 454
Clorpyralid 202
Cloud-water 388, 392
Co-deposition 348, 369, 374
CO_2 3, 429 et seq, 431, 432, 436
CO_2 enrichment in nursery 296
Coal wastes 305, 308
Cobalt 47, 253, 271
Cockchafers 136
Codes of Practice: 48, 49, 50
  herbicides in water 39
Coke: metal smelting 9
Collaborative International Pesticides
  Analytical Council (CIPAC) 56
Colliery spoil 308
Colonisation by squirrels 232
Combustion pollution 373, 374
Commodity chemicals 46, 151
Commodity substances 46, 151
Common names:
  active substances & products 22
Competition from weeds 181
Composts & composting 281, 288, 298
Composts and mycorrhizas 328
Concentrates:
  emulsifiable, soluble, suspension 26
Condensation nuclei 348
Conifer heart rot 149
Conifer litter 299
Conifer nursing mixtures 263
Consents:
  aerial application of pesticides 34
  under Control of Pesticide Regulations 44
Conservation and weeds 190
Container production 296
Containment: *Dendroctonus micans* 122
Contamination (heavy metal) 415
Control area: *Dendroctonus micans* 123
Control of alternate host 170, 171
Control of movement of logs 85
Control of Pesticides Regulations 27, 41
Control of substances hazardous to health
  (COSSH) 39, 41, 66
Control of surrounds 176, 177
Controlled droplet applicators 66, 67
Controlled genera: plant health 82
Controls on pesticides 20
Copper 266, 272, 275, 414, 417
Copper paints 220
Copper sulphate: early use 37
Copper-chrome arsenate 241, 242
Coppice nutrient cycle 324
Corsican pine nutrient cycle 318

COSSH 39, 41, 66
  assessments 66
Cost:
  pesticide development 28, 29
  squirrel control 234
  weed control 194, 195, 196
Creosote:
  Fomes rot control 150
  wood preservative 239
Critical levels for air-borne
  pollutants 379, 385
Critical load 384, 388, 410
Cronartium ribicola 171
Crop losses due to pests *etc*, worldwide 18
Crop production & climatic change 437
Crop Protection Products:
  Approval Schemes 37
Crop type under 'Field of use' 45
Crop vulnerability to squirrels 233, 234
Cryptococcus fagisuga 138
Culbin forest:
  Bupalus piniaria 107
  nutrient studies 317
Cultivation:
  & nutrient availability 279
  & nutrient loss 337
  & soil carbon 440
Cultural control:
  bracken 205
  disease 146, 167
  nursery insects 136
Curculio elephas 143
Cutworms 133, 136
Cyanides 149
Cyclic processes 315
2,4-D 57, 197, 199, 203, 209
Dalapon 199, 201
Damage:
  by pollutants 348
  by salt 421, 423
  by squirrels 227, 234
  due to heavy metals 415
  to plants - ozone 375
  to trees - fluorine 382
  to trees - methane 381
Damage prediction - grey squirrel 233
Damage tolerance 100
Damaging insects 98
Damping off 174
Dangerous working practices 73
Dazomet 216
DDT 37, 38, 107, 119
  in food chain 4

Debarking:
by squirrels   227, 228
chemical   220
Deer   224
Deer repellents   225
Deficiency symptoms: nutrients   250, 252
De-icing salt   420, 421
Dendroctonus micans:
pheromones   88
protected zone   82
search & control   102, 122
Denitrification   362
Deposition:
ozone   373
pollutants   392 *et seq*
Derelict woodland   184
Development names for pesticides   23
Diatom chronology for acidification   407
Dicamba   206
Die-back:
ash   427
oak   180
Diet and food requirement   6
Diflubenzuron:
aerial application   34
Bupalus piniaria   107
Panolis flammea   112
Dilution of susceptible varieties   172, 173
Diphenamid   198, 199, 216, 217
Diprion pini   131
Diquat   220
Diseases of poplar   167, 173
Disodium octaborate   150
Dispersing agent   26
Dissent from underdeveloped countries   5
DMCA: DMPZ   123
Douglas fir:
Adelges   142
leaf cast diseases   168
Drainage & N saturation   353
Drainage & nutrient loss   333, 337
Drepanopeziza   166
Drey poking - squirrels   233
Dried blood   261
Drift - spray applications   66
Drip (water from canopy)   388
Droplets & spray drift   66
Drought & forest decline   400
Drought & ozone   377
Dry deposition   348, 352, 363, 374, 392
Dunemann seedbeds   299
Dunnage: carrier of pests   85
Dust filtering   428
Dutch elm disease   13, 145, 156

Earth summit meeting, 1992   4
EC Directive 91/414/EC   19, 28
Economic appraisal:
fertilizer use   246
non-market benefits   461
Fomes control   152
pole-stage fertilizing   276
Edge-effects   397
Eight-toothed European spruce bark beetle:
see *Ips typographus*
Elatobium abietinum   141
Electric fencing   226
Electricity & biomass   452
Electrodyn application system   120
Electrolyte leakage   297
Elm bark beetles   88, 125, 126
Elm breeding for resistance to DED   160
Elm phloem necrosis   43
Emission sources:
nitrogen   350
sulphur   364, 368
Emulsifier   26
Endothia parasitica   43
Energy cropping   445 *et seq*
Engineering for operator protection   71
Entomopathegenic nematodes   118
Environmental review groups (UK)   346
Epicormic shoots   220
EPPO   144
Erica   203
Erwinia amylovora   170
Erwinia salicis   168
Ethics:
afforestation   15
stewardship & priorities   3, 4, 459
UK pesticide practice   55, 463
Ethics, Economics, Environment   459
Euproctis chrysorrhoea   115
Europe & forest decline   401
European Community Directives   19, 28, 40
European plant protection organisation   144
Eutrophication risk   340
Exotic pests   81
pesticides   29
pesticide approvals   28
Expenditure: fertilizers   245
Extended use approval   46
Extenders   27
Fallow nursery land   219
Farm & Garden Chemicals Act   38
Farm forestry : approved products   46
Farm woodlands   212
FCCC   454
Fencing   226

Fenitrothion:
Panolis flammea 111
damage in forest nurseries 291
residues 287
FEPA 38, 41
Fertilizers 245, 253, 254, 265
& nutrient loss 338, 339
in forest nurseries 285
Fields of use 44
Fillers 26
Fireblight 170
protected zone 82
'Fire brigade' response: new pests 84
Fire fighting 426
Fire sites 172
Fire trace vegetation control 222
Fish & acid rain 383
Fixed C valuation 442
Fluor-chrome arsenate 239
Fluorine 302, 381
Foam fire-suppressants 426
Fog 391
Fogging: *Bupalus piniaria* & DDT 107
Foliage analysis 251, 252
Foliage treatments: woody weeds 207
Fomes rot 149
Food & Environment Protection Act
(FEPA) 38, 41
Food chain & heavy metals 418
Food intake : human diet 6
Forest & soil conservation guidelines 412
Forest & water guidelines 384
Forest certification 463
Forest condition reports 399
Forest decline 363, 397 *et seq*
Europe 401
S. Wales coalfield 141, 366
Forest decline reversed 403, 448
Forest dying 398
Forest fires 426
Forest health 83, 383 *et seq*
Forest health surveys 399
Forest nurseries 174, 214, 238, 281
Forest nursery diseases 146
Forest nursery nutrition 281 *et seq*
Forest operations & water quality 411
Forest protection strategy 78
Forest roads - erosion 425
Forestry pesticides 31, 33
Formalin 216
Formulated products 25
Formulations 25
Fosamine-ammonium 206, 208, 209, 222
Fox-coloured sawfly 131
Foxes 237

Framework Convention on Climatic Change
(FCCC) 454
Free radicals 372
Frequency of weeding in the nursery 216
Frost susceptibility 293, 294, 370
Fuelwood 11, 12, 449
Fungal inhibitors against Fomes rot 150
lists and references 471, 481
Fungicides against Dutch elm disease 159
background data 481
Gafsa 256
gamma-HCH *see lindane*
Gassing 226
Gassing compounds: storage 74
GB woodland area 12, 13
General nature of risk 58
Genetic codes 3
Genetic engineering 94, 465
Panolis flammea NPV 108
restriction of access to seed 7
Geological carbon 433
Ghost swift moth 136
Gilpinia hercyniae 132
Global:
deforestation 10, 11
land use change 9, 10
mean surface temperature 434, 436
Glufosinate 202
Glyphosate 199, 201, 204, 209, 210
GMP 256
Gorse 209
Granules: water dispersible 27
Grass weeds 201
Gravel workings 310
Grazing: effect on woodland recovery 9
to control weeds 191
Great spruce bark beetle:
see *Dendroctonus micans*
Green crops 281, 287
Green revolution:
agriculture 7, 8, 17
forestry 30
Green spruce aphis 141
Greenhouse gases 430, 431
Grey mould 175
Grey squirrel 227, 231
Ground mineral phosphate 255
Groundwater Regulations 40, 47
Grounsel:
alternate hosts for rusts 170
resistance to simazine 219
Growth regulators 222
Growth retardants 222
Guidance notes 50
Gypsum 269

Gypsy moth 85, 114
Hardwood dust: carcinogen & sensitiser 62
Harvesting & blue stain 163
Hazard: definition 50
Hazel nuts & squirrels 236
Health & Safety in the Workplace 38, 39, 41
Health monitoring 70
Heat stress & protective clothing 72
Heather & check 203, 327, 329
Heather burning: nutrient loss 243
Heathland nurseries 281
Heathland nursery origins 329
Heathland soils: phosphate 257, 258
Heathland, upland & lowland 249
Heavy metals 413 *et seq*, 414, 417
Heavy metals in food chain 418
Helsinki conference on
 Protection of European Forests 4
Hepialus humuli 136
Herbaceous weeds 201
Herbicides:
 background data 482
 lists and references 472, 482
Heterobasidion 31, 149
High human populations 5, 6, 436
High input nutrition 280
High rates of N deposition 349 *et seq*
Holistic living 459
Honey fungus 155
Hoof & horn 261
Hoppers (squirrel bait) 232
Horse-chestnut scale 141
Hot-logging 162
Humus & calcium 270
Hydraulic fluid 425
Hydrogen cyanide 226
Hydroperoxy radical (HO$_2$) 373
Hydroxyl radical (OH) 373
Hylastes spp 116
Hylobius abietis 116
Hyper-accumulation of heavy metals 419
Imazapyr 204, 206, 209, 210
Import control:
 pests 42
 plants 42, 43, 82, 83
Inadvertent carriage of pests 85
Incremental spraying 67
Industrial sites: priority for recovery 8
Ineffective international control
 on pest movement 82
Infra-red record of forest condition 403
Injection of nutrients into trees 314
Insect pheromone traps 86

Insecticides:
 background data 480
 in British forest practice 103, 105
 lists and references 470, 480
Insects: 97 *et seq*
 silviculture 30, 31, 183, 245
Integrated:
 forest protection 77, 79
 weed control 183
Intensified cropping or land clearance 9, 10
Interception fraction 388
Interception of rainfall 386
Intergovernmental Panel on
 Climatic Change (IPCC) 454
International collaboration
 on pesticide analysis 56
Introduced pests 81
Ips cembrae 87, 128
Ips sexdentatus 128
Ips typographus 81, 127
 1997 outbreak 102
 pheromone trapping 87
 protected zone 82
Irish potato famine 7
Iron 253
Iron pyrites 414
Ironstone 311
Irrigation 8
Irrigation: control of cutworm 137
Isoxaben 202, 220
IUPAC chemical names 21
Judgements on hazard and risk 53
K 257, 267, 285
K & peat 267
Kairomones 86
Keithia disease 177
Kyoto Protocol 455
'Label' uses 45
Labelling standards 49
Labels - approved products 45, 58
Lachnellula willkommii 168
Laissez faire approach 4
Land clearance cycle 8
Land clearance or intensified cropping 9, 10
Land need per head 6
Land reclamation 304
Land-use change 9, 10
Landfill 312, 314
Landfill gas 381
Larch & willow rust 450
Larch bud moth 115
Larch canker 168
Large larch bark beetle 87, 126
Large pine sawfly 131,133
Large pine weevil 116

Late-season topdressings 294
LC50 60
LD50 59, 60
Leaching of S from canopy 372
Lead 47, 253, 302, 414
Leaf spot of willow 166
Leaf spot on cherry 178
Legislation 35 *et seq*
Leguminous plants 261
LERAP 65
Lime sulphur 37
Lime-induced chlorosis 271
Limestone 269
Limestone wastes 311
Liming in acid catchments 410
Limits to growth 7
Lindane:
 as plant dip against *Hylobius* 117, 118
 review by ACP 16, 57
 biological monitoring 62
Lithology & P 255
Litter/leaf-fall 324
Liverworts 220
Local Environment Risk Assessment
 for Pesticides (LERAP) 65
Lodgepole pine: *Panolis flammea* 108
Long-range trans-boundary air poll'n 349
Long-term maintenance of
 nursery soil fertility 287, 291
Lophodermium seditiosum 178
Loss:
 forest area: past & forecast 9, 10
 timber through Dutch elm disease 13
 due to Fomes rot 150
 through squirrels 235
Lymantria dispar 85, 102, 111, 114
Machine clearance 210
Magnesium deficiency - forest decline 401
Magnesium phosphide 226
Maleic hydrazide 220, 222
Malthus 7
Manganese 253
Marketing competence 71
Marketing new pesticides 27
Marssonina salicicola 166
Maximum exposure limits 49, 61
Maximum residues in food:
 wild nuts & berries 63
Mechanical vegetation control 184, 186
Medical checks on health 74
Megastigmus spermatrophus 142
Melampsora larici-populi 173
Melampsora pinitorqua 170
Melampsora spp 172
Melfluidide 222

Mercury 47, 302, 414
Meria laricis 176
Methane:
 as greenhouse gas 431, 440, 444, 445
 as pollutant 381
Methyl bromide 216
Mg 253, 266, 271
Mice - squirrel poisons 232
Mice in forest nurseries 238
Microsphaera alphitoides 176
Mineral oils 47, 217
Mineral pollutants 413 *et seq*
Mineral workings 305, 310
Minimising insecticide use 103
Mining wastes 307, 314, 414
Misuse of pesticides 75
Mixtures of tree species 263
Modelling:
 atmospheric gas cycles 349 *et seq*
 climatic change 436
 critical loads 408
 hydrological cycle 390
 N deposition 360
 nutrient requirements for restocking 279
 stream water drainage 355
 woodland nutrient cycles 316 *et seq*
Molybdenum 275, 302
Monitoring pest populations 85, 86
Monoculture and forest decline 398
Mor humus 270
Mounding 188
Mulching 189
Mull humus 270
Munzell colour charts 251
Mycorrhizal symbiosis 330
Mycorrhizas:
 in forest nurseries 285
 in the forest 327
 & heavy metals 416
Mycovirus against Dutch elm disease 160
Myelophilus 129
Myxomatosis 225
Myzus cerasi 136
N & sulphur interaction 369
N (nitrogen):
 as nutrient 260 *et seq*, 285, 317, 327
 as pollutant 349 *et seq*
 check 203
 depositions in forests 349, 353, 354
 fertilizer in forest stands 273, 274, 275
 in drainage water 355
 interaction - mycorrhizas & heather 327
 levels in large rivers 358
 saturation 353

N₂O (nitrous oxide):
  & N deposition — 352
  & climatic warming — 431
  as greenhouse gas — 446
Names: chemical — 21-23
Naphthyl-acetic acid — 220
Natural attractants — 86
Natural gas — 381
Nectria coccinea — 140, 163
Nectria galligena — 165
Needle cast of larch — 176
Needle cast of pine — 178
Needle-fusion — 271
Nematodes:
  entomopathegenic — 118
  nursery losses — 137
  pinewood — 83, 141
Neodiprion sertifer — 131
New pesticides — 27
NFFO — 452
Nickel — 47, 253, 302, 414, 417
Nicotine: poisons rules — 36
Nitric acid — 373
Nitrogen oxides (NOₓ) — 350, 373
No observed adverse effect level (NOAEL):
  pesticide residues in food — 63
Non-fossil fuel obligation — 452
Non-ionic wetters — 27
Non-market benefits — 461
Non-woodland trees — 427
Norway spruce scorch — 292
Notification of pesticides scheme — 37
Nuclear polyhedrosis virus (NPV) — 110, 131, 133
  Panolis flammea — 110
  Neodiprion sertifer — 131
Nursery Nutrition Committee — 282
Nursing mixtures (larch/spruce etc) — 263
Nutrient cycles — 279, 315 *et seq*
Nutrient deficiencies:
  foliage analysis — 251, 252
  symptoms on plants — 250
Nutrient flux models — 322
Nutrient loss & cultural operations — 333
Nutrient status: soil — 243
Oak:
  die-back — 180
  fertilizer response — 277
  powdery mildew — 176
  wilt — 83, 164
Occult precipitation — 388
Occupational exposure standards — 49, 61, 62

Off-label use: approved products — 45
Oil spills in the forest — 425
Older crop response to added N — 276
Open-cast restoration — 308
Open-top growth chambers — 376
Operator competence — 70
Operator safety — 68
Operophthera brumata — 116
Ophiostoma spp — 125, 145, 156, 164
Organic materials as nutrient sources — 297
Organic toxic wastes — 416
Organophosphorus pesticides:
  health monitoring — 74
Orgyria antiqua — 115
Orographic enhancement of precipitation — 389, 395
Otiorhynchus spp — 130, 138
Overall control of heather — 203
Oxadiazon — 220
Ozone:
  & climatic warming — 431
  & forest decline — 400
  as pollutant — 373 *et seq*
  critical loads — 384
  damage to plants — 375
P (phosphorus): — 10
  as nutrient — 254 *et seq*, 285
Pactrobutrazol — 222
Paints — 426
Panolis flammea — 87, 101, 108
Paraquat — 199, 201
Parasitic wasps — 118
Particles in air — 348
Patch weeding — 181
Peat:
  & carbon storage — 440, 441
  & K — 268
  & methane — 440
  drainage & nutrition — 280
  mineralisation — 267
  soils — 256
  soils & radioactive Cs — 420
Peniophora — 150, 152, 153
Pennines - pollution — 367
Permethrin — 120
Persistent organochlorine insecticides — 37
Personal protective clothing — 41, 71, 73
Pest movement restrictions — 42
Pesticide:
  availability - UK totals — 19, 20
  cost — 28
  early uses — 37
  maximum residues — 39
  nomenclature — 21 *et seq.*

Pesticides (contd)
residues in food 62, 63
residues in water 63, 64
resistance to 94
types of hazard 52
usage in forestry 32, 33, 105, 148, 193
wastes 42
Pesticides Safety Precautions Scheme 38
listed active substances 480 et seq
Pests Act 225
Petrochemical wastes 305
pH of nursery soils 283
Phaeocryptopus gaumannii 168
Pheasant feed areas - effect of squirrels 229
Pheromones 86, 110, 114, 127
Phlebiopsis gigantea 150, 152, 153
Phomopsis oblonga:
competitor with *Scolytus* 126, 159
Phosphine gas 226
Phosphorus as nutrient 254 et seq
Photolysis of N compounds 365
Phyllaphis fagi 136
Physical barrier:
against honey fungus 156
against *Hylobius* 117
Physiological impairment & ozone 378
Phytophthora diseases 173, 174, 175, 219
Phytoremediation 418
Phytosanitary certificate 83
Pine beauty moth 87, 101, 108
Pine looper moth see *Bupalus piniaria*
Pine shoot beetle 129
Pine shoot moth 88, 113
Pine stump removal 154
Pine twisting rust 170
Pine weevil 116
Pinewood nematode 83, 141
Pissodes spp 129
Plant conditioning 92
Plant dipping against *Hylobius* 117, 118
Plant health legislation 42, 43
Plant passports 43, 82
Plant protection products regulations 39
Plant symptoms of nutrient deficiency 250
Plantations & carbon storage 441
Planting areas, FC & private 193, 194
Plants' defences 92
Plastic sheet mulches 189
Poa annua and pH 285
Poison baits 224, 231, 232
Poisoning by pesticides:
humans 74, 76
wildlife 76
Poisonous plants 101

Poisons Lists 36
Poisons Rules 36
Pole-crop fertilizing 271
Pollution:
climate 399
damage 366
reduction 428
resistance 372
water 42
Poplar:
bacterial canker 167
for energy biomass 449
leaf rust 173
Population changes: squirrels 232, 235
Population pressures: human 5, 6, 436
Population trends 5, 6
Post-planting spraying: *Hylobius* 120
Potassium:
as nutrient 267
metaphosphate 256
Potentially toxic elements (PTEs) 413
Poverty 6
Power line vegetation 222
Power station wastes 312
Pppp 44
Pre-vernal flora 190
Precautionary principle 4, 466
Precipitation & climatic change 435, 436
Prescribed plant protection products 44
Pressure groups 53
Prevention better than cure 464
Pristophora erichsonii 132
Process modelling 466
Product evaluations:
European Community co-ordinated 57
UK 55
Products lists: for amateur use 23
herbicides, fungicides etc. 22
wood preservatives etc. 23
Proprietary brand names 23
Propyzamide 199, 202, 217, 218
plant passports 82
Protected zone: *Dendroctonus micans* 123
Protein foams 222
Prudence principle 466
Pseudomonas spp 166
Public safety and pesticides 35
Pulpwood & products 10, 11
Pulvinia regalis 141
Pupal surveys 101
Pyrethrin I - full name 21
Pyrolysis 453
Pythium spp 174

Qualitative judgement on pesticide use 54
Quantified hazards 53
Quantitative judgement
  on pesticide use 55, 57
Rabbit haemorrhagic disease 225
Radio-activity 419
Radio-tagging 233
Read the label 58
Reafforestation 244
Red lead as seed dressing 237
Red squirrels 227, 230
Reduced nitrogen 349
Reduced volume spraying 68
Renewable energy 445 et seq
Re-oxidation of reduced sulphur 372
Reporting incidents 55
Residence time (pollutants) 392
Resistance to pesticides 94
Respacing 211
Respiratory protective equipment 73
Respiratory standards 60
Restriction of movement of logs 85
Reviews
  by Advisory Committee on Pesticides 57
  of approvals 56
Rhabdocline pseudotsugae 168
Rhizina undulata 172
Rhizomorphs (honey fungus) 155
Rhizophagus grandis 124
  pheromones 88
Rhododendron 200, 210
Rhyaconia buoliana 88, 113
RIDDOR 39, 74
Rio de Janeiro meeting 4
Risk: definition 52
Risks to operators: plant dipping 120
River acidification 406
Roadside salt spray 422
Robinia as N fixer 262
Rock phosphate 256
Root development & ozone 378
Root transmission: Dutch elm disease 160
Rotary atomisers 66, 107, 111
Rotation length & carbon accumulation
  436, 437, 438, 441, 442
Rotation of stock 176, 177
Rothamsted Experimental Station 282
Roundwood supply 10, 11, 14
Rusts 170, 171, 172, 173, 178, 450
S (sulphur): & acid rain 363
  & climatic change 435
  & nitrogen interaction 369
  as nutrient 252, 253, 295
  leaching from canopy 372

Safe disposal of pesticide wastes 74
Safe use of pesticides 40, 51, 65
Safe use of toxic chemicals:
  Cook report 38
Safety 45 et seq
Salinity 420 et seq
Salt 420 et seq
Salt licks - effect on squirrels 230
Salt spray 420, 422, 424
Salt tolerance 423, 424
Sand & gravel workings 310
Sand beds 175
Sanitation fellings:
  cricket bat willow 170
  Dendroctonus 123
  Dutch elm disease 158
Sap-displacement
  for spruce transmission poles 239
Sawdust 298
Sawflies 131
Sawn wood supply 11, 14
Scab of willow 167
Scale of use:
  fertilizers 247
  fungicides 148
  herbicides 193
  insecticides 105, 106
  Scolytus spp 125, 126
  pheromones 88
Scorch due to fertilizers 292
Scots pine nutrient cycle 320, 325
Scottish Renewables Obligation (SRO) 452
Screefing 187
Sea level change 435, 436
Sea spray 420, 423
Sea water flooding 421
Search for pests 102
Season of use - warfarin 234
Seasonal restriction of cutting 172
Second rotation crops:
  nutrient responses 278 et seq
Sediment & soil erosion 411
Seed production and decline in beech 400
Seedbed grit cover and lime 284
  repeat low dose regime 218
Seedbed weeds:
  pre & post emergence 217, 218
Selenium 47, 302, 414
Semiochemicals 86, 88
Sensitisers 62, 74
Sequestered carbon 437, 438, 439, 448
Sewage sludge 300, 302, 305
Shielding with woodland 418
Short rotation coppice 213

Silent Spring 126
Silver leaf disease 165, 166
Silvicultural controls of disease 167
Silviculture & carbon storage 438 - 441
Simazine 199, 201, 216, 217, 218
Simazine residues in transplant lines 218
Simazine resistance 219
Site controller for spraying operations 65
Sitka spruce nutrient cycle 320, 326
Situation under 'Field of use' 45
Skin exposure to pesticides 62
Slags 414
Slow release fertilizers 296
Slow-release insecticides:
container stock 138
SO$_2$:
critical levels 385
pollution 364 *et seq*
Sodium alginate 222
Sodium arsenite 220
Sodium cyanide 226
Sodium nitrite 150
Soil:
acidity in nurseries 283
aluminium 403
compaction 306
cultivation & organic carbon 440
erosion & sediment 411, 425
nutrient status 243
organic carbon 290, 291
profile development 332
recognition of value 8
sterilisation 282
Solvent 26
Sooty bark disease of sycamore 171
South Wales coalfield 141, 366
Species response to pollutants 372
Spiritual values 462
Spontaneous biological control 132, 133, 134
Spot-weeding 181
Spray: drift 66, 67, 73
Spraying: against *Hylobius* 120
Spraying equipment 185
nozzles & quality 66, 67, 73
rotary atomiser 66, 67
incremental 67
Spreaders 27
Spring frosts 294
Spruce root aphis 141
Squirrels 227
SRO 452
Stabiliser 26
Starch barrier 220

Stem treatments - woody plants 207
Stem-flow 388
Sterile barrier: - Dutch elm disease 160
Sticking agents 27
Storage of pesticides 74
Store-keeper competence 71
Stream-water & aluminium 404
Strontium 253
Strophosomus sp 130
Stump removal:
honey fungus 156
pines - *Heterobasidion* 154
Stump treatments 207
Sulphur cycle 364
Superphosphate 255
Surface water and forestry 64
Surfactants 27
Surveys: *Dendroctonus* 123
threatening insects 101
pesticide use and practice 58, 75
Sustainability & development
in the UK 3, 4, 459
Sweet chestnut: fertilizer response 277
Symbiosis (mycorrhizal) 330
2,4,5-T 57, 197, 198, 199, 208, 210
Tailings 414
Tank-mixes: adjuvants 27
Tentsmuir: *Bupalus piniaria* 107
Termites 143
Therbuthylazine 201
Thinning and fertilizing 276
Threshold for damage by insects 100
Timber quality and fertilizers 324
Timber supply in Britain 12, 13, 14
Titanium 47, 253
Toleration of damage 100
Tomicus piniperda 129
Top-up spraying against *Hylobius* 120
Toxic chemicals in agriculture 37
Toxic wastes 413 *et seq*
Toxicity hazard ratings: LD$_{50}$ 59, 60
Trace elements 253, 271
Transmission poles 239
Transpiration water loss 389, 390
Transplants: herbicides 218
Traps, trapping (squirrel) 228, 231, 235
Tree guards 236
Tree species & drainage 334
Trees & soil profile development 332
Trichoderma viride on wounds 165, 166
Triclopyr 199, 202, 208, 208, 210
Triclopyr/dicamba/2,4-D 209, 222
Triple superphosphate 255
Trouble-shooting: careful use of names 24
Turf planting 249

UK Forestry Standard 77, 385, 462
UK timber supply & demand 12, 13, 14
UK water resource 406
Ultra-low volume spraying:
  diflubenzuron 112
  fenitrothion 110
Unforeseen hazards 39, 55, 74
Unquantified hazards 55
Urban ozone 374
Urban population trends 6
Urban woodland 353
Urbanisation 6
Urea:
  alternative to road salt 424
  as fertilizer 260
  commodity substance 46
  Heterobasidion control 150
  nutrient loss 340
Valuation of fixing carbon 442
Vanadium 47, 253
Vaporising oil 217
Vapourer moth 115
Vector-borne diseases 146, 156
Vertebrate damage 223
Verticillium wilt 175
Veterinary medicines 36
Vigilance 4, 81, 109, 117, 143, 466
Virox 130
Virus:
  biological control 91, 131
  Panolis flammea 110
Virus diseases 178
Volatilisation of herbicides 200
Voles 232, 233, 237
Volume Median Diameter (VMD)
  of spray droplets 66
Warfarin 227, 231, 233, 236
Waste paper 454
Wastes 312
Water:
  & aluminium 404
  acidity 386
  contribution to green revolution 8
  demand of willow coppice 450
  hardness as quality discriminant 408
  pollution 42, 204
  quality 383 *et seq*
  resources in UK 406
  stress and weeds 182
  yield 386

Water supplies: legislation 39
Water for irrigation 8
Water vapour as greenhouse gas 431
Watermark disease
  of cricket bat willow 168
Weathering of soil 332
Web-spinning larch sawfly 132
Weed competition 181
Weed control - costs 194, 195, 196
Weed suppression by trees 183
Weeds & heathland nurseries 215
Weeds - biological control 190
Weeds in water 65
Weevils: nursery losses 138
Wet deposition 351, 365, 392
Wet storage of logs 129, 162, 426
Wettable powders 27
Wetting agents/surfactants 27
White pine blister rust 171
White spirit 217
Whole-tree harvesting 278
Wildlife 69, 75
Wildlife and weeds 190, 194
Willow:
  diseases 166, 167, 169, 172
  energy biomass 448
  rusts 172, 450
Willowherb 219
  & heavy metals 418, 419
Winter hardiness 293, 370
Winter moth 116
Wood as energy source 449, 451
Wood chips:
  energy 451
  mulch 189
Wood distillation 453
Wood preservatives 239
Wood products & carbon storage 438
Wood supply & demand 10, 11
Woody weeds 207
Woolly beech aphid 136
Work study and weeding 196
World land use change 10
World population increase 3, 4, 5
Wound pathogens & protectants 165
Wounds as disease source 146
Xyloterus spp 130
Zeiraphera diniana 115
Zinc 47, 253, 302, 414, 416, 417
Ziram 225, 226